Polynuclear Aromatic Hydrocarbons:

Physical and Biological Chemistry

Marcus Cooke
Analytical Chemistry
Battelle's Columbus Laboratories

Anthony J. Dennis
Biomedical Sciences
Battelle's Columbus Laboratories

Gerald L. Fisher
Toxicology and Pharmacology
Battelle's Columbus Laboratories

Sixth International Symposium
Sponsored by:
U.S. Environmental Protection Agency
Battelle's Columbus Laboratories
The American Petroleum Institute

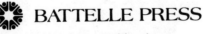

BATTELLE PRESS
Columbus • Richland

SPRINGER-VERLAG
New York Heidelberg Berlin

Library of Congress Cataloging in Publication Data

Main entry under title:

Polynuclear aromatic hydrocarbons:

Proceedings of the Sixth International Symposium on Polynuclear Aromatic Hydrocarbons held at the Battelle Columbus Laboratories, Oct. 27-29, 1981.
Includes bibliographical references and index.
 1. Hydrocarbons—Toxicology—Congresses.
 2. Carcinogens—Congresses
 3. Aromatic compounds—Toxicology—Congresses.
 4. Polycyclic compounds—Toxicology—Congresses.
 I. Cooke, Marcus, 1943-
 II. Dennis, Anthony J., 1948-
 III. Fisher, Gerald L., 1946-
 IV. International Symposium on Polynuclear Aromatic Hydrocarbons (6th : 1981 : Battelle Columbus Laboratories)
 V. United States. Environmental Protection Agency.
 VI. Battelle Memorial Institute. Columbus Laboratories.
 VII. Electric Power Research Institute.

RC268.7.H9P647   1982   616.99'4074   82-16439
ISBN 0-935470-13-1

Copyright © 1982 Battelle Memorial Institute

No part of this book may be reproduced in any form or by any electronic or mechanical means, including information storage and retrieval devices or systems, without prior written permission from the publisher, except that brief passages may be quoted for reviews.

Preface

Advances in the study of the chemistry and biology of polynuclear aromatic hydrocarbons continue to move at a pace which is significantly more rapid than any other such diverse class of carcinogenic or potentially carcinogenic compounds. The "PAH Symposium" continues to provide a forum for the dissemination and discussion of new techniques for the detection of both the pure compounds and for their effects on biological systems. The data presented at previous symposia and in the current volume represent a compilation of information which spans numerous fields of specialization and provides the scientific community with a source of uniquely condensed data.

The data presented in the current volume indicates that the mechanisms of production of the PAH are becoming more narrowly elucidated and that the sources of the various species of PAH are being better described. All of this information is leading to a global picture of the contribution of various processes to the burden of PAH in the industrialized western world and to its distribution in the environment. PAH have been described as arising from combustion of hydrocarbons from both stationary and mobile sources and have been found in both food crops and food processes.

The effects of various PAH on biological systems are addressed for whole animals and for cellular and subcellular systems. With each passing year the mechanisms of cell death, mutation and carcinogenesis produced by PAH become more clearly defined. The metabolic pathways by which these compounds are converted from biologically inert materials to active intermediates are becoming clear and with refinements in organic synthesis and separation, more of the purified intermediates are available for definitive biological testing in cells and animals.

It remains abundantly clear that the study and exchange of information concerning the chemistry and biological activity of the various PAH compounds is the key to defining the source of these materials in our environmnet and for ultimately shedding light on the biological mechanisms of PAH induced carcinogenesis.

The Sixth International Symposium on Polynuclear Aromatic Hydrocarbons proved to be an especially productive meeting. The compiled papers in this document are the written record of the discussions and research results presented during the conference. What this book cannot reveal is the synergism between scientific disciplines that produced a fertile exchange of ideas, solutions, and interesting new experiments. Battelle is committed to keeping this forum active by promoting the "PAH Symposium".

<u>The Editors</u>

Marcus Cooke
Anthony Dennis
Gerry Fisher

Contributors

L.D. ABRAMS
O.H. Materials Company
Scientific Services Division
P.O. Box 551
Findlay, OH 45840
USA

T. ALSBERG
Department of Analytical
  Chemistry, Arrhenius
  Laboratory
University of Stockholm
S-106 91 Stockholm
Sweden

P.A. ANDREWS
Department of Chemistry
McMaster University
Hamilton, Ontario
Canada, L8S 4M1

A. AUSTIN
U.S. Environmental
  Protection Agency
Research Triangle Park, NC 27711
USA

E. BALFANZ
Fraunhofer-Institut fur
  Toxikologie und Aerosolforschung
D-4400 Minister-Roxel
West Germany

J.J. BANCSI
Zenon Environmental
  Enterprises Ltd.
Hamilton, Ontario
Canada L8P 3V3

R. BARBELLA
Instituto di-Ricerche sulla
  Combustione   C.N.R.
Piazzale V. Tecchio
80125 Napoli
Italy

R.H. BARNES
Battelle Columbus Laboratories
505 King Avenue
Columbus, OH 43201
USA

P.C. BAUMANN
Columbia National Fishery Research
  Laboratory Research Station
Columbus, OH 43210
USA

F.A. BELAND
National Center for
  Toxicological Research
Jefferson, AR 72079
USA

P. BELTZ
Centre for Research on
  Environmental Quality
York University
Downsview, Ontario
Canada M3J 1P3

A. BENEDEK
Zenon Environmental
  Enterprises Ltd.
Hamilton, Ontario
Canada L8P 3V3

J.M. BENSON
Lovelace Inhalation Toxicology
  Research Institute
P.O. Box 5890
Albuquerque, NM 87185
USA

F. BERETTA
Instituto Chimica Industriale
  e Impianti Chimici
University di Napoli
Piazzale V. Tecchio
80125 Napoli
Italy

J.G.T. BERGSTROM
Studsvik Energiteknik AB
S-611 82 Nykoping
Sweden

D.R. BICKERS
Department of Dermatology
Cleveland Veterans Admin.
  Medical Center & Case
  Western Reserve University
Cleveland, OH 44106
USA

L. BLAU
Kernforschungszentrum
  Karlsruhe
Institut fur Radiochemi
Postfach 3640
7500 Karlsruhe
Federal Republic of Germany

G.M. BOOTH
Departments of Chemistry
  and Zoology
Brigham Young University
Provo, UT 84602
USA

A.S. BOPARAI
Chemical Engineering Division
Argonne National Laboratory
Argonne, IL 60439
USA

S.L. BOWIE
Organic Analytical Research
  Division
Center for Analytical Chemistry
National Bureau of Standards
Washington, DC 20234
USA

C. BRUCE
Torry Research Station
135 Abbey Road
Aberdeen AB9 8DG
Scotland

H. BRUNE
Advisory Board for
  Preventive Medicine
  and Environmental
  Protection LTD.
2000 Hamburg
Federal Republic of Germany

R.M. BUCHAN
Occupational Health and
  Safety Section
110 Veterinary Science
Colorado State University
Fort Collins, CO 80523
USA

J.J. BURBAUM
Department of Chemistry
Rensselaer Polytechnic Institute
Troy, NY 12181
USA

R. BURTON
U.S. Environmental
  Protection Agency
Research Triangle Park, NC
USA

M.A. BUTLER
Xerox Corporation
J.S. Wilson Center for
  Technology
Webster, NY 14580
USA

L.M. CALLE
Department of Chemistry
Ohio University
Athens, OH 45701
USA

D.A. CASCIANO
National Center for
  Toxicological Research
Jefferson, AR 72079
USA

J.E. CATON
Analytical Chemistry Division
Oak Ridge National Laboratory
P.O. Box X
Oak Ridge, TN 37830
USA

E. CAVALIERI
Eppley Institute for
  Research in Cancer
University of Nebraska
  Medical Center
Omaha, NE 68105
USA

F.D. CAZER
Comprehensive Cancer Center
410 West 12th Avenue
The Ohio State University
Columbus, OH 43210
USA

C.E. CERNIGLIA
National Center for
  Toxicological Research
Food & Drug Administration
Jefferson, AR 72079
USA

M.J.W. CHANG
Department of Health &
  Human Services
Food & Drug Administration
National Center for
  Toxicological Research
Jefferson, AR 72079
USA

R.L. CHANG
Department of Biochemistry
  and Drug Metabolism
Hoffmann-La Roche, Inc.
Nutley, NJ 07110
USA

D.J-C. CHEN
Genetics Group
Life Sciences Division
Los Alamos National Lab.
Los Alamos, NM 87545
USA

S.N. CHESLER
Organic Analytical Research
  Division
Center for Analytical Chemistry
National Bureau of Standards
Washington, DC 20234
USA

P-L. CHIU
Department of Pharmacology
School of Medicine
Uniformed Services University
  of the Health Sciences
Bethesda, MD 20814
USA

M.W. CHOU
National Center for
  Toxicological Research
Jefferson, AR 72079
USA

C.C. CHUANG
Battelle Columbus Laboratories
505 King Avenue
Columbus, OH 43201
USA

A. CIAJOLO
Instituto Chimica Industriale
  e Impianti Chimici
Universita di Napoli
Piazzale V. Tecchio
80125 Napoli
Italy

L. CLAXTON
U.S. Environmental
  Protection Agency
Genetic Toxicology Division
Research Triangle Park, NC 27711
USA

A.L. COLMSJO
University of Stockholm
Department of Analytical Chemistry
S-10691
Stockholm
Sweden

A.H. CONNEY
Department of Biochemistry
  & Drug Metabolism
Hoffmann-La Roche, Inc.
Nutley, NJ 07110
USA

C.S. COOPER
Chester Beatty Research Inst.
Institute of Cancer Research
Royal Cancer Hospital
Fulham Road
London SW3 6JB
England

J. CROOKS
Waters Associates, Inc.
Milford, MA 01757
USA

W.F. CUTHRELL
Organic Analytical Research
  Division
Center for Analytical Chemistry
National Bureau of Standards
Washington, DC 20234
USA

A. D'ALESSIO
Piazzale V. Tecchio
80125 Napoli
Italy

F.B. DANIEL
U.S. Environmental
  Protection Agency
Health Effects Research Lab.
26 W. St. Clair Street
Cincinnati, OH 45268
USA

R.C. DAVIS
Battelle Columbus Laboratories
505 King Avenue
Columbus, OH 43201
USA

R. DePAUS
Commission of the EC
JRC
Petten Establishment
P.O. Box 2
1755 ZG Petten
The Netherlands

G. DETTBARN
Biochemical Institute of
  Environmental Carcinogens
2070 Ahrensburg
Federal Republic of Germany

R. DEUTSCH-WENZEL
Advisory Board for
  Preventive Medicine
  and Environmental
  Protection Ltd.
2000 Hamburg
Federal Republic of Germany

J.L. DiCESARE
Perkin-Elmer Corporation
Norwalk, CT 06856
USA

J. DOMMEN
Department of Chemistry
Rensselaer Polytechnic Institute
Troy, NY 12181
USA

M.W. DONG
Perkin-Elmer Corporation
Norwalk, CT 06856
USA

M.A. DRUM
U.S. Environmental
  Protection Agency
Health Effects Research Lab.
26 W. St. Clair Street
Cincinnati, OH 45268
USA

J. DuBOIS
Commission of the EC
JRC
Petten Establishment
P.O. Box 2
1755 ZG Petten
The Netherlands

B.P. DUNN
Environmental Carcinogenesis Unit
British Columbia Cancer
  Research Centre
601 West 10th Avenue
Vancouver, BC
Canada V5Z 1L3

R. EASTERLING
U.S. Environmental
  Protection Agency
Research Triangle Park, NC 27711
USA

D.A. EASTMOND
Departments of Chemistry
  and Zoology
Brigham Young University
Provo, UT 84602
USA

W. EISENHUT
Bergbau-Forschung GmbH
Franz-Fischer-Weg 61
D-4300 Essen-Kray
Federal Republic of Germany

G. EKLUND
Studsvik Energiteknik AB
S-611 82
Nykoping
Sweden

K. EL-BAYOUMY
Division of Chemical
  Carcinogenesis
Naylor Dana Institute for
  Disease Prevention
American Health Foundation
Valhalla, NY 10595
USA

L.E. ELLIS
Department of Chemistry
Ohio University
Athens, OH 45701
USA

D.L. EVANS
Xerox Corporation
J.C. Wilson Center for
  Technology
Webster, NY 14580
USA

F.E. EVANS
National Center for
  Toxicological Research
Jefferson, AR 72079
USA

T.J. FACKLAM
Battelle Columbus Laboratories
505 King Avenue
Columbus, OH 4321
USA

P.F. FENNELLY
GCA/Technology Division
213 Burlington Road
Bedford, MA 01730
USA

H. FINKELMANN
Institut fur Physikalische
  Chemie der Technischen
  Universitat Clausthal
D-3392 Clausthal-Zellerfeld
West Germany

G. FISHER
Battelle Columbus Laboratories
505 King Avenue
Columbus, OH 43201
USA

R.J. FORDHAM
Commission of the EC/JRC
Petten Establishment
Box 2/1755 ZG Petten
The Netherlands

D.G. FOX
U.S. Forest Service
Rocky Mountain Forest &
　Ranger Experiment Station
240 W. Propsect Street
Fort Collins, CO 80526
USA

P.P. FU
Division of Carcinogenesis
　Toxicological Research
Jefferson, AR 72079
USA

J.E. FULFORD
SCIEX, Inc.
55 Glencameron Road/#202
Thornhill, Ontario
Canada L3T 1P2

W. FUNCKE
Fraunhofer-Institut fur
　Toxikologie und Aerosolforschung
D-4400 Munster-Roxel
West Germany

N.E. GEACINTOV
Department of Chemistry &
　Radiation & Solid State Univ.
New York University
New York, NY 10003
USA

A.T. GIAMMARISE
Xerox Corporation
J.S. Wilson Center for
　Technology
Webster, NY 14580
USA

Ph. GLAUDE
Commission of the EC
JRC
Petten Establishment
P.O. Box 2
1755 ZG Petten
The Netherlands

A. GOLD
Environmental Sciences
　and Engineering/201H
University of North Carolina
Chapel Hill, NC 27514
USA

W.R. GOWER
Department of Surgery
College of Medicine
The Ohio State University
410 W. 12th Avenue
Columbus, OH 43210
USA

D.E. GRAFF
Department of Physiology
College of Medicine
The Ohio State University
Columbus, OH 43210
USA

W.H. GRIEST
Analytical Chemistry Division
Oak Ridge National Laboratory
P.O. Box X
Oak Ridge, TN 37830
USA

G.D. GRIFFIN
Health & Safety Research Div.
Oak Ridge National Laboratory
Oak Ridge, TN 37830
USA

G. GRIMMER
Biochemical Institute of
　Environmental Carcinogens
2070 Ahrensburg
Federal Republic of Germany

P.L. GROVER
Chester Beatty Research Inst.
Institute of Cancer Research
Royal Cancer Hospital
Fulham Road
London SW3 6JB
England

P.M. GSCHWEND
School of Public &
  Environmental Affairs
  & Department of
  Chemistry
Indiana University
400 E. 7th Street
Bloomington, IN 47405
USA

H. GUSTEN
Kernforschungszentrum Karlsruhe
Institut fur Radiochemic
Postfach 3640
7500 Karlsruhe
Federal Republic of Germany

R.R. HALL
GCA/Technology Division
213 Burlington Road
Bedford, MA 01730
USA

R. HARDY
Torry Research Station
135 Abbey Road
Aberdeen AB9 8DG
Scotland

R.W. HART
Department of Health &
  Human Services
Food & Drug Administration
National Center for
  Toxicological Research
Jefferson, AR 72079
USA

D.A. HAUGEN
Division of Biological
  & Medical Research
Argonne National Laboratory
Argonne, IL 60439
USA

T.L. HAYES
Battelle Columbus Laboratories
505 King Avenue
Columbus, OH 43201
USA

S.S. HECHT
Division of Chemical
  Carcinogenesis
Naylor Dana Institute
  for Disease Prevention
American Health Foundation
Valhalla, NY 10595
USA

R.H. HEFLICH
National Center for
  Toxicological Research
Jefferson, AR 72079
USA

A. HEWER
Chester Beatty Res. Institute
Institute of Cancer Research
Royal Cancer Hospital
Fulham Road
London SW3 6JB
England

J.O. HILL
Lovelace Inhalation
  Toxicology Research
  Institute
P.O. Box 5890
Albuquerque, NM 87185
USA

R.A. HITES
School of Public &
  Environmental Affairs &
  Department of Chemistry
Indiana University
400 E. 7th Street
Bloomington, IN 47405
USA

D. HOFFMANN
Naylor Dana Institute for
  Disease Prevention
American Health Foundation
Valhalla, NY 10595
USA

A.P. HOLKO
Zenon Environmental Enterprises
Hamilton, Ontario
Canada L8P 3V3

M. HOYT
GCA/Technology Division
213 Burlington Road
Bedford, MA 01730
USA

D. HSIEH
Department of Environmental
 Toxicology
University of California
Davis, CA 95616
USA

G.T. HUNT
GCA/Technology Division
213 Burlington Road
Bedford, MA 07130
USA

M. INBASEKARAN
Comprehensive Cancer Center
410 W. 12th Avenue
The Ohio State University
Columbus, OH 43210
USA

W.A. IVANCIC
Battelle Columbus Laboratories
505 King Avenue
Columbus, OH 43201
USA

D.R. JAASMA
Department of Mechanical
 Engineering
Virginia Polytechnic Institute
 & State University
Blacksburg, VA 24061
USA

J. JACOB
Biochemisches Institut
 fur Umweltcarcinogene
2070 Ahrensburg
Federal Republic of Germany

D.M. JERINA
Laboratory of Bioorganic Chemistry
National Institute of Arthritis,
 Diabetes, and Digestive and
 Kidney Diseases
National Institutes of Health
Bethesda, MD 20205
USA

N.J. JOYCE
U.S. Environmental
 Protection Agency
Health Effects Research
 Laboratory
26 W. St. Clair Street
Cincinnati, OH 45268
USA

R. JUNGERS
U.S. Environmental
 Protection Agency
Research Triangle Park, NC 27711
USA

F.F. KADLUBAR
National Center for
 Toxicological Research
Jefferson, AR 72079
USA

C. KANDASWAMI
Department of Biochemistry
Memorial University of
 Newfoundland
St. John's, Newfoundland
Canada

W. KARCHER
Commission of the EC
Petten Establishment
P.O. Box 2
1755 ZG Petten
The Netherlands

M. KATZ
Centre for Research on
  Environmental Quality
York University
Downsview, Ontario
Canada M3J 1P3

P. KAUR
Environmental Health
  Research & Treating, Inc.
3217 Whitfield Avenue
Suite 11
Cincinnati, OH 45268
USA

R.J. KINDYA
GCA/Technology Division
213 Burlington Road
Bedford, MA 01730
USA

D.K. KIRIAZIDES
Xerox Corporation
J.C. Wilson Center for
  Technology
Webster, NY 14580
USA

J. KONIG
Fraunhofer-Institut fur
  Toxikologie und
  Aerosolforschung
D-4400 Munster-Roxel
West Germany

P.J. KOWALCZYK
Department of Chemistry
Rensselaer Polytechnic Institute
Troy, NY 12181
USA

S. KUMAR
Department of Chemistry
University of Missouri
St. Louis, MO 63121
USA

D.A. LANE
SCIEX, Inc.
55 Glencameron Rd./#202
Thornhill, Ontario
Canada L3T 1P2

E. LANGER
Bergbau-Forschung GmbH
Franz-Fischer-Weg 61
D-4300 Essen-Kray
Federal Republic of Germany

B. LARSSON
Food Laboratory
National Food Administration
P.O. Box 622
S-751 26 Uppsala
Sweden

D.W. LATER
Department of Chemistry
Brigham Young University
Provo, UT 84602
USA

R.J. LAUB
Department of Chemistry
The Ohio State University
Columbus, OH 43210
USA

E.J. LAVOIE
Naylor Dana Institute for
  Disease Prevention
American Health Foundation
Valhalla, NY 10595
USA

M.L. LEE
Departments of Chemistry
  and Zoology
Brigham Young University
Provo, UT 84602
USA

R.E. LEHR
Department of Chemistry
University of Oklahoma
Norman, OK 73019
USA

W. LEVIN
Department of Biochemistry
  and Drug Metabolism
Hoffmann-La Roche, Inc.
Nutley, NJ 07110
USA

S.P. LEVINE
O. H. Materials Company
Scientific Services Division
P.O. Box 551
Findlay, OH 45840
USA

J. LEWTAS
U.S. Environmental
  Protection Agency
Research Triangle Park, NC 27711
USA

K.F. LEWIS
Department of Biochemistry
New Jersey Medical-CMDNJ
Newark, NJ 07103
USA

I.E. LICHTENSTEIN
Johnson Matthey, Inc.
Malvern, PA 19355
USA

J.P. LOWE
Department of Chemistry
Pennsylvania State University
University Park, PA 16802
USA

C. LU
Biochemistry Department
McMaster University
Hamilton, Ontario
Canada L8S 4M1

R.G. LUTHY
Department of Civil Engineering
Carnegie-Mellon University
Pittsburgh, PA 15213
USA

C.R. MACKERER
Mobile Environmental and
  Health Sciences Laboratory
P.O. Box 1029
Princeton, NJ 08540
USA

P.R. MACKIE
Torry Research Station
135 Abbey Road
Aberdeen AB9 8DG
Scotland

M.C. MacLEOD
Biology Division
Oak Ridge National Laboratory
Oak Ridge, TN 37830
USA

A.D. MacNICOLL
Chester Beatty Res. Institute
Institute of Cancer Research
Royal Cancer Hospital
Fulham Road
London SW3 6JB
England

M. MALAIYANDI
Bureau of Chemical Hazards
Environmental Health
  Directorate
Health Protection Branch
Tunney's Pasture, Ottawa
Canada K1A OL2

D. MARSH
Xerox Corporation
J.C. Wilson Center for
  Technology
Webster, NY 14580
USA

T.O. MASON
Department of Physiological
  Chemistry
College of Medicine
The Ohio State University
Columbus, OH 43210
USA

T. MAST
Department of Environmental
  Toxicology
University of California
Davis, CA 95616
USA

W.E. MAY
Organic Analytical
  Research Division
Center for Analytical
  Chemistry
National Bureau of Standards
Washington, DC 20234
USA

D. MAYS
Battelle Columbus Laboratories
505 King Avenue
Columbus, OH 43201
USA

D.R. McCALLA
Biochemistry Department
McMaster University
Hamilton, Ontario
Canada L8S 4M1

B.E. McCARRY
Department of Chemistry
McMaster University
Hamilton, Ontario
Canada L8S 4M1

F.C. McELROY
Analytical & Information Div.
Exxon Research & Engineering
Linden, NJ 07036
USA

A.S. McGILL
Torry Research Station
135 Abbey Road
Aberdeen AB9 8DG
Scotland

L. McMILLAN
U.S. Environmnetal
  Protection Agency
Health Effects Research
  Laboratory
Cincinnati, OH 45268
USA

P. MELIUS
Department of Chemistry
Auburn University
Auburn, AL 36849
USA

O. MERESZ
Laboratory Services Branch
Ontario Ministry of the
  Environment
P.O. Box 213
Rexdale, Ontario
Canada M9W 5L1

R. MERMELSTEIN
Xerox Corporation
J.C. Wilson Center for
  Technology
Webster, NY 14580
USA

F. MESSIER
Department of Chemistry
McMaster University
Hamilton, Ontario
Canada L8S 4M1

C. MEYER
Bergbau-Forschung GmbH
Franz-Fischer-Weg 61
D-4300 Essen-Kray
Federal Republic of Germany

J. MICHL
Department of Chemistry
University of Utah
Salt Lake City, UT 84112
USA

K.J. MILLER
Department of Chemistry
Rensselaer Polytechnic Institute
Troy, NY 12181
USA

R.E. MILLER
Department of Physiology
College of Medicine
The Ohio State University
Columbus, OH 43210
USA

G.E. MILO
Department of Physiological
  Chemistry/College of Medicine
The Ohio State University
Columbus, OH 43210
USA

J. MISFELD
Institute of Mathematics
Technical University
3000 Hannover
Federal Republic of Germany

A. MOSBERG
Battelle Columbus Laboratories
505 King Avenue
Columbus, OH 43201
USA

H. MUKHTAR
Department of Dermatology
Cleveland Veterans Admin.
  Medical Center & Case
  Western Reserve University
Cleveland, OH 44106
USA

J. MUMFORD
U.S. Environmental
  Protection Agency
Health Effects Research Lab.
Research Triangle Park, NC 27711

B.P. MURPHY
Waters Associates, Inc.
Milford, MA 01757
USA

C.B. MURPHY
Xerox Corporation
J.C. Wilson Center for
  Technology
Webster, NY 14580
USA

D.J. MURPHY
Occusafe, Inc.
1040 S. Milwaukee
Wheeling, IL 60090
USA

R.W. MURRAY
Department of Chemistry
University of Missouri
St. Louis, MO 63121
USA

D.F.S. NATUSCH
Department of Chemistry
Colorado State University
Fort Collins, CO 80523
USA

K.-W NAUJACK
Biochemical Institute of
  Environmental Carcinogens
2070 Ahrensburg
Federal Republic of Germany

A. NELEN
Commission of the EC
JRC
Petten Establishment
P.O. Box 2
1755 ZG Petten
The Netherlands

S. NESNOW
Carcinogenesis & Metabolism
  Branch
U.S. Environmental
  Protection Agency
Research Triangle Park, NC 27711
USA

U.D. NEUE
Waters Associates, Inc.
Milford, MA 01757
USA

A.M. NEVILLE
Ludwig Institute for
  Cancer Research
Haddow Laboratories
Institute of Cancer Research
Sutton, Surrey
United Kingdom

M.G. NISHIOKA
Battelle Columbus Laboratories
505 King Avenue
Columbus, OH 43201
USA

P.O. O'BRIEN
Department of Biochemistry
Memorial University
  of Newfoundland
St. John's, Newfoundland
Canada

I.J. OCASIO
Department of Chemistry
Ohio University
Athens, OH 45701
USA

J.F. O'CONNELL
Department of Physiology
College of Medicine
The Ohio State University
Columbus, OH 43210
USA

K. OGAN
Perkin-Elmer Corporation
Norwalk, CT 06856
USA

M. O'HARE
Ludwig Institute for
  Cancer Research
Sutton, Surrey
United Kingdom

R.T. OKINAKA
Genetics Group
Life Sciences Division
Los Alamos National Lab.
Los Alamos, NM 87545
USA

H. OLSEN
Department of Environmental
  Toxicology
University of California
Davis, CA 95616
USA

C.E. OSTMAN
University of Stockholm
Department of Analytical
  Chemistry
S-10691 Stockholm
Sweden

K. PAL
Chester Beatty Research
  Institute
Institute of Cancer Research
Royal Cancer Hospital
Fulham Road
London SW3 6JB
England

A.G. PALMER III
O. H. Materials Company
Scientific Services Division
P.O. Box 551
Findlay, OH 45840
USA

E. PARSONS
Torry Research Station
135 Abbey Road
Aberdeen AB9 8DG
Scotland

M.J. PEAK
Division of Biological
  & Medical Research
Argonne National Laboratory
Argonne, IL 60439
USA

T.C. PEDERSON
Biomedical Science Department
General Motors Research Lab.
Warren, MI 48090-9055
USA

R.A. PELROY
Pacific Northwest Laboratory
Battelle Memorial Institute
Richland, WA 99352
USA

R. PENG
Department of Biochemistry
New Jersey Medical-CMDNJ
Newark, NJ 07103
USA

M.A. PEREIRA
U.S. Environmental
  Protection Agency
Health Effects Res. Laboratory
Cincinnati, OH 45268
USA

B.A. PETERSEN
Battelle Columbus Laboratories
505 King Avenue
Columbus, OH 43201
USA

T.L. PIERCE
Department of Mechanical
  Engineering
Virginia Polytechnic Institute
  & State University
Blacksburg, VA 24061
USA

J.F. PITT
Department of Physiology
College of Medicine
The Ohio State University
Columbus, OH 43210
USA

M.A. QUILLIAM
Department of Chemistry
McMaster University
Hamilton, Ontario
Canada L8S 4M1

U. RANNUG
Department of Analytical
  Chemistry
Arrhenius Laboratory
University of Stockholm
S-106 91 Stockholm
Sweden

R.R. REAGAN
Analytical Chemistry Division
Oak Ridge National Laboratory
P.O. Box X
Oak Ridge, TN 37830
USA

R.E. REBBERT
Organic Analytical Research
  Division
Center for Analytical
  Chemistry
National Bureau of Standards
Washington, DC 20234
USA

M. RIBICK
Columbia National Fishery
  Research Laboratory
Columbia, MO 65201
USA

R. RIGGIN
Battelle Columbus Laboratories
505 King Avenue
Columbus, OH 43201
USA

W.K. ROBBINS
Analytical & Information Division
Exxon Research & Engineering
Linden, NJ 07036
USA

W.L. ROBERTS
Department of Chemistry
The Ohio State University
Columbus, OH 43210
USA

E. ROGAN
Eppley Institute for
  Research in Cancer
University of Nebraska
  Medical Center
Omaha, NE 68105
USA

T. ROMANOWSKI
Fraunhofer-Institut fur
  Toxikologie und
  Aerosolforschung
D-4000 Munster-Roxel
West Germany

D.A. ROKOSH
Microbiology Section
Ontario Ministry of the
  Environment
P.O. Box 213
Rexdale, Ontario
Canada M9W 5L1

R.E. ROYER
Lovelace Inhalation Toxicology
 Research Institute
P.O. Box 5890
Albuquerque, NM 87185
USA

P.S. SABHARAWAL
Environmental Health Research
 and Testing, Inc.
3217 Whitfield Avenue
Suite 11
Cincinnati, OH 45268
USA

G. SAHLBERG
Food Laboratory
National Food Administration
P.O. Box 622
S-751 26 Uppsala
Sweden

T. SAKUMA
SCIEX, Inc.
55 Glencameron Road/#202
Thornhill, Ontario
Canada L3T 1P2

M.F. SALAMONE
Microbiology Section
Ontario Ministry of the
 Environment
P.O. Box 213
Resources Road
Rexdale, Ontario
Canada M9W 5L1

R. SANGAIAH
Environmental Sciences
 and Engineering
University of North Carolina
Chapel Hill, NC 27514
USA

A. SCHMOLDT
Pharmakologisches Institut
 der Universitat Hamburg
Federal Republic of Germany

C.A. SCHREINER
Mobil Environmental and
 Health Sciences Laboratory
P.O. Box 1029
Princeton, NJ 08540
USA

M.R. SCHURE
Department of Chemistry
Colorado State University
Fort Collins, CO 80523
USA

D.D. SCHURESKO
Health & Safety Res. Division
Oak Ridge National Laboratory
Oak Ridge, TN 37830
USA

W. SEGMULLER
Department of Chemistry
Rensselaer Polytechnic Inst.
Troy, NY 12181
USA

J. SEIBER
Department of Environmental
 Toxicology
University of California
Davis, CA 95616
USA

J.K. SELKIRK
Biology Division
Oak Ridge National Laboratory
Oak Ridge, TN 37830
USA

B.I. SHUSHAN
SCIEX, Inc.
55 Glencameron Road/#202
Thornhill, Ontario
Canada L3T 1P2

J-S. SIAK
Biomedical Science Department
General Motors Reseach Laboratory
Warren, MI 48090-9055
USA

B.D. SILVERMAN
IBM
Thomas J. Watson
  Research Center
Yorktown Heights, NY 10598
USA

P. SIMS
Chester Beatty Research
  Institute
Institute of Cancer Research
Royal Cancer Hospital
Fulham Road
London SW3 6JB
England

N.P. SINGH
Department of Health and
  Human Services
Food & Drug Administration
National Center for
  Toxicological Research
Jefferson, AR 72079
USA

L.M. SKEWES
Ford Motor Company
Scientific Research Staff
Analytical Sciences Department
Dearborn, MI 48121
USA

T.J. SLAGA
Biology Division
Oak Ridge National Laboratory
Oak Ridge, TN 37830
USA

C.A. SMITH
Department of Chemistry
The Ohio State University
Columbus, OH 43210
USA

W.D. SMITH
Zoology Department
The Ohio State University
Columbus, OH 43210
USA

W.J. SONNEFELD
Department of Chemistry
University of Maryland
College Park, MD 20742
USA

V.C. STAMOUDIS
Energy & Environmental
  Systems Division
Argonne National Laboratory
Argonne, IL 60439
USA

U. STEBNERG
Department of Analytical
  Chemistry
Arrhenius Laboratory
University of Stockholm
S-106 91 Stockholm
Sweden

D.L. STEWART
Biology Department
Pacific Northwest Laboratory
Richland, WA 99352
USA

G.F. STRNISTE
Genetics Group
Life Sciences Division
Los Alamos National lab.
Los Alamos, NM 87545
USA

V.V. SUBRAHMANYAM
Department of Biochemistry
Memorial University of
  Newfoundland
St. John's, Newfoundland
Canada

P.D. SULLIVAN
Department of Chemistry
Ohio University
Athens, OH 45701
USA

A. SUNDVALL
Department of Toxicology
  Genetics
Wallenberg Laboratory
University of Stockholm
S-106 91 Stockholm
Sweden

W. SYDOR, JR.
Department of Biochemistry
New Jersey Medical-CMDNJ
Newark, NJ 07103
USA

B. TAN
Department of Chemistry
University of Massachusetts
Amherst, MA 01003
USA

D.R. THAKKER
Laboratory of Bioorganic
  Chemistry
National Institute of
  Arthritis, Diabetes, and
  Digestive & Kidney Diseases
National Institute of Health
Bethesda, MD 20205
USA

B.A. TOMKINS
Analytical Chemistry Division
Oak Ridge National Laboratory
P.O. Box X
Oak Ridge, TN 37830
USA

G.E. TONEY
Environmental Sciences
  & Engineering
University of North Carolina
Chapel Hill, NC 27514
USA

S.H. TONEY
Northrop Services, Inc.
Environmental Sciences
Research Triangle Park, NC 27709

D.A. TRAYSER
Battelle Columbus Laboratories
505 King Avenue
Columbus, OH 43201
USA

L.L. TRIPLETT
Biology Division
Oak Ridge National Laboratory
Oak Ridge, TN 37830
USA

K. TRZCINSKI
Studsvik Energiteknik AB
S-611 82, Nykoping
Sweden

A. TURTURRO
Department of Health
  and Human Services
Food & Drug Administration
National Center for
  Toxicology Research
Jefferson, AR 72079
USA

L.E. UNRUH
National Center for
  Toxicology Research
Jefferson, AR 72079
USA

F.L. VANDEMARK
Perkin-Elmer Corporation
Norwalk, CT 06856
USA

D.L. VASSILAROS
Departments of Chemistry
  and Zoology
Brigham Young University
Provo, UT 84602
USA

K.P. VYAS
National Inst. of Arthritis,
  Diabetes, & Digestive and
  Kidney Diseases
Bethesda, MD 20205
USA

R.W. WALTERS
Department of Civil
  Engineering
Carnegie-Mellon University
Pittsburgh, PA 15213
USA

D.T. WANG
Laboratory Services Branch
Ontario Ministry of the
  Environment
P.O. Box 213
Rexdale, Ontario
Canada M9W 5L1

G.H. WEEKS
Department of Chemistry
University of Utah
Salt Lake City, UT 84112
USA

H.S. WEISS
Department of Physiology
College of Medicine
The Ohio State University
Columbus, OH 43210
USA

W.R. WEST
Departments of Chemistry
  and Zoology
Brigham Young University
Provo, UT 84602
USA

R. WESTERHOLM
Department of Analytical
  Chemistry
Arrhenius Laboratory
University of Stockholm
S-106 91 Stockholm
Sweden

C. WILLEY
Physical Sciences Department
Pacific Northwest Laboratory
Richland, WA 99352
USA

B.W. WILSON
Pacific Northwest Laboratory
Battelle Memorial Institute
Richland, WA 99352
USA

S.A. WISE
Organic Analytical Research
  Division
National Bureau of Standards
Washington, DC 20234
USA

D.T. WITIAK
Division of Medicinal
  Chemistry & Pharmacognosy
College of Pharmacy/The
  Comprehensive Cancer Center
The Ohio State University
Columbus, OH 43210
USA

T.K. WONG
Department of Pharmacology
School of Medicine
Uniformed Services University
  of the Health Sciences
Bethesda, MD 20814
USA

A.W. WOOD
Department of Biochemistry
  and Drug Metabolism
Hoffmann-La Roche, Inc.
Nutley, NJ 07110
USA

J. WOODROW
Department of Environmental
  Toxicology
University of California
Davis, CA 95616
USA

H. YAGI
Laboratory of Bioorganic Chemistry
National Institute of Arthritis,
  Diabetes, and Digestive and
  Kidney Diseases
National Institutes of Health
Bethesda, MD 20205
USA

C.S. YANG
Department of Biochemistry
New Jersey Medical-CMDNJ
Newark, NJ 07103
USA

S.K. YANG
Department of Pharmacology
School of Medicine
Uniformed Services University
  of the Health Sciences
Bethesda, MD 20814
USA

J. YEE
Department of Environmental
  Toxicology
University of California
Davis, CA 95616
USA

W.H. ZOLLER
Department of Chemistry
University of Maryland
College Park, MD 20742
USA

Acknowledgments

Coordinating a conference of this size, providing the technical guidance to ensure quality in the material presented, compiling and publishing the proceedings document is an exacting task. The diligence and dedication of many persons was required.

The sponsorship of the U.S. Environmental Protection Agency and the American Petroleum Institute gave the meeting relevance to modern society and made the conference possible. Dr. Charles Holdsworth (API) is gratefully acknowledged for his assistance in sponsoring the meeting. Several scientists in the U.S. Environmental Protection Agency's laboratory at the Research Triangle Park, North Carolina, provided technical leadership for the meeting. A special word of thanks is given to Dr. Larry D. Johnson (IERL/RTP) for his continuing assistance in organizing the meeting. The Environmental Sciences Research Laboratory (EPA/RTP) participated actively in program planning and execution. The meeting is deeply indebted to Dr. Roy L. Bennett, Dr. Alfred H. Ellison, Dr. Kenneth T. Knapp, and Dr. William E. Wilson for their advice and direction.

The Battelle conference coordination staff worked diligently for many months to provide the organization, planning, and communications necessary to make the meeting a success. Denise Cooley and Susan R. Armstrong deserve a special word of thanks for their hard work in putting together a meeting that provides an effective vehicle for researchers studying polynuclear aromatic hydrocarbons to come together and present their findings.

This manuscript is the product of many hours of diligent effort, not only by the scientists who assembled the manuscripts for the proceeds, but also the editorial staff who bring these documents together into a combined unit. Karen Rush was responsible for manuscript coordination at Battelle. Her care, attention to detail, and long hours of effort are greatly appreciated by everyone who uses this book.

Contents

POLYNUCLEAR AROMATIC HYDROCARBONS: EFFECTS OF
CHEMICAL STRUCTURE ON TUMORIGENICITY
    D. Hoffmann, E.J. Lavoie, and S.S. Hecht 1

THE BAY REGION THEORY: HISTORY AND CURRENT
PERSPECTIVES
    R.E. Lehr, A.W. Wood, W. Levin, A.H. Conney,
    D.R. Thakker, H. Yagi, and D.M. Jerina 21

COMPARATIVE REMOVAL OF POLYCYCLIC AROMATIC
HYDROCARBON-DNA ADDUCTS IN VIVO
    R.W. Hart, P.P. Fu, and M.J.W. Chang 39

EVALUATION OF EXTRACTION METHODS FOR CARBON
BLACK: POM ANALYSIS AND MUTAGENICITY ASSAY
    T. Alsberg, U. Rannug, U. Stenberg,
    and A. Sundvall 73

LASER EXCITED FLUORESCENCE AND CHROMATOGRAPHIC
TECHNIQUES FOR THE DETERMINATION OF PAH IN A
SPRAY OIL FLAME
    R. Barbella, F. Beretta, A. Ciajolo,
    and A. D'Alessio 83

HEPATIC TUMOR RATES AND POLYNUCLEAR AROMATIC
HYDROCARBON LEVELS IN TWO POPULATIONS OF
BROWN BULLHEAD (*ICTALURUS NEBULOSUS*)
    P.C. Baumann, W.D. Smith, and M. Ribick 93

METABOLISM OF PHENANTHRIDINE, AN AZA-ARENE
PRESENT IN LOW BTU GASIFIER EFFLUENTS
    J.M. Benson, R.E. Royer, and J.O. Hill 103

CHARACTERIZATION AND COMPARISON OF ORGANIC
EMISSIONS FROM COAL, OIL AND WOOD FIRED BOILERS
    J.G.T. Bergström, G. Eklund, and K. Trzcinski . . 109

METABOLISM OF BENZO(A)PYRENE BY SKIN MICROSOMES:
COMPARATIVE STUDIES IN C57BL/6N AND DBA/2N MICE
AND SPRAGUE-DAWLEY RATS
    D.R. Bickers, H. Mukhtar, and S.K. Yang 121

QUANTUM YIELDS OF THE PHOTODECOMPOSITION OF
POLYNUCLEAR AROMATIC HYDROCARBONS ADSORBED
ON SILICA GEL
   L. Blau and H. Güsten. 133

CARCINOGENICITY OF 3-METHYLCHOLANTHRENE
DERIVATIVES AND CYCLOPENTENO(CD)PYRENE
IN THE RAT MAMMARY GLAND
   E. Cavalieri and E. Rogan. 145

MICROBIAL OXIDATION OF 7-METHYLBENZ(A)ANTHRACENE
AND 7-HYDROXYMETHYLBENZ(A)ANTHRACENE
   C.E. Cerniglia, P.P. Fu, and S.K. Yang 157

INTERACTION OF BENZO(A)PYRENE AND CHRYSOTILE
   *M.J.W. Chang, N.P. Singh, A. Turturro,
   and R.W. Hart*. 167

COMPARISON OF BENZO(A)PYRENE METABOLISM AND
MUTATION INDUCTION IN CHO CELLS USING RAT
LIVER HOMOGENATE (S9) OR SYRIAN HAMSTER
EMBRYONIC CELL-MEDIATED ACTIVATION SYSTEMS
   D.J-C Chen, R.T. Okinaka, and G.F. Strniste. . . . 177

7-METHYLBENZO(A)PYRENE AND BENZO(A)PYRENE:
COMPARATIVE METABOLIC STUDY AND MUTAGENICITY
TESTING IN SALMONELLA TYPHIMURIUM TA100
   P-L. Chiu, T.K. Wong, P.P. Fu, and S.K. Yang . . . 183

A STRUCTURE-ACTIVITY RELATIONSHIP STUDY
OF MONOMETHYLBENZO(A)PYRENES BY THE USE
OF SALMONELLA TYPHIMURIUM TESTER STRAIN
TA100 AND BY ANALYSIS OF METABOLITE FORMATION
   P-L. Chiu and S.K. Yang. 193

SHPOL'SKII SPECTRA OF POLYCYCLIC AROMATIC
COMPOUNDS IN SAMPLES FROM CARBON BLACK
AND SOIL
   A.L. Colmsjö and C.E. Östman 201

THE POSSIBLE INVOLVEMENT OF POLYCYCLIC
HYDROCARBONS IN THE AETIOLOGY OF HUMAN
MAMMARY CANCER
   *C.S. Cooper, P.L. Grover, A. Hewer, M. O'Hare,
   A.D. MacNicoll, A.M. Neville, K. Pal, and
   P. Sims*. 211

THE DNA-BINDING OF A MONOHYDROXYMETHYL
METABOLITE FOLLOWING DMBA ADMINISTRATION
TO RATS
    F.B. Daniel, N.J. Joyce, and M.A. Drum 221

IDENTIFICATION AND QUANTIFICATION OF
PAH-DNA ADDUCTS BY LASER FLUORESCENCE
    *R.C. Davis, T.J. Facklam, W.A. Ivancic
and R.H. Barnes* 229

VERY HIGH-SPEED LIQUID CHROMATOGRAPHY
FOR PAH ANALYSIS: SYSTEM AND APPLICATIONS
    M.W. Dong, K. Ogan, and J.L. DiCesare 237

BINDING OF ORALLY ADMINISTERED
BENZO(A)PYRENE TO THE DNA OF MICE
OVER A DOSAGE RANGE OF 100,000
    B.P. Dunn 247

DETERMINATION OF PAH POLLUTION AT
COKE WORKS
    W. Eisenhut, E. Langer, and C. Meyer 255

COMPARATIVE METABOLISM *IN VITRO* OF
5-NITROACENAPHTHENE AND 1-NITRONAPHTHALENE
    K. El-Bayoumy and S.S. Hecht 263

USE OF MIXED PHASES FOR ENHANCED GAS-
CHROMATOGRAPHIC SEPARATION OF POLYCYCLIC
AROMATIC HYDROCARBONS. III. PHASE
TRANSITION BEHAVIOR, MASS-TRANSFER NON-
EQUILIBRIUM, AND ANALYTICAL PROPERTIES OF
A MESOGEN POLYMER SOLVENT WITH SILICONE
DILUENTS
    *H. Finkelmann, R.J. Laub, W.L. Roberts,
and C.A. Smith.* 275

IN VITRO METABOLISM OF 6-NITROBENZO(A)PYRENE:
IDENTIFICATION AND MUTAGENICITY OF ITS
METABOLITES
    *P.P. Fu, M.W. Chou, S.K. Yang, L.E. Unruh,
F.A. Beland, F.F. Kadlubar, D.A. Casciano,
R.H. Heflich, and F.E. Evans* 287

REAL-TIME ANALYSIS OF EXHAUST GASES
USING TRIPLE QUADRUPOLE MASS SPECTROMETRY
    J.E. Fulford, T. Sakuma, and D.A. Lane 297

DETECTION OF HIGH MOLECULAR POLYCYCLIC
AROMATIC HYDROCARBONS IN AIRBORNE
PARTICULATE MATTER USING MS, GC, AND GC/MS
    W. Funcke, T. Romanowski, J. Konig,
    and E. Balfanz. 305

REACTION PATHWAYS OF BENZO(A)PYRENE DIOL
EPOXIDE AND DNA. CONFORMATIONS OF ADDUCTS
    N.E. Geacintov. 311

IMPROVED METHODOLOGY FOR CARBON BLACK
EXTRACTION
    A.T. Giammarise, D.L. Evans, M.A. Butler,
    C.B. Murphy, D.K. Kiriazides, D. Marsh,
    and R. Mermelstein. 325

ANALYSIS OF BALANCE OF CARCINOGENIC IMPACT
FROM EMISSION CONDENSATES OF AUTOMOBILE
EXHAUST, COAL HEATING, AND USED ENGINE OIL
BY MOUSE-SKIN-PAINTING AS A CARCINOGEN-
SPECIFIC DETECTOR
    G. Grimmer, K.-W. Naujack, G. Dettbarn,
    H. Brune, R. Deutsch-Wenzel, and
    J. Misfeld. 335

ISOLATION AND IDENTIFICATION OF MUTAGENIC
PRIMARY AROMATIC AMINES FROM SYNTHETIC
FUEL MATERIALS
    D.A. Haugen, V.C. Stamoudis, M.J. Peak,
    and A.S. Boparai. 347

THE ULTIMATE FATES OF POLYCYCLIC
AROMATIC HYDROCARBONS IN MARINE
AND LACUSTRINE SEDIMENTS
    R.A. Hites and P.M. Gschwend. 357

THE POLYCYCLIC AROMATIC ENVIRONMENT OF
THE FLUIDIZED BED COAL COMBUSTION PROCESS-
AN INVESTIGATION OF CHEMICAL AND BIOLOGICAL
ACTIVITY
    G.T. Hunt, R.J. Kindya, R.R. Hall,
    P.F. Fennelly, and M. Hoyt. 367

COMPARISON OF THE METABOLIC PROFILES OF
PYRENE AND BENZ(A)ANTHRACENE IN RAT LIVER
AND LUNG BY GLASS CAPILLARY GAS CHROMATOGRAPHY/
MASS SPECTROMETRY
    J. Jacob, G. Grimmer, and A. Schmoldt 383

THE EFFECT OF ASBESTOS ON THE
BIOACTIVATION OF BENZO(A)PYRENE
  C. Kandaswami, V.V. Subrahmanyam, and
  P.J. O'Brien. 389

MOLECULAR SPECTRA OF POLYCYCLIC
AROMATIC HYDROCARBONS
  W. Karcher, R.J. Fordham, A. Nelen,
  R. DePaus, J. Dubois, and Ph. Glaude. 405

POLYCYCLIC AROMATIC HYDROCARBONS IN
LETTUCE. INFLUENCE OF A HIGHWAY AND
AN ALUMINUM SMELTER
  B. Larsson and G. Sahlberg. 417

IDENTIFICATION AND MUTAGENICITY OF
NITROGEN-CONTAINING POLYCYCLIC
AROMATIC COMPOUNDS IN SYNTHETIC FUELS
  D.W. Later, M.L. Lee, R.A. Pelroy,
  and B.W. Wilson 427

HIGH PERFORMANCE SEMI-PREPARATIVE LIQUID
CHROMATOGRAPHY AND LIQUID CHROMATOGRAPHY-
MASS SPECTROMETRY OF DIESEL ENGINE
EMISSION PARTICULATE EXTRACTS
  S.P. Levine, L.M. Skewes, L.D. Abrams,
  and A.G. Palmer, III. 439

THE RELATIVE CONTRIBUTION OF PNAs TO
THE MICROBIAL MUTAGENICITY OF RESPIRABLE
PARTICLES FROM URBAN AIR
  J. Lewtas, A. Austin, L. Claxton,
  R. Burton, and R. Jungers 449

EFFECT OF CATALYST ON PAH CONTENT OF
AUTOMOTIVE DIESEL EXHAUST PARTICULATE
  I.E. Lichtenstein 461

MEASUREMENT OF POTENTIALLY HAZARDOUS
POLYNUCLEAR AROMATIC HYDROCARBONS FROM
OCCUPATIONAL EXPOSURE DURING ROOFING
AND PAVING OPERATIONS
  M. Malaiyandi, A. Benedek, A.P. Holko,
  and J.J. Bancsi 471

THE POLYNUCLEAR AROMATIC HYDROCARBON
CONTENT OF SMOKED FOODS IN THE
UNITED KINGDOM
  A.S. McGill, P.R. Mackie, E. Parsons,
  C. Bruce, and R. Hardy. 491

MAGNETIC CIRCULAR DICHROISM AS AN
ANALYTICAL TOOL FOR SUBSTITUTED
POLYNUCLEAR AROMATIC HYDROCARBONS
  J. Michl and G.H. Weeks 501

BINDING OF BENZO(A)PYRENE-DIOL EPOXIDES
AND -TRIOL CARBONIUM IONS TO DNA
  K.J. Miller, J.J. Burbaum, and J. Dommen. 515

REPRESENTATIONS OF BINDING STIES FOR PAHs
IN DNA WITH COMPUTER GRAPHICS TECHNIQUES
  K.J. Miller, P.J. Kowalczyk, and W. Segmuller . . 529

NECESSITY FOR EXTRANUCLEAR METABOLISM?
AN ALTERNATE SCENARIO FOR BaP AND/OR
REDUCED A-RING 7,12-DMBA-INDUCED
TRANSFORMATION OF HUMAN FIBROBLASTS,
IN VITRO
  G.E. Milo, T.O. Mason, D.T. Witiak,
  M. Inbasekaran, F.D. Cazer, and W.R. Gower. . . . 537

PROPERTIES OF ^{3}H-LABELED 1-NITROPYRENE
DEPOSITED ONTO COAL FLY ASH
  A. Mosberg, G. Fisher, D. Mays, R. Riggin,
  M. Schure, and J. Mumford 551

AMBIENT PARTICULATE AND BENZO(A)PYRENE
CONCENTRATIONS FROM RESIDENTIAL WOOD
COMBUSTION, IN A MOUNTAIN COMMUNITY
  D.J. Murphy, R.M. Buchan, and D.G. Fox. 567

OXIDATION OF PHENANTHRENE WITH A
CARBONYL OXIDE
  R.W. Murray and S. Kumar. 575

COMPARISON OF THE SKIN TUMOR INITIATING
ACTIVITIES OF EMISSION EXTRACTS IN THE
SENCAR MOUSE
  S. Nesnow, L.L. Triplett, and T.J. Slaga. 585

AN INVESTIGATION OF FACTORS
INFLUENCING THE SEPARATION OF
POLYNUCLEAR AROMATIC HYDROCARBONS
BY LIQUID CHROMATOGRAPHY
    U.D. Neue, B.P. Murphy, and J. Crooks 597

COMPARISON OF NITRO-AROMATIC CONTENT
AND DIRECT-ACTING MUTAGENICITY OF
DIESEL EMISSIONS
    M.G. Nishioka, B.A. Petersen, and J. Lewtas. . . . 603

AN EVALUATION OF PAH CONTENT,
MUTAGENICITY AND CYTOTOXICITY
OF RICE STRAW SMOKE
    H. Olsen, J. Yee, T. Mast, J. Woodrow,
    G. Fisher, J. Seiber, and D. Hsieh 615

MUTAGENIC ACTIVATION AND INACTIVATION
OF NITRO-PAH COMPOUNDS BY MAMMALIAN ENZYMES
    T.C. Pederson and J-S. Siak. 623

COMPARISON OF BENZO(A)PYRENE AND DIESEL
PARTICULATE EXTRACT IN THE SISTER CHROMATID
EXCHANGE ASSAY *IN VIVO* AND *IN UTERO* AND THE
MICRONUCLEUS ASSAY
    M.A. Pereira, L. McMillan, P. Kaur,
    and P.S. Sabharawal. 633

ANALYSIS OF PAH IN DIESEL EXHAUST PARTICULATE
BY HIGH RESOLUTION CAPILLARY COLUMN GAS
CHROMATOGRAPHY/MASS SPECTROMETRY
    B.A. Petersen, C.C. Chuang, T.L. Hayes,
    and D.A. Trayser 641

SIMPLIFIED REAL-TIME PAH MEASUREMENT
TECHNIQUES
    T.L. Pierce and D.R. Jaasma. 655

THE METABOLISM OF NITRO-SUBSTITUTED
POLYCYCLIC AROMATIC HYDROCARBONS IN
SALMONELLA TYPHIMURIUM
    M.A. Quilliam, F. Messier, C. Lu, P.A. Andrews,
    B.E. McCarry, and D.R. McCalla 667

RATIONAL SELECTION OF SAMPLE PREPARATION
TECHNIQUES FOR THE MEASUREMENT OF
POLYNUCLEAR AROMATICS
    W.K. Robbins and F.C. McElroy. 673

THE EFFECT OF 5,6-BENZOFLAVONE ON
THE MUTAGENICITY OF PAH
    M.F. Salamone, P. Beltz, and M. Katz 687

SYNTHESIS OF PAH CONTAINING
CYCLOPENTA-FUSED RINGS
    R. Sangaiah, A. Gold, G.E. Toney, S.H. Toney,
    R. Easterling, L.D. Claxton, and S. Nesnow 695

MUTAGENESIS TESTING OF GASOLINE
ENGINE OILS
    C.A. Schreiner and C.R. Mackerer 705

THE EFFECT OF TEMPERATURE ON THE
ASSOCIATION OF POM WITH AIRBORNE
PARTICLES
    M.R. Schure and D.F.S. Natusch 713

ENHANCED BENZO(A)PYRENE METABOLISM IN
HAMSTER EMBRYONIC CELLS EXPOSED IN
CULTURE TO FOSSIL SYNFUEL PRODUCTS
    D.D. Schuresko, G.D. Griffin, M.C. MacLeod,
    and J.K. Selkirk 725

ANALYSIS OF PAH METABOLITES BY NOVEL
TRIPLE QUADRUPOLE MASS SPECTROMETRY
    B.I. Shushan, T. Sakuma, D.A. Rokosh,
    and M.F. Salamone 735

DIOL-EPOXIDE REACTIVITY OF METHYLATED
POLYCYCLIC AROMATIC HYDROCARBONS (PAH):
RANKING THE REACTIVITY OF THE POSITIONAL
MONOMETHYL ISOMERS
    B.D. Silverman and J.P. Lowe 743

ON-LINE MULTIDIMENTIONAL LIQUID
CHROMATOGRAPHIC DETERMINATION OF
POLYCYCLIC ORGANIC MATERIAL IN
COMPLEX SAMPLES
    W.J. Sonnefeld, W.H. Zoller, W.E. May,
    and S.A. Wise 755

ENRICHMENT OF PAH AND PAH DERIVATIVES
FROM AUTOMOBILE EXHAUSTS, BY MEANS OF
A CRYO-GRADIENT SAMPLING SYSTEM
    U. Stenberg, R. Westerholm. T. Alsberg,
    U. Rannug, and A. Sundvall 765

SUNLIGHT ACTIVATION OF SHALE OIL
BYPRODUCTS AS MEASURED BY GENOTOXIC
EFFECTS IN CULTURED CHINESE HAMSTER CELLS
  G.F. Strniste, D.H. Chen, and R.T. Okinaka. . . . 773

CHEMICAL AND ENZYMATIC OXIDATIONS
OF SUBSTITUTED BENZO(A)PYRENES
  P.D. Sullivan, L.E. Ellis, L.M. Calle,
  and I.J. Ocasio 779

EFFECTS OF BUTYLATED HYDROXYANISOLE
ON THE METABOLISM OF BENZO(A)PYRENE
  W. Sydor, Jr., K.F. Lewis, R. Peng,
  and C.S. Yang 791

BENZO(A)PYRENE METABOLISM IN HEPATIC S-9
FRACTIONS OF AROCLOR 1254-TREATED MULLET
(MUGIL CEPHALUS)
  B. Tan and P. Melius. 801

MULTICOMPONENT ISOLATION AND ANALYSIS
OF POLYNUCLEAR AROMATICS
  B.A. Tomkins, W.H. Griest, J.E. Caton,
  and R.R. Reagan 813

EFFECTS OF TRANSPLACENTALLY ADMINISTERED
POLYAROMATIC HYDROCARBONS ON THE GENOME
OF DEVELOPING AND ADULT RATS MEASURED BY
SISTER CHROMATID EXCHANGE
  A. Turturro, N.P. Singh, M.J.W. Chang,
  and R.W. Hart 825

THE APPLICATION OF HIGH RESOLUTION
PREPARATIVE LIQUID CHROMATOGRAPHY TO
THE POLYCYCLIC AROMATIC HYDROCARBONS
  F.L. Vandemark and J.L. DiCesare. 835

DETERMINATION AND BIOCONCENTRATION OF
POLYCYCLIC AROMATIC SULFUR HETEROCYCLES
IN AQUATIC BIOTA
  D.L. Vassilaros, D.A. Eastmond, W.R. West,
  G.M. Booth, and M.L. Lee. 845

STEREOSELECTIVITY IN THE METABOLISM,
MUTAGENICITY AND TUMORIGENICITY OF THE
POLYCYCLIC AROMATIC HYDROCARBON CHRYSENE
  K.P. Vyas, H. Yagi, D.R. Thakker, R.L. Chang,
  A.W. Wood, W. Levin, A.H. Conney, and
  D.M. Jerina 859

MODEL FOR PREDICTING ADSORPTION OF
PAH FROM AQUEOUS SYSTEMS
   R.W. Walters and R.G. Luthy 873

OCCURRENCE AND POTENTIAL UPTAKE OF
POLYNUCLEAR AROMATIC HYDROCARBONS OF
HIGHWAY TRAFFIC ORIGIN BY PROXIMALLY
GROWN FOOD CROPS
   D.T. Wang and O. Meresz 885

FACTORS MODIFYING THE EFFECT OF
AMBIENT TEMPERATURE ON TUMORIGENESIS
IN MICE
   H.S. Weiss, J.F. O'Connell, J.F. Pitt,
   R.E. Miller, and D.E. Graff 897

COMPARATIVE ANALYSIS OF POLYCYCLIC
AROMATIC SULFUR HETEROCYCLES ISOLATED
FROM FOUR SHALE OILS
   C. Willey, R.A. Pelroy, and D.L. Stewart 907

ANALYTICAL METHODS FOR THE DETERMINATION
OF POLYCYCLIC AROMATIC HYDROCARBONS ON
AIR PARTICULATE MATTER
   S.A. Wise, S.L. Bowie, S.N. Chesler,
   W.F. Cuthrell, W.E. May, and R.E. Rebbert 919

METABOLISM OF BAY REGION TRANS-DIHYDRODIOLS
TO VICINAL DIHYDRODIOL EPOXIDES
   S.K. Yang, M.W. Chou, and P.P. Fu 931

POLYNUCLEAR AROMATIC HYDROCARBONS: EFFECTS OF CHEMICAL STRUCTURE ON TUMORIGENICITY*

DIETRICH HOFFMANN, EDMOND J. LAVOIE, AND STEPHEN S. HECHT
Naylor Dana Institute for Disease Prevention, American Health Foundation, Valhalla, New York 10595.

This publication is dedicated to the founder of the American Health Foundation, Dr. Ernst L. Wynder, on the occasion of the 10th anniversary of the Naylor Dana Institute for Disease Prevention.

INTRODUCTION

The necessity of holding International Symposia on Polynuclear Aromatic Hydrocarbons (PAH) annually has often been questioned. Would not one meeting every second or third year suffice to permit free exchange of scientific information regarding the formation, environmental analysis, metabolic fate and carcinogenicity of PAH? We believe that there is no clear cut answer to this question especially in view of the many and varied aspects involved in such an annual symposium, but we prefer an affirmative answer for a number of scientific reasons.

While much of our knowledge about the environmental and biological aspects of PAH-carcinogenesis has emerged during recent decades, the carcinogenic effects of PAH have indirectly and directly been recorded since 1775 (1-4). Basic studies on the pyrosynthesis, analysis, metabolic activation and genotoxic effects of PAH have greatly influenced and advanced our understanding of environmental toxicity and of chemical carcinogenesis per se. Recent experiments with PAH have significantly enhanced our comprehension of the events occuring during various stages of carcinogenesis on the cellular level. Technological advances have made it possible for us to study methods and agents which can potentially inhibit PAH-induced carcinogenesis on the molecular level (5-9).

---

*Presented at the Sixth International Symposium on Polynuclear Aromatic Hydrocarbons, Columbus, Ohio, October 27-30, 1981.

PAH: EFFECTS OF CHEMICAL STRUCTURE ON TUMORIGENICITY

Since the first PAH-Symposium in 1975, a scientific basis for practical studies on the prevention of PAH carcinogenesis has developed, at least partly because the field of PAH-carcinogenesis has attracted many young scientists. If we can sustain the momentum and enthusiasm with which studies of PAH-carcinogenesis are pursued, our understanding of chemical carcinogenesis will certainly progress to such a degree that a reduction of the biological effects of these agents will be feasible. Toward this end, the annual PAH Symposia are needed.

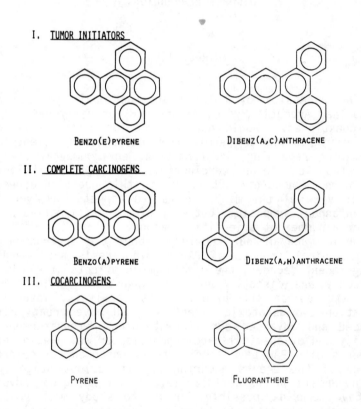

FIGURE 1. PAH with varying roles in tumor development.

PAH: EFFECTS OF CHEMICAL STRUCTURE ON TUMORIGENICITY

PAH as Tumorigenic Agents

About 20 years ago, Badger et al. demonstrated with ^{14}C-labelled precursors that practically all organic agents give rise to trace amounts of PAH upon incomplete combustion (6). At that time, it was realized that three-ring and higher aromatic hydrocarbons can be active as tumor initiators, as complete carcinogens, or as cocarcinogens. Several environmental PAH, such as dibenz(a,c)anthracene and benzo(e)pyrene are inactive as complete carcinogens, but are active as tumor initiators (10; Figure 1). Tumor initiators lead to tumor development only when the treated tissue is subsequently exposed to a promoter and/or when they are applied together with a cocarcinogen over a longer time period. Generally, neither promoters nor cocarcinogens induce tumors by themselves. A PAH is considered to be a complete carcinogen when its single or repeated applications induce benign and malignant tumors in epithelial cells and/or connective tissues.

A large number of methylated, as well as parent PAH with four, five or six condensed aromatic rings are complete carcinogens and exhibit their relative potency by induction of tumors in a clear dose-response relationship (11). When such compounds are applied to epithelial tissues, such as mouse skin, in "subthreshold doses", they are generally still sufficiently active as tumor initiators, even in concentrations as low as one-hundredth of the carcinogenic dose. The degree of tumor yield obtained with these initiators depends then largely on the potency of a tumor promoter or cocarcinogen as a stimulant of neoplastic development.

Tumor initiation by PAH, which underlies the effects in complete carcinogenesis, results from irreversible changes in cellular macromolecules. Such changes are caused by reactions with electrophiles formed during the metabolism of PAH. The development and progression of the latent or dormant tumor cell into differentiated tumors or undifferentiated forms of cancer is realized by promoting agents and/or cocarcinogens (12; Figure 2). At present, little is known about the cellular events involved in the stimulation of initiated cells by cocarcinogenic PAH. The complexity of the situation is illustrated in Figure 3 which shows that some PAH can act as cocarcinogens whereas others may have inhibitory effects.

PAH: EFFECTS OF CHEMICAL STRUCTURE ON TUMORIGENICITY

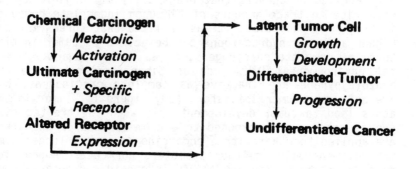

FIGURE 2. Sequence of complex events during chemical carcinogenesis. (Weisburger, 1976).

Mechanisms of Metabolic Activation of PAH

The activity of PAH as carcinogens or as tumor initiators is dependent on metabolic activation processes in the host and is counteracted, or at least diminished, by detoxification reactions. Upon resorption, the PAH undergo a complex spectrum of metabolic changes. The role of the various metabolites in activation and detoxification has been the subject of intensive studies in recent years.

Several research teams primarily in the U.S.A. and England have established the pathways of BaP metabolism as shown in Figure 4. One enantiomer of the anti-BaP-7,8-dihydrodiol-9,10-epoxide appears to be a major ultimate carcinogenic form of BaP. Interaction of this metabolite

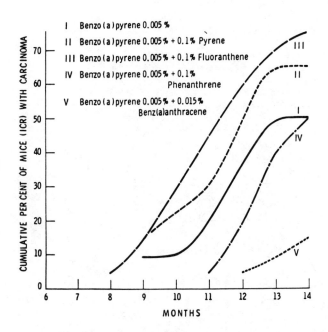

FIGURE 3. Cocarcinogenic and inhibiting effect of PAH. (Hoffmann and Wynder, 1963). See reference 34.

with DNA gives a deoxyguanosine adduct, as shown in Figure 5 (9). This adduct is a major product of DNA modification in numerous tissues exposed to BaP. These observations gave rise to the bay region theory of Jerina, Conney and co-workers, which predicts that bay region dihydrodiol epoxides of other PAH, if formed, could be major ultimate carcinogens (See Figure 6; 8). Extensive studies by several groups have supported this hypothesis for a number of PAH but it is becoming clear that there are complexities in PAH activation that are not explained by the bay region theory (13,14).

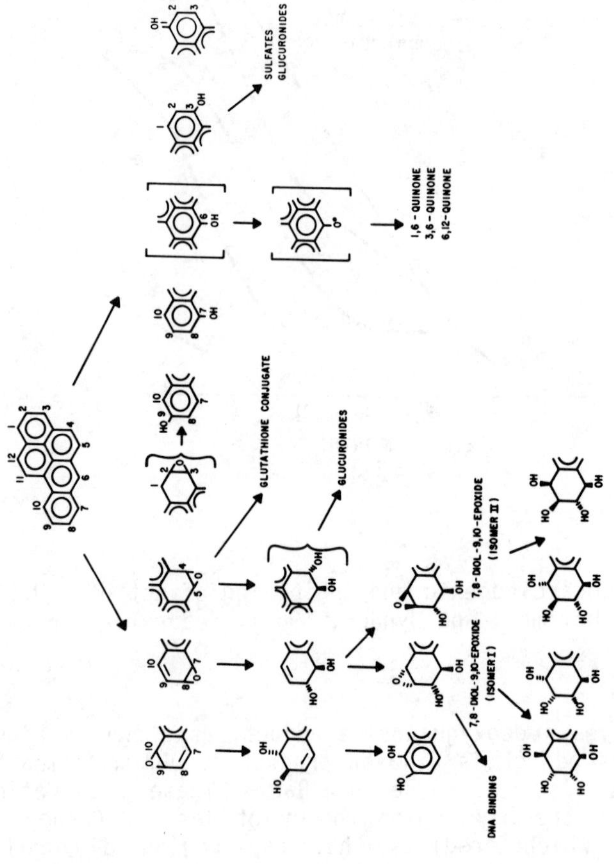

FIGURE 4. Metabolism of Benzo(a)pyrene. See reference 35.

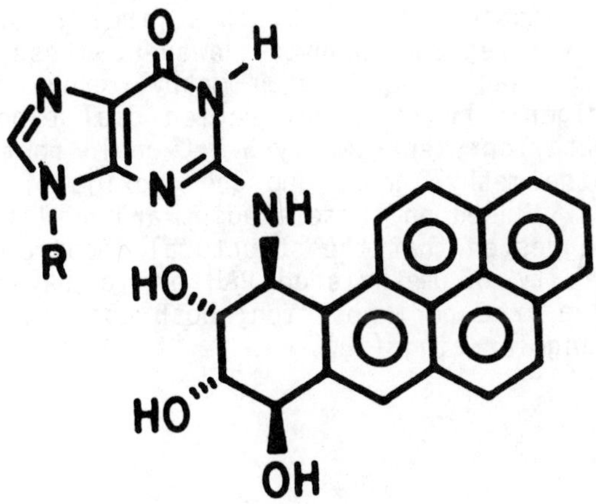

FIGURE 5. Structure of the guanosine-BaP adduct formed from the reaction of the <u>anti</u>-7,8-diol-9,10-epoxide of BaP with poly (G).

Our own studies on the structural aspects of PAH carcinogenicity started in 1972 with the observation that cigarette smoke contains chrysene and 5 of the 6 possible monomethylchrysenes (15). Bioassays on mouse skin of synthetic chrysene and monomethylchrysenes revealed that 5-methylchrysene (5-MeC) is a strongly carcinogenic alkyl-derivative and that its activity is comparable to that of BaP, while the other isomeric methylchrysenes and chrysene itself have only weak or marginal carcinogenic activity (Figure 7). Subsequently, we synthesized and bioassayed fluorinated 5-methylchrysenes and found that out of 7 tested fluoro-derivatives, 1-, 3- and 12-fluoro-5-MeC had less tumorigenic activity than 5-MeC (16; Table 1). HPLC traces of the metabolites formed from 1-F-5-MeC and 3-F-5-MeC showed only a minor peak corresponding in retention volume to 1,2-dihydrodiols of 5-McC (16; Figure 8).

Finally, we assayed the tumor initiating activity on mouse skin of the 5-MeC dihydrodiols (17). As shown in Figure 9, the 1,2-dihydrodiol was more active than 5-MeC whereas the 7,8-dihydrodiol was less tumorigenic than 5-MeC; the 9,10-dihydrodiol was inactive.

These results show that the 1,2-dihydrodiol is a major proximate tumorigen of 5-MeC. This finding is in agreement with the bay region theory. However, these studies also demonstrate that a bay region methyl group could enhance the tumorigenicity of a bay region diol expoxide. Among the monomethylchrysenes, only 5-MeC could form a diol epoxide with the methyl group and the epoxide ring in the same bay region. Based on these studies and on literature data, we have suggested that the structural requirements favoring tumorigenicity of methylated PAH are a bay-region methyl group and a free peri-position, both adjacent to an unsubstituted angular ring (18).

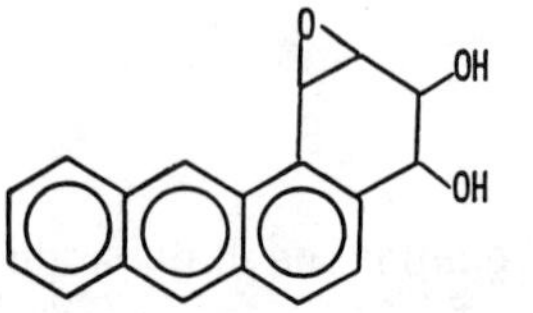

BENZ(A)ANTHRACENE-3,4-DIOL-1,2-EPOXIDE

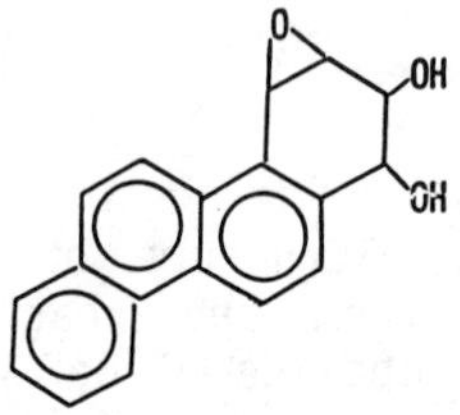

CHRYSENE-1,2-DIOL-3,4-EPOXIDE

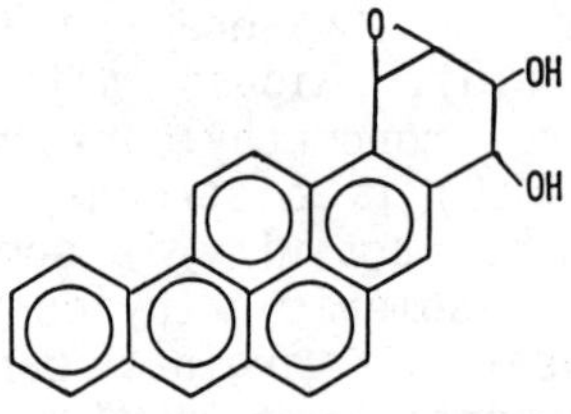

DIBENZO(A,I)PYRENE-3,4-DIOL-1,2-EPOXIDE

FIGURE 6. Diol epoxides predicted as most reactive form of 3 PAH.

TABLE 1 COMPLETE CARCINOGENICITY OF FLUORINATED DERIVATIVES OF 5-MeC

Compound	Percent of skin tumor-bearing animals[a]	No. of skin tumors per animal	Total No. of skin tumors	
			Benign	Malignant
1-F-5-MeC[b]	5[c]	0.1	1	0
3-F-5-MeC[b]	0[c]	0	0	0
6-F-5-MeC[b]	100	3.2	39	25
7-F-5-MeC[d]	70	1.2	11	13
9-F-5-MeC[d]	75	1.3	11	15
11-F-5-MeC[d]	85	1.3	15	11
12-F-5-MeC[d]	25[c]	0.3	2	3
5-MeC[b]	80	2.0	26	14
5-MeC[d]	80	1.2	12	12

[a]Each group consisted of 20 animals. [b]Assay I. [c]Significant; P<0.01. [d]Assay II. Hecht et al., 1979. See reference 16.

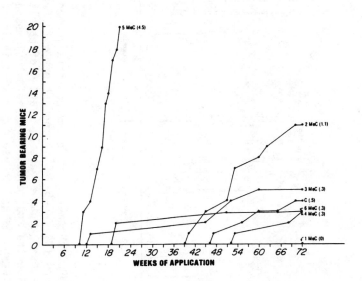

FIGURE 7. Bioassay of the methylchrysenes as complete carcinogens at a dose of 0.1 mg of each compound three times weekly. Numbers in parentheses represent tumors/animal.

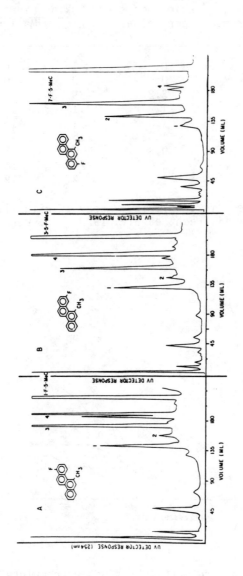

FIGURE 8. HPLC chromatograms of metabolites formed from (A) 1-F-5-MeC; (B) 3-F-5-MeC; and (C) 7-F-5-MeC in incubations with Aroclor-induced rat liver 9,000 x g supernatant. Peaks 1, 3 and 4 of panels A and B were identified by their UV spectra and relative retention volumes as 7,8-dihydrodiols, hydroxymethyl-chrysenes and chrysenols respectively of 1-F-5-MeC and 3-F-5-MeC. Peaks 2 and 3 of panel C were similarly identified as the 1,2-dihydrodiol and hydroxymethyl derivatives of 7-F 5-MeC. For details see reference 16.

Studies with methylated phenanthrenes have demonstrated that in addition to the presence of a bay-region methyl group, inhibition of metabolism at the 9,10-position is essential for tumorigenicity. Thus 4-methylphenanthrene satisfies the above structural requirements which favor tumorigenicity, but was inactive as a tumor initiator (19). Studies on the metabolism of 4-methylphenanthrene have shown that 9,10-dihydro-9,10-dihydroxy-4-methylphenanthrene is a major metabolite. Similar results have been observed in metabolism studies with phenanthrene (20,21).

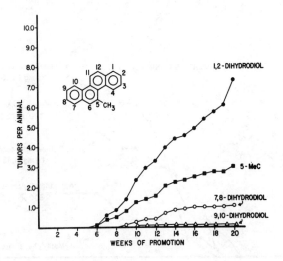

FIGURE 9. Tumor initiating activity of 5-MeC and three dihydrodiol metabolites. From reference 17.

Inhibition of 9,10-dihydrodiol formation has been observed for phenanthrenes which contained methyl groups at or adjacent to this K-region (19). 1,4-Dimethylphenanthrene and 4,10-dimethylphenanthrene when assayed as tumor initiators on mouse skin were found to be active (Figure 10). Both of these substituted phenanthrenes have a bay-region methyl group and a free peri-position, both adjacent to an unsubstituted angular ring in addition to a methyl substituent at or adjacent to the K-region of phenanthrene to inhibit 9,10-dihydrodiol formation. 4,9-Dimethylphenanthrene, which has a peri methyl substituent adjacent to the free angular ring, was inactive as a tumor initiator on mouse skin.

The results of comparative studies on the in vitro metabolism of dimethylphenanthrenes suggest that dihydrodiol precursors to bay-region diol epoxides are associated with the tumor-initiating activity of 1,4- and 4,10-dimethylphenanthrene. Since synthetic 1,2-dihydro-1,2-dihydroxyphenanthrene was not active as a tumor initiator on mouse skin (22), the presence of a bay region methyl group seems to be essential. The mechanism by which a bay-region methyl group alters tumorigenic potential has not been elucidated, but it appears to be an essential structural requirement for tumorigenicity of methylated phenanthrenes on mouse skin.

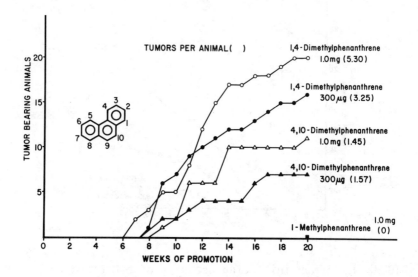

FIGURE 10. Tumor initiating activities on mouse skin of 1,4-dimethylphenanthrene, 4,10-dimethylphenanthrene, and 1-methylphenanthrene after initiating doses of 0.3 or 1.0 mg and promotion by 3 times weekly application of tetradecanoylphorbol acetate.

The inhibition of specific metabolic pathways as observed for methylated phenanthrenes may be a major structural requirement favoring the carcinogenicity of other PAH. Bioassays have shown for example that 4-,5-, and 6-methylbenzo(c)phenanthrene are the most tumorigenic isomers of methylbenzo(c)phenanthrene (23,24; Figure 11).

The high mutagenic and tumorigenic activity of 3,4-dihydro-3,4-dihydroxybenzo(c)phenanthrene, the precursor to a bay region dihydrodiol epoxide, has been demonstrated (25,-26). The 5,6-dihydrodiol of benzo(c)phenanthrene was inactive as a tumorigen. Inhibition of 5,6-dihydrodiol formation may, therefore, be directly associated with the tumorigenic potential of 4-, 5-, and 6-methylbenzo(c)phenanthrene.

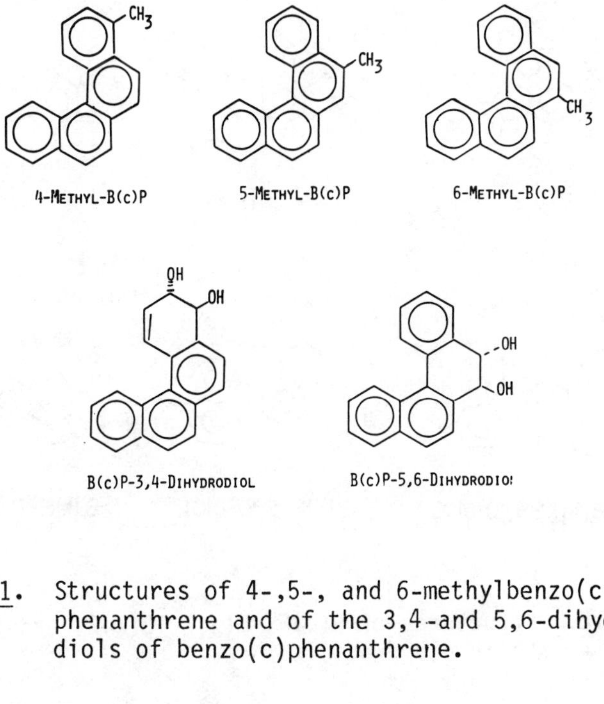

FIGURE 11. Structures of 4-,5-, and 6-methylbenzo(c)phenanthrene and of the 3,4-and 5,6-dihydrodiols of benzo(c)phenanthrene.

A group of widely occurring environmental PAH are the benzofluoranthenes (27-29). Benzo(b)fluoranthene [B(b)F] and benzo(j)fluoranthene [B(j)F] are strongly carcinogenic on mouse skin, while benzo(k)fluoranthene [B(k)F] is a weak tumor initiator and benzo(ghi)fluoranthene [B(ghi)F] is inactive as a complete carcinogen and as a tumor initiator. Studies on the metabolic activation of B(b)F have shown that the 9,10-dihydrodiol (Figure 12), the requisite precursor for the formation of a bay region dihydrodiol epoxide, is as active as B(b)F as a tumor initiator (30). However, this dihydrodiol has not been identified as a metabolite of B(b)F in studies on its in vitro metabolism in rat liver homogenates (31). The principal dihydrodiol metabolites of B(b)F have been identified as B(b)F-11,12-diol and B(b)F-1,2-diol (31). The possible involvement of B(b)F-9,10-diol-11,12-epoxide as an ultimate carcinogen of B(b)F is currently being investigated by DNA binding studies in mouse skin.

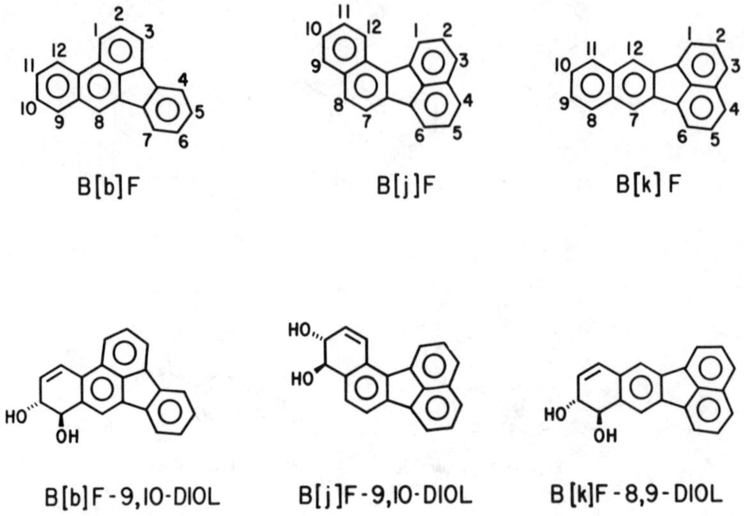

FIGURE 12. Structures of B(b)F, B(j)F, B(k)F, and three of their dihydrodiols.

B(j)F-9,10-diol, which could form a "pseudo bay region" diol epoxide, was more mutagenic than B(j)F toward S. typhimurium TA 100, in the presence of rat liver homogenate (32). It was also formed as a metabolite of B(j)F in vitro and can, therefore, be considered as one proximate mutagen of B(j)F. However, the tumorigenicity of B(j)F on mouse skin cannot be explained by B(j)F-9,10-diol alone, because it showed relatively low activity compared to B(j)F (30).
The lower tumor initiating activity of B(j)F-9,10-diol than that of B(j)F indicates that a bay region dihydrodiol

epoxide in a four sided bay region may not necessarily have high tumorigenic potential. This is in contrast to results with benzo(c)phenanthrene, which has a similar but not identical type of bay region (fjord; Figure 11). The 3,4-dihydrodiol of benzo(c)phenanthrene and the corresponding dihydrodiol epoxides were more tumorigenic than the parent hydrocarbon (25). These contrasting results in the B(j)F and benzo(c)phenanthrene systems are probably due to differences in carbonium ion stabilities, as well as to stereochemical factors. Our earlier studies have shown that a second mutagenic dihydrodiol, so far unidentified, is also formed in the metabolism of B(j)F in vitro (32). It is possible that this dihydrodiol may contribute to the tumorigenicity of B(j)F.

The major dihydrodiol metabolite of B(k)F which is formed in vitro is 8,9-dihydro-8,9-dihydroxy-B(k)F. While this dihydrodiol was active as a mutagen in S. typhimurium TA 100, it was inactive as a tumor initiator on mouse skin (30,32). The inactivity of B(k)F-8,9-diol as a tumor initiator was not surprising, in view of results with a structurally related linear diol epoxide, 8,9-dihydro-8,9-dihydroxybenz(a)anthracene, which was also inactive (33). Similar to B(j)F-9,10-diol, B(k)F-8,9-diol can be considered to be one proximate mutagen of B(k)F, but is clearly not a proximate tumorigen. It is possible that the 8,9-monoepoxide of B(k)F may be responsible for its tumor initiating activity.

ACKNOWLEDGEMENTS

Our studies are supported by National Cancer Institute Cancer Center Support (Core) Grant 5 R30 CA17613 and 1 R01 CA29580 and National Institute of Environmental Health Sciences Grant 5 R01 ES02030.

We thank Mrs. C. Hickey, Mrs. L. Landy and Mrs. B. Stadler for assistance in the preparation of this paper.

REFERENCES

1. Pott, P. (1775): Cancer scroti. In "The Chirurgical Works of Percival Pott, F.R.S." L. Hawes, W. Clarke and R. Collins, eds., London, pp 734-736.
2. Yamagiwa, K., Ichikawa, K. (1915): Experimentelle Studie über die Pathogenese der Epithelialgeschwülste Mitt. Med. Fak. Tokyo 15: 295-344.
3. Cook, J.W., Hewett, C.L. and Hieger, I. (1933): The isolation of a cancer producing hydrocarbon from coal tar - Parts I, II and III. J. Chem. Soc. Trans. 1: 395-405.
4. Berenblum, I. and Shubik, P. (1947): Role of croton oil applications associated with single painting of carcinogen in tumor induction of mouse skin. Brit. J. Cancer 1: 379-382.
5. Newman, M.S. (1976): Carcinogenic activity of benz-(a)anthracenes. In "Carcinogenesis", Vol. 1 Polynuclear Aromatic Hydrocarbons: Chemistry, Metabolism, Carcinogenesis. R.I. Freudenthal and P.W. Jones, eds. Raven Press, New York, pp 203-207.
6. Badger, G.M. (1962): Mode of formation of carcinogens in the human environment. Natl. Cancer Inst. Monogr. 9: 1-16.
7. Sims, P. and Grover, P.L. (1974) Epoxides in polycyclic aromatic hydrocarbon metabolism and carcinogenesis. Advan. Cancer Res. 20: 165-274.
8. Jerina, D.M., Lehr, R., Schaefer-Ridder, M., Yagi, H., Kale, J.M., Thjakker, D.R., Wood, A.W., Lu, A.Y.H., Ryan, D., West, S., Levin, W., and Conney, A.H. (1977): Bay region epoxides of dihydrodiols: A concept which explains the mutagenic and carcinogenic activity of benzo(a)pyrene and benz(a)anthracene. In "Origins of Human Cancer" (H. Hiatt, J.D. Watson, and I. Winsten, eds.) Cold Spring Harbor, New York, pp 639-658.
9. Weinstein, I.B., Jeffrey, A.M., Jenette, K.W., Blotstein, S.H., Harvey, R.G., Harris, C., Autrup, H., Kasai, H., and Nakanishi, K. (1976): Benzo(a)pyrene diol epoxides as intermediates in nucleic acid binding in vitro and in vivo. Science 193: 592-595.
10. Van Duuren, B.L., Sivak, A., Langseth, L., Goldschmidt, B.M., and Segal, A. (1968): Initiators and promoters in tobacco carcinogenesis. Natl. Cancer Inst. Monogr. 28: 173-180.

11. Dipple, A. (1976): Polynuclear Aromatic Carcinogens. In "Chemical Carcinogens" C.E. Searle, ed. Am. Chem. Soc. Monogr. 17: 245-314.
12. Weisburger, J.H. (1976): Bioassays and tests for chemical carcinogens. In "Chemical Carcinogens" C.E. Searle, ed. Am. Chem. Soc. Monogr. 173: 1-23.
13. Nordquist, M., Thakker, D.R., Yagi, H., Lehr, R.E., Wood, A.W., Levin, W., Conney, A.H., and Jerina, D.M. (1980): Evidence in support of the bay-region theory as a basis for the carcinogenic activity of polycyclic aromatic hydrocarbons. In "Molecular Basis of Environmental Toxicity" R.S. Bhatnager, ed., Ann Arbor Science Publishers, Ann Arbor, MI, pp. 329-357.
14. Sims, P. and Grover, P.L. (1981) The involvement of dihydrodiols and diol-epoxides in the metabolic activation of polycyclic hydrocarbons other than benzo[a]pyrene. In "Polycyclic Hydrocarbons and Cancer", Vol. 3, H.V. Gelboin and P.O.P. T'so, eds., Academic Press, New York, pp. 117-181.
15. Hecht, S.S, Bondinell, W.E., and Hoffmann, D. (1974): Chrysene and methylchrysenes: Presence in tobacco smoke and carcinogenicity. J. Natl. Cancer Inst. 53: 1121-1133.
16. Hecht, S.S., LaVoie, E., Mazzarese, R., Hirota, N., Ohmori, T. and Hoffmann, D. (1979): Comparative mutagenicity, tumor-initiating activity, carcinogenicity, and in vitro metabolism of fluorinated 5-methylchrysenes. J. Natl. Cancer Inst. 63: 855-861.
17. Hecht, S.S., Rivenson, A., and Hoffmann, D. (1980) Tumor initiating activity of dihydrodiols formed metabolically from 5-methylchrysene. Cancer Res. 40: 1396-1399.
18. Hecht, S.S., Amin, S., Rivenson, A., and Hoffmann, D. (1979) Tumor initiating activity of 5,11-dimethylchrysene and the structural requirements favoring carcinogenicity of methylated polynuclear aromatic hydrocarbons. Cancer Lett. 8: 65-70.
19. LaVoie, E.J., Tulley-Freiler, L., Bedenko, V., and Hoffmann, D. (1981): Comparison of the mutagenicity, tumor initiating activity and metabolism of methyl phenanthrenes. Cancer Res. 41: 3441-3447.
20. Boyland, E., Sims, P. (1962): Metabolism of polynuclear compounds. The metabolism of phenanthrene in rabbits and rats: dihydro-dihydroxy compounds and related glucosiduronic acids. Biochem. J. 84: 571-582.
21. Sims, P. (1970): Qualitative and quantitative studies on the metabolism of a series of aromatic hydro-

carbons by rat-liver preparations. Biochem. Pharmacol. 19: 795-818.
22. Wood, A.W., Chang, R.L., Levin, W., Ryan, D.E., Thomas, P.E., Mah, H.D., Karle, J.M., Yagi, H., Jerina, D.M., and Conney, A.H. (1979): Mutagenicity and tumorigenicity of phenanthrene and chrysene epoxides and diol expoxides. Cancer Res. 39: 4069-4077.
23. Dipple, A., Lawley, P.D., and Brookes, P. (1968): Theory of tumor initiation by chemical carcinogens: dependence of activation on structure of ultimate carcinogen. Europ. J. Cancer 4: 493-506.
24. Stevenson, J.L. and von Haam, E. (1965): Carcinogenicity of benz(a)anthracene and benzo(c)phenanthrene derivatives. Amer. Ind. Hygiene Assoc. J. 475-478.
25. Levin, W., Wood, A.W., Chang, R.L., Ittah, Y., Croisy-Delcy, M., Yagi, H., Jerina, D.M., and Conney, A.H. (1980): Exceptionally high tumor-initiating activity of benzo(c)phenanthrene bay-region diol-epoxides on mouse skin. Cancer Res. 40: 3910-3914.
26. Wood, A.W., Chang, R.L., Levin, W., Ryan, D.E., Thomas, P.E., Croisy-Delcey, M., Ittah, Y., Yagi, H., Jerina, D.M., and Conney, A.H. (1980): Mutagenicity of the dihydrodiols and bay-region diolepoxides of benzo(c)phenanthrene in bacterial and mammalian cells. Cancer Res. 40: 2876-2883.
27. Wynder, E.L. and Hoffmann, D. (1959): The carcinogenicity of benzofluoranthenes. Cancer 12: 1149-1159.
28. Wynder, E.L. and Hoffmann, D. (1967): Tobacco and Tobacco Smoke. Studies in Experimental Carcinogenesis. Academic Press, New York, N.Y. p 730.
29. Hoffmann, D. and Wynder, E.L. (1977): Organic particulate pollutants. Chemical analysis and bioassays for carcinogenicity. In "Air Pollution" 3rd Edition A.C. Stern, ed., Vol. II, Part B, Academic Press, New York, N.Y., pp 361-455.
30. LaVoie, E.J., Amin, S., Hecht, S.S., Furuya, K., and Hoffmann, D. (1982). Tumor initiating activity of dihydrodiols of benzo(b)fluoranthene, benzo(j)fluoranthene, and benzo(k)fluoranthene. Carcinogenesis 3: 49-52.
31. Amin, S., LaVoie, E.J., and Hecht, S.S. (1982) Identification of metabolites of benzo[b]fluoranthene. Carcinogenesis 3: 171-174.
32. LaVoie, E.J., Hecht, S.S., Amin, S., Bedenko, V., and Hoffmann, D. (1980): Identification of mutagenic dihydrodiols as metabolites of benzo(j)fluoranthene and benzo(k)fluoranthene. Cancer Res. 40: 4528-4532.

33. Wood, A.W., Levin, W., Chang, R.L., Lehr, R.E., Schaefer-Ridder, M., Karle, J.M., Jerina, D.M., and Conney, A.H. (1977): Tumorigenicity of five dihydrodiols of benzo(a)anthracene on mouse skin: Exceptional activity of benzo(a)anthracene 3,4-dihydrodiol. Proc. Natl. Acad. Sci. U.S.A., 74: 3176-3179.
34. Hoffmann, D. and Wynder, E.L. (1963). Studies on gasoline engine exhaust. J. Air Poll. 13: 322-327.
35. LaVoie, E.J. and Hecht, S.S. (1981). Chemical Carcinogens: In Vitro Metabolism and Activation. In: "Hazard Assessment of Chemicals, Current Developments", Vol. 1, J. Saxena, ed. Academic Press, New York, pp. 155-249.

THE BAY REGION THEORY: HISTORY AND CURRENT PERSPECTIVES.

ROLAND E. LEHR*, ALEXANDER W. WOOD**, WAYNE LEVIN**, ALLAN H. CONNEY**, DHIREN R. THAKKER***, HARUHIKO YAGI***, and DONALD M. JERINA***
*Department of Chemistry, University of Oklahoma, Norman, Oklahoma, 73019; **Department of Biochemistry and Drug Metabolism, Hoffmann-LaRoche Inc., Nutley, N.J., 07110; ***Laboratory of Bioorganic Chemistry, NIAMDD, NIH, Bethesda, Maryland, 20205.

INTRODUCTION

<u>History</u>

More than two centuries have passed since Pott's (1) observation of enhanced scrotal cancer among chimney sweeps. A complete understanding of the cancer-causing properties of the complex coal tar mixtures to which these workers were exposed remains elusive and progress has generally been slow. However, significant experiments and observations have periodically resulted in an increasing refinement of our concept of the basis for the carcinogenicity of coal tar and of cancer induction by chemicals.

<u>Identification of Polycyclic Aromatic Hydrocarbons (PAH) in Coal Tar</u>. By the mid 19th century, benzene, anthracene and other products of commercial interest had been isolated from coal tar and distillates. However, it was 1921 before PAH were postulated as the agents primarily responsible for the carcinogenicity of coal tar (2), 1930 before Hieger (3) demonstrated the carcinogenicity of dibenzo[a,h]anthracene, and 1933 before Cook, <u>et al</u>. (4) isolated benzo[a]pyrene (BaP) from coal tar and reported its high carcinogenicity. Subsequent research established that BaP was not the only carcinogenic PAH in coal tar, and since 1935 hundreds of PAH have been synthesized and tested for carcinogenicity as part of the effort to establish the structural factors involved in PAH carcinogenesis.

The bounteous legacy of tested compounds has provided abundant data against which to test various theories of PAH induced carcinogenicity. Numerous attempts to correlate physical and spectral properties of PAH with carcinogenic potency have ultimately been found lacking. The development of the simple Hückel method of quantum mechanics in 1931 (cf. 5) provided the basis for the Pullmans' application of quantum chemical calculations to PAH which resulted in the well known K-region theory in 1955 (6). While their predictions of carcinogenic activity based upon calculated reactivity parameters of PAH were encouraging, K-region arene oxides (7) have failed to show the tumorigenic properties expected of ultimate carcinogenic forms of PAH (8,9,10).

Metabolic Activation of PAH. The lack of success of the initial efforts to rationalize the carcinogenic properties of PAH is understandable, since they focused upon calculated or observed physical properties of the parent hydrocarbon. It is now generally acknowledged that metabolic activation of PAH to reactive intermediates provides the key to understanding their carcinogenicity. Studies of the metabolism of PAH were initiated as early as the mid-1930's, most notably by Boyland, with dihydrodiols and phenols being identified as typical metabolites of simple PAH such as naphthalene and anthracene (11,12). As early as 1950, PAH epoxides (arene oxides) were proposed by Boyland (13) as possible reactive metabolites of PAH that could explain the formation of dihydrodiols, mercapturic acids and other derivatives of PAH. However, it was 1968 before unequivocal evidence for the involvement of arene oxides as primary oxidative metabolites of PAH was obtained (14). Arene oxides are now recognized as the immediate metabolic precursors (15) of dihydrodiols, via hydration with epoxide hydrolase, as well as as major intermediates in the formation of phenols, via NIH shift (16) (Figure 1). However, non-K-region arene oxides, while reactive, have proven not to be the elusive ultimate carcinogenic forms of PAH. Borgen et al. (17), in a study of the metabolic activation of BaP and its derivatives, found that BaP 7,8-dihydrodiol was bound more extensively to DNA in the presence of liver microsomes than were BaP or several other of its metabolites. A diol oxide was postulated as the possible reactive species (15) and Sims et al. (18) provided preliminary evidence that further metabolism of BaP 7,8-dihydrodiol led to a 7,8-diol-9,10-epoxide, which is a metabolically activated form of BaP that binds to DNA and other cellular macromolecules. Very high mutagenicity of the diastereomeric BaP 7,8-diol-9,10-epoxides (Figure 2) and high tumorigenicity of the trans-diastereomer were quickly established (19,20) once the compounds were available by chemical synthesis (21,22). When combined with the extensive evidence that BaP is metabolized by mammalian enzymes to BaP 7,8-diol-9,10-epoxides and that

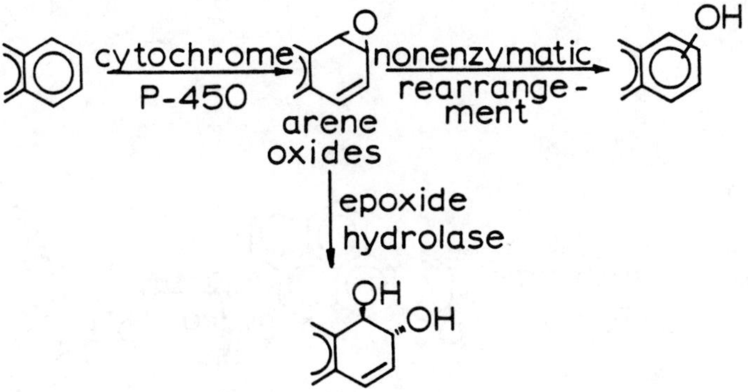

FIGURE 1. Arene oxides as metabolic precursors of dihydrodiols and phenols.

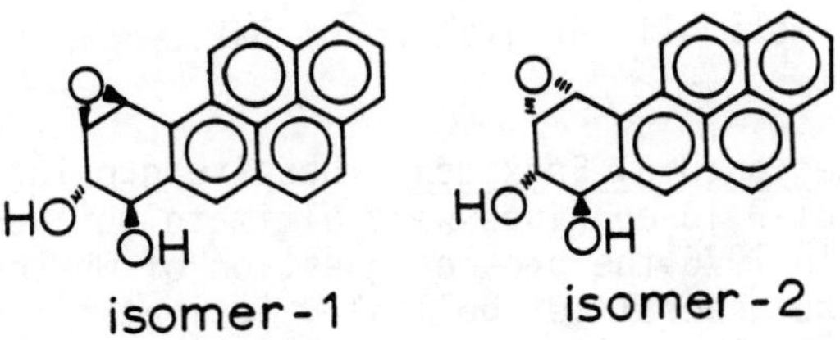

FIGURE 2. Diastereomeric BaP 7,8-diol-9,10-epoxides. Isomer-1 has the benzylic hydroxyl group at C-7 cis to the epoxide oxygen atom, whereas in isomer-2 they are trans.

the diol epoxides bind at C-10 to nucleophilic centers in DNA without further metabolism, convincing evidence for classifying the diol epoxides as ultimate mutagens and/or carcinogens had been secured (20). The scheme for metabolic activation of BaP is summarized in Figure 3.

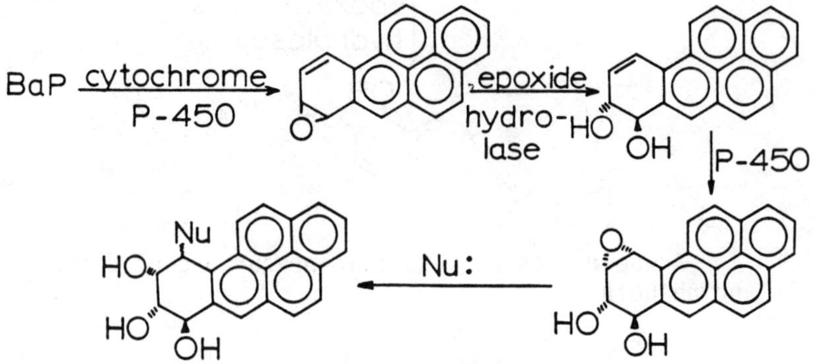

FIGURE 3. Metabolic activation of BaP.

Bay-Region Diol Epoxides. Despite growing evidence that BaP 7,8-diol-9,10-epoxides were ultimate mutagens and/or carcinogens, in 1976 the broader question of whether diol epoxides are general metabolically activated forms of PAH remained. Further, if diol epoxides were involved, was there a structural feature in the BaP diol epoxides that could enable one to predict which of the metabolically possible diol epoxides of a given PAH would be a likely ultimate carcinogenic form of the PAH? In 1976 two lines of reasoning led to the proposal that the critical structural feature in the BaP 7,8-diol-9,10-epoxides was the presence of the epoxide in the tetrahydrobenzo ring in a "bay region," and that, for a given PAH, similar diol epoxides (examples shown in Figure 4) would prove to be the most chemically reactive and biologically active.

First, existing data on the tumor-inhibiting effects of substituents on the angular ring of 7-methyl- and 7-fluoro-benzo[a]anthracenes were interpreted (23) to reflect the im-

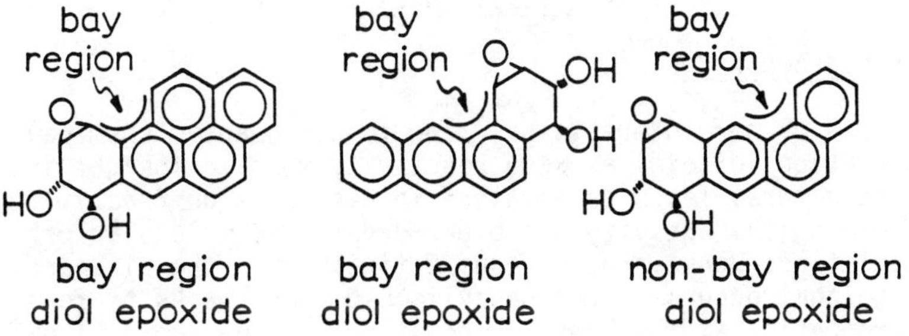

FIGURE 4. Bay- and non bay-region diol epoxides. The prototype of a "bay region" is the concave region between C4 and C5 of phenanthrene.

portance of metabolism on the angular ring, and analogy with BaP suggested the bay-region 1,2-epoxide (versus the non bay-region 3,4-epoxide). Second, quantum chemical calculations (24,25) predicted that, for a given PAH, ring-opening of the benzylic C-O bond of the oxirane to form carbonium ions should be much more facile when the epoxide forms part of a bay-region than when the epoxide is benzylic, but not in a bay-region. Furthermore, the diol epoxide calculated to be the most reactive for a given carcinogenic PAH was generally calculated to be much more reactive than its counterpart for noncarcinogens. Since electrophilicity is apparently an important factor in the carcinogenicity of organic compounds (26), it seemed reasonable that the calculations might also prove of value in predicting the relative mutagenicity and carcinogenicity of diol epoxides, at least for the positional isomers of a given PAH. During the past five years, considerable evidence in support of the predictions of the bay-region theory has been obtained. Compounds for which bay-region diol epoxides appear to be ultimate carcinogens are: BaP, benzo[a]anthracene, 7,12-dimethylbenzo[a]anthracene, 5-methylchrysene, benzo[c]phenanthrene (27) and 15,16-dihydro-11-methylcyclopenta[a]phenanthren-17-one (28). The evidence for most of these compounds has been recently reviewed (29), and the present article will deal primarily with a discussion of structure-activity relationships within the extensive series of diol epoxides that have been synthesized and studied in our

STRUCTURE-ACTIVITY RELATIONSHIPS

Tetrahydroepoxides

One of our interests has been to use quantum chemical calculations of diol epoxide reactivity to gain insight into the structural features involved in diol epoxide reactivity and biological activity. The perturbational molecular orbital calculations we have used focus entirely upon the pi-energy change that occurs as the benzylic C-O bond breaks to form a carbocation at the benzylic position, and neglect all effects due to the presence of oxygen atoms and their stereochemistry. Thus, tetrahydroepoxides in which the oxirane ring occupies a position analogous to that in a benzo ring diol epoxide will have a calculated value of $\Delta E_{deloc}/\beta$ identical to that of the diol epoxide. For example, (Figure 5) the $\Delta E_{deloc}/\beta$ value

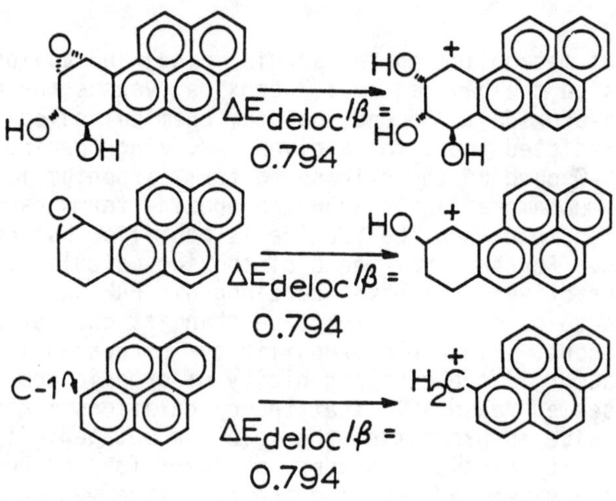

FIGURE 5. $\Delta E_{deloc}/\beta$ values are identical for benzylic carbocation formation from tetrahydro- and diol epoxides.

for the bay region BaP tetrahydro- and diol epoxides are identical (0.794) since the pi-energy change in both cases corresponds to the energy change attendant upon conjugating an empty p-orbital at C-1 of pyrene. In these calculations, large values of $\Delta E_{deloc}/\beta$ correspond to greater stabilization of the carbocation relative to the aromatic system from which it was generated. Since tetrahydroepoxides are frequently synthetically more accessible than diol epoxides and should also provide a test of the utility of PMO calculations, we studied the relationship between mutagenicity toward Salmonella typhimurim strain TA 100 and $\Delta E_{deloc}/\beta$ for an extensive series of tetrahydroepoxides (30). A good correlation between the logarithm of TA 100 mutagenicity and $\Delta E_{deloc}/\beta$ for the nine compounds was observed, with r = 0.74. When the correlation was confined to four- and five-ring epoxides by deleting the two phenanthrene tetrahydroepoxides, which were less mutagenic than expected based on the overall correlation, the r value improved to 0.96.

Diol Epoxides.

Solvolytic Reactivity. Subsequent to our application of PMO calculations to PAH carcinogenicity in 1976, a variety of other quantum mechanical computational methods have been reported. In part to determine if any of the calculations were superior in predicting solvolytic reactivity, bay region diol epoxides of seven PAH (phenanthrene, benzo[c]phenanthrene (BcPh), chrysene, BA, BaP, dibenzo[a,h]pyrene and dibenzo[a,i] pyrene) were solvolyzed (31). Generally good correlations (r = 0.95-0.98) were observed between the logarithms of the uncatalyzed rate constant (k_o) and $\Delta E_{deloc}/\beta$, with no significant differences between the various computational procedures in predicting the rates. Accordingly, in the following discussion, we have continued to use $\Delta E_{deloc}/\beta$ as the quantum chemical parameter.

Mutagenicity. Diol epoxides exist as diastereomeric pairs, depending upon whether the oxirane oxygen atom is cis (series-1) or trans (series-2) to the benzylic hydroxyl group (Figure 2), and the relative mutagenicity of members of the two series are analyzed separately in the following discussion. The relative mutagenicity data plotted in Figures 6 and 7 are taken from a recent article (32), with the exception of the data for the non bay-region BA diol epoxides, which was taken from an earlier source (33). Figures 6a and 7a show plots of the logarithm of mutagenicity versus $\Delta E_{deloc}/\beta$ for all diol epoxides in the two series. The least squares lines indicate an increase in the logarithm of mutagenicity with increased calculated reactivity. However, there is considerable scatter with some substantial deviations from the least squares lines. A careful analysis of the data reveals patterns to the deviations that enable some comments about structure/mutagenicity relationships to be made. First, it

THE BAY REGION THEORY

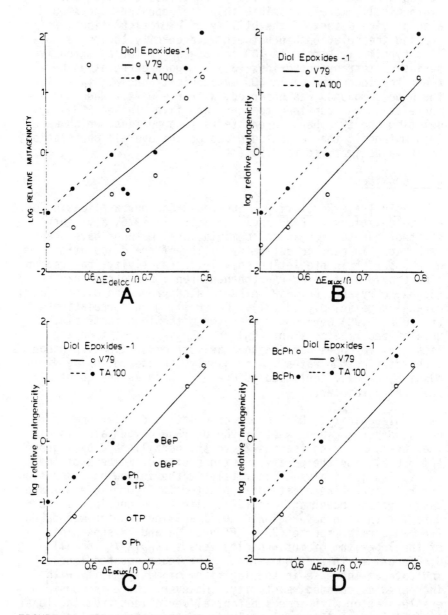

FIGURE 6. Plots of log relative mutagenicity vs. $\Delta E_{deloc}/\beta$ for diol epoxides-1. Abbreviations are: Ph, phenanthrene; TP, triphenylene; BeP, benzo[e]pyrene; BcPh, benzo[c]phenanthrene.

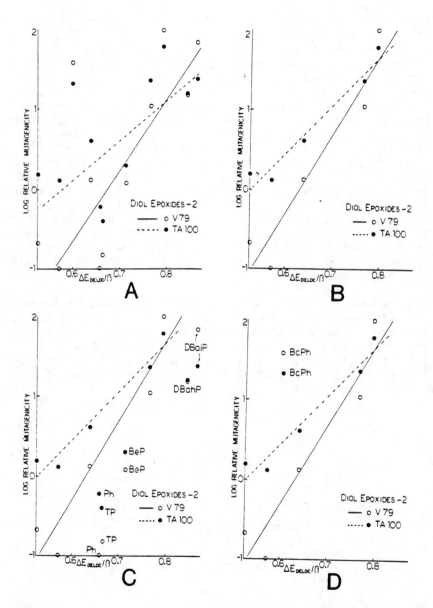

FIGURE 7. Plots of log relative mutagenicity vs. $\Delta E_{deloc}/\beta$ for diol epoxides-2. Abbreviations are same as for Figure 6, and additionally: DBahP, dibenzo[a,h]pyrene; DBaiP, dibenzo[a,i]pyrene.

is noteworthy that five diol epoxides in each series (the BaP, chrysene and BA bay-region diol epoxides and the BA diol epoxides on the non-angular ring) show a good correlation of log relative mutagenicity and $\Delta E_{deloc}/\beta$, both for V79 cells and for TA 100 cells, as shown in Figures 6b and 7b. Interestingly, for these compounds the conformational preference for each diastereomer stays the same; hydroxyl groups quasidiaxial in the isomer-1 series and quasidiequatorial in the isomer-2 series. It then becomes convenient to examine the other diol epoxides with respect to this "well behaved" series of five compounds. Diol epoxides that are less mutagenic than expected on this basis are shown in Figures 6c and 7c. For both diol epoxide series, the mutagenicity of the benzo[e]pyrene (BeP) and triphenylene (TP) benzo ring epoxides as well as that of the phenanthrene (Ph) bay-region diol epoxides is substantially attenuated relative to expectations based upon the "well behaved" series. For the diol epoxide-2 series, the dibenzo[a,h]- and [a,i]-pyrene bay region diol epoxides are also much less mutagenic than expected (data for the corresponding diol epoxides-1 is lacking due to the inavailability of the compounds).

We have previously noted the lower than expected mutagenicity and reactivity of the BeP and TP diol epoxides (34,35) that appears to be associated with the presence of two bay regions around the tetrahydrobenzo ring and the resultant locking of the hydroxyl groups into quasidiaxial conformations (Figure 8). For the phenanthrene bay-region diol epoxides,

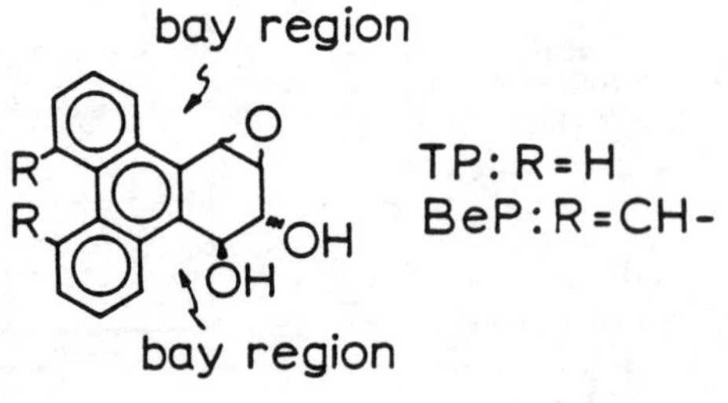

FIGURE 8. Benzo[e]pyrene and triphenylene diol epoxides.

the lower mutagenicity parallels that previously cited for the phenanthrene tetrahydroepoxides (30) and is unrelated to reactivity, since the solvolytic reactivity (k_0) of the phenanthrene diol epoxides correlates well with $\Delta E_{deloc}/\beta$. The basis for the reduced mutagenicity of the tetrahydro and diol epoxides derived from this three-ring aromatic hydrocarbon is presently unclear.

While the series-2 bay region dibenzo[a,h]- and [a,i] pyrene diol epoxides are highly mutagenic, with mutagenicities approaching those of the analogous BaP bay region diol epoxide, they do not exhibit the even higher mutagenicity levels that could have been expected based upon their calculated $\Delta E_{deloc}/\beta$ values (36). As with the phenanthrene bay-region diol epoxides, neither conformational nor reactivity differences can account for the lower mutagenicities, since the dibenzopyrene bay-region diol epoxides are more reactive (k_0) than their BaP counterpart, as expected on the basis of the $\Delta E_{deloc}/\beta$ calculations (36), and the conformational preferences on the benzo ring are the same as those of the "well behaved" series of diol epoxides. Interestingly, reduced mutagenicity is also observed for tetrahydroepoxides derived from these six-ring aromatic hydrocarbons, with the non bay-region tetrahydroepoxides of dibenzo[a,h]- and [a,i]pyrene also being significantly less mutagenic than expected (37).

Perhaps most surprising are the bay-region diol epoxides derived from benzo[c]phenanthrene (Figures 6d and 7d). They are significantly more mutagenic than expected (38), with diol epoxide-1 being about 25 and 200 times as mutagenic as expected in TA 100 and V79 cells, respectively, and diol epoxide-2 being about 6 and 60 times as mutagenic as expected in TA 100 and V79 cells, respectively, based on the correlations in Figures 6b and 7b. This high mutagenicity is unrelated to solvolysis rate, since the hydrolysis rate (k_0) for the diol epoxides is very low as expected on the basis of the relatively low calculated value of $\Delta E_{deloc}/\beta$ (0.600) (39). The chemistry and biological activity of these diol epoxides have recently be reviewed (31), so only a few points will be made here. First, because of the high extent of steric crowding in the bay-region (Figure 9), the hydroxyl groups for isomer-1 now prefer the quasidiequatorial conformation. This is the first example of an isomer-1 diol epoxide with a preferred quasidiequatorial conformation. Second, both diastereomeric diol epoxides are highly tumorigenic on mouse skin, with this being the first example of a diol epoxide-1 isomer that has shown significant tumorigenic activity. While conformational factors can be invoked to help explain the much

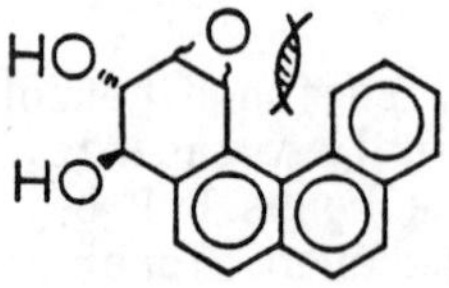

FIGURE 9. Benzo[c]phenanthrene bay-region diol epoxides.

higher than expected mutagenic and tumorigenic activity of diol epoxide-1 relative to diol epoxide-2, both isomers are much more mutagenic and tumorigenic than would have been expected based upon their calculated $\Delta E_{deloc}/\beta$ values and observed solvolysis rates (39). While the precise basis for the effect is not clear, the high mutagenicity and tumorigenicity of the benzo[c]phenanthrene diol epoxides may provide the first direct example of the effects of additional steric crowding in the bay region. Hecht and coworkers (40) have previously noted the apparent importance of methyl groups placed in such positions upon the tumorigenicity of PAH, especially 5-methylchrysene. Other hydrocarbons that appear likely to involve bay region diol epoxides as ultimate carcinogens, and in which a similarly situated methyl group appears to enhance the carcinogenicity of the molecule relative to the hydrogen-substituted PAH are: 7,12-dimethylbenzo[a]anthracene (41), 11-methylbenzo[a]pyrene (42), 1,4-dimethylphenanthrene (43) and 15,16-dihydro-11-methylcyclopenta[a]phenanthren-17-one (28).

CONCLUSIONS

Since evidence implicating bay-region diol epoxides as ultimate mutagens and carcinogens has been steadily accumulating, a thorough understanding of the structural factors that affect their chemical and biological behavior is needed. In this article, we have primarily examined the utility of a quantum chemical parameter, $\Delta E_{deloc}/\beta$, as a guide to understanding diol epoxide behavior. The parameter is successful in accounting for the relative solvolysis rates of several bay-region diol epoxides, but its ability to account for differences in mutagenicities of diol epoxides is more limited. *The most obvious conclusion to be reached is that factors in addition to pi-energy change can be important determinants in mutagenic activity.* One way to identify such factors is to note the structural variations in those diol epoxides whose mutagenicities deviate from expectations based upon the behavior of a series of diol epoxides whose mutagenicities nicely correlate with $\Delta E_{deloc}/\beta$. When this is done, a number of factors can be identified: size, conformational effects, and steric effects. Absolute configuration of the diol epoxides is also very important, but is beyond the scope of the present discussion, which has been confined to racemic diol epoxides. Diol epoxides derived from PAH with three aromatic rings (phenanthrene) and six aromatic rings (dibenzo[a,h]- and [a,i]pyrenes) are substantially less mutagenic than expected based upon the mutagenicities of analogous diol epoxides derived from four- and five-aromatic ring PAH in which the conformational preferences on the tetrahydrobenzo ring are the same. The effect is unrelated to chemical reactivity, since solvolysis rates for the compounds correlate well with $\Delta E_{deloc}/\beta$. Benzo[e]pyrene and triphenylene diol epoxides are also less mutagenic than expected, and this appears to be related to the fact that both benzylic positions on the benzo ring form parts of bay-regions, and the hydroxyl groups are consequently locked in quasidiaxial conformations. It should be noted that in this instance the solvolytic reactivity is also lowered (34). Finally, significantly enhanced mutagenicity is observed in the bay-region diol epoxides derived from one PAH: benzo[c]phenanthrene. In addition to interesting conformational effects, it appears likely that the close proximity of an angular benzo ring to the epoxide moiety is having an effect upon enhancing the mutagenicities of the compounds. Despite variations in the ability of the present quantum mechanical calculations to quantitatively predict mutagenicity for a series of diol epoxides, the bay-region theory has been highly successful in predicting ultimately carcinogenic metabolites for all of the carcinogenic alternant PAH thus far studied.

ACKNOWLEDGMENT

Partial support of this research by the National Cancer Institute (Grant No. 22985, awarded to R.E.L.) is gratefully acknowledged.

REFERENCES

1. Pott, P. (1963): Chirurgical observations relative to the cancer of the scrotum. London, 1775, Nat. Cancer Inst. Monograph 10:7-13.
2. Block, B. and Dreifuss, W. (1921): Experimental tar cancer, Schweiz. Med. Wochschr., 51:1033-1037.
3. Hieger, I. (1930): The spectra of cancer-producing tars and oils and of related substances, Biochemical J., 24:505-511.
4. Cook, J. W., Hewett, C. L., and Hieger, I. (1933): The isolation of a cancer-producing hydrocarbon from coal tar. Parts I, II and III, J. Chem. Soc., 395-405.
5. Hückel, E. and Gilde, H. G. (1972): Interview with Erich Hückel, J. Chem. Ed., 49:2-4.
6. Pullman, A. and Pullma, B. (1955): Electronic structure and carcinogenic activity of aromatic molecules, Adv. Cancer Res., 3:117-169.
7. Jerina, D. M., Yagi, H., and Daly, J. W. (1973): Arene oxides-oxepins, Heterocycles, 1:267-326.
8. Boyland, E. and Sims, P. (1967): The carcinogenic activities in mice of compounds related to benz[a]-anthracene, Int. J. Cancer, 2:500-504.
9. Miller, E. C. and Miller, J. A. (1967): Low carcinogenicity of the K-region epoxides of 7-methylbenz[a]-anthracene and benz[a]anthracene in the mouse and rat, Proc. Soc. Exp. Biol. Med., 124:915-919.
10. Levin, W., Wood, A. W., Yagi, H., Dansette, P. M., Jerina, D. M., and Conney, A. H. (1976): Carcinogenicity of benzo[a]pyrene 4,5-, 7,8- and 9,10-oxides on mouse skin, Proc. Natl. Acad. Sci. (USA), 73:243-247.
11. Boyland, E. and Levi, A. A. (1935): Metabolism of polycyclic compounds. I. Production of dihydroxydihydroanthracene from anthracene, Biochem. J., 29:2679-2683.
12. Booth, J. and Boyland, E. (1949): Metabolism of polycyclic compounds. Formation of 1,2-dihydroxy-1,2-dihydronaphthalenes, Biochem. J., 44:361-365.
13. Boyland, E. (1950): The biological significance of metabolism of polycyclic compounds, Biochemical Society Symposia, No. 5:40-54.
14. Jerina, D. M., Daly, J. W., Witkop, B., Zaltzman-Nirenberg, P. and Udenfriend, S. (1969): 1,2-Naphthalene oxide as an intermediate in the microsomal hydroxylation of naphthalene, Biochemistry, 9:147-156.
15. Jerina, D. M. and Daly, J. W. (1974): Arene oxides: a new aspect of drug metabolism, Science, 185:573-582.
16. Daly, J. W., Jerina, D. M. and Witkop, B. (1972): Arene oxides and NIH shift. Metabolism, toxicity and carcinogenicity of aromatic compounds, Experientia, 28:1129-1149.

17. Borgen, A., Darvey, J., Castagnoli, N., Crocker, T. T., Rasmussen, R. W. and Wang (1973): Metabolic conversion of benzo[a]pyrene by Syrian hamster liver microsomes and binding of metabolites to DNA, J. Med. Chem., 16:502-506.
18. Sims, P., Grover, P. L., Swaisland, A., Pal, A., and Hewer, A. (1974): Metabolic activation of benzo[a]pyrene proceeds by a diol epoxide, Nature, 252:326-328.
19. Levin, W., Wood, A. W., Wislocki, P. G., Chang, R. L., Kapitulnik, J., Mah, H. D., Yagi, H., Jerina, D. M., and Conney, A. H. (1978): Mutagenicity and carcinogenicity of benzo[a]pyrene and benzo[a]pyrene derivatives. In Polycyclic Hydrocarbons and Cancer, Vol. 1, edited by H. V. Gelboin and P. O. P. T'so, pp. 189-202, Academic Press, New York.
20. Conney, A. H. Levin, W., Wood, A. W., Yagi, H., Thakker, D. R., Lehr, R. E., and Jerina, D. M. (1980): Metabolism of polycyclic aromatic hydrocarbons to reactive intermediates with high biological activity. In Human Epidemiology and Animal Laboratory Correlations in Chemical Carcinogenesis, edited by F. Coulston and P. Shukik, pp. 153-183, Ablex Publishing Corp., Norwood, New Jersey.
21. Yagi, H., Hernandez, O., and Jerina, D. M. (1975): Synthesis of (±)-7β,8α-dihydroxy-9β,10β-epoxy-7,8,9,10-tetrahydrobenzo[a]pyrene, a potential metabolite of the carcinogen benzo[a]pyrene with stereochemistry related to the antileukemic triptolides, J. Am. Chem. Soc., 97:6881-6888.
22. Yagi, H. Thakker, D. R., Hernandez, O., Koreeda, M. and Jerina, D. M. (1977): Synthesis and reactions of the highly mutagenic 7,8-diol-9,10-epoxides of the carcinogen benzo[a]pyrene, J. Am. Chem. Soc., 99:1604-1611.
23. Jerina, D. M. and Daly, J. W. (1976): Oxidation at carbon. In Drug Metabolism-from Microbe to Man, edited by D. V. Parke and R. L. Smith, p. 13, Taylor and Francis, Ltd., London.
24. Jerina, D. M. and Lehr, R. E. (1978): The bay region theory. A quantum mechanical approach to aromatic hydrocarbon-induced carcinogenicity. In Microsomes and Drug Oxidations, edited by V. Ulbrick, I. Roots, A. Hildebrandt and R. W. Estabrook, p. 709, Pergamon Press, Elmsford, New York.
25. Jerina, D. M., Lehr, R. E., Yagi, H., Hernandez, O., Dansette, P. M., Wislocki, P. G., Wood, A. W., Chang, R. L., Levin, W. and Conney, A. H. (1976): Mutagenicity of benzo[a]pyrene and the description of a quantum mechanical model which predicts the ease of carbonium ion formation from diol epoxides. In In Vitro Metabolic Activation in Mutagenesis Testing, edited by F. J. de Serres,

J. R. Fouts, J. R. Bend and R. M. Philpot, p. 159, Elsevier/North Holland Biomedical Press, Amsterdam.
26. Miller, E. C. and Miller, J. A. (1981): Mechanisms of chemical carcinogenesis, Cancer, 47:1055-1064.
27. Levin, W., Wood, A. W., Chang, R. L., Ittah, Y., Croisy-Delcey, M., Yagi, H., Jerina, D. M. and Conney, A. H. (1980: Exceptionally high tumor initiating activity of benzo[c]phenanthrene bay-region diol epoxides on mouse skin, Cancer Res., 40:3910-3914.
28. Coombs, M. M., Kissonerghis, A. -M., Allen, J. A. and Vose, C. W. (1979): Identification of the proximate and ultimate forms of the carcinogen, 15,16-dihydro-11-methylcyclopenta[a]phenanthrene-17-one, Cancer Research, 39:4160-4165.
29. Nordqvist, M., Thakker, D. R., Yagi, J., Lehr, R. E., Wood, A. W., Levin, W., Conney, A. H. and Jerina, D. M. (1980): Evidence in support of the bay region theory as a basis for the carcinogenic activity of polycyclic aromatic hydrocarbons. In Molecular Basis of Environmental Toxicity, edited by R. S. Bhatnagar, pp. 329-357, Ann Arbor Science Publishers, Inc., Ann Arbor, Michigan.
30. Wood, A. W., Levin, W., Chang, R. L., Yagi, H., Thakker, D. R., Lehr, R. E., Jerina, D. M. and Conney, A. H. (1979): Bay region activation of carcinogenic polycyclic hydrocarbons. In Polynuclear Aromatic Hydrocarbons, Third International Symposium on Chemistry and Biology-Carcinogenesis and Mutagenesis, edited by P. W. Jones and P. Leber, pp. 531-551, Ann Arbor Science Publishers, Inc. Ann Arbor, Michigan.
31. Jerina, D. M., Sayer, J. M., Yagi, H., Croisy-Delcey, M., Ittah, Y. and Thakker, D. R. (in press): Highly tumorigenic bay-region diol epoxides from the weak carcinogen benzo[c]phenanthrene. In Biological Reactive Intermediates, Vol. 2, Chemical Mechanisms and Biological Effects, edited by R. Snyder, D. V. Parke, J. Kocsis, C. J. Jollow and G. G. Gibson, Plenum Publishing Corp., New York.
32. Levin, W., Wood, A., Chang, R., Ryan, D., Thomas, R., Yagi, H., Thakker, D., Vyas, K. Boyd, C., Chu, S. -Y., Conney, A., and Jerina, D. (in press): Oxidative metabolism of polycyclic aromatic hydrocarbons to ultimate carcinogens. In Drug Metabolism Reviews.
33. Lehr, R. E., Yagi, H. Thakker, D. R., Levin, W., Wood, A. W., Conney, A. H., and Jerina, D. M. (1978): The bay region theory of polycyclic aromatic hydrocarbon-induced carcinogenicity. In Carcinogenesis, Vol. 3: Polynuclear Hydrocarbons, edited by P. W. Jones and R. I. Freudenthal, pp. 231-241, Raven Press, New York.

34. Lehr, R. E., Kumar, S., Levin, W., Wood, A. W., Chang, R. L., Buening, M. K., Conney, A. H., Whalen, D. L., Thakker, D. R., Yagi, H., and Jerina, D. M. (1980): Benzo[e]pyrene dihydrodiols and diol epoxides: chemistry, mutagenicity and tumorgenicity. In Polynuclear Aromatic Hydrocarbons: Fourth International Symposium on Analyses, Chemistry, and Biology, edited by A. Bjørseth and A. J. Dennis, pp. 675-688, Battelle Press, Columbus, Ohio.
35. Wood, A. W., Chang, R. L., Huang, M. -T., Levin, W., Lehr, R. E., Kumar, S., Thakker, D. R., Yagi, H., Jerina, D. M., and Conney, A. H. (1980): Mutagenicity of benzo[e]pyrene and triphenylene tetrahydroepoxides and diol epoxides in bacterial and mammalian cells, Cancer Res., 49:1985-1989.
36. Wood, A. W., Chang, R. L., Levin, W., Ryan, D. E., Thomas, P. E., Lehr, R. E., Kumar, S., Sardella, D. J., Boger, E., Yagi, H., Sayer, J. M., Jerina, D. M. and Conney, A. H. (1981): Mutagenicity of the bay-region diol-epoxides and other benzo-ring derivatives of dibenzo[a,h]pyrene and dibenzo[a,i]pyrene, Cancer Research, 41:2589:2597.
37. Wood, A. W., unpublished results, these laboratories.
38. Wood, A. W., Chang, R. L., Levin, W., Ryan, D. E., Thomas, P. E., Croisy-Delcey, M., Ittah, Y., Yagi, H., Jerina, D. M., and Conney, A. H. (1980): Mutagenicity of the dihydrodiols and bay region diol-epoxides of benzo[c]phenanthrene in bacterial and mammalian cells, Cancer Res., 40:2876-2883.
39. Sayer, J. M., Yagi, H., Croisy-Delcey, M., and Jerina, D. M. (1981): Novel bay-region diol epoxides from benzo[c]phenanthrene, J. Am. Chem. Soc., 103:4970-4972.
40. Hecht, S. S., Ribenson, A., and Hoffmann, D. (1980): Tumor-initiating activity of dihydrodiols formed metabolically from 5-methylchrysene, Cancer Research, 40:1396-1399.
41. Newman, M. S. (1976): Carcinogenic activity of benz[a]anthracenes. In Polynuclear Aromatic Hydrocarbons: Chemistry, Metabolism and Carcinogenesis, edited by R. Freudenthal and P. W. Jones, p. 203, Raven Press, New York.
42. Iyer, R. P., Lyga, J. W., Secrist, J. A., Daub, G. H., and Slaga, T. J. (1980): Comparative tumor-initiating activity of methylated benzo[a]pyrene derivatives in mouse skin, Cancer Research, 40:1073.
43. LaVoie, E. J., Tulley-Freiler, L., Bendeko, V., and Hoffmann, D. (1981): Mutagenicity, tumor-initiating activity and metabolism of methylphenanthrenes, Cancer Research, 41:3441-3447.

COMPARATIVE REMOVAL OF POLYCYCLIC AROMATIC HYDROCARBON-DNA ADDUCTS IN VIVO

R.W. HART, P.P. FU, and M.J.W. CHANG. Health and Human Services/Food and Drug Administration/National Center for Toxicological Research, Jefferson, Arkansas 72079, U.S.A.

INTRODUCTION

Living systems are continuously exposed to a wide variety of agents which may damage their DNA. This damage has been implicated in carcinogenesis, mutagenesis, and aging, and repair of these lesions is considered an ameliorating factor for these effects. In order to rationally evaluate environmental health effects, it is important to examine the induction of DNA damage and its subsequent repair in vivo. Since we cannot use humans for these studies, animal and cell models are selected as alternatives. In order to extrapolate from animal data to human health effects, it is necessary to study the differences between species relative to the induction and repair of DNA damage.

There are two basic paradigms for research on the action and effect of toxic substances. One, which we can call the ontogenetic, seeks to account for or describe the effect of a toxic substance on cells and tissues of an organism during its lifetime as a function of dose, time or route of administration. The other, the evolutionary-comparative approach, looks for the differences in the genetically determined constitutive characteristics of species, or any genetically defined populations.

The ontogenetic approach dominates contemporary research in toxicology, for it produces important descriptive information. However, it cannot be sufficient by itself to solve the problem of risk assessment because it does not provide the information required for extrapolating animal data to humans. One reason for this is that the ontogenetic approach focuses primarily on effect rather than cause and difference.

The evolutionary-comparative paradigm is based on two postulates: first, a difference in the effect of a toxic substance in two basically similar organisms is due to a difference in the physiochemical environments of the essential macromolecules in the two organisms; and second, the parameters of the molecular environment within an organism are an expression of its genome. Most specifically, the hypothesis is that the eventual fate of the organism, in terms of its susceptibilities to toxic substances, is primarily determined by specific genetically determined constitutive properties, both molecular and organizational. In extrapolation of cell and animal data to human health effects the basic question becomes what does one species do differently than another, if anything, by way of protection, stabilization and repair of its essential molecules in order to maintain homeostasis in the face of environmental toxicants.

The direction and extent of differences between species to toxic substances will be based upon the molecular target of the substance under investigation and the rate and degree of evolutionary change of the target and its operational parameters as a function of evolutionary time (1,2). That such constitutive differences exists is suggested by the observation that despite a 50-fold difference in maximum achievable lifespan and a 10,000-fold difference in total number of cells at risk per unit time between placental mammals when different mammalian species are compared on a fraction-of-maximum lifespan basis, most of the age-related changes observed appear to occur at similar times in the life of an animal (3).

The capacity for coordinated and sometimes rapid-selective modification of the survival characteristic of a population implies that stability of a system, loss of which may be reflected in either aging or cancer, may be governed by a relatively small number of genetically controlled longevity - assurance mechanisms, or overall genetic stability systems, which are governed by a comparatively small number of regulatory loci (4,5). If the latter alternative is correct, then longevity may be under the same kind of genetic regulation as the systems that govern the size an organism might achieve without lost of homeostasis. Thus, a direct measurement of changes in the target molecule (DNA) as a function of species, damage category, and organ type should provide the information required for the rational extrapolation of animal data to human health effects for genotoxic agents (6).

The rapid evolution of genetic stability may be reflected in species differences in risk per cell per unit time for spontaneous malignant transformation without the concurrent input of large numbers of new genes. Under such an assumption the ultimate dose to the target molecule (DNA) would vary between species as would the ultimate phenotypic expression of such damage even under conditions where the level of insult to which different species were exposed was the same. The comparative metabolic activation of polycyclic aromatic hydrocarbons (PAHs) and their subsequent binding to and rate of removal from DNA can be used to study these concepts. This review covers interspecies and intertissue differences between metabolic activation and detoxification of PAHs, DNA binding through their active forms, and repair of these adducts.

METABOLISM

The total enzyme level and substrate specificity of the cytochrome P-450 system, which is the primary system by which PAHs are metabolically activated, vary among tissues and species. Induction of this system is dependent on the sex, age, and strain of the animals, as well as the inducer (i.e. 3-methylcholanthrene (3MC), phenobarbital and Arochlor 1254) and substrate studied. Since the cytochrome P-450 enzymes are present in multiple forms (7), the inducibility of each form may vary as a function of the condition of induction.

The primary metabolites resulting from cytochrome P-450 activation, arene oxides, may: 1) subsequently become hydrated [catalyzed by epoxide hydrolase (EH)] to form trans-dihydrodiols; 2) spontaneously rearrange to phenol(s); and 3) undergo addition of glutathione (catalyzed by cytoplasmic glutathione-S-epoxide-transferase) or other cellular nucleophiles (DNA, RNA, protein). The relative ratio between these reactions depends on the stability and reactivity of the arene oxide, and the relative amounts of the enzymes and nucleophiles present (8.9).

The resulting arene oxides are biologically more reactive than their corresponding parent PAHs in binding with both DNA and RNA. However, at least in the case of benzo[a]pyrene (BaP), 7,12-dimethylbenz[a]anthracene (DMBA) and 7-methylbenz[a]anthracene (7-MBA), the major DNA adduct of the parent hydrocarbon in vivo is not formed from the arene oxide intermediate. Thus, it is now believed that in most of the cases these oxides may not be the ultimate carcinogens.

In 1973 Borgen et al. (10) reported that BaP-trans-7,8-dihydrodiol, which is formed by the enzymatic hydrolysis of BaP-7,8-oxide, incubated in the presence of liver microsomes, had a 10-fold higher level of binding to DNA compared to BaP and other BaP metabolites. Later, it was found that BaP-trans-7,8-dihydrodiol-anti-diolepoxide was the metabolically formed species responsible for the binding of BaP to DNA and the carcinogenicity of BaP in vivo (9,11). Subsequently, Jerina et al. (12) proposed a bay-region theory, which predicted that if a PAH can enzymatically form more than one diolepoxide, the one with its epoxy ring located at the bay region will be the most biologically active among the isomeric diolepoxides. Many experimental results support this theory, although contrary results do exist (13). Since diolepoxides may represent the ultimate carcinogenic forms of the majority of PAHs it is important to determine the factors which govern the enzymatic formation of these metabolites, particularly the bay-region diolepoxides, since these factors may vary as a function of diet, age, stress, tissue and species in a predictable and quantitative fashion.

The enzymes cytochrome P-450 and epoxide hydrolase exhibit regioselectivity and stereoselectivity toward PAHs which may be isozyme specific and vary as a function of species. A striking example of this is the enzymatic formation of both syn- and anti-diolepoxides of BaP from metabolism of BaP with liver microsomes of rats pretreated with 3MC (9). The absolute configurations of 7,8-epoxide, trans-7,8-dihydrodiol, and syn- and anti-diolepoxides of BaP have been determined, and their stereostructures are as shown in Figure 1. The formation of (+)anti-diolepoxide is about 19-fold higher than that of

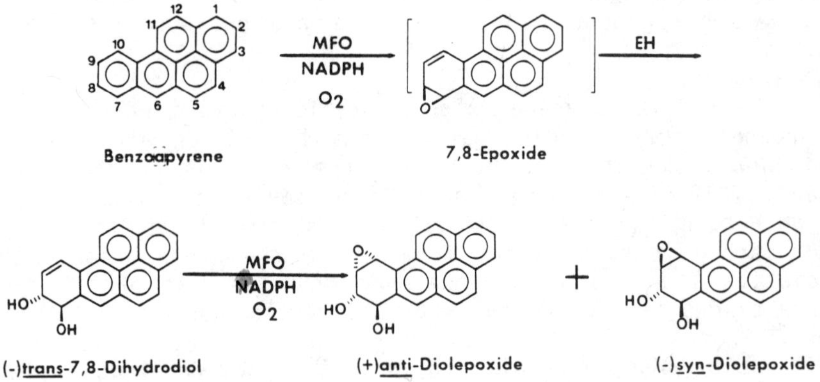

FIGURE 1. Metabolic activation pathway of BaP leading to the ultimate carcinogen anti- and syn-diolepoxides

(-)syn isomer (9). However, when the synthetically prepared

(+)BaP trans-7,8-dihydrodiol was metabolized under similar conditions, the (+) syn-diolepoxide was predominantly formed (11) (Figure 2). Enzymatic epoxidation of these dihydrodiols as catalyzed by cytochrome P-450, preferentially occurs on the same side of the molecular plane (9). This specific stereoselectivity has also been found in the metabolism of several other PAH dihydrodiols. Among the four BaP diolepoxides shown in Figures 1 and 2, (+)BaP-anti-diolepoxide, the enzymatically formed major diolepoxide, possesses the highest biological activity relative to DNA binding and tumor initiation (9,11). In addition to exhibiting differential DNA binding (14), these four isomers also lead to different types of DNA adducts which may show differential susceptibility to DNA repair or alter the fidelity of DNA replication to varying degree. Thus, it is apparent that the stereospecificities of the metabolizing enzymes is an important factor for determining the biological activities of a PAH on a species-specific basis.

While in vitro metabolism study can provide useful information and methodology for in vivo metabolism, it can never be considered as a substitute for the latter. Different metabolic patterns between the in vivo and in vitro metabolism have been reported (15,16) and metabolism of PAH could conceivably have marked species-specificity, strain-specificity, as well as organotropic effect. For these reasons, the study of PAH metabolism in vivo and between species under various environmental conditions is a necessity. Such an evolutionary-comparative approach to the pharmacokinetics and pharmacodynamics of PAH metabolism is required if we are to determine the cause(s) underlying species difference which will then permit quantitative extrapolation of risk between species.

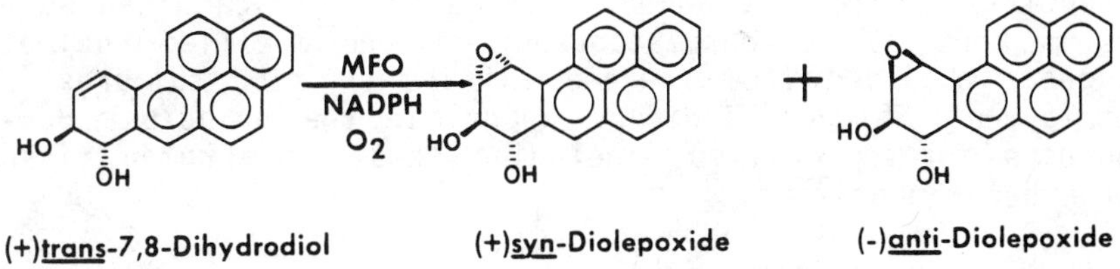

FIGURE 2. Formation of diolepoxides from metabolism of the synthetically prepared (+)BaP-trans-7,8-dihydrodiol.

DNA BINDING

PAHs interact with cellular DNA through their metabolically activated forms. Since activated metabolites were not available, initial attempts to demonstrate DNA adduct formation were based upon total binding in vivo of parent PAHs. As early as 1964, Brookes and Lawley (17) showed a good correlation between the carcinogenicity of a series of PAHs of widely differing carcinogenic potencies and their extent of covalent binding with DNA. The structures of the 7-bromomethylbenz[a]-anthracene-guanosine adducts were determined in 1971 (18). Structural characterizations of the PAH K-region-oxide--guanosine adducts and benzo[a]pyrene diolepoxide (BPDE) guanosine adducts were first reported in 1976 (19-21). Research in this area of PAH studies has since progressed rapidly both in vitro and in vivo; however, few comparative studies between species or tissues have yet been performed.

Since the level of PAH-modified DNA formed in vivo is low; many studies report the reaction of synthetically-prepared activated forms of PAHs, including oxides and diolepoxides, and directly-acting alkylating agents, with homopolymers or DNA in vitro. The modified DNA homopolymer is hydrolyzed and the modified nucleosides are separated from the unmodified nucleosides by column chromatography on Sephadex LH-20. Further separation and purification of the modified nucleosides can be achieved by high performance liquid chromatography. Structures of the purified nucleosides are then determined by their high resolution nmr spectra, high resolution electron impact or field desorption mass spectra of the adduct derivatives, UV-visible and fluorescence spectra, pKa determinations, nitrous acid treatment, circular dichroism, and chemical reactivities. By these means, not only the structures, but also in some cases, the conformation and absolute configurations of the adducts can be elucidated. The identified nucleoside adducts formed in vitro can then be used as markers for co-chromatography of the modified nucleoside adducts obtained from treated animals or cell cultures. By employment of this indirect approach, the principal BaP-DNA adducts formed in vivo, including those from human tissues, have been identified.

PAH-DNA Adducts Formed In Vitro

When 7-bromomethylbenz[a]anthracene (7-BrMBA) was reacted with salmon sperm DNA in sodium phosphate buffer at pH 7, three adducts were obtained and identified as N^2-dG, N^6-dA, and N^4-dC adducts (compounds 1-3 shown in Figure 3) (18). Similar types of adducts were obtained from 7-bromomethyl-12-methylbenz[a]-anthracene (7-BrM-12MBA) reacted with salmon sperm DNA (18). Although 7-BrM-12MBA is more carcinogenic than 7-BrMBA, it formed adducts with DNA to a lesser extent. Both compounds reacted more extensively with native DNA than with denatured DNA. The decrease of total binding with denatured DNA is at the expense of the binding with deoxyguanosine. The extent of binding to the N^6 position of adenine is similar in either native or denatured DNA. These results indicate that the DNA double helix assists the reaction of the exocyclic amino group of the guanosine with the PAH bromo derivatives, but that it has no effect upon deoxyadenosine substitution.

Reaction of (+) anti-BPDE with DNA has been reported to result in a large number of adducts mainly derived from deoxyguanosine and deoxyadenosine residues. Nine adducts have been identified from the reaction of (+) anti-BPDE with native calf thymus DNA or ØX174 DNA (22). Among these adducts, three are derived from deoxyguanosine (4-6), four from deoxyadenosine (10-13), and two from deoxycytidine (14) (see Figure 3). When the resolved (+) anti-BPDE enantiomer was reacted with double-stranded calf thymus DNA, the major adduct was the N^2-dG adduct 4, together with two N^6dA adducts 10 and 11 as minor adducts, while the other two N^4-dC adducts were found in trace amounts. The (-) anti-BPDE enantiomer reacted with native calf thymus DNA to form mainly two N^6-dA adducts, 12 and 13, and two other N^2-dG adducts, 5 and 6 (22) as minor adducts. The N^6-dA adducts formed from the (+) anti-BPDE enantiomer are in nearly equal amounts to those N^6-dA adducts formed from the (-) anti-BPDE enantiomer. However, the amounts of N^2-dG adduct(s) formed from the (+) and from the (-) enantiomers are quite different. The ratio of the N^2-dG adduct from the (+) anti-BPDE to that of adducts from the (-) enantiomer was 20:1, when either native calf thymus DNA or double-stranded ØX174 DNA was used. However, reaction of the racemic (+) anti-BPDE with single-stranded ØX174 DNA resulted in nearly equal amounts of the N^2-dG adducts. The total amount of N^2-dG adducts decreased, while the proportion of N^6-dA adducts increased to 40% compared to 10% formed in the reaction with double-stranded DNA. Thus, Meehan and Straub (22) suggested that the covalent binding of the two enantiomeric anti-BPDEs to form N^2-dG adducts in double-stranded DNA was dependent on the secondary structure of the polymer. As a result, the more native the polymer, the higher the percentage of the N^2-dG adducts formed (greater than 90% in the normal case); and the more denatured the polymer, the higher the proportion of N^6-dA adducts formed. The possible mechanism proposed by these authors for the dependence of DNA secondary structure is the stereoselective

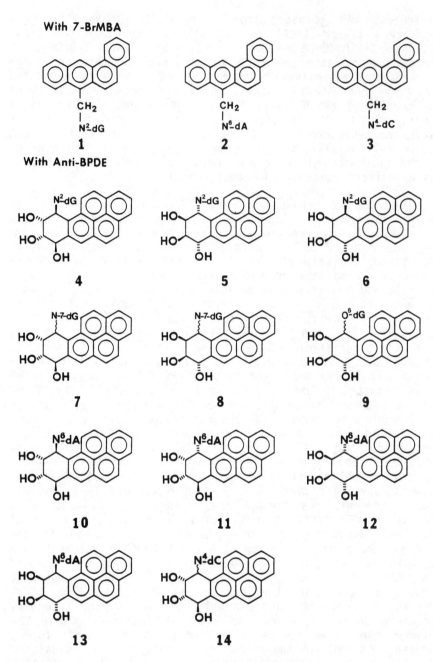

FIGURE 3. PAH-DNA adducts formed in vitro.

physical interactions, such as intercalation prior to the covalent binding. The differences in physical interactions may also account for, in part, the higher DNA binding efficiency of the anti-BPDE isomer than the syn-BPDE isomer. On the other hand, the N^6-dA adduct formation may proceed by a different mechanism from that of N^2-dG, since N^6-dA adducts were formed in increasing amounts from double-stranded to single-stranded DNA (22).

A separate report by Osborne et al. (23) provided somewhat different results. It was found that the (+) anti-BPDE enantiomer reacted with E. coli DNA only in 2-3 fold higher amounts than did (-) anti-BPDE enantiomer. Besides the above, two N-7-dG (7-8) and one O^6-dG (9) adducts were identified (Figure 3). While these two N-7-dG adducts were each formed from (+) and (-) anti-BPDE, the O^6-dG adduct was only obtained from the (-) anti-BPDE. The N-7-dG adduct of anti-BPDE was reported to be quite unstable (23). At neutral pH, this adduct was cleaved from the DNA spontaneously with a half-life of 3 hours. Therefore, the identification of this adduct was mainly by a chemical approach, not by standard spectroscopic techniques. The relative yields of adducts were reported as follows: from (+) anti-BPDE, 91% N^2-dG, 2% N-7-dG, 2% N^6-dA, and 5% unidentified guanine adducts; from (-) anti-BPDE, 45% N^2-dG, 31% O^6-dG, 18% N-7-dG, and 6% N^6-dA. These results again demonstrated that the (+) anti-BPDE exhibited high stereospecificity, forming virtually a single adduct; but the (-) isomer did not. This difference may be a result of preferential intercalation, as suggested by Meehan and Straub (22).

Geacintov et al. (24,25) however, have suggested that intercalation is not the factor facilitating the covalent binding of (+) anti-BPDE with DNA. Based on electric linear dichroism studies, fluorescence spectra, and stopped-flow kinetics, they proposed that non-covalent binding of BPDE at an intercalation site only facilitates its hydrolysis to tetrols. Presumably, physical interactions at the exterior site of the DNA helix could eventually lead to stereospecific covalent binding.

Due to the difficulty in preparing DMBA diolepoxides in sufficient amounts for DNA binding studies, DMBA-diolepoxide-deoxynucleoside adducts have not been fully characterized. Reaction of DMBA-trans-3,4-diol-1,2-epoxide with DNA has been reported by Jeffrey et al. (26), and by Vigny et al. (27). The isolated deoxynucleoside adduct was tentatively identified as the product of linkage between the exocyclic amino group of the guanosine residues to 1- or 2-position of the DMBA diolepoxide.

PAH Adducts Formed In Vivo and in Cell Culture

1. Benzo[a]pyrene

The first BaP adduct isolated from a cell system was the hydrolysis product of modified RNA isolated from cultured bovine bronchial mucosa treated with tritiated BaP (20). Subsequently, hydrolysis of DNA isolated from mouse skin treated with ^{3}H-BaP gave two major adducts; one identified as compound 4 the other was derived from 9-hydroxy-BaP-4,5-oxide (28,29). This and other in vivo studies on binding of BaP to mouse and rat skin indicated that the binding to DNA or RNA of BPDE is highly stereoselective, always resulting in the formation of N^2-dG, 4 as the major adduct (14, 29-31). Two additional deoxyguanosine and two deoxyadenosine adducts were also detected in mouse and rat skin DNA (30). The total DNA binding in rat skin was three times less than in mouse skin, but the quality and distribution of bound products was similar (30). Since rat skin is resistant to BaP carcinogenesis while mouse skin is susceptible (30), the difference in total BaP binding in the skin of these two species correlates with their respective susceptibilities to tumor induction. In hamster embryo cells the form of BaP-DNA adducts produced is dependent upon the duration of time of exposure (32). At six hours post exposure, most of the adducts resulted from reaction with ($\pm$) syn-BPDE, but after 24 hours, most adducts resulted from reaction with ($\pm$) anti-BPDE (32). Human cell cultures of bronchus, trachea, colon, esophagus, lung, bronchial mucosa, epidermal keratinocytes, and endometrium (33-37) all exhibit stereoselective metabolism of BaP to ($\pm$) anti-BPDE as a major and ($\pm$) syn-BPDE as a minor metabolite and stereoselective binding of DNA with BPDE. The absolute and relative amounts of BaP adducts formed vary as a function of duration of exposure, species, tissue and cell line. The factors which may account for these differences include: a) a multiplicity of metabolic pathways; b) the stability and reactivity of ($\pm$) synBPDE and ($\pm$) anti-BPDE; c) rate of excision; and d) growth condition of the culture systems.

2. 7,12-Dimethylbenz[a]anthracene

Isolation and hydrolysis of the skin DNA of male C57BL mice treated with ^{3}H-labeled DMBA resulted in three adducts whose fluorescence spectra were anthracene-like and similar to those of adducts obtained from reaction of DNA with DMBA-trans-3,4-diol-1,2-epoxide (27). Similar results were obtained when DMBA was applied to the skin of female or male Swiss mice (38,39), or incubated with rodent cell cultures (38,40,42). Thus, the deoxynucleoside adduct patterns formed in vivo supported the bay-region theory. The results of metabolism and DNA-binding in Syrian hamster embryo cells of DMBA and its 1-, 2-, 4-, 5-, 9-, and 11-fluoro derivatives by Hart and coworkers (42-44) also supported the bay-region theory.

The deoxynucleoside adduct pattern formed in vitro is, however, quite different from that obtained in vivo or in cell culture. Metabolism of DMBA with rat liver microsomes in the presence of calf thymus DNA, gave the deoxynucleoside adducts mainly derived from the K-region arene oxide (38). When a liver homogenate, prepared from male Sprague-Dawley rats pretreated with Aroclor 1254 (S9 fraction) was used for metabolizing DMBA, the DNA adduct pattern is concentration-dependent (45). At high DMBA/S9 ratios, adducts derived from the K-region arene oxide were prominent among the products. However, at low DMBA/S9 ratio, the adducts derived from diolepoxide occurred (45).

3. Benz[a]anthracene

Binding of BA with nucleic acids in vivo in mouse skin (46) and in hamster embryo cells (46-49) has been studied. Surprisingly, the major identified adducts from the DNA of mouse skin and hamster embryo cells were derived from the non-bay-region BA-trans-8,9-diol-anti-10,11-epoxide. The second major adduct was derived from the bay-region BA-trans-3,4-diol-anti-1,2-epoxide. It appears that both the non-bay-region and bay-region diolepoxides may contribute to the biological activity of BA (46). The results showed that the non-bay-region diolepoxide reacted in hamster embryo cells with guanosine and adenosine in RNA and with deoxyguanosine in DNA. However, the bay-region diolepoxide reacted mainly with deoxyguanosine. Both deoxyguanosine adducts derived from these two diolepoxides have been shown to be N^2-dG derivatives (48).

4. Other PAHs

DNA binding studies of 3MC (50) and 7-MBA in mouse skin in vivo and in hamster embryo cells (51) showed that the DNA binding adducts formed were derived from bay-region diolepoxide. However, the structures of these adducts have not yet been determined.

DNA binding adducts formed from Swiss S female mice treated with 7-BrMBA or 7-BrM-12-MBA resulted in three identified adducts in both cases (52). They are N^2-dG as major; N^6-dA and N^4-dC as minor.

DNA REPAIR

As described in the previous section, interaction of chemical electrophiles with DNA leads to covalently bound adducts. Formation of strand breaks, phosphotriesters, cross-links, apurinic and apyrimidinic sites, deamination and hydration may also occur as a result of such interactions. The level of DNA damage susceptible to misreplication is assumed to be inversely proportional to both the extent and rate of DNA repair prior to DNA replication. Thus, the greater the repair capacity, or the longer the period between the induction of damage and the replication of DNA, the lower the level of misreplication and mutation.

There appear to be at least four general categories of DNA repair: excision, post-replication, strand break, and photoreactivation. The first three of these systems repair DNA damage induced by PAHs. Excision repair, the most widely-examined form of repair, operates in four steps to remove bases which have been damaged by physical or chemical agents: a) a damage-specific endonuclease nicks the DNA backbone adjacent to the damaged site; b) an exonuclease removes the damaged nucleotides and others adjacent to them; c) a polymerase replaces these nucleotides with new ones using the opposite strand as a template; and d) a ligase seals the final phosphodiester bond. This type of excision repair is nucleotide-excision repair. A second type of excision repair, base excision repair, has an additional initial step and appears to operate on bases damaged by alkylating agents or via thermal denaturation. In this process, an N-glycosidase cleaves the altered purine base and the resulting apurinic site is nicked by apurinic-site endonuclease and then the exonuclease, polymerase and ligase complete the repair (6).

Post-replication repair is the repair process which operates after DNA replication has started. Strand-break repair rapidly restores broken single-strands and in some organisms, double-strands.

The methods for DNA repair <u>in vivo</u> have been reviewed by Brash and Hart (6).

Comparative Removal of PAH-DNA Adducts from Different Species and Strains

1. Benzo[a]pyrene

Studies on the excision repair of BaP-modified DNA have been performed in different cell lines including baby hamster kidney 21/C13 cells (BHK), secondary mouse embryo fibroblasts C57BL/6J (MEF), normal hamster tracheal epithelial cells (NHT), epithelioid human alveolar tumor cells A549, and 10T1/2 mouse embryo fibroblasts (10T1/2) (53-57). The BaP-modified DNA adducts formed are mainly dG-BaP adducts which arise from the reaction of DNA with syn- and anti-BPDE metabolites. The results of excision are summarized in Table 1. In two reports (53,54), the excision of N^2-dG-syn-BPDE and N^2-dG-anti-BPDE as a function of times was examined (Table 2). As shown in both Tables, the adducts were excised slowly, particularly in normal hamster tracheal epithelial cells (NHT) and in MEF cells. Because these experiments were done in different laboratories under different conditions and measurements of the excision were at different intervals, it is difficult to give a quantitative comparison of excision rate in different cell lines.

TABLE 1

EXCISION OF BaP-MODIFIED DNA ADDUCTS FROM DNA IN DIFFERENT CELL LINES

Cell Line	0 hr[a]	24 hr	48 hr	68 hr	72 hr	5 days
			BaP-DNA Adduct Excised (%)			
BHK[b]	0	28			~80	
MEF[b]	0	15			~40	
10T1/2[c]				~40		
A549[d]	0				45-60	
NHT[e]	0		~50			~60

[a] Duration of post-treatment incubation. [b] Data from ref. 53.
[c] Data from ref. 57. [d] Data from ref. 54. [e] Data from ref. 55.

TABLE 2

EXCISION OF N^2-dG-BaP ADDUCTS FROM DNA IN DIFFERENT CELL LINES

Cell Line	N^2-dG-syn-BPDE Excised (%)				N^2-dG-anti-BPDE Excised (%)			
	0 hr[a]	24 hr	48 hr	72 hr	0 hr	24 hr	48 hr	72 hr
BHK[b]	0	37.7	-	78.9	0	38.0	-	81.7
MEF[b]	0	13.3	39.8	46.0	0	14.8	32.0	32.9
A549[c]	0	-	-	60	0	-	-	45

[a] Duration of post-treatment incubation. [b] Data from ref. 53.
[c] Data from ref. 54.

Tentatively, the relative rates of excision decreased according to the order: BHK>A549>10T1/2 >MEF>NHT.

It was estimated that in a 72 hr period of post-treatment incubation, a BHK cell on the average excised 1.1×10^4 molecules of dG-BaP adducts, and an MEF cell excised 4.3×10^4 molecules (53). Although results indicated that excision of dG-BaP adducts is faster in BHK than in MEF, a similar value of 14-15% was calculated for the fraction of N^2-dG-BaP adducts removed from DNA per cell generation. This calculation is based on the information that one cell generation time of BHK is 12 hr and that of MEF is 22-24 hr (53,56).

Excision of N^2-dG-syn-BPDE and N^2-dG-anti-BPDE from DNA in different cell lines took place at different rates. As indicated in Table 2, while both MEF and A549 cells removed N^2-dG-syn-BPDE faster, the BHK cells removed both adducts at nearly equal rates. The faster excision of N^2-dG-syn-BPDE was also observed in 10T1/2 mouse embryo fibroblasts (57).

2. Benzo[a]pyrene-trans-7,8-diol-anti-9,10-epoxide (anti BPDE) and benzo[a]pyrene-trans-7,8-diol-syn-9,10-epoxide (syn-BPDE)

Formation and excision of N^2-dG-syn-BPDE and N^2-dG-anti-BPDE from syn- and anti-BPDE with DNA in A549 and in 10T1/2 cells were studied (57-59). In 10T1/2 cells, 50 to 70% of these adducts were excised from cellular DNA during a 68 hr post-treatment incubation (57). When a mixture of syn- and anti-BPDE was used for 2 hr, the extent of the modified DNA was 4-fold greater than that obtained after a 24 hr incubation with BaP (57). The higher initial adduct concentration observed with syn- and anti-BPDE may account for the faster excision rate of N^2-dG-BPDE adducts seen with BPDE exposure (50-70% excised) vs. BaP exposure (~40% excised) when measured at the same post-treatment time (68 hr) (57). Kinetics of excision of N^2-dG-anti-BPDE formed from exposure of anti-BPDE with human lung cells A549 and with normal human foreskin fibroblasts were studied (58,59). As shown in Tables 3 and 4, excision of this modified adduct is much faster in the A549 cells than in the human foreskin fibroblasts. In the human lung carcinoma cells A549, the percentage of excision reached 49% after 29 hr of post-treatment incubation, and ~73% at 54 hr. Also, the rate of excision had decreased after 72 hr, leaving about 20% of the adducts unexcised (58). The rate of excision in human foreskin fibroblasts was much slower. After eight days of post-treatment incubation, there was still about 30% of the adducts unexcised (59). However, such observations may be, as indicated above, misleading since the cell generation time in human foreskin fibroblasts is between 2-3 times longer.

TABLE 3

EXCISION OF N^2-dG-ANTI-BPDE FROM DNA IN HUMAN LUNG TUMOR CELLS A549 (58)

	Time After Post-Treatment (hr)			
	0	29	54	72
Conc. of adduct (μmol/mol of DNA-p)	28.5	14.1		
Excision (%)	0	49	~73	~80

TABLE 4

EXCISION OF N^2-dG-ANTI-BPDE FROM DNA IN NORMAL HUMAN FORESKIN FIBROBLASTS (59)

	Time After Post-Treatment (days)			
	0	2	4	8
Adduct Residues per 10^6 DNA Nucleotides	10.2	4.9	3.7	3.1
Excision (%)	0	56	62	70

As described by Feldman et al. (54,58), after 72 hr of post-treatment incubation the percentage of excision of the DNA lesions in the experiments with BaP is about 45-60% (see Tables 1 and 2), and the percentage of the DNA lesions in the experiments with anti-BPDE is about 80% (see Table 3). The different excision rates in these two systems were explained as due to differences in the initial distribution of the adducts in chromatin (58). The adduct concentration was similar in micrococcal nuclease resistant and sensitive DNA following 48 hr incubation with BaP; but the adducts formed from incubation of anti-BPDE for 30 min were 10 times higher in the micrococcal nuclease sensitive DNA. These authors proposed that "the adducts located at the chemically more reactive sites of the nuclease sensitive DNA of chromatin may be more accessible to repair enzymes and removed more completely" (58). Further, "the extrapolation of results of in vitro mutagenesis and transformation using ultimate metabolites to the in vivo situation, where in general, procarcinogens are metabolized in the target organ have to be done with caution." As initially pointed out by Wilkins and Hart (60), differential distribution and repairability of structurally identical lesions may mean different biological potency for the same lesion based upon its location within the genome.

3.   7-Bromomethylbenz[a]anthracene (7-BrMBA)

Excision of 7-BrMBA modified DNA adducts has been studied in different cell lines including human lymphocytes, HeLa cells, Chinese hamster V-79 cells, normal human fetal lung cells, several XP variant fibroblasts, mouse skin cells, and several strains of E. coli (61-65). The adducts formed in cell cultures are mainly N^2-dG, N^6-dA and N^4-dC adducts, as are those obtained from reaction of 7-BrMBA with DNA under cell free conditions. The excision of these 7-BrMBA-modified adducts in the cell lines studied thus far has shown that the N^6-dA adduct is always excised at a faster rate than the N^2-dG adduct. This difference has also been found in the other PAH-modified adducts, and thus is discussed separately.

The results of excision of 7-BrMBA modified adducts from DNA in human lymphocytes are summarized in Table 5. When 1 μM of 7-BrMBA was exposed to the cells, about 15 to 17% of the lesions were excised in a period of 12 hr (62). When the dose

TABLE 5

EXCISION OF 7-BROMOMETHYLBENZ[a]ANTHRACENE-MODIFIED ADDUCTS FROM DNA IN HUMAN LYMPHOCYTES

Cell	Conc. of 7-BrMBA (μM)	Time (hr)	Excision (%)	Reference
Normal	1	12	15.1	64
Normal	1	12	17.2	64
Normal	20	12	8.8	64
Normal	5	6	18.7	65
Normal	5	6	15.3	65
XP6	5	6	2.1	65
XP9	5	6	1.4	65

of 7-BrMBA was 20 μM, excision decreased by approximately 15% to 8.8%. However, because more adducts were formed when the higher dose of 7-BrMBA was used, the overall amount of excision was still slightly higher. In a separate report, when the dose of 7-BrMBA used was 5 μM (63), the total excision of the 7-BrMBA-modified DNA from the DNA of normal cells 6 hr after treatment was 15 to 18%. As might be expected in XP cells of complimentation group A, only 1 to 2% of the lesions were excised in the same period at the same dose (63).

As shown in Table 6, more detailed excision results were obtained by Dipple and Roberts (61) in HeLa cells and Chinese hamster V-79 cells. In both cell lines, the percentage of total excised adducts decreased after treatment with higher concentrations of 7-BrMBA. When 0.2 µM 7-BrMBA was used in both cell lines, comparison of the excision rate at different intervals indicates that the Chinese hamster V-79 cells possess higher excision capacity than the HeLa cells. The differences in excision capacity between these two cell lines may account for the lower sensitivity of the Chinese hamster cells to the toxic effects of 7-BrMBA (61).

TABLE 6

PERCENTAGE EXCISION OF 7-BROMOMETHYL[a]ANTHRACENE-MODIFIED ADDUCTS FROM DNA IN HeLa AND V-79 CELL LINES (61)

Time (hr)	HeLa		V-79	
	Conc. of 7-BrMBA			
	0.2 µM	0.6 µM	0.1 µM	0.2 µM
3	15	10	16	17
9	21	13	25	22
18	29	16	44	31
30	37	25	52	47

When Chinese hamster V-79 cells were exposed to 7-BrMBA in a low concentration (0.1 µM) for a period of 30 hr, almost 100% survival of the cells was found and all of the DNA was replicated during this stage while only 50% of the adducts were excised (61). These observations imply that complete excision of the DNA lesions from DNA is not a prerequisite for DNA replication and subsequent cell survival (61).

If the excision rates obtained from the non-dividing and dividing cell lines were comparable, the results shown in Tables 5 and 6 would suggest that among these three cell lines, normal human lymphocytes have the highest excision capacity, and the HeLa cells have the least. Any general conclusion as to species differences is difficult, however, since these comparisons are between transformed vs nontransformed and lymphocytes vs fibroblasts.

4. 7,12-Dimethylbenz[a]anthracene and Its Fluoro Derivatives

Sheikh et al. (66) studied the DNA binding in vivo of DMBA in two rat strains, Sprague-Dawley (SD) and Long-Evans (LE). The maximum binding in five different organs (kidney, lung, heart, mammary gland, and liver) and total excision in seven days were determined. The results are summarized in Table 7. The maximum binding levels were two to eight times higher in all organs of the LE strain when compared to those of SD rats. In LE rats, lung had the highest maximum binding,

TABLE 7

EXCISION OF DMBA-MODIFIED DNA ADDUCTS FROM DIFFERENT ORGANS OF FEMALE SPRAGUE-DAWLEY (SD) AND LONG-EVANS (LE) RAT STRAINS IN VIVO (66)

Organ	Maximum Binding (pmol DMBA/mg DNA)		Binding After 7 days (pmol DMBA/mg DNA)		Excision in 7 days (%)	
	SD	LE	SD	LE	SD	LE
Kidney	9.6	77.4	6.83	2.5	29	97
Lung	14.4	151.9	8.22	5.3	43	91
Heart	31.2	124.0	23.3	7.3	26	94
Mammary	16.6	113.7	15.0	11.3	9	90
Liver	44.6	91.8	11.4	4.3	75	95

followed by heart and mammary. In SD rats, liver had the highest binding. In both strains, kidney received the least maximum binding. Contrary to maximum binding, the extent of removal of adducts was consistently greater in the LE strain of rats when compared to that in SD rats. Seven days post-treatment all organs of the LE rats had lost at least 90% of their DMBAmodified adducts; whereas, with the exception of liver only 9 to 43% were lost in SD rats. Livers of both SD and LE rats excised relatively large amounts of adducts.

Consistent with these findings is the observation of Huggins et al. (67) which established that administration of

DMBA (10 mg/day/i.g.) to young female SD rats produced a 100% incidence of mammary cancer whereas the same dose produced a 30% incidence in LE animals. Thus, the results of Sheikh et al. may suggest that the relatively greater resistance of the LE strain to mammary carcinogenesis is a function of repair processes rather than absolute binding.

Similarly, the total maximum binding of 2-fluoro-DMBA was at levels only 3 to 10% of that found with DMBA in all tissues studied in SD rats (Table 8). The rates of excision of 2-fluoro-DMBA-modified adducts from the DNA of the heart, mammary and liver tissues of SD rats were higher than those observed for DMBA. The much lower maximum adduct formation and more efficient excision of the lesion may account for the non-carcinogenicity of 2-fluoro-DMBA.

TABLE 8

EXCISION OF 2-FLUORO-DMBA-MODIFIED DNA ADDUCTS FROM DIFFERENT ORGANS OF FEMALE SPRAGUE-DAWLEY (SD) RATS IN VIVO (66)

Organ	Maximum Binding (pmol DMBA/mg DNA)	Excision in 7 Days (%)
Kidney	0.37	-
Lung	0.57	-
Heart	1.86	61
Mammary	1.25	20
Liver	4.90	63

DNA repair in Syrian hamster embryo cells treated with DMBA and its weakly carcinogenic analog, 5-fluoro-DMBA was also studied by D'Ambrosio et al. (68). The amount of post-replication repair was estimated by measuring the percentage of pulse-labeled DNA sedimenting as high-molecular-weight DNA. It was found that at a concentration of 5 µg/ml, 15% of the pulse-chased DNA sedimented as high-molecular-weight DNA after DMBA treatment vs 26% after 5-fluoro-DMBA treatment and 28% after no treatment. Similar results were obtained when the concentration 1 µg/ml of substrates was used. 5-Fluoro-DMBA is about 100-fold less active as a carcinogenic initiator and mutagen than DMBA. It appears that these differences are probably not related to the total binding, but rather, to the levels of specific types of adducts entering DNA replication (68).

Dipple and Hayes found that DMBA-DNA adducts were intrinsically difficult to excise in the primary mouse embryo cell cultures (69). Even at a very low initial binding concentration (1 adduct per 2.5×10^6 nucleotides) only about 7% of the adducts were excised at 48 hr.

Tay and Russo studied DMBA-induced DNA binding and excision in Sprague-Dawley rat mammary epithelial cells (70). In these studies cells were cultured from young virgin (YV), old virgin (OV), and parous (P) rats. Over a dose range of 0.5-2.0 µg/ml, excision repair of the adducts from DNA in YV rats was 1.5 times higher than in OV cells and 2 times higher than in P cells at 48 hr. However, at doses of < 0.5 µg/ml, the excision levels in all three groups of rats were very similar at 48 hr despite the observation that the total amount of adducts formed in YV cells was found to be 1.5- to 2.0-fold higher than in OV or P cells. When both total adduct formation and excision rate are considered together, it is apparent that at a lower concentration of DMBA (< 0.5 µg/ml), OV and P cells exhibit higher DNA excision per unit damage (70). Therefore, Tay and Russo suggested, based on these results, that age and parity can decrease the DMBA-induced DNA adduct formation and increase the efficiency of excision of these lesions in mammary epithelial cells. These results may also explain the lower susceptibility of both OV and P rats to DMBA-induced mammary carcinogenesis (70).

5. 3-Methylcholanthrene

Comparative removal of 3MC-induced DNA adducts from lung and liver DNA in four different strains of mice in vivo was studied by Eastman and Bresnick (71). [^{3}H]-3MC was i.v. injected into A/J, C3H/HeJ, DBA/2J, and C57BL/6J mice and the adducts formed in lung and liver tissues were analyzed 4 hr, 7 days, and 28 days after injection. The results are shown in Table 9. The adducts in liver were completely removed by 28 days in all four strains. These results correlate with the findings that livers from these animals are resistant to carcinogenesis by 3MC. The susceptibility of A/J lung to 3MC-induced carcinogenesis was also consistant with the observation that after 28 days, the lung of A/J mice contained the highest adduct level (71).

TABLE 9

BINDING OF [^{3}H]-3MC TO DNA IN MICE (71)

Strain & tissue	in vivo exposure time	fmol 3MC bound per mg DNA[a]	fmol DNA adduct per mg DNA
A/J			
Lung	4 hr	113	44.9
	7 days	74	30.0
	28 days	106	15.9
Liver	4 hr	135	5.5
	7 days	96	5.5
	28 days	96	0
C3H/HeJ			
Lung	4 hr	76	17.6
	7 days	43	8.3
	28 days	43	6.1
Liver	4 hr	45	4.4
	7 days	20	0.6
	28 days	29	0
DBA/2J			
Lung	4 hr	68	16.3
	7 days	35	6.3
	28 days	24	1.1
Liver	4 hr	64	4.8
	7 days	50	1.3
	28 days	28	0
C57BL/6J			
Lung	4 hr	173	16.1
	7 days	103	2.7
	28 days	83	1.7
Liver	4 hr	139	9.7
	7 days	135	0
	28 days	107	0

[a] These values assume that all detected radioactivity represents 3MC adducts. The values represent the mean from 6 to 10 mice (3 to 5 determinations). A fairly large individual variation was observed with S.D. of ± 25% of the mean value.

Like 3MC, BaP was found to exhibit high excision capability in liver and kidney of A/J mice and relatively low excision capability in lung tissue. Thus, these results also appear to correlate with the high pulmonary neoplasia induction by both BaP and 3MC in mouse.

6. 7-Bromomethyl-12-methylbenz[a]anthracene

Despite the observation that both bromo compounds form similar types of DNA adducts, 7-BrM-12-MBA is more carcinogenic than 7-BrMBA (53,72); and induces lower levels of DNA adduct formation both in vivo and in vitro (53,61). Dipple and Schultz (72) studied the excision of both compounds in primary cultures of mouse embryo cells (ME cells) and in mouse L929 cell suspension cultures (L cells) in order to determine if a difference in excision repair existed. The results indicated that excision of both 7-BrM-12-MBA-modified and 7-BrMBA-modified adducts were faster in ME cells than in L cells, an average of about 2.8% and 1.5% of the initial adducts per hour were excised in the ME cells and L cells, respectively. However, the extent of excision of 7-BrM-12-MBA-modified and 7-BrMBA-modified adducts is similar. Thus, the greater carcinogenic potency of 7-BrM-12-MBA is not because its adducts are less susceptible to cellular repair systems (72).

Selectivity of Excision of N^2-deoxyguanosine and N^6-deoxyadenosine Adducts

As mentioned previously, PAH-modified DNA adducts are mainly those involving the linkage of the exocyclic amino groups of deoxyguanosine, deoxyadenosine and deoxycytidine to give N^2-dG, N^6-dA, and N^4-dC adducts, respectively. The N^2-dG adducts are always the predominant, if not the only, products and the N^6-dA adducts are formed in smaller amounts. The excision of PAH-modified N^2-dG and N^6-dA adducts has been found to occur at different rates. In most cases, the N^6-dA adduct is excised much faster than the N^2-dG adduct.

N^6-dA of 7-BrM-12-MBA formed upon topical application of 7-BrM-12-MBA to the back of Swiss S mice was excised faster than that of the N^2-dG adduct (53). On the relative excision rate of 7-BrMBA-modified N^2-dG and N^6-dA adducts (53,61,62,65), about 25-30% of the N^6-dA is excised in a period of 12 hr in non-dividing human lymphocytes, while only 13-16% of N^2-dG is excised (Table 10) (62). As shown in Table 10, repair of the N^6-dA adduct in HeLa cells is much faster than N^2-dG. Thirty hours after exposure of HeLa cells to 0.2 or 0.6 μM of 7-BrMBA, excision of the N^6-dA adduct was nearly complete and thus the excision rate difference was found to be about 4-fold for these two adducts in HeLa cells (61). Additionally, in V-79 cells

TABLE 10

SELECTIVE REMOVAL OF 7-BROMOMETHYLBENZ[a]ANTHRACENE INDUCED N^2-dG AND N^6-dA ADDUCTS IN CELL CULTURE

Cell Line	Ratio of N^6-dA/N^2-dG						
	0	3hr	9hr	12hr	18hr	24hr	30hr
Non-dividing Human Lymphocytes[a]	0.27			0.23			
	0.27		0.22				
HeLa[b]	0.15	0.10	0.08		0.04		0.03
	0.16	0.13	0.13		0.11		0.07
V-79[b]	0.19	0.16	0.14		0.11		0.10
	0.22	0.18	0.15		0.09		0.10
	0.23	0.21	0.15		0.09		0.13
	0.25	0.21	0.25		0.24		0.24
XP2BE[c]	0.24					0.22	
	0.20					0.19	
	0.20					0.19	
	0.14					0.05	
Normal human fetal lung cells	0.17					0.08	

[a] Data from ref. #62. [b] Data from ref. #61. [c] Data from ref. #65.

the half-life for excision of N^2-dG was about twice that for the excision of N^6-dA (61). The rates of loss of N^2-dG and N^6-dA are dose-dependent (83); with the lower the concentration, the greater the rate difference. When the concentration of 7-BrMBA was 1.8 μM in V-79 cells, the excision of N^2-dG and of N^6-dA was nearly equal (61). Consistent with the above in both normal human fetal lung cells and in XP4BE variant fibroblast cells, removal of N^6-dA is faster than N^2-dG (65). However, in the XP2BE cells, which are partially excision repair-deficient, the excision rates for N^6-dA and N^2-dG are almost equal (Table 10) (65), since XP2BE cells are not capable of excision of either form of damage.

A study on the removal of BaP adducts of DNA in hamster tracheal epithelial cells also indicated that four adenosine adducts, probably N^6-dA adducts, were removed almost completely in 24 hours, while the others, including N^2-dG adducts, were poorly repaired (55). The N^2-dG adducts derived from either anti- or syn-BPDE thus appear to be very persistent regardless of the stereoisomers studied (53,54,58,59).

The basis for the preferential excision of the N^6-dA to the N^2-dG adducts is not clear. Based on the results of susceptiblity to digestion by the single-strand-specific nuclease S_1, formaldehyde unwinding, heat denaturation, fluorescence quenching, and computer-generated model (73), it was suggested that the N^2-dG residue of anti-BPDE lies in the minor groove and causes little local destabilization of the DNA double helix. The covalently bound BPDE residue is on the exterior of the molecule, where the pyrene chromophore resides with a 35° maximum angle to the long axis of the DNA. Because of its orientation, the N^2-dG adduct of anti-BPDE may be less susceptible to excision. The crystal structure of N^6-dA of 7-BrM-12MBA was determined (74). Unlike the unsubstituted deoxyribonucleosides, this adduct adopts the syn conformation in the crystalline form. It was proposed that, if the N^6-dA residues of 7-BrM-12-MBA, 7-BrMBA, and syn- and anti-BPDE adopt the syn conformation in DNA, this would disrupt the normal base pairing, result in some local denaturation and thus lead to the preferential excision of these lesions by single-strand specific nucleases, such as S_1 (74).

CONCLUSION

These observations are immediately apparent: 1) a paucity of data exists comparing the effect of a single compound on either cells or individuals of different species, 2) differences in metabolic activation, detoxification and DNA repair of PAHs exist between organs, organisms and species as well as under different environmental conditions; however, binding of the same metabolite of a PAH to cellular DNA is similar regardless of the source of the DNA, and 3) the dependent variables in PAH metabolism and removal of the adducts from DNA which have been identified between organs, organisms and species have yet to be associated with an independent variable such as species maximum achievable lifespan or risk per cell per unit time for spontaneous malignant transformation. Until such studies are performed and correlations made, any attempt to extrapolate animal data to human health effects must remain as a qualitative exercise. Further, all studies reviewed had one or more important limitations: 1) few studies attempted to adjust for differences in cell cycle time between the different cell culture systems used; 2) when total radiolabel bound to total DNA was used as a measure, differences in DNA content per cell were oftentimes not adjusted for; 3) few in vivo studies took into consideration species and organ specific differences in cell killing which could in part account for different rates in loss of adducts; 4) chronicological time rather than time from maximum binding was often used in determining rate of repair in different tissues, species and between compounds; and 5) through both metabolism and repair may vary as a function of age, sex, etc. most strain/species comparative studies were not adjusted for percentage of species maximum lifespan expended at time of exposure. It is important to remember, however, that most of these studies were based on an ontogenetic rather than comparative-evolutionary approach to toxicology and are thus valid for the purpose intended but of only minimal use in the extrapolation of data between species and organs.

Therefore, at this time no universal correlation between either activation, detoxification, binding or repair of PAHs and PAH-induced damage can be made as it can between the excision repair of UV-induced cyclobutane-type pyrimidine dimers and species maximum achievable lifespan of the species from which the cultures were derived (75,76). That such a correlation might exist is not inconsistent with what little data is presently available. This is especially the case since data exists suggesting that PAH and UV-induced DNA damage may be repaired by similar if not identical excision repair systems.

Metabolic activation, detoxification and DNA repair of genetic damage induced by PAHs all exhibit different patterns between in vitro and in vivo and could conceivably have marked species-specificity, strain-specificity, as well as organotropic effects. The rate of metabolism of PAHs by these enzyme systems could vary widely depending on the species, strain, age and sex of the animal. The enzyme activities as well as the regio- and stereoselectivities of these enzymes in an experimental animal can also be altered by inhibitor, inducer, starvation, stress, hepatectomy, nutritional factors, hormonal status, castration, etc. The balance between activation and deactivation is also dependent on various enzyme systems that may be altered by these same environmental factors. Thus, sufficient genetically defined parameters exist that species susceptibility could be altered via organ, strain and species differences in these parameters. In order to identify such differences, however, more studies in noninduced animals will have to be performed.

Unlike metabolism and DNA repair, the binding of PAHs to DNA, once the directly acting metabolite is identified, appears to be somewhat independent of species and organ. Reaction of the activated metabolites of PAHs and bromomethylated PAHs with cellular nucleic acids both in vivo and in vitro always results in guanosine-adducts as the major products, adenosine-adducts as the second major, and cytidine-adducts the minor. Adducts derived from thymidine or uridine have not yet been identified. It is known that the order of the nucleophilicity of these nucleic bases is: G>A>C>U and dG>dA>dC>dT. Therefore, the ratio of the adducts formed in vivo and in vitro is in agreement with the nucleophilicity of these nucleic bases. However, the total guanosine-adduct formation decreases from native DNA to denatured DNA, while the total adenosine-adduct formation increases slightly. Therefore, the formation of guanosine-adducts in amounts larger than the other adducts in native DNA is partly ascribed to both the nucleophilicity of the guanosine base and the "driving force" by the conformation of the DNA double helix.

Binding of the PAH metabolites (and derivatives) to cellular nucleic acids is highly stereoselective. The (+) anti-BPDE exhibits the highest binding capability with double-stranded DNA than the (-) anti-BPDE diastereoisomer and the (+) and (-) syn-BPDE stereoisomers, particularly for the N^2-dG adduct formation. This implies that the binding capability of a chemical to double-stranded DNA is very much dependent upon the secondary structure of the native DNA. Nevertheless, the stereoselectivity is not observed for the N^6-dA formation, since total formation of this N^6-dA adduct is similar in native and denatured DNA.

It is amazing that all the studied diolepoxides and bromomethyl-PAHs attack preferentially at the exocyclic amino group of guanosine, adenosine, and cytosine residues of DNA and RNA. DMBA-5,6-oxide also attacks to the exocyclic amino group of poly G at neutral conditions. The biological activities of (+) anti-BPDE, (+) syn-BPDE, bay-region DMBA-diolepoxide, bay-region BA-diolepoxide, and non-bay-region BA-diolepoxide are known to be greatly different; but yet they all form N^2-dG adducts as the major DNA adducts in vivo and in vitro. Whether or how these N^2-dG adducts formed from diolepoxides of various biological activities exert different degrees of DNA genetic damage is of great importance for better understanding of PAH carcinogenesis. On the other hand, the minor adducts, such as N^6-dA and N^4-dC, cannot be overlooked in their role in tumor induction.

ACKNOWLEDGEMENT: We thank Dr. Frederick A. Beland for many stimulating discussions and Ms. Bobbye James and Ruth York for preparation of this manuscript.

REFERENCES

1. Trosko, J.E., Hart, R.W. (1976): DNA mutation frequencies in mammals. In: "Interdisciplinary Topics in Terontology," edited by R.G. Cutler, Vol. 9, pp. 168-197, Kurger, Basel.
2. Bourliere, F. (1962): Comparative longeviity of higher vertebrates. In: "Biological Aspects of Aging," edited by N. Shock, Columbia University Press, pp. 3-21, New York.
3. Burch, P.P. (1968). In: An Inquiry Concerning Growth, Disease, and Aging, Oliver and Boyd, Edinburg.
4. Sacher, G.A. (1975): Maturation and longevity in relation to cranial capacity in hominid evolution. In: Antecedents of Man and After I. Primates: Functional Morphology and Evolution, edited by R. Tuttle, The Hague: Mouton Publishers, pp. 417-441.
5. Cutler, R.G. (1975): Evolution of human longevity and the genetic complexity governing aging rats. Proc. Natl. Acad. Sci. U.S.A. 72:4664-4668.
6. Brash, D.E., and Hart, R.W. (1978): DNA damage and repair In Vivo. J. Env. Path. Tox., 2:79-114.
7. Lu, A.Y.H. (1979): Multiplicity of liver drug metabolizing enzymes. Drug Metabolism Rev., 10(2):187-208.
8. Sims, P., and Grover, P.L. (1974): Epoxides in polycyclic aromatic hydrocarbon metabolism and carcinogenesis. Adv. Cancer Res., 20:165-274.
9. Yang, S.K., Deutsch, J., and Gelboin, H.V. (1978): Benzo[a]pyrene metabolism: activation and detoxification. In: Polycyclic Hydrocarbons and Cancer, edited by H.V. Gelboin and P.O.P. Tso, Vol. 1, pp. 205-231, Academic Press, New York.
10. Borgen, A., Darvey, H., Castagnoli, N., Crocker, T.T., Rasmussen, R.E., and Wang, I.Y. (1973): Metabolic conversion of benzo[a]pyrene by Syrian hamster liver microsomes and binding of metabolites to deoxyribonucleic acid. J. Med. Chem., 16:502-506.
11. Polycyclic Hydrocarbons and Cancer (1978), edited by H.V. Gelboin and P.O.P. Ts'o, Vol. 1 and Vol. 2, Academic Press, New York.
12. Jerina, D.M., Lehr, R.D., Yagi, H., Hernandez, O., Dansette, P., Wislocki, P.G., Wood, A.W., Chang, R.L., Levin, W., and Conney, A.H. (1976): Mutagenicity of

benzo[a]pyrene derivatives and the descrip of a quantum mechanical model which predicts the base of carbonium ion formation from diol epoxides. In: In Vitro Metabolic Activa In Mutagenesis Testing, edited by F. J. de Serres, J.R. Fouts, J.R. Bend, and R.M. Philpot, pp. 159-177, Amsterdam, Elsevier.
13. Harvey, R.G. (1981): Activated metabolites of carcinogenic hydrocarbons. Acc. Chem. Res., 14:218-226.
14. Koreeda, N., Moore, P.D., Wislocki, P.G., Levin, W., Conney, A.H., Yagi, H., and Jerina, D.M. (1978): Binding of benzo[a]pyrene 7,8-diol-9,10-epoxides to DNA, RNA, and protein of mouse skin occurs with high stereoselectivity. Science, 199:778-781.
15. Bigger, C.A.H., Tomaszewski, J.E., and Dipple, A. (1980): Limitations of metabolic activation systems used with in vitro test for carcinogens. Science, 209:503-505.
16. Selkirk, J. (1977): Divergence of metabolic activation systems for short-term mutagenesis assays. Nature (London), 270:604-607.
17. Brookes, P. and Lawley, P.D. (1964): Evidence for the binding of polynuclear aromatic hydrocarbons to the nucleic acids of mouse skin. Relation between carcinogenic power of hydrocarbons and their binding to deoxyribonucleic acid. Nature (London), 202:781-784.
18. Dipple, A., Brookes, P., Mackintosh, D.S., and Rayman, M.P. (1971): Reaction of 7-bromomethyl-benz[a]anthracene with nucleic acids, polynucleotides, and nucleosides. Biochemistry, 10(23):4323-4330.
19. Jeffrey, A.M., Blobstein, S.H., Weinstein, I.B., Beland, F.A., Harvey, R.G., Kasai, H., and Nakanishi, K. (1976): Structure of 7,12-dimethylbenz[a]anthracene-guanosine adducts. Proc. Natl. Acad. Sci. USA, 73(7):2311-2315.
20. Weinstein, I.B., Jeffrey, A.M., Jennette, K.W., Blobstein, S.H., and Harvey, R.G. (1976): Benzo[a]pyrene diol epoxides as intermediates in nucleic acid binding in vitro and in vivo Science, 193:592-595.
21. Koreeda, M., Moore, P.D., Yagi, H., Yeh, H.J.C., and Jerina, D.M. (1976): Alkylation of polyguanylic acid at the 2-amino group and phosphate by the potent mutagen $(\pm)$-7β,8α-dihydroxy-9β,10β-epoxy-7,8,9,10-tetrahydrobenzo[a]pyrene. J. Am. Chem. Soc., 98:6720-6722.
22. Meehan, T., and Straub, K. (1979): Double-stranded DNA stereoselectively binds benzo[a]pyrene diol epoxides. Nature, 277:410-413.
23. Osborne, M.R., Jacobs, S., Harvey, R.G., and Brooks, P. (1981): Minor products from the reaction of (+) and (-) benzo[a]pyrene-anti-diolepoxide with DNA. Carcinogenesis, 2(6):553-558.

24. Ibanez, V., Geacintov, N.E., Gagliano, A.G., Brandimarte, S., and Harvey, R.G. (1980): Physical binding of tetraols derived from 7,8-dihydroxy-9,10-epoxy-benzo[a]pyrene to DNA. J. Am. Chem. Soc., 102:5661-5666.
25. Geacintov, N.E., Ibanez, V., and Yoshida, H. (1981): Stopped-flow kinetic studies on the interaction of benzo[a]pyrene diol epoxide with DNA. AACR Abstracts, 328.
26. Jeffrey, A.M., Kinoshita, T., Santella, R.M., Grunberger, D., Katz, L., and Weinstein, I.B. (1980): The Chemistry of Polycyclic Aromatic Hydrocarbon-DNA Adducts. In: Carcinogenesis: Fundamental Mechanisms and Environmental Effects, edited by B. Pullman, P.O.P. Ts'o, and H. Gelboin, pp. 565-579, D. Reidel Publishing Co., New York.
27. Vigny, P., Kindts, M., Cooper, C.S., Grover, P.L., and Sims, P. (1981): Fluorescence spectra of nucleoside-hydrocarbon adducts formed in mouse skin treated with 7,12-dimethylbenz[a]anthracene. Carcinogenesis, 2(2):115-119.
28. Vigny, P., Ginot, Y.M., Kindts, M., Cooper, C.S., Grover, P.L., and Sims, P. (1980): Fluorescence spectral evidence that benzo[a]pyrene is activated by metabolism in mouse skin to a diol-epoxide and a phenol-epoxide. Carcinogenesis, 1(11):945-949.
29. Boroujerdi, M., Kung, H.C., Wilson, A.G.E., and Anderson, M.W. (1981): Metabolism and DNA binding of benzo[a]pyrene in vivo in the rat. Cancer Res., 41:951-957.
30. Baer-Dubowska, W., and Alexandrov, K. (1981): The binding of benzo[a]pyrene to mouse and rat skin DNA. Cancer Lett., 13:47-52.
31. Ashurst, S.W., and Cohen, G.M. (1981): In vivo formation of benzo[a]pyrene diol epoxide-deoxyadenosine adducts in the skin of mice susceptible to benzo[a]pyrene-induced carcinogenesis. Int. J. Cancer, 27:357-364.
32. Baird, W.M., and Diamond, L. (1977): The nature of benzo[a]pyrene-DNA adducts formed in hamster embryo cells depends on the length of time of exposure to benzo[a]pyrene. Biochem. Biophys. Chem. Commun., 77(1):162-167.
33. Autrup, H., Harris, C.C., Trump, B.F., and Jeffrey, A.M. (1978): Metabolism of benzo[a]pyrene and identification of the major benzo[a]pyrene-DNA adducts in cultured human colon. Cancer Res., 38:3689-3696.
34. Grover, P.L., Hewer, A., Pal, K., and Sims, P. (1976): The involvement of a diol-epoxide in the metabolic activation of benzo[a]pyrene in human bronchial mucosa and in mouse skin. Int. J. Cancer, 18:1-6.
35. Shinohara, K., and Cerutti, P.A. (1977): Formation of benzo[a]pyrene-DNA adducts in peripheral human lung tissue. Cancer Lett., 3:303-309.

36. Theall, G., Eisinger, M., and Grunberger, D. (1981): Metabolism of benzo[a]pyrene and DNA adduct formation in cultured human epidermal keratinocytes. Carcinogenesis, 2:581-587.
37. Mass, M.J., Rodgers, N.T., and Kaufman, D.G. (1981): Benzo[a]pyrene metabolism in organ cultures of human endometrium. Chem.-Biol. Interact., 33:195-205.
38. Bigger, C.A.H., Tomaszewski, J.E. and Dipple, A. (1978): Differences between products of binding of 7,12-dimethylbenz[a]anthracene to DNA in mouse skin and in a rat liver microsomal system. Biochem. Biophys. Res. Comm. 80:229-235.
39. Cooper, C.S., Ribeiro, O., Hewer, A., Walsh, C., Grover, P.L. and Sims, P. (1980): Additional evidence for the involvement of the 3,4-diol 1,2-oxides in the metabolic activation of 7,12-dimethylbenz[a]anthracene in mouse skin. Chem.-Biol. Interact., 29:357-367.
40. Moschel, R.C., Baird, W.M., and Dipple, A. (1977): Metabolic activation of the carcinogen 7,12-dimethylbenz[a]anthracene for DNA binding. Biochem. Biophys. Res. Commun., 76:1092-1098.
41. Ivanovic, V., Geacintov, N.E., Jeffrey, A.M., Fu, P.P., Harvey, R.G., and Weinstein, I.B. (1978): Cell and microsome mediated binding of 7,12-dimethylbenz[a]anthracene to DNA studied by fluorescence spectroscopy. Cancer Lett., 4:131-140.
42. Daniel, F.B., Wong, L.K. Oravec, C.T., Cazer, F.D., Wang, C.L.A, D'Ambrosio, S.M., Hart, R.W., and Witiak, D.T. (1979): Biochemical studies on the metabolism and DNA-binding of DMBA and some of its monofluoro derivatives of varying carcinogenicity. In: Polynuclear Aromatic Hydrocarbons, edited by P.W. Jones and P. Leber, pp. 855-883, Ann Arbor Science Publishers, Michigan.
43. Daniel, F.B., Cazer, F.D., D'Ambrosio, S.M., Hart, R.W., Kim, W.H., and Witiak, D.T. (1979): Comparative metabolism of 7,12-dimethylbenz[a]anthracene and its weakly carciogenic 5-fluoro analog. Cancer Lett., 6:263-72.
44. Sheikh, Y.M., Inbasekaran, M.N., Daniel, F.B., Cazer, F.D., Hart, R.W., and Witiak, D.T. (1980): A study of the 7,12-dimethylbenz[a]anthracene (DMBA) bay region involvement in the production of carcinogen and mutagen metabolites. In: Polynuclear Aromatic Hydrocarbons: Chemistry and Biological Effects, edited by A. Bjorseth and A.J. Dennis, pp. 689-731, Battelle Press, Columbus, OH.
45. Bigger, C.A.H., Tomaszewski, J.E., Andrews, A.W., and Dipple, A. (1980): Evaluation of metabolic activation of

7,12-dimethylbenz[a]anthracene in vitro by Aroclor 1254-induced rat liver S-9 fraction. Cancer Res., 40:655-661.
46. Cooper, C.S., Ribeiro, O., Hewer, A., Walsh, C., Pal, K., Grover, P.L., and Sims, P. (1980): The involvement of a 'bay-region' and a non-'bay-region' diol-epoxide in the metabolic activation of benz[a]anthracene in mouse skin and in hamster embryo cells. Carcinogenesis, 1:233-243.
47. Cary, P.D., Turner, C.H., Cooper, C.S., Ribeiro, O., Grover, P.L., and Sims, P. (1980): Metabolic activation of benz[a]anthracene in hamster embryo cells: the structure of a guanosine-anti-BA8,9-diol10,11-oxide adduct. Carcinogenesis, 1:505-512.
48. Hemminki, K., Cooper, C.S., Ribeiro, O., Grover, P.L., and Sims, P. (1980): Reactions of bay-region and non-bay-region diolepoxides of benz[a]anthracene with DNA: evidence indicating that the major products are hydrocarbon-N^2-guanine adducts. Carcinogenesis, 1:277-286.
49. Cooper, C.S., Macnicoll, A.D., Ribeiro, O., Giovanni, G., Hewer, A., Walsh, C., Pal, K., Grover, P.L., and Sims, P. (1980): The involvement of a non-bay-region diol-epoxide in the metabolic activation of benz[a]anthracene in hamster embryo cells, Cancer Lett., 9:53-59.
50. Cooper, C.S., Vigny, P., Kindts, M., Grover, P.L., and Sims, P. (1980): Metabolic activation of 3-methylcholanthrene in mouse skin: fluorescence spectral evidence indicates the involvement of diol-epoxides formed in the 7,8,9,10-ring. Carcinogenesis, 1:855-860.
51. Vigny, P., Duquisne, M., Coulomb, H., Lacombe, C., Tierney, B., Grover, P.L., and Sims, P. (1977): Metabolic activation of polycyclic hydrocarbons: fluorescence spectral evidence is consistent with metabolism at the 1,2- and 3,4-double bonds of 7-methylbenz[a]anthracene. FEBS Lett., 75:9-12.
52. Rayman, M.P., and Dipple, A. (1973): Structure and activity in chemical carcinogenesis. Comparison of the reactions of 7-bromomethylbenz[a]anthracene and 7-bromomethyl-12-methylbenz[a]anthracene with mouse skin deoxyribonucleic acid in vivo., Biochemistry, 12:1538-1542.
53. Shinohara, K. and Cerutti, P.A. (1977): Excision repair of benzo[a]pyrene-deoxyguanosine adducts in baby hamster kidney 21/C13 cells and in secondary mouse embryo fibroblasts C57BL/6J, Proc. Natl. Acad. Sci. USA, 74:979-983.
54. Feldman, G., Remsen, J., Shinohara, K., and Cerutti, P.A. (1978): Excisability and persistence of benzo[a]pyrene

DNA adducts in epithelioid human lung cells, Nature, 274(24):796-798.
55. Eastman, A., Mossman, B.T., and Bresnick, E. (1981): Formation and removal of benzo[a]pyrene adducts of DNA in hamster tracheal epithelial cells, Cancer Res., 41:2605-2610.
56. Cerutti, P., Shinohara, K., and Remsen, J. (1977): Repair of DNA damage induced by ionizing radiation and benzo[a]pyrene in mammalian cells, J. Toxicol. Environ. Health, 2:1375-1386.
57. Brown, H.S., Jeffrey, A.M., and Weinstein, I.B. (1979): Formation of DNA adducts in 10T1/2 mouse embryo fibroblasts incubated with benzo[a]pyrene or dihydrodiol oxide derivatives, Cancer Res., 39:1673-1677.
58. Feldman, G., Remsen, J., Wang, T.V., and Cerutti, P. (1980): Formation and excision of covalent deoxyribonucleic acid adducts of benzo[a]pyrene 4,5-epoxides and benzo[a]pyrene diol epoxide I in human lung cells A549, Biochemistry, 19:1095-1101.
59. Yang, L.L., Maher, V.M., and McCormick, J.J. (1980): Error-free exicison of the cytotoxic, mutagenic N^2-deoxyguanosine DNA adduct formed in human fibroblasts by $(\pm)$-7β, 8α-dihydroxy-9α, 10α-epoxy7,8,9,10-tetrahydro-benzo[a]pyrene, Proc. Natl. Acad. Sci.USA, 77:5933-5937.
60. Wilkins, R.J., and Hart, R.W. (1974): Preferential DNA repair in human cells. Nature, 247:35-36.
61. Dipple, A. and Roberts, J.J. (1977): Excision of 7-bromomethylbenz[a]anthracene-DNA adducts in replicating mammalian cells, Biochemistry, 16:1499-1503.
62. Lieberman, M. and Dipple, A. (1972): Removal of bound carcinogen during DNA repair in nondividing human lymphocytes, Cancer Res., 32:1855-1860.
63. Slor, H. (1973): Induction of unscheduled DNA synthesis by the carcinogen 7-bromomethylbenz[a]anthracene and its removal from the DNA of normal and xeroderma pigmentosum lymphocytes, Mutation Res., 19:231-235.
64. Tarmy, E.M., Venitt, S., and Brookes, P. (1973): Mutagenicity of the carcinogen 7-bromomethylbenz[a]-anthracene: a quantitative study in repair-deficient strains of Escherichia coli, Mutation Res., 19:153-166.
65. McCaw, B.A., Dipple, A., Young, S., and Roberts, J.J. (1978): Excision of hydrocarbon-DNA adducts and consequent cell survival in normal and repair defective human cells, Chem.-Biol. Interact., 22:139-151.
66. Sheikh, Y.M., Joyce,N.J., Daniel, F.B., Oravec, C.T., Cazer, F.D., Raber, J., Mhaskar, D., Witiak, D.T., Hart, R.W., and D'Ambrosio, S. (1981): Strain differences in

organ selective DMBA-induced carcinogenicity: comparative binding of dimethylbenz[a]anthracene and its 2-fluoro analogue in Sprague-Dawley and Long-Evans rats. In: Polynuclear Aromatic Hydrocarbons, edited by A.J. Dennis and W.M. Cooke, pp. 625-639, Battelle Press, Columbus, Ohio.
67. Huggins, C., Grand, L.C. and Brillantes, F.P. (1961): Mammary cancer induced by a single feeding of polynuclear hydrocarbon and its suppression, Nature, 189:204-207.
68. D'Ambrosio, S.M., Daniel, F.B., Hart, R.W., Cazer, F.D., and Witiak, D.T. (1979): DNA repair in Syrian hamster embryo cells treated with 7,12-dimethylbenz[a]anthracene and its weakly carcinogenic 5-fluoro analog, Cancer Lett., 6:255-261.
69. Dipple, A. and Hayes, M.E. (1979): Differential excision of carcinogenic hydrocarbon-DNA adducts in mouse embryo cell cultures, Biochem. Biophys. Res. Commun., 91(4):1225-1231.
70. Tay, L.K. and Russo, J. (1981): 7,12-dimethylbenz[a]-anthracene-induced DNA binding and repair synthesis in susceptible and nonsusceptible mammary epithelial cells in culture, J. Natl. Cancer Inst., 67:155-161.
71. Eastman, A. and Bresnick, E. (1979): Persistent binding of 3-methylcholanthrene to mouse lung DNA and its correlation with susceptibility to pulmonary neoplasia, Cancer Res., 39:2400-2405.
72. Dipple, A. and Schultz, E. (1979): Excision of DNA damage arising from chemicals of different carcinogenic potencies, Cancer Lett., 7:103-108.
73. Beland, F.A. (1978): Computer-Generated graphic models of the N^2-substituted deoxyguanosine adducts of 2-acetylaminofluorene and benzo[a]pyrene and the O^6-substituted deoxyguanosine adduct of 1-naphthylamine in the DNA double helix, Chem.-Biol. Interact, 22:329-339.
74. Carrell, H.L., Glusker, J.P., Moschel, R.C., Hudgins, W.R., and Dipple, A., (1981): Crystal structure of a carcinogen: nucleoside adduct, Cancer Res., 41:2230-2234.
75. Hart, R.W., and Setlow, R.B. (1974): Correlation between deoxyribonucleic acid excision-repair and life-span in a number of mammalian species, Proc. Natl. Acad. Sci. USA, 71:2169-2173.
76. Francis, A.A., Lee, W.H., and Regan, J.D. (1981): The relationship of DNA excision repair of UV induced lesions to the maximum life span of mammals, Mech. Aging Dev., 16(2):181-190.

EVALUATION OF EXTRACTION METHODS FOR CARBON BLACK; POM ANALYSIS AND MUTAGENICITY ASSAY.

TOMAS ALSBERG*, ULF RANNUG**, ULF STENBERG*, ANNICA SUNDVALL**
*Department of Analytical Chemistry, Arrhenius Laboratory;
**Department of Toxicology Genetics, Wallenberg Laboratory, University of Stockholm, S-106 91 Stockholm, Sweden.

INTRODUCTION

The extraction yield of organic matter from particles is dependent on the particle surface, the nature of the adsorbed molecules and extraction method (1). For diesel particulates Soxhlet extraction with dichloromethane (DCM) seems to be the most commonly used method (2). Sanders (3) found o-dichlorobenzene superior to toluene and DCM for extraction of NO_2-pyrenes from carbon black. Agurell and Löfroth (4) found that extraction of carbon black with benzene gave extracts twice as mutagenic as acetone extracts.

The carbon black investigated here is suitable for methodology studies concerning correlation between mutagenicity testing and chemical analysis. Although the organic composition is simple enough to allow gas chromatographic analysis without clean up, most major PAH compounds, typical for automobile exhausts (5), are present. In addition, there are a number of sulfur heterocyclics and some oxygenated species associated with the particles.

MATERIALS AND METHODS

The carbon black was derived from combustion of natural gas and had a mean particle diameter of 200 nm. Its elemental composition was 97.4% C, 0.5% H, 0.1% N, 0.7% O and 1.1% S. When comparing the different extraction methods three portions of 1.7 g carbon black were extracted for 24 h and tested separately.

Chemistry

Vacuum sublimation and solvent extraction procedures are described elsewhere (1). To check possible artifact formation during Soxhlet extraction the following experiment was performed: The vacuum sublimation extract of 2 g carbon black, obtained by extracting 1 g at a time first for 5 min and then for 24 h to get a sample covering the complete Mw range of DCM extracts, was divided into two samples. One was the reference sample, the other was absorbed by an extraction thimble which was Soxhlet extracted 48 h with DCM.

Gas Chromatography (GC)   A 5-10% aliquot of all samples for mutagenicity testing was taken for GC analysis. Internal standards, added to these aliquots, were β,β'-binaphthyl (β,β'-BN) and p-quarterphenyl (p-QF). The gas chromatograph, a modified PYE Unicam GCV with flame ionization detector (FID), was connected to a Spectra Physics SP 4100 Integrator. The temperature of the glass capillary column, coated with SE-30 stationary phase was programmed from 70 to 300°C at 7 deg/min.

Gas Chromatography Mass Spectrometry (GC-MS).   A Carlo Erba Fractovap Series 2150 gas chromatograph was connected to a Jeol MS D300 double focussing mass Spectrometer with an Incos 2000 data system.

High Pressure Liquid Chromatography (HPLC).   The HPLC system consisted of two Constametric III pumps (flow rate range 0.033-3.3 ml/min), Gradient Master, Gradient Mixer and a Spectromonitor III UV detector all from Laboratory Data Control. Injection was done with a Rheodyne 7125 loop injector fitted with a 500 µl loop. The semipreparative column used for fractionation was made of stainless steel, 150 x 10 mm, and packed with 5 µm Spherisorb ODS. Water and acetonitrile (ACN), pro analysi quality from Fisher Scientific Company, were used as solvents and redistilled in glass.

When separating cyclopenteno[cd]pyrene (CpcdP) from a DCM extract the program given in Fig 4 was used. The CpcdP and benzo[ghi]fluoranthene (BghiF) fractions were rechromatographed at 30% H_2O in ACN. The compounds eluting before CpcdP and after BghiF were combined. When separating high molecular weight (Mw $\geq$ 276) POM from a benzene extract the separation was done under isocratic conditions, 5% H_2O in ACN. After collection of the effluent the volume was reduced by approximately 50% and water was added to give a water content of at least 50%. The POM was extracted twice with DCM, or cyclohexane (C-6) (CpcdP and BghiF fractions).

Open Column Chromatography.   Fractionation on silica gel was performed on an open glass column, 90 x 10 mm, packed with Merck silica gel 63-200 µm (pretreated at) 500°C and deactivated with 10% w/w water). The column was conditioned with 50 ml cyclohexane (C-6). The sample was eluted with 50 ml of C-6 followed by 30 ml of acetone. During elution the column was wrapped in aluminum foil for light protection.

Biology

Mutagenicity Tests   The plate incorporation assay was performed as described by Ames et al. (6) using Salmonella typhimurium TA 98. All samples were tested in the presence and absence of a metabolizing system (S9) (10% if not specified). The S9 fraction was prepared from Aroclor pretreated male Sprague Dawley rats according to Ames .The results are given as number of revertants per plate, Fig 5 (mean values of four plates + S.E.) or as number of revertants per mg extracted carbon black, determined by regression analysis. Blanks from all solvents and chromatographic systems used for fractionation were also tested.

RESULTS AND DISCUSSION

Comparison Between Extraction Methods

Chemistry.   A drawback of the vacuum sublimation procedure is that some compounds are degraded. The levels of CpcdP (Mw 226) for instance is approximately one third in sublimation extracts compared to DCM extracts, Fig 1. Oxygenated PAHs with more than one oxygen atom, Fig 3. are probably also degraded during vacuum sublimation. Of the four initially investigated solvents for Soxhlet extraction, the highest overall yield of POM was achieved with DCM, especially for those of high molecular weight (Mw $\geq$ 276). Later, benzene extraction was compared to DCM extraction. Benzene proved to be more efficient than DCM, especially for high molecular weight POM (Mw $\geq$ 276), Table 3.

Biology.   The results from the mutagenicity tests showed that with S9, DCM and acetone gave more than twice as mutagenic extracts as C-6, methanol and vacuum sublimation, Table 1. Without S9 a significant effect was seen only with DCM and acetone extraction. The mutagenicity of the benzene extracts was of the same order of magnitude as that of DCM.

EVALUATION OF EXTRACTION

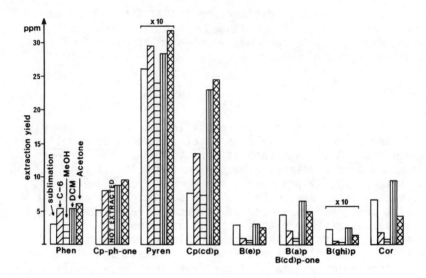

FIGURE 1. Concentrations of selected POM in extracts obtained by vacuum sublimation and Soxhlet extraction. Triplicate samples, with a typical Rel std dev of 15%.

TABLE 1

MUTAGENICITY OF EXTRACTS OBTAINED BY DIFFERENT EXTRACTION METHODS

	V.Subl[a]	DCM[a]	Acetone[a]	Methanol[a]	Cyclohexane[a]
-S9	0.03	0.45	0.20	0.03	0.05
+S9	0.80	2.3	2.3	0.72	0.93

a 1.7 g carbon black extracted for 24 h in triplicate.

Possible Causes for Differences in Mutagenic Effects.

Experiments were performed to explain the differences in mutagenicity between vacuum sublimation and DCM extracts.

Low Molecular Weight Compounds. DCM extracts contained low boiling compounds eluting from the GC-column between the solvent and phenanthrene (Phen) Fig. 2. Those compounds were absent in the vacuum sublimation extracts. Vacuum sublimation for 5 min however, extracts these compounds but such samples showed no mutagenic activity. The main peaks in this area in the chromatogram represent aliphatic hydrocarbons up to $C_{15}H_{32}$.

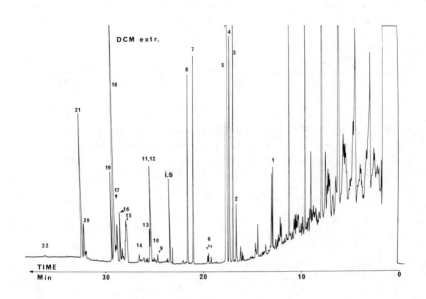

FIGURE 2. Capillary gas chromatogram of DCM extract. Peak identities: 1. Phen; 2. Mw 204 (Fig. 3); 3. fluoranthene; 4. 4,5-S-phen[a]; 5. pyrene; 6. Mw 216; 7. Bhif; 8. CpcdP; 9. benzofluoranthenes; 10. Mw 258[a]; 11. benzo[e]pyrene; 12. 6H-benzo[cd]pyrene-6-one (Fig. 3); 13. benzo[a]pyrene; 14. Mw 256 (Fig. 3); 15. Mw 272 (Fig. 3); 16. indeno[1,2,3-cd]pyrene; 17. 6,7-S-perylene[a]; 18. benzo[ghi]perylene; 19. anthanthrene; 20. 7,8-S-benzo[ghi]perylene[a]; 21. coronene; 22. Mw 324.
a=cf ref 7.

Oxygenated PAHs (oxy-PAH). The oxy-PAHs, were separated from a DCM extract on a silica column. It was found that the indirect mutagenic effect was exerted by compounds in the C-6 fraction, i.e. PAHs and S-PAHs. The acetone fraction from the column, containing the oxy-PAHs, accounted for the direct mutagenic effect in the extract.

EVALUATION OF EXTRACTION

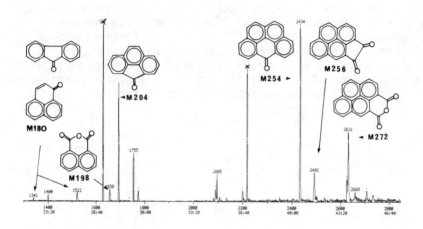

FIGURE 3. GC-MS Total Ion Chromatogram of acetone fraction from silica column. Electron impact, ionization voltage 70 eV. Tentative identification. Mass spectra can be found in refs 1(M 256, M 272), 8 (M 204, M 254, M 272), 5 (M 254).

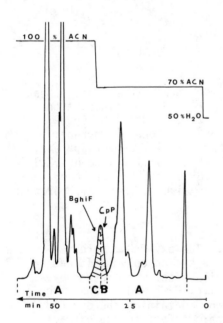

FIGURE 4. Semiprep HPLC-separation of CpcdP and BghiF from a DCM extract. Column: 10x150 mm, Spherisorb ODS 5 μm. Flow rate 1.5 ml/min. Injected volume 0.5 ml. Sample amount equivalent to 1 g carbon black. Fractions B and C were rechromatographed at 30% H_2O in ACN.

Since the oxy-PAH containing two or more oxygen atoms were found only in Soxhlet extracted samples and not in sublimation extracts, it is possible that they either are degraded during vacuum sublimation or formed during Soxhlet extraction. Treatment of a sublimation extract with DCM, as described in Materials and Methods, however, did not change the mutagenic effects of the sample, indicating no significant degradation or formation of mutagenic substances during Soxhlet extraction with DCM. This result is also supported by GC analysis. However, the possibility of the carbonaceous material acting as a catalyst for the formation of oxy-PAH during Soxhlet extraction was not investigated.

Contribution of CpcdP to the Total Mutagenicity. The low levels of CpcdP in sublimation extracts and the fact that CpcdP is a well known indirect acting mutagen, made it a possible candidate responsible for the difference in mutagenicity between DCM and sublimation extracts.

The result of the HPLC fractionation of CpcdP is shown in Tables 2 and 3. The mutagenic effect of the recombined HPLC-fractions is 80% of the effect of the unfractionated sample. This is in accordance with GC analysis which yielded an average over all POM recovery of 90%. Apparently the effects of the BghiF and CpcdP fractions are rather low when tested separately. However, when combining those fractions with the rest of the sample a somewhat higher mutagenic effect is seen than would be expected from the data of the separate fractions. Addition of the CpcdP fraction, to the remaining sample caused a 40% increase in the mutagenic

TABLE 2

MUTAGENIC EFFECTS OF HPLC-FRACTIONED DCM EXTRACT

	DCM ext[a] A	B[b]	C[b]	A+B	A+C	A+B+C
-S9	0.56	-	-	-	-	-
+S9	2.2	0.92	0.24	0.13	1.5	1.2	1.7

a. 3 g carbon black extracted 64 h, same sample as in Table 3.
b. B=CpcdP fraction, C=BghiF fraction, Fig 4 and Table 3.

effect, which indicates that CpcdP accounts for a substantial part of the mutagenicity in DCM extracts.

Mutagenicity of High and Low Molecular Weight POM. Benzene, DCM, and acetone extracts showed approximately the same mutagenicity and also the same levels of POM with Mw lower than 276. However, benzene extracts had higher levels of POM with Mw higher than 276 than DCM extracts, Table 3, which in turn had higher concentrations than acetone extracts, Fig. 1. One explanation for the similarity in mutagenicity could be that the high Mw POM does not significantly contribute to the total mutagenicity. In order to check if the high Mw POM was without mutagenic effect, a benzene extract was divided into a low and a high molecular fraction by HPLC, Table 3. Both these fractions were mutagenic in the presence of S9 (1.3 and 1.5 rev per mg carbon black, respectively). The recombined fractions gave a higher mutagenic effect (2.2 rev per mg carbon black) than

TABLE 3

POM CONCENTRATIONS IN HPLC FRACTIONED SAMPLES

Compound	DCM raw extract[a]	Fract. A	Fract. B	Fract. C	Benzene raw extr[a]	Low Mw	High Mw
1. Phen	6.1	5.5	–	–	7.2	3.5	–
2. CpPh-one	5.3	5.1	–	–	5.8	4.8	–
3. Fluo	25	24	–	–	27	23	–
4. S-Phen	26	24	–	–	27	23	–
5. Pyr	230	220	–	–	240	215	–
7. BghiF	16	–	4.5	9.5	17	15	–
8. CpcdP	14	–	10	2.0	16	13	–
11. BeP	9.7	8.8	–	–	12	9.2	–
12. BcdP-one							
13. BaP	2.3	2.2	–	–	3.5	2.7	–
17. S-Per	3.6	3.3	–	–	3.7	–	3.7
18. BghiP	21	20	–	–	47	–	43
19. Anth	3.1	2.4	–	–	13	–	9.4
20. S-BghiP	3.0	2.6	–	–	13	–	9.0
21. Cor	4.3	4.1	–	–	30	–	21

[a] 3 g carbon black was extracted for 64h

the individual fractions, showing that the high Mw fraction contributes to the mutagenicity although this was not detected in the screening of the above mentioned extracts (benzene, DCM, acetone).

S9 Dependence. The difficulties in correlating mutagenicity data to chemical data encountered in some of these experiments are to be expected, since the two systems inherently respond differently to different components. That is, differences in chemical composition may cover nonmutagenic compounds, as in the case of the low boiling non POM compounds, or compounds for which the mutagenicity assay is not optimized.

The latter aspect is exemplified in Fig. 5. The results show that, although fraction B (CpcdP) obviously contributes to the mutagenicity of the recombined sample, especially at high S9 levels the potency of this fraction can only be seen when tested alone at low S9 levels.

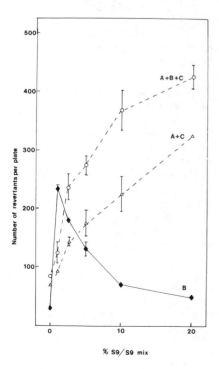

FIGURE 5. The mutagenicity of fraction B (CpcdP), A+B (sample - CpcdP) and A+B+C (recombined sample) in the presence of different amounts of S9 fraction.

ACKNOWLEDGEMENT

This work has been sponsored by the National Swedish Environment Protection Board.

REFERENCES

1. Stenberg, U.R., and Alsberg, E.T. (1981): Vacuum sublimation and solvent extraction of polycyclic aromatic compounds adsorbed on carbonaceous materials, Anal. Chem., 53 (in press).
2. Pepelko, W.E., Danner, R.M., and Clarke, N.A., editors (1980): Health Effects of Diesel Engine Emissions: Proceedings of an International Symposium, vol. 1. EPA-600/9-80-057a, U.S. Environmental Protection Agency, Health Effects Research Laboratory, Cincinatti OH 45268, 1980, 570 pp.
3. Sanders, R.D. (1981): Nitropyrenes: The isolation of trace mutagenic impurities from the toluene extract of an aftertreated carbon black. In: Chemical Analysis and Biological Fate: Polynuclear Aromatic Hydrocarbons, edited by M. Cooke and A.J. Dennis, pp 145-158, Battelle Press. Columbus Ohio.
4. Agurell, E., and Löfroth, G. (1981): Mutagenicity Testing of Carbon Black Extracts: Presence of PAH and non-PAH Mutagens, Poster Presentation at Nord EMS Meeting, May 24-26, Koge, Danmark.
5. Stenberg, U., Alsberg, T., Blomberg, L., and Wännman, T. (1979): Gas chromatographic separation of high-molecular polynuclear aromatic hydrocarbons in samples from different sources, using temperature-stable glass capillary columns. In: Polynuclear Aromatic Hydrocarbons, Third International Symposium on Chemistry and Biology-Carcinogenesis and Mutagenesis, Edited by P.W. Jones and P.L. Leber, pp 313-326, Ann Arbor Science, Ann Arbor, Mich. 48106
6. Ames, B.N., McCann, J., and Yamasaki, E. (1975): Methods for detecting carcinogens and mutagens with the Salmonella/mammalian-microsome mutagenicity test, Mutat. Res., 31:347-364.
7. Colmsjö, A.L., and Östman, C.E. (1982): Shpol'skii spectra of polycyclic aromatic compounds in samples from carbon black and soil. In: This book.
8. Gold, A. (1975): Carbon black adsorbates: Separation and identification of a carcinogen and some oxygenated polyaromatics, Anal. Chem., 47:1469-1472.

LASER EXCITED FLUORESCENCE AND CHROMATOGRAPHIC TECHNIQUES FOR THE DETERMINATION OF PAH IN A SPRAY OIL FLAME

R. BARBELLA, F. BERETTA, A. CIAJOLO, A. D'ALESSIO
Istituto di Ricerche sulla Combustione, C.N.R.; Istituto Chimica Industriale e Impianti Chimici, Università di Napoli; Piazzale V. Tecchio, 80125 Napoli, Italy.

INTRODUCTION

One of the major problems presented today to combustion technologists is to predict soot and PAH emissions in the exhaust gases starting from the input external design or operative variables.

Recent results showed that soot and PAH emissions follow different and, sometimes, opposite trends (1) and it is often impossible to generalize the experimental results obtained on specific combustion systems either in furnaces or in diesel engines. Consequently the formation and oxidation of soot and PAH have to be studied inside the combustion systems, in order to characterize the controlling processes which are responsible of the final appearance of these classes of pollutants in the discharge gases.

The development of an optical measurement technique, as an ancillatory tool for mapping PAH concentration in these conditions, is therefore high desiderable since it should make possible also a systematic study of the influence of different input parameters on PAH formation, with a reasonable amount of experimental time and resources.

The present paper reports selected results obtained in the study of an unconfined oil spray flame, combining sampling and chromatographic techniques with laser-excited fluorescence measurements. It provides detailed information on the evolution of PAH and also exploration of the analytical potential of the visible fluorescence as applied to practical flames.

Measurements of the PAH in flames have been carried out mainly on laminar-rich premixed flames generated by gaseous fuels (2,3,4,5). Therefore the high temperature pyrolytic processes leading the PAHs in absence of oxygen or oxygenated radicals were preferentially analyzed. Kern and Spengler (6) have measured the concentration of different compounds, including PAH, along the axis of a laminar hexane-air diffusion flame in the 700-1200°K temperature range. Their results show that PAH maximum concentration precedes the onset of soot formation and analogous results were obtained by Prado et al. (7) in turbulent kerosene and benzene diffusion flames.

Broadband fluorescence, excited in the visible by argon ion lasers, has been detected in laminar premixed and diffusion flames by different authors (5,8,9) and this effect has been attributed to PAH present in flames, although it was impossible to determine which individual compound is the primary contributor.

EXPERIMENTAL TECHNIQUES

The experiments were carried out on a special burner with a variable swirl generator, in which 6 kg/hr of light oil was introduced through a 60° semisolid commercial nozzle. The air was supplied through two coaxial ducts: a pure axial jet was produced in the 24 mm inner one, while a partially swirled flow was generated in the 74 mm external tube. The ratio of the internal to external flow rate was around 0.1 in this particular experiment. The flame was unconfined so that the surrounding air was entrained in its later development. More detailed descriptions of the burner and the possible flame configurations have been presented elsewhere (10,11).

The flame was sampled using a water-cooled stainless steel 2 mm i.d. probe which had a bronze sintered filter at 10 mm from the tip; a fiber glass filter and two traps placed on the sampling line ensured the complete collection of the sampled materials. Both the filters were extracted with DCM and the extracts were analyzed for PAH by reverse phase HPLC on chemically bonded C_{18} (octadecyl) stationary phase, using a 254 nm u.v. detector and a gradient program similar to that used by Katz and Ogan (12). Gas-chromatographic analysis was used in the fuel zone as a semi-quantitative method to determine the relative amount of the original fuel in the total collected material.

The light oil contained 70%, by weight, of paraffinic compounds from C_{10} to C_{30}, 20% of monoaromatic compounds and 5% of PAH, among which compounds with side chains were predominant. Therefore, the contribution of the PAH contained in the fuel was always considered in the measurements and it was found that it was also negligible in the initial preheating zone of the flame.

A systematic comparison between HPLC and GC analysis procedures was carried out in the luminous zone of the flame where the fuel concentration is negligible and a quite good agreement between the two methods was found.

On-line analysis of stable gases were also accomplished using a paramagnetic analyzer for O_2 and i.r. detectors for CO and CO_2.

Averaged flame temperatures were also measured by a Pt/Pt-13% Rh thermocouple, without any correction for the radiation and thermal losses.

A c.w. argon ion laser was employed for the fluorescence measurements, focusing the line at $\lambda = 514.5$ nm into a 3.10^{-6} cm^3 scattering volume inside the flame. The scattered light was collected at 90° on a f=300 mm grating monochromator and the signal was processed with a lock-in technique. Colour or broadband interference filters (peaked at $\lambda = 549.0$ nm) were employed in front of the monochromator in order to avoid the stray light caused by the intense scattering effects due to droplets and/or soot. The fluorescence profiles were corrected for the attenuation of the incident and scattered beams with additional chordal extinction measurements, and its signals were evaluated in absolute scale through the calibration of the signals in comparison with the Rayleigh scattering of propane at room temperature (13).

RESULTS

An overall description of the chemical processes which take place inside the flame is given by the axial profiles shown in Fig. 1, where the concentrations of CO, CO_2, O_2, PAH, soot, and total condensable material are reported together with the uncorrected temperature values.

The initial region is characterized by a concentration of condensable material composed of about 99% by the original fuel and for 1% by the PAH analyzed in the present paper; the temperature is around 800°K and the oxygen concentration is quite high.

The greatest part of the oxygen disappears between Z=50 and Z=100 mm above the nozzle and the temperature increases up to the final value of 1300°K; in this zone the condensable material decreases by more than one order of magnitude but it becomes richer in PAH, which accounts up to 15% of the total

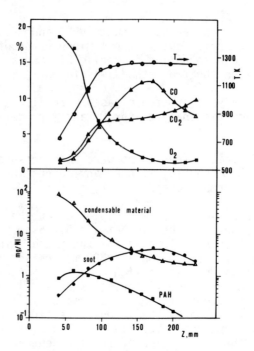

FIGURE 1. Axial profiles. Upper part; temperature and CO, CO_2, and O_2 percentage concentration; lower part: soot, PAH and condensable material mass concentration.

material. It is also important to note that soot begins to appear in this zone whereas the measured PAH are slightly decreased from their high initial values.

In the subsequent regions, up to Z=170 mm the temperature is constant along the axis, the CO concentration increases more rapidly than CO_2 concentration and the oxygen declines down to very low values. Soot concentrations go to a maximum in this region while PAH show the opposite trend. The final part of the flame is characterized by the lowering of the CO/CO_2 ratio and the destruction of soot particles and PAH.

Further informations on the chemistry of PAH and soot in this flame are given by the radial concentration traverses along the flame. Fig. 2 illustrates, as an example of the initial zones of the flame, the radial profiles of individual PAH at Z=40 mm above the nozzle in addition to the other properties reported also in Fig. 1. Temperature profiles and soot concentrations evidence off-axis peaks at R=30 mm where

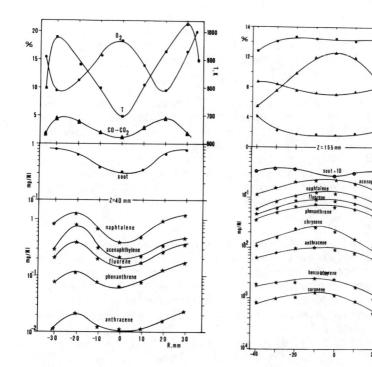

FIGURE 2. Radial profiles at Z=40 mm. Upper part: temperature and O_2, CO, and CO_2 concentration; middle part: soot mass concentration; lower part: individual PAH mass concentration.

FIGURE 3. Radial profiles at Z=155 mm. Upper part: temperature and O_2, CO, and CO_2 concentration; lower part: soot and individual PAH mass concentration (soot values has to be multiplied by 10).

the oxygen concentration reaches its minimum value. PAH have their maximum concentration nearer to the axis (R=20 mm). Compounds with two and three rings are prevailing among PAH, and naphthalene is the most abundant, whereas the species with more than three rings were not detected.

A traverse typical of a fully developed combustion zone is reported in Fig. 3 at Z=155 mm; oxygen has a low concentration in the central region and the radial temperature profile is nearly uniform. Also soot particles are almost uniformily distributed along the traverse and their concentration is much higher than in the initial zone of the flame. CO is prevailing over CO_2 in the central core and also PAH have their maxima on the axis. It is worth noting that two and three rings compounds are still the most abundant but PAH with four and seven rings have now reached an appreciable concentration.

The spectral distribution of fluorescence shows marked difference whether it was obtained on fuel isothermal sprays or inside the flame; in the last case, a broad band spectrum ranging from 400 to 600 nm and peaked at λ = 540 nm was obtained, showing Stokes and anti-Stokes features, in agreement with the results obtained in rich premixed flames (5). The spectrum obtained in cold conditions from the fuel did not evidence any anti-Stokes feature and was peaked at λ = 550 nm (13).

The radial and axial distribution of the fluorescence coefficients $Q_\lambda^F \cdot \Delta\lambda$ at λ = 549 nm with $\Delta\lambda$ = 2.65 nm, measured inside the flame is reported in Fig. 4, and they show a behaviour congruent with that presented by the PAH profile. In fact, they evidence pronounced external maxima in the initial regions, near to the spray nozzle, which become more prominent around Z=60 mm. For Z>80 mm the external peaks are decreasing and the distribution tends toward a flat profile with a very slight central maximum.

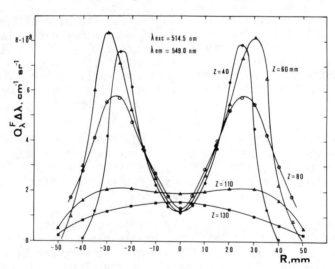

FIGURE 4. Radial profiles of fluorescence coefficient at different heights above the burner emitted at λ=549.0 nm excited with the 514.5 nm argon ion laser line; $\Delta\lambda$=2.65 nm.

DISCUSSION

The analysis of the results requires a short discussion on the structure of the flame itself. Its moderately swirled configuration generates an external recirculation region near the inlet air duct which appears as a luminuous toroidal region, extending for the first 20 mm above the nozzle. In the subsequent blue region between 30 and 80 mm only a limited amount of the smallest droplets are vaporized and burn in stoichiometric condition in a narrow external layer (11). Their combustion provides the energy necessary to the fuel vaporization and heat-up. Above 80 mm, the fuel vapors complete their mixing with the inlet air and are burned in fuel-rich conditions. The oxidation of the fuel is completed only in the last part of flame, beyond Z=130 mm, where the burnt gases mix with the ambient entrained air. Two-three rings PAH have their maximum concentration in the non-luminous preheating zone and decline in correspondence to the main reaction zone. Therefore, the main formation route of this class of compounds seems to be that of condensation and addition reactions using as "building bricks" the alkylated monoaromatic compounds contained in the fuel itself. This view is consistent with the results obtained by Smith (14) who found that in the pyrolysis of toluene below 1000°C the concentration of small radicals and unsaturated molecules was negligible with respect to that of aromatic reactive radicals.

Heavier PAH and soot are instead formed later, mainly in the higher temperature, rich combustion zone. Bittner and Howard (15) have recently analyzed, with great detail, the chemical processes which take place in the rich combustion of benzene. They underline that PAH and soot formation are controlled both by the concentration and reactivity of aromatic radicals and by the concentration of smaller radicals and species like CH_3, C_2H_2, H, and OH.

Soot is finally destroyed only in the last postcombustion zone in processes where OH radicals should play a very important role (16). It is worth noting the relative extension of the vaporization zone with respect to oxidation and hence the PAH formation rate is strongly dependent on the flow swirl level. Therefore, no general conclusions should be drawn by the analysis of a single flame.

A direct proportionality between fluorescence signals and PAH concentrations was found in a previous study on laminar pre-mixed flames (5). The present results obtained on a turbulent diffusion flame support this correlation.

Coe and Steinfeld (17) have recently found that vapors of acenaphthylene and fluoranthene at high temperatures emitted broadband fluorescence in the visible. Our measurements, at room temperature and in the liquid phase, indicate that only acenaphthalyene and, with a lower intensity, fluoranthene have a fluorescent emission among the PAH analyzed in this paper, for a $\lambda = 514.5$ nm excitation.

However, more information on the spectroscopy of PAH at high temperature and further experiments with different excitation wavelengths are necessary before a more definite correlation of this effect to a specific compound may be advanced.

CONCLUSIONS

PAH concentration profiles measured in turbulent spray oil flames show that two-three rings compounds are preferentially formed at low temperatures when most of the fuel is vaporized and preheated. Heavier PAH and soot are formed at higher temperatures in a rich zone beyond the main oxidation zone and are later oxidized by the entrained ambient air.

Laser excited fluorescence gives composite signatures of PAH but it allows following their formation and destruction zones. This method supplies a very useful diagnostic tool for a systematic study of the influence of external combustion parameters on PAH emissions.

REFERENCES

1. Longwell,J.P. (1981):Soot and PAH from practical combustion systems, Paper presented at NATO workshop on "Soot in combustion systems and its toxic properties: Le Bichenberg-Obernai, France, August 1981.
2. Thompkins,E.E. and Long,R.(1969):The flux of PAH and of insoluble material in premixed acetylene-oxygen flames. XII Symposium (Int.) on Combustion,p.625.
3. Homann,K.H.and Wagner,H.Gg.(1967):Some new aspects of the mechanism of carbon formation in premixed flames.XI Symposium(Int.) on Combustion,p.371,The Combustion Institute.
4. D'Alessio,A.,Di Lorenzo,A.,Sarofim,A.F.,Beretta,F.,Masi,S. and Venitozzi,C.(1975):Soot formation in CH_4/O_2 flames, XV Symposium(Int.) on Combustion,p.1427,The Combustion Institute
5. Di Lorenzo,A.,D'Alessio,A.,Cincotti,V.,Masi,S.,Menna,P. and Venitozzi,C.(1981):U.V. absorption ,laser excited fluorescence and direct sampling in the study of the formation of PAH in rich CH_4/O_2 flames. XVIII Symposium(Int.) on Combustion, pp.485-491, The Combustion Institute.
6. Kern,J. and Spengler,G.(1970):Untersuchungen an diffusionsflammen reaktionsprodukte in der achse einer hexan-flamme. Erdol und Kohle,Erdgas Petrochemie,23:813-817.
7. Prado,G.P.,Lee,M.L.,Hites,R.A.,Hoult,D.P.and Howard,J.B. (1977):Soot and hydrocarbons formation in a turbulent diffusion flame,XVI Symposium(Int.) on Combustion,p.649,The Combustion Institute.
8. Cincotti,V.,D'Alessio,A.,Menna,P. and Venitozzi,C.(1981): U.V.Absorption spectroscopy and laser excited fluorescence in the study of formation of high molecular mass compounds in the rich combustion of CH_4,La Rivista dei Combustibili, 35:59-68.
9. Haynes,B.S.,Jander,H. and Wagner,H.Gg.(1980):Optical studies of soot formation processes in premixed flames,Ber.Bunsenges. Phys.Chem.,84:585-593.
10. Beretta,F.Caveliere,A.,D'Alessio,A.,Noviello,C. and Scodellaro,C.(1980):Examination of the structure of practical oil flames with laser light scattering and absorption techniques La Rivista dei Combustibili,34:383-392.
11. Beretta,F.,Cavaliere,A.,Ciajolo,A.,D'Alessio,A.,Di Lorenzo, A.,Langella,C. and Noviello,C.(1981):Laser light scattering, emission/extinction spectroscopy and thermogravimetric ana-

lysis in the study of soot behaviour in oil spray flames,XVIII Symposium(Int)on Combustion,p.1091,The Combustion Institute.
12. Katz,E.,Ogan,K.(1980):Determination of PAH compounds in raw refinery waste water samples,<u>Chrom.Newsletter,</u> 8:18-20.
13. Beretta,F.,Cavaliere,A. and D'Alessio,A.(1981):Laser excited fluorescence measurements in spray oil flames for the detection of PAH and soot,<u>Comb.Sci.and Techn.</u>(in press).
14. Smith,R.D.(1979):Formation of radicals and complex organic compounds by high-temperature pyrolysis:The pyrolysis of toluene,<u>Comb. and Flame</u>,35:179-190.
15. Bittner,J.D. and Howard,J.B.(1981):Composition profiles and reaction mechanism in a near-sooting premixed benzene/oxygen/argon flame,XVIII Symposium (Int.) on Combustion pp.1105-1116, The Combustion Institute.
16. Neoh,K.G.,Howard,J.B. and Sarofim,A.F.(1981):Soot oxidation in flames, particulate carbon formation during combustion (Siegla,D.C. and Smith,G.W.,Eds.) Plenum Press,N.Y.
17. Coe,D.S. and Steinfeld,J. (1980):<u>Chem.Phys.Lett.</u> 76:185-189.

HEPATIC TUMOR RATES AND POLYNUCLEAR AROMATIC HYDROCARBON LEVELS IN TWO POPULATIONS OF BROWN BULLHEAD (ICTALURUS NEBULOSUS)

PAUL C. BAUMANN*, WILLIAM D. SMITH**, MICHAEL RIBICK***
*Columbia National Fishery Research Laboratory Research Station, Columbus, Ohio 43210; **Ohio State University, Zoology Department, Columbus, Ohio 43210; ***Columbia National Fishery Research Laboratory, Columbia, Missouri 65201. USFW, USDI.

INTRODUCTION

Increasing frequency of liver tumors (hepatomas) in wild populations of fish have been documented during the last decade. Greater than expected hepatoma rates have been reported in brown bullheads (Ictalurus nebulosus) from the Fox River, Illinois (12.2%); Atlantic hagfish (Myxine glutinosa) in Swedish estuaries (5.8%); English sole (Parophrys vetulus) from the Duwamish River estuary, Washington (32%); and tomcod (Microgadus tomcod) from the Hudson River, New York (25%) (5, 9, 15, 20). Hepatomas are of particular importance since they are a primary tumor (8, 11) and can be chemically induced (1, 19, 18).

In all of the instances of high rates of hepatomas previously mentioned, significant levels of contaminants were present in the environment. Hepatoma bearing fish from the Duwamish River, Hudson River, and Gullmar Fjord, Sweden all had high levels of PCBs. Analysis of the Fox River, Illinois revealed a wide variety of pollutants including mercury, lead, arsenic, toluene, benzene, naphthalene, benzanthracene, and chlorinated hydrocarbons. These data suggest that some industrial effluents may induce liver tumors in exposed populations of fish.

METHODS

Research Organism:

The brown bullhead was chosen as the indicator organism because it is fairly ubiquitous, pollution tolerant, occurs in abundance, and is known to be vulnerable to hepatic tumors. Its range extends across all of the eastern United States from southern Canada to Florida (6). Bullheads can survive in contaminated environments having low oxygen levels, resulting in exposure to pollutants and time for tumor formation to occur. Brown et al. (5) found that brown bullhead reached a 12.2% incidence of tumor in the Fox River, Illinois, but only a 1.98% incidence in Lake-of-the-Woods, Ontario.

HEPATIC TUMOR RATES AND PAH LEVELS

Study Sites:

The Black River near its junction with Lake Erie at Lorain, Ohio is a stream which is bordered by heavy industry. A study by Brass et al. (4) found a wide range of organic contaminants present in the sediments, including several polynuclear aromatic hydrocarbons (PAHs) at levels over 10 ppm. Two of these PAHs, dibenzanthracene and benz[a]pyrene are known to be higly carcinogenic to mammals. The basic difference between the Black River and control sites was the presence of industrial effluents containing PAHs. Lake-of-the-Woods is oligotrophic, and the Black River and Buckeye Lake are eutrophic with similar physical conditions (BOD, pH, DO) (4, 5, 22).

Buckeye Lake, Ohio is a shallow 1,335 hectare reservoir formed when the tributaries of the South Fork Licking River were dammed in 1825. The reservoir shoreline contains a large number of homes and cabins and a state park. No industrial pollution enters the lake (22), but it does receive large amounts of domestic effluent. Three municipal sewage plants, one of which does not meet EPA standards, discharge into the drainage basin, as do numerous septic systems, many of which have failed. Although 15-20% of the shoreline is cultivated farmland, pesticides are below detection limits (21).

Field Techniques:

Fish were collected by fyke net and held in tubs until processed. Sampling occurred in the Black River from April-June and in Buckeye Lake from July through August, 1980. Sediment samples were taken with a Peterson dredge in October 1980. Samples from two locations in the Black River (above and below the coke plant outfall) and one location in Buckeye Lake were made by combining two dredge hauls at each sample site.

Bullheads were measured live, then killed and dissected. The sex of the fish was recorded. Each fish was examined visually for external tumors, after which the liver was removed and also examined. Tumors were dissected and preserved in 10% neutrally buffered formaldehyde. All fish containing tumors were frozen for later residue analysis, and spines were removed for aging.

Preliminary aging provided a cut-off size for separating fish 3 yrs. and older from 2-yr.-olds. A subsample of 20 normal fish from each of these size categories was also frozen for residue analysis. Pectoral spines were removed from at least 50 fish in each size group at each location to provide age/length curves.

Laboratory Techniques:

A standard paraffin technique (12) was used for tissue preparation. Sections were stained with Harris hemotoxylin and Putts eosin. Tumor confirmation was performed by Dr. Charles Smith, U.S. Fish and Wildlife Service, Bozeman, Montana, and Dr. John Harshbarger, The Registry of Tumors in Lower Animals, Washington, D.C.

Age determination using pectoral spines followed the method of Marzolf (14). All complete rings were considered to be one annulus. Bullheads from which spines had not been removed were assigned an age according to their length, based on a linear regression of fish with known age and length.

Chemical Techniques:

Frozen composited fish tissue and dry sediment were saponified in 1.2 Molar solution of aqueous potassium hydroxide. The digestates were adjusted to pH 4 with glacial acetic acid and extracted three times with methylene chloride. The extracts were combined and concentrated by rotoevaporation. The concentrates were added to 8 g basic alumina columns. Elution with 30 ml hexane brought the aliphatics off first, the PAH fraction was then eluted with 100 ml of benzene. The aromatic fraction was concentrated and cleaned-up on a Bio Bead SX-12 gel permeation column, using methylene chloride. The cleaned-up fractions were gas-chromatographed on SE-52 fused silica capillary columns using hydrogen as a carrier gas. The temperature program for the column oven was 40°C to 260°C at 4°C/min. All identifications were confirmed by gas chromatography - mass spectrometry. The methods of analysis are described in greater detail by Vassilaros et al. (24).

RESULTS

Pathology:

Liver tumors in Black River fish appeared as small white to cream colored nodules embedded in or protruding above the surface of the liver. Harshbarger (10) described these tumors as cholangiomas, proliferations of bile duct tissue. A large number of mitotic figures occurred throughout the tumors, and there was invasion of surrounding normal tissue. The centers of the larger tumors contained acidophilic cells and large areas of necrosis. Lip and dermal tumors were also noted in both populations.

HEPATIC TUMOR RATES AND PAH LEVELS

Tumor Incidence:

Two-year-old brown bullhead from the Black River had a liver tumor rate of 1.2% while those 3 yrs. and older had a 33% rate. The tumor rate for fish 3 yrs. and older was statistically significantly higher ($P<0.01$) than that for 2-yr.-old fish. None of the brown bullhead examined from Buckeye Lake had visually observable hepatomas (sample size: 249 2-yr.-olds, 80 3-yr. and older).

Sediment Residues:

Two sediment samples were analyzed from the Black River. The highest concentrations of PAHs were found in sediment taken from the vicinity of the coke plant outfall (Table 1). These levels were 3 to 4 orders of magnitude greater than those in a Black River sediment from near a municipal discharge. The coke plant outfall sediment was a non-homogeneous mixture of clay soil, sand, and a black tar substance. This sample had concentrations of PAH, based on dry weight, ranging from 4800 ppb (coronene) to 390,000 ppb (phenanthrene). A sediment from Buckeye Lake is currently under investigation, but preliminary analyses indicate negligible levels of PAH relative to the Black River coke plant sediment.

Bullhead Residues:

A wide range of PAHs and other organics were present in Black River bullheads, while those from Buckeye Lake had a much lower contaminant burden (Table 1). Amounts of individual compounds were generally higher from Black River fish for the chemicals checked. Buckeye Lake bullhead contained no PAHs in amounts over 37 ppb (C-2 naphthalenes). However, along with total PCBs, Black River bullhead contained 4 PAHs in concentrations greater than 1 ppm: acenaphthylene, phenathrene, fluoranthene, and pyrene. Heavier weight PAHs were present, but only in lower amounts. Although the 3-year-old normal fish had higher residue levels than the tumorous fish of that age group, these results represent the parent compounds and not the reactive metabolites, the diols and phenol epoxides (23).

DISCUSSION

Sediment Residue Levels:

Table 1 shows that the PAHs prevalent in the Black River sediment are also present in the fish from that site. Lao et al. (13) detected PAHs in effluents from coke and coal tar production facilities. The profile of PAHs in these processes

TABLE 1

CONTAMINANT RESIDUE LEVELS (NG/G) FROM SEDIMENT AND COMPOSITES OF TWO-YEAR-OLD (2) AND THREE-YEAR-OLD-PLUS (3+) BROWN BULLHEAD BOTH WITH TUMORS (T) AND NORMAL (N)

Contaminant	Concentration					
		Black River			Buckeye Lake	
	Sediment	3+T	3+N	2N	3+N	2N
Naphthalene	31000	40	144	-	-	-
2-Methylnaphthalene	15000	114	322	3	0.4	14
Biphenyl	9700	21	119	1	-	3
C-2 Naphthalenes	-	815	721	78	-	37
Acenaphthylene	40000	775	2378	57	-	-
Acenaphthene	36000	88	258	26	-	-
Dibenzothiophene	22000	321	697	171	5	16
Phenanthrene	390000	2140	5724	1669	3	19
Methyldibenzothiophenes	NR	390	676	7	-	3
Fluoranthene	220000	583	1938	558	7	28
Phenanthro[4,5-bcd]thiophene	4800	60	117	35	-	-
Pyrene	140000	424	1089	401	7	36
Benz[a]anthracene	51000	4	33	3	-	0.1
Chrysene	51000	38	83	19	0.7	29
Benzofluoranthenes	75000	1	32	-	-	1
Benzo[e]pyrene	28000	4	21	1	-	0.1
Benzo[a]pyrene	43000	7	18	-	-	3
Perylene	12000	3	8	-	7	2
Indeno[1,2,3-cd]pyrene	26000	-	-	-	-	-
Diben[a,h]+[A,C]anthracene	9400	-	-	-	-	-
Benzo[b]chrysene	8000	-	-	-	-	-
Benzo[ghi]perylene	24000	-	-	-	-	-
Coronene	4800	-	-	-	-	-
Anthanthrene	13000	-	-	-	-	-
Total PCB	NA	1900	1300	1000	100	100

(-) indicates not detected
NA: indicates not analyzed for
NR: indicates not reported

is very similar to that found in the sediment, especially for phenanthrene thru ananthrene (Table 1). They did not report any heterocyclic PAHs, i.e. dibenzothiophenes and cogeners. The black tar material probably accounts for the high levels of PAHs in the sediment. Coal tar itself is approximately 44.4% PAH by weight (16). The Black River fish did not contain PAH's heavier than perylene (Table 1), however the profile for detected PAHs was similar to that of the sediment, especially for phenanthrene, fluoranthene, and pyrene. Apparently the residues in fish can be attributed to the uptake from the water above or from the polluted sediment. However, the mode of assimilation of the sediment adsorbed PAHs by fish or other aquatic organisms is not well understood.

Bullhead Residue Levels:

The high tumor rate for bullheads in the Black River correlates with their high body burden of PAHs. Bullheads from Lake-of-the-Woods, Ontario had only a 1.98% overall cancer rate (5), while no bullhead from Buckeye Lake contained tumors. Phenanthrene and benz[a]pyrene reached higher levels in bullhead from the Black River than in any of the species tested from such polluted areas as the Hersey River, Michigan, and the Buffalo River, New York (Table 2) (2, 3). Black River bullheads also had higher body burden of benzanthracene than any species except carp x goldfish hybrids from the Buffalo River. Neoplasms reported in fish from the Buffalo River include gonadal tumors (carp x goldfish hybrids), dermal lesions (sheepshead and brown bullhead), and papillomas (white sucker).

Both methylcholanthrene and benzo[a]pyrene have been used by Ermer (7) to induce epitheliomas in the fish Gasterosteus aculetus and Rhodeus amorus. The chemicals were painted on the skin (0.5 mg doses) twice a week for 3 to 7 months. However tumors were not induced on carp by this same method. Also 7-12-dimethylbenz[a]anthracene (DMBA) and 3-methylcholanthrene did not induce tumors in guppies (Peopcilea reticulata) or zebra fish (Danio rerio) when injected with 40 mg and 20 mg doses respectively (19). Neither did the addition of DMBA to the diet (120 mg/100 g dry diet) for 56 weeks induce tumors. However fish liver microsomes can metabolize PAHs into mutagenic and carcinogenic compounds (17, 19).

Of the compounds found in Black River bullhead, benz[a]anthracene, benzofluoranthenes, and benzo[a]pyrene (BaP) have been identified as animal carcinogens by the International Agency for Research on Cancer (IARC). Benzo[e]pyrene is listed as a suspected animal carcinogen by the AIRC and seven of the others have caused at least benign tumors in mammalian experiments.

TABLE 2

RESIDUE LEVELS OF PHENANTHRENE, BENZANTHRACENE, AND BENZ[a]PYRENE IN FISH FROM THREE CONTAMINATED RIVERS

Chemical	Concentration (ppb) in fish		
	Hersey R.*	Buffalo R.**	Black R.*
Phenanthrene	37.65 T	15.7 C	2,140 B
	28.47 X	773.1 CG	5,724 B
		26.3 B	2,703 B
			1,669 B
Benzanthracene	0.17 T	11.9 C	4 B
	0.13 S	127.5 CG	33 B
		20.48 B	22 B
			3 B
Benz[a]pyrene	0.07 T	1.05 B	0.7 B
	0.08 S	2.46 CG	18.0 B
		0.93 B	7.0 B
			ND

T = Brown trout (Salmo trutta)
S = White sucker (Catostomus commersoni)
C = Carp (Cyprinus carpio)
CG = Carp x goldfish hybrid (Cyprinus carpio x Carassius aratus)
B = Brown bullhead (Ictalurus nebulosus)
ND = Not detectable
* = Single composite (3)
** = Means of individual fish analyses with n=4 (carp and carp x goldfish) and n=1 (bullhead) (2)

Most of the heavier weight, more carcinogenic compounds such as BaP were found at relatively low levels in the bullhead as compared to lighter weight PAHs or sediment levels (Table 1). Other studies have also reported low BaP concentrations in fish as compared to their environment (2, 3, 25). However Varanasi and Gmur (23) found that both English sole and starry flounder metabolized BaP extensively and rapidly, and suggested that low BaP levels in fish tissues may be the result of this rapid biotransformation rather than low bioavailability. They also noted that English sole, which metabolized BaP more extensively than starry flounder, also had a higher liver tumor rate in the Duwanish River Estuary.

Therefore body burdens of BaP and similar parent PAHs may not reflect the levels present or amounts processed through time of such reactive metabolites as diol or phenol epoxides (23). Since the Black River differs from control sites principally in PAH levels, it is our conclusion that they are the most likely causal factor for the elevated cholangioma rate seen in the brown bullhead population from that river.

ACKNOWLEDGMENTS

This work was supported by the U.S. Fish & Wildlife Service. Residue analyses were performed at Alpine West Laboratories, Provo, Utah.

REFERENCES

1. Ahokas, J. T., O. Pelkonen, and N. T. Karki. (1977): Characterization of benzo[a]pyrene hydroxylase of trout liver. Cancer Res., 37:3737-3743.
2. Black, J. J., P. P. Dymerski, and W. F. Zapisek. (1980): Environmental Carcinogenesis Studies in the Western New York Great Lakes Aquatic Environment. National Science Foundation and National Cancer Inst., N.I.H. 15 pp. (unpublished report).
3. Black, J. J., T. F. Hart, Jr., and E. Evans. (1981): HPLC studies of PAH pollution in a Michigan trout stream. In: Chemical Analysis and Biological Fate: Polynuclear Aromatic Hydrocarbons. Edited by M. Cooke and A. J. Dennis. Fifth International Symposium. pp. 343-355. Battelle Press, Columbus, Ohio.
4. Brass, J. H., W. C. Elbert, M. A. Feige, E. M. Glick, and A. W. Lington. (1974): United States Steel, Lorain, Ohio, Works, Black River Survey: Analysis for hexane organic extractables and polynuclear aromatic hydrocarbons. United States Environmental Protection Agency, Cincinnati, Ohio.
5. Brown, R. E., J. J. Hazdra, L. Keith, I. Greenspan, and J. B. G. Kwapinski. (1973): Frequency of fish tumors in a polluted watershed as compared to non-polluted Canadian waters. Cancer Res., 33:189-198.
6. Clay, W. M. (1975): The Fishes of Kentucky. Kentucky Department of Fish and Wildlife Resources. Frankfort, Kentucky.

7. Ermer, M. (1970): Versuche mit cancerogenen mittelin bein kurzlebigen fischarten. Zool. Arten. Anz., 184:175-193.
8. Falkmer, S., S. O. Endin, Y. Ostberg, A. Mattisson, M. L. Johansson, S. Jobeck, and R. Fange. (1976): Tumor pathology of the hagfish (Myxine glutinosa) and the river lamprey (Lampetra fluviatilus). Prog. in Exp. Tumor Res., 20:217-250.
9. Falkmer, S., S. Marklund, P. E. Mattsson, and C. Rappe. (1977): Hepatomas and other neoplasms in the Atlantic hagfish (Myxine glutinosa), a histopathologic and chemical study. Annals of the New York Academy of Sciences, 298:342-355.
10. Harshbarger, J. C. (1981): Registry of Tumors in Lower Animals, Nat'l. Museum of Natural History, Smithsonian Inst., Washington, DC (personal communication).
11. Hueper, W. C. and W. W. Payne. (1961): Observations on the occurrence of hepatoma in rainbow trout. J. Nat'l. Cancer Inst., 27:1123-1144.
12. Humason, G. L. (1967): Animal Tissue Techniques. W. H. Freeman & Co., San Francisco, California.
13. Lao, R. C., R. S. Thomas, and J. L. Monkman. (1975): Computerized gas chromatographic - mass spectrometer analysis of polycyclic aromatic hydrocarbons in environmental samples. J. of Chromatography, 112:681-700.
14. Marzolf, R. C. (1955): Use of pectoral spines and vertebrae for determining age and rate of growth of the channel catfish. J. of Wildlife Management, 19(2):243-249.
15. McCain, B. B., K. V. Pierce, S. R. Wellings, and B. S. Miller. (1977): Hepatomas in marine fish from an urban estuary. Bull. of Environ. Contam. and Toxicol., 18(11):1-2.
16. Neff, J. M. (1979): Polycyclic Aromatic Hydrocarbons in the Aquatic Environment: Sources, Fate, and Biological Effects. Applied Science Publishers LTD, London. 262 pp.
17. Payne, J. F., I. Martins, and A. Rahimtula. (1978): Crankcase oils: Are they a major mutagenic burden in aquatic environments? Science, 200:329-330.
18. Pierce, K. V., B. B. McCain, and S. R. Wellings. (1978): Pathology of hepatomas and other liver abnormalities in English sole (Parophyrys vetulus) from the Duwamish River estuary, Seattle, Washington. J. Nat'l. Cancer Inst., 50(6):1445-1449.
19. Pliss, G. B. and V. V. Khudoley. (1975): Tumor induction by carcinogenic agents in aquarium fish. J. Nat'l. Cancer Inst., 55:129-136.

20. Smith, C. E., T. H. Peck, R. H. Klauda, and J. B. McLaren. (1979): Hepatomas in Atlantic tomcod (Microgadus tomcod) collected in the Hudson River estuary, New York. J. Fish Diseases, 2:313-319.
21. Tobin, R. L. and J. D. Youger. (1977): Limnology of Selected Lakes in Ohio 1975. United States Geological Survey Water Resources Investigations. pp. 77-105. Columbus, Ohio.
22. United States Environmental Protection Agency. (1975): Region V Working Paper. # 396. Report of Buckeye Lake, Fairfield, Licking, and Perry Counties, Columbus, Ohio.
23. Varanasi, V. and D. J. Gmur. (1980): In vivo metabolism of naphthalene and benz[a]pyrene by flatfish. In: Chemical Analysis and Biological Fate: Polynuclear Aromatic Hydrocarbons. Edited by M. Cooke and A. J. Dennis. Fifth International Symposium. pp. 367-376. Battelle Press, Columbus, Ohio.
24. Vassilaros, D. L., P. W. Stoker, G. M. Booth, and M. L. Lee. (1981): Capillary gas chromatographic determination of polycyclic aromatic compounds in vertebrate fish tissue. Analytical Chemistry. (in press).
25. Veldre, I. A., A. R. Itra, and L. P. Paalme. (1979): Levels of benzo[a]pyrene in oil shale wastes, some bodies of water in the estonian S.S.R. and in water organisms. Environ. Perspectives, 30:211-216.

METABOLISM OF PHENANTHRIDINE, AN AZA-ARENE PRESENT IN LOW BTU
GASIFIER EFFLUENTS

JANET M. BENSON, ROBERT E. ROYER AND JOSEPH O. HILL
Lovelace Inhalation Toxicology Research Institute
P. O. Box 5890
Albuquerque, NM 87185

INTRODUCTION

Aza-arenes are potentially toxic components of process and waste materials produced during low Btu gasification of coal. They have also been identified as components of urban air particles, crude petroleum distillates, shale oil and effluents produced during coal combustion (1). Because of their distribution, they may represent an inhalation hazard in both the occupational and general environment. Certain aza-arenes, including quinoline are carcinogenic (2) and mutagenic. Quinoline and benzoquinoline derivatives are active in Ames Salmonella/mammalian microsome mutagenicity assay, but require metabolism by microsomal enzymes to be mutagenic (1). In addition, they induce rat liver aryl hydrocarbon hydroxylase, indicating that this microsomal enzyme may be involved in aza-arene metabolism.

We have initiated studies on the comparative metabolism of phenanthridine (3,4 benzoquinoline) by rat lung and liver microsomes. Phenanthridine was chosen because it has been identified as a component of a mutagenic fraction of tar produced in low Btu coal gasification and because it has greater mutagenic activity than other aza-arenes identified in that fraction.

MATERIALS AND METHODS

Phenanthridine was purchased from Tridon Chemical Inc., Hauppauge, New York. Aroclor 1254 was obtained from Analabs, Inc., New Haven, CT. Benzo(a)pyrene was purchased from Aldrich Chemical Company, Milwaukee, WI. Phenanthridone was synthesized according to the method of Hawbecker et al (3) and its identity was confirmed by gas chromatography/mass spectrometry.

For preparation of microsomal enzyme-induced rat liver microsomes, laboratory-reared adult male Fischer-344 rats were injected intraperitoneally with Aroclor 1254 (500 mg/kg in corn oil). For isolation of microsomal enzyme-induced lung microsomes, male Fischer-344 rats were administered benzo(a)pyrene (BaP 10 mg/kg in 0.2% gelatin-saline suspension) by intratracheal instillation. Both Aroclor and BaP treated animals and untreated animals were sacrificed 24 hours later. Lungs and livers from these animals were homogenized in 1.15% KCl. Homogenates were centrifuged at 10,000 xg at 5°C for 20 minutes. Supernatant fractions were collected and respun at 100,000 xg at 5°C for 1 hour. Resulting supernatant fractions were discarded and microsome-containing pellets were resuspended in a small volume of 1.15% KCl. Incubation of phenanthridine and phenanthridone with lung and liver microsomes was carried out according to the method described by Selkirk et al. (4).

High pressure liquid chromatographic analyses were conducted using C-18 reversed column (30 cm x 0.4 cm i.d.) eluted with a linear 30 to 100% acetonitrile-H_2O gradient. Eluted compounds were detected by means of their absorbance at 254 nm. Ultraviolet absorbance spectra were obtained using Perking Elmer Model 559 scanning UV-visible spectrophotometer. Fluorescence spectra were obtained using a Farrand Mark I spectrofluorometer. For gas chromatography/mass spectrometric (GC/MS) analyses, a Finnigan Model 4023 GC/MS data system (GC/MS/DS) with a NBS mass spectral library was used to obtain spectra of standards and the isolated metabolite.

Mutagenic activity of phenanthridine and phenanthridone was assessed using the Ames Salmonella/mammalian microsome assay (5). Compounds were tested for activity in tester strain TA-98, with and without metabolic activation. Cytotoxic activity was assessed in canine pulmonary alveolar macrophages by a method developed by Hill and described by Benson et al. (6).

RESULTS

The HPLC chromatographic profiles of phenanthridine metabolites produced by enzyme-induced lung and liver microsomes are shown in Figure 1 a, b. Similar profiles were obtained when uninduced microsomes were used.

Metabolite E had approximately the same retention time as did the standard phenanthridine (16.5 min.). The UV absorbance spectra of metabolite E and phenanthridone were similar, with absorption wavelengths of 232 (max), 258, 269, 306, 321 and 334 nm. Fluorescence spectra were also similar with emission maxima at 365 nm (λ excitation = 240 nm). Mass spectra showed a molecular ion for both compounds at 195 m/e and fragment ions at 167, 139, 113, 87 and 63 m/e.

Phenanthridone was directly mutagenic (14 revertants/µg) and mutagenic activity was enhanced upon incubation with rat liver S-9 (34 revertants/µg). By comparison, phenanthridine was not directly mutagenic, but exhibited activity (6.8 revertants/µg) when incubated with rat liver S-9.

Results on the effects of phenanthridine and phenanthridone on canine pulmonary alveolar macrophages indicate that phenanthridine has a median effective concentration (EC_{50}) of 20 µg/mL of incubation medium, whereas the EC_{50} for phenanthridone is 60 µg/mL. Therefore, phenanthridine is about three times as cytotoxic to PAM as is phenanthridone.

In subsequent studies, when phenanthridone was incubated with uninduced rat lung and liver microsomes, no metabolism was detected.

DISCUSSION

Our results indicate that phenanthridine is metabolized by both rat lung and liver microsomes to several more polar metabolites. The pattern of metabolites produced by lung and liver as analyzed by HPLC are qualitatively similar, but differences exist in the relative amounts of metabolites produced. The greatest difference is in the amount of metabolite A produced by liver and lung microsomes, with liver microsomes producing greater than 4 times more A than lung. These differences may reflect differences in pattern of metabolites formed by lung and liver or may be a consequence of using different enzyme inducers in the two organs.

The phenanthridine metabolite with an HPLC retention of 16.5 min. (metabolite E) has been identified as phenanthridone. This identification was made on the basis of agreement between the HPLC retention times, UV absorbance fluorescence spectra and mass spectra of the isolated metabolite and the phenanthridone standard. These results are consistent with the findings of Tsuboi (7) who reported the metabolism of phenanthridine to phenanthridone in vitro by a mixture of crude liver homogenate and phenanthridine dehydrogenase.

It is interesting to note that microsomal metabolism of phenanthridine resulted in a compound with enhanced mutagenic activity but lower cytotoxic activity toward both Salmonella TA-98 and toward canine pulmonary alveolar macrophages. Results of Ames test indicate that phenanthridone may be further metabolized by microsomal enzymes to more polar metabolites. However, incubation of phenanthridone with uninduced rat lung and liver microsomes produced no metabolism of phenanthridone. It is possible that enzymes in Salmonella may be metabolizing phenanthridone to mutagenic metabolites, and be responsible in part for the observed mutagenic activity of this compound.

Results of these studies show that phenanthridine is metabolized by both rat lung and liver to mutagenic and potentially carcinogenic derivatives. Overall health effects which may be expected upon inhalation of phenanthridine depend on target organ specificities of the compound and its metabolites and upon the sites of deposition in the airways, its retention in the lung and distribution to other tissues upon inhalation.

ACKNOWLEDGEMENTS

We would like to thank Dr. C. R. Clark for directing the Salmonella mutagenicity assay and discussion of results.
Research performed under U. S. Department of Energy Contract Number DE-AC04-76EV01013.

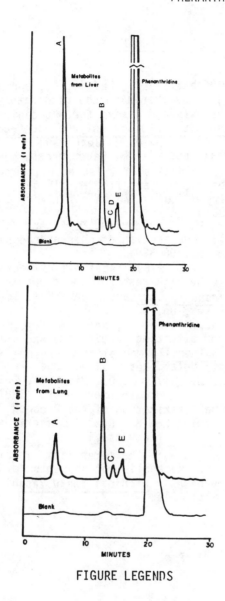

FIGURE LEGENDS

FIGURE 1a. High pressure liquid chromatogram of phenanthridine and metabolites produced by induced rat liver microsomes. b. Chromatogram of phenanthridine and metabolites produced by induced rat lung microsome.

REFERENCES

1. Dong, M., Schmeltz, I., LaVoie, E., and Hoffmann, D. (1978): Aza-arenes in the respiratory environment: Analysis and assays for mutagenicity. In: Carcinogenesis, vol. 3: Polynuclear Aromatic Hydrocarbons. Edited by P. W. Jones and R. I. Freudenthal, pp. 97-108, Raven Press, New York.
2. Hirao, K., Shinohara, Y., Tsuda, H., Fuykushiwa, S., Takahashi, M., and Ito, N. (1976): Carcinogenic activity of quinoline on rat liver, Cancer Res., 36:329-335.
3. Hawbecker, B. L., Radovich, D. A. and Tillotson, L. G. (1976): Preparation of phenanthridone. J. Chem. Ed., 53:398-399.
4. Selkirk, J. K. (1978): Analysis of benzo(a)pyrene metabolism by high-ressure liquid chromatography. In: Advances in Chromatography, Vol. 16, New York: Marcel Decker, pp. 1-36.
5. Ames, B. N., McCann, J., and Yamasaki, E. (1975): Methods for detecting carcinogens and mutagens with the Salmonella/mammalian microsome mutagenicity test, Mutat. Res., 31:347-364.
6. Benson, J. M., Hill, J. O., Mitchell, C. E., Newton, G. J., and Carpenter, P. L. (1981): Toxicological Characterization of the Process Stream of an Experimental Low Btu Coal Gasifier. Arch. Env. Contam. Toxicol. (In press.)
7. Tsuboi, A. (1959): The metabolism of quinoline derivatives. Metabolism of phenanthridine and its quaternary bases. Nichidai Igaku Zasshi, 18:2995-3003, CA 61:15194f.

CHARACTERIZATION AND COMPARISON OF ORGANIC EMISSIONS FROM COAL, OIL AND WOOD FIRED BOILERS

J G T BERGSTRÖM, G EKLUND AND K TRZCINISKI
Studsvik Energiteknik AB, S-611 82 Nyköping, Sweden

INTRODUCTION

Organic compounds such as PAH are present in the combustion flue gases. Studsvik has a program to quantify the emissions from different fuels and combustion processes. We want to determine the influence of design and operational factors on these emissions. The program will result in control parameters and devices to reduce the emissions.

As a part of our program we have been sponsored to collect samples from the flue gases in several boilers. The intent had been to find if the same type and amounts of organic compounds are emitted from efficient combustion of different fuels.

The same sampling and analytical methods are used for all the boilers. We present a comprehensive evaluation of organic emissions and have not focused on any special type of organic compounds due to their health or environmental effects.

TESTED BOILERS

We present results from samples collected in seven different boilers.

Plant A is a steam boiler (600 t/h steam) equipped with a electrostatic precipitator. During the tests the load was only 54% (312 MW_{th}). The boiler load was depending of the energy required for district heating.

Plant B was a hot water boiler without any flue gas cleaning system. The maximal load was 50 MW_{th}. During the tests the capacity was limited to 68% (37 MW_{th}) for the same reason as for Plant A.

COAL, OIL, AND WOOD FIRED BOILERS

Plant C was a small hot water boiler with a capacity of 5.8 MW_{th}. It was operated almost at maximal load during the tests. The boiler was not equipped with any flue gas cleaning system.

Plant D was a pulverized coal fired power plant operating at full load producing 270 MW_e (600 MW_{th}). The boiler was equipped with a high efficiency electrostatic precipitator. The unit fired a mixture of low sulphur (0.8%) coals.

Plant F was a coal fired industrial steam boiler (40 t/h steam). The low sulphur (0.9%) coal was fired on a moving grate (Wanderrost). During the tests the load was 78% (25 MW_{th}). The boiler was equipped with multiple cyclones.

Plant G was a coal fired steam boiler. The steam was heat exchanged and used for district heating. The low sulphur coal (0.7%) was fired on a moving grate similar to Plant F. During the tests the boiler was operated at three different loads; 16 MW_{th}, 23 MW_{th} and 29 MW_{th}. The samples were collected in a flue gas stream cleaned in a bag house filter installed for test purposes.

Plant J was a small fluidized bed coal combustor arranged mainly for demonstration but generating hot water for district heating. The capacity was 4.7 MW_{th} and the samples were collected at 87% load. The FBC was equipped with a bag house filter.

Plant W was a wood fired hot water boiler. The boiler capacity was 30 MW_{th}. The wood chips were fired in a cyclone furnace in front of the boiler. During the tests the fuel was oak chips with a moisture content of about 30%. The boiler was equipped with an electrostatic precipitator.

During the collection of samples for analyses of organic emissions we made measurements to prove the combustion conditions and the efficiency of the flue gas cleaning equipment.

Table 1 shows the flue gas condition in the different boilers during the tests.

TABLE 1

FLUE GAS CONDITIONS DURING TESTS

Boiler	A	B	C	D	F	G	J	W
Gas flow m^3/MJ^1	0.31	0.29	0.29	0.35	0.51	0.50	0.35	0.37
Gas temperature $°C$	158	151	220	140	155	112	164	195
Moisture vol-%	11	11	11	7	5	5	7	15
CO_2 vol-%	13.0	13.7	13.6	14.1	9.3	9.8	12.9	14.5
CO ppm	40	3	16	<1	15	11	2000	50-5000
SO_2 ppm	1050	520	-	603	394	377	333	-
NO_x ppm	190	255	150	479	49	132	130	85
Particulate mg/m^3sd	10	10	154	80	78	1	2200	206

[1] Calculated as standard m^3 dry (sd) gas and MJ heat input with the fuel.

The CO_2 content in the flue gases indicates only the amount of excess air present at the sampling point in the stack. In some boilers there was a significant leakage of air into the flue gas in the air preheater and the precipitators.

There is a difference in combustion conditions shown in CO-content for Plant J, W and the other boilers. Also particulate emission is high from Plant J depending on problems in the bag house filter.

MATERIALS AND METHODS

Sampling equipment and procedure

Sampling. Two different kinds of sampling equipment were used. Light hydrocarbons (LHC) were collected (1) in "U"-formed glass tubes height 240 mm, OD 6 mm, ID 2 mm, equipped with Swagelok quick connects on both ends. The lowest part of the trap was occupied by a 50 mm portion of the trapping material, plugged with silane treated glass-wool on each side. Porapak Q 80/100 mesh, was choosen for trapping of light aliphatics and 20% SP-2100/0.1% Carbowa 20M on Chrom W for trapping of light aromatics. When sampling, the traps were immersed in liquid nitrogen and the stack gas was drawn with a flow rate of 40 ml/min by a personal sampling pump with a digital flow meter. The sample volumes were 50 to 300 ml depending on the concentration of LHCs in the stack gas.

The sampling system for semivolatile and high boiling organic compounds was a modified version of SASS train. The equipment (2) was developed by Studsvik to provide for obtaining a isokinetic sampling by an all-glass system. The Studsvik sampler is shown schematically in Figure 1.

The interchangeable probe extended into an oven module which was kept at $160^{\circ}C$ during the sampling. The oven contained a variable range cyclone and a high purity tissue quartz thimble filter. At a flow rate of 4 m^3/h sd the cut-size of the cyclone was adjusted to 1.5 μm. After leaving the filter the flue gas was drawn through a high efficiency cooler, a condensate collector and through an XAD-2 sorbent trap. The coolant was maintained at temperature below $-10^{\circ}C$ so temperature of gas entering XAD-2 adsorbent bed did not exceed $5-10^{\circ}C$. At the conclusion of the sampling run the probe and the cooler were washed with acetone which was then treated as a part of the sample. The flue gas volumes were between 10 and 20 m^3sd in each sample.

Materials. Technical grade cyclohexane and acetone were purified by distillation. Dichloromethane (Rathburne HPLC Grade) was used without further purification. Sodium sulphate (Merck Pa) was purified by heating to 450° for two hours. Potassium hydroxide (EKA pa) was used without purification. Quartz fiber filter was purified by heating to 530°. Amberlite XAD-2 (Rohm-Haas) was purified by successive washing with

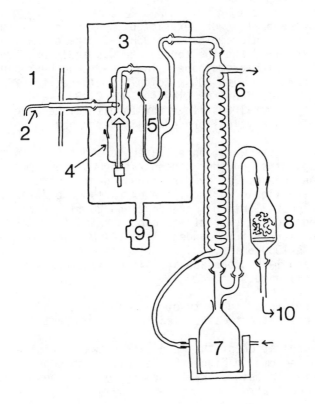

FIGURE 1. Studsvik sampling system. 1. Stack 2. Glass probe 3. Heated oven 160° 4. Glass cyclone 5. Quartz fiber thimble 6. Cooler 7. Condensate collector 8. XAD-2 adsorbent bed 9. Heater 10. To the pump and volume meter

water, methanol and acetone. The resin was then packed in the sampling ampoule and Soxhlet extracted with acetone and dichloromethane for 48 h with intermediate changes of the solvent. All glassware used for sampling was carefully rinsed and then heated to 450° over night.

Sample treatment. The glassware containing the samples was transported to the laboratory as soon as possible.

The condensates were extracted in a separatory funnel with 2x50 ml dichloromethane at pH 1-2. The pH was then adjusted to 11-12 and the condensate was again extracted with 2x50 ml dichloromethane. The XAD-2 adsorbent bed was Soxhlet extracted with 400 ml dichloromethane for 24 h. The particulate fractions for plants A, B, C, D, F and G were analyzed by Central Institute for Industrial Research (Oslo, Norway). The analytical procedure is described in ref (3).

The extracts were concentrated to about 5 ml in a Kurdena-Danish apparatus. Further concentration was performed in a centrifuge tube at 30° by directing a gentle stream of nitrogen onto the liquid surface. The extracts were dried with sodium sulphate. Internal standard and 20 ml of cyclohexane were added and the extract was concentrated to 10 ml. In some cases the extract from condensate and XAD-2 adsorbent were combined. A 1000 ppm solution of 1-chlorooctane, 1-chlorododecane and 1-chlorohexadecane in cyclohexane was used as an internal standard.

The volume of the extract was made up to 50 ml with cyclohexane. This extract was called the main extract.

The separation of the main extract was performed according to the scheme in Figure 2. The aim of this procedure was to separate the main extract into three fractions; acidic, basic and neutral.

Analytical procedure

Light hydrocarbons. The thermal desorption of the traps was achieved by heating at $230^{\circ}C$. A HP-5880 GC/FID was used for the analysis. The analytical columns were 0.5 m, ID 2 mm, Al_2O_3 for light aliphatics and 12.5 m, fused silica WCOT, SP-2100 (Hewlett-Packard) for analysis of light aromatics. The precision of the method was investigated and the relative standard deviation was found to be 3.4% for seven runs.

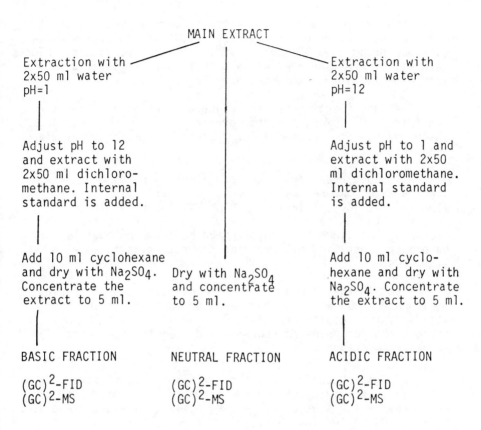

FIGURE 2. Treatment of extracts

High boiling hydrocarbons. A HP-5880A gas chromatograph equipped with flame ionization detector, split/splitless injector, recording integrator and autosampler was used. Chlorinated compounds were analyzed on a Varian 3920 GC equipped with split/splitless injector and a Ni-63 electron capture detector. For mass spectral identification a Finnigan 1020 GC-MS was used. The GC was equipped with a split/splitless injector. The capillary column was directly connected to the mass spectrometer.

A glass capillary column (purchased from GC Labor, Jaeggi, Switzerland) used in this work was 50 m, 0.3 mm ID coated with SE-54.

Three microliters of the extracts were injected without split at 50°. After an initial isothermal period of 5 min the oven was programmed 3°/min to 270°, injector and detector blocks were set of 270°. Helium was used as a carrier gas at a flow rate of 3 ml/min at ambient temperature.

The mass spectrometer interface was set at 250° and the manifold at 80°. Mass spectra were recorded in the mass range 50-400 dalton at a scan rate of 2s, with automatic repetetive scanning.

All substances have been identified by GC-MS. The mass spectra were compared with those of standard mass spectra library. The identification of PAHs have been confirmed by comparing the GC retention times with those of prepared PAH standard. The isomers of chlorinated benzens and phenols have been identified by comparing the retention times with standards.

Peaks in the chromatograms were quantified by comparing their peak areas with the peak area of the nearest internal standard. The results were not adjusted with relative response factors. Polychlorinated benzenes and phenols were quantified relative to standards.

The losses caused by concentration of the extracts and the extraction efficiency of the XAD adsorbent have been studied. The recovery for substances less volatile than naphthalene was found to be better than 90%.

To investigate the extraction yield and precision of the method a solution of several standard components was prepared in acetone. 400 µl of the solution was added to 2000 ml distilled water. The concentration of each component was 20 ppb in the water phase. The mixture of standard substances was devided into four 500 ml aliquots, and each aliquot was subjected to the procedure outlined in Figure 2. The recovery was determined by comparison with the original standard solution. The results are shown in Table 2.

TABLE 2

Compound	Neutral fraction		Acidic fraction		Basic fraction	
	Yield %	Rel stand deviation %	Yield %	Rel stand deviation %	Yield %	Rel stand deviation %
Chlorobenzene	102	12	-		-	
Xylene	74	8				
Decane	40	2	-		-	
Dichlorobenzene	101	1	-		-	
Naphthalene	95	4	-		-	
Trichlorobenzene	97	1	-		-	
1-chlorodecane	61	16	-		-	
Tetrachlorobenzene	105	2	-		-	
Tetradecane	60	3	-		-	
Fluorene	105	1	-		-	
1-chlortetradecane	101	4	-		-	
Pyrene	103	2	-		-	
3,6-dimethylphenanthrene	101	2	-		-	
Chrysene	96	4	-		-	
1,1-binaphthyl	97	3	-		-	
Benzo(a)pyrene	97	8	-		-	
Benzo(ghi)perylene	97	8	-		-	
Phenol	3	17	70	6	-	
2-chlorophenol	-		93	3	-	
Cresoles	-		72	6	-	
2,3-dichlorophenol	-		101	3	-	
Quinoline					99	5
Azafluorene					106	7

RESULTS

The same types of compounds were found in the extracts from all the boilers. The amount was, however, very different depending on combustion conditions. The detection limits were 100 ng/m^3 for LHC and between 10 and 100 ng/m^3 for the high boiling organics depending on gas volume in the sample and compound.

A comparison of the total amount of organics in the sample related to heat input in the boiler is given in Table 3.

TABLE 3

AMOUNT OF ORGANICS IN THE SAMPLES AND CARBON MONOXIDE IN FLUE GAS

Boiler	A	B	C	D	F	G	J	W
CO mg/MJ	15	1	6	<1	10	6	900	210
Total organics mg/MJ	1	0.08	0.08	0.1	0.05	0.07	9	1

Carbon monoxide is given as an indicator for combustion efficiency. The level of emission from the coal fired FBC-boiler is quite different than from the others. Almost all types of identified compounds were present on a much higher level in the extracts from the boiler J than from the others.

The n-alkanes and phthalates were present on almost the same level in all samples. The same was found for benzoic acid and acetic acid. The benzoic acid was mainly found in XAD-adsorbent and acetic acid in the condensate.

The oxidation of the alkyl benzenes by nitrogen oxides are commonly known. Hanson (4) confirms that benzoic acid has been found in XAD-2 which was exposed to NO and NO_2. However, it is impossible to estimate to what extent benzoic acid is a product of XAD degradation or is present in the flue gas.

Table 4 gives an overview of the major components in the samples.

TABLE 4

SOME ORGANIC COMPOUNDS IN FLUE GASES

Compound	Pulverized coal D	Wander-rost G	Heavy fuel oil B	Coal deficient combustion J	Wood chips W
1. Ethene	180	5	100	5	41
2. Benzene	5	5	0.5	2000	n a
3. Toluene	10	10	0.5	600	n a
4. Naphtalene	~1	5	0.5	7000	59
5. Acenaphtene	<	<	<	270	0.4
6. Acenaphtylene	<	<	<	1400	26
7. Fluorene	<	<	<	200	0.10
8. Phenanthrene	0.2	0.2	<	600	16
9. Anthracene	<	<	<	100	1.7
10. Pyrene	<	0.07	<	200	21
11. Fluoranthene	0.1	0.07	<	200	12
12. Benz(a)anthracene	<	<	<	30	0.9
13. Chrysene	<	<	<	10	0.9
14. Benzo(ghi)fluoranthene	<	<	<	20	5
15-16. Benz(bok)fluoranthene	<	<	<	25	3
17. Benzo(a)pyrene	<	<	<	12	0.4
18. Indeno(123-cd)pyrene	<	<	<	5	1.6
19. Dibenzo(ah)anthracene	<	<	<	1	<
20. Benzo(ghi)perylene	<	<	<	4	3.6
30. Phenol	4	<0.1	<1	700	n a
31. 2-chlorophenol	<	<	<	TO	n a
33. 4-chlorophenol	<	<	<	10	n a
35-36. 2.4-/2.6-chlorophenol	<	<	<	3	n a
40. Acetic acid	200	100	160	200	n a
41. Benzoic acid	80	25	12	40	n a
42. Benzaldehyde	2	1	1	750	0.8
43. Acetophenone	5	3	1	-	1.0
44. Benzonitrile	<	<	<	500	0.20
45. Quinoline	<	<	<	200	<0.10

All figures in $\mu g/m^3$ sd, n a = not analyzed, < = less than detection limit.

The PAHs associated with particulate emissions were identified in the two particle fractions. The concentration of the extractable organics were low. It was only in the samples from boiler J that a firm conclusion about distribution of the compounds between two particle fractions can be presented. The concentrations of the PAHs associated with the fine particle fraction were three to fifty times higher than those of the coarser particles.

Distribution of the PAHs between particulate and vapor phase was quantified for boiler J. As expected the low boiling PAHs were mainly found in vapor phase and high boiling PAHs are associated with particles.

ACKNOWLEDGEMENTS

This research was performed under contracts from the Swedish Board for Energy Source Development and the COAL-HEALTH-ENVIRONMENT Project. We acknowledge the cooperation and assistance received from the staff of the different plants tested. We appreciate the analytical cooperation with the Central Institute for Industrial Research, Oslo, Norway. We are grateful for the assistance received from the staff of the Development for Chemical Technology at Studsvik.

REFERENCES

1. Trzcinski, K: Bestämning av lätta kolväten (in Swedish). Determination of light hydrocarbons in gas samples.
2. Bergström, J: Emissioner av PAH. Utformning av provtagningsutrustning (in Swedish). Emissions of PAH. Development of a sampling system.
3. Kveseth, K: Organisk analyse av utslipp fra kull- og oliefyrte kraftverk (in Norwegian). Organic analysis of emissions from coal and oil fired power plants.
4. Hansson R L (1981): Evaluation of Tenax GC and XAD-2 as polymer adsorbents for sampling fossil fuel combustion products containing nitrogen oxides. <u>Env. Sci. Tec.</u>, 15(6), 701.

METABOLISM OF BENZO(A)PYRENE BY SKIN MICROSOMES: COMPARATIVE STUDIES IN C57BL/6N AND DBA/2N MICE AND SPRAGUE-DAWLEY RATS.

DAVID R. BICKERS*, HASAN MUKHTAR*, AND SHEN K. YANG**
*Department of Dermatology, Cleveland Veterans Administration Medical Center & Case Western Reserve University, Cleveland, Ohio, 44106, **Department of Pharmacology, School of Medicine, Uniformed Services University of the Health Sciences, Bethesda, Maryland, 20814.

INTRODUCTION

Cancer of the skin is the most common form of malignant neoplasm in the human population. Of the approximately one million new cancers diagnosed annually in the United States, almost 1/3 occur in the skin (1). In general, cutaneous cancers are relatively slow-growing, non-aggressive malignancies with little tendency to metastasize to other body organs. However, a small percentage of human skin cancers (1-3%) are highly aggressive and potentially fatal tumors.

Epidemiologic data have clearly shown that ultraviolet radiation is a major cause of human skin cancer (2). Those wavelengths between 290-320 nm, the so-called sunburn spectrum or ultraviolet B, are oncogenic in the skin of experimental animals (3). Certain chemicals are present in the environment which also can cause human skin cancer (4). Among these are coal tar products (5) which contain the polycyclic aromatic hydrocarbons such as benzo(a)pyrene (BaP). BaP is a ubiquitous pollutant chemical generated whenever fossil fuels are incompletely combusted and is present in significant amounts in polluted urban air and in a number of occupational environments (6).

The skin, because of its essential function as a barrier between the body and its environment, is in direct and continuing contact with numerous physical and chemical agents that are potentially oncogenic. Of the chemical carcinogens perhaps the best studied insofar as the skin is concerned is BaP. The known biological activities of BaP, many of which have been shown to occur in skin or in cells cultured from skin, include cytotoxicity, tumorigenicity and covalent binding to macromolecules including DNA, RNA and proteins (7). Most if not all of these biological effects of BaP appear to require metabolism of the compound into reactive moieties that ultimately mediate the toxic response (8).

The metabolism of BaP in the skin has been known to occur for many years. In 1948, Weigert first showed that application of BaP to the skin of experimental animals resulted in the formation of a metabolite with fluorescent properties slightly different from the parent compound (9). Subsequently it has become clear that BaP is metabolized in the skin by a cytochrome P-450-dependent membrane-bound enzyme system commonly known as aryl hydrocarbon hydroxylase (AHH) (10). Products of this reaction are converted to diol-epoxides of BaP which bind to cutaneous macromolecules thereby, perhaps, initiating tumor formation (11).

The relationship of AHH activity and/or inducibility to the tumorigenicity of the polycyclic aromatic hydrocarbons (PAH) has been proposed and studies by Nebert and his co-workers have suggested that the responsiveness of AHH to PAH in inbred mouse strains is a genetic characteristic inherited as an autosomal dominant trait and that AHH responsiveness and skin tumorigenicity are directly related (12). Thus induction of skin tumors by topically applied PAH occurs much more frequently in the AHH-responsive C57BL/6N mouse strain than in the AHH-non-responsive DBA/2N mouse strain. Despite numerous studies of AHH activity and inducibility in cutaneous tissue, little information is available concerning the patterns of BaP metabolism in the skin and whether such differences correlate in any way with tumor susceptibility. The present study was designed to assess patterns of BaP metabolism that occur in microsomes prepared from an AHH-responsive (C57BL/6N) and an AHH-non-responsive (DBA/2N) inbred mouse strain. For comparative purposes the Sprague-Dawley rat was also studied. BaP metabolism was assessed using reverse phase high performance liquid chromatography (HPLC). Data were obtained in control animals as well as in animals pretreated with topical application of the environmental pollutant chemical known as the polychlorinated biphenyls (Aroclor 1254) and the PAH carcinogen 3-methylcholanthrene (3-MC).

MATERIALS AND METHODS

Chemicals

(^{14}C)BaP was purchased from New England Nuclear, Boston, Mass. and was purified by a silica gel (Bio-Sil A, 100-200 mesh, BioRad Laboratories) column with hexane as the eluting solvent and subsequently by reverse-phase HPLC using a DuPont Zorbax ODS column (6.2 mm x 25 cm) eluted with methanol:water (9:1, v/v). ^{14}C-BaP was diluted with unlabelled compound to a specific activity of 22.5 mCi/mmol. All solvents used were of HPLC grade and were purchased from Fisher Scientific Co., Fair Lawn, NJ. Gold Label BaP and 3-MC were obtained from Aldrich Chemical Co. All other chemicals were obtained in the purest form available. The polychlorinated biphenyl Aroclor 1254 was a gift from the Monsanto Corp., St. Louis, MO.

Animals and Treatment

5-6 week-old female C57BL/6N and DBA/2N mice were obtained from the NCI, Frederick Cancer Research Center. For studies of neonatal Sprague-Dawley rats, pregnant animals were obtained from Holtzman Rat Farm, Madison, Wisconsin. Newborn rats were allowed to suckle until the 4th day after birth when they were treated with topical application of the inducers. The animals received 1 mg/10 grams body weight of the polychlorinated biphenyl Aroclor 1254 or 3-MC applied in 100µl acetone on the back of the animal. Control animals received solvent alone.

Preparation of Microsomes

Twenty four hours after a single application of the inducer, the animals were sacrificed by decapitation and skin microsomes were prepared according to procedures established in this laboratory (13). Microsomal suspensions were always incubated in the metabolic assay system on the day of preparation.

In Vitro Incubation System

Skin microsomes prepared from the neonatal rats or from C57BL/6N and DBA/2N adult mice were used as the enzyme source. The incubation mixture contained 1.5 mg microsomal protein, 0.05 mmol Tris-HCl (pH 7.5), 3 mmol $MgCl_2$ and 1 mg NADPH in a final volume of 0.96 ml. The reaction was initiated by the addition of 80 nmol BaP(^{14}C-BaP) in 40 µl methanol. The samples were incubated for 30 min in the dark at 37°C in a Dubnoff metabolic shaker. The reaction was terminated by adding 1 ml of acetone followed by 2 ml of ethyl acetate. The mixture was vortexed for 1 min to extract any unreacted BaP as well as the metabolites into the organic phase. Organic and aqueous layers were separated by centrifugation at 1500 r.p.m. for 5 min. The radioactivity in the aqueous phase was less than 0.5% of total radioactivity. The organic phase was then dehydrated over anhydrous $MgSO_4$, dried under a stream of N_2, and dissolved in 50 µl of tetrahydrofuran for HPLC analysis. All extractions were performed under yellow light.

Fluorescent Assay of AHH

AHH activity was determined by a modification of the method of Nebert and Gelboin (14) as previously described (15). The quantitation of phenolic BaP metabolistes was based on comparison of fluorescence to a standard solution of 3-OHBaP. Protein was determined according to Lowry et al.(16) using bovine serum albumin as reference standard.

HPLC Analysis of Formation of Metabolites

A Waters Associates model 204 liquid chromatograph, fitted with a Whatman ODS-2 column (4.6 mm x 25 cm) was used for the analysis of radiolabeled metabolite mixtures of BaP. An unlabeled metabolite mixture prepared from a larger-scale in vitro incubation of BaP with rat liver microsomes was added as a source of UV marker for all radiolabeled samples. The column was eluted at ambient temperature with a 10-min linear gradient of methanol:water (1:1,v/v) to methanol at a solvent flow rate of 0.8 ml/min.

The eluates were monitored at 254 nm, fractions of approximately 0.2 ml were collected dropwise, and the radioactivity of each fraction was determined by liquid scintillation spectrometry. The counting efficiencies of the early eluted fractions, containing a higher percentage of water, were about 2% lower than those of the fractions eluted with methanol. The conversion of liquid scintillation counting data to concentration was based on the counting efficiency of the fractions eluted with methanol.

RESULTS AND DISCUSSION

AHH Induction

The levels of AHH activity, as measured by the fluorescence assay, in skin microsomes prepared from neonatal rat, C57BL/6N and DBA/2N mouse following the topical application of Aroclor 1254 of 3-MC are presented in Table 1. In both mouse strains and in the rat Aroclor 1254 was a more potent inducer of the cutaneous enzyme than was 3-MC. Aroclor 1254 treatment resulted in greater induction of the enzyme in neonatal rat (1500%) and in the C57BL/6N mouse (1685%) than in the DBA/2N mouse (102%). 3-MC on the other hand was 3-4 times more effective in inducing AHH activity in neonatal rat skin (1056%) than in the skin of either mouse strain (320% for C57BL/6N mice and 61% for DBA/2N mice).

TABLE 1

AHH ACTIVITY IN NEONATAL RAT, C57BL/6N AND DBA/2N MOUSE SKIN: EFFECT OF TOPICAL APPLICATION OF AROCLOR 1254 AND 3-MC

Treatment[a]	AHH (pmoles OHBP/mg protein/30 minutes)		
	Neonatal Rat	C57BL/6N Mouse	DBA/2N Mouse
Control	68	46	133
Aroclor 1254	1086[b]	821[b]	268[b]
3-MC	786[b]	193[b]	214[b]

[a] Animals were treated with a single topical application (100 mg/kg) of Aroclor 1254 or 3-MC 24 hours before sacrifice. Controls received solvent alone.

[b] Results significantly different from respective controls ($p < 0.05$).

HPLC Analysis of BaP Metabolism

The HPLC separation of (^{14}C)BaP metabolites obtained by incubation of (^{14}C)BaP with Aroclor 1254-induced C57BL/6N mouse skin microsomes is shown in figure 1. The overall pattern of metabolism of BaP obtained from neonatal Sprague-Dawley rat and the DBA/2N mouse, whether from control, Aroclor 1254, or 3-MC treated skin was qualitatively similar to that depicted in figure 1. Under the conditions of our HPLC system there was no significant radioactivity eluted prior to 12 minutes.

The quantitation of BaP metabolism by skin microsomes from the mouse strains and from the rat are shown in tables 2-4. Metabolism in control skin was substantially higher in DBA/2N mouse than in C57BL/6N mouse or in the neonatal rat. This is consistent with the AHH activity levels shown in table 1. The formation of diols, phenols and quinones by microsomes prepared from the animals increased substantially

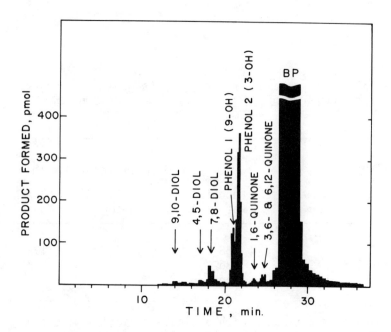

FIGURE 1. BaP metabolism by skin microsomes prepared from C57BL/6N mice pre-treated with Aroclor 1254.

following pre-treatment with both Aroclor 1254 and 3-MC. In the C57BL/6N mouse, total metabolism of BaP was 12 and 4 times higher, respectively, in Aroclor 1254 and 3-MC induced microsomes as compared to untreated controls (table 2). On the other hand, in the DBA/2N mouse, 3-MC had no stimulatory effect on the metabolism of BaP whereas Aroclor 1254 treatment resulted in a 3-fold increase in metabolism compared to controls (table 3). There were significant increases in the formation of diols, phenols and quinones in both Aroclor 1254 and 3-MC treated C57BL/6N mouse and neonatal rat and in Aroclor treated DBA/2N mouse (tables 2,3,4). Maximal trans-7,8-diol formation occured in Aroclor 1254-induced C57BL/6N mouse with a 1200% increase as compared to untreated controls. Following topical application of 3-MC, trans-7,8-diol formation increased 320% in the skin of C57BL/6N mouse and 608% in the neonatal rat. In DBA/2N mouse, Aroclor treatment resulted in a 58% increase in trans-7,8-diol formation whereas 3-MC was ineffective in this regard.

TABLE 2

EFFECT OF TOPICAL APPLICATION OF AROCLOR 1254 AND 3-MC ON SKIN MICROSOMAL METABOLISM OF ^{14}C-BaP IN C57BL/6N MICE

Metabolite[a]	C57BL/6N Mouse		
	Control	Aroclor 1254[b]	3-MC[b]
trans - 9,10-Diol	3	24	10
trans - 4,5-Diol	10	29	30
trans - 7,8-Diol	10	119	42
Phenol 1 (mainly 9-OHBP)	18	289	88
Phenol 2 (mainly 3-OHBP)	34	780	277
Quinones (1,6;3,6; 6,12 Quinones)	30	127	66
Others	13	19	16
Total metabolites	118	1387	477

[a] pmol metabolite formed/30 min/mg protein. Eighty nmol ^{14}C-BaP (S.A.=22.5mCi/mmol) were incubated with skin microsomes (1.5 mg protein) in a final volume of 1.0 ml.

[b] animals were treated with a single topical application of the inducer 24 hours prior to sacrifice. Controls received solvent alone.

These studies re-emphasize the complexity of carcinogen metabolism in the skin. Furthermore, they illustrate the futility of attempting to predict tumor susceptibility based on patterns of carcinogen metabolism in cutaneous tissue. This lack of correlation could be due to several factors: 1) the current initiation-promotion method of induction of tumors that is almost universally used is not the sole manner in which tumors are induced. 2) the production of dihydrodiols such as 7,8-diol that are precursors of bay-region diol-epoxides may be necessary but not sufficient to evoke tumors. In addition, the bay-region theory may not satisfactorily account for all possible mechanisms of tumor induction by polyaromatic hydrocarbons such as BaP. 3) there may be competing pathways in a tissue at any given time for the metabolism of a compound. It is already known that the

TABLE 3

EFFECT OF TOPICAL APPLICATION OF AROCLOR 1254 AND 3-MC ON SKIN MICROSOMAL METABOLISM OF ^{14}C-BaP IN DBA/2N MICE

Metabolite[a]	Control	DBA/2N Mouse Aroclor 1254[b]	3-MC[b]
trans - 9,10-Diol	8	15	6
trans - 4,5-Diol	11	42	16
trans - 7,8-Diol	20	48	22
Phenol 1 (mainly 9-OHBP)	54	142	44
Phenol 2 (mainly 3-OHBP)	134	334	114
Quinones (1,6;3,6; 6,12 Quinones)	30	156	38
Others	9	51	13
Total Metabolites	266	788	253

a pmol metabolite formed/30 min/mg protein. Eighty nmol ^{14}C-BaP (S.A.=22.5 mCi/mmol) were incubated with skin microsomes (1.5 mg protein) in a final volume of 1.0 ml.

b animals were treated with a single topical application of the inducer 24 hours prior to sacrifice. Controls received solvent alone.

skin contains in addition to AHH, enzymes such as epoxide hydrolase and glutathione transferase. The latter could be especially effective in binding potentially oncogenic electrophilic metabolites in so-called "tumor-resistant" strains of animals. 4). It is also quite likely that the patterns of metabolism of BaP that occur in an in vitro incubation system differ substantially from those that occur in vivo. The studies described here clearly show that skin, a tissue frequently considered to be metabolically inert, is capable of metabolizing polycyclic aromatic hydrocarbons into a spectrum of chemical moieties most of which are inert but some of which may be precursors of the ultimate carcinogenic species of BaP. It is also of interest that Aroclor 1254, a polychlorinated biphenyl with little or no proven oncogenicity for skin was a more potent inducer of cutaneous BaP metabolism than was 3-MC a known potent skin carcinogen. This further illustrates the point that enzyme inducibility by a chemical does not necessarily correlate with its ability to induce tumors.

TABLE 4

EFFECT OF TOPICAL APPLICATION OF AROCLOR 1254 AND 3-MC ON SKIN MICROSOMAL METABOLISM OF ^{14}C-BaP IN SPRAGUE-DAWLEY RATS

Metabolite[a]	Sprague-Dawley Rat		
	Control	Aroclor 1254	3-MC
trans - 9,10-Diol	5	64	38
trans - 4,5-Diol	10	63	50
trans - 7,8-Diol	12	127	85
Phenol 1 (mainly 9-OHBP)	34	342	266
Phenol 2 (mainly 3-OHBP)	59	656	478
Quinones (1,6;3,6;6,12 Quinones)	17	109	167
Others	2	26	15
Total metabolites	140	1420	1100

a pmol metabolite formed/30 min/mg protein. Eighty nmol ^{14}C-BaP (S.A.=22.5 mCi/mmol) were incubated with skin microsomes (1.5 mg protein) in a final volume of 1.0 ml.

b animals were treated with a single topical application of the inducer 24 hours prior to sacrifice. Controls received solvent alone.

In summary, the data reported here demonstrate that the skin is an active site of carcinogen metabolism, that there are quantitative and qualitative differences in the pattern of BaP metabolism by skin microsomal enzymes and that it is currently not possible to relate patterns of metabolism of BaP to tumor susceptibility of C57BL/6N and DBA/2N mice and Sprague-Dawley rats.

ACKNOWLEDGMENTS

Supported in part by NIH grant ED-1900 and NIOSH grant OH-01149 and funds from Veterans Administration.

REFERENCES

1. Haenszel, W. (1963): Variations in skin cancer incidence within the United States, Natl. Cancer Inst. Mono., 10: 225-243.
2. Emmett, E.A. (1973): Ultraviolet radiation as a cause of skin tumors, CRC Crit. Rev. Toxicol., 2:211-225.
3. Epstein, J.H. (1970): Ultraviolet carcinogenesis. In: Photophysiology, edited by A.C. Giese, pp 235-273, Academic Press, New York.
4. Hueper, W.C. (1963): Chemically induced skin cancers in man, Natl. Cancer Inst. Monogr. 10:377-391.
5. Henry, S.A. (1947): Occupational cutaneous cancer attributable to certain chemicals in industry. Br. Med. Bull., 4:389-401.
6. Guerin, M.R. (1978): Energy sources of polycyclic hydrocarbons. In: Polycyclic Hydrocarbons and Cancer, Vol. 1 Edited by H.V. Gelboin and P.O.P. Ts'o, pp 3-42, Academic Press, New York.
7. Theall, G., Eisinger, M., Grunberger, D. (1981): Metabolism of benzo(a)pyrene and DNA adduct formation in cultured human keratinocytes. Carcinogenesis, 2:581-587.
8. Miller, E.C. (1978): Carcinogenesis by chemicals: an overview. G.H.A. Clowes Memorial Lecture. Cancer Res. 30:559-576.
9. Weigert, F., Calcutt, G., and Powell, A.V. (1948): The course of the metabolism of benzpyrene in the skin of the mouse. Br. J. Cancer 2:405-409.
10. Levin, W., Conney, A.H., Alvares, A.P., Merkatz, I., and Kappas, A. (1973): Induction of benzo(a)pyrene hydroxylase in human skin. Science 176:419-420.
11. Koreeda, N., Moore, P.D., Wislocki, P.G., Levin, W., Conney, A.H., Yagi, H., and Jerina, D.M. (1978): Binding of benzo(a)pyrene 7,8-diol-9,10-epoxides to DNA, RNA and protein of mouse skin occurs with high stereoselectivity, Science 199:778-781
12. Nebert, D.W. and Jensen, N.M. (1979): The Ah Locus: genetic regulation of the metabolism of carcinogens, drugs and other environmental chemicals by cytochrome P-450-mediated monooxygenases. CRC Crit. Rev. Biochem. 4:401-438.
13. Bickers, D.R. (1980): The skin as a site of drug and chemical metabolism. In: Current Concepts in Cutaneous Toxicity, edited by V.A. Drill and P. Lazar, pp 95-126, Academic Press, New York.
14. Nebert, D.W. and Gelboin, H.V. (1968): Substrate inducible aryl hydrocarbon hydroxylase in mammalian cell culture, J. Biol. Chem. 243:6250-6261.

15. Bickers, D.R., Dutta-Choudhury, T., and Mukhtar, H. (1982): Epidermis: a site of drug metabolism in neonatal rat skin: Studies on cytochrome P-450 content, the mixed function oxidase and epoxide hydrolase activity. <u>Mol. Pharmacol.</u> (in press).
16. Lowry, O.H., Rosebrough, N.J., Farr, A.L., Randall, R.J. (1951): Protein measurement with Folinphenol reagent. <u>J. Biol. Chem.</u>, 193:265-275.

QUANTUM YIELDS OF THE PHOTODECOMPOSITION OF POLYNUCLEAR AROMATIC HYDROCARBONS ADSORBED ON SILICA GEL

LOTHAR BLAU and HANS GÜSTEN
Kernforschungszentrum Karlsruhe, Institut für Radiochemie,
Postfach 3640, 7500 Karlsruhe, Federal Republic of Germany

INTRODUCTION

The ecotoxicological assessment of polynuclear aromatic hydrocarbons (PAH) as a basis for defining standards of air quality calls for the knowledge of the residence times of PAHs in the atmosphere. Due to their low vapor pressure virtually all the PAHs released into the atmosphere are associated with aerosols and airborne particulate matter and they are transported together with this matter through the atmosphere (1, 2). Studies of particle size distributions (3, 4, 5) show that PAHs are primarily associated with aerosol particles in the sub-micron range. Particles of this size can have atmospheric lifetimes of days or weeks which imply long-range transport of the aerosols. During the last years evidence has been accumulated in laboratory studies of an enhanced photocatalytic degradation of PAHs on solid surfaces (6, 7, 8, 9). Thus, it is likely that PAHs, during the atmospheric transport on airborne aerosols, are subject to a photochemically induced heterogeneous oxidation process by solar light.

In this paper we report on the procedure and quantitative data from simulated laboratory experiments involving the heterogeneous photodegradation of several PAHs on silica gel as a model aerosol. The measured quantum yields of the heterogeneous photodegradation are then used to estimate the atmospheric residence times of the PAHs.

EXPERIMENTAL METHODS

MATERIALS

Silica gel 18 - 32 with a particle size of 18 - 32 μm was supplied by Woelm GmbH, Eschwege (Germany). Using the BET method the surface area was determined to be 420 m² · g^{-1}. After drying the silica gel at 200 °C the PAHs were adsorbed from benzene solutions. Benzo(b)chrysene, benzo(b)fluoranthene, benzo(j)fluoranthene and benzo(k)fluoranthene were obtained with 99.5 percent purity as polycyclic aromatic hydrocarbon reference materials from the Commission of the European Communities (Brussels, Belgium). The other

QUANTUM YIELDS OF THE PHOTODECOMPOSITION

PAHs such as anthracene, phenanthrene, benzo(a)anthracene, fluoranthene and benzo(a)pyrene were purified by column chromatography on neutral Al_2O_3 in benzene and final recrystallization from ethanol. A pure sample of benzo(a)fluoranthene was kindly supplied by Dr. P. Studt.

APPARATUS AND PROCEDURE

The photodecomposition of the PAHs adsorbed on silica gel was carried out in a circulating tubular reactor vessel (1500 cm³) incorporated in the horizontal photoreactor shown in Fig. 1

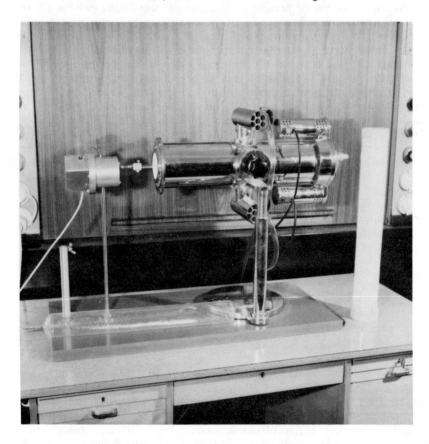

FIGURE 1. PHOTOREACTOR FOR THE HETEROGENEOUS PHOTODECOMPOSITION OF PAH

A charge of 2 g of silica gel adsorbed with a PAH was placed into the reaction vessel and the interior rotated by a drilling machine in a light converter cylinder made of quartz (55 cm, 8 cm i. d.). This light converter cylinder was coated with a fluorescent luminophor (Y-triumsulfate doped with Bi-ions) and allowed to illuminate the contents along the whole length of the reaction vessel (50 cm). This rotating bed reactor secures that the content is tumbled continuously through the illuminated volume, thereby providing maximum contact between the silica gel and the incident light. The luminophor converts the 254 nm emission of two low-pressure mercury arcs to the desired wavelength range of 300 to 400 nm. In this range of wavelengths the PAHs display absorption bands which can absorb tropospherical solar light. Since the reaction vessel is made of Pyrex glass it serves also as a cut-off filter for the short wavelength light below 290 nm, i.e. together with the light converter cylinder it simulates the spectral distribution of solar light. The two low-pressure mercury lamps are located inside the horizontal photoreactor - made of high-grade steel - in an array of 8 straight U-shaped tubes of 50 cm length operating with cold-cathodes and a high-voltage transformer. The average photon flux inside the reaction vessel was found to be 2×10^{19} photons/l · min^{-1}. It was determined by using a solution of potassium ferrioxalate as the chemical actinometer. The photoreactor is cooled by forced air from an air turbine. The temperature inside the reactor can be kept at 30 -35 °C even over very long periods of operation. The light output of the photoreactor at constant temperature is very constant even over very long irradiation times. This was monitored with anthracene on silica gel 18 - 32 or with (E)-α-(2,5-dimethyl-3-furylethylidene) (isopropylidene) succinic anhydride (11) on silica gel 18 - 32 as the relative chemical actinometer (12).

The photodecomposition of a PAH was followed by taking small samples of approximately 50 mg of silica gel and after elution with methanol, monitoring the disappearance of PAH by absorption or fluorescence spectroscopy. The relative quantum yields determined from the time dependence of the photodecomposition are transformed into real quantum yields by independent measurement of the photodecomposition of anthracene adsorbed on silica gel 18 - 32, using the immersion technique of Leermakers et al. (13). The general technique and method for determining the quantum yield is described in (14). The quantum yield for the photodecomposition of anthracene adsorbed on silica gel in a cyclohexane slurry was found to be 0.085.

BACKGROUND

Due to limitations of space the theoretical background for measuring quantum yields in the adsorbed state will be published elsewhere (15). A short excerpt of the theoretical framework will be briefly outlined here.

QUANTUM YIELDS OF THE PHOTODECOMPOSITION

When the incident light is weakly absorbed by the PAH the rate of the photodecomposition can be described by a first-order rate expression

$$-\frac{d(PAH)}{dt} = K \cdot Q' \cdot I_a \cdot (PAH), \qquad (1)$$

i.e., the photodecomposition rate at a given concentration (PAH) is proportional to the product of the quantum yield Q, the absorbed light energy I_a and a proportional factor K. Thus, I_a is the fraction of the total incident photon flux which is absorbed by a PAH in the wavelength region of 300 - 400 nm. K is a factor for correcting the relative quantum yield to become a real quantum yield. In a first approximation Eq. 1 can be written as

$$Q' = \frac{\Delta c}{I_a(PAH) \cdot \Delta t} \qquad (2)$$

with $\Delta c = c(t_1) - c(t_2)$ the change of concentration of (PAH) between the irradiation times t_1 and t_2, (PAH) the mean concentration of $c(t_1)$ and $c(t_2)$, and I_a (PAH) the absorbed light energy. Thus, Eq. 2 gives relative quantum yields for different PAHs which then are transformed into real quantum yields by using the independently measured Q value for the photodecomposition of anthracene on silica gel using the immersion technique. When the real quantum yields are known, the differential equation 1 is solved on a computer using the Runge-Kutta method. In Fig. 2 the theoretical photodecomposition curves according to Eq. 1 of anthracene and phenanthrene adsorbed on silica gel 18 - 32 and the measured data points demonstrate that the analytical framework allows determining the quantum yields of PAHs in the adsorbed state.

This approach is only valid when
1. the quantum yield of the photodecomposition is independent of the wavelength of the incident light, i.e., in the 300 - 400 nm wavelength range,
2. the absorption spectra and the molar extinction coefficients of the PAHs in the adsorbed state on silica gel are comparable to those in solution.

The first condition is very likely, the second one has been proved by Leermakers et al. (13).

RESULTS AND DISCUSSION

It has been argued that PAHs in the adsorbed state spon-

QUANTUM YIELDS OF THE PHOTODECOMPOSITION

taneously oxidize in the absence of light (16). The rate of the non-photochemical decomposition of the PAHs investigated is, if at all, lower by many orders of magnitude than the photochemical decomposition in our rotating bed reactor. This confirms the recent findings of Butler and Crossley (17) that PAHs adsorbed on soot particles do not react significantly in the dark for periods of up to 230 days.

We investigated the photodecomposition of the following PAHs: anthracene (A), phenanthrene (P), benzo(a)anthracene (BaA), benzo(b)chrysene (BbC), fluoranthene (F), benzo(a)fluoranthene (BaF), benzo(b)fluoranthene (BbF), benzo(j)fluoranthene (BjF), benzo(k)fluoranthene (BkF) and benzo(a)pyrene (BaP). All but two, benzo(b)chrysene and benzo(a)fluoranthene, are on the EPA Priority Pollutant List. The relative and real quantum yields according to Eqs. 1 and 2 for the photodegradation of the ten PAHs adsorbed on silica gel are compiled in Table 1. In addition to the initial concentration c_0 of the PAHs their wavelength maxima λ_{max} within the window of the fluorescence emission of the light converter as well as the molar extinction coefficients ε_{max} are summarized. The standard deviation σ for Q was calculated from the mean values of Q'.

The reproducibility of the relative Q' values of anthracene adsorbed on silica gel has been determined in nine independent irradiations under identical conditions, i.e., same concentration and light intensity. The relative quantum yield is Q' = 2.1 ± 0.6 x 10^{-5}. The most critical task is to maintain a regular distribution of the silica gel in the rotating reaction vessel under irradiation.

Since we are not yet able to determine the accurate light intensity within the photoreactor by a chemical actinometer as the secondary standard, the relative Q' values are converted into real quantum yields by separate measurement of the quantum yield using the immersion technique (13). Leermakers et al. (18) have studied the electronic absorption spectra and the photochemistry of organic compounds adsorbed on silica gel in the form of a silica gel-solvent slurry. The immersion technique avoids light scattering associated with non-transparent systems. The photodecomposition of anthracene was measured in cyclohexane at a low anthracene concentration of 5 x 10^{-5} M in order to suppress other possible photoreactions of anthracene such as photodimerization. Although mechanistic details are difficult to sort out, is is likely that the disappearance of anthracene adsorbed on silica gel in the cyclohexane slurry is due to photooxidation. Oxygen is well in excess of a factor of 100 compared to the anthracene concentration, i.e. in the slurry photooxidation is considered to be the favored process like in the "dry adsorbed" state (6, 7, 8). The quantum yield of the photodecomposition of anthracene adsorbed on silica gel in a cyclohexane slurry was determind to be Q = 0.085. With this value all relative quantum yields Q' in Table 1 are

QUANTUM YIELDS OF THE PHOTODECOMPOSITION

converted into real quantum yields Q. The measured quantum yields are then used to calculate the theoretical curve for the photodegradation of a PAH with increasing irradiation time by solving the differential equation 1. Fig. 2 demonstrates that despite the great differences in the irradiation times the theoretical curve for the photodegradation of anthracene and phenanthrene adsorbed on the same silica gel is well covered by the experimental data.

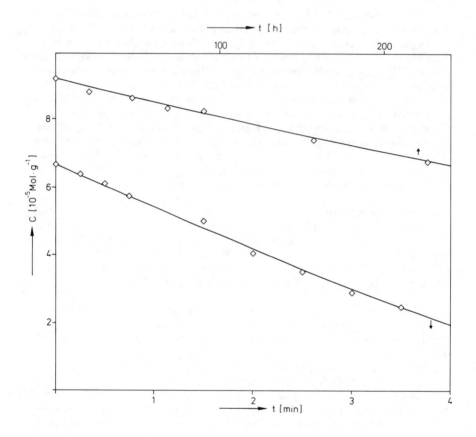

FIGURE 2. THEORETICAL CURVE FOR THE PHOTODEGRADATION OF ANTHRACENE AND PHENANTHRENE ADSORBED ON SILICA GEL. LOWER CURVE: ANTHRACENE; UPPER CURVE: PHENANTHRENE. ◇EXPERIMENTAL DATA

TABLE 1

QUANTUM YIELD OF THE HETEROGENEOUS PHOTODEGRADATION OF PAHS ON SILICA GEL AT ROOM TEMPERATURE

PAH	$c_0 \times 10^5$ $[mol \cdot g^{-1}]$	τ max* $[nm]$	ε max $\times 10^{-4}$* $[l \cdot mol^{-1} \cdot cm^{-1}]$	rel. Q'	$Q \times 10^5$	$\sigma \times 10^5$
Anthracene	6,64	355	0.87	2.1×10^{-5}	8500	-
Phenanthrene	9.21	331	0.03	2.4×10^{-8}	9.7	4.4
Benzo(a)anthracene	2.10	341	0.74	2.1×10^{-7}	85	61
Benzo(b)chrysene	1.15	361	0.94	5.0×10^{-8}	20	7
Fluoranthene	4.15	357	0.88	3.9×10^{-9}	1.6	0.4
Benzo(a)fluoranthene	1.28	420	0.78	4.7×10^{-7}	190	120
Benzo(b)fluoranthene	2.18	350	1.17	3.9×10^{-9}	1.6	0.9
Benzo(k)fluoranthene	2.00	401	2.26	1.4×10^{-8}	5.7	2.4
Benzo(j)fluoranthene	1.60	383	1.31	8.1×10^{-7}	33	29
Benzo(a)pyrene	0.91	383	3.00	1.3×10^{-8}	5.3	3.6

*in n-heptane

QUANTUM YIELDS OF THE PHOTODECOMPOSITION

In order to test the validity of the theoretical framework for determining quantum yields in the adsorbed state we measured the quantum yields of the photodecomposition of various PAHs from Table 1 on the same silica gel at an approximate 100 fold lower concentration.

TABLE 2

RELATIVE QUANTUM YIELDS OF THE PHOTODEGRADATION OF PAHS ON SILICA GEL AS A FUNCTION OF THE PAH CONCENTRATION

PAH	$c_0 \times 10^7$ $[mol \cdot g^{-1}]$	rel. Q'
Benzo(a)anthracene	1.75	2.7×10^{-7}
Benzo(b)chrysene	2.26	9.5×10^{-8}
Benzo(b)fluoranthene	3.20	1.2×10^{-8}
Benzo(k)fluoranthene	2.60	4.5×10^{-9}
Benzo(a)pyrene	2.45	7.0×10^{-8}

The relative quantum yields at different concentrations vary by up to a factor of three (see Tables 1 and 2). The low concentrations in Table 2 correspond approximately to "environmental" concentrations of airborne PAHs. Airborne PAH concentrations are about 10 - 50 µg/g particulate matter (19, 20).

Table 1 demonstrates the large difference in the quantum yields for the photodegradation of the various PAHs. The sequence of increasing quantum yields follows roughly the qualitative investigations by Katz et al. (10) with anthracene being the most photosensitive PAH.

Knowing the quantum yield of the photodecomposition of a PAH one can estimate with the equation

$$k_1 = 1/\tau = \int I_{(\nu, z)} \cdot \sigma \cdot Q \, d\nu \qquad (3)$$

the mean photodegradation lifetime τ in the troposphere from the product of the solar flux intensity I over the absorbing wavelength of the PAHs, the photoabsorption cross section σ in this wavelength range and the quantum yield of the photodecomposition. Taking the solar flux intensity I for a zenith angle of z = 60 ° from the data of Leighton (21) and calculating the absorption cross section from the absorption spectra of the PAHs in n-heptane (22) in the wavelength range of 300 - 400 nm we obtain the tropospheric lifetime of the PAHs. A solar flux intensity at a zenith angle of 60 ° corresponds approximately to the daily mean value in Central Europe in the middle of the summer (23). With a typical ground level wind velocity of 4 m/sec we then obtain the average distance over which an adsorbed PAH molecule would be transported from its source prior to photodecomposition, provided there are no fallout or washout processes during the long-range transport (see Table 3).

TABLE 3

MEAN TROPOSPHERIC LIFETIME τ AND DISTANCE R OF PAHS TRANSPORTED FROM SOURCES IN CENTRAL EUROPE

PAH	τ [h]	R [km]
A	0.02	0.3
P	1100	15000
BaA	2.5	35
BbC	6.5	90
F	110	1600
BaF	0.25	3.3
BbF	80	1100
BjF	3	40
BkF	14	200
BaP	10	140

It is interesting to note that while the quantum yields of the photodecomposition of anthracene and phenanthrene vary by a factor of 10^3, the distance transported from the source differs by a factor of more than 10^4. This illustrates that beyond the quantum yield the absorption cross section over the range of the solar spectrum plays an important role for the rate of photodecomposition of a PAH. The data of Table 3 indicate further that PAHs in the adsorbed state may travel over distances of up to 10000 km and more. The large differences in the rate of photodecomposition might also explain why in geographic regions subject to atmospheric long-range transport the local phenanthrene concentration is higher by a factor of up to 20 compared to that of anthracene. When the air mass trajectories from England or Central Europe are directed to Norway the airborne phenanthrene concentration is higher by a factor of 10 to 20 than that of anthracene (24). Norwegian drinking water resources contain much more phenanthrene than anthracene (25) though their emission concentrations are roughly the same. While the phenanthrene to anthracene ratio in the Ruhr valley near Essen is 6:1, in the remote Sauerland area ratios of 25:1 have been assayed (26). These results can be rationalized by the rapid photodecomposition of airborne anthracene. The calculated tropospheric lifetimes of the PAHs in Table 3 are minimum values, since the solar flux intensity is representative only for an eight hour day. The lifetime calculation does not take this into consideration. During the winter the photodecomposition rate of the airborne PAHs can be lower by a factor of up to ten due to the low zenith angle of the sun (27). For this reason a more than ten fold higher PAH concentration is generally found in airborne dust during the winter.

Since among the mineral components SiO_2 is present at the highest concentration in atmospheric aerosols (28) the silica gel here used as a model aerosol might be representative for an airborne aerosol. The photodegradation of PAHs on natural aerosols such as sand or fly ash will be reported elsewhere.

REFERENCES

1. Natusch, D.F.S. and Tomkins, B. A. (1978): A theoretical consideration of the adsorption of polynuclear aromatic hydrocarbon vapor onto fly ash in a coal-fired power plant. In: Carcinogenesis, Vol. 3: Polynuclear Aromatic Hydrocarbons, edited by P. W. Jones and R. I. Freudenthal, pp. 145-153, Raven Press, New York.
2. Cautreels, W. and van Cauwenberghe, K (1978): Experiments on the distribution of organic pollutants between airborne particulate matter and the corresponding gas phase, Atmos. Environ., 12:1133-1141.
3. Pierce, R. C. and Katz, M. (1975): Dependency of polynuclear aromatic hydrocarbon content on size distribution of atmospheric aerosols, Environ. Sci. Technol., 9:347-353.
4. Butler, J. D. and Crossley, P. (1979): An appraisal of relative airborne sub-urban concentrations of polycyclic aromatic hydrocarbons monitored indoors and outdoors, Sci. Total Environ., 11:53-58.
5. Pedersen, P. S., Ingwersen, J., Nielsen, T., and Larsen, E. (1980): Effect on fuel, lubricant, and engine operating parameters on the emission of polycyclic aromatic hydrocarbons, Environ. Sci. Technol., 14:71-79.
6. "Particulate Polycyclic Organic Matter", Committee on Biologic Effects of Atmospheric Pollutants, National Academy of Sciences, Washington, D.C. (1972).

7. Lane, D. A. and Katz, M. (1977): The photomodification of benzo(a)pyrene, benzo(b)fluoranthene, and benzo(k)fluoranthene under simulated atmospheric conditions. In: Advances in Environmental Science and Technology, Vol. 9, edited by I. H. Suffet, pp. 137-154, Wiley-Interscience, New York.
8. Fox, M. A. and Olive, S. (1979): Photooxidation of anthracene on atmospheric particulate matter, Science, 205:582-583.
9. Lotz, F., Nitz, S., and Korte, F. (1979): Photomineralisierung adsorbierter organischer chemikalien im mikromaßstab, Chemosphere, 10:763-768.
10. Katz, M., Chan, C., Tosine, H., and Sakuma, T. (1979): Relative rates of photochemical and biological oxidation (in vitro) of polynuclear aromatic hydrocarbons. In: Polynuclear Aromatic Hydrocarbons, edited by P. W. Jones and P. Leber, pp. 171-189, Ann Arbor Science Publ., Inc., Ann Arbor.
11. Heller, H. G. and Langan, J. R. (1981): Photochromic heterocyclic fulgides. Part 3. The use of (E)-α-(2,5-dimethyl-3-furyl-ethylidene) (isopropylidene) succinic anhydride as a simple convenient chemical actinometer, J. Chem. Soc. Perkin II, 341-344.
12. Güsten, H. and Bozicević, Z.: to be published.
13. Leermakers, P. A., Thomas, H. T., Weis, L. D., and James, F. C. (1966): Spectra and photochemistry of molecules adsorbed on silica gel, Amer. Chem. Soc., 88:5075-5083.
14. Schulte-Frohlinde, D., Blume, H., and Güsten, H. (1962): Photochemical cis-trans-isomerisation of substituted stilbenes, J. Phys. Chem., 66:2486-2491.
15. Blau, L., Heinrich, G., and Güsten, H.: to be published.
16. Korfmacher, W. A., Natusch, D.F.S., Taylor, D. R., Mamantov, G., and Wehry, E. L. (1980): Oxidative transformation of polycyclic aromatic hydrocarbons adsorbed on coal fly ash, Science, 207:763-765.
17. Butler, J. D. and Crossley, P. (1981): Reactivity of polycyclic aromatic hydrocarbons adsorbed on soot particles, Atmos. Environ., 15:91-94.
18. Nicholls, C. H. and Leermakers, A. (1971): Photochemical and spectroscopic properties of organic molecules in adsorbed or other perturbing polar environments, Advan. Photochem., Vol. 8, pp. 315-336.
19. Katz, M., Sakuma, T., and Ho, A. (1978): Chromatographic and spectral analysis of polynuclear aromatic hydrocarbons-quantitative distribution in air of Ontario cities, Environ. Sci. Technol., 12:909-915.

20. Gordon, R. J. (1976): Distribution of airborne polycyclic aromatic hydrocarbons throughout Los Angeles, Environ. Sci. Technol., 10:370-371.
21. Leighton, P. A. (1961): Photochemistry of Air Pollution. Academic Press, New York, p. 29.
22. Heinrich, G. and Güsten, H. (1980): Fluorescence spectroscopic properties of carcinogenic and airborne polynuclear aromatic hydrocarbons. In: Polynuclear Aromatic Hydrocarbons: Chemistry and Biological Effects, edited by A. Bjørseth and A. J. Dennis, pp. 983-1003, Battelle Press, Columbus, Ohio.
23. Schulze, R. (1970): Strahlenklima der Erde, Steinkopff-Verlag, p. 71, Darmstadt, Germany.
24. Bjørseth, A., Lunde, G., and Lindskog, A. (1979): Long-range transport of polynuclear aromatic hydrocarbons, Atmos. Environ., 13:45-53.
25. Olufsen, B. (1980): Polynuclear aromatic hydrocarbons in Norwegian drinking water resources. In: Polynuclear Aromatic Hydrocarbons: Chemistry and Biological Effects, edited by A. Bjørseth and A. J. Dennis, pp. 333-343, Battelle Press, Columbus, Ohio.
26. Hettche, O. (1972): Polyzyklische aromaten als schädliche fremdstoffe in der außenluft. VDI-Zeitschrift, 114:346-347.
27. Penzhorn, R. -D., Filby, W. G., and Güsten, H. (1974): Die photochemische abbaurate des schwegeldioxids in der unteren atmosphäre mitteleuropas., Naturforsch., 29a:1449-1453.
28. Reiter, R., Sládkovic, R., and Pötzl, K. (1978): Chemische komponenten des reinluftaerosols in abhangigkeit von luftmassencharakter und meterologischen bedingungen, Ber. Bunsenges. Phys. Chem., 82:1188-93.

CARCINOGENICITY OF 3-METHYLCHOLANTHRENE DERIVATIVES AND CYCLOPENTENO[cd]PYRENE IN THE RAT MAMMARY GLAND

ERCOLE CAVALIERI AND ELEANOR ROGAN
Eppley Institute for Research in Cancer, University of Nebraska Medical Center, Omaha, Nebraska 68105.

INTRODUCTION

Multiple Mechanisms of Activation in PAH Carcinogenesis

Current data in polycyclic aromatic hydrocarbon (PAH) carcinogenesis indicate multiple mechanisms of activation in tumor initiation (1). It is plausible to assume that a predominant mechanism for each hydrocarbon triggers tumor formation. However, the mode of activation may change, depending on the nature of the activating enzymes present in the target tissues of various animals. In addition to bay-region diol-epoxides, other electrophilic intermediates, such as radical cations, simple arene oxides, and some benzylic carbonium ions of alkylated PAH generated by enzymatic hydroxylation and subsequent formation of reactive benzylic esters have been postulated as ultimate carcinogenic forms of PAH (1,2).

A number of experimental results suggest that the bay-region diol-epoxides are not the sole active ultimate carcinogenic electrophiles of certain PAH or play no role in the carcinogenesis of some other PAH. Recently, fluorinated derivatives of benzo[a]pyrene (BP), i.e., BP-7-F, BP-8-F, BP-9-F and BP-10-F, synthesized with the precise objective of blocking formation of the bay-region diol-epoxide, were tested for carcinogenicity by s.c. injection in Sprague-Dawley rats (Miller, unpublished results). Although less potent than the parent compound, BP-7-F, BP-8-F and BP-10-F induced sarcomas. In other tumorigenesis experiments BP-7-CH_3 was found to be slightly less active than the parent compound BP in the induction of sarcomas in rats by s.c. injection (2). With a methyl group at the 10 position the carcinogenicity is almost nil in rats by s.c. injection (2), but it is only moderately reduced when compared to BP in an initiation-promotion experiment on mouse skin (3). It does not seem likely that the carcinogenicity of BP-7-CH_3 and particularly BP-10-CH_3 derives from formation of the bay-region diol-epoxides. Furthermore, the carcinogenicity of 10-azabenzo[a]pyrene (4), in which nitrogen replaces the carbon atom at the 10 position, cannot be explained by formation of the bay-region diol epoxide. Similarly, this mechanism of activation cannot be invoked to account for the activity of 2-fluoro-7-methylbenz[a]anthracene, 4-fluoro-7,12-dimethylbenz[a]anthracene (DMBA-4-F) and DMBA-4-CH$_3$ (2).

In dose-response studies by repeated application (5-7) and initiation-promotion (8-12) in mouse skin, the carcinogenic activity of BP and the putative proximate metabolite BP-7,8-dihydrodiol is similar. Furthermore, BP-7,8-dihydrodiol does not elicit any carcinogenic response when directly applied to rat mammary gland, while the parent compound BP is active (12). Since it would be expected that the proximate BP-7,8-dihydrodiol is more active than BP, these data pose a serious question about the BP diol-epoxide as sole carcinogenic metabolite for BP. Attribution of the relatively low tumorigenicity of BP-7,8-dihydrodiol to its polarity, which prevents access to critical target cells (13), is speculative and not supported by any solid experimental facts.

Anthanthrene, which lacks a bay-region, is not carcinogenic, but its 6- and 6,12-methyl derivatives induce tumors in rats by s.c. injection (14) and in mouse skin by repeated application and initiation-promotion (12). Cyclopenteno[cd]pyrene (CPEP), which cannot form a diol-epoxide, is a moderately potent carcinogen by repeated application on mouse skin, whereas its 3,4-dihydro derivative, cyclopentano[cd]pyrene, is inactive (15). These results suggest that the ultimate carcinogenic metabolite for CPEP is the 3,4-oxide derivative.

The aforementioned results provide clear evidence that activation pathways other than epoxidation at the bay-region are involved in the tumor-initiating process.

Peroxidases as Possible Activating Enzymes of PAH by One-Electron Oxidation

The hypothesis that one-electron oxidation of PAH to radical cations might be an important mechanism of activation leading to carcinogenesis is based on several lines of evidence (16-21). It is plausible to assume that the ability of PAH radical cations to bind covalently with nucleophilic groups of biological macromolecules depends on several known and unknown factors which include (a) ease of formation of these reactive intermediates, (b) geometry of the hydrocarbon, (c) close spatial relationship between the activating enzyme, hydrocarbon and cellular target, (d) high charge localization of the PAH radical cation at the most positive carbon atom(s) and (e) proper orientation of the PAH with the cellular target in order that the most reactive position of the PAH radical cation is close to the nucleophilic group. The ease of formation of radical cations (a, above) is dependent on the ionization potential (IP) of the hydrocarbon, since these intermediates are obtained by removal of an electron from the PAH. IP of unsubstituted and alkyl-substituted alternant PAH and their carcinogenic activities have been reported (1). If we choose a tentative IP cut-off at 7.35 eV, we can observe that there are relatively few carcinogenic compounds above this value and their potency is not high. Among the well-investigated carcinogenic PAH with IP above 7.35 eV are 5-methylchrysene, dibenz[a,h]anthracene and 7-methylbenz[a]anthracene on the borderline, while the most potent carcinogens, such as BP, DMBA and 3-methylcholanthrene (MC) have an IP lower than 7.35 eV.

Activation of N-hydroxy-2-acetylaminofluorene (22,23) and a number of PAH (21) by one-electron oxidation was observed with horseradish peroxidase (HRP), as well as with rat mammary cell peroxidase (24) for the former compound. The HRP-catalyzed binding to DNA is low or negligible for hydrocarbons with IP above approximately 7.35 eV, while it is high for those below this IP value (unpublished results). These data and the mammary carcinogenicity results of several PAH (see below) led us to speculate that the only PAH capable of biological activation by one-electron oxidation are those with IP below approximately 7.35 eV. The low IP is a necessary, but not sufficient, condition for PAH carcinogenesis, since other parameters such as charge localization of the PAH radical cation, proper orientation of the PAH toward the cellular target, etc. play a role. Recently, we have found that the level of rat mammary peroxidase is increased by treatment with the monooxygenase inducers β-naphthoflavone and possibly MC. I.p. injection of β-naphthoflavone caused a 10-fold increase in mammary peroxidase and a 2-fold increase in mammary aryl hydrocarbon hydroxylase (25).

Rat Mammary Gland: A Target Organ for PAH Activation by One-Electron Oxidation

Thus far, the only PAH carcinogenic by direct application on mammary gland are those we have hypothesized to be activated by one-electron oxidation because they have low IP and sufficient charge localization in their radical cation: DMBA, MC, BP, BP-6-CH_3 and 7-methylbenz[a]anthracene (12,26). Compounds such as dibenz[a,h]anthracene, 5-methylchrysene and BP-7,8-dihydrodiol, which have relatively high IP, are not carcinogenic by direct application in mammary gland (12), but are potent carcinogens in mouse skin (12,27). All of these results can be explained by postulating that one-electron oxidation might be the predominant and probably selective pathway of activation in mammary gland for unsubstituted and alkyl-substituted PAH. To substantiate this hypothesis, the carcinogenicity of selected MC derivatives and CPEP has been tested by direct application in rat mammary gland.

RESULTS AND DISCUSSION

Carcinogenicity of 3-Methylcholanthrene Derivatives and Cyclopenteno[cd]pyrene in Mammary Gland

MC activated by one-electron oxidation in chemical systems (28,29) reacts with nucleophiles specifically at position 1. With this in mind, we have compared the carcinogenicity of MC to MC substituted at positions 1 and 2. The derivatives were MC-1-OH, MC-2-OH, MC-1-one, MC-2-one, MC-1-CH$_3$ and MC-2-CH$_3$. Substituents at position 1 block or render difficult the reaction of MC radical cation with nucleophiles at that position, while substituents at position 2 do not interfere when nucleophilic attack occurs at C-1, with the exception of MC-2-one in which the carbonyl group considerably deactivates the 1-carbon atom (29). In this experiment we have also tested the carcinogenicity of CPEP, which is active in mouse skin (15). Groups of 20 female Sprague-Dawley rats (Harlan Sprague-Dawley, Indianapolis, IN), 49-56 days old, were treated at two dose levels of 24 or 8 µmole of compound. The PAH used were MC, MC-2-OH, MC-2-one, MC-1-one, MC-2-CH$_3$, MC-1-CH$_3$ and CPEP. One group was treated with DMBA at the low dose as positive control. The rats were lightly anesthesized with ether and the abdominal skin was everted after surgical midline incision to expose the fourth right mammary gland. The crystalline PAH was dispersed on the gland and the skin flipped back and closed with sutures. Regular palpation of the treated area determined tumor onset and progression. Animals were sacrificed when tumor size was greater than 1.5 cm in diameter. Surviving rats were killed 34 weeks after treatment. Histological examination showed that all growths were benign and malignant mammary and mesenchyme tumors (Table 1).

The MC-treated groups had a tumor incidence of 100% and 90% in the fourth mammary gland region at the high and low dose, respectively. The rats developed a 61% incidence of mammary tumors at the high dose, while this percentage decreased to 40% at the low dose. The MC-2-OH-treated groups showed a

TABLE 1

CARCINOGENICITY OF MC DERIVATIVES AND CPEP IN SPRAGUE-DAWLEY RAT MAMMARY GLAND

Compound (dose, µmol)	Effective[a] no. animals	Tumor-bearing animals (%)	Mammary tumors		Mesenchyme tumors	
			No. malignant tumors no. rats	No. benign tumors no. rats	No. malignant tumors no. rats	No. benign tumors no. rats
MC (24)	13	100	10/8	0	14/10	1/1
MC (8)	20	90	9/8	0	23/16	0
MC-2-OH (24)	20	100	7/7	1/1	24/19	1/1
MC-2-OH (8)	20	100	1/1	0	22/20	0
MC-1-OH (24)	20	25	3/3	0	3/3	0
MC-1-OH (8)	20	0	0	0	0	0
MC-2-one (24)	20	0	0	0	0	0
MC-2-one (8)	20	0	0	0	0	0
MC-1-one (24)	20	0	0	0	0	0
MC-1-one (8)	20	0	0	0	0	0
MC-2-CH$_3$ (24)	20	40	0	0	9/6	2/2
MC-2-CH$_3$ (8)	20	30	0	0	6/6	0
MC-1-CH$_3$ (24)	20	0	0	0	0	0
MC-1-CH$_3$ (8)	20	0	0	0	0	0
CPEP (24)	20	0	0	0	0	0
CPEP (8)	20	0	0	0	0	0
DMBA (8)	20	80	13/10	0	5/4	6/6

[a] Histologically verified

tumor incidence of 100% at both doses; however the percentage of mammary tumors was less than in the MC groups, i.e., 35% and 5% at the high and low dose, respectively. MC-1-OH was active only at the high dose with a 25% tumor incidence. MC-2-CH_3 induced only mesenchyme tumors at both doses, suggesting that this compound is a weak mammary carcinogen. MC-1-one, MC-2-one, MC-1-CH_3 and CPEP were all inactive. DMBA, used as positive control at the low dose, induced tumors in 80% of the rats. Except for the carcinogenicity of MC-1-OH, the results for MC and derivatives are consistent with MC activation by one-electron oxidation. We have hypothesized that the weak activity of MC-1-OH might be attributed to the formation of MC-1-$OCOCH_3$ in mammary gland (see below). CPEP, which requires epoxidation to be carcinogenic in mouse skin (15), was inactive in mammary gland, suggesting that this pathway of activation is inefficient in this target organ.

Mutagenicity of 1-Acetoxy-3-Methylcholanthrene

It has been postulated that one of the possible mechanisms of activation of N-hydroxy-2-acetylaminofluorene in mammary gland involves acetylation to yield the alkylating compound N-acetoxy-2-acetylaminofluorene (30). We have postulated that MC-1-OH could follow a similar pathway of activation. To gain some evidence in this regard, we have compared the mutagenicities of MC-1-$OCOCH_3$ and MC-2-$OCOCH_3$ in S. typhimurium TA98, using BP-6-CH_2OCOCH_3 as a positive control (31). Mutagenicity tests were performed as described by Ames et al. (32). As expected, the alkylating MC-1-$OCOCH_3$ was mutagenic, whereas MC-2-$OCOCH_3$ was not (Figure 1). These results indicate that the stability of the carbonium ion at C-1 in MC is sufficiently high to make the corresponding acetate ester an alkylating agent. To gain more evidence that acetylation is the mechanism of activation for MC-1-OH, a comparison of the carcinogenicity in mammary gland of MC-1-OH versus the corresponding acetate ester will be pursued.

CONCLUSIONS

At present the only PAH carcinogenic in mammary gland are those we have postulated to be activated by one-electron oxidation, because they have low IP, namely, DMBA, MC, MC-2-OH, MC-2-CH_3, BP, BP-6-CH_3 and 7-methylbenz[a]anthracene. The inactivity of 5-methylchrysene and dibenz[a,h]anthracene (12 in mammary gland suggests that diol epoxides are not formed. This interpretation of the results is substantiated by the lack of carcinogenicity of BP-7,8-dihydrodiol (12) and CPEP,

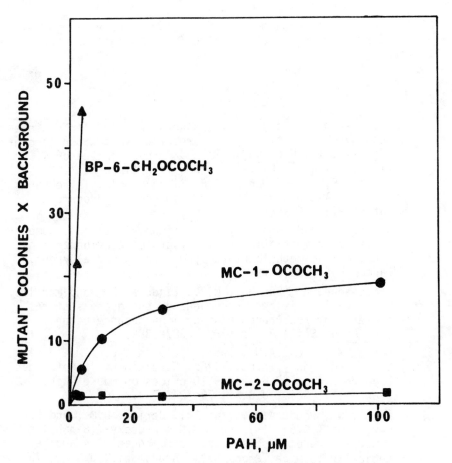

FIGURE 1. Mutation of S. typhimurium strain TA98 by MC-1-OCOCH$_3$, MC-2-OCOCH$_3$ and BP-6-CH$_2$OCOCH$_3$.

both of which require a simple epoxidation to be activated. Thus oxygenation of PAH by monooxygenase enzymes to form epoxides or diol epoxides does not seem to be involved in the carcinogenesis of these compounds in this target organ.

ACKNOWLEDGEMENTS

We wish to thank Dr. S. Salmasi for the histopathological examinations, Dr. R. Balasubramanian, Dr. A. Hakam, Ms. A. Munhall and Ms. B. Walker for valuable assistance and Ms.

M. Susman for editorial aid. This work was supported by Public Health Service contract NO1 CP05620 from the National Cancer Institute, NIH.

REFERENCES

1. Cavalieri, E., Rogan, E., and Roth, R. (1981): Multiple mechanisms of activation in aromatic hydrocarbon carcinogenesis. In: Free Radicals and Cancer, edited by R. Floyd, Marcel Dekker, Inc., New York (in press).
2. Harvey, R.G., and Dunne, F.B. (1978): Multiple regions of metabolic activation of carcinogenic hydrocarbons, Nature 273, 566-568.
3. Hecht, S.S., Hirota, N., and Hoffman, D. (1978): Comparative tumor initiating activity of 10-methylbenzo(a)pyrene, 7,10-dimethylbenzo(a)pyrene and benzo(a)pyrene, Cancer Lett., 5, 179-183.
4. Lacassagne, A., Buu Hoi, N.P., Zajdela, F., and Marille, P. (1964): Activité cancérogène de quelques isostères azotés d'hydrocarbures pentacycliques cancérogènes. C.R. Acad. Sci., Paris, 258, 3387-3389.
5. Levin, W., Wood, A.W., Yagi, H., Dansette, P.M., Jerina, D.M., and Conney, A.H. (1976): Carcinogenicity of benzo[a]pyrene 4,5-, 7,8- and 9,10-oxides on mouse skin, Proc. Natl. Sci., U.S.A., 73, 243-247.
6. Levin, W., Wood, A.W., Yagi, H., Jerina, D.M., and Conney, A.H. (1976): (+)-trans-7,8-dihydroxy-7,8-dihydrobenzo[a]pyrene: A potent skin carcinogen when applied topically to mice, Proc. Natl. Acad. Sci., U.S.A., 73, 3867-3871.
7. Levin, W., Wood, A.W., Wislocki, P.G., Kapitulnik, J., Yagi, H., Jerina, D.M., and Conney, A.H. (1977): Carcinogenicity of benzo-ring derivatives of benzo[a]pyrene on mouse skin, Cancer Res., 37, 3356-3361.
8. Chouroulinkov, I., Gentil, A., Grover, P.L., and Sims, P. (1976): Tumor-initiating activities on mouse skin of dihydrodiols derived from benzo[a]pyrene, Brit. J. Cancer, 34, 523-532.
9. Slaga, T.J., Viaje, A., Berry, D.L., and Bracken, W. (1976): Skin tumor initiating ability of benzo[a]pyrene 4,5-, 7,8-diol,9,10-epoxides and 7,8-diol, Cancer Lett., 2, 115-121.

10. Levin, W., Wood, A.W., Chang, R.L., Slaga, T.J., Yagi, H., Jerina, D.M., and Conney, A.H. (1977): Marked differences in the tumor-initiating activity of optically pure (+)- and (-)-trans-7,8-dihydrobenzo[a]pyrene on mouse skin, Cancer Res., 37, 2721-2725.
11. Slaga, T.J., Bracken, W.M., Viaje, A., Levin, W., Yagi, M., Jerina, D.M., and Conney, A.H. (1977): Comparison of the tumor-initiating activities of benzo[a]pyrene arene oxides and diol-epoxides, Cancer Res., 37, 4130-4133.
12. Cavalieri, E., Sinha, D., and Rogan, E. (1980): Rat mammary gland versus mouse skin: Different mechanisms of activation of aromatic hydrocarbons. In: Polynuclear Aromatic Hydrocarbons. Chemistry and Biological Effects, edited by A.J. Bjorseth and A.J. Dennis, pp. 215-231, Battelle Press, Columbus, Ohio.
13. Kouri, R.E., Wood, A.W., Levin, W., Rude, T.H., Yagi, H., Mah, H.D., Jerina, D.M., and Conney, A.H. (1980): Carcinogenicity of benzo[a]pyrene and thirteen of its derivatives in C3H/fCum mice, J. Natl. Cancer Inst., 64, 617-623.
14. Lacassagne, A., Buu Hoi, N.P., and Zajdela, F. (1958): Relation entre structure moléculaire et activité cancérogène dans trois séries d'hydrocarbures aromatiques hexacycliques, C. R. Acad. Sci., Paris, 246, 1477-1480.
15. Cavalieri, E., Rogan, E., Toth, B., and Munhall, A. (1981): Carcinogenicity of the environmental pollutants cyclopenteno[cd]pyrene and cyclopentano[cd]pyrene in mouse skin, Carcinogenesis, 2, 277-281.
16. Wilk, M., Bez, W., and Rochlitz, J. (1966): Neue Reaktionen der carcinogenen Kohlenwasserstoffe 3,4-Benzpyrene, 9,10-Dimethyl-1,2-benzanthracene und 20-Methylcholanthren, Tetrahedron, 22, 2599-2608.
17. Fried, J. (1974): One-electron oxidation of polycyclic aromatics as a model for the metabolic activation of carcinogenic hydrocarbons. In: Chemical Carcinogenesis, Part A, edited by P.O.P. Ts'o and J. DiPaolo, pp. 197-215, Marcel Dekker, New York.
18. Cavalieri, E., Roth, R., and Rogan, E.G. (1976): Metabolic activation of aromatic hydrocarbons by one-electron oxidation in relation to the mechanism of tumor initiation. In: Carcinogenesis, vol. 1. Polynuclear Aromatic Hydrocarbons: Chemistry, Metabolism and Carcinogenesis, edited by R.T. Freudenthal and P.W. Jones, pp. 181-190, Raven Press, New York.

19. Rogan, E., Roth, R., Katomski, P., Benderson, J., and Cavalieri, E.L. (1978): Binding of benzo[a]pyrene at the 1,3,6 positions to nucleic acids in vivo on mouse skin and in vitro with rat liver microsomes and nuclei, Chem.-Biol. Interact., 22, 35-51.
20. Cavalieri, E., Roth, R., Rogan, E., Grandjean, C., and Althoff, J. (1978): Mechanisms of tumor initiation by polycyclic aromatic hydrocarbons. In: Carcinogenesis, vol. 3. Polynuclear Aromatic Hydrocarbons, edited by P.W. Jones and R.I. Freudenthal, pp. 273-284, Raven Press, New York.
21. Rogan, E.G., Katomski, P.A., Roth, R.W., and Cavalieri, E.L. (1979): Horseradish peroxidase/hydrogen peroxide-catalyzed binding of aromatic hydrocarbons to DNA, J. Biol. Chem., 254, 7055-7059.
22. Bartsch, H., and Hecker, E. (1971): On the metabolic activation of carcinogen N-hydroxy-N-2-acetylaminofluorene. III. Oxidation with horseradish peroxidase to yield 2-nitrosofluorene and N-acetoxy-N-2-acetylaminofluorene, Biochim. Biophys. Acta, 237, 567-578.
23. Floyd, R.A., and Soong, L.M. (1977): Obligatory free radical intermediate in the oxidative activation of the carcinogen N-hydroxy-2-acetylaminofluorene, Biochim. Biophys. Acta, 498, 244-249.
24. Reigh, D.L., Stuart, M., and Floyd, R.A. (1978): Activation of the carcinogen N-hydroxy-2-acetylaminofluorene by rat mammary peroxidase, Experientia, 34, 107-108.
25. Rogan, E., and Cavalieri, E. (1982): Effect of monooxygenase inducers on rat mammary peroxidase and aryl hydrocarbon hydroxylase. Submitted.
26. Dao, T.L., King, C., and Tominga, T. (1971): Isolation, identification and biological study of compounds derived from 3-methylcholanthrene by irradiation in dimethyl sulfoxide, Cancer Res., 31, 1492-1495.
27. Hecht, S.S., Mazzarese, R., Amin, S., LaVoie, E., and Hoffmann, D. (1979): On the metabolic activation of 5-methylchrysene. In: Polynuclear Aromatic Hydrocarbons, edited by P.W. Jones and P. Leber, pp. 733-752, Ann Arbor Science Publishers, Inc., Ann Arbor, MI.
28. Cavalieri, E., and Roth, R. (1976): Reaction of methylbenzanthracenes and pyridine by one-electron oxidation: A model for metabolic activation and binding of carcinogenic aromatic hydrocarbons, J. Org. Chem., 41, 2679-2684.

29. Rogan, E., Roth, R., and Cavalieri, E. (1980): Manganic acetate and horseradish peroxidase/hydrogen peroxide: In vitro models of activation of aromatic hydrocarbons by one-electron oxidation. In: Polynuclear Aromatic Hydrocarbons. Chemistry and Biological Effects, edited by A. Bjorseth and A.J. Dennis, pp. 259-266, Battelle Press, Columbus, Ohio.
30. Malejka-Giganti, Rydell, R.I., and Gutman, H.R. (1977): Mammary carcinogenesis in the rat by topical application of fluorenylhydroxamic acids and their acetates. Cancer Res., 27, 111-117.
31. Cavalieri, E., Roth, R., and Rogan, E. (1979): Hydroxylation and conjugation at the benzylic carbon atom: A possible mechanism of carcinogenic activation for some methylsubstituted aromatic hydrocarabons. In: Polynuclear Aromatic Hydrocarbons, edited by P.W. Jones and P. Leber, pp. 517-529, Ann Arbor Science Publishers, Inc., Ann Arbor, MI.
32. Ames, B.N., McCann, J., and Yamasaki, E.: Methods for detecting carcinogens and mutagens with the Salmonella/mammalian microsome mutagenicity test. Mut. Res., 31, 347-364, 1975.

MICROBIAL OXIDATION OF 7-METHYLBENZ(A)ANTHRACENE AND 7-HYDROXYMETHYLBENZ(A)ANTHRACENE

C.E. CERNIGLIA*, P.P. FU*, and S.K. YANG**
*National Center for Toxicological Research, Food and Drug Administration, Jefferson, Arkansas 72079 and **Department of Pharmacology, School of Medicine, Uniformed Services University of the Health Sciences, Bethesda, Maryland 20814

INTRODUCTION

There has been increasing concern over the fate of polycyclic aromatic hydrocarbons (PAHs) in terrestrial and aquatic ecosystems since many of these compounds exhibit cytotoxic, mutagenic and carcinogenic properties (1). PAHs are biologically inactive per se and require metabolic activation by mammalian microsomal enzyme systems to reactive intermediates which can lead to their carcinogenic properties (13).

Recent reports on the microbial metabolism of aromatic hydrocarbons have suggested a similarity between fungal and mammalian enzyme systems in the metabolism of aromatic substrates (2). Previous investigations have indicated that the filamentous fungus, Cunninghamella elegans is able to convert benzo[a]pyrene (3-6), benz[a]anthracene (9), and 3-methylcholanthrene (7) to metabolites that have been implicated as proximate carcinogenic forms in higher organisms. However, little is known about the ability of fungi to metabolize methyl-substituted benz[a]anthracenes.

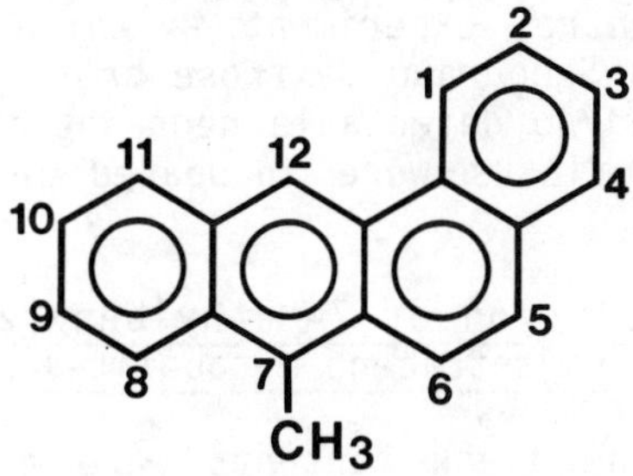

FIGURE 1: Structure of 7-Methylbenz[a]anthracene

7-Methylbenz[a]anthracene (7-MBA) (Figure 1) is the most tumorigenic and biologically active of the 12 monomethyl derivatives of benz[a]anthracene (8,10-12,14,17,18). Studies on the mammalian metabolism of 7-MBA have indicated that this compound is metabolized to form mainly 5,6- 8,9-, 3,4-, 10,11- and 1,2-dihydrodiols (15,16,19,24). The 7-MBA-trans 3,4-dihydrodiol has been implicated as the proximate carcinogen of 7-MBA since upon further metabolism, it is more mutagenic to Salmonella typhimurium strain TA98 and Chinese hamster V79 cells than 7-MBA and the other 7-MBA dihydrodiols (11). In addition, a trans-3,4-dihydrodiol-7-MBA 1,2-epoxide, appears to be the reactive intermediate which binds to mouse skin DNA (19,20). However, it has also been reported that a non-K-region dihydrodiol-epoxide, trans-8,9-dihydrodiol 7-MBA-10,11-epoxide, is mutagenic towards S. typhimurium strain TA100 (12).

In this study we report on the fungal metabolism of 7-MBA and 7-hydroxymethylbenz[a]anthracene (7-OHMBA). This paper describes the isolation and identification of the metabolites formed from 7-MBA and 7-OHMBA by C. elegans. The results show that C. elegans oxidizes these compounds in a manner that is similar to those observed in higher organisms.

MATERIALS AND METHODS

Microorganism and Growth Conditions

C. elegans ATCC 36112 was grown in 125 ml Erlenmeyer flasks which contained 30 ml of Sabouraud dextrose broth. The flasks were incubated at 25°C for 48 h on a rotary shaker operating at 150 rev/min. After 48 h, cells were collected by filtration and transferred to 30 ml of fresh medium. 7-MBA or 7-OHMBA (3 mg in 0.3 ml of dimethylformamide) was added to each flask. A control experiment in which 7-MBA or 7-OHMBA was incubated with Sabouraud dextrose broth in the absence of the organism showed no detectable degradation of either 7-MBA or 7-OHMBA. All flasks were incubated in the dark as described above.

Detection and Isolation of 7-Methylbenz[a]anthracene and 7-Hydroxymethylbenz[a]antracene Metabolites

After 24 h the flask contents were extracted with six volumes of ethyl acetate. The combined organic extracts were dried over anhydrous sodium sulfate and the solvent was removed under reduced pressure at 40°C in the dark. The residue was dissolved in methanol and analyzed by high performance liquid chromatography (hplc). A Beckman model 332 hplc and model 155-10 variable wavelength absorbance detector operated at 254 nm was used to separate the metabolites of 7-MBA and 7-OHMBA. A C_{18} Ultrasphere ODS column (25 cm x 4.6 mm id) was used and the separation was achieved with a programmed 30 min linear gradient of methanol/water (1:1, v/v) to methanol at a solvent flow rate of 1.2 ml/min. UV-visible absorption spectra were determined on a Beckman model 25 recording spectrophotometer. Mass spectra were obtained with a Finnigan model 4023 mass spectrometer at 70 eV ionizing voltage with a solid probe. Direct probe mass spectrometry were performed on samples which were dissolved in 5 μl of methanol and dried in glass sample cups. Spectra were recorded as the probe temperature was increased ballistically from $30°C$ to $300°C$ monitoring the ion source temperature at $270°C$.

Chemicals

7-MBA and 7-OHMBA were synthesized in this laboratory according to published procedures (22,24). Solvents for HPLC analysis were purchased from Burdick and Jackson Laboratories; all other chemicals were of reagent grade in the highest available purity.

RESULTS

The high pressure liquid chromatogram of the 7-MBA metabolites obtained by the oxidation of 7-MBA by C. elegans is shown in Figure 2. The metabolites and recovered substrate contained in peaks A through K were collected and their ultraviolet-visible absorption and mass spectra were determined (Table 1). Component K is the parent hydrocarbon, 7-MBA. Components C and E were major metabolites in the fungal oxidation of 7-MBA. Mass spectral evidence indicated that each component had molecular ions at m/e 292 and a characteristic fragment ion at m/e 274 (M^+ 18) indicative of a loss of a water molecule which suggests that both compounds are dihydrodiols. Further evidence was provided by the absorption spectra (Table 1) which indicated that compounds C and E are 7-OHMBA-trans-8,9-dihydrodiol and 7-OHMBA-trans-3,4-dihydrodiol, respectively. The trans configuration was assigned since each of the dihydrodiols did not form a vicinal cis-acetonide in anhydrous acetone/cupric sulfate (23). Furthermore, all dihydrodiols formed from the fungal metabolism of PAHs have been shown to be trans-stereoisomers (2). To further confirm that metabolites C and E (Figure 2) were dihydrodiols derived from 7-OHMBA, C. elegans was incubated with 7-OHMBA in a manner identical to the 7-MBA incubation. Hplc analysis of the ethyl acetate soluble metabolites formed by the incubation of 7-OHMBA with C. elegans indicated that components C and E formed from 7-MBA were also formed from 7-OHMBA (Figure 3).

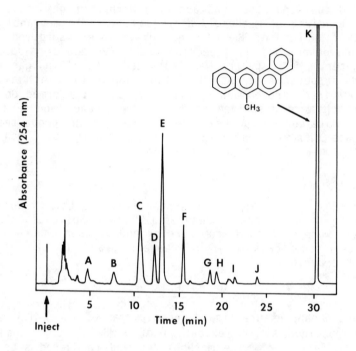

FIGURE 2: HPLC elution profile of metabolites formed from 7-methylbenz[a]anthracene by C. elegans.

Metabolite J (Figure 2) had an hplc retention time, mass spectrum m/e at 258 (M^+) and absorption spectrum identical to those given by authentic 7-OHMBA. Components F and I had identical retention times, mass and UV absorption spectra (Table 1) to synthetic 7-MBA-trans-8,9-dihydodiol and 7-MBA-trans-3,4-dihydrodiol (20), respectively.

Metabolite D (Figure 2) showed molecular ions at m/e 310(M^+) 292 (M^+-H_2O) and 274 (M^+-2H_2O) which suggests that it is a tetraol.

TABLE 1

HPLC RETENTION TIMES, ABSORPTION AND MASS SPECTRAL DATA OF THE 7-METHYLBENZ[a]ANTHRACENE METABOLITES ISOLATED FROM THE EXTRACTS OF CUNNINGHAMELLA ELEGANS CULTURES

HPLC Peak	HPLC Retention Time (min)	UV-Visible Absorption maxima (nm)	Mass Spectra m/e (%)	Assignment
A	4.8	260,275,288,322, 343,365	310(5)292(43)274(100) 258(53)244(100)203(19)	7-MBA-tetraol(?)
B	7.5	260,267,310,325, 355,375	292(12)274(23)258(15) 244(60)	7-MBA-triol(?)
C	10.5	266,293,304,317	292(40)274(28)246(23) 245(100)228(23)215(27) 202(33)	7-OHMBA-8,9 diol
D	12.5	254,260,287,293, 305	310(48)292(5)274(30) 258(18)246(50)245(100) 202(52)	7-MBA-8,9-10,11 tetraol
E	13.0	259,363,383,408	292(100)274(21)257(10) 246(26)245(34)228(17) 217(56)215(33)202(29)	7-OHMBA-3,4-diol
F	15.5	266,293,307,320	276(100)258(44)245(12) 232(33)230(57)229(26) 215(45)204(24)202(26)	7-MBA-8,9-diol
G	18.5	260,320,350,365, 400	274(55)258(88)245(87)	7-OHMBA phenol or dihydroxy 7-MBA
H	19.5	272,283,294,343, 360,377,390	274(96)257(25)245(100)	7-OHMBA phenol or dihydroxy 7-MBA
I	21.5	262,350,368,390, 413	276(100)258(55)245(32) 232(18)230(52)229(24) 215(39)204(20)202(18)	7-MBA-3,4-diol
J	24.0	257,267,277,288, 315,330,347,362, 385	258(98)241(26)229(100) 226(21)	7-OHMBA
K	31.0	255,268,282,288, 317,332,350,365, 387	242(100)241(52)239(22)	7-MBA

Based on its absorption spectrum, (Table 1) compound D was tentatively assigned as 8,9,10,11-tetraol of 7-MBA. Tetraols have previously been isolated from the fungal oxidation of PAHs (5,6,9).

Metabolites G and H each gave a mass spectrum which had molecular ions at m/e 274 (Table 1) which suggests that these compounds are either dihydroxy derivatives of 7-MBA or monohydroxylated derivatives of 7-OHMBA. Owing to the small amounts formed, no attempt was made to further characterize compounds G and H. Metabolites A and B (Figure 2) are very polar metabolites of 7-MBA. The mass spectral fragmentation patterns indicate that they are polyhydroxylated derivatives of 7-MBA, however, insufficient material was available for further structural analysis.

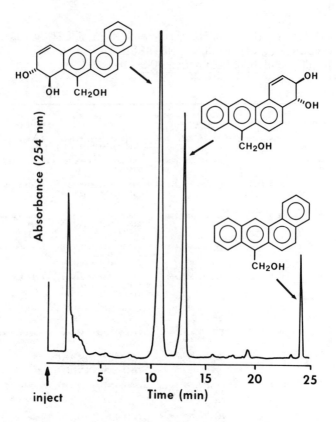

FIGURE 3. HPLC elution profile of metabolites formed from 7-hydroxymethylbenz[a]anthracene.

DISCUSSION

The results presented show that the filamentous fungus, C. elegans, can oxidize 7-MBA and 7-OHMBA to detoxified products as well as metabolites which have been implicated as carcinogens. To our knowledge, this represents the first study on the fungal metabolism of a monomethyl and a hydroxymethylated derivative of benz[a]anthracene. The proposed metabolic pathways for the fungal oxidation of 7-MBA and 7-OHMBA to dihydrodiols are shown in Figure 4. C. elegans oxidized 7-MBA and 7-OHMBA in the 8,9 and 3,4 positions of the ring to form dihydrodiols with a trans-configuration. Similar to our studies on the fungal metabolism of benzo[a]pyrene (3-6), benz[a]anthracene (9) and 3-methylcholanthrene (7), we did not detect any metabolites derived from the 5,6 positions of the ring. The K-region 7-MBA-trans-5,6-dihydrodiol is a major metabolite in the mammalian metabolism of 7-MBA (15,16,19,24). This data suggests that their may be differences in the fungal cytochrome P-450 from that reported for mammalian hepatic microsomal and nuclear cytochrome P-450. However, it should be noted that in this experimental system, the inability to detect a K-region dihydrodiol may be due to the K-region epoxide being bound to cellular macromolecules or conjugated to glutathione.

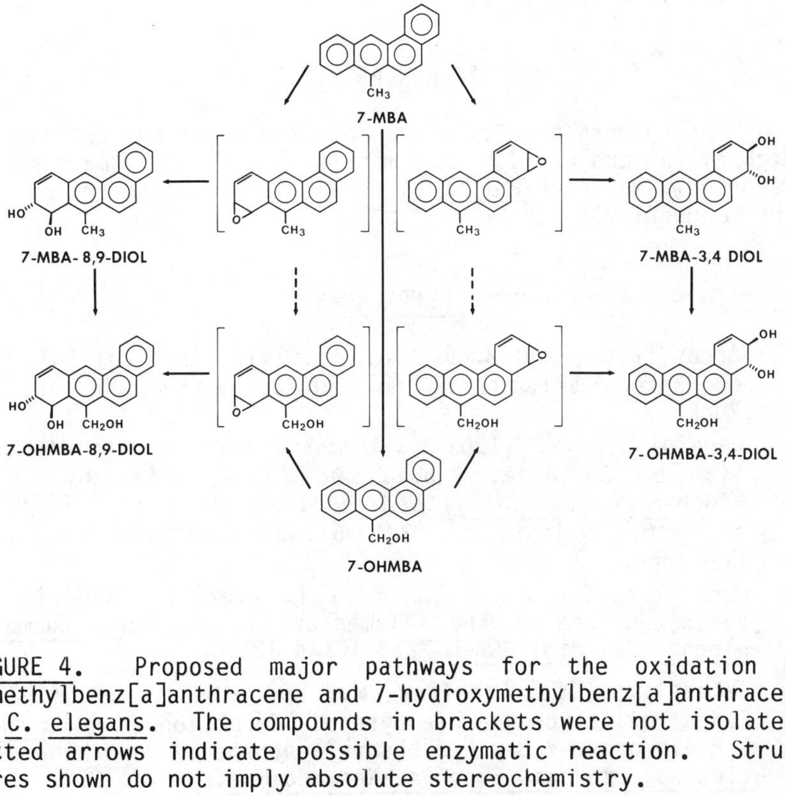

FIGURE 4. Proposed major pathways for the oxidation of 7-methylbenz[a]anthracene and 7-hydroxymethylbenz[a]anthracene by C. elegans. The compounds in brackets were not isolated. Dotted arrows indicate possible enzymatic reaction. Structures shown do not imply absolute stereochemistry.

The conversion of 7-MBA by C. elegans to the 3,4-dihydrodiol is of particular interest since this compound has been shown to be more biologically active than 7-MBA and the other dihydrodiols. In addition, fluorescence spectral studies of the 7-MBA bound to DNA in mouse skin or hamster embryo cells indicated that the 1,2,3,4 positions of the molecule are involved in the metabolic activation of 7-MBA (8,11,12,14,-17,21). Although no cytotoxic or mutagenic effects were noted in the fungi used in this report after incubation with either 7-MBA or 7-OHMBA, studies on the binding of polycyclic aromatic hydrocarbons to fungal DNA are now in progress to see if similar PAH interactions with cellular macromolecules occur in microbial systems and also the environmental consequences and significance of these reactions.

Finally, the results show that the fungal metabolism of 7-MBA and 7-OHMBA is qualitatively similar to that reported for mammalian enzyme systems.

ACKNOWLEDGMENTS

The authors greatly appreciate the technical assistance of Michael Fox and Dr. J.P. Freeman in obtaining the mass spectra reported. We thank Ruth York for assistance in preparation of the manuscript.

REFERENCES

1. Arcos, J.C., and Argus, M.F. (1974): Chemical Induction of Cancer, Vol. II A, pp. 30-33, Academic Press, New York.
2. Cerniglia, C.E. (1981): Aromatic Hydrocarbons: Metabolism by Bacteria, Fungi, and Algae. In: Reviews in Biochemical Toxicology, edited by E.Hodgson, J.R. Bend, and R.M. Philpot, pp. 321-361, Elsevier/North Holland, New York.
3. Cerniglia, C.E., and Gibson, D.T. (1979): Oxidation of benzo[a]pyrene by the filamentous fungus Cunninghamella elegans, J. Biol. Chem. 254:12174-12180.
4. Cerniglia, C.E., Mahaffey, W., and Gibson, D.T. (1980): Fungal oxidation benzo(a)pyrene. Formation of (-)-trans-7,8-dihydroxy-7,8-dihydrobenzo[a]pyrene by Cunninghamella elegans, Biochem. Biophys. Res. Commun. 94:226-232.
5. Cerniglia, C.E., and Gibson, D.T. (1980): Fungal oxidation of benzo[a]pyrene and (+)-trans-7,8-dihydroxy-7,8-dihydrobenzo[a]pyrene: Evidence for the formation of a benzo[a]pyrene 7,8-diol-9,10-epoxide, J. Biol. Chem. 255:5159-5163.

6. Cerniglia, C.E., and Gibson, D.T. (1980): Fungal oxidation of (+)-trans-9,10-dihydroxy-9,10-dihydrobenzo[a]pyrene: Formation of diastereomeric benzo[a]pyrene 9,10-diol-7,8-epoxide, Proc. Natl. Acad. Sci. USA, 77:4554-4558.
7. Cerniglia, C.E., Dodge, R.H., and Gibson, D.T. (1981): Fungal oxidation of 3-methylcholanthrene: Formation of proximate carcinogenic metabolites of 3-methylcholanthrene, Chem.-Biol. Interact. (in press).
8. Chouroulinkov, I., Gentil, A., Tierney, B., Grover, P.L., and Sims, P. (1977): The metabolic activation of 7-methylbenz[a]anthracene in mouse skin: high tumor-initiating activity of the 3,4-dihydrodiol, Cancer Lett. 3:247-253.
9. Dodge, R.H., and Gibson, D.T. (1980): Fungal metabolism of benzo[a]anthracene. Abst. 81st Annual American Society for Microbiology Meeting, p. 138.
10. Dunning, W.F., and Curtis, M.R. (1960): Relative carcinogenic activity of monomethyl derivatives of benz[a]anthracene in Fischer line 344 rats, J. Natl. Cancer Inst. 25:387-391.
11. Malaveille, C., Tierney, B., Grover, P.L., Sims, P., and Bartsch, H. (1977): High microsome-mediated mutagenicity of the 3,4-dihydrodiols of 7-methylbenz[a]anthracene in S. typhimurium TA98, Biochem. Biophys. Res. Commun. 75:427-433.
12. Marquardt, H., Baker, S., Tierney, B., Grover, P.L., and Sims, P. (1977): The metabolic activation of 7-methylbenz[a]anthracene: the induction of malignant transformation and mutation in mammalian cells by non-K-region dihydrodiols, Int. J. Cancer. 19:828-833.
13. Miller, E.C., and Miller, J.A. (1974): Biochemical Mechanisms of Chemical Carcinogenesis. In: The Molecular Biology of Cancer, edited by H. Bush, pp. 377-403, Academic Press, New York.
14. Newman, M.S. (1976): Carcinogenic Activity of Benz[a]anthracene. In: Carcinogenesis, Vol. 1, Polynuclear Aromatic Hydrocarbons: Chemistry, Metabolism, and Carcinogenesis, R.I. Freudenthal and P.W. Jones, Eds., pp. 203-207, Raven Press, New York.
15. Sims, P. (1967): The metabolism of 7- and 12-methylbenz[a]anthracene and their derivatives, Biochem. J. 105:591-598.

16. Sims, P. (1970): Studies on the metabolism of 7-methylbenz[a]anthracene and 7,12-dimethylbenz[a]anthracene and its hydroxy methyl derivatives in rat liver and adrenal homogenates, Biochem. Pharmacol. 19:2261-2275.
17. Slaga, T.J., Gleason, G.L., Mills, G., Ewald, L., Fu, P.P., Lee, H.M., and Harvey, R.G. (1980): Comparison of the skin tumor initiating activities of dihydrodiols and diol-epoxides of various polycyclic aromatic hydrocarbons, Cancer Res. 40:1981-1984.
18. Stevenson, J.L., and Von Haam, E. (1965): Carcinogenicity of benz[a]anthracene and benzo[c]phenanthrene derivatives, Amer. Ind. Hyg. Assoc. J. 26:475-478.
19. Tierney, B., Hewer, A., Walsh, C., Grover, P.L., and Sims, P. (1977): The metabolic activation of 7-methylbenz[a]anthracene in mouse skin, Chem.-Biol. Interact. 18:179-193.
20. Tierney, B., Abercrombie, B., Walsh, C., Hewer, A., Grover, P.L., and Sims, P. (1978): The preparation of dihydrodiols from 7-methylbenz[a]anthracene, Chem.-Biol. Interact. 21:289-298.
21. Vigny, P., Duquesne, M., Coulomb, H., Lacombe, C., Tierney, B., Grover, P.L., and Sims, P. (1977): Metabolic activation of polycyclic hydrocarbons: fluorescence spectral evidence is consistent with metabolism at the 1,2- and 3,4-double bonds of 7-methylbenz[a]anthracene, FEBS Lett. 75:9-12.
22. Wood, J.L. and Fieser, L.F. (1940): Sulfhydryl cysteine derivatives of 1,2-benzanthracene and 3,4-benzopyrene, J. Am. Chem. Soc. 62:2674-2681.
23. Yang, S.K., McCourt, D.W., Gelboin, H.V., Miller, J.R., and Roller, P.P. (1977): Stereochemistry of the hydrolysis products and their acetonides of two stereoisomeric benzo[a]pyrene 7,8-diol 9,10-epoxides, J. Am. Chem. Soc. 99:5124-530.
24. Yang, S.K., Chou, M.W., and Fu, P.P. (1980): Metabolism of 6-, 7-, 8-, and 12-methylbenz[a]anthracenes and Hydroxy Methylbenz[a]anthracenes. In: Polynuclelar Aromatic Hydrocarbons: Chemistry and Biologic Effects, edited by A. Bjorseth and A.J. Dennis, pp. 645-662, Battelle Press, Columbus, Ohio.

INTERACTION OF BENZO(A)PYRENE AND CHRYSOTILE

M.J.W. CHANG, N.P. SINGH, A. TURTURRO and R.W. HART
Department of Health and Human Services, Food and Drug Administration, National Center for Toxicological Research, Jefferson, Arkansas, 72079

INTRODUCTION

Benzo(a)pyrene (BaP), an ubiquitous environmental contaminant and a carcinogenic component of cigarette smoke, has been implicated in human respiratory neoplasms (1), tumor induction in experimental animals (2,3) and transformation of human fibroblast cells in vitro (4). BaP has also been suggested as a cocarcinogen in asbestos related neoplasms (5).

Various studies have shown that in the presence of chrysotile, BaP uptake across both an artificial membrane and the rat liver microsomal membrane was enhanced (6-8). McLemore et al. (9) reported different degrees of increase in the induction of aryl hydrocarbon hydroxylase (AHH) in human mitogen-stimulated lymphocyte cultures when amosite asbestos and benz(a)anthracene or cigarette tars were used in different combination. Additionally, Daniel et al. (10) observed that in the presence of chrysotile, the cytotoxicity of BaP to a normal human fibroblast cell line was enhanced and the binding of radioactive BaP to DNA was also increased, while the cell free metabolism of BaP in the presence of chrysotile was not affected. In another cell-free metabolic study of BaP, Kandaswami and O'Brien (11) by measuring the activity of AHH in the presence of asbestos fibers concluded the metabolism was slightly impaired.

In order to understand the interactions of polycyclic aromatic hydrocarbons (PAHs) and asbestos, we started a series of studies using BaP, chrysotile and a human foreskin fibroblast cell line as the model system.

MATERIALS AND METHODS

Adsorption of BaP on Chrysotile

(^{3}H)-BaP (S.A. = 52.9 Ci/mmol, New England Nuclear) was diluted with nonradioactive florisil (silica gel) purified BaP and dissolved in dimethylsulfoxide (DMSO) to give a final concentration of 0.6 mg/ml. NIEHS intermediate sized chrysotile was heated in an oven at $70°C$ for 24 h before it was suspended in a growth medium without fetal bovine serum (FBS, Gibco) at a concentration of 2 mg/ml. Tissue culture growth media with and without FBS were used to prepare the test chrysotile suspensions at a final concentration of 10 µg/ml containing 3 µg/ml (^{3}H)-BaP. Three sets of experiments were carried out with each time duplicate samples of 0.5 ml of the mixture were taken for radioactivity assay before incubation in a water bath at 37°C. At different time intervals, the chrysotile was pelletted by a brief centrifugation at room temperature. Aliquots of 0.5 ml of the supernatant were removed for radioactivity measurement. Radioactivity was determined with a Packard 3380 Liquid Scintillation Spectrophotometer after adding 0.5 ml of distilled water and 10 ml of Aquassure (New England Nuclear) to the samples.

Desorption of Chrysotile-Adsorbed BaP

The BaP coated chrysotile pellet obtained from the above experiment was transferred to a freshly prepared tissue culture medium with 10% FBS and incubated in a 37°C shaker water bath. At different time intervals the suspension was briefly centrifuged, and duplicate samples of 0.5 ml of the supernatant were sampled and assayed for the radioactivity as described above. The experiment was repeated once.

Adsorption of ^{14}C-Thymidine on Chrysotile

The study was performed similarly as the adsorption of BaP on chrysotile. ^{14}C-thymidine (S.A. = 20 mCi/mmol, New England Nuclear) 0.02 µCi/ml was added in a serum free chrysotile suspension (3 doses of chrysotile were studied: 10, 20 and 100 µg/ml). Each time, triplicate of 0.1 ml aliquot was sampled and assayed.

Tissure Culture

Fibroblasts were cultured in Eagle's minimal essential medium containing Earle's balanced salt solution supplemented with 1.5 X essential amino acids and 1.5 X BEM vitamins (Gibco Laboratories, Grand Island, NY). The medium was completed with the addition of 10% fetal bovine serum. Cells were grown in T-75 flask at 37°C under 95% air and 5% CO_2 in a Forma Scientific UN-I-TROL CO_2 incubator. Cultures were harvested at 100% confluency by 0.01% trypsine in 0.01% ethylenediaminetetra acetic acid (EDTA) solution.

BaP Uptake With/Without Chrysotile

Human fibroblasts were seeded at 0.25 x 10^6 cells per 100 mm petri dish. Three dishes per treatment per time point were used. 24 h later, media were removed and 9 ml each of freshly prepared ^{3}H-BaP (3.3 µg/ml) and BaP + chrysotile (11 µg/ml) in growth media with/without 10% FBS were added. Duplicate 0.1 ml samples were taken pre- and 24 h post-treatment and assayed for their radioactivities. One ml of pure FBS or complete growth medium was then added to the culture dish and the radioactivity in the medium was continuously assayed each 24 h interval through 96 h post treatment.

^{14}C-Thymidine Incorporation

There were four experimental groups: (1) control; (2) BaP treated (3 µg/ml); (3) chrysotile treated (10 µg/ml); and (4) BaP plus chrysotile treated at the same concentrations as groups 2 and 3. Cells were seeded at a concentration of 0.25×10^6 cells per 100 mm petri dish and permitted to grow for 24 h. Three dishes per sample per time point were used. The first 24 h treatment was carried out without 10% FBS; the treatment was left in contact with the cells during the whole experiment.

Effects of BaP Plus Chrysotile on Population Growth

Human fibroblasts were seeded at a concentration of 0.25×10^6 in a 100 mm petri dish and permitted to grow for 24 h before the treatment. 9 ml of 3.3 µg/ml BaP with or without 10% FBS, in the presence or absence of 11 µg/ml chrysotile were used to treat the cultures. 24 h later, the medium was supplimented with 1 ml of whole serum or 1 ml of complete medium. At different time intervals (24, 27, 48, 72 and 96 h post treatment), cells were harvested by trypsinization. After washing with cold 1/10 x SSC (0.15 M NaCl and 0.015 M sodium citrate), cell pellets were dissolved in 0.2 N NaOH. Aliquots of the solution were used in duplicate for protein determination by Bio Rad microassay (12). Total protein in µg/dish was used as a measure of population growth. There were consistently three dishes per sample per time point.

RESULTS

Adsorption/Desorption on Chrysotile

As shown in Fig. 1A, adsorption of BaP starts almost immediately; reaching a plateau at about 2 h after mixing. At the concentration studied (10 µg/ml chrysotile and 3 µg/ml ^{3}H-BaP), approximately 85% of the BaP is adsorbed to chrysotile within 21 h of incubation.

From Fig. 1B, it appears that the adsorption of BaP on chrysotile fibers is not a very strong one. As soon as the BaP covered fibers were transferred to a complete culture growth medium (with 10% FBS) at 37°C, about 35% of the BaP was eluted to the medium within 2 min. By 130 min post mixing and incubation at 37°C, 95% of the BaP was eluted off the fibers.

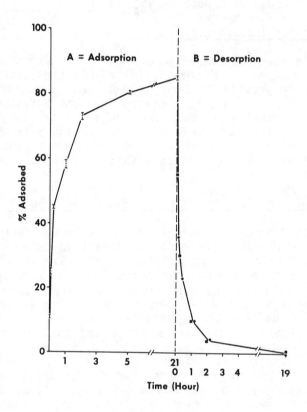

FIGURE 1. BaP (3 µg/ml) on a chrysotile (10 µg/ml): A-adsorption in serum free media, 3 sets of experiments were carried out at different times; B-desorption in complete media (with 10% FBA), experiment was repeated once.

Table 1 gives the results of 3 sets of experiments. At a 10-fold difference in the concentration of the adsorbing material and after 41 h of incubation, there was negligible adsorption of thymidine on chrysotile.

TABLE 1

ADSORPTION OF ^{14}C-THYMIDINE ON CHRYSOTILE[a]

Chrysotile[b] (µg/ml)	Time (min)				
	0[c]	5	15	60	2460
10	100.1 ± 0.0	100.2 ± 1.6		100.6 ± 2.0	99.7 ± 0.9
20	101.9 ± 0.5	100.6 ± 1.1	101.0 ± 2.1	-	101.4 ± 0.3
100	100.1 ± 1.0	99.6 ± 0.6	99.6 ± 1.8	-	100.2 ± 0.8

[a] % ± S.D. radioactivity left in the supernatant after minutes of incubation at room temperature.
[b] Tissue culture growth medium without FBS was used to suspend the mineral fibers.
[c] ^{14}C-thymidine (S.A. = 20 mCi/mmol) 0.02 µCi/ml was added to the chrysotile suspension and the mixture was centrifuged immediately.

BaP Uptake by Human Fibroblasts

BaP uptake by human fibroblasts was studied in the presence and absence of chrysotile under the conditions with and without 10% FBS in the culture media. The results are given in Table 2. In the presence of 10% serum the uptake in 24 h had no significant difference between BaP alone and in combination with chrysotile. But when serum was deleted from the culture medium, the presence of chrysotile enhanced the uptake of BaP by 70%. The serum-free effect alone increased BaP

TABLE 2

BaP UPTAKE (% ± S.D.) BY HUMAN FIBROBLAST

	BaP	BaP + CH	't' test
plus serum	10.1 ± 2.8	8.2 ± 0.6	NS
minus serum	16.0 ± 2.3	27.4 ± 2.2	.001< P <.01
serum effect	.02< P <.05	P <.001	

Human fibroblast 2.5×10^5/dish were seeded for 24 h; treatment (3.3 µg/ml ^{3}H-BaP (0.1 µCi/dish) + 11 µg/ml chrysotile) was also permitted for 24 h.

uptake by 60%. In the presence of chrysotile, the serum-free effect on a 24 h BaP uptake was recorded as a 170% enhancement.

Cytotoxicity of BaP and BaP + Chrysotile

The cultures were divided into four groups at 24 hr after seeding and treated in the presence of 5 µl DMSO per ml serum-free media for 24 h as described in the methods section. Survival was monitored every 24 h by the trypan blue exclusion assay. Duplicate experiments showed that BaP treatment did not significantly enhance the level of cell killing observed by chrysotile alone (data not shown). When total protein content per dish was used as a measure of population growth, it was found that the presence of chrysotile severely retarded the growth (Table 3A and 3B). In contrast, when measured by ^{14}C-thymidine incorporation the additive effect of BaP + chrysotile versus chrysotile alone was not statistically different (Table 4). However, thymidine incorporation in the chrysotile treated cultures was strongly enhanced.

DISCUSSION

By using minimally toxic doses of BaP (3.3 µg/ml) (13) and chrysotile (10 µg/ml) (9), we investigated the interaction of these two agents on a nontransformed human fibroblast cell line. In the absence of serum, the uptake of BaP was significantly enhanced by the presence of chrysotile (Table 2). This phenomenon may relate to the observation that BaP adsorbs onto chrysotile in the absence of serum (Fig. 1A) and can subsequently easily be eluted from the fibers by the addition of 10% serum (Fig. 1B). Previously we have reported that chrysotile has a strong affinity to macromolecules such as bovine serum albumin and nucleic acids (14). We propose here that the desorption of BaP from chrysotile may be due to its replacement by such macromolecules found in both serum and in plasma membrane by competitive binding. This situation may be analogous to the detachment of BaP from its riding vehicle, chrysotile, when it comes into contact with the macromolecules such as glycoproteins under cell free conditions. The demonstration of a fiber facilitated uptake of BaP by an artificial membrane system (6,7) or by a microsomal membrane preparation (7,8) was carried out with fibers onto which an unknown quantity of BaP was precoated by soaking the fibers in a BaP solution with subsequent evaporation of the organic solvent. The precoating is not a good simulation of what may occur naturally in the environment. Hence, in reality, whether an enhanced

TABLE 3A

CHRYSOTILE EFFECTS ON POPULATION GROWTH OF BaP-TREATED HUMAN FIBROBLAST: NO SERUM

Hour post treatment	Protein μg/dish ± S.D.		't' test
	BaP	BaP + Ch	
24	79.7 ± 12.1	79.2 ± 2.3	NS
27	73.0 ± 3.3	63.8 ± 3.2	.05< P <.1
48	130.8 ± 4.9	104.8 ± 8.2	.001< P <.01
72	349.3 ± 18.0	308.8 ± 6.8	.02< P <.05
96	477.7 ± 11.1 485.6 ± 17.1	415.7 ± 119.2 401.7 ± 22.8	NS .001< P <.01

2.5×10^5 cells/dish were seeded for 24 h before treatment of 3.3 μg/ml BaP + 11 μg/ml chrysotile, serum was added at 24 h post treatment.

TABLE 3B

CHRYSOTILE EFFECTS ON POPULATION GROWTH OF BaP-TREATED HUMAN FIBROBLASTS: WITH SERUM

Hour post treatment	Protein μg/dish ± S.D.		't' test
	BaP	BaP + Ch	
24	99.8 ± 9.8	91.8 ± 7.5	NS
27	101.5 ± 11.8	88.5 ± 4.4	NS
48	202.4 ± 12.6	124.6 ± 8.7	P <.001
72	422.1 ± 31.7	306.9 ± 15.4	.001< P <.01
96	514.7 ± 58.4 524.4 ± 11.6	437.3 ± 21.8 474.4 ± 15.2	.05 < P <.1 .01 < P <.02

2.5×10^5 cells/dish were seeded for 24 h before treatment of 3.3 μg/ml BaP + 11 μg/ml chrysotile, 1 ml/dish more complete medium was added at 24 h post treatment.

BaP uptake exist is questionable since the forced adsorption of BaP onto the fiber produces an artificial condition.

When radioactive thymidine incorporation was measured as an indicator of short-term cytotoxicity, no additive effect was observed between BaP and chrysotile at the concentrations used (Table 4). However, when a viable cell count was performed by dye exclusion, a slight enhancement of cytotoxicity

TABLE 4

^{14}C-THYMIDINE INCORPORATION^a (CPM/CELL $\pm$ S.D.) x 10^5

	Cell harvested at hours after treatment^b			
	24	48	72	96
Control	194.9 $\pm$ 25.5^c	541.4 $\pm$ 98.2	341.4 $\pm$ 14.0	29.2 $\pm$ 6.9
	244.9 $\pm$ 27.8^c	340.6 $\pm$ 24.7	298.9 $\pm$ 7.6	30.6 $\pm$ 3.3
BaP	202.3 $\pm$ 54.4	522.0 $\pm$ 119.2	358.7 $\pm$ 25.3	27.4 $\pm$ 0.9
	207.6 $\pm$ 22.6	369.8 $\pm$ 26.4	378.6 $\pm$ 15.7	59.4 $\pm$ 9.1
Chrysotile	599.1 $\pm$ 154.6	782.1 $\pm$ 63.8	431.7 $\pm$ 21.1	136.7 $\pm$ 32.9
	462.3 $\pm$ 57.1	657.8 $\pm$ 42.0	700.2 $\pm$ 72.9	232.4 $\pm$ 26.5
BaP + chrysotile	442.5 $\pm$ 104.6	722.9 $\pm$ 80.7	458.8 $\pm$ 10.5	115.1 $\pm$ 4.9
	504.8 $\pm$ 50.8	621.8 $\pm$ 57.1	606.3 $\pm$ 27.0	212.1 $\pm$ 6.4

^{a} 2.5 x 10^5 human fibroblasts/dish were seeded for 24 h before treatment of 3.3 µg/ml BaP and/or 11 µg/ml chrysotile; serum was added at 24 h post treatment.
^{b} 2 µCi ^{14}C-thymidine was added per dish at 1 h before cell harvesting.
^{c} Two sets of experiment were carried out at different time with identical experimental design.

was observed with additional BaP treatment for up to 72 h post asbestos treatment. By 96 h, the enhancement of cell killing was no longer detectable. The 10 µg/ml chrysotile treatment was the predominant cause of cell killing and it strongly increased thymidine incorporation; whereas the addition of BaP at the concentration used did not statistically enhance the cytotoxicity. Again, when protein content per dish was measured as an index for cytokinetic effects, we found the cotreatment of chrysotile significantly retarded the population growth rate of the culture (Table 3A and 3B). The mechanism of enhanced thymidine incorporation in the chrysotile treated culture is not clear. Since we observed a negligible adsorption of thymidine onto the fibers (Table 1), a faciliated thymidine uptake sufficient to modify the size of the thymidine pool is unlikely.

The extent of polyaromatic hydrocarbon binding to nucleic acid, especially to DNA, has been proposed as an indicative index of transforming potential. Our preliminary data of higher RNA binding of BaP in the presence of chrysotile suggested that the carcinogenic effect of BaP can be further potentiated by co-exposure to certain classes of mineral fibers. Currently we are investigating whether this enhanced RNA binding is simply due to more substrate (BaP) available for the activating enzyme system (resulted from enhanced BaP uptake by the presence of chrysotile) or an altered activating enzyme system subsequent to the chrysotile treatment.

REFERENCES

1. Hammond, E.C., Selikoff, I.J., Lawther, P.L., and Seidman, H. (1976): Inhalation of benzo(a)pyrene and cancer in man, Ann. N.Y. Acad. Sci., 271:116-124.
2. Roe, F.J.C., and Waters, M.A. (1967): Induction of hepatoma in mice by carcinogens of the polycyclic hydrocarbon type, Nature, 214:299-300.
3. Vesselinovitch, S.D., Kyriazis, A.P., Mihailovich, N., and Rao, K.V.N. (1975): Conditions modifying development of tumors in mice at various sites by benzo(a)pyrene. Cancer Res., 35:2948-2953.
4. Tejwani, R., Witiak, D.T., Inbasekaran, M.N., Cazer, F.D., and Milo, G.E. (1981): Characteristics of benzo-(a)pyrene and A-ring reduced 7,12-dimethylbenz(a)anthracene induced neoplastic transformation of human cells in vitro, Cancer Lett., 13:119-127.
5. Selikoff, I.J., Hammond, E.C., and Churg, J. (1968): Asbestos exposure, smoking and neoplasia, JAMA, 204:106-113.
6. Lakowicz, J.R., and Hylden, J.L. (1978): Asbestos-mediated membrane uptake of benzo(a)pyrene observed by fluorescence spectroscopy, Nature, 275:446-448.
7. Kandaswami, C., and O'Brien, P.J. (1980): Effects of asbestos on membrane transport and metabolism of benzo(a)pyrene, Biochem. Biophys. Res. Comm., 97:794-801.
8. Lakowicz, J.R., and Bevan, D.R. (1980): Benzo(a)pyrene uptake into rat liver microsomes: effects of adsorption of benzo(a)pyrene to asbestos and non-fibrous mineral particulates, Chem.-Biol. Interact., 29:129-138.
9. McLemore, T.L., Jenkins, W.T., Arnott, M.S., and Wray, N.P. (1979): Aryl hydrocarbon hydroxylase induction in mitogen stimulated lymphocytes by benzanthracene or cigarette tars adsorbed to asbestos fibers, Cancer Lett., 7:171-177.

10. Daniel, F.B., Beach, C.A., and Hart, R.W. (1980): Asbestos-induced changes in the metabolism of polycyclic aromatic hydrocarbons in human fibroblast cell cultures. In: <u>The In Vitro Effects of Mineral Dusts</u>, edited by R.C. Brown, I.P. Gormley, M. Chamberlain, and R. Davies, pp. 255-262, Academic Press, New York.
11. Kandaswami, C., and O'Brien, P.J. (1981): Pulmonary metabolism of benzo(a)pyrene: Effect of asbestos, <u>Biochem. Pharmacol.</u>, 30:811-814.
12. Bradford, M.M. (1976): A rapid and sensitive method for the quantitation of microgram quantities of protein utilizing the principle of protein-dye binding, <u>Anal. Biochem.</u>, 72:248-254.
13. Marquardt, H., Grover, P.L., and Sims, P. (1976): <u>In vitro</u> malignant transformation of mouse fibroblasts by non-k-region dihydrodiols derived from 7-methylbenz(a)anthracene, 7,12-dimethylbenz(a)anthracene, and benzo(a)pyrene, <u>Cancer Res.</u> 36:2059-2064.
14. Chang, M.J.W., Joseph, L.B., Stephens, R.E., and Hart, R.W. (1981): Modulation of biological processes by mineral fiber adsorption of macromolecules <u>in vitro</u>. In: <u>Proceeding First National Conference 'Health Risks in the Arts, Crafts and Trades'</u> (in press).

COMPARISON OF BENZO[a]PYRENE METABOLISM AND MUTATION INDUCTION IN CHO CELLS USING RAT LIVER HOMOGENATE (S9) OR SYRIAN HAMSTER EMBRYONIC CELL-MEDIATED ACTIVATION SYSTEMS

DAVID JEN-CHI CHEN, RICHARD T. OKINAKA and GARY F. STRNISTE, Genetics Group, Life Sciences Division, Los Alamos National Laboratory, Los Alamos, New Mexico 87545

INTRODUCTION

Many carcinogens/mutagens are chemically non-reactive and have to be metabolically activated by cellular enzymes before they can exert their biological effects (1). Many cell types are not capable of activating such procarcinogens. Among these are Chinese hamster ovary (CHO) cells, which are routinely used for studies on mutagenesis. Mutagenesis in CHO cells has been studied by the addition of an enzymatically active liver homogenate (S9) fraction (2). However, the metabolism of procarcinogens, such as benzo[a]pyrene [BaP], by rat liver homogenate differs from that in intact cellular activation systems. Consequently, BaP-induced mutation frequencies in mammalian cells may vary when different activation systems are used (3,4).

In this study, we are attempting to compare BaP metabolism and conjugation in rat liver homogenate (S9 preparation) and in Syrian hamster embryonic (SHE) cells. Furthermore, a CHO mutation assay incorporating either of the activation systems is being used to measure the mutation induction frequency.

MATERIALS AND METHODS

Cell Culture and Mutagenicity Assay

Chinese hamster ovary cells (CHO-AA8-4) were cultured under conditions described elsewhere (5). The protocols for S9 activation and determining cytotoxicity and mutagenicity at the hypoxanthine-guanine phosphoribosyl transferase (HGPRT) locus has been previously reported (2,6). Primary Syrian hamster embryo (SHE) cells were obtained using the protocol of Pienta et al. (7). For activation by SHE cells, CHO cells (3×10^5 per 60 mm dish) were co-incubated with lethal irradiated (x-ray, 4000 r) SHE cells (2×10^6/60 mm dish) and BaP in αMEM plus 10% fetal calf serum. After 48 h incubation, cytotoxicity and mutagenicity were determined. The plating efficiencies for non-treated CHO cells were regularly between 90 and 100%. The observed mutant frequency is the ratio of mutant colonies per dish to the number of viable cells plated per dish.

Assay of BaP Metabolites

For metabolic activation, S9 protein (0.5 mg/ml), cofactors (2) and 1 µg/ml [^{3}H]-BaP were incubated at 37°C. At noted times, one ml samples were removed and twice extracted with 2.5 volumes of ethyl acetate. The samples were eluted with a water:acetonitrile gradient through an Altex reverse phase ultrasphere-ODS column (4.5 x 150 mm) using a Beckman model 334 high performance liquid chromatography (HPLC) system. The flow rate was 1 ml per min; fractions were collected every 15 sec. For the analysis of SHE cell metabolism, cells at 2×10^6 per 60 mm dish were incubated with 1 µg/ml [^{3}H]-BaP. The resulting metabolites were extracted into ethyl acetate after 24 and 48 h of incubation and were then analyzed by HPLC.

Separation of Conjugated BaP Metabolites on Alumina Columns

BaP metabolites remaining in the aqueous phase after ethyl acetate extraction were separated by a modification of techniques developed by Autrup (8). The aqueous phases were dried, resuspended in 70% ethanol, and applied to alumina columns. The columns were eluted first with 25 ml of ethanol followed by elutions with 25 ml of water, 50 ml of $(NH_4)_2HPO_4$ (pH 3) and finally with 25 ml of 25% formic acid. Five ml fractions were collected and aliquots were assayed for radioactivity.

RESULTS AND DISCUSSION

Using ethyl acetate extraction methods, the S9 fraction shows rapid water solubilization of BaP (> 80% after 2 h incubation) whereas the SHE cells show less than 50% solubilization even after prolonged incubation (24 h) (data not shown). Elution profiles of BaP metabolites from HPLC are shown in Figure 1. The results indicate that BaP metabolites formed in the S9 preparation differ quantitatively and qualitatively from those formed in SHE cells. For SHE cells (Fig. 1, bottom panel), the major BaP metabolite is the 9,10-diol. The 4,5- (or 7,8-) diols and the 7,8,9,10-tetrols (hydrolysis products of the

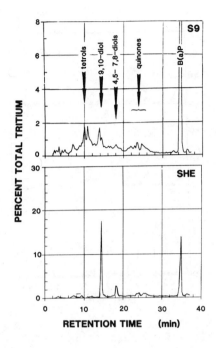

FIGURE 1. HPLC elution patterns of BaP metabolites formed by rat hepatic S9 fraction (top) and SHE cells (bottom). The S9 reactions containing [^{3}H]-BaP were extracted with ethyl acetate at various times up to 2 h after the reaction was started. Dishes containing SHE cells were incubated with [3H]-BaP for 48 h. The resulting metabolites were extracted into ethyl acetate at various times. Samples of standards of BaP metabolites were gifts from the NCI chemical repository.

7,8-diol-9,10-epoxide) are also present in lesser amounts in the medium. In addition to the variety of metabolites mentioned above quinones (primarily the 1,6- and 3,6-) and very hydrophilic derivatives of BaP are produced in the S9 preparation (Fig. 1, top panel).

The pattern of BaP metabolites produced by x-ray treated (4000 r) SHE cells does not change qualitatively as a function of passage number (up to passage 10, data not shown). This pattern in nonirradiated cells, however, does change with the disappearance of the 9,10-diol and the appearance of more hydrophilic species including tetrols (data not shown).

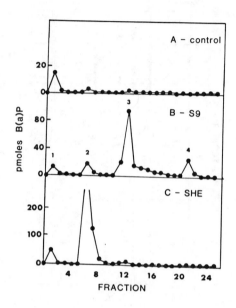

FIGURE 2. Separation of conjugated BaP metabolites on alumina columns (8). The first peak contains primarily non-conjugated BaP and some sulfate conjugates. The metabolites in the 2nd, 3rd, and 4th peaks have been presumptively identified as conjugated sulfates-, glucuronides, and glutathiones, respectively. Panel A: BaP with no activation; panel B: S9 activation (2 h); and panel C: SHE cell activation (48 h).

Chromatographic separations of conjugated BaP species using alumina columns and a 4-step gradient (8) indicate that the two metabolic systems appear to differ significantly in their ability to conjugate BaP to sulfates, glucuronic acid, and glutathione. As shown in Fig. 2 (panel C), SHE cells conjugate BaP to products which have been presumably identified as being sulfate containing. The S9 reaction appears to contain all three forms of conjugated products (Fig. 2, panel B). These data suggest that the S9 preparations possess a greater potential in removing cytotoxic and mutagenic species by conjugation mechanisms. This notion is consistent with our previous results which showed that at a fixed BaP concentration increasing S9 concentrations are followed by the detoxification of cytotoxic and mutagenic events(6).

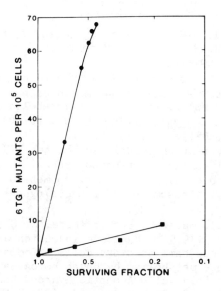

FIGURE 3. 6TG^R mutants induced in CHO cells by BaP activated by rat liver S9 (■) or SHE cells (●) as a function of the surviving fraction of cells.

Figure 3 shows the increase in frequency of induced mutations as a function of cell survival after treatment with BaP activated by S9 preparation or irradiated SHE cells. At 50% survival SHE cell mediated BaP induced mutation frequency is approximately 20-fold higher than that observed with rat liver S9 activation. The lower mutation induction efficiency with S9 activation may be due to its extensive conjugating capacity (Fig. 2). We have also observed that the rate of induction of mutations in CHO is independent of passage number of the x-ray treated feeder layer SHE cells (data not shown). This is consistent with the similar HPLC patterns observed for these x-ray treated cells.

ACKNOWLEDGMENTS

We thank Ms. E. Wilmoth, Ms. J. Bingham and Ms. B. Espinoza for their technical assistance. This work was funded by the Department of Energy and the Environmental Protection Agency.

REFERENCES

1. Miller, A. J. (1970): Carcinogenesis by Chemicals: An overview. G. H. A. Clowes Memorial Lectures. Cancer Res., 30: 559-576.
2. Okinaka, R. T., Barnhart, B. J., and Chen, D. J. (1981): Comparison between sister chromatid exchange and mutagenicity following exogenous metabolic activation of promutagens. Mutation Research, 91: 57-61.
3. Jenssen, D., Beije, B., and Ramel, C. (1979): Mutagenicity testing of Chinese hamster V79 cells treated in the in vitro liver perfusion system. Comparative investigation of different in vitro metabolizing systems with dimethylnitrosamine and benzo(a)pyrene. Chem.- Biol. Interactions, 27: 27-39.
4. Carver, J. H., Salazar, E. P., Knize, M. G., Orwig, and Felton, J. S. (1979): Mutation induction at multiple gene loci in Chinese hamster ovary cells: Comparisons of benzo(a)pyrene metabolism by organ homogenates and intact cells. In: Polycyclic Aromatic Hydrocarbons, edited by M. Cooke and A. J. Dennis, pp. 177-192, Battelle Press, Columbus, Ohio.
5. Strniste, G. F. and Chen, D. J. (1981): Cytotoxic and mutagenic properties of shale oil byproducts. I. Activation of retort process waters with near ultraviolet light. Environ. Mutagenesis, 3: 221-231.
6. Chen, D. J., Okinaka, R. T., Strniste, G. F., and Barnhart, B. J. (1981): Induction of 6-thioguanine resistant mutations by rat liver homogenate (S9)-activated promutagens in human embryonic skin fibroblasts. Mutation Res., (in press).
7. Pienta, R. J., Poiley, J. A. and Lebherz, W. B. (1978): Further evaluation of a hamster embryo cell carcinogenesis bioassay. In: Cancer prevention and detection, H. E. Nieburgs, V. E. O. Valli and A. Kay, Eds., part 1, Vol. 5, pp. 1993-2011, Marcel Dekker, Inc., New York.
8. Autrup, H. (1979): Separation of water-soluble metabolites from cultured human colon. Biochem. pharmacol., 28: 1727-1730.

7-METHYLBENZO[A]PYRENE AND BENZO[A]PYRENE: COMPARATIVE
METABOLIC STUDY AND MUTAGENICITY TESTING IN SALMONELLA
TYPHIMURIUM TA100

PEI-LU CHIU[*], THOMAS K. WONG[*1], PETER P. FU[**], AND SHEN K. YANG[*]
*Department of Pharmacology, School of Medicine, Uniformed Services University of the Health Sciences, Bethesda, Maryland 20814 and **Division of Carcinogenesis, National Center for Toxicological Research, Jefferson, Arkansas 72079 U.S.A.

INTRODUCTION

A methyl substituent at the 7-position of benzo[a]pyrene (BaP) may significantly alter the metabolism at the 7,8,9,10 benzo-ring. Although BaP is metabolically activated to the potent ultimate carcinogen 7,8-dihydrodiol 9,10-epoxides (7, 11), the metabolic activation pathway of 7-methylbenzo[a]-pyrene (7-MBaP) is unknown. Conflicting results exist regarding the relative carcinogenicity of BaP and 7-MBaP. Iyer et al. (5) report that the 7-MBaP has no skin tumor-initiating activity in the 2-stage carcinogenicity testing system in female Sencar mice, whereas Harvey and Dunne (4) indicate that 7-MBaP appears to be as good a sarcomagenic agent as BaP in male Long Evans rats. It was suggested that the metabolic activation of 7-MBaP may occur through pathways other than the in vivo formation of the bay region 7,8-dihydrodiol 9,10-epoxides (4). The aim of this study is to ascertain whether 7-MBaP is metabolically activated at the 7,8,9,10 benzo-ring.

MATERIALS AND METHODS

Materials

BaP 4,5-epoxide, and the BaP trans-7,8-dihydrodiol anti- and syn- 9,10-epoxides were obtained from the Chemical Repository of the National Cancer Institute. 7-MBaP and the 7,8-dihydrodiols of BaP and 7-MBaP were synthesized according to established procedure (3). All chemicals were analyzed by re-

---

[1]present address: Biometry Branch, National Institute of Environmental Health Sciences, Research Triangle Park, North Carolina 27709 U.S.A.

versed-phase HPLC and showed high purity except that the 7-MBaP trans-7,8-dihydrodiol 9,10-epoxides (a mixture of anti- and syn- isomers) were contaminated with approximately 10% of 7-MBaP trans-7,8-dihydrodiol.

Mutagenicity Test

Tester strain TA100 of Salmonella typhimurium was obtained from Dr. B. Ames of the University of California, Berkeley, California. The bacterial mutagenicity assay was conducted and S9 mix was prepared under conditions recommended by Ames et al (2). For assays without S9 mix, each of the testing chemicals (dissolved in DMSO just prior to the testing) in 0.1 ml DMSO was mixed with 0.1 ml of testing bacteria (3 x 10^7 cells in nutrient broth) and 2 ml top agar in a test tube. This mixture was poured on a minimum glucose agar plate. In assays requiring metabolic activation, 0.5 ml of S9 mix containing 30 µl S9 and an NADPH-regenerating system were added (2). The mixture was incubated in a 37°C dry bath for 10 min before the addition of top agar. When 1,1,1-trichloropropene oxide (TCPO) was used, it was added to S9 mix to give a final concentration of 0.2 mM. The number of histidine-independent revertants determined in triplicate was counted after incubating the agar plates in a 37°C incubator for 40 to 48 hrs.

In Vitro Incubations and HPLC Analysis of Metabolites

Liver microsomes were prepared from aroclor 1254 (PCBs)-pretreated male Sprague-Dawley rats (50 mg/kg body weight) (10) This microsomal preparation is referred to as PCB-microsomes. 7-MBaP was incubated with PCB-microsomes under the conditions as previously described (10). Metabolites were analyzed on a Waters Associates model 204 liquid chromatograph fitted with a DuPont Zorbax SIL column (6.2 mm x 25 cm) and were eluted with a 20-min convex gradient (Waters Associates model 660 solvent programer, setting #10) from 40% tetrahydrofuran (THF) in hexane to THF at a solvent flow rate of 2 ml/min. BaP trans-7,8-dihydrodiol and 7-MBaP trans-7,8-dihydrodiol were each incubated in a 25-ml incubation mixture containing 2 µmol of substrate (dissolved in 1 ml acetone), 1.26 mmol of Tris-HCl buffer (pH 7.5), 75 µmol of $MgCl_2$, 50 µmol of $NADP^+$, 50 µmol of glucose-6-phosphate (G-6-P), 2.5 units of G-6-P dehydrogenase (type II, Sigma), and 25 mg protein equivalent of PCB-microsomes. After 30 min of incubation at 37°C, reactions were stopped by adding 25 ml acetone and 10 µg of 7,8,9,10-tetrahydro-BaP (in 20 µl methanol) were added to each incubation flask to serve as an internal standard for chromatography. After extraction with 50 ml of ethyl acetate, the organic layer was dried and the residue was dissolved in 150 µl of methanol for HPLC analysis. In each run, 15 µl of the sample were injected onto a DuPont Zorbax ODS column (4.6 mm x 25 cm). Metabolites were eluted with a 20-min convex gradient (setting #10) from 65 to 95% methanol in water at a solvent flow rate of 1.2 ml/min. The areas under the chromatographic peaks were analyzed by a Hewlett Packard model 3388A integrator. The amount of each metabolite was calculated relative to the area of the internal standard.

RESULTS AND DISCUSSION

Metabolites which were identified from the incubation of 7-MBaP with PCB-microsomes include 4 dihydrodiols, 2 phenols, 2 quinones, 1 hydroquinone, and 7-hydroxymethyl-BaP (Fig. 1). The 3-hydroxy-7-MBaP is the predominant metabolite. Among the dihydrodiols, 7-MBaP 9,10-dihydrodiol is in the largest amount; both 4,5- and 7,8- dihydrodiols of 7-MBaP are minor metabolites. Under similar conditions (1), 4,5-, 7,8-, and 9,10-dihydrodiols of BaP are formed from BaP at much higher amounts than those of 7-MBaP shown in Fig. 1. Thus a methyl substituent at the 7-position of BaP does not hinder the metabolic formation of 7,8-dihydrodiol, but it does inhibit the formation of all dihydrodiols. Similar results were also reported by Jeffrey et al (6).

In the Ames system, the mutagenicities of BaP and 7-MBaP are comparable (Fig. 2A). In the presence of 0.2 mM TCPO which effectively abolishes the formation of all dihydrodiols, the mutagenicity of BaP is reduced by 10-15%, but the mutagenicity of 7-MBaP remains approximately the same (Fig. 2B). Both BaP 4,5-epoxide and BaP 7,8-dihydrodiol 9,10-epoxides are potent mutagens in the S. typhimurium TA100 system (8,9). The results in Figs. 2A and 2B suggest that the mutagenicities of BaP and 7-MBaP are primarily caused by epoxides and/or other metabolites that are formed enzymatically from phenolic intermediates. The mutagenicities apparently are not caused by the metabolic formation of the bay region 7,8-dihydrodiol 9,10-epoxides because 7,8-dihydrodiol was not formed due to the presence of TCPO. Under our experimental conditions, aryl hydrocarbon hydroxylase (AHH) activity of PCB-microsomes was not inhibited by the presence of TCPO.

Trans-7,8-dihydrodiols of both BaP and 7-MBaP have comparable mutagenic activities at doses up to 1 µg/plate (Fig. 2C). BaP trans-7,8-dihydrodiol is more toxic to the bacteria at higher doses. This is due to the highly toxic property of BaP 7,8-dihydrodiol 9,10-epoxides (9) which are formed from the metabolism of BaP 7,8-dihydrodiol. These results suggest that the mutagenicity of 7-MBaP 7,8-dihydrodiol may be caused by the metabolic formation of the 7,8-dihydrodiol 9,10-epoxides and/or by some other metabolites.

To further understand the causes of the mutagenicity of 7-MBaP trans-7,8-dihydrodiol, the metabolites of 7-MBaP trans-7,8-dihydrodiol with PCB-microsomes were compared with those of

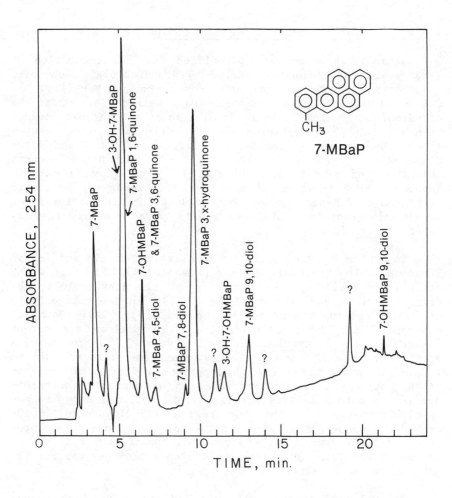

FIGURE 1. Normal-phase HPLC separation of 7-MBaP metabolites. The chromatographic conditions are described in Materials and Methods.

BaP trans-7,8-dihydrodiol (Fig. 3). Five metabolites (A1-A5 in Fig. 3A) formed from the incubation of 7-MBaP trans-7,8-dihydrodiol have uv absorption spectra characteristic of a pyrene chromophore. Chromatographic peaks A2, A3, and A4 cochromatographed with the hydrolysis products (tetrols) of chemically synthesized 7-MBaP trans-7,8-dihydrodiol 9,10-epoxides (Fig. 3A). Under the same incubation conditions, four 7,8,9,10-tetrols (B1-B4 in Fig. 3B) were formed from the metabolism of BaP 7,8-dihydrodiol. When the peak areas of the internal standard in Figs. 3A and 3B were normalized, the amount of tetrols formed from 7-MBaP trans-7,8-dihydrodiol was about half of that formed from BaP trans-7,8-dihydrodiol. Because the tetrol formation is the indicator of the metabolic formation of 7,8-dihydrodiol 9,10-epoxides, the results in Figs. 3A and 3B indicate that the formation of 7,8-dihydrodiol 9,10-epoxides from 7-MBaP trans-7,8-dihydrodiol is only about half of that from BaP trans-7,8-dihydrodiol. Even though less 7,8-dihydrodiol 9,10-epoxide is formed from 7-MBaP 7,8-dihydrodiol, the mutagenicity of 7-MBaP 7,8-dihydrodiol may still be due to a higher mutagenic potency of the 7,8-dihydrodiol 9,10-epoxides formed.

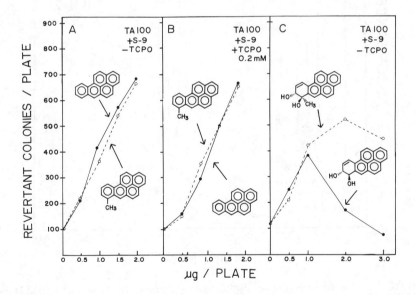

FIGURE 2. Mutagenic activities of BaP, 7-MBaP, BaP trans-7,8-dihydrodiol, and 7-MBaP trans-7,8-dihydrodiol in S. typhimurium TA100. In (B), TCPO was used at 0.2 mM in addition to the S9 mix.

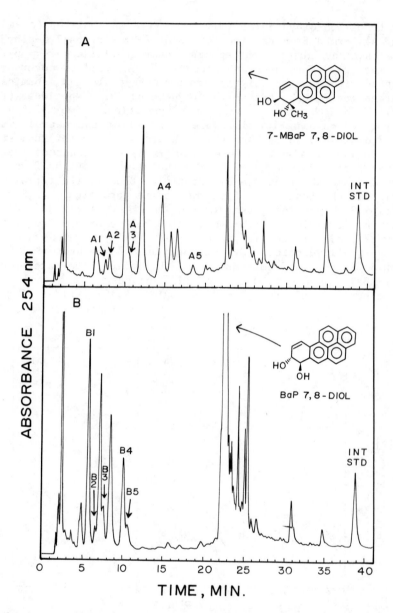

FIGURE 3. Reversed-phase HPLC separation of the metabolites of (A) 7-MBaP trans-7,8-dihydrodiol and of (B) BaP trans-7,8-dihydrodiol. The metabolites were eluted under conditions described in Materials and Methods. The peak areas of the internal standards (INT STD) have been normalized so that the relative amount of tetrols formed can be readily recognized in this figure.

The mutagenic potency of 7-MBaP 7,8-dihydrodiol 9,10-epoxides was determined by Ames test and was compared with those of BaP 7,8-dihydrodiol anti- and syn- 9,10-epoxides and BaP 4,5-epoxide (Fig. 4). At 0.1 µg/plate, the mutagenicity of 7-MBaP 7,8-dihydrodiol 9,10-epoxides is only 20% that of BaP 7,8-dihydrodiol anti-9,10-epoxide; 25% that of BaP 7,8-dihydrodiol syn-9,10-epoxide; and 65% that of BaP 4,5-epoxide. Although the 7-MBaP 7,8-dihydrodiol 9,10-epoxides were contaminated with 10% of 7-MBaP 7,8-dihydrodiol, the mutagenicity of 7-MBaP 7,8-dihydrodiol 9,10-epoxides should not be significantly affected by this contamination because 7-MBaP trans-7,8-dihydrodiol is not mutagenic in the absence of metabolic activation.

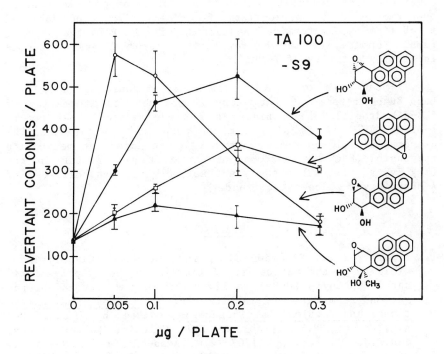

FIGURE 4. Mutagenic activities of BaP trans-7,8-dihydrodiol anti-9,10-epoxide, BaP 7,8-dihydrodiol syn-9,10-epoxide, BaP 4,5-epoxide, and 7-MBaP 7,8-dihydrodiol 9,10-epoxides (anti- and syn- isomers) in S. typhimurium TA100 in the absence of metabolic activation.

The above experimental results suggest that the presence of a methyl group at the 7-position of BaP greatly reduces the metabolism at the 7,8,9,10 benzo-ring. Our results indicate: (i) trans-7,8-dihydrodiol is a minor metabolite of 7-MBaP, (ii) 7,8-dihydrodiol 9,10-epoxides are minor metabolites of 7-MBaP 7,8-dihydrodiol, and (iii) 7-MBaP 7,8-dihydrodiol 9,10-epoxides are weak mutagens. Based on these observations, it is concluded that the 7-MBaP trans-7,8-dihydrodiol 9,10-epoxides are not the major metabolites responsible for the mutagenic activity of 7-MBaP.

The results of our metabolic study may be used to interpret the lack of tumor-initiating activity of 7-MBaP in female Sencar mice (5). Female Sencar mice are sensitive to epidermal carcinogenesis in the two-stage initiation and promotion sytem. The skin of female Sencar mice may be insensitive to the carcinogenic effect caused by metabolites formed at positions other than the 7,8,9,10 benzo-ring of 7-MBaP. Thus the lack of metabolism at the terminal benzo-ring of 7-MBaP by the skin of female Sencar mice may be responsible for the lack of tumor-initiating activity of 7-MBaP.

7-MBaP is a relatively potent sarcomagenic agent in male Long Evans rats (4). Our study showed that the formation of 7,8-dihydrodiol 9,10-epoxides is a minor metabolic pathway of 7-MBaP. If this is also a minor pathway in vivo, the 7,8-dihydrodiol 9,10-epoxides must not be an important metabolite responsible for the sarcomagenic activity of 7-MBaP in the Long Evans rats. Thus the exact sarcomagenic metabolite(s) of 7-MBaP is yet to be established.

REFERENCES

1. Alvares, A.P., Eiseman, J.L., and Yang, S.K. (1980): Inducibility of the metabolism of benzo[a]pyrene by polychlorinated biphenyls (PCBs) in liver and lung microsomal preparations from rats and rabbits. In: Microsomes, Drug Oxidations, and Chemical Carcinogenesis, edited by M.J. Coon, A.H. Conney, R.W. Estabrook, H.V. Gelboin, J.R. Gillette, and P.J. O'Brien, pp. 1207-1210, Academic Press, New York.
2. Ames, B.N., McCann, J., and Yamasaki, E. (1975): Methods for detecting carcinogens and mutagens with the Salmonella/mammalian-microsome mutagenicity test, Mut. Res., 31:347-364.
3. Fu, P.P., Lai, C.-C., and Yang, S.K. (1980): Synthesis of cis- and trans- 7,8-dihydrodiols of 7-methylbenzo[a]pyrene, J. Org. Chem., 46:220-222.
4. Harvey, R.G., and Dunne, F.B. (1978): Multiple regions of

metabolic activation of carcinogenic hydrocarbons, Nature, 273:566-568.
5. Iyer, R.P., Lyga, J.W., Secrist,III, J.A., Daub, G.H., and Slaga, T.J. (1980): Comparative tumor-initiating activity of methylated benzo[a]pyrene derivatives in mouse skin, Cancer Res., 40:1073-1076.
6. Jeffrey, A.M., Kinoshita, T., Santella, R.M., Grunberger, D., Katz, L., and Weinstein, I.B. (1980): The chemistry of polycyclic aromatic hydrocarbon-DNA adducts. In: Carcinogenesis: Fundamental Mechanisms and Environmental Effects, edited by B. Pullman, P.O.P. Ts'o and H. Gelboin, pp. 565-579, D. Reidel Publishing Co., Dordrecht-Holland.
7. Sims, P., Grover, P.L., Swaisland, A., Pal, K., and Hewer, A. (1974): Metabolic activation proceeds by a diol-epoxide, Nature, 252:326-328.
8. Wislocki, P.G., Wood, A.W., Chang, R.L., Levin, W., Jerina, D.M., and Conney, A.H. (1976): Mutagenicity and cytoxicity of benzo[a]pyrene arene oxides, phenols, quinones, and dihydrodiols in bacterial and mammalian cells, cancer Res., 36:3350-3357.
9. Wislocki, P.G., Wood, A.W., Chang, R.L., Levin, W., Yagi, H., Hernandez, O., Jerina, D.M., and Conney, A.H. (1976): High mutagenicity and toxicity of a diol epoxide derived from benzo[a]pyrene, Biochem. Biophys. Res. Commun., 68(3): 1006-1012.
10. Wong, T.K., Chiu, P.-L., Fu, P.P., and Yang, S.K. (1981): Metabolic study of 7-methylbenzo[a]pyrene with rat liver microsomes: Separation by reversed-phase and normal-phase high performance liquid chromatography and characterization of metabolites, Chem.-Biol. Interact., 38:153-166.
11. Yang, S.K., McCourt, D.W., Roller, P.P., and Gelboin, H.V. (1976): Enzymatic conversion of Benzo[a]pyrene leading predominantly to the diol-epoxide r-7,t-8-dihydroxy-t-9,10-oxy-7,8,9,10-tetrahydrobenzo[a]pyrene through a single enantiomer of r-7,t-8-dihydroxy-7,8-dihydrobenzo[a]pyrene, Proc. Natl. Acad. Sci. USA, 73:2594-2598.

A STRUCTURE-ACTIVITY RELATIONSHIP STUDY OF MONOMETHYLBENZO[A]-
PYRENES BY THE USE OF SALMONELLA TYPHIMURIUM TESTER STRAIN
TA100 AND BY ANALYSIS OF METABOLITE FORMATION

PEI-LU CHIU AND SHEN K. YANG
Department of Pharmacology, School of Medicine, Uniformed Services University of the Health Sciences, Bethesda, Maryland 20814
U.S.A.

INTRODUCTION

Monomethylbenzo[a]pyrenes (MBaPs) have been used to elucidate the site(s) responsible for the metabolic activation of BaP (2,4). It was reported that 2-, 3-, 4-, 5-, 6-, 7-, 11-, and 12-MBaPs have equal or higher sarcomagenic activities as compared with that of BaP (5,7,9,11,12). 8-MBaP and 10-MBaP were found to be inactive (8). Iyer et al. (4) found that methylation at the 7-, 8-, 9-, and 10- positions of BaP abolished the tumor-initiating activities (Fig. 1). These results are consistent with the findings that the predominant carcinogenic metabolite is formed at the 7,8,9,10 positions of BaP. To date very little data exist regarding the metabolic pathways of MBaPs. A methyl substituent at the 7,8,9,10 positions of BaP may sterically hinder the metabolic formation and the biological activity of 7,8-dihydrodiol 9,10-epoxide. On the other hand, methyl substitution at other positions may favor the metabolic formation and/or biological activity of 7,8-dihydrodiol 9,10-epoxide. Thus we initiated metabolic studies of twelve MBaPs, and mutagenicity tests with S. typhimurium tester strain TA100 for identifying biologically reactive metabolite(s). This linked approach, for example, has enabled us to answer key questions regarding the biological activities of 7-MBaP (3,13). This paper reports the mutagenic activity testings of twelve MBaPs and the high perfor-

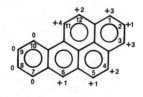

(BaP +2)

FIGURE 1. Relative tumor-initiating activities of monomethylbenzo[a]pyrenes (8). The activity of BaP is ranked +2.

mance liquid chromatographic (HPLC) analysis of metabolites formed from 1-MBaP and 4-MBaP by rat hepatic enzymes.

MATERIALS AND METHODS

Materials

7-MBaP and 10-MBaP were synthesized through 9,10-dihydrobenzo[a]pyrene-7(8H)-one and 7,8-dihydrobenzo[a]pyrene-(9H)10-one respectively according to established procedures (6,13). Other MBaPs were obtained from the Chemical Repository of the National Cancer Institute. The purity of each MBaP was checked by reversed-phase HPLC and showed >96% pure.

Mutagenicity Testing

Salmonella typhimurium strain TA100 was obtained from Dr. B. Ames of the University of California, Berkeley, California. S9 mixture containing an NADPH-regenerating system was prepared according to the method of Ames et al (1). Testing bacteria were grown in nutrient broth in a 37°C water bath shaker for 6 hrs prior to test. In the absence of metabolic activation, the testing sample in 0.1 ml DMSO was mixed with 0.1 ml testing bacteria (5 x 10^7 cells), and 2 ml top agar. The mixture was poured on a minimum glucose agar plate, and the plate was incubated in a 37°C incubator for 40-48 hrs before the revertant colonies on the plate were counted. The number of revertant colonies represents the average of triplicate plates. To test compounds that require metabolic activation, 0.5 ml S9 mixture was added to the testing chemical and bacteria. The mixture was then incubated at 37°C in a dry bath for 10 min before the addition of top agar. When 1,1,1-trichloropropene oxide (TCPO) was used, it was added to S9 mix to give a final concentration of 0.2 mM. This concentration of TCPO did not inhibit the aryl hydrocarbon hydroxylase activity, nor was it mutagenic to the testing bacteria.

In Vitro Metabolism and HPLC Analysis of 1-MBaP and 4-MBaP

Substrate (1-MBaP or 4-MBaP, 0.8 µmol in 0.2 ml DMSO) was added to 48 ml S9 mix containing 5 mmol of sodium phosphate buffer (pH 7.4), 0.4 mmol of $MgCl_2$, 1.65 mmol of KCl, 0.25 mmol of glucose-6-phosphate, 0.2 mmol of $NADP^+$, and 3 ml of S9. The mixture was incubated at 37°C in a water bath shaker for 10 min. At the end of incubation, the reaction was stopped by adding 50 ml of acetone. Ten µg of 7,8,9,10-tetrahydro-BaP were added to serve as an internal standard for chromatography. Ethyl acetate (100 ml) was then added to extract the substrate and metabolites. The solvent of the organic phase was removed under reduced pressure and the residue was dissolved in 0.15 ml of methanol. The metabolites were injected onto a µBondapak C_{18} column (3.9 mm x 30 cm, Waters Associates) fitted on a Spectra-Physics model 3500B liquid chromatograph. The metabolites were eluted with a 40-min linear gradient from 65% methanol in water to methanol at a flow rate of 1 ml/min. The eluted metabolites were tentatively identified by comparing their uv absorption spectra with those of BaP derivatives.

RESULT AND DISCUSSION

None of the twelve MBaPs are mutagenic in the absence of metabolic activation. When activated by hepatic S9 fractions, the mutagenic activities of the MBaPs are as follows (in a decreasing order): 4-MBaP > 9-MBaP > 6-MBaP ≃ 11-MBaP > BaP ≃ 8-MBaP ≃ 7-MBaP > 2-MBaP ≃ 3-MBaP ≃ 5-MBaP ≃ 10-MBaP > 12-MBaP ≃ 1-MBaP. It is apparent that the relative mutagenic activities shown in Fig. 2 are not consistent with the relative tumor-initiating activities reported by Iyer et al. (8). It has been reported that the carcinogenic activity of BaP is mainly due to the metabolic formation of the 7,8-dihydrodiol 9,10-epoxide (10), whereas the bacterial mutagenic activity of BaP is primarily due to the metabolic formation of the K-region 4,5-epoxide (14). Thus the lack of correlation between the bacterial mutagenic activity and tumor-initiating activity of the MBaPs may be due to different biological responses of these two testing systems to different reactive metabolites.

Both 1-MBaP and 4-MBaP are potent tumor-initiating agents on mouse skin (8). In contrast, when compared to BaP, 4-MBaP is metabolized to become a more potent bacterial mutagen, whereas 1-MBaP is metabolized to become a weaker bacterial mutagen (Fig. 2). In the presence of 0.2 mM TCPO, the mutagenic activity of 1-MBaP remained unchanged, whereas a reduction of approximately 20% was observed for 4-MBaP (Table 1). The formation of dihydrodiol was completely abolished when 0.2 mM TCPO was incubated in the testing mixture (Fig. 3). Because 7,8-dihydrodiol 9,10-epoxides cannot be formed in the presence of TCPO, the mutagenic activities of 1-MBaP and 4-MBaP must be primarily caused by epoxides, phenols, and/or metabolites formed by further oxygenation. The reduction in mutagenic activity of 4-MBaP by TCPO suggests that the metabolites derived by further metabolism of dihydrodiols also contributed to the mutagenic activity of 4-MBaP when tested in the absence of TCPO.

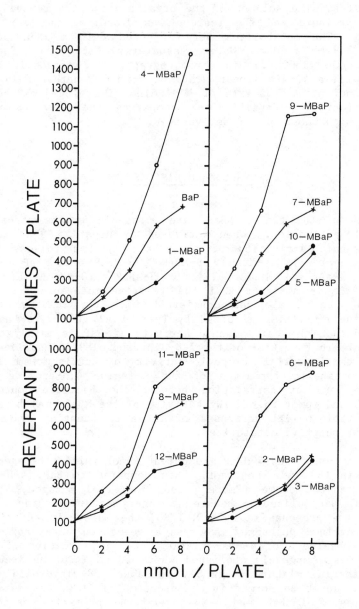

FIGURE 2. A comparison of mutagenicities of BaP and twelve isomeric monomethylbenzo[a]pyrenes in <u>Salmonella typhimurium</u> tester strain TA100. The mutagenicity tests were carried out under conditions described in <u>Materials and Mathods</u>.

TABLE 1

MUTAGENIC ACTIVITIES OF 1-MBaP and 4-MBaP IN THE PRESENCE AND ABSENCE OF TCPO IN SALMONELLA TYPHIMURIUM TA100[a]

nmol/plate	His^+ revertants / plate			
	1-MBaP		4-MBaP	
	-TCPO	+TCPO	-TCPO	+TCPO
0	84	91	84	91
2	114	147	187	117
4	148	161	375	211
6	230	230	651	457
8	282	254	1023	810

[a]TCPO was used at 0.2 mM.

BaP 4,5-epoxide was suggested to be the major mutagenic metabolite of BaP in the S. typhimurium testing system (14). Although 4-MBaP is the most potent mutagen (Fig. 2), the K-region 4,5-epoxide may not be formed from 4-MBaP because of the methyl substitution. We compared the metabolites formed from 1-MBaP and 4-MBaP by hepatic S9 fractions in the presence and absence of TCPO (Fig. 3).

The metabolites formed from incubation of 1-MBaP by hepatic S9 fractions are 3-hydroxy-1-MBaP, 1-MBaP 3,6-quinone, 1-hydroxymethyl-BaP, 1-MBaP 7,8-, 4,5-, and 9,10- dihydrodiols (Fig. 3A, solid curve). Dihydrodiols were not detected when the incubation was carried out in the presence of 0.2 mM TCPO (Fig. 3A, dotted curve) but the other metabolites were not significantly affected.

The metabolic formation of 3-hydroxy-4-MBaP was drastically inhibited because of the peri 4-methyl group (Fig. 3B, solid curve). Major metabolites include 4-MBaP 7,8- and 9,10- dihydrodiols, 4-MBaP 1,6-quinone, 9- and 7- hydroxy-4-MBaP, and 4-hydroxymethy-BaP. The K-region 4,5-dihydrodiol was not detected. Again, in the presence of TCPO, dihydrodiols were completely abolished with concomitant increased formation of 9-hydroxy-4-MBaP (Fig. 3B, dotted curve). The absence of 4,5-dihydrodiol as a metabolite of 4-MBaP suggests that a methyl substitution at the K-region double bond of BaP can sterically block the oxidative metabolism at the methyl-substituted double bond. Methyl substitution at the K-region of benz[a]anthracene also blocks the metabolic formation of K-region dihydrodiols (15).

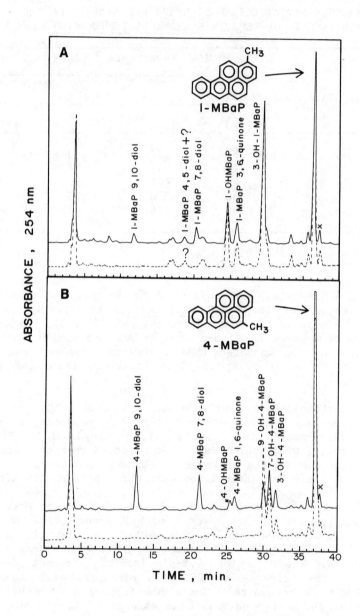

FIGURE 3. Reversed-phase HPLC separation of metabolites from 1-MBaP (A) and 4-MBaP (B) in the presence (———) and absence (-----) of 0.2 mM TCPO. Internal standard is indicated by X.

The results indicate that the major MBaP metabolites responsible for the mutagenic activity in S. typhimurium may include epoxides, dihydrodiols, phenols, and/or their further enzymatic products. The mutagenic activity of BaP in S. typhimurium tester strains may be primarily due to the metabolic formation of K-region 4,5-epoxide (14). However, the K-region 4,5-epoxide is not responsible for the mutagenic activity of 4-MBaP because it is not formed metabolically. 7,8-Dihydrodiol is a major metabolite for both 1-MBaP and 4-MBaP. Thus it is possible that 7,8-dihydrodiol 9,10-epoxides are the ultimate carcinogenic metabolites of 1-MBaP and 4-MBaP.

The results in this report indicate that no correlation can be found between the bacterial mutagenic activity and tumor-initiating activity of BaP and the twelve MBaPs. Further metabolic study and biotesting of the parent compounds and their metabolites must be carried out to reveal the molecular basis for the differences in carcinogenicity of the MBaPs.

REFERENCES

1. Ames, B.N., McCann, J., and Yamasaki, E. (1975): Methods for detecting carcinogens and mutagens with the Salmonella/mammalian-microsome mutagenicity test, Mut. Res., 31:347-364.
2. Arcos, J.C., and Argus, M.F. (1974): Chemical Induction of Cancer, Vol. IIA: Structural Bases and Biological Mechanisms. Academic Press, Inc. New York, pp. 34-39.
3. Chiu, P.-L., Wong, T.K., Fu, P.P., and Yang, S.K. (1982): 7-Methybenzo[a]pyrene and benzo[a]pyrene: Comparative metabolic study and mutagenicity testing in Salmonella typhimurium TA100, In: This volume.
4. Dipple, A. (1976): Polynuclear aromatic carcinogens. In: Chemical Carcinogens, edited by C.E. Searle, pp. 245-314, American Chemical Society, Washington, D.C.
5. Dunlap, C.E., and Warren, S. (1941): Chemical configuration and carcinogenesis, Cancer Res., 1:953-954.
6. Fu, P.P., Lai, C.-C., and Yang, S.K. (1980): Synthesis of cis- and trans-7,8-dihydrodiols of 7-methylbenzo[a]pyrene. J. Org. Chem., 46:220-222.
7. Harvey, R.G., and Dunne, F.B. (1978): Mutiple regions of metabolic activation of carcinogenic hydrocarbons. Nature, 273: 566-568.
8. Iyer, R.P., Lyga, J.W., Secrist,III, J.A., Daub, G.H., and Slaga, T.J. (1980): Comparative tumor-initiating activity of methylated benzo[a]pyrene derivatives in mouse skin, Cancer Res., 40:1073-1076.

9. Lacassagne, A., Zajdela, F., Buu-Hoi, N.P., Chalvet, O., and Daub, G.H. (1968): High carcinogenic activity of mono-, di-, and trimethyl-benzo[a]pyrene, Int. J. Cancer, 3:238-243.
10. Levin, W., Wood, A.W., Yagi, H., Jerina, D.M., and Conney, A.H. (1976): (±)-trans-7,8-dihydroxy-7,8-dihydrobenzo[a]pyrene: a potent skin carcinogen when applied topically to mice, Proc. Natl. Acad. Sci, USA, 73:3867-3871.
11. Schurch, O., and Winsterstein, A. (1935): Uber die Krebserregende Wirkung aromatischer Kohlenwasserstoffe, Ztschr. Physiol. Chem., 236:79-91.
12. Shear, M., and Perrault, A. (1939): Studies in carcinogenesis VII. compounds related to 3:4-benzpyrene, Amer. J. Cancer, 36:211-228.
13. Wong, T.K., Chiu, P.-L., Fu, P.P., and Yang, S.K. (1981): Metabolic study of 7-methylbenzo[a]pyrene with rat liver microsomes: separation by reversed-phase and normal-phase high performance liquid chromatography and characterization of metabolites, Chem.-Biol. Interact., 36:153-166.
14. Wood, A.W., Levin, W., Lu, A.Y.H., West, S.B., Yagi, H., Mah, H.D., Jerina, D.M., and Conney, A.H. (1977): Structural requirement for the metabolic activation of benzo[a]pyrene to mutagenic products: effects of modifications in the 4,5-, 7,8-, and 9,10- positions, Mol. Pharmacol., 13:1116-1125.
15. Yang, S.K., Chou, M.W., and Fu, P.P. (1981): Microsomal oxidations of methyl-substituted and unsubstituted aromatic carbons of monomethylbenz[a]anthracenes, In: Polynuclear Aromatic Hydrocarbons, edited by M. Cooke, and A.J. Dennis, pp. 253-264, Battelle Press, Columbus, Ohio.

SHPOL'SKII SPECTRA OF POLYCYCIC AROMATIC COMPOUNDS IN SAMPLES FROM CARBON BLACK AND SOIL.

ANDERS L. COLMSJÖ AND CONNY E ÖSTMAN
University of Stockholm, Dept of Analytical Chemistry, S-10691 Stockholm, Sweden.

INTRODUCTION

Low temperature fluorescence techniques utilizing the Shpol'skii effect have for a number of years been applied to the analysis of polycyclic aromatic hydrocarbons (1-7). The Shpol'skii effect, which is characterized by high accuracy and sensitivity in fingerprint identification of certain compounds has by now been well documented and reference spectra of more than one hundred polycyclic aromatic compounds (PACs) are accessible (8). The use of the Shpol'skii effect in the analysis of heterocyclic aromatic compounds has hitherto not been well reported. This is due partially to the lack of standard substances and partially to the fact that many heterocyclic compounds exhibit poorly resolved spectra or have low quantum efficiency. This paper deals with the characterisation of a number of hitherto unreported spectra that have appeared in carbon black and soil samples.

MATERIALS AND METHODS

A home-built low temperature fluorescence equipment described elsewhere (9) was used for recording the spectra. The soil and carbon black samples were treated by Soxhlet extraction and a clean up procedure. After clean up, the samples were injected on a HPLC equipped with a reversed phase partition system (ODS). Eluted fractions were collected by the aid of a UV-detector response, evaporated and dissolved in n-hexane.

RESULTS

When registering quasilinear low temperature fluorescence spectra from eluted HPLC-fractions, a large number of well resolved peaks can be observed, which emanate from compounds not found amongst the most commonly registered PAHs. As the standard library of reference spectra grew larger and finally contained all condensed benzene ring systems of interest up to five rings, it became evident that these spectra emanated from substituted parent compounds or heterocyclic aromatic compounds.

Methylated PAHs

At first a special study was made of methyl substituted PAHs which provided an explanation of some unidentified spectra and also gave valuable information about the relation between the Shpol'skii spectra from a parent compound and its methyl derivatives. For instance, fig 1 shows a reference Shpol'skii spectrum of 1-methylcoronene that was synthesized. The typical red-shift of a few nanometers and an enhancement of the 0-0' transition compared with the parent compound (8) can clearly be discerned. This spectrum has previously been reported as emanating from an unidentified compound present in automobile exhausts (15) and has in this work also been shown to be present in carbon black. The necessity for a good reference library of quasilinear fluorescence spectra has been stressed by the fact that two earlier commersially available batches of 4-methylpyrene were shown to be the 2-methyl isomer of pyrene. This was revealed after achieving all three possible isomers expertly synthesized by Arne Berg in Århus, Denmark. The three spectra are quite different and can easily be distinguished (8). Identification of the isomers was also made by using the PMO theory together with unspecific methylation of the pyrene molecule (11).

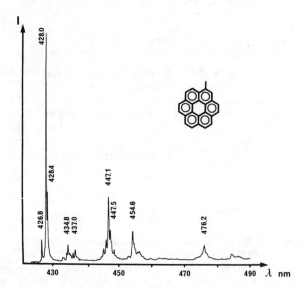

FIGURE 1. Fluorescence spectrum of 1-methylcoronene in n-hexane at 63 K. The compound was detected in automobile exhausts and in carbon black.

Nitrogen-containing heterocyclic aromatic compounds

The quasilinear fluorescence from compounds containing at least one nitrogen atom in the aromatic ring system are, except from benzoacridines, rarely reported. This is due mostly to the lack of reference compounds. Nevertheless, one very characteristic spectrum of a compound obtained from the soil collected close to the edge of an airport runway in Stockholm, emanates from 1-azapyrene, fig 2. The spectrum shows a slightly blue-shifted 0-0' transition compared to that of pyrene (8). The typical triplett structure of the emission spectrum, derived from three fluorescence spectra slightly shifted in energy compared to each other, emanates from three site distributions within the solute-solvent crystal structure. This compound was the origin of the dominating peak when the PAC fraction of the soil sample was analysed by glass-capillary gas-chromatography with a nitrogen-specific detector. The compound was not observed in the chromatogram derived with a flame ionization detector because of interfering PACs.

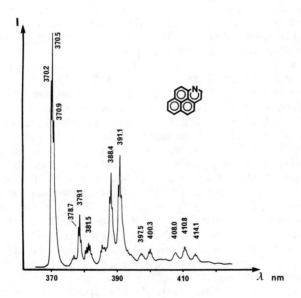

FIGURE 2. Fluorescence spectrum of 1-azapyrene in n-hexane at 63 K. The compound was detected in a soil sample emanating from the edge of an airport runway.

Sulfur-containing heterocyclic aromatic compounds

A large number of spectra emanating from unidentified compounds in HPLC-fractions of samples of various types could not, in spite of the investigations of substituted PAHs and nitrogen-containing aromatic compounds, be finally interpreted. Many other types of compounds, showing quasilinear fluorescence, obviously passed through all clean up procedures and showed retentions on reversed phase HPLC and GC similar to "pure" PAHs. Usually the very complex compositions of compounds in the HPLC-fractions from samples makes correlation studies between GC-MS and fluorescence analysis impossible. Nevertheless, from a sample consisting of selected carbon black a number of compounds, exhibiting very well resolved Shpol'skii spectra at 63 K, could be isolated.

Elementary analysis of the carbon black sample showed that one per cent of the weight consisted of sulfur. Mass spectrometrical analysis of the compounds exhibiting hitherto unreported Shpol'skii spectra, gave the result that almost all of these compounds showed a M-45 fragment and an enhanced M+2 peak which is characteristic for sulfur-containing heterocyclic aromatic compounds (S-PAC), (12). After a correlation study of the mass spectra, room temperature and cryofluorescence spectra, UV-spectra, chromatographic retention data on HPLC and GC and published data in the field (13, 14), a number of hitherto unknown Shpol'skii spectra could be ascertained (10). According to a theory of formation, proposed by Karcher <u>et. al.</u> (13), sulfur-substituted analogs to PAHs are formed simultaneously if sulfur is present in the formation process. This would indicate that sulfur-substituted analogs to pyrene, benzo(a)pyrene, benzo(e)pyrene, benzo(ghi)perylene, anthanthrene and coronene should be present, since these PAHs together with cyclopenta(cd)pyrene and benzo(ghi)fluoranthene are the major components of the PAC fraction, fig 3. This was indeed found to be the case, which was established in this investigation primarily for the reason that the peri-condensed thiophenes found exhibited well-resolved quasilinear fluorescence spectra at 63 K, (in contrast to kata-condensed thiophenes which show poor fluorescence, such as dibenzo(b,d)thiophene or dinaphto(2,1,1',2')thiophene). If the spectra of these sulfur-containing heterocyclic compounds are to be compared with those of "pure" PAHs, it is notable that the spectra of the sulfur-containing heterocyclic compounds can in most cases be observed as blue-shifted spectra of their PAH analog (-S- replaced with -C=C-) and not as red-shifted spectra of their parent compounds. Fig 3 shows that the blue-shift of an S-PAC with respect to its PAH analog is approximately 12-18 nm, with one exception.

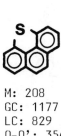
M: 208
GC: 1177
LC: 829
0-0': 356.2

M: 202
GC: 1213
LC: 898
0-0': 371.4

Δλ

15.2

M: 258
GC: 1652
LC: 2060
0-0': 390.2

M: 252
GC: 1696
LC: 1798
0-0': 402.4

12.2

M: 258
GC: 1647
LC: 1548
0-0': 385.9

M: 252
GC: 1696
LC: 1798
0-0': 402.4

16.5

M: 258
GC: 1662
LC: 1500
0-0': 358.2

M: 252
GC: 1689
LC: 1506
0-0': 376.2

18.0

FIGURE 3. Sulfur containing heterocyclic aromatic compounds detected in a carbon black sample and their PAH analogs. Fig. explanation; M: molecular weight; GC: retention time on an unpolar methyl silicone gum capillary column temperature pro-

SHPOL'SKII SPECTRA

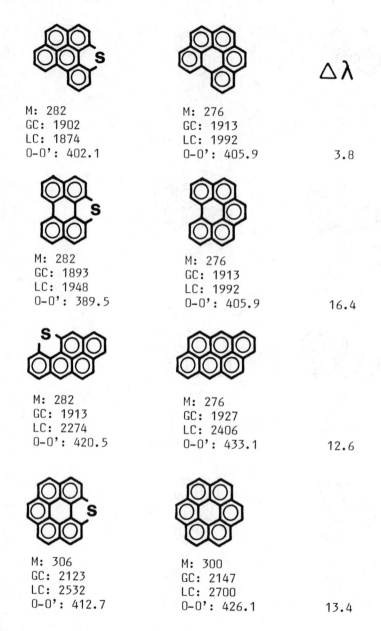

FIGURE 3. (continued) grammed from 68°C (1 min) to 290 °C (7°C/min); LC: retention time on an HC-ODS column gradient eluted from 50%/50% (AN/H$_2$O) (1 min) to 100% AN (30 min), flow rate 1.2 ml/min.; 0-0': emission wavelength of the 0-0'-transition (nm).

The only compound that did not show the characteristic mass fragmentation of a sulfur-containing aromatic compound was the compound assumed to be 7,8 -thiabenzo(ghi)perylene giving the Shpol'skii spectrum shown in fig 4. This compound, together with 1,12-thiaperylene, also exhibited well resolved quasilinear phosphorescence spectra. The assumed structure of the compound in fig 4 was later also verified by the synthesis of 7,8 -thiabenzo(ghi)perylene.

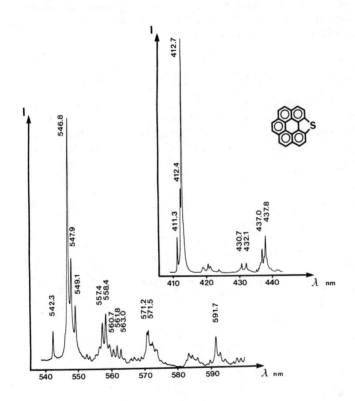

FIGURE 4. Fluorescence (uppermost spectrum) and phosphorescence spectrum of 7,8 - thiabenzo(ghi)perylene in n-hexane at 63 K. The compound was detected in carbon black.

DISCUSSION

Quasilinear fluorescence spectroscopy has so far proven to be a valuable analytical method for the identification of certain polycyclic aromatic compounds in low concentrations. Isomeric compounds that would otherwise be difficult to distinguish can be discerned by this low temperature fluorescence technique. The technique is still used exclusively in a few laboratories the reason for this being found in the fact that the cryo-cell must be home-made, since a commercial instrument is still lacking and in the fact that the availability of reference spectra has been limited.

ACKNOWLEDGEMENTS

We thank Ms Beryl Holm for reviewing the manuscript, Ms Ulrika Örn for synthesising the 1-methylcoronene and Dr W Karcher for supplying the 4,5-thiaphenanthrene, 1,12-thiabenzo(a)anthracene and the dinaphto(2,1,1',2')thiophene.

REFERENCES

1. Colin, J., Vion, G., Lamotte, M. and Joussot-Dubien, J. (1981): Quantitative trace analysis of polycyclic aromatic hydrocarbons by high-performance liquid-chromatography and low temperature fluorimetry in Shpol'skii matrices. In: J Chromatogr. 204:135.
2. Colmsjö, A. and Östman, C. (1980): Selectivity properties in Shpol'skii fluorescence of polynuclear aromatic hydrocarbons. In: Anal. Chem. 52:2093
3. Yang, Y., D'Silva, A, Fassel, V. and Iles, M. (1980): Direct determination of polynuclear aromatic hydrocarbons in coal liquids and shale oil by laser excited Shpol'skii spectrometry. In: Anal. Chem. 52:1350
4. Yang, Y., D'Silva, A. and Fassel, V. (1981): Laser excited Shpol'skii spectroscopy for the selective excitation and determination of polynuclear aromatic hydrocarbons. In: Anal. Chem. 53:894
5. Ewald, M., Moinet, A., Bellocq, J., Wehrung, P and Albrecht, P. (1978): Identification de marquers géochemiques aromatiques de la série de hopane, extraits de sédiments marins d'origine autochtone, par leur spectre de fluorescence quasi linéaire dans l'heptane a 4 K (effect Shpol'skii). In: Géochimie organique des sédiment marins profonds, Orgon IV, Golfe d'Aden, mer d'Oman, Edition CNRS, 405.
6. Keshina, A., Smirnov, G., Shabad, L., Pritch, V., Jaskulla, N. and Hüningen, E. (1978): Polycyclic aromatic hydrocarbon occurancy in car exhaust in ECE-tests. In: Hyg and San. 1:44
7. Colmsjö, A. (1981): The practice and utility of the Shpol'skii effect in chemical analysis. Thesis at University of Stockholm, Dept of Analytical Chemistry, S-106 91 Stockholm, Sweden.
8. Colmsjö, A. and Östman, C. (1981): Atlas of Shpol'skii Spectra and other low temperature spectra of POM. Printed at the University of Stockholm, Dept of Analytical Chemistry S-106 91 Stockholm, Sweden.
9. Colmsjö, A. and Stenberg, U. (1977): Temperature effects on resolution of quasilinear fluorescence spectra of some polyaromatic hydrocarbons. In: Chemica Scripta, 11:220
10. Colmsjö, A. and Östman, C. (1981): The utility of the Shpol'skii effect in analysis of sulfur containing heterocyclic aromatic compounds. (In press).
11. Dewar, M. and Dougherty, R. (1975): The PMO theory of organic chemistry. Plenum press, New York and London.
12. Gallegos, E. (1975): CHS^+ sulfur compound analysis by gas chromatography-mass spectrometry. In: Anal. Chem. 47:1150

13. Karcher, W., Depaus, R., van Eijk, J. and Jacob, J. (1979): Separation and identification of sulfur containing polycyclic aromatic hydrocarbons (thiophene derivates) from some PAH. In: Polynuclear Aromatic Hydrocarbons - 3rd Int. Symp. on <u>chemistry and biology - Carcinogenesis and Mutagenesis.</u> Edited by P. Jones and P. Leber. pp 341-356. Ann Arbor Science Publichers Inc., Ann Arbor, Mich.
14. Karcher, W., Nelen, A., Depaus, R., van Eijk, J., Glaude, P. and Jacob, J. (1981): New results in the detection and mutagenic testing of heterocyclic polynuclear aromatic hydrocarbons. In: <u>Polynuclear Aromatic Hydrocarbons - 5th Int. Symp. on chemistry and Biology - Carcinogenesis and Mutagenesis.</u> Edited by M. Cooke and A. Dennis, Battelle Press, Columbus, Ohio.
15. Colmsjö, A. and Stenberg, U. (1979): Identification of Polynuclear Aromatic Hydrocarbons by Shpol'skii Low Temperature Fluorescence. In: <u>Anal. Chem.</u>, 51: 145

THE POSSIBLE INVOLVEMENT OF POLYCYCLIC HYDROCARBONS IN THE
AETIOLOGY OF HUMAN MAMMARY CANCER

C.S. COOPER, P.L. GROVER, A. HEWER, M.O'HARE*, A.D. MacNICOLL,
A.M. NEVILLE*, K. PAL and P. SIMS
Chester Beatty Research Institute, Institute of Cancer
Research: Royal Cancer Hospital, Fulham Road, London SW3 6JB;
*Ludwig Institute for Cancer Research, Haddow Laboratories,
Institute of Cancer Research, Sutton, Surrey, U.K.

INTRODUCTION

Although evidence for the involvement of environmental and hormonal factors in the aetiology of human mammary cancer has accumulated, the types of agent involved have not been identified. Results obtained with experimental animals have shown that mammary cancer can be induced in rats by some polycyclic aromatic hydrocarbons (PAHs) (1) and there appears to be no fundamental reason why PAHs should not also be implicated in the initiation of human mammary tumours since they are environmental and dietary contaminants.

In other tissues, such as mouse skin, that are known to be susceptible to the carcinogenic action of PAHs, the hydrocarbons are activated by the formation of non-K-region dihydrodiols that can be further metabolised to form reactive vicinal diol-epoxides (2). We report here that cultures of epithelial cell aggregates and of fibroblasts prepared from rat and human mammary tissues possess the ability to metabolise 7,12-dimethylbenz(a)anthracene (DMBA), benzo(a)pyrene (BP) and benz(a)anthracene (BA) to form non-K-region dihydrodiols and to activate them to products that react covalently with cellular macromolecules.

MATERIALS AND METHODS

Materials

^{3}H-Labelled DMBA (22-38Ci/mmol), BP (13-27Ci/mmol) and BA (8Ci/mmol) were prepared by tritium exchange (Radiochemical Centre, Amersham, Bucks, UK). Dihydrodiol and diol-epoxide derivatives of the unlabelled hydrocarbons were prepared as described (3).

Metabolism and Activation of Hydrocarbons by Cells in Culture

Epithelial cell aggregates and fibroblasts were prepared as described (3-5) from samples of non-neoplastic Wistar rat and human mammary tissue. Cells were cultured at 37° in 150 cm^2 flasks containing Dulbecco's MEM (50 ml) supplemented with foetal calf serum (10% v/v) and filled with 10% CO_2 in air. ^{3}H-Labelled DMBA, BP or BA was added to the culture media and after 24h, the media were removed and the levels of dihydrodiols, total ether-soluble metabolites and water-soluble metabolites present in the media were assessed exactly as described (3). DNA and protein were isolated from the cells (3) and the extents of covalent binding of hydrocarbon to protein and DNA were determined (3).

Samples of enzymic hydrolysates of DNA isolated from cells after incubation with ^{3}H-labelled DMBA were cochromatographed on columns of Sephadex LH20 in admixture with nucleoside-hydrocarbon adducts, obtained from chemical reactions of trans--10,11-dihydro-10,11-dihydroxy-7,12-dimethylbenz(a)anthracene 8,9-oxide (DMBA 10,11-diol 8,9-oxide) with DNA, that were used as fluorescent markers. Columns were eluted with water-methanol gradients and the eluant fractions were examined for the presence of radioactive and fluorescent (excitation 317 nm, emission 372 nm) materials.

RESULTS

Metabolism and Activation by Rat Mammary Cells

The Wistar rats used as a source of mammary tissue are susceptible to the induction of mammary tumours by DMBA but are not as susceptible as, for example, the Sprague-Dawley strain (P.S. Rudland, personal communication). The results obtained in experiments on the metabolism of ^{3}H-labelled BP, DMBA and BA by rat mammary epithelial cells and fibroblasts show that all three hydrocarbons are metabolised by both types of cell to form ether-soluble and water-soluble metabolites (Table 1 and unpublished observations). The ether-soluble metabolites were further examined by high pressure liquid chromatography and found to include several ^{3}H-labelled metabolites whose chromatgraphic characteristics appeared to be indistinguishable from those of authentic reference dihydrodiols derived from the parent hydrocarbons. The radioactive dihydrodiol metabolites included the 7,8-diol of BP, the 3,4-diol of DMBA and the 3,4-diol and 8,9-diol of BA, the dihydrodiols that are believed to be intermediates in the metabolic activation of the respective parent hydrocarbons.

TABLE 1

METABOLISM AND ACTIVATION OF HYDROCARBONS BY CULTURED RAT MAMMARY TISSUE[a]

Hydrocarbon	Cell type[b]	Ether-soluble metabolites (nmol/mg protein)	Dihydrodiols adducts (pmol/mg protein)			Hydrocarbon-DNA (pmol/mg protein)
			3,4-diol			
DMBA	E	1.4	23			0.6
	F	1.4	23			0.7
			7,8-diol			
BP	E	1.3	90			0.45
	F	1.3	110			0.41
			3,4-diol	8,9-diol		
BA	E	1.4	24	246		<0.05
	F	1.8	10	201		<0.05

[a] The metabolism and activation of DMBA, BP and BA by epithelial cells and fibroblasts that were prepared from rat mammary tissue was examined. Details of the conditions of maintenance of the mammary cells, their treatment with hydrocarbon (8 ug/ml of culture medium), the separation of the metabolites and the analysis of the DNA-hydrocarbon adducts are presented elsewhere (3,5). Each value is the average of the values obtained with three preparations of tissue.

[b] E, epithelial cells; F, fibroblasts

Both epithelial cell aggregates and fibroblasts can convert DMBA, BP and BA into products that bind covalently to protein (results not shown). Similarly both cell types can convert the potent carcinogens DMBA and BP into products that react with cellular DNA but, in contrast, little or no covalent binding of the weak carcinogen BA to DNA was detected (Table 1). The deoxyribonucleoside-hydrocarbon adducts present in hydrolysates of DNA isolated from epithelial cells and fibroblasts that had been incubated in culture with ^{3}H-labelled DMBA were examined by Sephadex LH20 column chromatography. For both cell types, the major adducts (products II-IV, Figures 1A and 1B) were eluted before the major fluorescent adduct formed in reactions of DMBA-10,11-diol 8,9-oxide with DNA (product VI, Figure 1A) and in the regions expected for adducts formed in reactions of DNA with DMBA 3,4-diol 1,2-oxide (3), the diolepoxide thought to be involved in the metabolic activation of DMBA in other tissues.

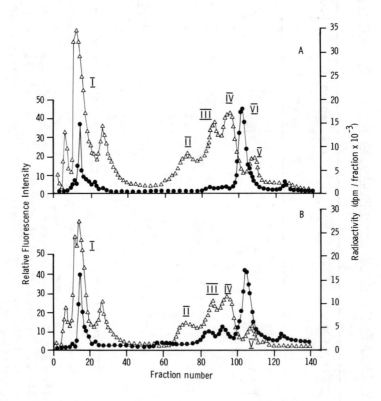

FIGURE 1. Sephadex LH20 column elution profiles obtained when hydrolysates of DNA from rat mammary epithelial cell aggregates

(A) and fibroblasts (B) that had both been treated in culture with ^{3}H-labelled DMBA were chromatographed in admixture with hydrolysates of DNA that had been reacted with DMBA-10,11-diol 8,9-oxide and that were used to provide fluorescent markers. Eluted fractions were examined for the presence of fluorescent (●——●) and radioactive (△——△) materials.

Metabolism and Activation by Human Mammary Cells

The metabolism and the activation of PAHs by epithelial cell aggregates and fibroblasts prepared from non-neoplastic human breast tissue from three mammoplasty patients have been examined. The cells prepared from patient 1 were treated with DMBA whilst those from patients 2 and 3 were treated with BP. The epithelial cell aggregates and fibroblasts from patient 1 metabolised DMBA to form ether-soluble metabolites, including dihydrodiols, water-soluble metabolites and products that reacted with cellular macromolecules (Table 2 and unpublished observations) and notably there were no appreciable differences in the levels of ether-soluble metabolites formed, in the amounts of the 3,4-diol of DMBA formed and in the covalent binding of hydrocarbon to DNA observed for the two cell types. Similarly, both epithelial cells and fibroblasts prepared from patients 2 and 3 metabolised and activated BP but higher levels of ether-soluble metabolites, of the 7,8-diol of BP and of covalent binding of hydrocarbon to DNA were observed in experiments with epithelial cells than in analogous experiments with fibroblasts.

The hydrocarbon-DNA adducts formed when cells from patient 1 were treated with DMBA were examined further by subjecting hydrolysates of DNA from the cells to chromatography on columns of Sephadex LH20. The results (Figure 2) show that the profile of adducts formed in epithelial cells is similar to that for adducts formed in fibroblasts. As observed in experiments in which the activation of DMBA in rat mammary tissue in culture was examined (Figures 1A and 1B), some of the adducts (products II-IV, Figures 2A and 2B) were eluted in the region expected for adducts formed in reactions of DMBA-3,4-diol 1,2-oxide with DNA. However, the overall profiles of adducts from rat and human mammary tissues that had been treated with DMBA were different; one of the adducts formed in rat mammary tissue (product II, Figures 1A and 1B) was not detected in analogous experiments with human mammary tissue whilst one of the major adducts formed in human mammary tissue (product V, Figures 2A and 2B) was present only as a minor adduct in hydrolysates of DNA from rat mammary tissue (product V, Figures 1A and 1B).

These differences could be explained if some aspects of the mechanism of activation of DMBA in mammary tissue from rats and humans are different and, in this respect, it is notable that the profiles of nucleoside-hydrocarbon adducts observed in experiments in which the activation of BP in rat mammary tissue in culture was examined were also different from those obtained in analogous experiments carried out with human mammary tissue (results not shown).

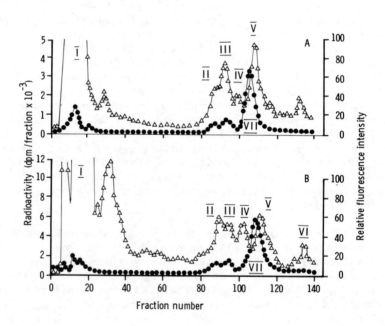

FIGURE 2. Sephadex LH20 column elution profiles obtained when hydrolysates of DNA from human mammary epithelial cell aggregates (A) and fibroblasts (B) that had both been treated in culture with ^{3}H-labelled DMBA were chromatographed in admixture with hydrolysates of DNA that had been reacted with DMBA 10,11-diol 8,9-oxide and that were used to provide fluorescent markers. Eluted fractions were examined for the presence of fluorescent (●——●) and radioactive (△——△) materials.

TABLE 2

METABOLISM AND ACTIVATION OF HYDROCARBONS BY CULTURED HUMAN MAMMARY TISSUE[a]

Hydrocarbon	Patient No.	Cell type[b]	Ether-soluble metabolites (nmol/mg protein)	Dihydrodiols (pmol/mg protein)	Hydrocarbon-DNA adducts (pmol/mg protein)
DMBA	1	E	0.34	3,4-diol 7.3	0.18
		F	0.19	6.8	0.17
BP	2	E	1.7	7,8-diol 84	0.27
		F	0.5	27	<0.05
BP	3	E	2.6	7,8-diol 118	0.63
		F	0.6	31	<0.05

a   The metabolism and activation of DMBA and BP by epithelial cells and fibroblasts that were prepared from non-neoplastic human mammary tissue was examined. Details of the conditions of maintenance of the mammary cells, their treatment with hydrocarbon (0.8 ug/ml of culture medium), the separation of the metabolites and the analysis of the DNA-hydrocarbon adducts are presented elsewhere (3).
b   E, epithelial cells; F, fibroblasts.

DISCUSSION

The results presented here show that epithelial cell aggregates and fibroblasts isolated from non-neoplastic rat and human mammary tissue can metabolise and activate PAHs. The comparison between the abilities of epithelial cell aggregates and fibroblasts to activate PAHs is of particular interest since it is the epithelial cells that are thought to give rise to mammary tumours (6). In rat mammary tissue, this specificity does not appear to be reflected in the different abilities of the two cell types to activate PAHs because both exhibited the same abilities to activate the hydrocarbons. Similarly there were no differences in the abilities of human epithelial cell aggregates and fibroblasts prepared from patient 1 to activate DMBA. However, in contrast, only the epithelial cells prepared from patients 2 and 3 could efficiently activate BP so it is possible that the differing abilities of the two cell types to activate carcinogens may account in part, in some cases, for the specificity of tumour induction.

Thus it is quite possible that PAHs present in the diet or in the environment may be activated *in vivo* by human mammary epithelial cells and might therefore contribute to the incidence of mammary cancer. Since PAHs are fat soluble, it is conceivable that activation of PAHs is facilitated by the accumulation of hydrocarbon in the adipose tissue that surrounds the mammary ducts and, in this connection, it is notable that high levels of 3-methylcholanthrene were detected in the mammary glands and in other fatty tissues when rats were treated orally with this hydrocarbon (7). An implication of the results reported here, and of others, is that cells present in mammary tissue may well be able to activate PAHs *in vivo*.

In conclusion it would appear that although the potential of mammary tissue to activate PAH carcinogens has been established, further work will be necessary to define the precise role of these environmental carcinogens in the aetiology of human breast cancer.

ACKNOWLEDGMENTS

This study was supported in part by grants to the Chester Beatty Research Institute, Institute of Cancer Research: Royal Cancer Hospital from the Medical Research Council and the Cancer Research Campaign, and in part by PHS Grant No. CA21959 awarded by the National Cancer Institute, DHHS.

REFERENCES

1. Huggins, C., and Yang, N. C. (1962): Induction and extinction of mammary cancer, Science, 137:257-262.
2. Grover, P. L., Hewer, A., Pal, K., and Sims, P. (1976): The involvement of a diol-epoxide in the metabolic activation of benzo(a)pyrene in human bronchial mucosa and in mouse skin, Int. J. Cancer, 18:1-6.
3. Grover, P. L., MacNicoll, A. D., Sims, P., Easty, G. C., and Neville, A. M. (1980): Polycyclic hydrocarbon activation and metabolism in epithelial cell aggregates prepared from human mammary tissue, Int. J. Cancer, 26:467-475.
4. MacNicoll, A. D., Easty, G. C., Neville, A. M., Grover, P. L., and Sims, P. (1980): Metabolism and activation of carcinogenic polycyclic hydrocarbons by human mammary cells, Biochem. Biophys. Res. Commun., 95:1599-1606.
5. Cooper, C. S., Pal, K., Hewer, A., Grover, P. L., and Sims, P. L. (1981): The metabolism and activation of polycyclic aromatic hydrocarbons in epithelial cell aggregates and fibroblasts prepared from rat mammary tissue (submitted for publication).
6. Wellings, S. R., and Jensen, H. M. (1973): On the origin and progression of ductal carcinoma in human breast, J. Nat. Cancer Inst., 50:1111-1118.
7. Dao, T. L., Bock, F. G., and Crouch, S. (1959): Level of 3-methylcholanthrene in mammary glands of rats after intragastric instillation of carcinogens, Proc. Soc. Exp. Biol. Med., 102:635-638.

THE DNA-BINDING OF A MONOHYDROXYMETHYL METABOLITE FOLLOWING DMBA ADMINISTRATION TO RATS

F. B. DANIEL, N. J. JOYCE, AND M. A. DRUM
U.S. Environmental Protection Agency, Health Effects Research Laboratory, 26 W. St. Clair Street, Cincinnati, Ohio 45268.

INTRODUCTION

7,12-Dimethylbenz(a)anthracene (DMBA, Figure 1) is a potent carcinogen toward the female Sprague-Dawley (S.D.) rat mammary gland (1). It has been proposed, in agreement with the "bay-region" theory (2), that DMBA binds to DNA subsequent to the metabolic formation of a reactive diol-epoxide in the A-ring (Figure 1). In contrast to previous observations in cultured rodent cells (3,4) and mouse skin *in vivo* (5), we now report the observation of an additional, previously unobserved type of hydrocarbon-deoxynucleoside adduct in rat liver *in vivo* following intraperitoneal (i.p.) administration of DMBA. Chromatographic and spectrofluorometric evidence are presented to show that this adduct results from the formation and DNA binding of a 7-hydroxymethyl-12-methylbenz(a)anthracene (7OHM-12MBA, Figure 1) A-ring diol-epoxide.

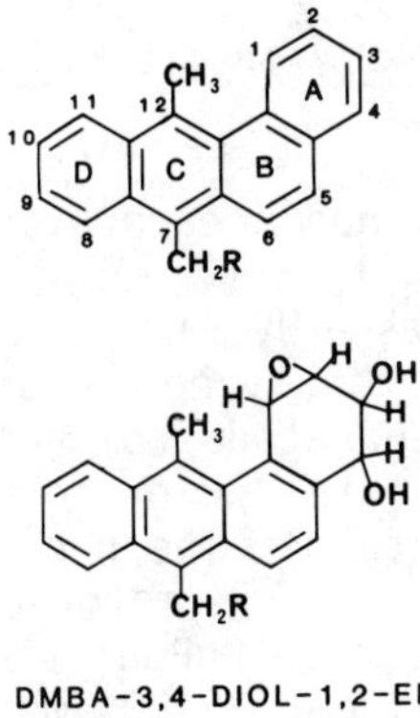

DMBA R=H

7OHM-12MBA R=OH

DMBA-3,4-DIOL-1,2-EPOXIDE R=H

7OHM-12MBA-3,4-DIOL-1,2-EPOXIDE R=OH

FIGURE 1. Line structures of DMBA, 7OHM-12MBA, and their "bay-region" (A-ring) diol-epoxides.

MATERIALS AND METHODS

DMBA was purchased from Eastman Chemical Co., NY, generally tritiated DMBA from Amersham Corp., IL. Samples of cold and [^{3}H]7OHM-12MBA were supplied by Dr. David Longfellow of the National Cancer Institute, MD. All preparations were checked by analytical high pressure liquid chromatography (hplc) before usage, radiochemical and chemical purities were either greater than 96% or were purified to that level.

Female S.D. rats from Charles River, MA, were housed in controlled rooms (22-23°C, 12 hours light) with food and water ad libitum. Animals were given 20 umoles of the appropriate hydrocarbon by i.p. injection at 50 days of age, and were killed 48 hr later, the livers removed, rinsed, weighed, frozen in liquid N_2 and stored at -70°C. All work was done under yellow (Westinghouse FG040) lamps.

DNA was isolated and purified by chloroform: isoamyl alcohol and phenol extractions and enzyme (pronase, RNase A, a-amylase) treatment. Binding levels were calculated by quantitation of DNA using 3,5-diaminobenzoic acid and radioactivity by liquid scintillation analysis. The DNA was hydrolyzed enzymatically to deoxyribonucleosides (3). The system for preparation of DMBA-DN adducts from Syrian hamster embryo cells has been described previously (3). Fluorescence spectra were recorded on an Aminco SPF500 instrument in methanol at room temperature.

The Sephadex LH-20 system has been previously described (3). A small amount of 4-(p-nitrobenzyl)pyridine (PNP) was added to each sample to serve as a UV marker. One hundred 2 ml fractions were collected and analyzed. Hydrocarbon-deoxyribonucleoside adducts were separated from unmodified deoxyribonucleosides in the hydrolysates prior to hplc by extraction with water-saturated n-butanol. The n-butanol phase was dried under N_2 and redissolved in 35% methanol in water for hplc. The hplc system consisted of a DuPont Zorbax-ODS column (4.6 x 250 mm) eluted with a linear 35-80% methanol in water gradient at a flow rate of 1 ml/min. Ninety 0.5 ml fractions were collected and analyzed.

RESULTS

When hydrolysates of liver DNA from [^{3}H]DMBA treated rats, were chromatographed on Sephadex LH-20 the profile in Figure 2 was obtained. Four peaks of radioactivity are

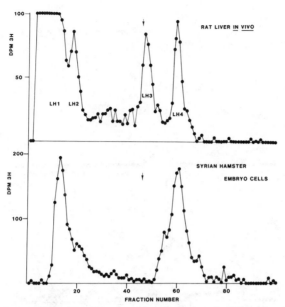

FIGURE 2. Sephadex LH-20 chromatograms of DMBA-deoxyribo-nucleoside adducts. PNP elutes at arrow.

consistently observed in these chromatograms, early eluting activity (LH1 and LH2) has been observed by other investigators (4,5,7,8) but has remained uncharacterized and varies as a percentage of the total activity isolated with the DNA from one experiment to another. Two peaks of activity elute in the mid-region of the chromatogram, LH3 and LH4. LH4 is chromatographically identical to the major DMBA-deoxyribo-nucleoside (DMBA-DN) adduct observed when DNA hydrolysates from Syrian hamster embryo cells, previously treated for 24 hours with 0.5 uM [^{3}H]DMBA, were analysed on this system (Figure 2) (3). In contrast the other peak, LH3, eluting in this system with the UV standard, PNP, has not been found in cultured rodent cells in our laboratory (3) nor in those of other investigators (4).

To further resolve these adducts, DNA hydrolysates from the liver of DMBA treated rats, were extracted with water-saturated n-butanol and the organic phase analyzed by hplc on Zorbax-ODS columns, the resulting chromatogram is shown in Figure 3 (lower). Further experiments (data not shown) have confirmed that hplc peaks III and IV correspond to Sephadex LH-20 peaks, LH3 and LH4 respectively.

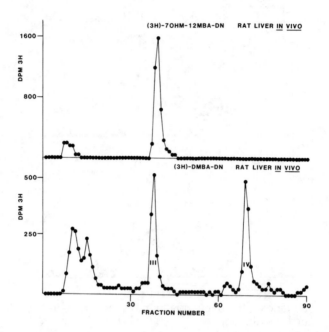

FIGURE 3. Zorbax-ODS hplc chromatograms of hydrocarbon-deoxyribonucleosides from S.D. rat livers.

The upper chromatogram depicted in Figure 3 is the hplc profile resulting from the organic extract of a liver DNA hydrolysate from a female S.D. rat treated with [^{3}H]7OHM-12MBA. The resulting 7OHM-12MBA-deoxyribonucleoside (7OHM-12MBA-DN) adduct elutes in the same fractions as adduct III from [^{3}H]DMBA treated rat liver DNA.

Quantities of III and IV sufficient for spectrofluorometric analysis were obtained from a pool of liver DNA hydrolysates which were partially purified on Sephadex LH-20 and further purified by hplc. Figure 4 shows the fluorescence spectra of III, IV and 9,10-dimethylanthracene, in methanol.

DISCUSSION

DMBA is a well studied carcinogen toward the female S.D. rat mammary gland (1). It has been shown to bind to cellular DNA in many systems (4,5,7,8) through a diol-epoxide of the A-

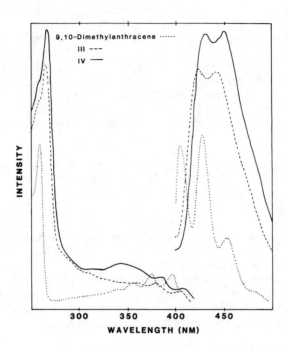

FIGURE 4. Fluorescence excitation and emission spectra.

ring in agreement with the "bay-region" theory (2). It has been proposed that the formation of carcinogen-DNA adducts are initiation events in the process of chemical carcinogenesis (6). If so it is important not only to confirm that a particular chemical binds to DNA but also identify the actual metabolite(s) involved in the binding and explore their significance in the carcinogenic process.

In this study Sephadex LH-20 chromatography of liver DNA hydrolysates, from S.D. female rats treated with [^{3}H]DMBA, reveals the presence of a major adduct, LH4, which has also been observed in Syrian hamster embryo cells treated with the hydrocarbon (Figure 2) (3). When LH4 is further purified by hplc the major component, IV, has a fluorescence spectrum (Figure 4) in excellent agreement with that reported for DMBA-DN adducts containing a hydrocarbon with a saturated A-ring (4,7,8).

However in this study the Sephadex LH-20 chromatograms from rat liver also show the presence of another major adduct, LH3, which is not found in the hamster embryo cell hydrolysates. When LH3 is further purified by hplc the major component, III, has a fluorescence spectrum similar to IV (Figure 4) but shifted 5-7 nm to shorter wavelengths relative to IV, which is in excellent agreement with that reported by Moschel et al. (7) for 7OHM-12MBA-DN adducts. Also included in Figure 4 for comparative purposes is the spectrum of 9,10-dimethylanthracene, a model fluorochrome for an A-ring saturated DMBA.

Further evidence which indicates that this previously unreported adduct is a 7OHM-12MBA-DN adduct is shown in Figure 3, which illustrates the hplc profiles obtained when n-butanol extracts of liver DNA hydrolysates from (lower) [^{3}H]DMBA treated and (upper) [^{3}H]7OHM-12MBA treated animals were analysed. The 7OHM-12MBA-DN adduct obviously elutes in the same fractions as does DMBA-DN adduct III. While we have not rigorously eliminated the possibilities that III may be the isomeric 7-methyl-12-hydroxymethylbenz(a)anthracene (7M-12O-HMBA)-DN adduct or a mixture of 7OHM-12MBA and 7M-12OHMBA adducts, it seems clear that it is a monohydroxymethyl-methylbenz(a)anthracene-deoxyribonucleoside adduct.

The significance of these observations to the carcinogenicity of DMBA are uncertain at this time. 7OHM-12MBA has been demonstrated to be a weaker carcinogen than DMBA in the mouse skin initiation-promotion assay (9) and in the newborn mouse (10), but the 3,4-dihydrodiols of both DMBA and 7OHM-12MBA have been found to be more potent than DMBA (9,10). This leaves a question as to the biological effect a monohydroxymethyl-methylbenz(a)anthracene adduct has relative to a DMBA adduct.

It should be noted here that except when DMBA is administered during regeneration (11) it is not oncogenic in the rat liver. We have not chromatographed the mammary DMBA-DN adducts from this study, however, we have observed adducts LH3 and LH4 (Sephadex LH-20), and III and IV (hplc) in a number of other female S.D. rat tissues including the mammary gland, target tissue, when DMBA was administered intravenously, these findings are the subject of a manuscript in preparation.

ACKNOWLEDGEMENTS

We gratefully acknowledge the skillful preparation of the manuscript by Ms. F. Jean Roe.

REFERENCES

1. Huggins, C.B. (1979): Experimental Leukemia and Mammary Cancer. University of Chicago Press, Chicago, 212 pp.
2. Jerina, D.M. and Daly, J.W. (1977): Oxidation of carbon. In: Drug Metabolism, edited by D.V. Parke and R. L. Smith, pp. 13-32, Taylor and Francis, London.
3. Daniel, F.B., Wong, L.A., Oravec, C.T., Cazer, F.D., Wang, C., D'Ambrosio, S.M., Hart, R.W., and Witiak, D.T. (1979): Biochemical studies on the metabolism and DNA-binding of DMBA and some of its monofluoro derivatives of varying carcinogenicity. In: Polynuclear Aromatic Hydrocarbons, edited by P.W. Jones and P. Leber, pp. 855-883, Ann Arbor Press, Ann Arbor.
4. Dipple, A., Tomaszewski, J.E., Moschel, R.C., Bigger, C.A.H., Nebzydoski, J.A. and Egan, M. (1979): Comparison of metabolism-mediated binding of 7-hydroxymethyl-12-methylbenz(a)anthracene and 7,12-dimethylbenz(a)anthracene, Cancer Res., 39:1154-1158.
5. MacNicoll, A.D., Burden, P.M., Ribeiro, O., Hewer, A., Grover, P.L. and Sims, P. (1979): The formation of dihydrodiols by the chemical or enzymic oxidation of 7-hydroxymethyl-12-methylbenz(a)anthracene and the possible role of hydroxymethyl dihydrodiols in the metabolic activation of 7,12-dimethylbenz(a)anthracene, Chem. Biol. Interactions, 26:121-132.
6. Brookes, P. (1977): Role of covalent binding in carcinogenicity. In: Biological Reactive Intermediates, edited by O.J. Jallow, J.J. Kocsis, R. Snyder, and H. Vainio, pp. 470-480, Plenum Press, New York.
7. Moschel, R.C., Hudgins, W.R. and Dipple, A. (1979): Fluorescence of hydrocarbon-deoxynucleoside adducts, Chem. Biol. Interactions, 27:69-79.
8. Vigny, P., Kinds, M., Cooper, C.S., Grover, P.L. and Sims, P. (1981): Fluorescence spectra of nucleoside-hydrocarbon adducts from mouse skin treated with 7,12-dimethylbenz(a)anthracene, Carcinogenesis, 2:115-119.

9. Wislocki, P.G., Gadek, K.M., Chou, M.W., Yang, S.K. and Lu, A.Y.H. (1980): Carcinogenicity and mutagenicity of the 3,4-dihydrodiols and other metabolites of 7,12-dimethylbenz(a)anthracene and its hydroxymethyl derivatives, Cancer Res., 40:3661-3664.
10. Wislocki, P.G., Juliana, M.M., MacDonald, J.S., Chou, M.W., Yang, S.K. and Lu, A.Y.H. (1981): Tumorigenicity of 7,12-dimethylbenz(a)anthracene, its hydroxymethylated derivatives and selected dihydrodiols in the newborn mouse, Carcinogenesis, 2:511-514.
11. Marquardt, H., Sternberg, S.S. and Phillips, F.S. (1970): 7,12-Dimethylbenz(a)anthracene and hepatic neoplasia in regenerating rat liver, Chem. Biol. Interactions, 2:401-403.

IDENTIFICATION AND QUANTIFICATION OF PAH-DNA ADDUCTS BY LASER FLUORESCENCE.

ROBERT C. DAVIS, THOMAS J. FACKLAM, WILLIAM A. IVANCIC AND RUSSELL H. BARNES
Battelle's Columbus Laboratories, 505 King Avenue, Columbus, Ohio 43201.

INTRODUCTION

Quantitative information about the levels of DNA adducts formed in vivo is very important in understanding and predicting the effects of exposure to carcinogenic and mutagenic xenobiotics. Unfortunately, experimental results for animals is very limited and data relating to human populations, virtually nonexistent. In this paper, we present some initial experimental results which demonstrate the utility and sensitivity of laser fluorescence analysis of DNA adducts.

There is need for molecular dosimetry in two contexts. We will refer to the first as laboratory studies: establishing dose-response relations; exploring synergism, the effect of the presence of one compound on the incorporation of others; incorporation and repair kinetics and mechanisms; distribution of DNA adducts among tissues; etc. Most of these studies have depended on radiotracers. The sensitivity of these measurements is limited by the fact that the specific activity of a radiolabeled compound depends on the number of radioactive nuclei incorporated per molecule and the half life. On the other hand, a fluorescent molecule can be repetitively excited and fluoresce, with limits being imposed by the intensity of the exciting source and the possible photocomposition. A molecule with one ^3H incorporated has a specific activity of 30 Ci/mmole, more or less the limit for most tritiated compounds. Using tritiated dimethylbenzanthracene at 12 Ci/mmole, Shoyab (1) was able to detect adducts at a level of 125 per mammalian haploid genome, corresponding to 1 cpm/μgram DNA. The limited availability of radiolabeled mutagens and carcinogens makes these experiments difficult or impossible for many, if not most, important compounds. If one wants to do a double-label experiment to simultaneously detect adducts formed from two compounds and ^{14}C is used, its specific activity limit is lower because its half life is 463 times longer than ^3H. Radiotracer studies with more natural xenobiotic sources such as combustion particulates would be very much more difficult. On the other hand, fluorescent spectra will generally differ, enabling one to analyze a mixture in terms of its fluorescent components. We will present data demonstrating that we have already achieved sensitivity at least as good as that achievable by radiotracer techniques.

The other application of DNA-adduct dosimetry is in the context of screening populations, especially humans, who may have been exposed to carcinogenic or mutagenic compounds. Methods presently used include measurement of physical damage to genetic material (e.g., chromosomal aberrations or DNA strand breakage) or induction of mutations. These approaches all suffer from limited sensitivity and frequently from their expense and problems caused by the subjectivity involved in measurements. None of these techniques gives causal information. On the other hand, preparation of DNA by standard methods would yield a sample on which laser fluorescence could determine not only the levels of polynuclear aromatic adducts, but information about their nature and therefore the nature of the precursor xenobiotic(s).

INSTRUMENT

The approach we used for our high-sensitivity fluorescence measurements is very straightforward in concept, although its success depends on some subtle details. Figure 1 is a schematic depiction of the instrument used. The nitrogen/dye laser combination gives tunability throughout the visible region of the spectrum. Frequency doubling allows converting visible output into ultraviolet output, down to 220 nm, when needed. The nitrogen/dye laser gives intense pulses about 8 nsec in duration. Additionally, the coherent nature of the laser beam enables it to be focused to less than 50 μm in diameter. In these studies the sample was contained in a standard fluorescence cuvette, with vertical illumination, so that the fluorescent column could be imaged on the entrace slit of the monochromator. In the future, smaller sample sizes will be used, either 5-10 μl in a capillary or we could use the technique of Diebold and Zare (2) in which a small (∼4 μl) droplet was maintained in the gap between a stainless steel tube which was slightly separated from a solid rod, thereby serving as a windowless fluorescence cell. It should be noted that Diebold and Zare (2) were able to detect 6×10^{-12} M solutions or 7 femtograms of aflatoxins in a 4 μl volume using a different type of laser fluorescence technique.

A photomultiplier detector and its associated amplification system was used in a photon counting mode in

FLUORESCENCE ANALYSIS OF DNA ADDUCTS

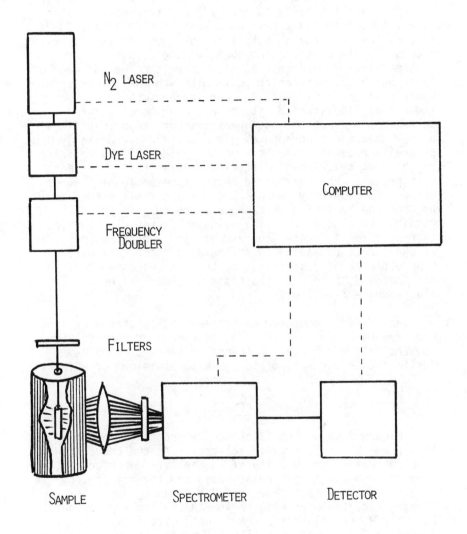

FIGURE 1. Schematic representation of a laser fluorescence spectrometer.

order to maximize sensitivity. Its output was gated in order to reduce background noise and other nonfluorescent artifacts. Since the fluorescent lifetimes of polynuclear aromatics are typically 10 nsec (3) and the optimum repetition rate for the laser system used is approximately 30 Hz, most of the time the output of the photomultiplier corresponds only to noise. Consequently, the output of the photomultiplier more than 60 nsec after the leading edge of laser pulse was not registered. Although the intensity of Raman scattering is typically considered to be very much lower than fluorescence, it becomes overwhelming when solvent is present in concentrations more than 10^{12} times higher than the fluorophore. This artifact, especially severe in aqueous solvents, can be avoided by capitalizing on the fact that Raman scattering is a virtually instantaneous process when considered on the time scale of fluorescence. Consequently, we discarded all signals within 25 nsec of the beginning of each laser pulse. It should be possible to reduce this time by using faster electronics, thereby observing more fluorescence and improving the limits of DNA adduct detection. It should be noted that gating has the potential of discriminating the fluorescence of different compounds solely on the basis of their different fluorescence lifetimes.

An Apple II computer controlled the spectrometer function and data acquisition. The spectrometer consists of commercially available components so that it could be duplicated in other laboratories with a minimum of effort.

RESULTS

Figure 2 shows the fluorescence spectra of benzo[a]pyrene in simple organic solvents. As the concentration decreases, the quality of the spectra deteriorates, as seen in both the signal:noise ratio and the ability to resolve spectral features. At 10^{-9} M benzo[a]pyrene fluorescence features at both 405 and 430 nm are obvious. At 10^{-10} M, the 430nm feature is only marginally identifiable. At 10^{-11} M, the 430 nm feature is indistinguishable and the signal:noise ratio is only 4:1.

Even though we can make these measurements, the question remains as to whether there are problems inherent in the use of aqueous systems, buffers, DNA or impurities. To address these questions; we measured the fluorescence of increasingly dilute levels of 9-aminoacridine intercalated into the DNA double helix. 9-aminoacridine was chosen

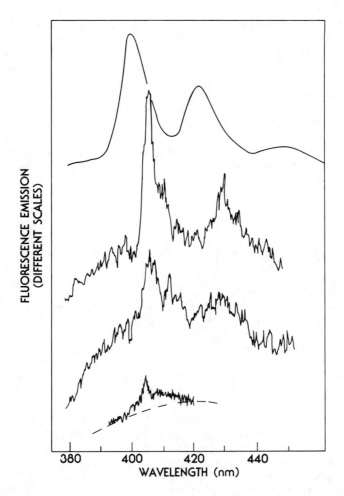

FIGURE 2. Fluorescence emission spectra of benzo[a]pyrene solutions excited at 295 nm. From top to bottom: 10^{-6} M in methanol; 10^{-9} M, 10^{-10} M and 10^{-11} M in cyclohexane. The top spectrum was measured on a (commercial) Aminco SPF-500 fluorimeter and the lower three were measured by laser fluorescence system. The only spectral difference between methanol and cyclohexane is that the latter has lower background fluorescence (indicated by the dashed line on the lowest spectrum).

because it is a fluorescent, water-soluble mutagen, and is more than 99 percent intercalated under the conditions used (4). Although interaction with the DNA is noncovalent, the fluorescence measurement is unaffected. All the problems associated with measurements of covalent adducts are present in this system. Cacodylic acid, 9-aminoacridine, calf thymus DNA, and other reagents from commercial sources were used without further purification. Figure 3 shows that our sensitivity for measuring the fluorescence of the 9-aminoacridine-DNA complex is comparable to that for benzo[a]pyrene. At the lowest concentration used, the ratio of 9-aminoacridine to DNA corresponds to $\sim$ 50 molecules per mammalian haploid genome (5), less than half the value measured in the most sensitive radiotracer experiments we are aware of (1).

DISCUSSION

The differences between the fluorescence spectra of 9-aminoacridine and benzo[a]pyrene demonstrate the utility of fluorescence spectra. If one is willing to make a multi-parameter fluorescence measurement (_e.g._, fluorescence emission as a function of exciting wavelength, time after exciting pulse, whether DNA is native or denatured, _etc._), then one's ability to discriminate compounds becomes extraordinarily powerful. The question of how many compounds (or classes of compounds) can be identified simultaneously by fluorescence at a given level of experimental sophistication is not answered at this time.

Fluorescence-based techniques clearly cannot detect all DNA adducts at levels comparable to those achieved for 9-aminoacridine and benzo[a]pyrene. For example, adducts formed by simply alkylating compounds have absorption (and therefore excitation) spectra very similar to that of DNA itself (6). This makes it difficult to detect the adducts in the presence of low-level DNA or protein impurity fluorescence when excitation has to be in the mid-ultraviolet, and the adduct is present in very much lower concentration than other species. But, it should be remembered that the only criteria for suitability for quantifying DNA adducts are their high fluorescent yield and excitation at wavelengths longer than the DNA adsorption. Classes of detectable compounds include polycyclic aromatic hydrocarbons, aromatic amines,

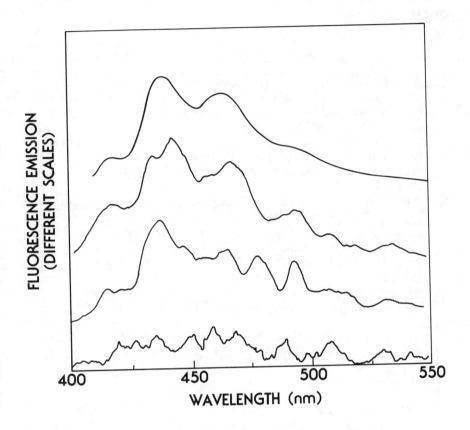

FIGURE 3. Fluorescence spectra of 9-aminoacridine in 10^{-3} M (nucleotide) calf thymus DNA, aqueous cacodylate buffer pH = 7. Concentrations of 9-aminoacridine are 10^{-5} M, 5×10^{-7} M, 4×10^{-9} M and 9×10^{-12} M for the spectra from top to bottom. Excitation was at 391 nm. The top spectrum was recorded on a (commercial) Amino SPF-500 fluorimeter and the lower three on our laser fluorescence system. The control for 10^{-3} M calf thymus DNA with no added 9-aminoacridine was flat, with a noise level comparable to that of the lowest trace.

combustion products (e.g., nitropyrenes), mycotoxins and aminoazo dyes. We feel that this approach will have substantial applications to a number of different types of studies of the incorporation of these compounds into DNA as covalent adducts.

REFERENCES

1. Shoyab, M. (1978): Dose-dependent preferential binding of polycyclic aromatic hydrocarbons to reiterated DNA of murine skin cells in culture, Proc. Nat. Acad. Sci. U.S., 75:5841-5845.
2. Diebold, G. J., and Zare, R. N. (1977): Laser fluorimetry: subpicogram detection of aflatoxins using high-pressure liquid chromatography, Science, 196:1439-1441.
3. Schmillen, A., and Legler, R. (1967): Luminescence of Organic Substances. Springer-Verlag, Berlin, pp. 8-164.
4. Bloomfield, V. A., Crothers, D. M., and Tinoco, Jr., I. (1974): Physical Chemistry of Nucleic Acids, pp 429-470. Harper and Row, New York, and references therein.
5. Sober, H., editor (1970): Handbook of Biochemistry, second edition. Chemical Rubber Co., Cleveland, OH, H 112-113.
6. Singer, B. (1972): Reaction of guanosine with ethylating agents, Biochemistry, 11:3939-3947.

VERY HIGH-SPEED LIQUID CHROMATOGRAPHY FOR PAH ANALYSIS: SYSTEM AND APPLICATIONS

MICHAEL W. DONG, KENNETH OGAN, JOSEPH L. DiCESARE
Perkin-Elmer Corporation, Norwalk, Connecticut 06856

INTRODUCTION

Recent advances in column and instrument technology have made possible the development of a new generation of high-speed liquid chromatography (LC) (1). Using small-dimensioned columns packed with small particles and specialized equipment, performance levels of 300 - 450 plates/second are attainable (2), thus permitting many useful analyses to be performed in 1 - 2 minutes with over 10,000 plates (1,3,4). Comparing to conventional LC systems using 10 µm column, the new High-Speed LC system permits analysis times to be reduced 4 - 5 fold without sacrificing resolution. Reductions in method development time are equally impressive. In addition to offering higher speed and efficiencies, the new high-speed system also yields higher mass sensitivities and lowers solvent consumption.

In PAH analyses, mixtures containing compounds with one to seven condensed rings can be separated in about six minutes. Alternately, for detailed analysis of very complex mixtures, higher-resolution in the range of 30,000 to 100,000 plates can be attained by connecting columns in series. In this paper, the column and instrumental requirements of high-speed LC are reviewed, and several practical applications of this new technique in PAH analysis are described.

MATERIALS AND METHODS

Apparatus

<u>Instrument.</u> A Perkin-Elmer Series 3B Liquid Chromatograph equipped with a Model LC-85 variable wavelength UV/Vis detector (190 - 600 nm), and a Sigma 15 data station were used. A high-speed chromatography package designed to minimize extra-column band-broadening effects was utilized in connection with the chromatographic system. The package consists of a Rheodyne Model 7125 injector with a 6 µl loop, a 2.4 µl microcell for the LC-85 variable wavelength UV detector, and about 80 cm length of 0.007" i.d. connecting

tubing in the sample flowpath. A 250 mm X 9.5 mm i.d. column tube filled with 55 - 100 μm high surface silica served as a presaturator column (5). A specially designed high-speed gradient mixer with a void volume of 0.5 ml was used for rapid gradient analysis (1). A Perkin-Elmer Model 650-10S Fluorescence Spectrophotometer was also used.

Columns. Columns used in this study were Perkin-Elmer high-speed columns packed with 3-μm and 5-μm octadecylsilyl bonded-phase packings. For general method development and analysis, the 5-μm column (125 mm X 4.6 mm) with about 10,000 plates was used. For high resolution analysis, the 3-μm column (100 X 4.6 mm) with 15,000 plates was employed. Details on the columns and packings specifications were available elsewhere (1,4). For the separation of certain PAH isomers, a Supelco LC-PAH column (150 mm X 4.6 mm) was utilized.

Chemicals

PAH standard materials used in this study were of the highest purity grade available and were obtained from the following companies: Aldrich Chemical Company, Inc. (Milwaukee, WI), Applied Science (State College, PA), Chem. Service, Inc. (West Checter, PA), Eastman Kodak Co. (Rochester, NY), K & K (Plainville, NY), Nanogen International (Watsonville, CA), and Pfalz & Bauer, Inc. (Stamford, CT). HPLC grade methanol, acetonitrile and tetrahydrofuran from Fisher Scientific (Pittsburgh, PA) were used for the mobile phase. Water was further purified using a mixed bed ion-exchange resin and an activated charcoal filter. The creosote oil was purchased from the open market in Connecticut in 1980. The coal liquefaction sample was obtained from sources in the industry. No information on the process, origin, and pretreatment procedure is available.

Sample Clean-Up

PAH fractions from the creosote oil and the coal oil sample were prepared by the coupled-column techniques described previously (6,7).

RESULTS AND DISCUSSION

The capabilities and advantages of high-speed LC has been described in detail previously (1-4). Briefly, the increase of performance level is due to the following factors.

a.) small particle supports (e.g. 3 μm) which generate the high efficiencies at high mobile phase velocities in relatively short column lengths.

b.) small-diameter columns and their low void volumes which result in low mobile phase consumption and high mass sensitivity, and

c.) an instrumental system having minimized band-broadening effects which permits the high column efficiencies to be maintained.

The capabilities of the High-Speed LC system described above for rapid analysis is shown in Figure 1 in which 18 PAH from the 2-ring naphthalene to the 7-ring coronene are

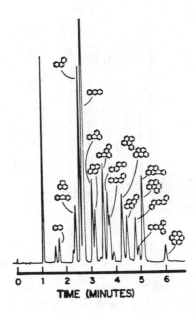

FIGURE 1. Rapid Gradient Analysis of 18 PAH Test Mix. LC conditions: PE/HS-5 C18 column; mobile phase: CH_3CN/H_2O programmed from 60 to 100% in 5 minutes at 4 ml/min, pressure 25.5 MPa; ambient temperature; LC-85 detector at 254 nm with 2.4 μl flow cell.

separated in six minutes. The gradient conditions are 65% acetonitrile/water to 100% CH_3CN in 5 minutes at a flow rate of 4 ml/min. The column used is a 5-µm C-18 column with a void volume of 1.2 ml. Since the column equilibration time is only 1.3 minutes under these conditions, the total time between sample injections is only eight minutes. Although not completely resolving several PAH isomers, this method offers a rapid analysis which is useful for screening a large number of samples.

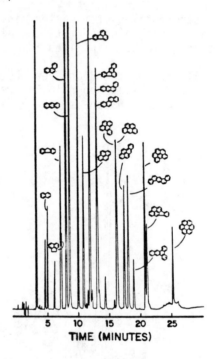

FIGURE 2. High Resolution Analysis of 18 PAH Test Mix. LC conditions: two PE/HS-3 C18 columns coupled in series; mobile phase:CH_3CN/H_2O programmed from 65 to 90% in 20 minutes at 1.8 ml/min; pressure 31 MPa; ambient temperature; LC-85 detector at 254 nm with 2.4 µl flow cell.

If higher resolution is desired, the 5-μm column (10,000 plates/column) can be substituted with a 3-μm column (15,000 plates/column). For even higher efficiencies, in still relatively short analysis time, two or three 3-μm columns can be connected in series to yield 30,000 - 45,000 plates. Figure 2 shows the separation of the same 18 PAH mixture using two 3-μm column in series. Analysis time is 25 minutes; however, the higher resolution attained is quite apparent. Using this two column system, very complex mixtures can be analyzed quite effectively as demonstrated by the separation of close to 100 components in a creosote oil sample (Figure 3). The attainment of over 100,000 plates using similar techniques has been described previously (4).

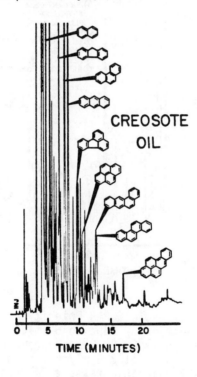

FIGURE 3. Analysis of PAH Fraction of Creosote Oil. LC conditions: same as in Figure 2. Compound identification by retention times only.

For the separation of all 16 PAHs listed in the EPA priority pollutant list, columns with special chromatographic selectivity is required (8-9). Using a 250 mm length column packed with 10 μm particles, analysis times ranging from 40 - 80 minutes are quite typical (10). However, by changing the column to a 5-μm high-speed column filled with similar packing material, all 16 PAH priority pollutants can be resolved quite adequately in only eight minutes (Figure 4).

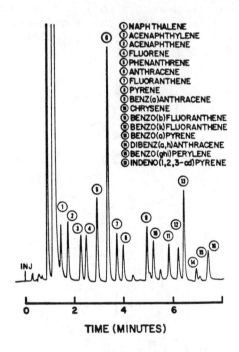

FIGURE 4. Separation of 16 PAH Priority Pollutants. LC conditions: Supelco 5-μm LC-PAH column; mobile phase: CH_3CN/H_2O programmed from 55 to 100% in 4 minutes at 3.5 ml/min; hold initial conditions for 1 min; pressure 24 MPa; ambient temperature; LC-85 detector at 254 nm with 2.4 μl flow cell.

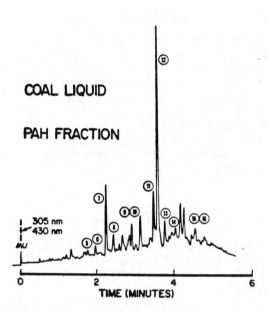

FIGURE 5. Analysis of PAH Fraction of a Coal Liquefaction Sample. LC conditions: PE/HS-3 C18 column; mobile phase CH_3CN/H_2O: 60 to 100% in 5 minutes at 3 ml/min; pressure 30 MPa; PE Model 650-10S fluorescence Spectrophotometer (excitation at 305 nm, emission at 430 nm).

Figure 5 shows the rapid analysis of a coal liquid sample using a fluorescence detector. For analyzing PAH in these complex samples, the sensitivity and spectral selectivity of a fluorescence spectrophotometry (11) can be used quite effectively to reduce sample clean-up requirements. Detection limits of sub-picograms range are possible for many PAH (10,12).

For the coal oil sample, sample clean-up using the coupled column technique is automated and rapid (20 minutes) (6-7). Final analysis is efficient and takes only six minutes (Figure 5). The total analysis time for PAH in coal oils and similar samples can therefore be completed in less than 30 minutes.

CONCLUSION

The use of short high-speed columns packed with 3- or 5-μm particles with the proper equipment permits a 4 - 5 fold reduction in analysis time without sacrificing resolution. For routine PAH separations, analysis can be completed in six to eight minutes. Resolution in the range of 30,000 - 100,000 plates can be obtained by connecting these columns in series. Instruments with small extra-column band-broadening effect, however, must be used to preserve the very high performance levels of these columns.

REFERENCES

1. DiCesare, J.L., Dong, M.W., and Ettre, L.S. (1981): Very-high-speed liquid column chromatography: The system and selected applications, Chromatographia, 14: 257-268.
2. DiCesare, J.L., Dong, M.W., and Atwood, J.G. (1981): Very-high-speed liquid chromatography II: Some instrumental factors influencing performance, Advances in Chromatography, edited by A. Zlatkis, Chromatography Symposium, Dept. of Chemistry, University of Houston, Houston, pp. 357-374.
3. Dong, M.W., and DiCesare, J.L. (1981): Very-high-speed liquid chromatography III: Quantitative analysis of parabens in cosmetic products, J. Chromatogr. Sci. (in press).
4. DiCesare, J.L., Dong, M.W., and Vandemark, F.L. (1981): Instrumentation and columns for very-high-speed LC, Amer. Lab., 13: 52-64.
5. Atwood, A.G., Schmidt, G.J., and Slavin, W. (1979): Improvement in liquid chromatography column life and method flexibility by saturating the mobile phase with silica, J. Chromatogr., 171: 109-115.
6. Ogan, K., and Katz, E. (1981): Analysis of complex samples by coupled-column chromatography, Anal. Chem., (in press).
7. Katz, E., and Ogan, K. (1981): The use of coupled-column and high resolution liquid chromatography in the analysis of petroleum and coal liquid samples, Chemical Analysis and Biological Fate: Polynuclear Aromatic Hydrocarbons, edited by M. Cooke and A.J. Dennis, Battelle Press, Columbus, OH, pp. 169-178.

8. Polynuclear Aromatic Hydrocarbons, Method 610 (1979): Federal Register, Vol. 44, No. 233, pp. 66514-69517.
9. Ogan, K., and Katz, E. (1980): Retention characteristics of several bonded phase liquid chromatography columns for some polycyclic aromatic hydrocarbons, J. Chromatogr., 188: 115-127.
10. Ogan, D., Katz, E., and Slavin, W. (1979): Determination of polycyclic aromatic hydrocarbons in aqueous samples by reversed-phase liquid chromatography, Anal. Chem., 51: 1315-1320.
11. Ogan, K., Katz, E., and Porro, T.J. (1979): The role of spectral selectivity in fluorescence detection for liquid chromatography, J. Chromatogr. Sci., 17: 597-600.
12. Das, B.S., and Thomas, G.H. (1978): Fluorescence detection in high-performance liquid chromatographic determination of polycyclic aromatic hydrocarbons, Anal. Chem., 50: 967-973.

BINDING OF ORALLY ADMINISTERED BENZO(A)PYRENE TO THE DNA OF MICE OVER A DOSAGE RANGE OF 100,000.

BRUCE P. DUNN
Environmental Carcinogenesis Unit, British Columbia Cancer Research Centre, 601 West 10th Avenue, Vancouver, B.C., Canada V5Z 1L3.

INTRODUCTION

There have been numerous reports of the presence of polycyclic aromatic hydrocarbons (PAH) in foods (1,2). In general, these appear to be a result of either cooking practices which cause deposition of soot (e.g., charbroiling), of preserving practices (e.g., smoking), or of pre-existing environmental contamination of foodstuffs. Since foods containing PAH are often contaminated for identifiable and preventable reasons (1,2), there is a possibility of legally regulating the levels of PAH in foods. Decisions on the regulation of carcinogens in foodstuffs depend critically on the perceived risk of these compounds to human health. One of the major problems in risk estimation is in determining the hazard of low doses of a carcinogen, such as are present in human foods, using animal carcinogenicity data obtained at high doses. This communication examines the dose dependency of one of the early stages in carcinogenesis, namely, the binding of the carcinogen to the DNA of target organs. The results have implications for the choice of model used in predicting human risks from PAH carcinogens in foods.

MATERIALS AND METHODS

Carcinogen

Generally labelled benzo(a)pyrene, ^{3}H-G-BaP (30-65 Ci/mmole) and 6-position labelled benzo(a)pyrene, ^{3}H-6-BaP (25 Ci/mmole) were purchased from the Amersham Corp. and were used without further purification. Radioactive and non-radioactive benzo(a)pyrene in corn oil were mixed to give the desired specific activity.

Administration of Carcinogen

Mice were random-bred Swiss male mice weighing 30-35 g. Mice were maintained on a 12 hour light/dark cycle. For experiments, mice were starved for 6 hours before force-feeding, starting one hour into the light phase of their cycle. After starvation, mice were force-fed with BaP in preservative-free corn oil (0.2 ml) by stomach intubation under light ether anesthesia. Animals were maintained for one hour without access to water or food, then maintained for another 17 hours with water but without food. Bedding was also removed during starvation prior to and after carcinogen administration, as it may be eaten in the absence of food.

Isolation and Hydrolysis of DNA

Eighteen hours after force-feeding, animals were sacrificed by cervical dislocation, and stomach, intestine (distal 10 cm adjacent to colon), colon and liver samples excised. DNA was isolated by digestion with RNase and pronase, followed by extraction with phenol and then chloroform/isoamyl alcohol. Isolated DNA was reprecipitated two times with ethanol, then redissolved and determined spectrophotometrically.

An equal volume of 1 N HCl was added to DNA samples, and the DNA hydrolyzed by heating at $85°$ for 1 hour. These conditions have been reported to result in the hydrolysis of benzo(a)pyrene-guanine adducts and the release of the BaP moiety of the adducts as a BaP tetrol (3). The acid digest was extracted two times with 1.5 volumes of ethyl acetate, and radioactivity in the ethyl acetate phase (BaP tetrols resulting from hydrolysis) and the aqueous phase (non-hydrolyzable adducts) determined by scintillation counting. Radioactivity was expressed as moles BaP bound/g DNA, assuming no loss of tritium label from the BaP. In some cases, DNA was also hydrolyzed enzymatically, and the hydrolysate chromatographed on a small column of Sephadex LH-20 (4). Nucleosides and hydrophilic BaP adducts were eluted from the column with water, then hydrophobic BaP-nucleoside adducts were eluted with methanol.

RESULTS

An experiment designed to explore the degree and nature of the covalent binding of orally administered benzo(a)pyrene to the DNA of various organs in the mouse was performed. Mice were fed 1 microgram of either generally tritiated benzo(a)-pyrene (^{3}H-G-BaP) or 6-position tritiated BaP (^{3}H-6-BaP), dissolved without the addition of non-radioactive BaP in 0.2 ml corn oil. After 18 hours, DNA was isolated from stomach, intestine, colon and liver of the experimental animals. DNA from each individual organ was isolated and purified separately.

Figure 1 indicates the covalent binding of BaP to DNA as determined by acid hydrolysis and ethyl acetate extraction. There is generally a close agreement between the amount of binding measured using ^{3}H-G-BaP and ^{3}H-6-BaP. The level of binding was lowest in the stomach and highest in the intestine and liver. In all organs, acid-hydrolyzable adducts represented only a minor fraction of the total amount of radioactivity associated with DNA, namely, 15-20% in liver and colon samples and 25-30% in stomach and intestine samples.

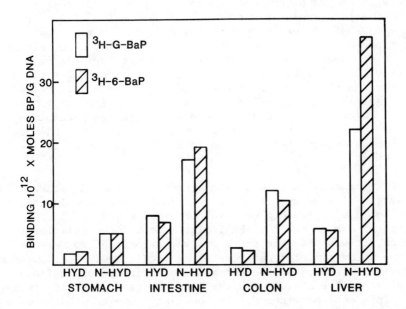

FIGURE 1. Hydrolyzable (HYD) and non-hydrolyzable (N-HYD) adducts formed after intragastric administration of ^{3}H-G-BaP (open bars) or ^{3}H-6-BaP (hatched bars). Data are mean values for determinations on 18 mice for each isotopically labelled form of the carcinogen. Standard errors (not shown) were generally less than ±15%.

A comparison was made between acid hydrolysis and enzyme hydrolysis in a parallel experiment. DNA from pooled liver samples was degraded by enzyme hydrolysis (4) and hydrophobic BaP-nucleoside adducts separated from hydrophilic adducts and unmodified nucleosides by chromatography on Sephadex LH-20. Other aliquots of the same DNA were degraded by acid hydrolysis and fractionated by ethyl acetate extraction as described above. For DNA from animals treated with ^{3}H-G-BaP, acid hydrolysis/

ethyl acetate extraction indicated 15.8% hydrolyzable adducts, while enzyme hydrolysis/Sephadex LH-20 chromatography indicated 14.3% hydrophobic adducts. Corresponding figures for the DNA from animals treated with ^{3}H-6-BaP were 19.4 and 19.2%, respectively.

Acid hydrolysis of DNA followed by ethyl acetate extraction is quick, simple, and is easily performed on multiple samples. It is also capable of being used with DNA samples containing less radioactivity than is required for analysis by enzyme hydrolysis and Sephadex LH-20 chromatography. For these reasons, acid hydrolysis was used in investigating the nature of the BaP-DNA adducts formed when mice were administered a wide range of carcinogen doses. Figure 2 shows the relationship between the amount of ^{3}H-G-BaP administered to animals and the amount of hydrolyzable and non-hydrolyzable adducts formed in DNA from the stomach. The amount of both forms of adducts was approximately linearly related to dose over a range of 10^{-7} to 10^{-3} g. Over the entire dose range, the relative percentage of hydrolyzable adducts remained relatively constant, ranging from approximately 25-30% of the total DNA-associated radioactivity.

Similar results were obtained from liver samples taken from the same series of experimental animals (Figure 3). In this case, because of the higher level of binding and the larger amounts of DNA which could be isolated from liver relative to stomach, it was possible to extend the experiment to measure binding at an administered dose of as little as 10^{-8} g. Over a dose range of 100,000, the formation of hydrolyzable and non-hydrolyzable BaP-DNA adducts was linearly related to the amount of administered carcinogen. In contrast, however, to the data from stomach DNA, the relative proportion of hydrolyzable adducts was not constant with dose, increasing from a minimum of 17% at the lowest dose to a maximum of 27% at the highest dose.

In a parallel experiment, the removal of BaP adducts over a 7-day period was investigated. After one week, no detectable radioactivity was found in either the hydrolyzable or non-hydrolyzable fraction of stomach DNA in animals administered doses ranging from 10^{-7} to 10^{-3} g. In this experiment, this implies a repair or removal of greater than 80% for hydrolyzable adducts and greater than 95% for non-hydrolyzable adducts. In the liver, however, significant amounts of radioactivity remained detectable after 7 days. Approximately 20 to 40% of the original level of non-hydrolyzable adducts remained after this time period, while the repair or removal of hydrolyzable adducts was more efficient, with only 15-30% of the original radioactivity remaining. Although there was some variability in the efficiency of repair or removal with different doses, there was no clearcut trend to the data, and no evidence that repair efficiency was dose-dependent over a range from 10^{-8} to 10^{-3} g BaP.

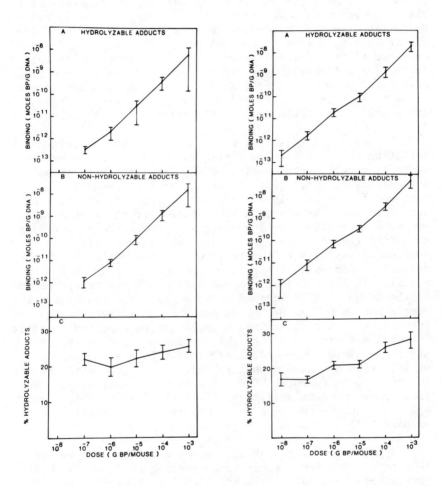

FIGURES 2 AND 3. Binding of ^{3}H-G-BaP to the DNA of stomach (left, Fig. 2) and liver (right, Fig. 3). Data are mean ±S.D. for groups of 9 or 10 animals per point which were administered either 10 ng ^{3}H-G-BaP (Fig. 3, dose of 10^{-8} g) or 100 ng ^{3}H-G-BaP with or without addition of non-radioactive BaP to adjust the specific activity to give the desired dose.

DISCUSSION

There was insufficient radioactivity associated with DNA in the current experiments to properly explore the relationship between the "non-hydrolyzable" adducts from acid hydrolysis and the "hydrophilic" adducts from the more commonly employed enzyme hydrolysis procedure. It seems likely, however, that these two procedures are measuring the same or closely related products. Reaction products, such as those between BaP 7,8-dihydrodiol-9,10-epoxide and DNA bases, have been implicated as major adducts formed in hepatocytes and other cells (5). In the current experiments, the majority of the radioactivity associated with DNA in four different organs does not appear to be in the form of such adducts, which are acid-hydrolyzable and result in hydrophobic products after enzyme hydrolysis.

It seems unlikely that DNA-associated radioactivity resistant to acid hydrolysis results simply from the exchange of tritium onto normal nucleosides. Firstly, there is no substantial or consistent difference in the amount or proportion of "non-hydrolyzable" adducts when a different form of tritiated BaP, ^{3}H-6-BaP, is substituted for the more commonly employed generally labelled material. For exchange reactions, such differences would be expected, due to the differing degrees of lability of the tritium at different sites on the molecule. Secondly, the bulk of non-hydrolyzable radioactivity in liver DNA is removed during a 7-day repair period in the animal, while radioactivity in isolated DNA is stable over a period of months. This suggests an active repair process in living cells, which would not be expected to exist for radioactivity on normal DNA nucleosides. Finally, although they are commonly ignored or eliminated by pre-analysis purification steps, hydrophilic adducts are frequently reported in experiments in which DNA is degraded enzymatically. These include experiments utilizing ^{14}C-labelled polycyclic aromatic hydrocarbons, where there is no possibility of tritium exchange (6).

Typical levels of BaP in various foods range from 0.1 to 10 ng/g (1,2). For a mouse consuming 5 g food per day, this corresponds to a daily dose of 0.5-50 ng of carcinogen. The lowest doses of BaP used in the current study, 10 and 100 ng, thus approximate levels which might be encountered as food contaminants. The highest dose of carcinogen used, 1 mg, is somewhat greater than the lowest dose which has been reported to induce stomach tumors in mice (0.2 mg) (7). The data on BaP binding to DNA thus span the range from those doses typical of foods to those doses which are capable of inducing tumors in a substantial proportion of treated animals. Over this entire range of 100,000 fold (liver) or 10,000 fold (stomach), the degree of binding of orally administered benzo(a)pyrene is directly related to dose. There is neither evidence of a threshold effect nor of increased binding at high doses due to induction of metabolizing enzymes. In addition to the relative linearity of the dose-response curves, the relative proportion of hydrolyzable and non-hydrolyzable adducts changes either only modestly (liver) or not at all (stomach) as the dose is varied. These data are similar to those found by Pereira et al., who obtained a linear dose-response relationship for the binding of topically applied benzo(a)pyrene to mouse epidermal DNA (6). In addition, in the liver, the efficiency of the subsequent removal of BaP-DNA adducts over a period of one week does not change as the administered carcinogen dose is varied over a range of 100,000. These results suggest that, unless some stage in carcinogenesis occurring subsequent to DNA binding is dose-dependent, the carcinogenic risk of polycyclic aromatic hydrocarbons is likely to be directly proportional to dose, even at very low doses. This in turn suggests that, in the absence of other information, linear non-threshold models may be the most appropriate, conservative choice for estimating the risks of polycyclic aromatic hydrocarbons in human foods.

ACKNOWLEDGMENTS

The support of Health and Welfare Canada, in the form of a Research Scholar Award to the author, is gratefully acknowledged.

REFERENCES

1. Lo, M.T., and Sandi, E. (1978): Polycyclic aromatic hydrocarbons (polynuclears) in foods, Residue Rev., 69: 35-86.
2. Howard, J.W., and Fazio, T. (1980): Review of polycyclic aromatic hydrocarbons in foods, J. Assoc. Off. Anal. Chem., 63:1077-1104.
3. Koreeda, M., Moore, P.D., Wislocki, P.G., Levin, W., Conney, A.H., Yagi, H., and Jerina, D.W. (1978): Binding of benzo(a)pyrene 7,8-diol-9,10-epoxides to DNA, RNA, and protein of mouse skin occurs with high stereoselectivity, Science, 199:778-781.

4. Baird, W.M., and Brookes, P. (1973): Isolation of the hydrocarbon-deoxynucleoside products from the DNA of mouse embryo cells treated in culture with 7-methylbenz-(a)anthracene-^{3}H, Cancer Res., 33:2378-2385.
5. Ashurst, S.W., and Cohen, G.M. (1980): A benzo(a)-pyrene-7,8-dihydrodiol-9,10-epoxide is the major metabolite involved in the binding of benzo(a)pyrene to DNA in isolated viable rat hepatocytes, Chem.-Biol. Interact., 29:117-127.
6. Pereira, M.A., Burns, F.J., and Albert, R.E. (1979): Dose response for benzo(a)pyrene adducts in mouse epidermal DNA, Cancer Res., 39:2556-2559.
7. Pierce, W.E.H. (1961): Tumor-promotion by lime oil in the mouse forestomach, Nature, 189:497-498.

DETERMINATION OF PAH POLLUTION AT COKE WORKS

WERNER EISENHUT, ERNST LANGER, CLAUS MEYER
Bergbau-Forschung GmbH, Franz-Fischer-Weg 61,
D-4300 Essen-Kray, Federal Republic of Germany.

INTRODUCTION

Polynuclear Aromatic Hydrocarbons (PAH) arise from each kind of pyrolytic decomposition of organic matter, as _e.g._ carbonization of coal. For better understanding the resulting pollution problems, the technique of coke manufacture should be briefly described.

First, coal is fed into the coke oven by a charging car. Then, the charging holes are closed by lids. Heating up to 1000°C or higher is obtained indirectly by heated gas. During this carbonization period, coal decomposes into coke and volatile matter, which are composed of a lot of chemical substances including PAH. The volatiles are sucked off continuously from the oven, pass through one or two stand pipes and other installations and after cooling yield by-products such as coke oven gas and tar. When carbonization is finished after roughly 20 hours, the coke oven doors are opened and the coke is pushed out by a machine, passes the coke guide car and falls into the quenching car. By this car, it is brought to the quenching tower and quenched with water.

Because coke ovens and by-product plants, during carbonization, form a more or less closed system, most of the PAH formed are condensed within this system after cooling and yield tar. However, when opening the coke oven, such as when filling holes, if stand pipes and doors are not sufficiently sealed, volatile matter can be emitted. The discharged PAH are condensed into the solid state, where they are mainly associated with particles. Emissions may also arise from charging and pushing.

Determination of PAH pollution at coke works includes measuring of:
- work place exposure,
- emissions from different sources,
- pollution in the neighbourhood.

Several PAH investigations have been carried out by various scientists. Their differing results may depend on a number of factors such as the coal rank, charging method, carbonizing temperatures and times, design, and above all, the age and state of maintenance of the coke oven plant. A further factor influencing PAH emissions is the efficiency of emission reducing equipment. Because all these factors are not yet sufficiently studied, Bergbau-Forschung GmbH started in 1977 a detailed PAH measuring program with the aid of the Department of Commerce of Northrhine-Westfalia. First results will be given in this paper.

SAMPLING AND ANALYSIS

Because PAH are in the solid state and associated with particulate matter, sampling is the same as in case of particulates. For determination of work place exposure, sampling is carried out with personal samplers. We use samplers of the TBF-50 (1) and the Casella (2) type. For determination of PAH in emissions, the individual source, e.g. the door (Figure 1) is covered with a special suction device.

FIGURE 1. Suction device for collecting emissions from coke oven doors.

Pollution in the neighbourhood should be collected by high volume samplers, but until now we did not yet carry out this measurement.

Analysis begins with extraction of PAH. As the solvent, we normally use benzene. However, because of the toxicity of benzene, we worked additionally with cyclohexane and in the future we will only use this solvent.

The extract is fractionated by chromatographic methods. First, we used thin-layer chromatography (TLC) and determined PAH by fluorescence spectrometry. Then we worked with gas chromatography (GC) and FID. Hereby, an internal standard is used and 12 individual PAHs are identified by their relative retention times. As parallel runs gave good agreement between TLC and GC, we prefer the second method for our further work.

WORK PLACE EXPOSURE

After measuring dust concentration at different work places and different plants, we obtained first results on PAH exposure by determining the benzene soluble material (BSM).

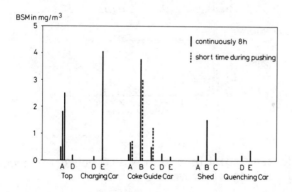

FIGURE 2. Benzene soluble material at different work places and different plants.

As Figure 2 shows, there is a great difference between individual results, giving BSM concentrations from 0.1 to 4.0 mg/m³. When it is assumed that, according to general experience, 1% of BSM is benz(a)pyrene (BaP), BaP is estimated to be 1 to 40 µg/m^3.

In a subsequent series of tests, we took samples at oven top and at working platforms of both the pusher side and coke side. Sampling was carried out parallel with the TBF-50 and the Casella sampler, which had been installed stationary. Analysis was made on BSM and BaP. As Figure 3 shows, BSM varies from 0.5 to 2.5 mg/m^3 on top and from 0.1 to 1.0 mg/m^3 on working platforms. BaP is between 5 and 15 µg/m^3 on top and between 0.04 and 3.3 µg/m^3 on working platforms. Depending on different sampling techniques, the Casella instrument gives higher values than the TBF-50 f instrument does. This shows that the sampling method should be taken into account when values from different authors are compared.

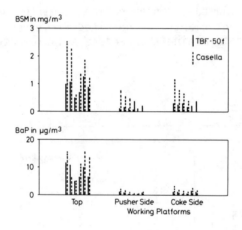

FIGURE 3. Benzene soluble material and benzo(a)pyrene at oven top and working platforms.

Tests with portable personal samplers are planned, but have not yet been made. It is well known, that this kind of measurement gives considerably lower exposure values, because workers usually stay in places where no dust and no fumes are to be seen.

EMISSIONS FROM DIFFERENT SOURCES

Since 1973, we have measured dust emissions from coke charging. The measuring includes determination of the benzene soluble materials. We found BSM emissions from 0.2 to 5.5 g per ton of coke, the mean being 1.3 g/t. When BaP is estimated again to be 1 %, 13 mg per ton of coke are emitted. It is to be pointed out that in Germany all charging cars are equipped with emission control facilities, either of the wet scrubber or of the gas exhausting type.

PAH emissions from pushing have not yet been measured but they seem to be small when coke is well carbonized. Emissions from filling holes and stand pipes are found to be nearly zero when the holes are luted and the pipes are water sealed.

The major sources of PAH emissions from coke works are coke oven doors. For diminishing emissions, Bergbau-Forschung GmbH carries out a door research and development program with Ruhrkohle AG and several engineering companies, which is financially supported by the German Federal Bureau of Environment and the Commission of European Communities. First measurements of PAH at four doors of 6m height have shown that BaP emissions, which are 16 and 82 mg per ton of coke in case of conventional doors with rigid sealing, can be reduced to 5 and 13 mg/t when doors are equipped with spring loaded seals.

Within this series of measurements, it was the first time that we analysed 12 individual PAHs. The development of emission during one carbonization period and the concument determined emission factors of fluoranthene, pyrene, benzo(a)anthracene, chrysene, benzo(e)pyrene, benzo(a)pyrene, perylene, benzo(b)fluoranthene plus benzo(j)fluoranthene, benzo(k)fluoranthene, indeno (1,2.3 - cd)pyrene, dibenzo(a,h)-anthracene and benzo(g,h,i)perylene are shown in Figure 4. However, even if the values are very important, they should only be taken as a first investigation.

DETERMINATION OF PAH POLLUTION AT COKE WORKS

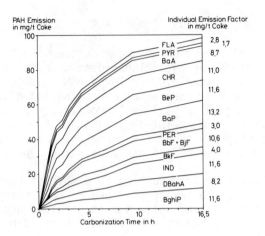

FIGURE 4. PAH emission from coke oven doors of 6m height.

DISCUSSION

The results show, that still a lot of work is to be done. The differences we obtained with the TBF-50 f and the Casella sampler prove the findings of P.W. Jones (3) who pointed out that "Sampling is probably the most critical stage ...; in the absence of a reproducible and quantitative sampling procedure, sophisticated and sensitive analytical procedures are of no value". Nevertheless, work place exposure is found to be in a wide range like the results from other investigators, on which we published at the PAH-Colloquium in Hannover (4). It is obvious, that even with modern emission control facilities at coke works it will be difficult to meet the US limit for coke ovens which is 0.15 mg BSM per m^3.

Emissions from coke oven doors are found to be less than 100 mg BaP/ton of coke, and from other sources they are estimated to be considerably lower. This shows that an emission factor of 1.8 g/t, which is frequently used, e.g., World Health Organization (5) and is reported to be published by M.W. Smith (6) - where we did not find this factor - does not characterize the magnitude of emissions at least of German and other coke works with usual emission control facilities. We estimate that maximally 200 mg BaP are emitted per ton of coke.

REFERENCES

1. VDI-Richtlinie 2463/3: Messen der Massenkonzentration von Partikeln in der Außenluft. TBF-50 f Filterverfahren.
2. Available from Casella London Ltd., Regent House, Britannia Walk, London, N1 7ND.
3. Jones, P.W. (1980): Measurement of PAH emissions from stationary sources-an overview. VDI-Berichte Nr. 358, pp. 23-38.
4. Eisenhut, W. und G. Zimmermeyer (1980): Messung von PAH-emissionen aus Rokereien. VDI-Berichte Nr. 358, pp. 113-119.
5. Environmental Carcinogens Selected Methods of Analyses (1979). Jarc Publication No. 29, WHO, International Agency for Research on Cancer, Lyon, pp. 32.
6. Smith, M.W. (1970): Evaluation of coke oven emissions. Presented to the 78th General Meeting of the American Iron and Steel Institute, New York City, May 28th.

COMPARATIVE METABOLISM IN VITRO OF 5-NITROACENAPHTHENE AND 1-NITRONAPHTHALENE

KARAM EL-BAYOUMY AND STEPHEN S. HECHT
Division of Chemical Carcinogenesis, Naylor Dana Institute for Disease Prevention, American Health Foundation, Valhalla, New York 10595.

INTRODUCTION

Nitropolynuclear aromatic hydrocarbons (NPAH) are an important class of compounds due to their mutagenic and carcinogenic activities and their possible presence in the environment (1-5). Our interest in two members of this class, 5-nitroacenaphthene and 1-nitronaphthalene, was stimulated by the results of comparative carcinogenicity and mutagenicity assays (6-8).

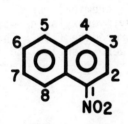
1-NITRONAPHTHALENE

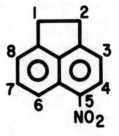
5-NITROACENAPHTHENE

5-Nitroacenaphthene is mutagenic toward S. typhimurium with and without activation and induces tumors in rats, mice, and hamsters (6,7,9-11). 1-Nitronaphthalene is less mutagenic than 5-nitroacenaphthene and appears to be non-carcinogenic (8). Thus the presence of a 2 carbon bridge (C_1-C_2) in 5-nitroacenaphthene seems to be a key structural factor in its biological activities.

METABOLISM OF 5-NITROACENAPHTHENE AND 1-NITRONAPHTHALENE

In the present study, we have determined and compared the major metabolic pathways of both compounds under conditions similar to those employed in the mutagenicity assays. The results indicate that upon incubation with rat liver 9000xg supernatant from Aroclor pretreated rats in the presence of cofactors, 5-nitroacenaphthene is both oxidized and reduced while 1-nitronaphthalene is oxidized (ring hydroxylation) with no detectable nitro reduction. The data in this report could explain the comparative mutagenic response of both compounds toward S. typhimurium under identical conditions.

MATERIALS AND METHODS

Chemicals

The syntheses of 5-nitroacenaphthene and its metabolites were carried out as described previously (12). 1-Nitronaphthalene was obtained from the NCI chemical repository (No. 360 Lot # MI-3-72). 1-Naphthylhydroxylamine was synthesized according to a literature procedure (13). The commercial sources for other chemicals were identical to our previous studies (12).

Mutagenicity Assays

All samples tested for mutagenicity were at least 99% pure by HPLC analysis. Studies were performed as previously described (12,14) using S. typhimurium TA98 and TA100, both kindly provided by Dr. B. Ames of the University of California, Berkeley. The procedure of Ames et al. (15) was employed in performing these assays. Since the response of TA98 and TA100 were similar, the only assays reported here are for TA98.

Metabolism Studies

Incubations were typically carried out in 50 ml Erlenmeyer flasks containing 20 ml of S-9 mix and 2.0 mg of substrate which had been dissolved in 200 μl DMSO. The ratio of S-9 mix to substrate was the same as that which gave the maximum response in the mutagenicity assay of 5-nitroacenaphthene i.e. 200 μl/plate S-9 mix per 20 μg of 5-nitroacenaphthene. Control experiments were performed as des-

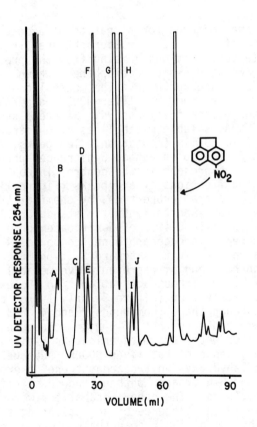

Figure 1. High pressure liquid chromatogram of metabolites formed from incubation of 5-nitroacenaphthene with rat liver 9000xg supernatant for 60 min. Peaks were identified as described in reference 12: A (14, Figure 2); B (15), C (12), D (6), E (13), F (7), G (4), H (5), I (2), J (3).

cribed above except that heat denatured S-9 mix was used. The preparations of 9000xg supernatant, S-9 mix, and HPLC analysis of metabolites were described previously (12).

RESULTS

Conventional methods for nitration of acenaphthene give a mixture of 3- and 5-nitroacenaphthene (16). To avoid possible complications arising from the presence of the 3-isomer, we used 5-nitroacenaphthene which was synthesized from 5-bromoacenaphthene. The metabolic studies were carried out with the same rat liver 9000xg supernatant as employed in the mutagenicity assays. Figure 1 shows a HPLC trace of the EtOAc soluble metabolites of 5-nitroacenaphthene formed in a 60 min incubation. Ten peaks, A-J, were observed in addition to unchanged 5-nitroacenaphthene. Peaks A-J were not present in the control incubations. The structural identifications of all peaks (Figure 2) were established by spectral properties, by chemical transformations, and by comparison to synthetic standards (12). The major primary metabolites of 5-nitroacenaphthene were 1- and 2-hydroxy-5-nitroacenaphthene (compounds 4 and 5).

The mutagenic activities toward S. typhimurium TA98 of 5-nitroacenaphthene and of representative metabolites were determined with and without activation. In the experiments conducted with activation (Figure 3) only 1-oxo- and 1-hydroxy-5-nitroacenaphthene (compounds 2 and 4) had mutagenic activities comparable to that of 5-nitroacenaphthene. The other metabolites were less mutagenic than 5-nitroacenaphthene. In the absence of rat liver 9000xg supernatant (Figure 4), only 1-hydroxy-5-nitroacenaphthene, 1-oxo-5-nitroacenaphthene and 5-nitroacenaphthene showed significant mutagenic activities. The other metabolites were less mutagenic than in the presence of an activating system.

The metabolism of 1-nitronaphthalene was carried out under the same conditions employed for 5-nitroacenaphthene. Figure 5 shows a HPLC trace of the EtOAc soluble metabolites of 1-nitronaphthalene formed in a 60 min incubation. Two major peaks were identified tentatively as nitronaphthol and a dihydrodiol derived from 1-nitronaphthalene. We did not detect any products arising from nitro reduction i.e. neither 1-naphthylhydroxylamine nor

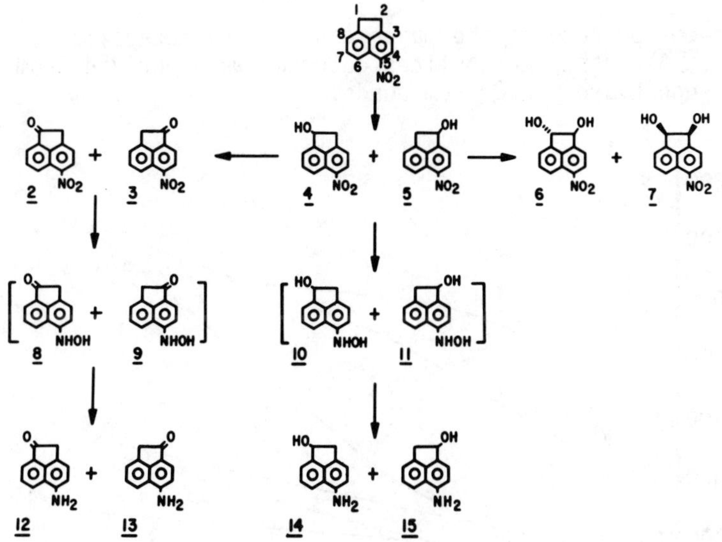

Figure 2. The metabolic pathways for 5-nitroacenaphthene (Compound 1) in vitro.

1-naphthylamine was found. In order to establish whether or not 1-naphthylhydroxylamine was formed metabolically as a transient intermediate, we studied the metabolism of 1-naphthylhydroxylamine under the same conditions. Its major metabolite was 1-naphthylamine; the latter was not detected in control experiments with heat deactivated S-9.

DISCUSSION

The results of this study demonstrate that the rat liver 9000xg supernatant, isolated from animals that had been treated with Aroclor, is capable of catalyzing both the oxidation and reduction of 5-nitroacenaphthene. Oxidation to give 1-hydroxy-5-nitroacenaphthene and 2-hydroxy-

5-nitroacenaphthene is the major and primary metabolic process. All other metabolites detected were derived from both bridge hydroxylated compounds.

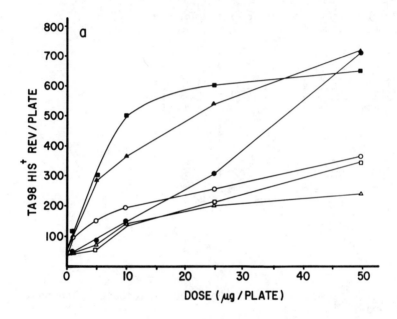

Figure 3. Comparative mutagenicity toward S. typhimurium TA98 of 5-nitroacenaphthene (Compound 1) ■——■ , 1-hydroxy-5-nitroacenaphthene (Compound 4) ●——●, 1-oxo-5-nitroacenaphthene (Compound 2) ▲——▲, cis-1,2-dihydroxy-5-nitroacenaphthene (Compound 7), o——o, 1-oxo-5-aminoacenaphthene (Compound 12) □——□, and 1-hydroxy-5-aminoacenaphthene (Compound 14) △——△ in the presence of rat liver S9 fraction.

Reduction of the nitro group in compounds 2 - 5 gave the keto-amino and hydroxy-amino derivatives (compounds 12 - 15) presumably via the hydroxylamines (compounds 8 - 11). We did not detect products resulting from further reduction of the dihydroxynitro metabolites (compounds 6 and 7). We also did not detect 5-aminoacenaphthene. These results indicate the relative ease of oxidation compared to reduction of 5-nitroacenaphthene. Analogous results have been reported in a study of the metabolism of 2,4-dinitrotoluene by rat primary hepatocytes (17).

The results of the mutagenicity assays indicate that 1-hydroxy-5-nitroacenaphthene and 1-oxo-5-nitroacenaphthene are the proximate mutagens of 5-nitroacenaphthene. Their mutagenic activities are probably expressed through the corresponding nitroso or hydroxylamine derivatives since both compounds were mutagenic without the activating system and S. typhimurium is known to possess a family of nitroreductase systems (18).

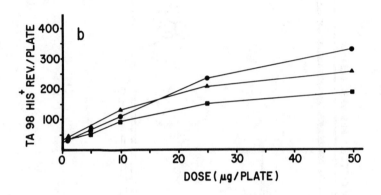

Figure 4. Comparative mutagenicity toward S. typhimurium TA98 of 5-nitroacenaphthene (Compound 1) ■——■, 1-hydroxy--5-nitroacenaphthene (Compound 4) ●——●, and 1-oxo-5-nitroacenaphthene (Compound 2) ▲——▲ in absence of rat liver S9 fraction. Compounds 7, 12, and 14 were not mutagenic in absence of S9 fractions.

We did not test 2-hydroxy- and 2-oxo-5-nitroacenaphthene and their corresponding amino derivatives but their mutagenic activities would be expected to be similar to those of the 1-substituted compounds. cis-1,2-Dihydroxy-5-nitroacenaphthene (compound 7) can be regarded as a detoxification product since it was less mutagenic than 5-nitroacenaphthene, with activation, and was inactive in the absence of rat liver 9000xg supernatant. The lack of mutagenic activity of both 1-keto- and 1-hydroxy-5-aminoacenaphthene (compounds 12 and 14) in the absence of rat liver 9000xg supernatant is typical for aromatic amines.

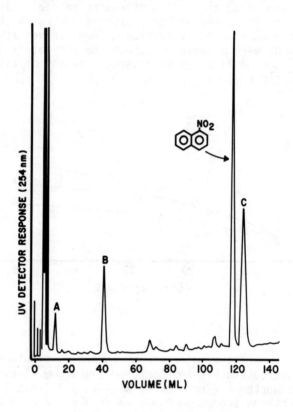

Figure 5. High pressure liquid chromatogram of metabolites formed from incubation of 1-nitronaphthalene with rat liver 9000xg supernatant for 60 min.

Rat liver 9000xg supernatant apparently catalyzes primarily ring oxidation, in the case of 1-nitronaphthalene. The major metabolites appear to be dihydrodiols (peaks A and B) and a nitronaphthol isomer (peak C) based on their mass spectral and chemical properties. We did not detect any products, such as 1-naphthylamine, that would have been formed by nitro reduction although the incubations were performed under conditions identical to those used for 5-nitroacenaphthene. It is possible however that secondary metabolites which are products of nitro reduction could have been formed, but were not extracted by ethyl acetate. The metabolites formed from 1-nitronaphthalene could be considered as detoxification products.

It can be concluded from this study that nitro reduction is an important pathway in the metabolic activation of 5-nitroacenaphthene to a mutagen. 1-Nitronaphthalene apparently is not reduced as readily as 5-nitroacenaphthene under the same conditions. These results could explain the relatively high mutagenic activity of 5-nitroacenaphthene compared to 1-nitronaphthalene.

ACKNOWLEDGEMENTS

This study was supported by the National Institute of Environmental Health Sciences Grant ES 2477-01.

REFERENCES

1. Hoffmann, D. and Rathkamp, G. (1970): Quantitative determination of nitrobenzenes in cigarette smoke. Anal. Chem., 42: 1643-1647.
2. Mermelstein, R., Kiriazides, D.K., Buher, M., Sanders, D.R., McCoy, E.C., and Rosenkranz, H.S. (1980): Nitropyrenes: isolation, identification, and reduction of mutagenic impurities in carbon black and toners. Sciene, 209: 1039-1042.
3. Pederson, T.C. and Siak, J-S. (1981): Involvement of nitro-substituted polycyclic aromatic hydrocarbons in the mutagenic activity extracted from diesel exhaust. The Toxicologist, 1: 74, Abstr. No. 268.
4. Pitts, J.N., van Cauwenberge, K.A., Grosjean, D., Schmid, J.T., Fitz, D.R., Belser, W.L., Knudson, G.B., and Hyndes, P.M. (1978): Atmospheric reactions of polycyclic aromatic hydrocarbons: facile formation of

mutagenic nitro derivatives. Science, 202: 515-518.
5. Wang, Y.Y., Rappaport, S.M., Sawyer, R.F., Talcott, R.E., and Wei, E. (1978): Direct-acting mutagens in automobile exhaust. Cancer Lett., 5: 39-47.
6. McCann, J., Choi, E., Yamasaki, E., and Ames, B.N. (1975): Detection of carcinogens as mutagens in the Salmonella/microsome test: Assay of 300 Chemicals. Proc. Natl. Acad. Sci. (U.S.A.), 72: 5135-5139.
7. National Cancer Institute (Bethesda, MD., U.S.A.) Report 78 (1979), DHEW/Pub/NIH-78-1373, NCI-CG-TR-118; Order No. PB-287347. Bioassay of 5-nitroacenaphthene for possible carcinogenicity.
8. National Cancer Institute (Bethesda, MD., U.S.A.) Technical Report Series No. 64 (1978). DHEW/Pub/NIH-78-1314, NCI-CG-TR-64. Bioassay of 1-nitronaphthalene for possible carcinogenicity.
9. International Agency for Research on Cancer (1978). Some Aromatic Amines and Related Nitro-Compounds - Hair Dyes, Coloring Agents and Miscellaneous Industrial Chemicals, IARC Monographs on the Evaluation of the Carcinogenic Risk of Chemicals to Man, Vol. 16, Lyon, France.
10. Tokiwa, H., Nakagawa, R., and Ohnishi, Y. (1981): Mutagenic assay of aromatic nitro compounds with Salmonella typhimurium. Mutat. Res., 91: 321-325.
11. Yahagi, T., Shimizu, H., Nagao, M., Takemura, N., and Sugimura, T. (1975): Mutagenicity of 5-nitroacenaphthene in Salmonella. Gann, 66: 581-582.
12. El-Bayoumy, K. and Hecht, S.S. (1981): Metabolism in vitro of 5-nitroacenaphthene. Cancer Res., (in press).
13. Willstätter, R. and Kubli, H. (1908): Über die Reduktion von Nitroverbindungen nach der Methode von Zinn. Chem. Ber., 4: 1936-1940.
14. El-Bayoumy, K., LaVoie, E.J., Hecht, S.S., Fow, E.A., and Hoffmann, D. (1981): The influence of methyl substitution on the mutagenicity of nitronaphthalenes and nitrobiphenyls. Mutat. Res., 81: 143-153.
15. Ames, B.N., McCann, J., and Yamasaki, E. (1975): Methods for detecting carcinogens and mutagens with the Salmonella/mammalian-microsome mutagenicity test. Mutat. Res., 31: 346-364.
16. Jones, L.A., Joyner, C.T., Kim, H.K., and Kyff, R.A. (1970): Acenaphthene I. the preparation of derivatives of 4,5-diaminonaphthalic anhydride. Can. J. Chem., 48: 3132-3135.

17. Bond, J.A. and Rickert, D.E. (1981): Metabolism of 2,4-dinitro[^{14}C]toluene by freshly isolated Fischer-344 rat primary hepatocytes. Drug Metab. Disp., 9: 10-14.
18. Speck, W.T., Bulmer, J.L., Rosenkranz, E.J., and Rosenkranz, H.S. (1981): Effect of genotype on mutagenicity of niridazole in nitroreductase-deficient bacteria. Cancer Res., 41: 2305-2307.

USE OF MIXED PHASES FOR ENHANCED GAS-CHROMATOGRAPHIC SEPARATION OF POLYCYCLIC AROMATIC HYDROCARBONS. III. PHASE TRANSITION BEHAVIOR, MASS-TRANSFER NON-EQUILIBRIUM, AND ANALYTICAL PROPERTIES OF A MESOGEN POLYMER SOLVENT WITH SILICONE DILUENTS

H. FINKELMANN †, R. J. LAUB*, W. L. ROBERTS*, AND C. A. SMITH*
†Institut für Physikalische Chemie der Technischen Universität Clausthal, D-3392 Clausthal-Zellerfeld, West Germany;
*Department of Chemistry, The Ohio State University, Columbus Ohio 43210.

ABSTRACT

Previous studies of the gas-chromatographic separation of polycyclic aromatic hydrocarbons have shown that open-tubular (capillary) columns containing liquid crystal solvents in admixture with silicone gums exhibit both high efficiency and high selectivity. Detailed here, for the first time, are selected analytical and physicochemical properties of a representative example of a unique class of phases which comprise silicone polymers with mesogenic side-chains. The retentions and relative separations achieved with the polymer resemble those found with a blend of 20% (w/w) BBBT with SE-52. However, the polymer exhibits a higher useful temperature range, lower volatility, and higher column efficiency. Relative retentions (alpha values) can differ by as much as a factor of three over a five-degree temperature range, and blending the mesogenic polymer with silicone diluents lowers the nematic transition temperature by as much as eighty degrees. The properties of mesogenic silicones thus provide an unprecedented spectrum of adjustable parameters for the quantitative control of solute retention and retention order, that is, a means of optimization of separations.

INTRODUCTION

Mesogenic (liquid-crystal) solvents enable the gas chromatographic separation of polycyclic aromatic hydrocarbons on the basis, almost entirely, of solute geometry (1). Use of such highly selective phases is generally avoided in preference to those of high efficiency, however, since, broadly speaking, those of the former type give rise to band-broadening (i.e., poor efficiency) due in large measure to unfavorable kinetics of mass transfer. While the trade-off between high efficiency and high selectivity has been overcome in this laboratory by the use of blends of phases (1-3), there remains the inconvenience in practice of fabrication of columns containing mixed solvents (or attachment in series of lengths of pure-phase columns) for each new separation. A solution to the difficulty, as voiced by us previously (3), appears to be combination of the two desired properties, namely, efficiency and selectivity, in a single pure phase. We therefore chose to investigate mesogenic polysiloxanes (MPMS); we report here preliminary studies of the utility of such solvents and offer comparison of these with those of our previous work which comprised binary phases.

EXPERIMENTAL

PAH standards used in this work were obtained from Chem Service, artificial mixtures of which were made with spectro-grade benzene from MCB. The silicone gums SE-30 and SE-52, and N,N'-bis-(p-butoxybenzylidene)-bis-p-toluidine (BBBT) liquid crystal (mp 159°C; 188° → nematic → 303°C isotropic) were obtained from Alltech Associates. Synthesis of the liquid crystalline polysiloxane was carried out by addition of a vinyl-substituted mesogen to a poly(hydrogenmethylsiloxane): the appropriate vinyl derivative and the polysiloxane were taken up in tetrahydrofuran to which was added 100 ppm of a metal catalyst. The solution was left standing for 12 hr at 50°C, following which the resultant mesogenic polymer was reprecipitated several times from methanol and dried under vacuum. The glass-capillary gc columns were fabricated in the usual manner (2,3), were 10 m in length by 0.2 mm i.d., and were coated with films of phase of 0.3 μm. The gas chromatograph was a Carlo Erba model 4160 which was used with split-mode injection (hydrogen carrier).

RESULTS AND DISCUSSION

Fig. 1 provides illustration at three temperatures of the separation of several impurities (primed) surrounding phenanthrene (1) and anthracene (2) solutes with a column containing 20% (w/w) BBBT + 80% SE-52. The change in elution order of, for example, nos. 6' - 9' is quite remarkable over the small temperature range considered, but is in fact consistent with our findings generally of the properties of liquid crystal stationary phases (3). Shown in Fig. 2 are the respective solute van't Hoff plots (here, log k' against $10^3 T^{-1}$), where linearity is extant both in the smectic and nematic regions but where significant changes in elution are clearly apparent at the transition temperature. Since the smectic form of the mesogen is a poor solvent for the aromatic hydrocarbon solutes, i.e., contributes little to their retentions, the effective con-

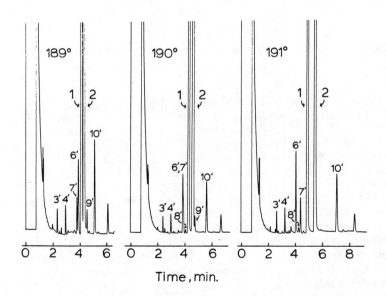

FIGURE 1. Chromatograms of solutes surrounding phenanthrene (no. 1) and anthracene (no. 2) at indicated temperatures with an open-tubular column containing 20% (w/w) BBBT + 80% SE-52 stationary phases.

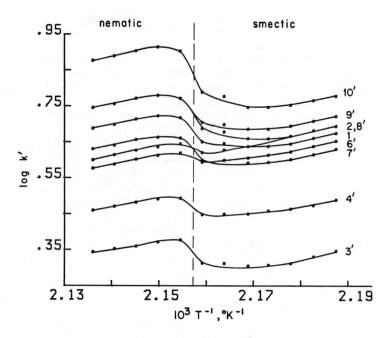

FIGURE 2. Plots of log k' against $10^3 T^{-1}$ for solutes of Fig. 1.

centration of BBBT in the binary phase is reduced from 20% to zero on passing below 188°C. In the intermediate range 188° to 191°, in contrast, some nematic character (presumed to be proportional to T) is present, hence the solute elution order is altered. Utilization of the column over this temperature span thus effectively mimicks changing physically the composition of the solvent. [The situation is therefore analogous to (but opposite in sense from) simulation of constant column temperature with binary or ternary blends of phases (4,5).] The effect of temperature upon resolution of phenanthrene from anthracene is dramatized in Fig. 3, where k' is plotted against T: the relative separation of the solutes changes abruptly from virtual unity at 180° to ca. 1.5 at 195°C, just as if two columns (one of SE-52 and the other of BBBT) had been employed. The inconvenience of fabrication of new columns for each separations task encountered can thereby be overcome at least in part by careful control of the system temperature with blends of phases of these types.

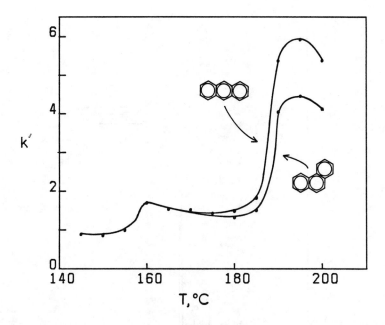

FIGURE 3. Plots of k' against T for phenanthrene and anthracene solutes with an open-tubular column containing pure BBBT.

We illustrate next the utility of a single (pure) phase with which the same ends can be achieved. Fig. 4 shows the separation at 85°C of biphenyl (1), 2,6-dimethylnaphthalene (4), and 1,3-dimethylnaphthalene (5) with a 10-m column containing SE-30. 2-Ethylnaphthalene (2), 1-ethylnaphthalene (3), 1,4-dimethylnaphthalene (6), 2,3-dimethylnaphthalene (7), and diphenylmethane (8), however, are not resolved at this temperature. Fig. 5 illustrates the resolution achieved with a column containing the mesogenic polysiloxane at 100°C. Fig. 6, that for the same phase at 70°C, shows complete separation of all components. The elution order in the latter two figures obviously differs considerably from that achieved with SE-30. Further, the order is changed with the former phase (Figs. 5,6) simply by passing from the nematic to the isotropic region of this mesogen ($T_{n \to i}$ = 97°C). Plots of plate height H against linear carrier velocity (van Deemter curves), Fig. 7, moreover reveal that the efficiency achieved in the nematic

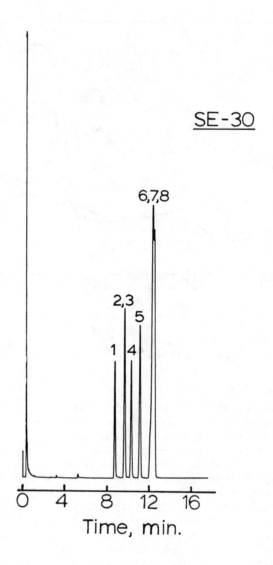

FIGURE 4. Chromatogram of naphthalenes and diphenylmethane with an open-tubular column containing pure SE-30. T = 85°C.

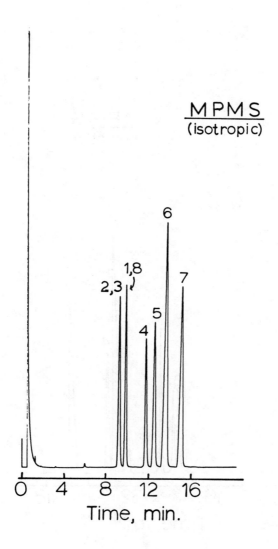

FIGURE 5. Chromatogram of the solutes of Fig. 4 with the pure mesogenic polysiloxane at 100°C.

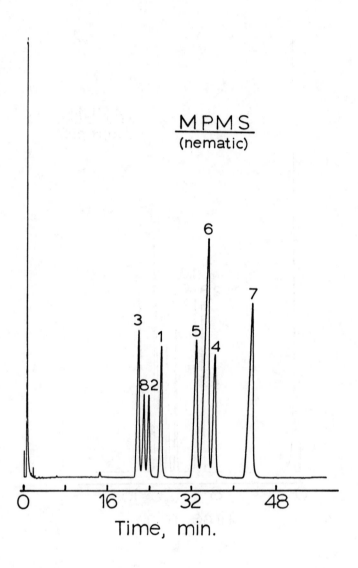

FIGURE 6. As in Fig. 5; T = 70°C.

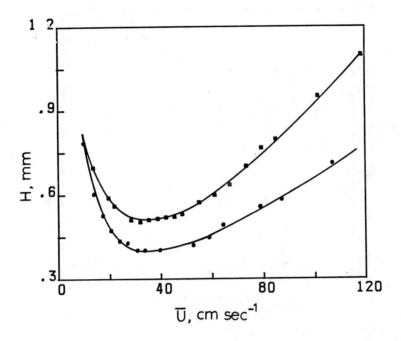

FIGURE 7. van Deemter plots for naphthalene (90°C; squares) and 2-methylnaphthalene (110°C; circles) solutes for the column of Fig. 5.

region (90°; naphthalene solute) is somewhat lower than that in the isotropic region (110°C; 2-methylnaphthalene solute) (temperatures chosen to yield a capacity factor of 7 for both solutes) which is undoubtedly a consequence of the well-known poor mass-transfer characteristics of nematic solvents. However, the chromatographic efficiency of the polymer, even when in the nematic state, far exceeds that of discrete mesogens such as BBBT (cf., for example, Figs. 1a-c of ref. 2). Finally, Fig. 8 provides illustration of the effect of dilution of the mesogen polymer with SE-52: the nematic/isotropic transition temperature is reduced geometrically as the weight fraction of the former is decreased, a result most likely due to the formation of a biphasic domain between the nematic and isotropic states of the liquid crystal (6).

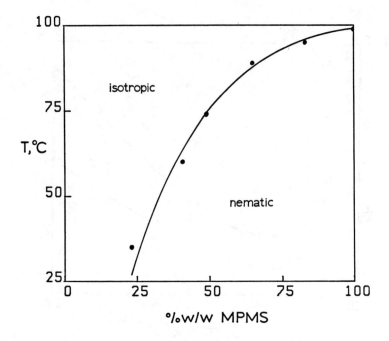

FIGURE 8. Plot of nematic → isotropic transition temperature for the mesogenic polysiloxane as a function of dilution in SE-52. Curve generated by following the separation of 1 + 2 methylnaphthalenes.

We assert by way of summary that incorporation of mesogenic monomers onto polysiloxane backbones yields stationary phases with concomitant high efficiency and high selectivity, and that these are comparable to discrete mesogens diluted in silicone matrices. Furthermore, mesogenic polymers with higher nematic (as well, of course, as chiral) ranges would clearly be of some utility in the analysis of solutes of low volatility. Synthesis of compounds of these types is currently under study in this laboratory, the results of which we hope soon to report.

ACKNOWLEDGMENTS

We gratefully acknowledge support received in part from the Department of Energy and from the National Science Foundation.

REFERENCES

1. Laub, R.J. and Roberts, W.L. (1980): In: <u>Polynuclear Aromatic Hydrocarbons: Chemistry and Biological Effects</u>, A. Bjørseth and A.J. Dennis, Eds., Battelle Press, Columbus, Ohio, p. 25.
2. Laub, R.J., Roberts, W.L., and Smith, C.A. (1980): <u>J. High Resolut. Chromatogr. Chromatogr. Commun.</u>, 3:355.
3. Laub, R.J., Roberts, W.L., and Smith, C.A. (1981): In: <u>Polynuclear Aromatic Hydrocarbons: Chemical Analysis and Biological Fate</u>, W.M. Cooke and A.J. Dennis, Eds., Battelle Press, Columbus, Ohio, p. 287.
4. McCrea, P.F. and Purnell, J.H. (1968): Temperature independent retention in gas chromatography, <u>Nature</u>, 219:261-262.
5. McCrea, P.F. and Purnell, J.H. (1969): Gas chromatographic column systems exhibiting temperature independence of solute retention, <u>Anal. Chem.</u>, 41(14):1922-1929.
6. Casagrande, C. DeLoche, B., Dubault, A., and Veyssie, M. (1980): Pseudo clearing temperature in binary polymer-nematic solutions, <u>Phys. Rev. Letters</u>, 45:1645-1648.

IN VITRO METABOLISM OF 6-NITROBENZO[A]PYRENE: IDENTIFICATION AND MUTAGENICITY OF ITS METABOLITES

PETER P. FU, MING W. CHOU, SHEN K. YANG[*], LEONARD E. UNRUH, FREDERICK A. BELAND, FRED F. KADLUBAR, DANIEL A. CASCIANO, ROBERT H. HEFLICH, FREDERICK E. EVANS
National Center for Toxicological Research, Jefferson, Arkansas 72079 and *Department of Pharmacology, School of Medicine, Uniformed Services University of the Health Sciences, Bethesda, Maryland 20814, USA.

INTRODUCTION

Nitro-polycyclic aromatic hydrocarbons (nitro-PAHs) are potent bacterial mutagens found in diverse sources including photocopier toners and urban air (9,16,18). Two processes have been suggested for their presence as atmospheric contaminants (2,5,13,14,16-18). They can result from incomplete combustion and, as such, may be responsible for the mutagenic activity associated with automobile exhaust. In addition, nitro-PAHs may be formed from the reaction of PAHs with $(NO)_x$, both of which are components of urban atmospheres.

6-Nitrobenzo[a]pyrene ($6-NO_2$-BaP), a representative nitro-PAH, has been detected as an air pollutant (7) and can be formed in model atmospheres containing trace quantities of benzo[a]pyrene (BaP), nitrogen oxide and nitric acid (13). Although this nitro-PAH exhibits direct-acting bacterial mutagenic activity, its mutagenicity is greater in the presence of a mammalian liver homogenate (S9) metabolizing system (13,17,19). The in vitro metabolic activation pathways for $6-NO_2$-BaP were unknown and are the subject of the present communication. Our investigation had two objectives: first, to determine if activation occurred through hydrocarbon ring metabolism, and if so, what effect did 6-nitro substitution have upon the metabolic pathways when compared to its parent hydrocarbon, BaP; and second, to determine if nitroreduction played a significant role in the mutagenic activity as has been observed with other nitro-PAHs, such as 2-nitrofluorene and 1-nitropyrene (11).

MATERIALS AND METHODS

All experiments were conducted under low-UV yellow lights.

METABOLISM OF 6-NITROBENZO[A]PYRENE

Chemicals

6-NO_2-BaP was synthesized by direct nitration of BaP according to the procedure of Fieser and Hershberg (4).

In Vitro Incubations

Male Sprague-Dawley rats (80-100 g), obtained from our breeding colony, received intraperitoneal injections of 3-methylcholanthrene (MC; 25 mg/kg body weight) or phenobarbital (PB; 75 mg/kg body weight) on 3 consecutive days before sacrifice. Liver microsomes were prepared by a modified (3) Nebert-Gelboin (12) procedure. Protein concentrations were determined by the method of Lowry et al. (10), using bovine serum albumin as the standard.

Incubation mixtures contained 50 mmol Tris-HCl, pH 7.5; 3 mmol $MgCl_2$; 1 mmol $NADP^+$ (Sigma, sodium salt); 2 mmol glucose-6-phosphate (Sigma, monosodiuum salt); 100 units glucose-6-phosphate dehydrogenase (Sigma, Type II); 1 g microsomal protein and 40 μmol 6-NO_2-BaP (dissolved in 40 ml acetone) in a total incubation volume of 1 l. Incubations were conducted with shaking for 60 min at 37° and then quenched by the addition of 1 l acetone. The metabolites and residual substrate were partitioned into 2 l ethyl acetate, and the organic phase was dried with anhydrous sodium sulfate. The solvent was evaporated under reduced pressure and the residue was dissolved in 1 ml methanol for analysis by high performance liquid chromatography (HPLC).

Instrumentation

Reversed-phase HPLC was performed with a Beckman system consisting of two model 100A pumps, a model 210 injector, a model 420 solvent programmer and a Waters Associates model 440 absorbance (254 nm) detector. Metabolites were separated by using a DuPont Zorbax ODS column (6.2 mm x 25 cm) and eluting with a 30-min linear gradient of 75-90% methanol in water at a flow rate of 1.6 ml/min.

UV-visible spectra were obtained with a Beckman model 25 spectrometer. Mass spectra were recorded with a Finnigan model 4000 system. The samples were introduced with a solid probe and ionized at 70 eV with an ionizer temperature of 250°. Fourier transform ^{1}H-nmr spectra were obtained with a Bruker WM 500 spectrometer.

Bacterial Mutagenicity Assays

Histidine auxotrophic Salmonella typhimurium tester strains TA98 and TA100 were obtained from Bruce N. Ames (Univ. Calif., Berkeley, CA) and nitroreductase-deficient (15) strains TA98-NR and TA100-NR were provided by Herbert S. Rosenkranz (Case Western Reserve, Cleveland, OH). Reversion to histidine prototrophy was determined in triplicate as described by Ames et al. (1). Each substrate was tested with and without S9 activation at 2.5, 5, 10 and 17.5 nmoles compound per plate. The S9 mixture was prepared from liver homogenates of Aroclor 1254-pretreated Sprague-Dawley rats as described by Ames et al. (1).

RESULTS

In Vitro Incubation of 6-NO_2-BaP

HPLC analysis of ethyl acetate extracts from incubations of 6-NO_2-BaP with hepatic microsomes from control, PB- or MC-pretreated male Sprague-Dawley rats indicated a linear decrease in substrate for at least 30 min. Microsomes from PB-pretreated

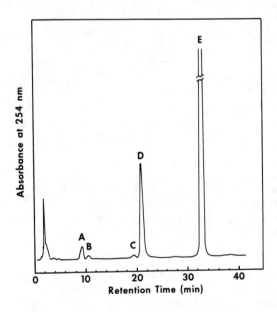

FIGURE 1. HPLC profile of ethyl acetate extractable metabolites obtained from a 1 l incubation (60 min) of 6-NO_2-BaP with hepatic microsomes from MC-pretreated Sprague-Dawley rats.

rats were 1.5 times more active than control microsomes in metabolizing 6-NO_2-BaP, whereas MC-pretreatment resulted in a four-fold increase over controls. In each case, HPLC analysis indicated formation of the same 4 major UV absorbing metabolite peaks. A representative chromatogram, from an incubation with microsomes from MC-pretreated rats is shown in Fig. 1. The metabolites contained in peaks labeled A through D were isolated in sufficient quantity for structural identification by nmr, UV and mass spectral analysis. Peak E is the substrate, 6-NO_2-BaP.

The mass spectrum of the material from peak D had a molecular ion (M^+) at m/z 313 and major fragments at m/z 283, 282, 268, 267, 254, 239 and 226 which suggested the presence of a phenolic metabolite. Support for this assignment came from examination of the metabolite's UV-visible spectra (Figure 2A) at neutral and basic pH's. As is characteristic with phenols,

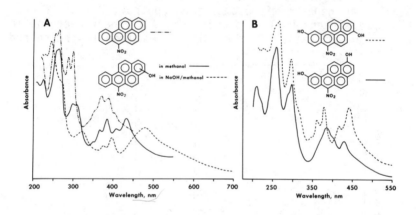

FIGURE 2. UV-visible spectra of [A] 6-NO_2-BaP in methanol (-.-.-) and the metabolites of peak D in methanol (——) and 0.1 N NaOH/methanol (- - -); [B] the metabolites of peak A (- - -) and peak B (——) in methanol.

a red shift was observed when the spectrum obtained under alkaline conditions was compared to the spectrum measured at neutral pH. ^{1}H nmr spectral analysis indicated that peak D actually contained two isomeric phenolic components in a 4:1 ratio. The ^{1}H nmr resonance assignments (Table 1) of the major component were determined by extensive homonuclear decoupling experiments including decoupling the long range peri $^4J_{11,12}$

and bay $^5J_{10,11}$ couplings. These data, plus the absence of peri coupling to H4, were uniquely consistent with substitution at C3. In addition, the upfield shifts of H2 and H1 by 0.41 and 0.16 ppm, respectively, and the downfield shift of H4 by 0.25 ppm when compared with those of 6-NO$_2$-BaP were consistent with ortho, meta, and peri substituent effects of a hydroxyl group (8,20). These data indicated the major metabolite of peak D was 3-hydroxy-6-NO$_2$-BaP. The resolution at 500 MHz was also sufficient to enable measurement of some of the resonances of the minor component in the mixture (Table 1). The combined nmr, UV-visible and mass spectral data were consistent with this metabolite being 1-hydroxy-6-NO$_2$-BaP.

TABLE 1

500 MHz PROTON NMR SPECTRAL DATA FOR 6-NITROBENZO[A]PYRENE AND ITS METABOLITES.[a]

Compound	Chemical Shift, δ										
	H_1	H_2	H_3	H_4	H_5	H_7	H_8	H_9	H_{10}	H_{11}	H_{12}
6-NO$_2$-BaP	8.56	8.21	8.42	8.35	7.95	8.17	8.06	8.06	9.39	9.36	8.67
1-Hydroxy-6-NO$_2$-BaP[b]	—	7.71	8.25	8.20	7.70	c	c	c	9.33	9.25	8.92
3-Hydroxy-6-NO$_2$-BaP[b]	8.40	7.80	—	8.60	7.83	8.11	7.98	7.98	9.28	9.09	8.51
6-NO$_2$-BaP-3,9-hydroquinone	8.32	7.77	—	8.48	7.76	8.02	7.63	—	8.51	8.84	8.41
6-NO$_2$-BaP-1,9-hydroquinone	—	7.67	8.17	8.07	7.62	8.02	7.64	—	8.56	9.01	8.82

Compound	Coupling Constant, Hz						
	$J_{1,2}$	$J_{2,3}$	$J_{4,5}$	$J_{7,8}$	$J_{8,9}$	$J_{9,10}$	$J_{11,12}$
6-NO$_2$-BaP	7.9	7.6	9.2	8.5	7.0	8.7	9.2
1-Hydroxy-6-NO$_2$-BaP	—	8.1	9.6	c	c	c	9.6
3-Hydroxy-6-NO$_2$-BaP	8.6	—	9.8	8.7	6.8	9.0	9.0
6-NO$_2$-BaP-3,9-hydroquinone	8.5	—	9.6	9.1	—	—	9.2
6-NO$_2$-BaP-1,9-hydroquinone	—	8.1	9.6	9.2	—	—	9.6

a Samples were dissolved in acetone-d$_6$. Chemical shifts are reported on the δ scale by assigning the acetone-d$_5$ resonance to 2.06 ppm.
b Analysis of the two phenolic metabolites was performed on a 1:4 mixture of 1- and 3-hydroxy-6-NO$_2$-BaP, respectively.
c Only a partial analysis of 1-hydroxy-6-NO$_2$-BaP was possible because the analysis was performed on a mixture.

The mass spectra of the metabolites contained in peaks A and B were similar to one another and had molecular ions at m/z 329 and major fragments at 298, 284, 270 and 255. These data are consistent with a dihydroxy (i.e., hydroquinone) derivative of 6-NO_2-BaP. Further support for this interpretation was obtained from their UV-visible spectra which indicated a fully aromatic structure as opposed to a quinone type compound. The exact sites of hydroxyl substitution in both metabolites were established by comparison of their 500 MHz ^{1}H nmr spectra to the spectra of 1- and 3-hydroxy-6-NO_2-BaP, 6-NO_2-BaP, and 9-hydroxy-BaP. The chemical shift data was in accord with hydroxyl substitution at C9 in both metabolites. Furthermore, the loss of an <u>ortho</u> coupling constant for H10 and H8 also supported this conclusion. These data indicate that peak A is 6-NO_2-BaP-3,9-hydroquinone while peak B is 6-NO_2-BaP-1,9-hydroquinone.

The UV-visible spectrum of the metabolite in peak C suggested that it was a quinone as opposed to phenolic product. Mass spectral analysis confirmed this; the spectrum is identical in all aspects with that obtained with an authentic sample

TABLE 2

MUTAGENICITY OF BaP, 3-HYDROXY-BaP, 6-NO_2-BaP, AND 1- AND 3-HYDROXY-6-NO_2-BaP IN S. TYPHIMURIUM STRAINS TA98 AND TA100

MUTAGEN[a]	S9 ACTIVATION	TA98	TA100
NONE	−	26±10	130±17.1
	+	32±1.7	163±16.6
BaP	−	32±2.1	84±3.8
	+	324±17.4	861±11.5
3-Hydroxy-BaP	−	111±10.5	197±13.6
	+	286±30.5	321±26.6
6-NO2-BaP	−	36±2.5	199±19.5
	+	982±179.5	1086±226.8
3-Hydroxy-6-NO2-BaP	−	163±14.7	238±11.5
	+	2332±448.6	1760±182.3

[a] Mutagen concentration was 10 nmol/plate

of BaP-3,6-quinone. Final substantiation for this assignment was provided by an exact correspondence in HPLC retention times for Peak C and BaP-3,6-quinone.

Mutagenicity of 1- and 3-Hydroxy-6-NO_2-BaP

Since they could not be resolved by HPLC, the mutagenicity of a mixture of 1- and 3-hydroxy-6-NO_2-BaP (1:4 molar ratio) was compared to 6-NO_2-BaP, 3-hydroxy-BaP, and BaP in Salmonella typhimurium tester strains TA98 and TA100 at 4 different concentrations. The data in Table 2 are from linear regions of the dose response curves. In the absence of S9 activation, 3-hydroxy-BaP and the mixture of 1- and 3-hydroxy-6-NO_2-BaP showed slight mutagenicity in strains TA98 while the other two compounds were inactive (< twice background). None of the hydrocarbons were mutagenic in strain TA100 without S9 activation. In the presence of an S9 activation system all four compounds were active in both strains with the mixture of 1- and 3-hydroxy-6-NO_2-BaP demonstrating the greatest mutagenicity. They had approximately 2.5 and 1.5 times the mutagenic activity observed with 6-NO_2-BaP in strains TA98 and TA100, respectively. By contrast, 3-hydroxy-BaP had a lower mutagenicity than that detected for BaP. 6-NO_2-BaP and the mixture of 1- and 3-hydroxy-6-NO_2-BaP were also tested for mutagenicity in nitroreductase-deficient strains, TA98-NR and TA100-NR. The results obtained were not significantly different than those observed with strains TA98 and TA100.

DISCUSSION

In this study we have shown that the environmental contaminant, 6-NO_2-BaP, is metabolized by a rat hepatic microsomal system to five products identified as 1- and 3-hydroxy-6-NO_2-BaP, 6-NO_2-BaP-1,9- and 3,9-hydroquinones, and BaP-3,6-quinone. The primary hydroxylated products, 1- and 3-hydroxy-6-NO_2-BaP were the major metabolites and accounted for an estimated 70% of the total metabolites. A comparison of the mutagenicity of BaP, 3-hydroxy-BaP, 6-NO_2-BaP and 1- and 3-hydroxy-6-NO_2-BaP in Salmonella typhimurium tester strains TA98 and TA100 indicated that the mixture of 1- and 3-hydroxy-6-NO_2-BaP was the most active and suggested that these metabolites are major proximal mutagens of 6-NO_2-BaP. Thus, although ring hydroxylation of PAHs is normally a detoxification process, ring hydroxylation at C1 and/or C3 position of 6-NO_2-BaP must be considered as an activation system.

In addition to the primary oxidation products, two secondary metabolites, 6-NO_2-BaP-1,9- and 3,6-hydroquinone were also detected. Although further studies will be necessary, the failure to detect 9-hydroxy-6-NO_2-BaP as a microsomal metabolite suggests that these hydroquinones result from further oxidation of 1- and 3-hydroxy-6-NO_2-BaP, respectively. A fifth product, BaP-3,6-quinone, was also found in the incubation mixture. The biological pathway for the formation of this quinone is not readily evident although it may result from nitro reduction of 3-hydroxy-6-NO_2-BaP and subsequent hydrolysis of the quinone imine intermediate. Alternatively it may be possible that it results from chemical oxidation of 6-NO_2-BaP.

The most abundant metabolite formed during the metabolism of BaP is 3-hydroxy-BaP and there is also a substantial degree of oxidation in the 4,5-, 7,8- and 9,10-regions (6). By comparison, while 3-hydroxylation is still the dominant pathway with 6-NO_2-BaP, the nitro substitution at C6 effectively blocks metabolism at regions peri to it (i.e., 4,5- and 7,8-) and at the 9,10-positions. It should also be noted that with both BaP and 6-NO_2-BaP there was no detectable hepatic oxidation of the 11,12-double bond.

BaP-7,8-diol-9,10-epoxides and BaP-4,5-oxide are mutagenic metabolites of BaP which do not result from metabolism of BaP phenols. Therefore, it is apparent that although both BaP and 6-NO_2-BaP are metabolized to bacterial mutagens, they have different pathways. The activation sequence for 1- and 3-hydroxy-6-NO_2-BaP is presently unknown but since the same mutagenicity was observed in the nitroreductase deficient strains as was found in the normal tester strains it appears that nitroreduction may not be important. It should be noted, however, the Salmonella probably have a number of nitroreductases (11) and the deficient strains used in this investigation may not be lacking in the enzymes necessary to reduce 6-NO_2-BaP or its phenolic metabolites. Therefore, in order to determine the activation pathways, metabolism and mutagenicity studies with 1- and 3-hydroxy-6-NO_2-BaP are presently being conducted.

ACKNOWLEDGEMENTS

We thank Dr. J. P. Freeman for obtaining the mass spectra, Ms. G. L. White in conducting the Ames' tests, and Ms. L. Amspaugh for assistance in preparation of the manuscript.

REFERENCES

1. Ames, B.N., McCann, J., and Yamasaki, E. (1975): Methods for detecting carcinogens and mutagens with the Salmonella/ mammalian-microsome mutagenicity test. Mutation Res., 31, 347-364.
2. Chiu, C.W., Lee, L.H., Wang, C.Y., and Bryan, G.T. (1978): Mutagenicity of some commerically available nitro compounds for Salmonella typhimurium. Mutation Res., 58:11-22.
3. Chou, M.W., and Yang, S.K. (1979): Combined reversed-phase and normal-phase high-performance liquid chromatography in the purification and identification of 7,12-dimethylbenz- [a]anthrancene metabolites. J. Chromatography, 185:635-654.
4. Fieser, L.F., and Hershberg, E.B. (1939): The orientation of 3,4-benzpyrene in substitution reactions. J. Am. Chem. Soc., 61:1565-1574.
5. Fitch, W.L., and Smith, D.H. (1979): Analysis of absorption properties and absorbed species on commerical polymeric carbons. Environ. Sci. Technol., 13:341-346.
6. Holder, G., Yagi, H., Dansette, P., Jerina, D.M., Levin, W., Lu, A.Y.H., and Conney, A.H. (1974): Effects of inducers and epoxide hydrase on the metabolism of benzo[a]pyrene by liver microsomes and a reconstituted system: Analysis by high pressure liquid chromatography. Proc. Nat. Acad. Sci. USA, 71:4356-4360.
7. Jager, J. (1978): Detection and characterization of nitro derivatives of some polycyclic aromatic hydrocarbons by fluorescence quenching after thin-layer chromatography: Application to air pollution analysis. J. Chromatography, 152:575-578.
8. Lee, H., Shyamasundar, N., and Harvey, R.G. (1981): Isomeric phenols of benzo[e]pyrene. J. Org. Chem., 46: 2889-2895.
9. Lofroth, G., Hefner, E., Alfheim, I., Moller, M. (1980): Mutagenicity in photocopies. Science, 209:1037-1039.
10. Lowry, O.H., Rosebrough, N.J., Farr, A.L., and Randall, R.J. (1951): Protein measurements with the Folin phenol reagent. J. Biol. Chem. , 193:265-275.
11. McCoy, E.C., Rosenkranz, H.S., and Mermelstein, R. (1981): Evidence for the existence of a family of bacterial nitroreductases capable of activating nitrated polycyclics to mutagens. Environ. Mut., 3:421-427.
12. Nebert, D.W., and Gelboin, H.V. (1968): Substrate-inducible microsomal aryl hydroxylase in mammalian cell culture. I. Assay and properties of induced enzyme. J. Biol. Chem., 243:6242-6249.

13. Pitts, J.N., Jr., van Cauwenberghe, K.A., Grosjean, D., Schmid, J.P., Fitz, D.R., Belser, W.L., Jr., Knudson, G.B., and Hynds, P.M. (1978): Atmospheric reactions of polycyclic aromatic hydrocarbons: Facile formation of mutagenic nitro derivatives. Science, 202:515-519.
14. Pitts, J.N., Jr. (1979): Photochemical and biological implications of the atmospheric reactions of amines and benzo(a)pyrene. Phil. Trans. R. Soc. Lond., 290:551-576.
15. Rosenkranz, H.S., and Speck, W.T. (1975): Mutagenicity of metronidazole: Activation by mammalian liver microsomes. Biochem. Biophys. Res. Commun., 66:520-525.
16. Rosenkranz, H.S., McCoy, E.C., Sanders, D.R., Butler, M., Kiriazides, D.K., Mermelstein, R. (1980): Nitropyrenes: isolation, identification, and reduction of mutagenic impurities in carbon black and toners. Science, 209: 1039-1041.
17. Tokiwa, H., Nakagawa, R., and Ohnishi, Y. (1981): Mutagenic assay of aromatic compounds with Salmonella typhimurium. Mutation Res., 91:321-325.
18. Wang, C.Y., Lee, M.S., King, C.M., and Warner, P.O. (1980): Evidence for nitroaromatics as direct-acting mutagens of airborne particulates, Chemosphere, 9:83-87.
19. Wang, Y.Y., Rappaport, S.M., Sawyer, R.F., Talcott, R.E., and Wei, E.T. (1978): Direct-acting mutagens in automobile exhaust. Cancer Lett., 5:39-47.
20. Yagi, H., Holder, G.M., Dansette, P.M., Hernandez, O., Yeh, H.J., LeMahieu, R.A., and Jerina, D.M. (1976): Synthesis and spectral properties of the isomeric hydroxybenzo[a]-pyrenes. J. Org. Chem., 41:977-985.

REAL-TIME ANALYSIS OF EXHAUST GASES USING TRIPLE QUADRUPOLE MASS SPECTROMETRY

J.E. FULFORD, T. SAKUMA and D.A. LANE
SCIEX INC., 55 Glencameron Rd., #202, Thornhill, Ontario,
L3T 1P2, Canada.

INTRODUCTION

The projected increase in the use of fossil fuels, especially coal and diesel fuel, has stimulated considerable interest in the toxic components (such as the PAH and nitro-PAH) of the emissions from the combustion of these fuels. The most common methodology for the analysis of the gaseous and particulate components of these emissions entails long and complicated chemical workup procedures followed by GC or GC/MS analysis.

In this paper we present a more direct and rapid means for the analysis of the gaseous and particulate emissions of combustion or hot gas streams.

Qualitative and quantitative analyses of hot gaseous components can be accomplished in <u>real-time</u> by transferring the gas directly into an atmospheric pressure chemical ion source coupled to a mass spectrometer, as has been demonstrated previously for the emissions of a coal-fired cement kiln (1). The use of APCI/MS/MS (atmospheric pressure chemical ionization mass spectrometry/mass spectrometry) for the direct real-time analysis of the gaseous emissions from a light duty diesel engine and for targeted compound analysis of the extracts of the diesel particulate matter is reported.

METHODS AND MATERIALS

The instrument involved in the analysis - the TAGA™ 6000 triple quadrupole mass spectrometer system - has been described elsewhere (1).

Direct Gas Analysis

For the direct analysis of the gas-phase components, the diesel exhaust from the tailpipe of a Datsun pickup truck was drawn (2 L/s) through a heated (150°C - 180°C) teflon (22 mm O.D.) sampling line. A small portion (0.2 L/s) of the total exhaust flow was diluted ten-fold with heated (50°C) ambient air and transferred directly into the atmospheric pressure ion source of the MS/MS system as shown in Figure 1.

APCI/MS/MS OF DIESEL ENGINE EMISSIONS

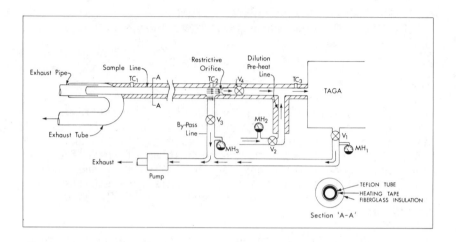

FIGURE 1. Schematic Diagram of Exhaust Sampling Line.

Emission characteristics of the engine were compared by determining the ratios of single analyzer scans (with only one quadrupole mass selective) on a mass to mass basis at various engine speeds. Three other modes of operation were also used:

(1) CID (collision induced dissociation) spectra were recorded by mass selecting a single m/z (parent ion) in the first quadrupole (Q1), colliding the selected parent with Argon target gas in Q2 and scanning the spectrum of product (daughter) ions with Q3.

(2) Spectra of all of the parent ions which yield a common daughter ion were obtained by mass selecting a single m/z with Q3 while scanning Q1 and promoting dissociation of the parent ions in the Q2 collision zone.

(3) Spectra of parent ions which lose common neutral fragments (constant neutral loss scan) were recorded by parallel scans of Q1 and Q3 with a constant mass difference while CID occurred in the collision zone.

Extract Analysis

Methylene chloride extracts of diesel exhaust particulate matter (2) were evaporated to dryness, and taken up in benzene. Aliquots containing 25 μg of extract were deposited on a direct insertion probe and after the solvent had evaporated, the sample was desorbed into the APCI source using a computer controlled temperature program.

Quantitation for the nitropyrene isomers (parent ion m/z=247) was achieved by integrating the response from the NO_2^- (m/z=46) daughter ion.

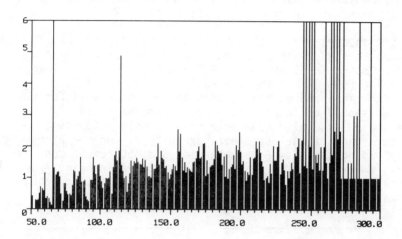

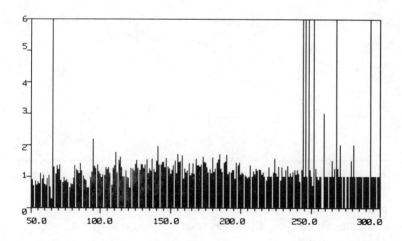

FIGURE 2. Single analyzer scan.
Ratio: (a) full idle/low idle; (b) half idle/low idle.

RESULTS AND DISCUSSION

Direct Gas Analysis

Speed-dependent emission characteristics of the diesel engine are shown in Figures 2a and b, in which the ratios of single analyzer scans at full idle/low idle and half idle/low idle, demonstrate that higher molecular weight species were more abundant at higher engine idle speed.

CID spectra of six standard nitro- and dinitro-PAH revealed that NO_2^- was a common daughter ion of abundant yield as shown in Figure 3 for 2,7-dinitrofluorene. Since the m/z=46 (NO_2^-) is both abundant and characteristic of nitro-PAH, it appeared that a constant daughter scan for the NO_2^- ion should be an effective means for the rapid screening of the nitro-PAH in a given sample. Constant daughter scans of the hot diesel exhaust gas for the parent ions yielding the m/z=46 daughter ion did not reveal conclusively, however, the presence of any nitro-PAH ranging from the nitronaphthalene isomers up to the nitrobenzo(a)pyrene isomers.

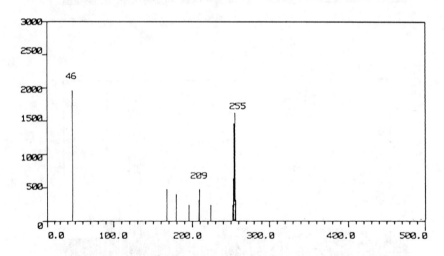

FIGURE 3. Negative ion CID spectrum of the (M-H)⁻ of 2,7-dinitrofluorene.

These compounds are more likely to be adsorbed on the soot particles even at the temperature (150°C) of the vehicle exhaust system. There was, however, evidence for the presence of lower molecular weight nitroaromatics such as the nitrobenzenes.

CID spectra of selected parent ions from the diesel emissions permitted tentative assignments for the m/z=163 (MH^+ of benzenepentanal), the m/z=114 (MH^+ of 1-acetylpyrrolidine) and the m/z=127 (MH^+ of 3-methylpyrazole-4-carboxylic acid) ions.

The applicability of APCI/MS/MS to the real-time analysis of hot gas streams was demonstrated by monitoring the m/z=97, 111, 125 and 139 parent ions as the engine speed was varied (Figure 4). These compounds have been tentatively identified as hexenals, heptenals, octenals and nonenals on the basis of the abundant m/z=43 ion corresponding to the CH_3CO^+ fragment ion.

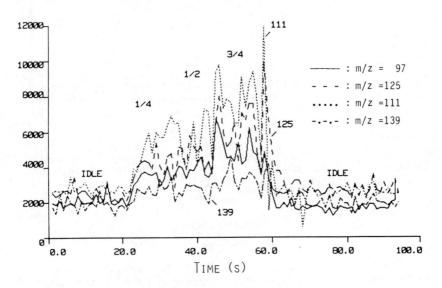

FIGURE 4. Effect of Engine Speed on Aldehyde Formation.

Extract Analysis

Calibration for nitropyrene in the extract from diesel exhaust particulate matter was achieved by spiking the raw extract with a known amount of nitropyrene and establishing a calibration curve as shown in Figure 5. The concentration of nitropyrene in the extract was determined to be 430 ppm, and is in agreement with a separate determination by more traditional methods (3). The extrapolated detection limit for nitropyrene in the extract samples was less than 400 ppb using the MS/MS technique.

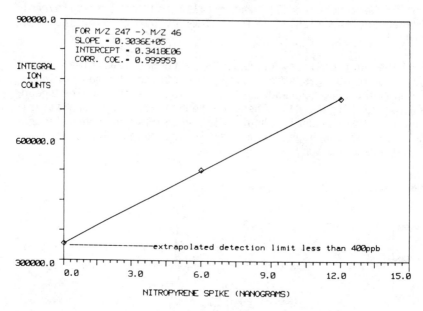

FIGURE 5. Calibration Curve for Nitropyrene in Extracts of Diesel Exhaust Particulate Matter.

CONCLUSIONS

The application of APCI/MS/MS to the rapid analysis of the two major constituents (gas phase and particulate matter) of diesel engine exhaust has been demonstrated. A number of aldehydes have been observed, in the vapor component of diesel engine exhaust, however, the nitro-PAH were observed only in the extracts of the particulate matter. Further work is required, to determine the extent of the loss of nitro-PAH vapors inherent in the gas sampling process. Since APCI/MS/MS permits the direct analysis of crude extracts, the requirement for separation and derivatization procedures has been eliminated. Consequently, the sample preparation time has been greatly shortened. The actual instrument time per sample is about 20 minutes and a quantitative result with a sensitivity comparable to, or better, than GC/MS techniques is realized.

REFERENCES

1. Sakuma, T., Davidson, W.R., Lane, D.A., Fulford, J.E., and Quan, E.S.K. (1981): The rapid analysis of gaseous PAH and other combustion related compounds in hot gas streams by APCI/MS and APCI/MS/MS. In: <u>Chemical Analysis and Biological Fate: Polynuclear Aromatic Hydrocarbons</u>, M. Cooke and A.J. Dennis, Eds., pp. 179-188, Battelle Press, Columbus, Ohio.
2. Henderson, T.R., Royer, R.E., and Clark, C.R. (1981): MS/MS characterization of diesel emissions. Proceedings of the 29th Annual Conference on Mass Spectrometry and Allied Topics (in press).

ACKNOWLEDGEMENTS

The authors gratefully acknowledge Nissan Automobile Company (Canada) for the loan of the test vehicle and Dr. T.R. Henderson of Lovelace Inhalation Toxicology Research Institute Inc. for sample collection and extraction.

DETECTION OF HIGH MOLECULAR POLYCYCLIC AROMATIC HYDROCARBONS IN AIRBORNE PARTICULATE MATTER, USING MS, GC, AND GC/MS

WERNER FUNCKE, THOMAS ROMANOWSKI, JOHANN KÖNIG, ECKHARD BALFANZ, Fraunhofer-Institut für Toxikologie und Aerosolforschung, D-4400 Münster-Roxel, West Germany

INTRODUCTION

Polycyclic aromatic hydrocarbons (PAH), which produce experimental cancer in animals and may also contribute to the occurrence of lung cancer in men, have been identified in many samples of different environmental matter. Most of these PAH analyses were confined to substances of a molecular weight (m.w.) up to 300 (coronene), although the general existence of higher molecular PAH has long been known and a potential carcinogenic property was anticipated (e.g., 1) for them.

Up to now, only a few authors have published details about PAH of higher m.w. in environmental samples. They were using thin layer chromatography (2) or high performance liquid chromatography (3) sometimes coupled with a mass spectrometer (4). However, for such complex mixtures as PAH fractions extracted from environmental matter, capillary gas chromatography (GC) combined with mass spectrometry (MS) should be the method of choice. Unfortunately, in the past, PAH could be investigated by GC/MS only up to a m.w. of 326 (5).

In this paper, we demonstrate the existence of high molecular PAH in airborne particulate matter by low voltage-mass spectrometry (LV-MS) using a direct inlet technique. Furthermore, employing a modified GC/MS system equipped with fused silica capillary columns, we have succeeded in seperating from airborne particulate matter a PAH fraction in the m.w. range of 300 to 402. Nearly 200 PAH could be characterized by their EI mass spectra.

MATERIALS AND METHODS

Airborne particulate matter was sampled in Duisburg, West Germany, during the winter months of 1979/80. The extraction of soluble components from the glass fiber filter deposits and the fractionation procedure have been described elsewhere (6). By means of column chromatography (column: 80 cm x 2,5 cm i.d.,

Sephadex LH 20, 2-propanol, flow rate: 0.6 ml/min) as the final isolation step, a pure PAH fraction with components of a m.w. higher than 300 was obtained by collecting the eluate from 1190 ml to 2000 ml.

For the analysis of the unseparated PAH fraction merely by mass spectrometry (Finnigan, model 4021), a direct insertion probe with a glass capillary was utilized. A gas chromatograph (Varian, model 3700) equipped with a flame ionization detector was used for seperating the PAH fraction. The original GC/MS system (Finnigan 4021) was modified as follows: the GC injector was replaced by an adapted Schomburg split/splitless injector (7); the GC/MS interface consisted of a temperature controlled oven of about 25 cm length, through which the fused silica capillary column extends directly into the ion source. In this way, we set up a GC/MS interface with an inert inner surface virtually free of dead volume. The high speed vacuum system, consisting of two turbo molecular pumps with a capacity of 200 and 500 1/sec, each connected with a mechanical forepump (15 m^3/h), maintained a vacuum of about 4×10^{-7} torr in the analyzer during operation. Further experimental details see Figs. 1 to 4.

RESULTS AND DISCUSSION

To obtain an indication of the molecular weights expected in the PAH fraction, the sample was introduced directly into the ion source. This technique yielded a mass spectrum as shown in Fig. 1 after cumulation of 119 scans.

In general, molecular ions constitute the base peaks of the EI-mass spectra of unsubstituted PAH. As a result of the stability of these ions the intensities of the fragment ions obtained by using an ionizing energy of only 20 eV lie below the detection limit under the conditions described above. The even-numbered ions in Fig. 1 (e.g., m/e 302, 326, 328, 340, 350, 352 with an upper limit of 402) are assumed to be the molecular ions of the corresponding PAH.
The high molecular PAH fraction investigated by LV-MS technique was separated by GC; a typical chromatogram is shown in Fig. 2.

The hardware modifications mentioned above eliminate most of the deficiencies for analyzing high molecular PAH by GC/MS. The chromatographic resolution as well as the peak intensities of the reconstructed ion current (RIC) are very similar to those obtained by GC (cf. Fig. 2; experimental conditions see legend of Fig. 3). Besides some data processing programmes, mass chromatography has been very useful for finding minor components

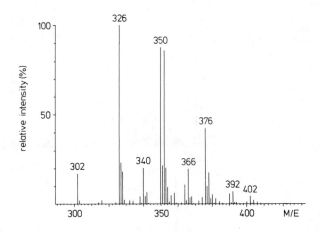

FIGURE 1

Low voltage mass spectrum of the high molecular PAH fraction derived from a direct insertion probe by accumulating the scans from number 61 to 179. MS conditions: solid probe temperature 3 min hold at 80°C, programmed from 80°C to 250°C with 30°C/min, 10 min hold at 250°C; mass range 290-450 amu; scan time 5 sec; ionizing energy 20 eV.

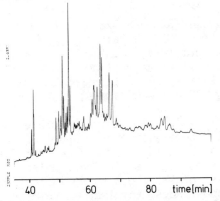

FIGURE 2

Capillary gas chromatogram (FID) of the high molecular PAH fraction. GC conditions: 15m x 0.32 mm i.d. fused silica column coated with 0.25 μm SE 54; oven temperature 5 min 200°C, with 2°C/min at 300°C, hold at 300°C; split 1:30; carrier gas helium at 0.8 bar.

not seen as peaks in the RIC or for detecting compounds in peak clusters. To illustrate this technique, Fig. 3 is given as an example for the PAH with a m.w. of 326 and 328. The molecular ion peaks of about 20 substances can be detected in a range of about 450 scans.

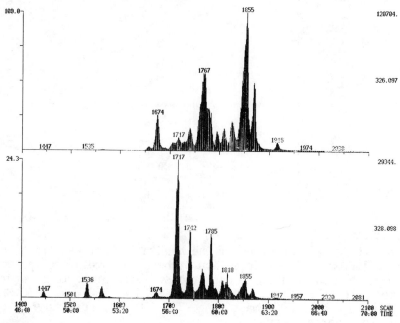

FIGURE 3

Mass chromatograms (326 and 328 amu) of high molecular PAH derived from the GC/MS analysis (GC conditions: column see Fig. 2; oven temperature 5 min 200°C, with 1.5°C/min at 300°C, hold at 300°C; carrier gas helium 0.5 bar; MS conditions: interface temperature 280°C; ion source temperature 300°C; ionizing energy 70 eV; 4400 scans; scan time 2 sec.

Table 1 lists 190 PAH, which have been obtained from the RIC, along with their scan number and their m.w.. Fig. 4 shows two typical mass spectra derived from RIC; without exception all EI mass spectra obtained so far have molecular ions, which produce the base peaks. There is no indication for the existence of alkylated substances. For lack of reference material, the identity of the PAH detected so far could not yet be established.

Summing up the results, the previous conclusion derived from LV-MS investigations (Fig. 1) is well confirmed by the GC/MS measurements. With regard to Table 1, the Sephadex LH 20 fraction investigated here turns out to be as complex as the PAH fraction in the m.w. range between 202 and 300 which is usually investigated in airborne particulate matter or other environmental samples. Special research efforts are now in progress to further explore this new fraction of high m.w. PAH and its influence on the carcinogenic potential of airborne particulate matter in animal experiments.

HIGH MOLECULAR WEIGHT PAH

TABLE 1

LIST OF PAH ALONG WITH THEIR SCAN NUMBER (CF. PARTLY FIG. 3) AND WITH THEIR MOLECULAR WEIGHT

Compound No.	Scan No.	M.W.	Compound No.	Scan No.	M.W.	Compound No.	Scan No.	M.W.
1	1275	308	65	1808	328	129	2276	352
2	1284	308	66	1810	326	130	2280	350
3	1289	302	67	1811	332	131	2286	376
4	1305	308	68	1818	328	132	2294	378
5	1316	308	69	1825	326	133	2314	356
6	1326	302	70	1826	340	134	2320	350
7	1332	308	71	1829	334	135	2325	366
8	1343	308	72	1847	332	136	2328	378
9	1346	300	73	1847	342	137	2354	378
10	1349	302	74	1855	326	138	2354	356
11	1379	302	75	1856	348	139	2355	350
12	1379	318	76	1870	326	140	2360	366
13	1390	308	77	1873	340	141	2383	350
14	1390	318	78	1879	340	142	2391	378
15	1401	302	79	1880	326	143	2396	366
16	1405	318	80	1886	348	144	2413	378
17	1415	308	81	1888	342	145	2426	366
18	1440	318	82	1890	354	146	2428	366
19	1445	316	83	1905	340	147	2461	378
20	1446	328	84	1917	340	148	2463	362
21	1466	316	85	1917	342	149	2486	364
22	1478	316	86	1929	354	150	2505	378
23	1493	314	87	1930	340	151	2514	364
24	1498	316	88	1933	342	152	2537	364
25	1507	316	89	1943	340	153	2555	374
26	1511	316	90	1943	352	154	2564	378
27	1526	316	91	1955	354	155	2569	364
28	1532	314	92	1956	342	156	2594	362
29	1536	328	93	1956	338	157	2607	376
30	1545	316	94	1965	340	158	2623	364
31	1552	314	95	1974	352	159	2625	362
32	1560	314	96	1977	340	160	2630	376
33	1564	328	97	1983	340	161	2671	378
34	1570	314	98	1995	354	162	2671	384
35	1579	328	99	2007	358	163	2690	376
36	1585	314	100	2013	349	164	2703	378
37	1657	326	101	2032	340	165	2723	376
38	1670	334	102	2037	338	166	2730	278
39	1672	342	103	2056	352	167	2732	384
40	1674	326	104	2064	358	168	2763	374
41	1681	334	105	2085	352	169	2776	382
42	1685	326	106	2089	350	170	2799	378
43	1690	326	107	2102	350	171	2805	392
44	1690	334	108	2113	352	172	2817	374
45	1706	326	109	2119	358	173	2832	382
46	1706	342	110	2127	352	174	2841	376
47	1707	334	111	2143	350	175	2854	378
48	1709	326	112	2158	352	176	2878	376
49	1717	328	113	2158	358	177	2890	374
50	1729	326	114	2164	350	178	2926	376
51	1737	334	115	2179	358	179	2946	374
52	1738	342	116	2186	352	180	2971	376
53	1740	326	117	2189	350	181	2981	402
54	1741	328	118	2192	352	182	3037	376
55	1761	332	119	2204	358	183	3071	390
56	1769	326	120	2220	352	184	3149	374
57	1773	342	121	2236	350	185	3180	390
58	1778	326	122	2249	350	186	3320	402
59	1781	342	123	2256	352	187	3785	402
60	1783	332	124	2256	356	188	3817	400
61	1785	328	125	2260	366	189	4099	400
62	1795	326	126	2260	378	190	4325	400
63	1796	332	127	2272	356			
64	1807	342	128	2275	378			

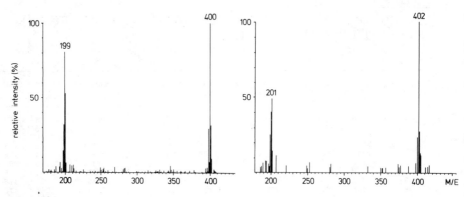

FIGURE 4

EI mass spectra (70 eV) of a PAH with a molecular weight of 400 and 402 obtained from the GC/MS analysis, scan number 4099 and 3320

REFERENCES

1. Arcos, J.C. and Argus, M.F. (1968): Molecular geometry and carcinogenic activity of aromatic compounds - New perspectives. Advan. Cancer Res., 11: 305-471.
2. Matsushita, H., Esumi, Y., Yamada, K. (1970): Identification of polynuclear hydrocarbons in air pollutants. Jpn. Anal., 19: 951-966.
3. Peaden, P.A., Lee, M.L., Hirata, Y., and Novotny, M. (1980): High-performance liquid chromatographic separation of high-molecular-weight polycyclic aromatic compounds in carbon black. Anal. Chem., 52: 2268-2271.
4. Bjørseth, A. (1980): Measurements of PAH content in workplace atmospheres. VDI-Berichte, 385: 81-93.
5. Stenberg, U., Alsberg, T., Blomberg, L., and Wänman, T. (1979): Gas chromatographic separation of high-molecular polynuclear aromatic hydrocarbons in samples from different sources, using temperature-stable glass capillary columns. In: Polynuclear Aromatic Hydrocarbons, edited by P.W. Jones and P. Leber, pp. 313-326, Ann Arbor Science Publishers, Inc., Ann Arbor.
6. König, J., Funcke, W., Balfanz, E., Grosch, B. and Pott, F. (1980): Testing a high volume air sampler for quantitative collection of polycyclic aromatic hydrocarbons. Atmospheric Environ., 14: 609-613.
7. Schomburg, G., Husmann, H., and Weeke, F. (1974): Preparation, performance and special applications of glass capillary columns. J. Chromatogr. 99: 63-79.

REACTION PATHWAYS OF BENZO(a)PYRENE DIOL EPOXIDE AND DNA. CONFORMATIONS OF ADDUCTS.

NICHOLAS E. GEACINTOV
Department of Chemistry and Radiation and Solid State Laboratory, New York University, New York, New York 10003.

INTRODUCTION

The diol epoxide $7\beta,8\alpha$-dihydroxy-$9\alpha,10\alpha$-epoxy-7,8,9,10-tetrahydrobenzo(a)pyrene (BaPDE) is biologically the most potent metabolite of the carcinogen benzo(a)pyrene. BaPDE binds covalently to cellular macromolecules including DNA (1,2) and it is important to understand the details of the reaction mechanisms and the properties of the covalent DNA adducts which are formed.

Utilizing a variety of spectroscopic methods we have been studying the conformations of the adducts formed and the reaction mechanisms of racemic BaPDE with DNA under controlled laboratory conditions (usually 25°C, pH 7, 5mM sodium cacodylate buffer).

SPECTROSCOPIC TECHNIQUES

Absorption spectra

The spectroscopic properties of BaPDE, of its tetraol hydrolysis product BaPT (7,8,9,10-tetrahydroxytetrahydrobenzo(a)pyrene), and of all the complexes formed between BaPDE, BaPT and DNA, are characterized by the pyrenyl chromophore. In aqueous solutions this chromophore exhibits absorption maxima at 313, 327 and 343-344 nm. Upon binding to DNA by an intercalation mechanism, in which the planar pyrene moeity is believed to be sanwiched between adjacent base pairs, a red shift of ~10nm occurs in the absorption spectrum of the pyrene system. A prominent new absorption maximum at 353-354nm appears (3,4) upon intercalative binding. At external DNA binding sites the spectral shift is at most 2-3nm.

Fluorescence spectroscopy

The fluorescence of the intercalated pyrenyl group is strongly quenched. Thus, there is no maximum in the fluorescence excitation spectrum (the emission monochromator wavelengths of a spectrofluorometer is held fixed, while the excitation wavelength is scanned over the absorption spectral region of the chromophore) corresponding to the 353nm absorption maximum of the intercalated molecules.

The fluorescence of the pyrenyl chromophore at other (non-intercalation) binding sites is readily observable (3,5). Thus, an analysis of the absorption spectra and fluorescence properties of carcinogen-DNA adducts can provide important information on their structures.

The diol epoxide BaPDE is non-fluorescent, while its tetraol hydrolysis product BaPD exhibits an easily detectable fluorescence in aqueous solutions. Thus, the hydrolysis of BaPDE to BaPT in aqueous solutions can be conveniently followed by monitoring the increase of the fluorescence intensity as a function of time (6,7).

Electric Linear Dichroism

In this technique a 2-5ms reactangular voltage pulse is applied to an aqueous solution containing the DNA complexes. The electric field E gives rise to a partial orientation of the DNA and the carcinogen attached to it. The change in the absorbance due to this orientation is measured using polarized light. The linear dichroism is defined by $\Delta A = A_{\parallel} - A_{\perp}$ where $A_{\parallel}$ is the absorbance when the electric vector of the light is polarized parallel to the direction of the applied electric field, and $A_{\perp}$ is the perpendicular component.

The ΔA for the DNA bases (250-300nm wavelength region) is negative, because the planes of the bases tend to align with their normals parallel to E, and the in-plane transition moments are thus oriented perpendicular to E. The pyrenyl chromophore linear dichroism is measured in the 310-360 nm spectral region. Since the transition moment vector is oriented along the long axis of the pyrene chromophore, it can be shown that $\Delta A < 0$ (tends to be negative) for intercalated pyrenyl residues (3), while $\Delta A > 0$ when the orientation of the pyrenyl plane tends to be perpendicular to the axis of the DNA helix (8). The sign of the electric linear dichroism ΔA thus also provides important information on the conformation of carcinogen-DNA complexes.

HYDROLYSIS OF BaPDE vs. COVALENT BINDING TO DNA

When a BaPDE and a DNA solution are mixed rapidly in a stopped-flow experiment, absorbance at 353nm due to intercalated BaPDE appears within less than 5ms (4) of mixing. A similar result is obtained when BaPT is mixed with DNA.

It has been found that DNA markedly catalyzes the hydrolysis of BaPDE to BaPT. The pseudo-first order hydrolysis rate constant k_H for the process

$$\text{BaPDE} \xrightarrow[\text{[DNA]}]{k_H} \text{BaPT}$$

is markedly dependent on DNA concentration. In the absence of DNA in 5mM sodium cacodylate buffer solution, at pH 7 and 25°C, $k_H = (9 \pm 0.5) \times 10^{-3} s^{-1}$. This constant increases with increasing DNA concentration and for DNA concentrations $> 5 \times 10^{-4}$M (in nucleotides) reaches a limiting constant value of 0.68 ± 0.04 s^{-1} (9). It is interesting to note that the extent of noncovalent binding of BaPDE, as monitored by the absorbance at 353nm as a function of DNA concentration in a stopped-flow experiment, displays the same DNA concentration dependence as k_H.

On the basis of these results it therefore appears that the DNA-catalyzed hydrolysis of BaPDE to a tetraol involves an intermediate intercalation complex. In this non-covalent complex it may well be that an intermediate C-10 carbonium ion is stabilized by hydrogen bonding of the negatively charged epoxide oxygen atom, thus facilitating the addition of a water molecule at the C-10 position.

Hydrolysis constitutes the major reaction pathway of the diol epoxide. About 90-95% of the total BaPDE molecules added to a DNA solution are hydrolyzed to BaPT, while only 5-10% bind covalently to the DNA bases (7). As is shown below, the covalently bound molecules are not intercalated.

The reaction pathways of BaPDE in the presence of DNA are summarized in Figure 1.

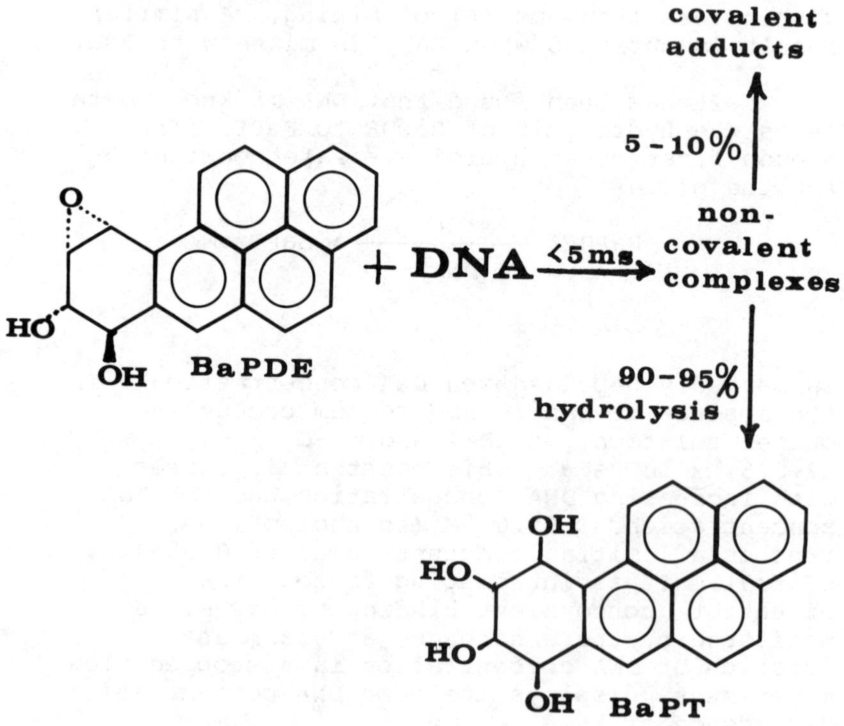

FIGURE 1. Reaction pathways of BaPDE in DNA solutions.

ABSORPTION AND FLUORESCENCE PROPERTIES OF NON-COVALENT AND COVALENT DNA-ADDUCTS

Non-Covalent Tetraol Complexes

Non-covalent BaPT-DNA adducts are stable and thus can be readily studied. A typical absorption spectrum is shown in Figure 2b. A characteristic shoulder at 353nm due to the intercalated species is apparent. In addition there is a second, external binding site (3). The absorption spectrum of free BaPT, exhibiting a maximum at 343nm, is shown for comparison. The fluorescence excitation spectrum, also shown in Figure 2b, indicates that the intercalated BaPT molecules do not contribute to the fluorescence of these complexes, since there is no fluorescence excitation maximum corresponding to the 353nm absorbance. The fluorescence of BaPT molecules at external binding sites is not quenched (3).

Non-Covalent BaPDE-DNA Complexes.

While at pH 7 BaPDE in the presence of DNA is hydrolyzed rapidly to BaPT, at a pH of 8.5 the lifetime of BaPDE is one hour or more (4). Thus, the absorption spectrum of non-covalent BaPDE-DNA adducts can be easily determined at higher pH values as long as the spectrum is taken within a few minutes after mixing. Such an absorption spectrum is shown in Figure 2a, and is compared to an absorption spectrum of free BaPDE. An absorption band appears at 354nm, and gradually decays to the BaPT-DNA adduct spectrum shown in Figure 2b. This shows that BaPDE also intercalates into DNA and that the association constant K for BaPDE is larger than for BaPT.

Covalent BaPDE-DNA Adducts

By extracting fully equilibrated BaPDE-DNA solutions (no unreacted BaPDE molecules remaining) with organic solvents, the non-covalently bound molecules can be removed. The residual

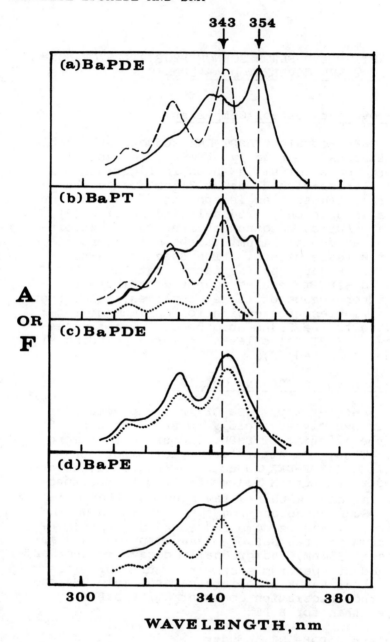

FIGURE 2. Absorbance of DNA complexes (—), free BaPDE or BaPT (---), and fluorescence excitation spectra, F, (····) of complexes. (a) and (b): non-covalent, (c) and (d): covalent DNA adducts.

absorption spectrum of such an extracted solution is shown in Figure 2c, and represents the absorption spectrum of the covalent adducts. The absorption maximum occurs at ~343-345nm and thus does not resemble the spectrum of an intercalated species. Furthermore, the fluorescence excitatation spectrum coincides with the absorption spectrum. The fluorescence of the covalently bound pyrene chromophore is not quenched, as would be expected if it were intercalated.

Covalent BaPE Adducts

We have recently (10) studied the properties of covalent adducts derived from the reaction of 9,10-epoxy-7,8,9,10-tetrahydrobenzo(a)pyrene (BaPE), whose structure is shown in Figure 3, with DNA.

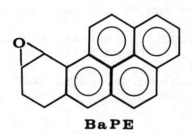

BaPE

FIGURE 3. Structure of BaPE

The BaPE molecule is similar to BaPDE but lacks the two hydroxyl groups in the 7 and 8 positions of BaPDE. Kinoshita et al. (11) have shown that chemically the binding of BaPE to DNA is more heterogeneous than in the case of BaPDE.

The absorption and fluorescence excitation spectra of covalent BaPE-DNA adducts are shown in Figure 2d. The absorption spectrum displays a broad maximum at 354nm and is thus quite different from the one displayed by the covalent BaPDE-DNA adducts. The fluorescence excitation spectrum does not resemble the absorption spectrum because the maximum is at ∼343nm, which does not coincide with the 354nm absorption maximum (Figure 2d). These data thus also indicate that the covalent binding of BaPE to DNA is more heterogeneous than in the case of BaPDE. There are two different types of binding sites, one of which displays the properties of an intercalation site (non-fluorescent, 354nm absorption maximum), while the other one resembles an exterior site (fluorescence excitation and thus inferred absorption maximum at ∼343nm). The relative proportion of this latter, probably exterior binding site, is relatively small, since there is no evidence of a 343nm maximum in the absorption spectrum of the covalent BaPE-DNA adduct.

These results indicate that the OH groups in BaPDE have a profound influence on the mode of covalent binding of this diol epoxide to DNA. The lack of these OH groups in BaPE gives rise to a greater heterogeneity of binding sites (10,11).

CONFORMATION OF ADDUCTS: ELECTRIC LINEAR DICHROISM STUDIES.

Some of the conclusions reached from an analysis of the absorption and fluorescence properties of the adducts can be confirmed from a detailed examination of the electric linear dichroism (ELD) spectra. Representative ELD spectra are shown in Figure 4.

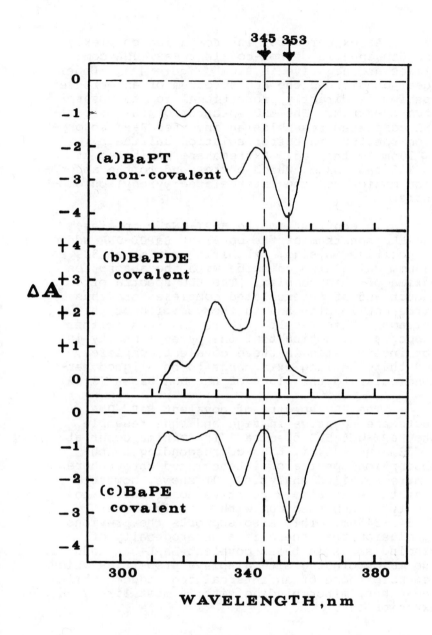

FIGURE 4. Electric linear dichroism spectra (ΔA) for different DNA complexes. Vertical scale in arbitrary units.

As expected for intercalation complexes, the ELD spectra of non-covalent BaPT-DNA complexes are negative in sign (Figure 4a). Under ideal conditions the ELD spectrum of an oriented species is directly proportional to its absorption spectrum. The ELD spectrum of the tetraol-DNA complexes resembles an inverted BaPT absorption spectrum in buffer solution shifted by ∼ 10nm to the red. Extrema are exhibited at ∼ 335 and 353nm, which correspond to the absorption maxima for the intercalated pyrenyl chromophore.

In contrast to the BaPT-DNA complexes, the ELD spectrum of the covalent BaPDE-DNA adduct is positive in sign, exhibiting maxima at 329 and 345nm, which also coincide with the absorption maxima of this complex. The ELD spectra of this adduct and of intercalated complexes are thus different in sign and in the position of the extrema. This result (Figure 2b) thus further supports the notion that the pyrenyl moeity is not intercalated in these covalent complexes, and that its long axis tends to be aligned perpendicular to the plane of the DNA bases (8,12).

The ELD spectra of covalent BaPE-DNA adducts are negative in sign and thus resemble the BaPT-DNA ELD spectra. The minima occur at ∼ 337 and ∼354nm, thus corresponding to the absorption spectra of intercalated chromophores. A more detailed analysis (10) shows, however, that there is also a positive contribution to the total ELD spectrum with a maximum near ∼ 343-345nm. This also supports the previous conclusion that there is a heterogeneity of binding sites in these covalent BaPE-DNA adducts. The major binding site displays properties which resemble those of an intercalated adduct, while the other, minor binding site is most likely an exterior one.

CONCLUSIONS AND SUMMARY

When a BaPDE solution and a DNA solution are mixed with one another, non-covalent complexes form within less than 5ms of mixing. A new red-shifted absorption maximum characteristic of an intercalation complex appears at 353-354nm. The major reaction pathway of non-covalently bound BaPDE is hydrolysis to the tetraol, a reaction which is markedly catalyzed by DNA. Only 5-10% of the molecules react with the DNA bases to form covalent bonds; the conformation of this adduct is quite different from that of an intercalated complex and it appears that the long axis of the pyrenyl chromophore is more perpendicular than parallel to the planes of the bases. In contrast, covalent adducts derived from the reaction of BaPE with DNA exhibit predominantly an intercalation-like conformation. It must be stressed, however, that such an intercalation-like conformation displayed by covalent adducts may be quite different from those of classical intercalation complexes (e.g. ethidium bromide-DNA or BaPT-DNA non-covalent complexes), since covalent binding of the carcinogen may give rise to significant local alterations in the DNA structure.

Finally, it is evident that a rather simple analysis of the absorption and fluorescence properties of pyrene-type carcinogen-DNA complexes can provide significant information about the conformations of the adducts. Such conclusions can be further confirmed using linear dichroism techniques.

ACKNOWLEDGEMENTS.

The author wishes to acknowledge the contributions of all of his co-workers whose names appear in the cited publications. The racemic samples of BaPDE and BaPE were synthesized in Dr. R.G. Harvey's laboratory at the University of Chicago. This investigation was supported by PHS grant number CA20851 awarded by the National Cancer Institute, DHHS, and in part by a Department of Energy Contract DE-ACO2-78EV04959, and Contract No. E(11-1)2386 at the Radiation and Solid State Laboratory at New York University.

REFERENCES

1. Gelboin, H.V. (1980): Benzo(a)pyrene metabolism, activation, and carcinogenesis: role and regulation of mixed function oxidases and related enzymes. *Physiol. Rev,.* 60: 1107-1166.
2. Harvey, R.G. (1981): Activated metabolites of carcinogenic hydrocarbons. *Accts. Chem. Res.*, 14: 218-227.
3. Ibanez, V., Geacintov, N.E., Gagliano, A.G., Brandimarte, S. and Harvey, R.G. (1980): Physical binding of tetraols derived from 7,8-dihydroxy-9,10-epoxybenzo-(a)pyrene to DNA. *J. Am. Chem. Soc.*, 102: 5661-5666.
4. Geacintov, N.E., Yoshida, H., Ibanez, V. and Harvey, R.G. (1981): Noncovalent intercalative binding of 7,8-dihydroxy-9,10-epoxybenzo(a)pyrene to DNA, *Biochem. Biophys. Res. Commun.*, 100: 1569-1577.
5. Prusik, T., Geacintov, N.E., Tobiasz, C. Ivanovic, V. and Weinstein, I.B. (1979): Fluorescence study of the physico-chemical properties of a benzo-(a)pyrene 7,8-dihydrodiol-9,10-oxide derivative bound covalently to DNA, *Photochem. Photobiol.*, 29: 223-232.
6. Kootstra, A., Slaga, T.J. and Olins, D.E. (1979): Studies on the binding of B(a)P diol epoxide to DNA and chromatin. In *Polynuclear Aromatic Hydrocarbons*, Edited by P.W. Jones and P. Leber, pp. 819-834, Ann Arbor Science Publishers, Inc., Ann Arbor, MI.
7. Geacintov, N.E., Ibanez, V., Gagliano, A.G., Yoshida, H., and Harvey, R.G. (1980): Kinetics of hydrolysis to tetraols and binding of benzo(a)-pyrene-7,8-Dihydrodiol-9,10-oxide and its tetraol derivatives to DNA. Conformation of adducts, *Biochem. Biophys. Res. Commun.*, 92:1335-1342.

8. Geacintov, N.E., Gagliano, A., Ivanovic, V. and Weinstein, I.B. (1978): Electric linear dichroism study on the orientation of benzo(a)pyrene-7,8-dihydrodiol-9,10-oxide covalently bound to DNA, Biochemistry, 17: 5256-5262.
9. Geacintov, N.E., Yoshida, H., Ibanez, V. and Harvey, R.G. (1982): Non-covalent binding of benzo(a)pyrene-7,8-dihydrodiol-9,10-oxide to DNA and its catalytic effect on the hydrolysis of the diol epoxide to tetraol. Submitted.
10. Geacintov, N.E., Gagliano, A.G., Ibanez, V. and Harvey, R.G. (1982): Spectroscopic characterizations and comparisons of the structures of the covalent adducts derived from the reactions of 7,8-dihydroxy-7,8,9,10-tetrahydrobenzo(a)-pyrene-9,10-oxide, and the 9,10-epoxides of tetrahydro benzo(a)pyrene and benzo(e)pyrene with DNA, Carcinogenesis, in press.
11. Kinoshita, T., Lee, H.M., Harvey, R.G., and Jeffrey, A.M. (1982): Structures of covalent adducts derived from the reactions of the 9,10-epoxide of 7,8,9,10-tetrahydrobenzo(a)pyrene and 9,10,11,12-tetrahydrobenzo(e)pyrene with DNA, Carcinogenesis, in press.
12. Lefkowitz, S.M., Brenner, H.C., Astorian, D.G., and Clarke, R.H. (1979): Optically detected magnetic resonance study of benzo(a)pyrene-7,8-dihydrodiol-9,10-oxide covalently bound to DNA, FEBS Lett., 105: 77-80.

IMPROVED METHODOLOGY FOR CARBON BLACK EXTRACTION

A. T. GIAMMARISE, D. L. EVANS, M. A. BUTLER, C. B. MURPHY,
D. K. KIRIAZIDES, D. MARSH AND R. MERMELSTEIN
Xerox Corporation, J. C. Wilson Center for Technology,
Webster, New York 14580.

INTRODUCTION

The isolation and identification of polynuclear aromatic hydrocarbons (PAHs), and especially their derivatives, have been the subject of considerable recent research. There is a need for a precise, accurate analytical methodology that can reliably and routinely quantitate these compounds at low levels in complex matrices. It was recently reported by Lofroth et al. (1) and Rosenkranz et al. (2) that extracts of selected xerographic copies and toners produced positive responses in the Ames Salmonella mutagenicity assay. This activity was traced to various nitropyrenes (NPs) which were found to be present as trace impurities in the carbon black used as the toner colorant. The NP level in the carbon black has been reduced to an extremely low level since that time (2,3), making measurement of NPs an even more difficult task. Toners produced from the modified carbon black and copies made using such toners do not result in a mutagenic response confirming that the problem was solved.

While the extraction of carbon blacks has been under examination for many years, a majority of the research has been conducted on the PAH and sulfur analog content of untreated carbon blacks (4,5). Pitts (6) and subsequently Tokiwa et al. (7) showed that non-mutagenic PAHs adsorbed onto filters and exposed to oxides of nitrogen and sulfur in the presence of trace acid formed mutagenic products. Further analyses for the determination of NPs have been directed predominantly toward fly ash (8) and diesel fuel particulates (9-12). The state-of-the-art on NP extraction and analysis from after-treated carbon blacks is expressed by Sanders (3).

In the present paper, a simple, reproducible, analytical procedure is described that can be routinely applied for quality control for the measurement of sub ppm NP content of pelletized after-treated carbon black.

The variables examined and optimized were: extraction thimble flow, time and solvents. Recovery of NP in "spiked" samples, and distribution of NPs in the carbon black aggregate distribution were examined. The Ames Salmonella assay was selected as the final detector and results related to the additive mutagenicity of the NPs. However, the methodology optimization procedure was performed by monitoring the extraction process for NP content by HPLC.

MATERIALS AND METHODS

Ten grams of carbon black were placed in Ace Glass (Vineland, N.J.) glass thimbles of porosity "B" and inserted into the Soxhlet extractor. To the 250ml boiling flask, 140ml of solvent was added. The solvents employed were: methylene chloride, benzene, toluene, monochlorobenzene (MCB) and o-dichlorobenzene (DCB). The solvents were HPLC grade obtained from either Fischer Scientific (Pittsburgh, Pa.) or Burdick and Jackson (Muskegan, Mich.). The boiling was regulated to give a Soxhlet cycle time of 3.5 to 4 minutes. Extraction rates were measured by collecting the extraction solvent and replacing it with fresh solvent at intervals up to 96 hours.

If any carbon black was carried over into the boiling flask, it was removed by filtration. The extracts were reduced in volume in two steps using a Büchi Rotovapor R (VWR Scientific, Rochester, N.Y.), with a water bath (50°C for toluene or 55°C for MCB at a pressure of 15 torr). The final reduction to one ml was obtained using a 50ml flask with a 2ml projection. The final volume was adjusted to either one or two ml with solvent using volumetric flasks. Analyses were made with a Hewlet-Packard (Avondale, PA) Model 1084B HPLC using a UV-visible detector operating at 400nm. Two DuPont Zorbax CN columns (250mm x 46mm) were used. The elution conditions were: Linear gradient, Hexane/Isopropanol, 5% IPA to 20% IPA in 12 minutes, 20% IPA to 60% IPA in 6 minutes, hold isocratic 60% for 4 minutes, reverse gradient to 5% IPA in 4 minutes, hold isocratic 4 minutes before reinjection at a flow rate of 2ml/min. The columns were reconditioned daily by overnight elution with 80/20 tetrahydrofuran/isopropanol at one ml/minute. Nitropyrene standards used for quantification were the same as those used by Rosenkranz, et al. (2). Two Black Pearls® L carbon blacks obtained from the Cabot Corporation, Boston, Mass., were used. They are identified as H, an obsolete black with high levels of NP impurities, and L, a current production black with trace levels of NP impurities.

After many extractions and measurements of NPs where the standard deviation of the measured NPs was greater than 30%, the sample handling and extraction procedure was examined in detail to obtain an understanding of the sources of variability.

Initial focus was on thimble flow. It was found that the flow rate through the empty glass thimbles, measured in terms of time for 40ml of solvent to pass through the frit, varied significantly. Figure 1 shows the flows measured for 33 thimbles. The flow ranged from 30 to 320 seconds. The interval ranging from 60 to 75 seconds was chosen because: (1) this time interval was well below the soxhlet cycling time of 210-240 seconds and (2) it represented the largest single narrow flow range population.

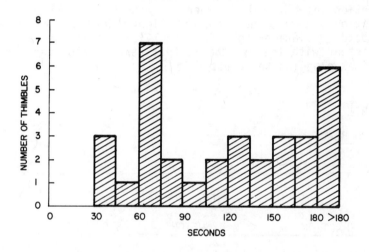

FIGURE 1. Empty thimble drain time-40 ml toulene.

The flow of solvent through the carbon black filled glass thimbles also varied. Measurements in this case were made with 20ml of solvent because of volume constraints. Flows were found to vary from 89 to 240 seconds even when the previously selected matched thimbles were used. On occasion, solvent overflowed the thimble during extraction.

In an attempt to avoid frit plugging thereby minimizing the tendency for solvent flow variations, the use of a narrow particle size range of the carbon black was examined.

Sieve analysis was performed on six 100g. samples of carbon black L and the results are given in Table 1. The fraction >250um but <500um was 51% of the sample and was selected for extraction. It was separately demonstrated that the quantity of extractable nitropyrenes was independent of aggregate particle size. Table 2 shows the results using MCB and toluene as solvents. The >500um fraction was rough crushed and sieved to obtain the >250um but <500um particles. Where MCB was used as the extraction solvent, sieving of the pelletized carbon black does not affect the extractable NPs. However, toluene, a less effective extraction solvent for NPs, showed sensitivity to sieving. The propensity for overflow was eliminated by sieving. Further, in extractions with toluene, employing the slow flowing thimbles, the relative standard deviation was 47% (n=10). Whereas, with toluene under the same conditions, using the selected flow thimbles, the relative standard deviation was 22% (n=14). These results are consistent with the hypothesis that reduced flow was a result of fines plugging the glass frit.

TABLE 1

CARBON BLACK L

	SIEVE ANALYSIS	
FRACTION	% BY WEIGHT	STD. DEV.
>1000um	5.2	1.1%
> 500um	12.9	2.0%
> 250um	51.3	2.5%
> 125um	24.4	2.8%
< 125um	6.1	2.6%
	$\Sigma = 99.9$	

TABLE 2

NITROPYRENE CONTENT AS A FUNCTION OF AGGREGATE PARTICLE SIZE[a]

SOLVENT	SIZE, um			
	125[b]	250[c]	500[d]	UNSIEVED
TOLUENE	--	0.177	--	0.136
MCB	0.283	0.290	0.273	0.300

a) Values in PPM of 1,6-DNP+1,8-DNP for all using Matched glass thimbles 60-75" for all.
b) Size >125um but <250um
c) Size >250um but <500um
d) Rough crushed to >250 but < 500um

Using carbon black H, benzene, toluene, MCB and DCB were found to be effective extraction solvents for NP. The efficiency of these solvents for NP extraction is shown in Figure 2. From these data, toluene and MCB were considered worthy of further examination. Extraction rate studies with carbon black L were undertaken with both solvents and the results are shown in Figure 3. The results indicated that MCB extracted 91.2% of the NPs in 24 hours, while toluene extracted only ~40% in the same time, and MCB extracted 97.7% in 48 hours while toluene extracted only 60.6%. To be on the safe side, MCB was chosen for extraction for 48 hours. It is noteworthy that carbon black "H" contained about 70 ppm extractable NP compared to approximately 0.50 ppm NP in carbon black "L." However, in both cases, MCB was the best extractant.

Figure 4 is a typical HPLC chromatogram of the NP standards in the order they elute under the previously cited HPLC conditions.

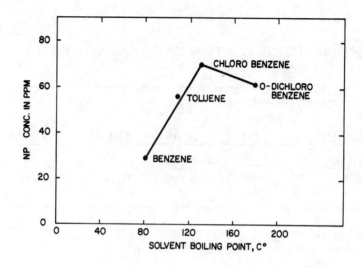

FIGURE 2. Carbon black H; 24 hour extraction; glass thimbles; 10 g sample.

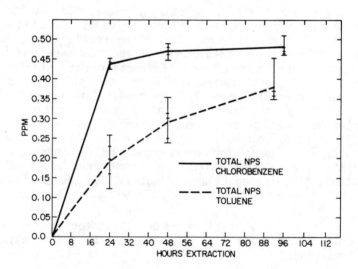

FIGURE 3. Carbon black L; ppm NPs vs time.

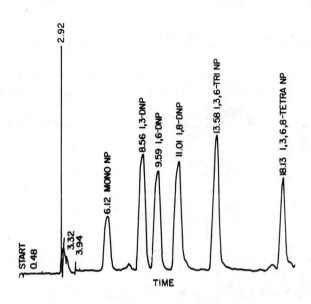

FIGURE 4. HPLC chromatogram of NP standards.

The recovery of nitropyrenes in the absence and presence of carbon black and in carbon black extracts was evaluated. As shown in Table 3, a known NP mixture containing the six NPs at levels which were representative of an authentic extract was added to a flask, and distilled to one ml. It was analyzed and 104% NPs were recovered. The same NP mixture added to the previous experiment was put into a Soxhlet flask and carried through the procedure of refluxing for 48 hours without carbon black in the extraction thimble. This was analyzed and the total NP recovery was 97.5%. In the third experiment, the same standard NP mixture as before was put into a Soxhlet flask. Carbon-black "L" was put into an extraction thimble refluxed for 48 hours. The extraction solvent was evaporated to one ml. and analyzed. Recovery was 97.0%. A previously extracted extract was put into a Soxhlet flask. NP standards as above were added to the extract. This mixture was carried through the distillation procedure and analyzed. The recovery was 111%.

The results of NP measurements from 29 separate measurements of carbon black "L" are summarized in Table 4. The relative standard deviation of 10% for total NP is much better than the 30% relative standard deviation obtained prior to method refinements.

TABLE 3

NITROPYRENE RECOVERY WITH AND WITHOUT CARBON BLACK L

	WITHOUT CARBON BLACK L SPIKED SAMPLES		WITH CARBON BLACK L SPIKED SAMPLES
EVAPORATION ONLY	104.0% -- 2	$\bar{X}$ s n	97%(a) 7.8 7
EXTRACTION AND EVAPORATION	97.5% 0.5 4	$\bar{X}$ s n	111.0%(b) -- 2

(a) For the carbon black, a value of 0.47ppm (n=29) was used.
(b) An extract removed with a prior extraction used instead of standards.

TABLE 4

RESULTS OF CHLOROBENZENE EXTRACTIONS OF CARBON BLACK L

	$\bar{X}$ (ppm)	Sx (ppm)	%
MONO NP	0.067	0.006	14.2
1,3-DNP	0.069	0.003	14.5
1,6-DNP	0.133	0.007	28.1
1,8-DNP	0.158	0.005	33.5
1,3,6-TRI NP	0.026	0.113	5.5
1,3,6,8-TETRA NP	< 0.020	--	< 4.2
TOTAL NP	0.473	0.050	100

SUMMARY

An improved, accurate, and precise technique for the extraction and analysis of NPs has been demonstrated. The method uses preselected glass Soxhlet extraction thimbles to achieve uniform solvent flow. Sieving of the carbon black results in uniform solvent flow through the black and avoids the experienced difficulty of pore clogging and solvent overflowing the thimble. The distribution of NPs was independent of aggregate size of the black.

It has also been shown that the quantitative extraction of NPs from a pelletized carbon black can be successfully accomplished using either toluene or MCB as extractant. However, MCB is superior to toluene due to its efficiency and is, thus, the preferred extraction solvent.

The use of matched fast-flowing thimbles, application to sieved pelletized carbon blacks, and efficient extraction with MCB are believed to constitute the basis for a useful, extraction technique for precise carbon black analyses. Major extraction variables are controlled as evidenced by the low standard deviations presented in Table 4. Mutagenic data (13) also support this viewpoint. Continuing efforts in our laboratory with other carbon black types appear to confirm the general applicability of this technique.

REFERENCES

1. Lofroth, G., Heffner, E., Alfheim, I., and Moller, M., (1980): Mutagenicity in photocopies, Science, 209: 1037-1039.
2. Rosenkranz, H. S., McCoy, E.C., Sanders, D. R., Butler, M., Kiriazides, D. K., and Mermelstein, R., (1980): Nitropyrenes: isolation, identification, and reduction of mutagenic impurities in carbon black and toners, Science, 209: 1039-1041.
3. Sanders, D. R. (1981): Nitropyrenes in aftertreated carbon black. In: Polynuclear Aromatic Hydrocarbon: Chemical Analysis and Biological Fate, edited by M. Cooke and A. J. Dennis, pp. 145-158, Battelle Press, Columbus.
4. Lee, M. L., Hites, R. A. (1976): Characterization of sulfur-containing polycyclic aromatic compounds in carbon blacks, Anal. Chem., 48: 1890-1893.

5. Peaden, P. A., Lee, M. L., Hirata, Y., Novotny, M. (1980): High-performance liquid chromatographic separation of high-molecular-weight polycyclic aromatic compounds in carbon black, <u>Anal. Chem.</u>, 52: 2268-2271.
6. Pitts, J. N., Jr. (1979): Photochemical and biological implications of the atmospheric reactions of amines and benzo(a)pyrene, Phil. Trans. Roy. Soc. London A290: 551-576.
7. Tokiwa, H., Nakagawa, R., Morita, K., and Ohnishi, Y. (1981): Mutagenicity of nitro derivatives induced by exposure of aromatic compounds to nitrogen dioxide, <u>Mutation Res.</u>, 85:195-205.
8. Hansen, L. D., Fisher, G. L., Chrisp, C. E., Eatough, D. J. (1981): Chemical properties of bacterial mutagens in stack collected coal fly ash. In: <u>Polynuclear Aromatic Hydrocarbons: Chemical Analysis and Biological Fate</u>, edited by M. Cooke and A. J. Dennis, pp. 507-516, Battelle Press, Columbus.
9. Schuetzle, D., Lee, F. S., Prater, T. J., Tejada, S. B. (1981): The identification of polynuclear aromatic hydrocarbon derivatives in mutagenic fractions of diesel particulate extracts, <u>Int. J. Environ. Anal. Chem.</u>, 9: 93-144.
10. Gibson, T. L., Ricci, A. I., Williams, R. L., (1981): Measurement of polynuclear aromatic hydrocarbons, their derivatives, and their reactivity in diesel automobile exhaust. In <u>Polynuclear Aromatic Hydrocarbons: Chemical Analysis and Biological Fate</u>, edited by M. Cooke and A. J. Dennis, pp. 707-717.
11. Petersen, B. A., Chuang, C. C., Mongard, W. L., and Trayser, D. A. (1981): Identification of mutagenic compounds in extracts of diesel exhaust particulates, Presented at the <u>74th Annual Meeting of the Air Pollution Control Association</u>, Philadelphia, June 21-26.
12. Pederson, T., and Siak, J. S. (1981): Dinitropyrenes: Their probable presence in diesel particulate extracts and consequent effect on mutagen activations by NADPH-Dependent S9 enzymes. In <u>EPA 1981 Diesel Emissions Symposium</u>, October 5-7, Raleigh, N.C.
13. Butler, M. A., Kiriazides, D. K. and Mermelstein, R., Mutagenicity measurement of carbon black extracts. Manuscript in preparation.

ANALYSIS OF BALANCE OF CARCINOGENIC IMPACT FROM EMISSION CONDENSATES OF AUTOMOBILE EXHAUST, COAL HEATING, AND USED ENGINE OIL BY MOUSE-SKIN-PAINTING AS A CARCINOGEN-SPECIFIC DETECTOR

G. GRIMMER*, K.-W. NAUJACK*, G. DETTBARN*, H. BRUNE**,
R. DEUTSCH-WENZEL**, J. MISFELD***
* Biochemical Institute of Environmental Carcinogens, 2070 Ahrensburg, FRG; ** Advisory Board for Preventive Medicine and Environmental Protection Ltd., 2000 Hamburg, FRG; *** Institute of Mathematics, Technical University, 3000 Hannover, FRG.

INTRODUCTION

The purpose of this investigation was to identify the most effective substances responsible for the carcinogenic impact of emission condensates of automobile exhaust, coal heating, and crankcase oil using mouse-skin-painting as a carcinogen-specific-bioassay.

Since all these matrices contain polycyclic aromatic compounds (PAC) the questions arise:
  (a) What portion of the carcinogenicity originates from PAC in these matrices?
  (b) What portion of these total carcinogenicity originates from benzo[a]pyrene?

The carcinogenic activity of automobile exhaust condensate (AEC), observed after long term application to the dorsal skin of mice, has been described repeatedly (1,2,3,4). We have confirmed these results with exhaust condensate using the same topical application technique. The incidence of carcinoma depends on the amount of the carcinogen administered which is demonstrated in the first table.

This clear cut dose-response-relation is the precondition to determine the effect resulting from fractions of the total material. A comparison of the tumorigenic effect of single fractions with that of the unseparated sample allows establishing a balance of the carcinogenic effect.

TABLE 1

CORRELATION OF THE AMOUNT OF AUTOMOBILE EXHAUST CONDENSATE
(285 µg BaP/g) AND THE FREQUENCY OF CARCINOMA

Amount (mg/year)	Animals with Carcinomas
53	1.3%
106	15.0%
158	29.7%
316	60.0%
438	71.8%

Dropping onto the skin of mice (80 mice per dose).

The aforementioned three matrices have been fractionated by liquid-liquid distribution and chromatography on Sephadex LH 20 using different solvent systems into several fractions containing polycyclic aromatic hydrocarbons (PAH), sulfur-containing polycyclic aromatic compounds (S-PAC), azaarenes (N-PAC), and a fraction free of these compounds. The separation scheme for smoke condensate from coal stoves is shown in the next figure.

Let me first demonstrate the results of the fractionation and the dose-incidence relation in case of AEC from a gasoline engine which is illustrated in Figure 1.

AUTOMOBILE EXHAUST CONDENSATE

Material

The material from fractionation (about 550 g condensate, trapped in a water-cooled steel tube heating exchanger and a micron-glass-fibre filter) (5) originates from a passenger car (1.5 liter, 50 HP) during Europa-tests (ECE reglement 15) which simulates city traffic on a chassis dynamometer.

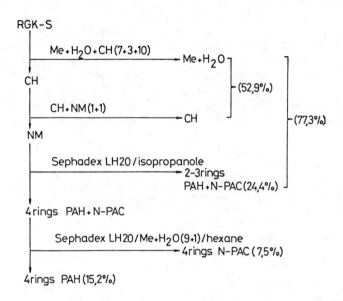

FIGURE 1. Fractionation of smoke condensate of coal combustion.

Fractionation

In the case of automobile exhaust, the experiments were started on the working hypothesis that the carcinogenic activity originates chiefly from PAH. Consequently, the exhaust condensate was separated into a PAH-containing and a PAH-free fraction which was achieved by two steps. Using new material, in a third step, the concentrated PAH fraction was separated into PAH with 2 and 3 rings, and PAH with more than 3 rings. The three separation steps were based on the following criteria:

(a) hydrophobic-hydrophilic properties
(b) aromatic-aliphatic properties
(c) number of aromatic rings; i.e. PAH with
2 and 3 rings; PAH with 4,5,6 and 7 rings.

Hydrophobic-Hydrophilic Substances. The first fractionation step was a distribution between methanol, water and cyclohexane (9+1+10). The hydrophilic phase was extracted three times with cyclohexane by which all PAH are extracted from the methanol water phase except for a small residue (about 1-2%). The next table (Table 2) demonstrates that AEC and its hydrophobic substances induce comparable incidence rates of carcinoma. Only a minor proportion of the carcinogenic effect of AEC cannot be attributed to PAH (4,5,6).

TABLE 2

CARCINOGENIC ACTIVITY OF AEC (TOPICAL APPLICATION). SEPARATION INTO HYDROPHOBIC AND HYDROPHILIC COMPONENTS.

Weight (%)	Designation	PAH Content	Tumor Yield Dose 1 (%)	Dose 2 (%)
100	Exhaust condensate	+	15	60
67	Methanol-water phase	-	1	3
33	Cyclohexane phase	+	14	61

Dose 1 and 2 correspond to 1.05 and 3.15 mg AEC, respectively, administered twice weekly. The stated tumor yield is medium value (probit).

Aromatic-Aliphatic Substances. The results of the separation into aromatic and aliphatic compounds by five times extraction of the cyclohexane phase with nitromethane are given in Table 3.

PAH with 2 and 3 Rings-PAH with 4 to 7 Rings. In the third fractionation step, the aromatic compounds were separated according to their molecular size by means of Sephadex LH 20 chromatography using isopropanol. Aromatic compounds with 2 and 3 rings were separated from those with more than 3 rings. The fractions were dosed proportionally (Table 4).

TABLE 3

CARCINOGENIC ACTIVITY OF AEC. SEPARATION OF THE HYDROPHOBIC COMPONENTS INTO AROMATIC AND ALIPHATIC SUBSTANCES

Weight (%)	Designation	PAH Content	Tumor Yield Dose 1 (%)	Tumor Yield Dose 2 (%)
100	AEC	+	15	60
15.3	Cyclohexane phase	-	1	0
17.7	Nitromethane phase	+	15	60

Dose 1 and 2 correspond to 1.05 and 3.15 mg AEC, respectively, administered twice weekly. AEC-mean value.

In this bioassay, another AEC was used than in the first and second step. Since the PAH-content of this AEC is higher (e.g., 0.514 mg BaP/g), the dose-incidence-relation on the base of weight differs from the first and second step. However, the figure clearly demonstrates that the carcinogenic activity of the total AEC can be attributed to the PAH with more than 3 rings.

The PAH-free fraction, which contains more than 80% of the original AEC, exhibits only a small carcinogenic effect. It should be noted that the 2- and 3-ring PAH together with their alkyl derivatives provoke only 1 tumor in 80 mice in both dose groups.

The probit analysis of these results shows:
- (1) The portion of the PAH containing more than 3 rings, accounts for about 85% of the total carcinogenicity of AEC.
- (2) The portion of benzo[a]pyrene of AEC accounts for about 10% of the total carcinogenicity.
- (3) Regarding the small effect of the PAH-free residue, no hints for a co-carcinogenic activity were obtained.

TABLE 5

COMPARISON OF THE CARCINOGENIC IMPACT OF USED LUBRICATING OIL AND FRACTIONS THEREOF APPLIED TO THE SKIN OF MICE. (65 CFLP female mice/group, a total of 1235 mice, 2 × 0.1 ml test material per week, 104 weeks total, solvent: acetone + cyclohexane, 3+1)

Proportion to total oil (%)	Test Material	Individual dose (mg)	Ratio of doses	Tumor Incidence	Animals bearing local tumors (%)
100.0	Used lubricating oil sample	0.625	1	3	4.6
		1.875	3	17	26.6
		5.625	9	43	69.4
91.6	PAH-free fraction	0.5725	1	1	1.5
		1.7175	3	2	3.0
		5.1525	9	1	1.5
7.3	PAH-fraction with 2 and 3 rings	0.0456	1	4	6.2
		0.1369	3	2	3.1
		0.4106	9	6	9.2
1.1	PAH-fraction with more than 3 rings	0.0069	1	0	0
		0.0206	3	9	13.9
		0.0619	9	33	53.2
100.0	Reconstituted oil from all fractions	0.625	1	3	4.9
		1.875	3	6	9.5
		5.625	9	35	53.9
	Benzo(a)pyrene	0.003846		26	40.0
		0.007692		50	78.1
		0.015385		60	93.8
	Solvent			1	1.5

Concentration of benzo(a)pyrene in used oil: 217 μg BaP/g oil.

USED LUBRICATING OIL

Material

For the animal experiments, an oil was used which had been aged in a gasoline-driven car on the chassis dynamometer under standard conditions simulating city traffic (1500 cm^3, 75 HP, distance 10,000 km; oil content: 217 mg BaP/kg oil; fuel with lead content of 0.4 g/l type CEC European Reference Fuel RF-01-T).

To determine the portion of PAH on the total biological activity, the used oil was separated into PAH-containing and PAH-free fractions as described for AEC. A comparison of the tumorigenic effect of single fractions with that of an unseparated sample of used crankcase oil is demonstrated in Table 5.

The PAH-fraction containing PAH with more than 3 rings is by far the most potent fraction. In contrast to that, the PAH-free portion of the oil, about 92% of its total weight, provokes only a very small tumor incidence. A small number of tumors were produced by PAH with 2 and 3 rings, which is in contrast to the results of the AEC.

The probit analysis of the results shows:
(1) The portion of the PAH containing more than 3 rings accounts for about 70% of the total carcinogenicity in case of crankcase oil.
(2) The portion of benzo[a]pyrene (217 mg BaP/kg) accounts for 18% of the total carcinogenicity of the used oil.
(3) Regarding the reduced carcinogenicity of the oil, reconstituted from all fractions, it seems possible that some of the carcinogenic substances are volatile and were lost during evaporation of the solvents from the oil fractions.

SMOKE CONDENSATE OF BRIQUET-FIRED STOVE

Material

For the animal experiments, a smoke condensate from a briquet-fired residential stove was used. As described in the case of automobile exhaust, the material was trapped by

TABLE 4

COMPARISON OF THE CARCINOGENIC IMPACT OF AUTOMOBILE EXHAUST CONDENSATE WITH GASOLINE ENGINE AND FRACTIONS THEREOF. (80 CFLP female mice/group, a total of 1120 mice, 2 x 0.1 ml test material per week, 104 weeks total, solvent: DMSO + acetone, 3+1)

Proportion to total condensate (%)	Test Material	Individual dose (mg)	Ratio of doses	Tumor Incidence	Animals bearing local tumors (%)
100.0	Automobile exhaust condensate	0.292 0.875 2.626	1 3 9	5 24 54	6.4 30.4 69.2
82.8	PAH-free fraction	0.969 2.907	4 12	3 9	4.0 11.4
13.0	PAH-fraction with 2 and 3 rings	0.152 0.455	4 12	1 1	1.3 1.3
3.5	PAH-fraction with more than 3 rings	0.0204 0.0613	2 6	6 42	7.6 56.8
0.0514	Benzo(a)pyrene	0.0039 0.0077 0.0154		15 34 54	23.4 53.1 84.4
	Solvent (DMSO+acetone, 3+1) untreated animals			0 0	0 0

Ratio of doses: The dose ratio corresponds to that of the total exhaust condensate from which the fraction was obtained.

a collecting system consisting of glass cooler and a glass fibre particulate filter (Draeger Werke AG, Lübeck, FRG; type MB 50, collecting area about 1 m^2). Hard coal briquets (three experiments, 3.66 kg each) have been combusted under standard conditions (DIN reglement; combustion period 240 min; Oranier Durchbrand stove, type 4312 DB; bitumen bound briquets). After evaporation of the solvents, the smoke condensate, consisting of filter-extraction, precipitation on the glass cooler, and the condensed water, weighed about 85 g, which contained 60.2 mg benzo[a]pyrene.

The smoke condensate was separated using the aforementioned scheme. In addition to that, a fraction of nitrogen containing polycyclic aromatic compounds (N-PAC) with more than 3 rings was separated from the smoke condensate. For the animal experiments, the PAH-free fraction has been combined with the fraction containing the 2- and 3-ring components.

Results

The results of the topical application of the smoke condensate and fractions thereof onto the skin of mice are given in Table 6. Since histological and mathematical evaluations have not yet been completed, all findings reported in the following refer to macroscopical interpretation. From our experience with previous epicutaneous experiments, however, a good correlation was found between macroscopically observed and histologically confirmed findings.

The table shows:
(1) The contribution of the PAC consisting of more than 3 rings to the total carcinogenicity of smoke condensate is very high. Proportionated doses provided, the carcinogenicity is comparable to that of the total condensate. Therefore, 15 mg of the PAC-fraction (more than 3 rings) are comparable to 100 mg of the total smoke condensate.
(2) Comparing the carcinogenicity of the N-PAC, fraction with that of the PAH-fraction, the effect of the N-PAC accounts only for less than 20% of the total activity.
(3) It should be noted, that the carcinogenic activity of the PAH-free-fraction and of the fraction containing PAH with 2 and 3 rings (together about 80% by weight of the

TABLE 6

COMPARISON OF THE CARCINOGENIC IMPACT OF SMOKE CONDENSATE OF BRIQUET-FIRED STOVE AND FRACTIONS THEREOF APPLIED TO THE SKIN OF MICE. (65 CFLP female mice/group, a total of 1170 mice, 2 x 0.1 ml test material per week, 104 weeks total, solvent: DMSO + acetone, 3+1)

Proportion to total condensate (%)	Test material	Individual dose (mg)	Ratio of doses	Tumor Incidence	Animals bearing local tumors (%)
100.0	Smoke condensate	0.205	1	23	35.0
		0.616	3	54	83.0
		1.849	9	58	89.0
77.3	PAH-free fraction and PAH-fraction with 2 and 3 rings	0.635	4	4	6.0
		1.905	12	18	28.0
15.2	PAH-fraction with more than 3 rings	0.062	2	49	75.5
		0.187	6	54	83.0
7.5	N-PAC-fraction with more than 3 rings	0.031	2	1	1.5
		0.092	6	18	28.0
22.7	PAH- and N-PAC- fraction with more than 3 rings	0.093	2	53	81.5
		0.280	6	55	85.0
ca.100	reconstituted smoke condensate from all fractions	0.411	2	50	77.0
		1.232	6	55	85.0
0.070	Benzo(a)pyrene	0.003365		46	71.0
		0.006730		54	83.0
		0.013462		57	88.0
	Solvent			0	0
	untreated animals			1	1.5

smoke condensate) counts only for a small effect.
(4) It may be stated, that the contribution of benzo[a]pyrene contained in the smoke condensate accounts for 5 to 8%.

REFERENCES

1. Kotin, P., Falk, H.L., and Thomas, M. (1954): Aromatic hydrocarbons. II. Presence in particulate phase of gasoline engine exhaust and carcinogenicity. <u>Archi. Industr. Hyg.</u>, 9:164-177.
2. Wynder, E.L. and Hoffmann, D. (1962): A study of air pollution carcinogenesis. III. Carcinogenic activity of gasoline engine exhaust condensate. <u>Cancer</u>, 15:103-108.
3. Brune, H. (1977): Experimental results with percutaneous applications of automobile exhaust condensates in mice. IARC Sci. Publ. No. 16, Lyon 1977, 41-47.
4. Brune, H., Habs, M., and Schmähl, D. (1978): The tumor-producing effect of automobile exhaust condensate and fractions thereof, II, <u>J. Environm. Path. and Tox.</u>, 1:737-746.
5. Grimmer, G. and Böhnke, H. (1978): The tumor-producing effect of automobile exhaust condensate and fractions thereof, I., <u>J Environm. Path and Tox.</u>, 1:661-667.
6. Misfeld, J. and Timm, J. (1978): The tumor-producing effect of automobile exhaust condensate and fractions thereof, III, <u>J. Environm. Path. and Tox.</u>, 1:747-772.

ISOLATION AND IDENTIFICATION OF MUTAGENIC PRIMARY AROMATIC AMINES FROM SYNTHETIC FUEL MATERIALS.

DAVID A. HAUGEN*, VASSILIS C. STAMOUDIS**, MEYRICK J. PEAK*, AND AMRIT S. BOPARAI***
*Division of Biological and Medical Research, **Energy and Environmental Systems Division, ***Chemical Engineering Division, Argonne National Laboratory, Argonne, Illinois 60439.

INTRODUCTION

Organic bases present in synthetic fuel materials are responsible for a major portion of their mutagenic activity (1-5). The principal bases are primary aromatic amines (PAA) and aza-arenes (AA). As determined by thin-layer (2) and column (6) chromatography, as well as by other less direct approaches (7,8), the PAA are responsible for most of the mutagenic activity of the basic fractions.

The objective of our study was to improve fractionation and identification procedures for mutagenic aromatic bases. We report here the application of a cation exchange high-performance liquid chromatographic (HPLC) method which, combined with GC/MS, provides qualitative and quantitative information superior to that provided by previous methods. Our data for materials from coal gasification and liquefaction processes strongly support the results from other laboratories (2,6) indicating that 3- and 4-ring PAA are the principal mutagenic bases derived from these processes.

MATERIALS AND METHODS

Materials Analyzed

Three samples were analyzed: (a) condensate from the low temperature reactor of the HYGAS coal gasification pilot plant operated by the Institute of Gas Technology, Chicago, IL, (b) recirculation oil from the same plant, and (c) coal oil A, a coal liquefaction product obtained from the EPA/DOE Fossil Fuels Research Materials Facility (Oak Ridge National Laboratory), the Comparative Research Material designated CRM-1.

Solutions of the oils in CH_2Cl_2 were extracted successively with 1 N NaOH and 1 N HCl. The acidic extracts were neutralized, and the organic bases were back-extracted with CH_2Cl_2.

The solutions were dried, and the solvent was evaporated at 60°C and 4-6 torr. The relative amount of residual anilines and pyridines depended on the evaporation time.

Separation of Aromatic Bases by Cation Exchange HPLC

The organic bases obtained as described above were dissolved in CH_3CN, and aqueous citric acid was added so that the final concentrations of CH_3CN, citric acid, and sample were 40% (v/v), 100 m$\underline{M}$, and 4 mg/ml, respectively. The mixture was fractionated on a 0.9 x 50-cm Partisil 10 SCX column (Whatman) using a model GPC/ALC-204 chromatograph (Waters) equipped with a silica precolumn between the pump and the injector, and a cation exchange guard column between the injector and the primary column. The mobile phase was 40% (v/v) aqueous CH_3CN containing 10 m$\underline{M}$ sodium citrate at the pH values specified. Bases were eluted at 3.5 ml/min with a 115-min concave gradient from pH 3.4 to 8.0, and 1-min fractions were collected for assay of mutagenic activity. Standardization and characteristics of the system are described elsewhere (9).

For fractions subsequently analyzed by GC and GC/MS, benzo[f]quinoline was added as an internal standard, the CH_3CN was evaporated under a stream in N_2 at 30°C, NaOH was added to 1 $\underline{N}$, the bases were extracted with CH_2Cl_2 with 80-90% yield, and the extract was concentrated to 20 μl.

GC and GC/MS Analysis

GC/MS was performed using a Hewlett-Packard (HP) model 5984A instrument operated in the electron impact mode, and equipped with an HP model 5840 GC and an HP model 5934A data system. GC was performed using an HP model 5880A instrument equipped with a flame ionization detector. Both GC instruments were equipped with 50-m x 0.31-mm flexible, fused silica, wall-coated OV-101 capillary columns from HP. The carrier gas was helium, and the temperature gradient was 2°/min to 270°C. Isomeric PAA and methyl-AA were distinguished by comparison of their mass spectra and relative GC retention times as determined by chromatography of numerous standards.

Mutagenesis Assay of HPLC Fractions

HPLC fractions were assayed for mutagenicity by the plate incorporation method (10) using Salmonella typhimurium strain TA98, and optimum concentrations of liver S9 from Arochlor 1254-treated rats. For a given experiment, a fixed amount (0.02-0.1 ml) of each fraction was assayed in duplicate. Dose-response relationships for the crude basic materials were determined under conditions identical to those used for the fractions. Linear dose-response relationships were also determined for fractions found to have high mutagenicity. The slopes of the dose-response relationships were consistent with the data obtained using a single, fixed volume of each fraction.

RESULTS

GC Analysis of Crude Basic Fractions

GC and GC/MS analysis of the crude basic fractions revealed the presence of at least 200 components, principally 1- to 4-ring PAA, 1- to 5-ring AA, and their respective alkyl derivatives. The ratio of AA to PAA was about 4:1.

Cation Exchange HPLC

Figure 1 illustrates the separation of standard PAA and AA by cation exchange HPLC, and also shows the effect of loading on the resolution of the basic fraction of CRM-1. Although high loading causes diminished resolution, it has been useful for collection of sufficient material for subsequent subfractionation of mutagenic fractions. In a previous report (9) we demonstrated that for this system, 1- to 5-ring PAA and AA are eluted in order of increasing basicity. PAA (weaker bases) are generally eluted before the AA, and positional isomers in each class are often well resolved (see examples in Figs. 1 and 2). The method is highly reproducible. Under identical conditions, retention times for standards at low load changed less than 1 min in 3 months.

Figure 2 illustrates liquid chromatography of 2.5 mg of the basic fraction of CRM-1. Most of the mutagenic activity was eluted in the earlier part of the chromatogram where PAA, but not AA elute. The general chromatographic profile for UV absorbance in Figure 3 was similar to those for the other samples, as was the position of the principal band of S9-dependent mutagenic activity relative to standard PAA. For each sample the limited mutagenesis data (single volume of each fraction) indicated that about 70% of the mutagenic activity was recovered.

GC and GC/MS Analysis of HPLC Fractions

Figure 3 illustrates the GC separation of three mutagenic HPLC fractions from the CRM-1 sample (HPLC shown in Figure 2). As for all samples examined, individual HPLC fractions contained only 10-20 principal components, and were thus highly purified relative to the crude basic mixture.

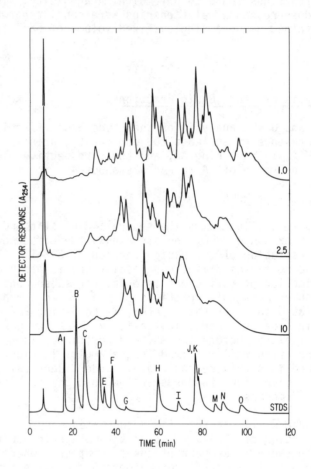

FIGURE 1. Liquid chromatograms of standard bases and the basic fraction from CRM-1. Numbers at the ends of the chromatograms indicate the load (mg). The detector sensitivity was inversely proportional to the load. The standard mixture contained: A, 1-aminopyrene; B, 1-aminoanthracene; C, 2-aminoanthracene; D, 4-aminobiphenyl; E, 2-aminonaphthalene; F, 2-aminofluorene; G, aniline; H, benzo[f]quinoline; I, quinoline; J, acridine; K, isoquinoline; L, pyridine; M, 3-methylpyridine; N, 2-methylpyridine; O, 2,4-dimethylpyridine.

For the aminophthalenes, the gas chromatograms (Figure 3) demonstrate that as predicted by HPLC data (Figure 2) fractions 35, 37, and 41 contained respectively, 1-aminonaphthalene (but a negligible amount of 2-aminonaphthalene), relatively little of either aminonaphthalene, and 2-aminonaphthalene (but a negligible amount of 1-aminonaphthalene). The coelution of C_3-anilines and aminonaphthalenes in the HPLC system is consistent with the behavior of standard PAA (9). The abrupt appearance and disappearance of different isomers of aminonaphthalenes and C_3-anilines in the HPLC fractions (Figure 3) shows that the HPLC system satisfactorily separates individual components of complex mixtures even at a semipreparative load which moderately diminishes resolution (Figure 1). Figure 3 also illustrates the HPLC separation of other less abundant components. Similar, but more detailed qualitative and quantitative data are presented elsewhere (9) for a different sample.

Table 1 summarizes the composite GC and GC/MS data for analysis of mutagenic HPLC fractions isolated from three samples. The data are semiquantitative because detailed GC data

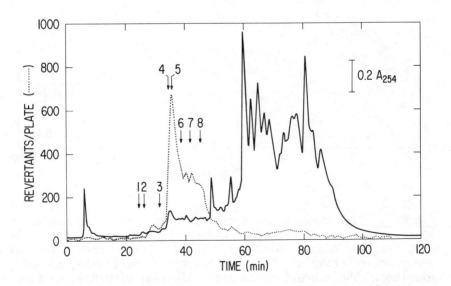

FIGURE 2. HPLC separation of the basic fraction of CRM-1. The numbered arrows indicate the retention times for standard PAA: 1, 2-aminobiphenyl; 2, 1-aminoanthracene; 3, 2-aminoanthracene; 4, 1-aminonapthalene; 5, 3-methyl-2-aminonaphthalene; 6, 4-aminobiphenyl; 7, 2-aminonaphthalene; 8, 2-aminofluorene.

is available for only 6 to 8 representative 1-min fractions in each mutagenic band. Only the CRM-1 sample contained significant amounts of (C_0-C_2)-aminoindans. For each mutagenic band, (C_0-C_2)-aminonaphthalenes comprised at least 80% of the PAA of 2 or more rings. It may be significant that the relative concentrations of (C_0-C_1)-amino(anthracenes/phenanthrenes) had

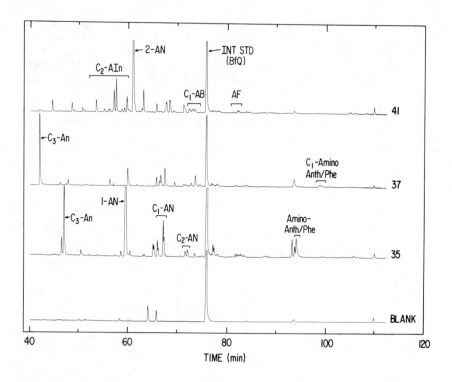

FIGURE 3. Gas chromatograms of HPLC fractions. The numbers at the right end of the upper three chromatograms refer to the HPLC fraction number (minutes), Fig. 2. These three fractions are representative of the seven fractions, (including no. 39) examined. The abbreviations are: AIn (aminoindans), AN (aminonaphthalenes), AB (aminobiphenyls), BfQ (benzo[f]quinoline), AF, (aminofluorenes), An (aniline), Amino-Anth/Phe (aminoanthracene/phenanthrene). The blank was a portion of HPLC mobile phase treated as the fractions. Components eluted before 40 min were either monocyclic bases, or were contaminants also appearing in the blank.

the same order as the relative specific mutagenicities of the original basic fractions.

In preliminary experiments, we have subfractionated broader 5-min HPLC fractions of the CRM-1 bases (10-mg load, Figure 1) by reverse phase HPLC. Reverse phase HPLC of the most mutagenic 5-min fraction from cation exchange HPLC resolved several major peaks. Mutagenesis assays and GC/MS analysis of the reverse phase HPLC fractions revealed that although one of the major peaks contained 1- and 2-aminonaphthalene, the mutagenic

TABLE 1

COMPOSITION OF MUTAGENIC BANDS ISOLATED BY HPLC

Component	Relative Concentration[a]		
	LTR	RO	CRM-1
(C_0-C_2)-Aminoindans	v. low	v. low	medium
1-Aminonaphthalene[b]	medium	high	high
2-Aminonaphthalene[b]	v. high	v. high	high
C_1-Aminonaphthalenes	high	v. high	high
C_2-Aminonaphthalenes	high	medium	medium
Aminobiphenyls	low	low	low
C_1-Aminobiphenyls	v. low	v. low	low
Aminofluorenes	low	v. low	low
Aminoacenaphthenes	medium	medium	low
C_1-Aminoacenaphthenes	medium	low	v. low
Amino(anthracenes or phenanthrenes)	high	low	medium
C_1-Amino(anthracenes or phenanthrenes)	low	v. low	low
Amino(pyrene or fluoranthene)	-	-	v. low[c]

a The samples were: low temperature reactor (LTR) organic condensate and recirculation oil (RO) from the HYGAS pilot plant, and CRM-1, a coal liquefaction product. The relative concentrations are v. high (>15%), high (7-5%), medium (2.5-7%), low (0.5-2.5%), and v. low (<0.5%).
b These were identified using internal and external standards in the HPLC and GC systems; others are tentatively identified from their masses and relative GC retention times.
c Detected only for CRM-1 by overloading the ion exchange HPLC system and subfractionating by reverse phase HPLC.

activity was associated primarily with other fractions containing C_2-aminonaphthalenes, C_1-amino(anthracenes/phenanthrenes), and unsubstituted 4-ring PAA.

DISCUSSION

We have applied a unique combination of complementary HPLC, GC/MS, and bioassay techniques to the semipreparative qualitative and quantitative analysis of mutagenic organic bases in synthetic fuel materials. The HPLC system (a) separates PAA from AA according to their basicity, (b) resolves positional isomers within each class, (c) provides fractions that can be assayed for mutagenic activity without work-up, and (d) provides relatively simple fractions from highly complex mixtures. The data provide direct evidence that PAA are the principal mutagens in basic fractions of materials derived from coal gasification and liquefaction processes.

Present data indicate that a relatively small number of 3- and 4-ring PAA appear to be principally responsible for the mutagenic activity in the coal gasification and liquefaction samples. Complementary qualitative information from the HPLC and GC/MS systems provides clues to the identity of PAA isomers, but specific assignments for the isomers are limited by the availability of reference chemicals. We have not yet eliminated the possibility that biologically important PAA of higher molecular weight are present in the samples. The importance of 3- and 4-ring PAA is in general agreement with results for other coal-derived materials for which more complex PAA-containing fractions were isolated by TLC or by conventional column chromotography (2,6).

Based on GC quantification, our present data indicate that the specific mutagenicities of the most active HPLC fractions are considerably greater than those of the unsubstituted 3- and 4-ring PAA we have examined. It is known that methylation of 2-aminonaphthalene and 4-aminobiphenyl adjacent to the amino group increases their carcinogenicity and mutagenicity (11,12). Similarly, methylation of polycyclic aromatic hydrocarbons at certain sites also increases their mutagenic and carcinogenic activity (11,13,14). Consistent with these observations, it is possible that methyl-PAA are of major importance for the mutagenic activity in the samples we have examined.

As an extension of the present exploratory approach, we envision the GC/MS analysis of broader HPLC fractions to routinely compare the concentrations of various sub-classes of

PAA as a function of sample type, process conditions, or biological activity.

ACKNOWLEDGMENTS

We thank K. M. Suhrbier, S. S. Dornfeld, D. Venters, R. Woznikaitis, and D. Castelli for expert technical assistance. This work was supported by the U. S. Department of Energy under contract No. W-31-109-ENG-38.

REFERENCES

1. Epler, J.L., Young, J.A., Hardigree, A.A., Rao, T.K., Guerin, M.R., Rubin, I.B., and Clark, B.R. (1978): Analytical and biological analysis of test materials from synthetic fuel technologies: Mutagenicity of crude oil determined by the Salmonella typhimurium/microsomal activation system, Mutat. Res., 57:265-276.
2. Wilson, B.W., Pelroy, R., and Cresto, T.T. (1980): Identification of primary aromatic amines in mutagenically active subfractions from coal liquefaction materials, Mutat. Res., 79:193-202.
3. Pelroy, R.A., and Petersen, M.R. (1979): Use of Ames test in evaluation of shale oil fractions, Environ. Health Perspect., 30:191-203.
4. Hsie, A.W., Brimer, P.A., O'Neill, J.P., Epler, J.L., Guerin, M.R., and Hsie, M.H. (1980): Mutagenicity of alkaline constituents of a coal-derived crude oil in mammalian cells, Mutat. Res., 78:79-84.
5. Stamoudis, V.C., Bourne, S., Haugen, D.A., Peak, M.J., Reilly, C.A., Jr., Stetter, J.R., and Wilzbach, K. (1980): Chemical and biological characterization of high-BTU coal gasification (the HYGAS process): Chemical characterization of mutagenic fractions. In: Coal Conversion and the Environment: Chemical, Biological, and Ecological Considerations, Proceedings of the 20th Annual Hanford Life Sciences Symposium, DOE Report, CONF-80-1039 (in press).
6. Guerin, M.R., Ho, C.-H., Rao, T.K., Clark, B.R., and Epler, J.L. (1980): Polycyclic aromatic primary amines as determinant chemical mutagens in petroleum substitutes, Environ. Res., 23:42-53.
7. Pelroy, R.A., and Gandolfi, A.J. (1980): Use of a mixed function amine oxidase for metabolic activation in the Ames/Salmonella assay system, Mutat. Res., 72:329-334.

8. Haugen, D.A., Peak, M.J., and Reilly, C.A., Jr. (1980): Chemical and biological characterization of high BTU coal gasification (the HYGAS process): Nitrous acid treatment for detection of mutagenic primary aromatic amines: non-specific reactions. In: Coal Conversion and the Environment: Chemical, Biological, and Ecological Considerations, Proceedings of the 20th Annual Hanford Life Sciences Symposium, DOE Report, CONF-80--1039 (in press).
9. Haugen, D.A., Peak, M.J., Suhrbier, K.M., and Stamoudis, V.C. (1980): Isolation of mutagenic aromatic amines from a coal conversion oil by cation exchange chromatography, Anal. Chem., (in press).
10. Ames, B.N., McCann, J., and Yamasaki, E. (1975): Methods for detecting carcinogens and mutagens with the Salmonella/mammalian-microsome mutagenicity test, Mutat. Res., 31:347-364.
11. Arcos, J.C., and Argus, M.F. (1974): Chemical Induction of Cancer: Structural Bases and Biological Mechanisms, vol. IIA and IIB. Academic Press, New York, 379 pp.
12. Hecht, S.S., El-Bayoumy, K., Tulley, L., and LaVoie, (1979): Structure-mutagenicity relationships of N-oxidized derivatives of aniline, o-toluidine, 2'-methyl-4-aminobiphenyl, and 3,2'-dimethyl-4-aminobiphenyl, J. Med.Chem., 22:981-987.
13. LaVoie, E.J., Tulley-Freiler, L., Bedenko, V., and Hoffmann, D. (1981): Mutagenicity, tumor-initiating activity, and metabolism of methylphenanthrenes, Cancer Res., 41:3441-3447.
14. Hecht, S.S., Loy, M., and Hoffman, D. (1976): On the structure and carcinogenicity of the methylchrysenes. In: Carcinogenesis, vol. 1: Polynuclear Aromatic Hydrocarbons: Chemistry, Metabolism, and Carcinogenesis, edited by R.I. Freudenthal and P.W. Jones, pp. 325-340, Raven Press, New York.

THE ULTIMATE FATES OF POLYCYCLIC AROMATIC HYDROCARBONS IN MARINE AND LACUSTRINE SEDIMENTS

RONALD A. HITES AND PHILIP M. GSCHWEND
School of Public and Environmental Affairs and Department of Chemistry, Indiana University, 400 East Seventh Street, Bloomington, Indiana 47405.

INTRODUCTION

Polycyclic aromatic hydrocarbons (PAH) are generated on land by the combustion of various fossil fuels and wood in both stationary and mobile sources. Some of the resulting PAH-laden particles are transported by the wind to distant locations, at which point, they are removed from the atmosphere by rain and dry fallout. Those PAH deposited on the surface of lakes and the sea are removed to and incorporated in the sediments where they have been found by numerous investigators (1, 2). Other PAH-laden particles settle to the surface near their point of origin and are delivered to natural waters by erosion and runoff, followed by the resuspension and transport of contaminated sediments. Existing data support this scenario because high PAH levels are found at locations close to centers of human activity while low concentrations occur in regions remote from significant anthropogenic influence (3).

In an effort to quantify the deposition rates for PAH in both remote and urban locales, we have determined the PAH abundances in several sediment cores from the northeastern United States (see Fig. 1) and have calculated the corresponding PAH fluxes to these sites. By assessing flux information (rather than concentrations), many of the differences between sites are taken into account, thereby allowing more meaningful intercomparisons. We were particularly interested in PAH fluxes to lakes located on islands and to remote high altitude lakes since these sites should most accurately reflect the atmospheric deposition of these combustion-derived pollutants. This background flux could then be compared to PAH inputs found nearer to urban centers, thereby quantitatively showing the relative importance of long-range aeolian versus short-range runoff delivery of PAH. Additionally, the cores provide a record of the historic variation of PAH inputs to these sites.

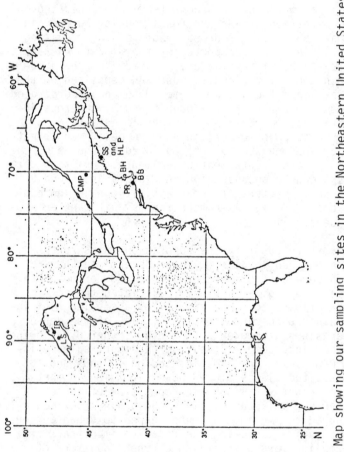

FIGURE 1. Map showing our sampling sites in the Northeastern United States: LS, Lake Superior; IR, Isle Royale; CMP, Coburn Mountain Pond; SS, Somes Sound; HLP, Hadlock Lower Pond; BH, outer Boston Harbor; PR, Pettaquams- cutt River (2); BB, Buzzards Bay (1).

RESULTS AND DISCUSSION

The PAH were observed to have very similar depth profiles at the remote sites of this study (see Fig. 2). In the cases where core subsampling resolution is fine enough, surface sediment concentrations are somewhat less than core sections corresponding to deposition approximately 30 years ago. It possibly reflects the transition from home heating using coal to that using oil and gas which occurred at about that time. Also in each case, the PAH concentrations become nearly undetectable in sections reflecting inputs of 100 years ago. Thus, it appears that the PAH depositional profile is fairly ubiquitous, and we suggest that this profile may, in fact, serve as a sedimentary time marker much as other indicators such as Cs-137, DDT, and PCBs have been used. The application of this aspect of our PAH profiles lends support to the sedimentation rate estimates based on Pb-210 for these sites.

Combining the observed PAH concentrations with the sedimentation rates and in situ dry densities, we calculated fluxes (ng/cm yr) of individual PAH to the five remote and three urban sites of this study. Table 1 shows the results of these flux calculations for core subsections reflecting PAH deposition to remote sites at present, in the interval including 1950, and at the turn of the century (1900). The first point to notice is that the average fluxes for most individual PAH (except anthracene) to remote northeastern United States sites are 0.8-3 ng/cm yr at present. Where core subsampling resolution permits, we observe that fluxes of approximately 30 years ago were as much as 2 to 3 times greater and that fluxes around 1900 were 5 to 10 times lower than at present. This historical PAH record clearly shows that man's activities over the last century resulted in an influx of PAH to the environment. These data support the conclusion that coal-derived energy is much more PAH-polluting than energy from oil and gas. Also, if the PAH flux reflects the strength of the same combustion sources as those producing acid rain, then our data indicate this pollution effect has slowed over the last two decades. Finally, we note that PAH, many of which are known carcinogens, are being delivered to some very remote sites of the northeastern United States.

The three easternmost sites (SS, HLP, CMP) are also the three southernmost (see Fig. 1); these locations showed higher input rates than their Great Lakes region counterparts. This may be because they are located downwind of a more intense source region. There is no apparent trend in the fluxes with respect to the water depths at which the

TABLE 1 PAH fluxes (in ng/cm yr) to sediments from 5 remote sites in the northeastern United States for 3 age intervals: present, approximately 1950, and 1900; and to sediments from 3 urban sites for 1940 to the present. Buzzards Bay data from Hites et al. (1), and Pettaquamscutt River PAH data from Hites et al. (2).

Remote Sites	interval	in situ dens	phen	anth
Lake Superior (LS) 0.02 cm/yr	1955-now	0.55	0.3	0.03
	1930-1955	0.55	0.2	0.02
	1870-1920	0.55	0.06	0.004
Isle Royale (IR) 0.09 cm/yr	1974-now	0.33	0.4	0.01
	1951-1955	0.32	1	0.05
Somes Sound (SS) 0.1 cm/yr	1960-now	0.43	2	0.2
	1940-1960	0.52	4	0.4
	1880-1940	0.51	0.4	<0.02
Hadlock Lower Pond (HLP) 0.07 cm/yr	1950-now	0.12	2	0.1
	1920-1950	0.11	0.3	0.04
Coburn Mtn. Pond (CMP) 0.3 cm/yr	1975-now	0.036	2	0.2
	1943-1947	0.057	8	0.5
	1898-1901	0.057	0.8	<0.07
AVERAGES	present	-	1	0.1
	∼1950	-	3	0.2
	∼1900	-	0.4	0.03
Urban Sites				
Boston Outer Harbor 0.1 cm/yr	1900-now	0.93	24	5.6
Buzzards Bay, Mass 0.3 cm/yr	1940-now	0.3	18	2
Pettaquamscutt River 0.3 cm/yr	1940-now	0.16	46	5
AVERAGES		-	30	4

*includes all $C_{20}H_{12}$ isomers except perylene

TABLE 1 (Continued)

meth phen	fluo	pyr	b(a)a	chry+ tri	b(e)p	b(a)p
0.3	1	0.6	0.3	1	0.9	0.4
0.2	0.9	0.5	0.3	1	0.9	0.3
0.08	0.2	0.1	0.07	0.3	0.2	0.07
0.5	0.4	0.3	0.2	0.8	0.8	0.2
2.5	1	0.7	0.2	0.7	0.9	0.2
1	5	4	2	2	2	2
4	8	5	3	4	2	2
<0.2	0.4	0.4	<0.05	<0.2	0.2	0.2
2	4	3	0.6	1	0.8	0.6
-	0.3	0.2	0.04	0.1	0.1	0.06
4	4	3	0.7	2	2	0.7
12	11	8	3	6	7	4
<1	<0.5	<0.3	<0.2	<0.9	0.4	0.1
1.5	3	2	0.8	1.5	1.5	0.8
4.5	4	3	1.5	2.5	3	1.5
0.4	0.4	0.3	0.1	0.2	0.5	0.1
17	37	39	19	23	14	17
-	53	48	37	37	-	140*
36	93	93	42	42	-	130*
25	55	55	30	35	~25	~30

FATES OF PAH

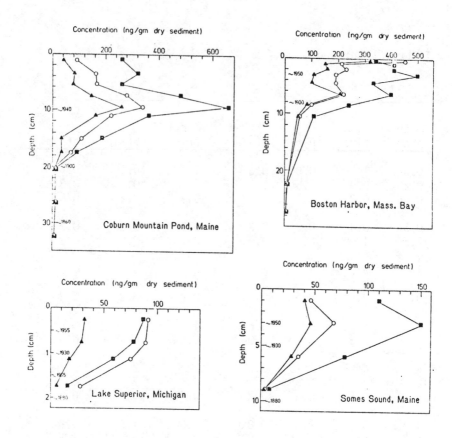

FIGURE 2. PAH profiles in four sediment cores.

SYMBOLS:

▲ benzo(a)pyrene

○ chrysene and triphenylene

■ pyrene

cores were acquired. This supports the tacit assumption that transport of the PAH through the water column is rapid and does not result in substantial removals of PAH.

It is noteworthy that we found similar PAH fluxes at our Siskiwit Lake-Isle Royale site and at our Lake Superior site. Swain (4) found that polychlorinated biphenyl (PCBs) concentrations in Siskiwit Lake water were more than ten times higher than those of Lake Superior surface water samples and that several precipitation samples collected on Isle Royale averaged four to five times more PCBs than those from the Duluth-Lake Superior area. These results indicate that PAH and PCBs are transported and accumulated differently in the environment. This may have been predicted since PAH are introduced into the atmosphere in association with particles and remain so associated, while PCBs are transported to remote locations chiefly in the vapor phase (5). We are pursuing this issue with further studies of rainfall collected at Isle Royale.

For comparison, we have calculated similar flux estimates for three sites located much closer to urban centers, both from our data for outer Boston Harbor and from literature data. These results are also shown in Table 1. Due to the overwhelming effect of bioturbation in the Boston Harbor core, we may only calculate the time averaged flux over the last 100 years at this site. These locations all show much greater PAH fluxes than our remote locations. As suggested by Windsor and Hites (3) and by Hites _et al._ (1), such locations probably receive most of their PAH contamination via runoff delivery. This delivery mechanism completely overwhelms the background atmospheric deposition rates reflected by our remote sites.

It is interesting to compare our observed flux results with fluxes estimated from dry and wet deposition of the atmospheric PAH burden. The dry depositional flux may be calculated as the product of the atmospheric PAH concentration and the dry depositional velocity. Lunde and Bjorseth (6) found individual major PAH at 0.1 to 7 ng m^{-3} in air over Norway, the highest concentrations corresponding to air masses coming from the direction of the United Kingdom and France. A few atmospheric samples have been collected on the Great Lakes (U.S.A.) and were found to have mean individual PAH concentrations of 0.5 to 1.1 ng m^{-3} (7). The dry depositional velocity is uncertain, but is between 0.01 and 0.5 cm/sec (7). Taking an atmospheric individual PAH concentration of 1 ng m^{-3}, the dry depositional flux is between 0.3 and 15 ng/cm yr. The wet depositional flux may be calculated as the product of the PAH concentration in rain times the annual rainfall. Eisenreich _et al._ (7) reported 1-3 ng/l

for major individual PAH in rain collected on the Great
Lakes. Lake Superior receives an annual rainfall of about 80
cm, while Maine receives about 100 cm/yr. Thus the wet
depositional flux may be estimated to be between 0.08 and 0.3
ng/cm yr. This relatively low wet to dry flux ratio may be
reasonable since it is known that atmospheric particles in
the submicron size range, on which PAH are believed to be
carried in the atmosphere (8), are not efficiently scavenged
by falling raindrops (9). Summing the wet and dry fluxes, we
have between 0.4 and 15 ng/cm yr. This brackets the flux
values we observed at the remote sites.

We can also compare our fluxes with inputs of PAH _into_
the atmosphere (as opposed to the content of the atmosphere). The United States' benzo(a)pyrene emission rate was
estimated to be about 1300 tons/yr in 1960 (10). If all of
this uniformly falls out back onto the U.S., we calculate a
benzo(a)pyrene flux of about 15 ng/cm yr. Compare this with
our background flux for this compound at remote sites of
0.2-2 ng/cm yr.

CONCLUSION

Polycyclic aromatic hydrocarbons, produced during anthropogenic combustion processes, appear to be delivered
uniformly to remote sites of the northeastern United States
at a flux of 0.8-3 ng/cm yr. Our deposition data indicate
that current PAH fluxes are 5 - 10 times greater than those
of 80 years ago, but that these inputs have diminished by a
factor of 2 during the last two decades. Sites nearer to
urban centers have much higher PAH fluxes (an average of 35
ng/cm yr), probably due to runoff delivery of PAH-associated
sediments. Our flux determinations may be interpreted in
terms of the United States' changing fossil fuel usage; the
data may indicate increased PAH impact on the environment if
increased use is made of coal. Additionally, our improved
understanding of the cycling of PAH in the environment may be
useful for predicting the fates of other organic compounds,
although caution is warranted due to the apparent geochemical
fractionation of PCBs and PAH near Isle Royale.

ACKNOWLEDGEMENTS

We thank Dr. S. Eisenreich (U. Minnesota) for sediment
samples from Lake Superior and Drs. S. Norton and R. Davis
(U. Maine, Orono) and P. Meyers (U. Michigan) for Coburn
Mountain Pond sediment extracts. This work was supported by
grants from the National Science Foundation (OCE-77-20252 and
OCE-80-05997).

REFERENCES

1. Hites, R. A., Laflamme, R. E., and Farrington, J. W. (1977): Polycyclic aromatic hydrocarbons in marine/aquatic sediments: Their ubiquity, Adv. Chem. Series, 185: 289-311.
2. Hites, R. A., Laflamme, R. E., Windsor, J. G., Jr., Farrington, J. W., and Deuser, W. G. (1980): Polycyclic aromatic hydrocarbons in an anoxic sediment core from the Pettaquamscutt River (Rhode Island, U.S.A.), Geochim. Cosmochim. Acta, 44: 873-878.
3. Windsor, J. G., Jr. and Hites, R. A. (1979): Polycyclic aromatic hydrocarbons in Gulf of Maine sediments and Nova Scotia soils, Geochim. Cosmochim. Acta, 43: 27-33.
4. Swain, W. R. (1978): Chlorinated organic residues in fish, water, and precipitation from the vicinity of Isle Royale, Lake Superior, J. Great Lakes Res., 4: 398-407.
5. Bidleman, T. F., Rice, C. P., and Olney, C. E. (1977): High molecular weight chlorinated hydrocarbons in the air and sea: Rates and mechanisms of air/sea transfer. In: Marine Pollutant Transfer, edited by H. L. Windom and R. A. Duce, pp. 223-251, Lexington Books, Lexington, Massachusetts.
6. Lunde, G. and Bjorseth, A. (1977): Polycyclic aromatic hydrocarbons in long-range transported aerosols, Nature, 268: 518-519.
7. Eisenreich, S. J., Looney, B. B., and Thornton, J. D. (1981): Airborne organic contaminants in the Great Lakes ecosystem, Environ. Sci. Technol., 15: 30-38.
8. Katz, M. and Pierce, R. C. (1976): Quantitative distribution of polynuclear aromatic hydrocarbons in relation to particle size of urban particulates. In: Carcinogensis, edited by R. Freudenthal and P. W. Jones, pp. 413-429, Vol 1., Raven Press, New York.
9. Lodge, J. P., Jr., Waggoner, A. P., Klodt, D. T., and Crain, C. N. (1981): Non-health effects of airborne particulate matter, Atmos. Environ., 15: 431-482.
10. National Academy of Sciences (1972): Particulate Polycyclic Organic Matter. National Academy of Sciences, Washington, D.C., 361 pp.

THE POLYCYCLIC AROMATIC ENVIRONMENT OF THE FLUIDIZED BED COAL COMBUSTION PROCESS - AN INVESTIGATION OF CHEMICAL AND BIOLOGICAL ACTIVITY

GARY T. HUNT, RORBERT J. KINDYA, ROBERT R. HALL, PAUL F. FENNELLY, MARILYN HOYT
GCA/Technology Division, 213 Burlington Rd., Bedford, MA 01730.

INTRODUCTION

The Fluidized-Bed Coal Combustion (FBC) technology represents a viable energy alternative to conventional combustion processes. The design of the combustion process offers an energy source that has several distinct advantages over conventional systems, including reduced nitric oxide emissions via lower operating temperatures, sulfur oxide control via combustion in a lime bed, and a significant reduction in capital and operating costs over scrubber equipped conventional systems. Reduced operating temperatures characteristic of the FBC process, however, have prompted some concern over possible increased emissions of potentially hazardous organic compounds (1). Particular attention has been focused on components historically exhibiting carcinogenic or mutagenic behavior in mammalian systems. A recent survey of organic emissions from conventional fossil fuel combustion systems concluded that reported emissions are restricted to the polynuclear aromatic hydrocarbon (PAH) category (2).

Known mutagens previously isolated from experimental FBC combustor particulate catches have included a variety of PAH homologues with molecular weights ranging from 128 to 252 (4). The same investigators reported that particulate PAH emissions may vary according to fuel type burned in the combustor (3).

An equivalent concern has also been demonstrated for the biological activity of these emissions, as well. Kubitschek and Williams have reported mutagenic activity in fly ash samples from an experimental 6-inch diameter pressurized FBC operated at the Argonne National Laboratory (5). This reponse was seen to vary in specific activity depending on temperature of sample collection as well as the operating state of the combustor. Mutagenic activity of similar samples was observed to increase two orders of magnitude during startup as compared to steady-state operation.

CHEMICAL AND BIOLOGICAL ACTIVITY OF FBC COMBUSTION PROCESS

Hanson, et al. (4) have reported mutagenic activity in stack particulates and bag filter catches collected at the Morgantown experimental FBC unit. Similar results have been reported for flue gas particulates collected at the Exxon (PFBC) unit (6). Despite these efforts both the chemical and biological characteristics of FBC emissions are limited. A recent survey on the environmental aspects of this process concurs with this observation (7).

This presentation is based on a comprehensive environmental assessment of the FBC technology recently completed at GCA/Technology Division. Data points were collected on both particulate and vapor phase organic emissions from each of three FBC units. A variety of multimedia samples were collected from the EPA sponsored experimental Exxon Miniplant pressurized combustor, Electric Power Research Institute-Babcock and Wilcox sponsored (AFBC) unit at Alliance, Ohio and the DOE sponsored Georgetown University FBB (AFBC). The industrial sized design of the latter unit, represents a reliable and yet economical application of this novel coal combustion technology to industrial users.

A collaborative study on the chemical and biological characteristics of FBC particulate emissions will be discussed. Comprehensive organic analyses were conducted on both particulate and vapor phase organics using the Level 1 screening protocols developed by EPA/IERL-RTP (8). Particular emphasis will be placed on fused ring polynuclear aromatics and heterocycles associated with process particulate emissions.

Biological screening protocols included the use of the Ames testing procedure as outlined in the EPA/IERL-RTP protocols (9). Further investigative efforts included the use of both capillary GC/MS and HPLC fluorescence in combination with a biological fractionation scheme to further evalute and define any apparent causal relationships between chemical and biological activity. A comparison of these results with the existing chemical data base on both fluidized-bed and conventional coal combustion processes will be made. Approximate concentrations of organics adsorbed on particulate catches from each of the three FBC units will be presented including any apparent correlations between mutagenic and chemical activity. Particulate emissions (ng/J) of potential mutagens will be discussed and compared to the existing FBC and conventional coal combustion data

CHEMICAL AND BIOLOGICAL ACTIVITY OF FBC COMBUSTION PROCESS

base with particular emphasis on polynuclear aromatic hydrocarbons (PAH).

MATERIALS AND METHODS

Sample Collection

Combustor flue gas emissions were collected at each of three FBC units using the Level 1 sampling protocol developed at IERL/RTP. This includes the use of the Source Assessment Sampling System (SASS) including a series of three cyclones and a glass fiber filter for final particulate collection and an XAD-2 sorbent trap module for the collection of vapor phase organics. In addition, bulk particulate samples were collected from each of the control devices noted in Table 1. Representative composite samples were collected during several sequential days of operation for each of the devices shown. A summary of pertinent unit operational parameters is also noted in Table 1. The following numbers and types of samples were collected:

- Two composite samples of baghouse filter ash from the Georgetown University FBC unit in Washington, D.C.

- A sample of baghouse filter ash from the Exxon PFBC unit in New Jersey.

- An ash sample from the electrostatic precipitator while in operation at the Exxon PFBC unit in New Jersey.

- Samples of composited cyclone ash collected by the secondary and tertiary combustor cyclones at the Exxon unit.

- Two composited samples of ash from the primary combustor cyclones of the Alliance unit.

Chemical Analysis

Sample Preparation. Representative samples of available particulates were extracted with methylene chloride in a Soxhlet apparatus as specified by the Level 1 EPA/IERL protocols (8). Typically, 100 to 500 gram quantities of bulk particulates were extracted when available. Due to the limited availability of sample, 5 to 10 gram quantities were extracted of both the baghouse hopper and ESP catch taken from the Exxon PFBC unit. Quantities of SASS cyclone were more limited, and for this reason analyses were often limited to bioassay and HPLC screening procedures. Only the Alliance unit contained large quantities of SASS cyclone due to the flue sampling location preceding any particulate control devices. Extracts were reduced in volume via rotary evaporation techniques. Aliquots were removed for gravimetric determinations and HPLC/fluorescence screening procedures.

TABLE 1

SUMMARY OF FBC COMBUSTOR OPERATING CONDITIONS

	PARAMETER	EXXON[a]	ALLIANCE[b]	GEORGETOWN[c]
1.	PRESSURE (ATM)	7.0 (PFBC)	1.0 (AFBC)	1.0 (AFBC)
2.	BED TEMP., (°C)	894	840	868
3.	FLUE GAS TEMP. (°C)[d]	150	450 (BEFORE CYCLONE)	177 (AFTER BAG FILTER)
4.	COAL TYPE	BITUMINOUS CHAMPION (1.7% S)	BITUMINOUS PITTSBURGH NO. 8 (2.4% S)	BITUMINOUS (3.3% S)
5.	ENERGY INPUT, J/HR (BTU/HR)	2.4×10^9 (2.3×10^6)	2.5×10^{10} (2.4×10^7)	1.3×10^{11} (1.2×10^8)
6.	% O_2 FLUE GAS	4.1	3.2	11.0
7.	PARTICULATE CONTROL			
	A. CYCLONES			
	• TYPE	2° 3°	1°	e
	• SIZING (μm)	20% ≤3 35% ≤3 50% ≤17 50% ≤5	60% ≤44	-
	• COLLECTION TEMP. (°C)	750 750	450	-
	• CATCH (kg/HR)	5.8 1.4	180	-
	B. BAG FILTER			
	• SIZING (μm)	86% ≤3	f	6% ≤3 15% ≤10
	• COLLECTION TEMP. (°C)	180	200	200
	• CATCH (kg/HR)	0.23	88	648
	C. ELECTROSTATIC PRECIPITATOR (ESP)			
	• SIZING (μm)	95% ≤4 50% ≤2		
	• COLLECTION TEMP. (°C)	190	-	-
	• CATCH (kg/HR)	0.15	-	-

[a] AVERAGE VALUES RECORDED DURING 3-DAY SAMPLING PROGRAM 5/2-5/4, 1979.
[b] AVERAGE VALUES RECORDED DURING 4-DAY SAMPLING PROGRAM 12/10-12/14, 1979.
[c] AVERAGE VALUES RECORDED DURING SAMPLING PERIOD 8/26-9/5, 1980.
[d] AT SAMPLE COLLECTION POINT.
[e] CYCLONES NOT IN OPERATION DURING SAMPLING PROGRAM. BAGHOUSE BEHAVED LIKE CYCLONE.
[f] NO PARTICULATE SIZING AVAILABLE FOR THIS SAMPLE.

Gravimetry

Typically, 1.0 ml aliquots of each particulate extract were removed and transferred to a tared pan. Samples were weighed to constant weight per the protocols noted earlier (8). Results are reported on a µg/g (ppm) basis for samples collected.

HPLC/Fluorescence

Samples selected for HPLC screening procedures included those suspected of containing polynuclear aromatics based on available Level 1 data and those samples which had demonstrated active mutagenic response in the Ames test to be described later in this section. Aliquots of selected samples were filtered and diluted with methanol. A 30 µl portion was screened using the instrumental conditions listed in Table 2.

TABLE 2

SUMMARY OF INSTRUMENT OPERATING CONDITIONS

HPLC	FLUORESCENCE	GC/MS
DUPONT 850 HPLC WITH GRADIENT ELUTION	PERKIN-ELMER 650-10S DETECTOR	HEWLETT-PACKARD 5985 GC/MS/DS
		SE-54 CAPILLARY COLUMN (30m)
COLUMN, ZORBAX ODS C_{18}, 25.0 cm x 4.7 mm	WAVELENGTH (nm) EXCITATION 300 EMISSION 400	FLOW RATE, 2.5 ml/min UHP HELIUM COLUMN TEMPERATURE,
FLOW RATE, 1.5 ml/min	SLIT WIDTH, 15 nm	50°C HELD FOR 2 MIN, 10°/min TO 260° AND HOLD
SOLVENT, METHANOL/WATER	OUTPUT, 1 VOLT TO SPECTRA PHYSICS MINI-GRATOR	INJECTION, SPLITLESS FOR 30 SEC
STEP GRADIENT		
METHANOL (%) TIME (MIN)	CHART SPEED, 40 cm/hr	VOLUME, 1.0 µl
1. 80-85 20		INJ. TEMP., 275°
2. 85-90 15		MS CONDITIONS
3. 90-HOLD 10		ELECTRON ENERGY, 70 eV
4. 90-95 10		SOURCE TEMP., 200°
OVEN TEMPERATURE, 40°C		

Gas Chromatography/Mass Spectrometry

Samples selected for capillary GC/MS analysis included those exhibiting both significant fluorescent activity in the HPLC screening procedure (see Figure 2) and a positive mutagenic response in the Ames test. 1.0 ml aliquots of selected sample extracts (CH_2Cl_2) were removed and analyzed by GC/MS using the instrument operating conditions

CHEMICAL AND BIOLOGICAL ACTIVITY OF FBC COMBUSTION PROCESS

noted in Table 2. Extracts were spiked with d_{10}-anthracene to serve as an internal standard. Measurable components in the total ion chromatogram were verified using computerized spectral matching in conjunction with manual verification techniques.

In addition, Selected Ion Monitoring (SIM) GC/MS scans for a variety of polynuclear aromatics were conducted on aliquots of the Exxon bag filter and ESP catch extracts. GC/MS operating conditions are identical to those shown in Table 2. Quantitative measurements in these instances were based on comparison to standard reference materials.

Assay Methodology

Ames testing of Exxon Miniplant and B&W/EPRI-Alliance samples was performed for GCA by Battelle Columbus Laboratories, Columbus, Ohio. Assay of samples from the Georgetown FBB was performed by the EPA-Health Effects Research Laboratory (HERL) at Research Triangle Park, N.C. Performance of all the assays followed the protocol recommended by Ames (10) et al., and subsequently specified by EPA-IERL biological testing (9). A listing of all solid samples assayed including the particulate catches noted earlier is shown in Table 3.

RESULTS AND DISCUSSION

Gravimetry

Gravimetric measurements representing organic extractables (CH_2Cl_2) with B.P. >300°C were recorded for a number of control device catches and particulates as illustrated in Figure 1. The highest concentrations are noted for the bag filter and ESP catches collected at the Exxon PFBC unit. Gravimetric extractables were undetected (less than blank) for both the secondary and tertiary combustor cyclones catches on the same unit.

Gravimetric measurements are next lowest for the Georgetown bag filter which functioned as a cyclone during the sampling program. The Alliance catches contained the lowest concentrations of extractables in all instances. The Alliance cyclone extractables were consistent with those observed for the Exxon cyclones. Similar gravimetric values were noted for the Alliance SASS cyclone catches (not shown), as well (12). Values for the <3μ catch averaged 16 μg/g, while the >3μ catches contained organic extractables less than blank values.

CHEMICAL AND BIOLOGICAL ACTIVITY OF FBC COMBUSTION PROCESS

TABLE 3

SUMMARY OF AMES TEST RESULTS FOR FLUIDIZED-BED COMBUSTION PARTICULATE AND SOLID SAMPLES

UNIT	SAMPLES TESTED	AMES RESULTS[a]	LRPC[b] (μg)
EXXON MINIPLANT PFBC WITH SORBENT REGENERATION	COMBUSTOR FLUE GAS PARTICLES, FINE FRACTION (<3 μm)	+ TA1538, TA98	5000
	COMBUSTOR FLUE GAS PARTICLES, COARSE FRACTION (>3 μm)	+ TA1537, TA1538, TA98	1000
	SOLIDS CAPTURED BY SECOND CYCLONE	-	
	SOLIDS CAPTURED BY THIRD CYCLONE	-	
	SOLIDS CAPTURED BY ELECTROSTATIC PRECIPITATOR (ESP)	+ TA1537, TA1538, TA98	1000
	SPENT COMBUSTION BED MATERIAL	-	
	REGENERATOR FLUE GAS PARTICLES CAPTURED BY REGENERATOR CYCLONE[c]	-	
	SOLIDS CAPTURED BY FABRIC FILTER	+ TA98[d]	40
B&W-EPRI/ALLIANCE AFBC	COMBUSTOR FLUE GAS PARTICLES, FINE FRACTION (<3 μm)	-	-
	COMBUSTOR FLUE GAS PARTICLES, COARSE FRACTION (>3 μm)	-	-
	SOLIDS CAPTURED BY CYCLONE	-	-
	SPENT COMBUSTION BED MATERIAL	-	-

[a]AMES TEST RESULTS ARE LISTED AS POSITIVE OR NEGATIVE (+ OR -); POSITIVE IN THE TESTER STRAIN INDICATED.
[b]LRPC = LOWEST RECORDED POSITIVE CONCENTRATION--THAT IS, THE LOWEST DOSE/PLATE WHICH RESULTED IN A POSITIVE RESPONSE IN ANY STRAIN. HIGHEST MUTAGENIC RESPONSE WAS OBSERVED IN TA-98 FOR ALL SAMPLES.
[c]THIS SAMPLE IS UNIQUE; ONLY THE EXXON MINIPLANT OPERATED WITH A SORBENT REGENERATOR.
[d]BAGHOUSE HOPPER ASH (CH_2Cl_2-EXTRACT) WAS TESTED IN TA98 ALONE AS PART OF A LEVEL 2 INVESTIGATION.

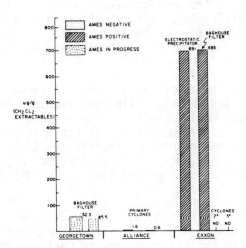

FIGURE 1. FBC combustor particulates--a comparison of organic extractables (gravimetric) and mutagenic activity in control device catches.

CHEMICAL AND BIOLOGICAL ACTIVITY OF FBC COMBUSTION PROCESS

No results are currently available from mutagenicity testing of samples from the Georgetown University FBB. Preliminary, unpublished results from testing of Georgetown baghouse hopper ash CH_2Cl_2-extract indicate mutagenic response similar to that observed in testing a comparable sample from the Exxon Miniplant (13). Review of the types of species identified in chemical analysis of this Georgetown sample noted in Table 4 reveals numerous known mutagens; so mutagenic activity in S. typhimurium is not surprising. Whether these compounds are responsible for the preliminary results observed is unknown at this time. More data are required before any conclusions may be advanced.

Results summarized in Table 3 reveal that baghouse hopper ash obtained from the Exxon Miniplant has the highest mutagenic activity of any particulate or solid sample tested to date. Because of this activity and the availability of a relatively large quantity (grams) of this material, the Exxon baghouse ash sample was fractionated using a high pressure liquid chromatographic technique which separated the components eluting prior to coronene (approximately MW 300) from those with higher molecular weights. This fractionation technique was attempted in an effort to further isolate and hopefully identify those species responsible for the observed mutagenicity of this sample.

Following HPLC separation, the two resultant fractions were assayed for mutagenicity in TA-98. Fraction No. 1 (MW <300) was almost 10 times more active than Fraction No. 2 (MW >300). In addition, Fraction No. 1 displayed higher specific activity than did the unfractionated sample. Table 3 shows a positive mutagenic response for baghouse hopper ash in TA-98 at a dose/plate of 40 µg. Positive mutagenic response for Fraction No. 1 was observed at the lowest dose tested, 12.5 µg.

The pattern of response was also different between HPLC Fraction No. 1 and the unfractionated sample. Addition of the S-9 activation mix resulted in lower mutagenic response in testing of neat baghouse ash, but activity was enhanced by addition of the microsomal mix in testing of the fractionated sample. This is not unexpected considering the variety of polynuclear aromatics suspected to be in this subsample (see Figure 2 and Table 4).

It must be emphasized that the positive mutagenic response observed for these FBC effluent streams is not a result unique to FBC processes nor to coal combustion emissions in general. In addition, any evaluation of environmental impacts would require further investigation beyond the screening protocols employed in this study. The data presented here corroborate the work of other investigators testing FBC effluents. The mutagenic response of these samples is similar to that observed previously at the small experimental FBC unit situated at the the Exxon Miniplant (6) and the Morgantown Energy Technology Center (11).

TABLE 4

ORGANIC EXTRACTABLES (CH_2Cl_2) ISOLATED FROM FBC FABRIC FILTER BAG CATCHES

GEORGETOWN - AFBC[a] (BITUMINOUS)		
COMPONENT	CONCENTRATION (NG/G)	COLLECTION RATE[b] (10^{-5} NG/J)
DIMETHYL BENZENE	300	150
BENZALDEHYDE	900	450
BENZONITRILE	1,500	750
BENZOFURAN	900	450
NAPHTHALENE[d]	11,000	5,500
METHYLNAPHTHALENE[d] (ISOMER)	3,600	1,800
BIPHENYL[e]	1,100	550
1,4 NAPHTHALENE DIONE	250	125
DIMETHYLNAPHTHALENE[d] (ISOMERS)	480	240
2-METHYL BIPHENYL[e]	400	200
DIBENZOFURAN[d,e]	1,900	950
METHYLDIBENZOFURAN[d,e]	450	225
XANTHENE[d]	230	115
9-FLUORENONE[d,e]	330	165
PHENANTHRENE/ANTHRACENE[d,e]	630	315
2-PHENYL NAPHTHALENE[d,e]	TRACE[f]	--
METHYLPHENANTHRENE[d] (ISOMER) OR METHYLANTHRACENE (ISOMER)	TRACE[f]	--

EXXON - PFBC (BITUMINOUS)		
COMPONENT	CONCENTRATION (NG/G)	COLLECTION RATE[c] (10^{-5} NG/J)
BENZALDEHYDE	210	2.0
BENZOTHIAZOLE	300	2.8
NAPHTHALENE[d]	300	2.8
METHYL NAPHTHALENE[d] (ISOMERS)	360	3.4
DIBENZOFURAN[d,e]	130	1.2
9-FLUORENONE[d,e]	350	3.3
PHENANTHRENE/ANTHRACENE[d,e]	1,200	11
BENZOCINNOLINE[d,e]	1,300	12
FLUORANTHENE[d,e]	880	8.2
PYRENE[d,e]	710	6.6
CHRYSENE[d]	580	5.4
BENZ(A)ANTHRACENE/TRIPHENYLENE[d]	1,000	9.3
BENZOFLUORANTHENE ISOMERS[d]	730	6.8
BENZO(A)PYRENE/PERYLENE/BENZO(E)PYRENE[d]	250	2.3

[a] DAY 2 AND 4 FILTER CATCH SAMPLE COMBINED.
[b] VALUE BASED ON A PARTICULATE COLLECTION RATE OF 5000 NG/J IN THIS CONTROL DEVICE.
[c] EMISSION VALUE BASED ON A PARTICULATE COLLECTION RATE OF 9.3 NG/J IN THIS CONTROL DEVICE. BAG FILTER NOT IN PLACE DURING ROUTINE OPERATION.
[d] COMPONENTS ISOLATED PREVIOUSLY BY HANSON, ET AL. (4) FROM FBC BAG FILTER CATCHES COLLECTED DURING COMBUSTION OF MONTANA ROSEBUD SUB-BITUMINOUS COAL.
[e] COMPONENTS ISOLATED PREVIOUSLY BY HANSON, ET AL. (4) FROM FBC BAG FILTER CATCHES COLLECTED DURING COMBUSTION OF KENTUCKY BITUMINOUS COAL.
[f] AN ESTIMATED CONCENTRATION OF <100 NG/G.

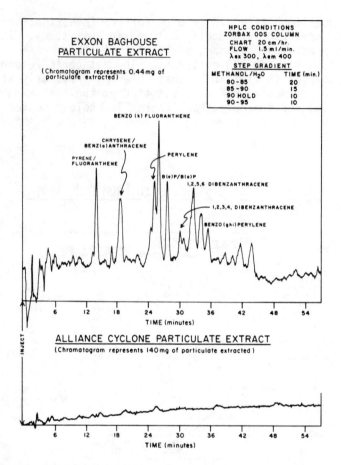

FIGURE 2. HPLC/fluorescence screening technique--representative chromatograms comparing FBC particulate catches.

HPLC/Fluorescence

HPLC/fluorescence screening procedures were used to direct all subsequent chemical and biological testing. Representative chromatograms are shown in Figure 2. The chromatogram of the Alliance cyclone catch shown here is representative of all the Alliance particulate extracts investigated, including both those collected in SASS cyclones and combustor control devices. Similar results were recorded for both of the Exxon cyclone samples, as well. The Exxon bag filter extract chromatogram is exemplary of the Exxon ESP catch and to a lesser extent the Georgetown bag filter.

The combined chemical and biological screening protocols were successful in prioritizing samples for further chemical analyses. Samples exhibiting little or no fluorescent activity in the HPLC analyses such as the Alliance SASS and bulk particulate catches consistently demonstrated a negative response in the Ames screening protocol. Interestingly enough the same samples contained little gravimetric extractables (CH_2Cl_2), as noted in Figure 1.

These results are consistent with our assumed correlation between fluorescent and mutagenic activity. Previous characterization studies of FBC particulate samples had reported a variety of known mutagens, the majority being PAHs typically associated with biologically active particulate samples (4). Our findings are consistent with these earlier investigators. As noted in Table 4 the Exxon combustor particulate catch warranted some further investigation. Owing to its marked biological and fluorescent activity (Figure 2) the Exxon fabric filter catch was investigated further.

Gas Chromatography/Mass Spectrometry (GC/MS)

Results of GC/MS analysis conducted on particulate samples designated in the screening protocols are shown in Table 4. This included analyses of CH_2Cl_2 extracts from the Georgetown fabric filter and Exxon fabric filter and ESP catches. As noted in Table 4 the majority of components isolated from both the Georgetown and Exxon fabric filters are consistent with the existing literature on FBC units burning bituminous coal (4). Components isolated from the Georgetown filter catch are particularly significant. This listing includes a number of known mutagens providing a potential causal relationship for the positive assay reported in preliminary Ames testing on this sample (13). The predominance of oxygenated PAHs suggests that the percent O_2 of the flue gas may be a contributing factor to the combustor chemistry. During the 4-day sampling program at Georgetown measured O_2 values averaged 11.0 percent. The Exxon and Alliance units conversely averaged 4.1 and 3.0, respectively.

Additional factors such as combustion efficiency in atmospheric FBC units may also contribute to combustor particulate chemistry. The Georgetown unit combustion efficiency measured 85 percent during testing while the Alliance unit ranged from 91 to 94 percent. These suggestions concur with observations on the operating state of the combustor noted by Kubitschek and Williams (5).

The chemistry of the Exxon fabric filter catch is consistent with the HPLC screening measurements shown in Figure 2. The predominance of PAH homologues is consistent with the mutagen screening results noted earlier, particularly the Ames testing conducted on the HPLC fractionated sample. Calculated emissions factors (ng/J) for a number of PAHs in the Exxon ESP and fabric filter catches are shown in Table 5. Comparison of average values for fluoranthene and pyrene with representative emissions factors for conventional coal units (2) is illustrated in Figure 3. While the FBC data base is somewhat limited, particulate emissions are significantly lower than those reported for conventional coal combustion systems. Conversely, as shown in Table 4 organic collection rates (ng/J) for the Georgetown fabric filter are significantly higher than those calculated for the Exxon fabric filter. Again the predominance here of oxygenated PAHs suggests some influence from the FBC combustor environment.

The diversity in PAHs isolated from fabric filter catches suggests factors other than fuel type such as percent O_2 in the flue gas and combustor efficiency may contribute to organic emissions from FBC units.

SUMMARY

Particulate PAH emission rates (ng/J) for two of three units tested appears to be lower than those typically reported for conventional coal combustion processes. Results of chemical and biological analyses of FBC fabric filter catches warrant some further study due to the occurrences of a variety of fused ring r-heterocyclic and oxygenated PAHs not routinely reported in FBC combustor particulates.

Such observations may again be related to the higher percent O_2 recorded during sampling at the Georgetown combustor flue gas. Some further investigation of this relationship is warranted particularly since the Ames testing results on the other two units have indicated several instances where homologous PAHS are not solely responsible for noted mutagenic activity.

TABLE 5

PAHS IDENTIFIED (SIM GC/MS) IN EXXON CONTROL DEVICE PARTICULATE CATCHES

COMPONENT	CONCENTRATION (NG/G)		PNA EMISSIONS (10^{-5} NG/J)	
	ESP	BAG FILTER	ESP[a]	BAG FILTER[b]
ANTHRACENE/PHENANTHRENE	1830	10	12	0.095
2-METHYLANTHRACENE*	759	281	5	2.6
9-METHYLANTHRACENE*	1000	ND[c]	6.5	-
FLUORANTHENE*	460	427	3.0	4.1
PYRENE	500	318	3.2	3.1
CHRYSENE*	10	3400	0.065	32

[a] EMISSION VALUES BASED ON A PARTICULATE COLLECTION RATE OF 64.5 NG/J IN THIS CONTROL DEVICE. THESE VALUES SHOULD REPRESENT ACTUAL EMISSIONS WHEN THE PORTABLE CONTROL DEVICE IS NOT IN PLACE (REF. 14).

[b] EMISSION VALUES BASED ON A PARTICULATE COLLECTION RATE OF 93.3 NG/J IN THIS CONTROL DEVICE. THESE VALUES SHOULD REPRESENT ACTUAL EMISSIONS SINCE THE PORTABLE CONTROL DEVICE IS TYPICALLY NOT IN PLACE. (REF. 14).

[c] ND <10 NG/G.

*KNOWN MUTAGENS PREVIOUSLY ISOLATED FROM FBC PARTICULATE CATCHES (REF. 3).

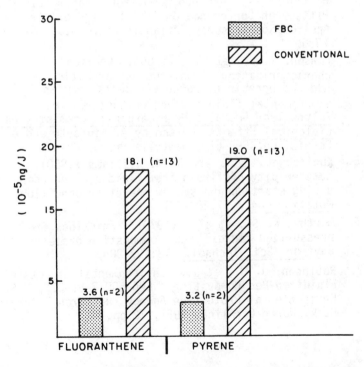

FIGURE 3. Particulate emissions (ng/J) of select PAHS--A comparison of FBC and conventional coal combustion processes

Results reported here provide further evidence that the Level 1 collaborative chemical and biological screening protocols in conjunction with HPLC/fluorescence techniques provide a useful tool for prioritizing combustion source particulate samples for further investigation. HPLC procedures, furthermore can provide a convenient means of fractionating particulate extracts to further isolate chemical mutagens as reported for the Exxon fabric filter catch.

REFERENCES

1. Fennelly, P. F., et al. (1977): Coal burns cleaner in a fluid bed, Environ. Sci. Technol., 11:244-248.
2. Zelenski, S. G., et al. Inventory of organic emissions from fossil fuel combustion for power generation. EPRI-EA-1394, TSP 78-820.
3. Hanson, R. L., et al. Chemical characterization of mutagenic extracts of effluent particles from an experimental fluid bed combustor. Paper No. 788, Pittsburgh Conference on Analytical Chemistry and Applied Spectroscopy, Atlantic City, N.J. March 1980.
4. Hanson, R. L., et al. (1980): Chemical characterization of polynuclear aromatic hydrocarbons in airborne effluents from an experimental fluidized-bed combustor. In: Polynuclear Aromatic Hydrocarbons: Chemistry and Biological Effects, edited by A. Bjorseth and A. Dennis, pp. 599-616, Battelle Press, Ohio.
5. Kubitschek, H. E. and D. M. Williams (1980). Mutagenicity of flyash from fluidized-bed combustor during start-up and steady operating conditions, Mutation Res., 77:287-291.
6. Murthy, K. S., et al. (1979): Emissions from pressurized fluidized-bed combustion processes, Environ. Sci. Technol., 13:197-204.
7. Robinson, J. M., et al. Environmental Aspects of Fluidized-Bed Combustion. EPA-600/7-81-075, U.S. Environmental Protection Agency, Research Triangle Park, North Carolina, 1981, 72 pp.

8. Lentzen, D. E., et al. IERL-RTP Procedures Manual: Level 1 Environmental Assessment. (Second Edition). EPA-600/7-78-201, U.S. Environmental Protection Agency, Research Triangle Park, North Carolina, 1978, 279 pp.
9. Duke, K. M. et al. IERL-RTP Procedures Manual: Level 1 Environmental Assessment Biological Tests for Pilot Studies. Prepared for the U.S. Environmental Protection Agency by Battelle Columbus Laboratories, Columbus, OH. EPA-600/7-77-043, April 1977.
10. Ames, B. N., et al. (1975): Methods for detecting carcinogens and mutagens with the salmonella/mammalian-microsome mutagenicity test, Mutatation Res., 31:347-364.
11. Clark, C. R., et al. Mutagenicity of Effluents Associated with the Fluidized-Bed Combustion of Coal. Inhalation Toxicology Research Institute Annual Report 1977-1978, pp. 274-284, Lovelace Biomedical and Environmental Research Institute, Albuquerque, NM. Prepared for the U.S. Department of Energy under Contract No. EY-76-C-04-1013, December 1978.
12. Kindya, R. J., et al. Environmental Assessment: Source Test and Evaluation Report B and W. Alliance Atmospheric Fluidized-Bed Combustor. EPA-600/7-81-076, U.S. Environmental Protection Agency, Research Triangle Park, North Carolina, 1981. 104 pp.
13. Dr. Joellen Lewtas and Ms. Judy Mumford of EPA-HERL, private communication.
14. Kindya, R. J., et al. Environmental Assessment: Source Test and Evaluation Report Exxon Miniplant Pressurized Fluidized-Bed Combustor with Sorbent Regeneration. EPA-600/7-81-077, U.S. Environmental Protection Agency, Research Triangle Park, North Carolina, 1981. 208 pp.

COMPARISON OF THE METABOLIC PROFILES OF PYRENE AND BENZ(A)ANTHRACENE IN RAT LIVER AND LUNG BY GLASS CAPILLARY GAS CHROMATOGRAPHY/MASS SPECTROMETRY.

J. JACOB*, G. GRIMMER*, A. SCHMOLDT**
* Biochemisches Institut für Umweltcarcinogene, 2070 Ahrensburg, FRG; ** Pharmakologisches Institut der Universität Hamburg, FRG.

INTRODUCTION

The initial event for the carcinogenesis of various polycyclic aromatic hydrocarbons (PAH) is their metabolic activation to highly reactive compounds, the so-called ultimate carcinogens. The formation of these products is catalyzed by microsomal monooxygenases and epoxide hydrolases the activity of which are by far highest in liver. Hence, the metabolism of PAH has been predominantly investigated in this tissue.

However, the carcinogenic effect of environmentally relevant PAH in man is manifested in the lung. This paper therefore deals with the comparison of the metabolic profiles formed from two very abundant PAH (pyrene and benz(a)anthracene (BaA)) by rat liver and lung microsomes. For this purpose, untreated, polychlorinated biphenyl- (PCB) as well as 5,6-benzoflavone- (BNF) treated animals were used.

MATERIAL AND METHODS

Pretreatment of the animals as well as preparation of microsomes and the incubation technique have been published previously (1,2,3). The cleanup and the enrichment of metabolites from incubation experiments, the condition of gas-liquid chromatography and mass spectrometry, and the identification of metabolites have been reported in detail elsewhere (4). A number of BaA phenols, dihydrodiols and dihydrodiol epoxides were kindly supplied by NCI Repository.

RESULTS

Pyrene

Pyrene is converted preferentially into 1-hydroxypyrene by liver microsomes of untreated rats (5,6) apart from small amounts of the K-region dihydrodiol. Only traces of metabolites formed by twofold oxidation were obtained (Table 1).

TABLE 1

METABOLITES FORMED FROM PYRENE BY RAT LIVER AND LUNG MICROSOMES OF NORMAL AND PRETREATED (PCB;BNF) RATS.

Metabolite[+]	Liver			Lung		
	normal	PCB	BNF	normal	PCB	BNF
1-OH	13.3	22.8	37.9	0.3	-	1.6
4,5-Diol	4.4	31.6	6.9	0.3	0.5	0.4
Diphenol I	1.0	10.6	2.6	-	-	-
Diphenol II	1.1	5.0	1.8	-	-	-
non-K-region-Triol	0.6	-	1.2	-	0.6	0.9
unknown I[++]	-	-	-	-	-	0.3
unknown II[++]	0.3	-	-	-	-	-

[+] Data are given in nmol/mg microsomal protein formed in 15 min.
[++] unknown I and II showed rel. retention times related to pyrene of 1.500 and 1.589, respectively.

Pretreatment with BNF results in an increased metabolic rate which is also based on the 1-phenol formation (Figure 1). In contrast to this 'P448-inducer', K-region oxidation is stimulated after pretreatment with PCB which induces additionally the P450-system. Moreover, a considerable increase of the diphenol I formation is found (factor 10). The total metabolic rate in the rat lung is only about 1/30 if compared with the liver. In normal rats no significant difference between 1-phenol formation and K-region oxidation is found. However, after BNF-treatment, the formation of 1-hydroxypyrene is increased, whereas the rate of K-region oxidation remains constant. After PCB-treatment only the K-region diol and a non-K-region triol are found as significant metabolites.

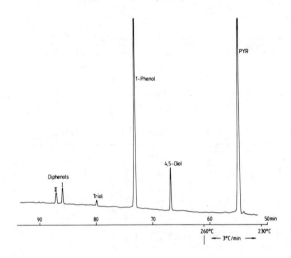

FIGURE 1. Profile of pyrene metabolites formed by liver microsomes of BNF-pretreated rats.

Benz(a)anthracene

In accordance with previous results (1-4), BaA is oxidized preferentially at the 10,11-position by liver microsomes of normal rats. PCB as well as BNF induce mainly the oxidation at the K-region (5,6-) and at the 8,9-position. Additionally, the formation of secondary metabolites (triols and tetrols) is increased (Table 2). In case of BNF-treatment the phenol formation is significantly increased (Fig. 2), whereas tetrols could not be observed.

In lung microsomes of untreated rats the oxidation at the 8,9-position predominates (formation of 8-OH-BaA and 8,9-dihydrodiol). After PCB-treatment only induction of the 5,6-oxidation could be observed. However, BNF-treatment resulted in an increase of the phenol formation (1-; 8- and 11-OH-BaA), whereas no 5,6-oxidation products were obtained.

TABLE 2

METABOLITES FORMED FROM BENZ(A)ANTHRACENE BY RAT LIVER AND LUNG MICROSOMES OF NORMAL AND PRETREATED (PCB;BNF) RATS.

Metabolite[+]	Liver			Lung		
	normal	PCB	BNF	normal	PCB	BNF
1-OH	-	0.1	-	0.9	0.8	0.6
4-OH	-	-	1.7	-	-	-
5-OH	-	-	1.0	-	-	-
8-OH	-	-	0.5	2.0	-	1.4
11-OH	-	-	0.6	0.3	-	1.2
5,6-Diol	0.9	24.3	10.8	-	2.0	-
8,9-Diol	1.0	18.1	11.4	1.4	1.0	0.9
10,11-Diol	8.2	-	3.0	-	-	-
unknown Diol	-	0.6	2.0	-	-	-
Triol I (1.220)[++]	-	0.9	0.9	-	-	-
Triol II (1.247)[++]	-	4.0	-	-	-	-
Triol III (1.157)	-	-	1.5	-	-	1.4
Triol IV (1.236)	-	1.4	2.5	-	-	-
Triol V (1.265)[++]	-	0.3	0.3	-	-	-
Triol VI + Tetrol I (1.230)	-	1.4	-	-	-	-
Triol VII + Tetrol II (1.336)	-	0.6	-	-	-	-
Tetrol III (1.255)[++]	-	0.2	-	-	-	-
Tetrol IV (1.440)	-	2.2	-	-	-	-
unknown I (1.136)	-	-	1.2	0.9	1.2	1.1
unknown II (1.200)	-	-	2.3	-	-	-

[+] Data are given in nmole/mg microsomal protein formed in 15 min.; values in parentheses are GC retention times related to BaA.
[++] Triol I and II are K-region triols. Triol V and Tetrol III showed the same GC and MS data as authentic samples of 2,3,4-triol and 1,2,3,4-tetrol, respectively.

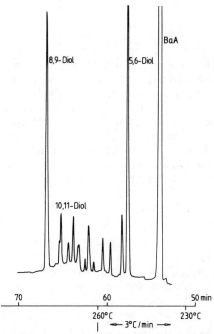

FIGURE 2. Profile of benz(a)anthracene metabolites formed by liver microsomes of BNF-pretreated rats.

DISCUSSION

The metabolism of benzo(a)pyrene in rat lungs is regulated by the P448-system (7), whereas also cytochrome P450 is involved in the liver. This is confirmed by our results for pyrene as well as for BaA in so far as the P448-inducer BNF does not induce the K-region oxidation in the lung. It is, however, in contrast to this, that PCB induce the K-region-oxidation in this tissue in case of BaA and, less significant, in case of pyrene. It can therefore be speculated that PCB may induce another isocytochrome than the P448 induced by BNF.

BaA is oxidized mainly at the 8,9-position in the lung whereas the 10,11-dihydrodiol is the predominant metabolite in the liver. The formation of the bay-region dihydrodiol epoxide of BaA observed in the liver after induction with PCB or BNF could not be observed in the lung in which metabolic rates were lower anyway.

ACKNOWLEDGEMENTS

The present studies were carried out in accordance with the environmental plan of the Federal Environment Office by the order of the Federal Ministry of the Interior. The authors are greatly indebted to the NCI Chemical Repository which supplied them with benz(a)anthracene metabolite reference materials. For their excellent technical assistance they would like to thank Mrs. S. Schinz and Mr. G. Raab. Some of these results are part of the thesis of M. Anderer.

REFERENCES

1. Jacob, J., Schmoldt, A., and Grimmer, G. (1981): Time course of oxidative benz(a)anthracene metabolism by liver microsomes of normal and PCB-treated rats, Carcinogenesis, 2:395-401.
2. Schmoldt, A., Jacob, J., and Grimmer, G. (1981): Dose-dependent induction of rat liver microsomal aryl hydrocarbon monooxygenase by benzo(k)fluoranthene, Cancer Lett., 13: 249-257.
3. Jacob, J., Schmoldt, A., and Grimmer, G. (1981): The influence of polycyclic aromatic hydrocarbons as inducers of monooxygenases on the metabolic profile of benz(a)anthracene in rat liver microsomes, Cancer Lett. (in press).
4. Jacob, J., Schmoldt, A., and Grimmer, G. (1981): Glass-capillary-gas-chromatography/mass spectrometry data of mono- and polyhydroxylated benz(a)-anthracene, Hoppe Seyler's Z. Physiol. Chem., 362: 1021-1030.
5. Harper, K.H. (1957): The metabolism of pyrene, Brit. J. Cancer, 11: 499-507.
6. Boyland, E. and Sims, P. (1964): Metabolism of polycyclic compounds. The metabolism of pyrene in rats and rabbits, Biochem. J., 90: 391-398.
7. Vadi, H., Jernström, B., and Orrenius, S. (1976): Recent studies on benz(a)pyrene metabolism in rat liver and lung. Carcinogenesis Vol. 1. Polynuclear Aromatic Hydrocarbons: Chemistry, Metabolism, and Carcinogenesis, edited by R.I. Freudenthal and P.W. Jones, pp. 45-61, Raven Press, N.Y.

THE EFFECT OF ASBESTOS ON THE BIOACTIVATION OF BENZO[A]PYRENE

C. KANDASWAMI, V.V. SUBRAHMANYAM AND P.J. O'BRIEN
Department of Biochemistry, Memorial University of
Newfoundland, St. John's, Newfoundland, Canada

INTRODUCTION

One neoplastic transformation of the respiratory tract epithelium is termed as bronchogenic carcinoma (1). The incidence of this type of neoplasm is usually considered to be a consequence of exposure to multiple chemical components (2). The best known example is cigarette smoke which, on the basis of several epidemiological studies, is regarded as the main etiologic agent in the induction of pulmonary carcinoma. However, the genesis of lung cancer seems to be related to the combined exposure of cigarette smoke and other factors such as particulates (3). In this context, exposure to various agents encountered in the work environment is also of crucial importance. These factors may act as causative agents or cofactors in the production of respiratory tract carcinoma. This is best illustrated in the case of asbestos mine workers (4), uranium miners (5), and coke oven workers (6). The risk of bronchogenic carcinoma posed by cigarette smoking can be greatly augmented by asbestos exposure (1). This action could explain the synergistic effect of asbestos inhalation and cigarette smoking, as exemplified by an increased incidence of pulmonary carcinoma in asbestos workers who smoke cigarettes (1). Experimental studies (1) also show that simultaneous exposure to polycyclic aromatic hydrocarbons, known to be present in cigarette smoke and asbestos, results in synergistic or cocarcinogenic effects. Asbestos appears to possess only weak carcinogenic activity (3). However, it can strikingly intensify the action of carcinogenic hydrocarbons introduced into the respiratory tract. Information is scarce on the mechanism of this synergistic interaction. Several explanations may be offered to account for the observed cocarcinogenic effect. The adsorption of the hydrocarbon on the surface of the asbestos fiber might facilitate greater penetration and enhanced tissue retention of the carcinogen (7). Experimental studies (8,9) have indicated that asbestos could readily adsorb benzo[a]pyrene in vitro and decrease the excretion of this hydrocarbon in experimental animals. If the fibers were to retard the metabolic disposition of benzo[a]pyrene, this action would prolong the duration of its contact with the site of metabolic activation. Alternatively, asbestos might modify the bioactivation of benzo[a]pyrene, resulting in increased binding of benzo[a]pyrene metabolites to cellular macromolecules. The objective of the present study is to examine whether asbestos fibers could modify the in vitro metabolism and macromolecular binding of benzo[a]pyrene.

MATERIALS AND METHODS

[G-^{3}H]-benzo[a]pyrene was obtained from Amersham Corporation. UICC standard reference samples of asbestos were used. The type of chrysotile employed in these studies was Canadian chrysotile B. The sources of chemicals have been indicated in earlier communications (10,11).

Male Sprague-Dawley rats weighing between 200 and 250 g were used in these experiments. Animals were pre-treated by daily i.p. injections of 3-methylcholanthrene (in corn oil, 40 mg/kg body weight) for 4 days. They had free access to laboratory chow and drinking water. Microsomal fractions from liver and lung homogenates were isolated by differential centrifugation as described earlier (10,11). Pulmonary alveolar macrophages from guinea pigs were isolated as described by Maxwell et al. (12). Aryl hydrocarbon hydroxylase activity was determined as described previously (10,11). A similar incubation system was set up for the isolation of benzo[a]pyrene metabolites. Microsomes were usually incubated with 20 μM [^{3}H]-benzo[a]pyrene (in dimethyl sulfoxide) and NADPH for 15 minutes at 37°C. The products were extracted with ethyl acetate:acetone (2:1, v/v) and the metabolites were separated by thin layer chromatography on silica gel G plates, following standard procedures (13,14). The different metabolites identified were scraped into scintillations vials to which aquasol was added and after shaking radioactivity was determined. All these operations were performed following established procedures (13,14).

Mouse embryo fibroblasts (3T6) were grown in Dulbecco's modified Eagle's medium, containing 5% fetal calf serum, at 37°C in a 5% CO_2 atmosphere for 2 days. Thereafter the cells were harvested and used for metabolic studies. The cells were incubated with 20 μM [^{3}H]-benzo[a]pyrene and metabolites were isolated as described above.

The irreversible binding of [^{3}H]benzo[a]pyrene metabolites to the macromolecular fraction was performed as described by Hill and Shih (15). In some experiments, the binding of metabolites to the microsomal protein was carried out as described by Sivarajah et al. (16).

RESULTS

Effect of Asbestos on Aryl Hydrocarbon Hydroxylase

Asbestos causes a marked inhibition of microsomal aryl hydrocarbon hydroxylase activity. Chrysotile and crocidolite appear to be very effective inhibitors of this activity in rabbit and rat liver microsomes, as illustrated in Table 1. Maximal enzyme inhibition was observed with chrysotile in all the microsomal fractions examined. In these experiments, benzo[a]pyrene was preincubated with the fibers at 37°C for 15 minutes. No enzyme inhibition was observed when the fibers and benzo[a]pyrene were concurrently introduced to the microsomal incubation system. However, the hydroxylase activity was greatly impaired when microsomes were pretreated with asbestos at 37°C for 1 hours.

TABLE 1

EFFECT OF ASBESTOS ON ARYL HYDROCARBON HYDROXYLASE ACTIVITY OF LIVER MICROSOMES[a]

Addition	Specific Activity (pmoles of phenol formed /min/mg of protein)		
	Rabbit	Rat[b]	Rat (3-MC)[c]
None	36.0	90.0	500.0
Crocidolite	15.9	44.1	380.0
Canadian chrysotile B	11.2	40.5	260.0
Amosite	22.0	55.8	420.0
Anthophyllite	22.9	63.9	315.0

a Benzo[a]pyrene was pretreated with asbestos (50 µg/ml). The values represent averages of three separate experiments.
b Control rat liver microsomes were used
c 3-methylcholanthrene-treated rat liver microsomes were used.

Treatment of rat lung microsomes with asbestos resulted in a drastic diminution of aryl hydrocarbon hydroxylase activity (Table 2). In the aforementioned studies, the assay for the hydroxylase activity was performed in the presence of the metal-chelating agent, EDTA. When liver microsomes were treated with asbestos in the absence of EDTA, the inhibition of aryl hydrocarbon hydroxylase was less pronounced (Table 3). If asbestos-associated trace metals were to impair this activity, one would expect a greater enzyme inhibition in the absence of the metal-chelating agent.

TABLE 2

EFFECT OF ASBESTOS ON ARYL HYDROCARBON HYDROXYLASE ACTIVITY OF LUNG MICROSOMES[a]

Addition	Specific Activity (pmoles of phenol formed /min/mg of protein)
None	15.0
Crocidolite	5.4
Canadian chrysotile B	6.3
Amosite	6.2
Anthophyllite	6.5

[a] Asbestos (1.0 mg) was preincubated with the microsomal fraction for 1 hour at 37°C. The values represent averages of three separate experiments.

TABLE 3

EFFECT OF ASBESTOS ON ARYL HYDROCARBON HYDROXYLASE ACTIVITY IN LIVER MICROSOMES[a]

Addition	Specific Activity (pmoles of phenol/min/mg of protein)	
	-EDTA	+EDTA
None	500.0	515.0
Chrysotile	375.6	263.5
Crocidolite	450.0	277.6
Amosite	284.5	196.6
Anthophyllite	326.5	243.5

a   3-Methylcholanthrene-pretreated rat liver microsomes were used. The concentration of asbestos was 1.0 mg/ml.

Effect of Asbestos on the Microsomal Generation of Benzo[a]pyrene Metabolites

The effect of asbestos fibers on the production of benzo[a]pyrene metabolites in lung microsomes is shown in Table 4. In this table, the major benzo[a]pyrene metabolites are grouped into phenolic compounds and quinones, dihydrodiols, and other more polar compounds. Crocidolite was less effective than the other types of asbestos in decreasing benzo[a]pyrene metabolites. Crocidolite, while reducing the relative proportions of polar compounds and phenols, did not affect the diol fraction. The other three asbestos types diminished the formation of both phenols and diols. The phenolic compounds are formed from the epoxides non-enzymically while the dihydrodiols are generated enzymically. The decrease in both these groups of compounds indicates that asbestos inhibits epoxide-metabolizing enzymes besides aryl hydrocarbon hydroxylase.

TABLE 4

EFFECT OF ASBESTOS ON THE PRODUCTION OF BENZO[A]PYRENE METABOLITES BY LUNG MICROSOMES[a]

Condition	Products formed (%)		
	Polar	Diols	Phenols and Quinones
Control	11.08	3.79	17.77
Chrysotile	0.89	0.49	2.01
Amosite	0.74	0.64	4.06
Anthophyllite	0.91	0.67	2.02
Crocidolite	5.9	5.8	10.52

[a] 3-Methylcholanthrene-pretreated rat lung microsomes were used. The microsomes were pretreated with the asbestos sample (1.0 mg/ml) for 1 hour. The figures represent percentages of the total amount of extracted radioactivity.

The Effect of Trace Metals on the Microsomal Production of Benzo[a]pyrene Metabolites

The common trace metals associated with asbestos fibers include nickel, chromium, maganese, cobalt, and iron (17). The effects of some of these metals on the metabolism of benzo[a]pyrene by liver microsomes were examined. A perusal of Table 5 shows that the rate of benzo[a]pyrene metabolism is considerably affected by Ni^{2+}, Cr^{3+}, and Fe^{3+}. These metal ions diminish the relative proportions of polar metabolites, phenols, and quinones. Cd^{2+}, Fe^{2+}, and Co^{2+} slightly decrease benzo[a]pyrene metabolism at a concentration of 10 μM. Mn^{2+} is more effective than these three metals in reducing the extent of diol formation. Chrysotile decreases the relative amounts of diols, phenols, and quinones. Chrysotile diminished the formation of benzo[a]pyrene metabolites by 31% in rat liver microsomes.

TABLE 5

EFFECT OF ASBESTOS AND METAL IONS ON THE MICROSOMAL GENERATION OF BENZO[A]PYRENE METABOLITES[a]

	Products formed (%)		
Addition	Polar	Diols	Phenols and Quinones
Control	6.89	7.72	50.12
Chrysotile	5.04	4.82	30.99
$NiSO_4$	4.90	2.74	27.91
$Cr_2(SO_4)_3$	4.54	2.09	29.64
$MnCl_2$	5.26	3.26	35.77
$CdCl_2$	7.38	6.98	44.47
$FeSO_4$	11.60	9.47	38.60
$CoCl_2$	8.77	7.97	43.16
$FeCl_3$	5.71	3.28	26.41

a 3-Methylcholanthrene-pretreated rat liver microsomes were used. The concentration of each metal ion was 10 µM and that of asbestos was 1.0 mg/ml.

The effect of both metal ions and asbestos upon benzo[a]-pyrene metabolism was also examined (Table 6). The metabolism of benzo[a]pyrene in the presence of the fibers is reduced 10 to 15% on treatment of the fibers with Cr^{3+} or Fe^{3+}. In this case, the inhibitory effect is more on the relative amounts of phenols and quinones. In the case of Co^{2+}, there is a slight stimulation of benzo[a]pyrene metabolism in the presence of chrysotile. The effect of trace metals on the formation of benzo[a]pyrene metabolites in rat lung microsomes was also examined. In this case, Ni^{2+}, Cr^{3+}, Fe^{2+}, and Fe^{3+} were found to depress benzo[a]pyrene metabolism. All these results show that trace metals, when added to asbestos, do not alter the modifying effect of the fibers on benzo[a]pyrene metabolism. Cobalt, however, seems to differ from other metals in this respect.

TABLE 6

EFFECT OF ASBESTOS AND METAL IONS ON THE MICROSOMAL PRODUCTION OF BENZO[A]PYRENE METABOLITES[a]

	Products formed (%)		
Addition	Polar	Diols	Phenols and Quinones
Chrysotile	5.04	4.82	30.99
Chrysotile + $Cr_2(SO_4)_3$	4.65	5.29	22.12
Chrysotile + $MnCl_2$	9.68	8.79	38.06
Chrysotile + $CdCl_2$	6.71	9.08	42.90
Chrysotile + $FeSO_4$	4.62	8.71	40.92
Chrysotile + $CoCl_2$	8.65	9.18	59.52
Chrysotile + $FeCl_3$	5.01	4.17	19.36

[a] 3-Methylcholanthrene-pretreated rat liver microsomes were used. The concentration of each metal ion was 10 μM and that of asbestos was 1.0 mg/ml.

The Effect of Asbestos on the Macromolecular Binding of Benzo[a]pyrene Metabolites in Microsomes

Chrysotile causes a 38% stimulation of the in vitro covalent binding of [^{3}H]-benzo[a]pyrene metabolites to the protein fraction in rat liver microsomes (Table 7). Even though Ni^{2+}, Cr^{3+}, and Fe^{3+} modify benzo[a]pyrene metabolism, only Cr^{3+} shows an increase in the microsomal protein binding of benzo[a]pyrene metabolites.

Metabolism and Binding of Benzo[a]pyrene in Mouse Embryo Fibroblasts

In one experiment, growing cells were harvested and exposed to Canadian chrysotile for 2 hours at 37°C, then incubated with [^{3}H]-benzo[a]pyrene for 30 minutes, and the metabolites were isolated. There was a 20% decrease in metabolism in chrysotile-treated cells. Analysis of the organic soluble metabolites revealed a 25% drop in the phenolic metabolites and quinones as shown in Table 8. Interestingly,

TABLE 7

THE EFFECT OF ASBESTOS ON THE IRREVERSIBLE BINDING OF BENZO[A]PYRENE METABOLITES TO MICROSOMAL PROTEIN[a]

Condition	Product bound (nmols/mg of protein)
Control	1.70
Chrysotile-treated	2.36

a 3-Methylcholanthrene-pretreated rat liver microsomes were used. Microsomes were treated with asbestos (1.0 mg/ml) for 1 hour at 37°C and then incubated with 20 µM [^{3}H]-benzo[a]pyrene for 15 minutes. The total metabolism of benzo[a]pyrene was 4.2 nmols/min/mg of protein.

TABLE 8

BENZO[A]PYRENE METABOLISM BY MOUSE EMBRYO FIBROBLASTS (3T3)[a]

Conditions	Products formed (%)		
	Polar	Diol	Phenols and Quinones
Control	1.29	3.03	16.66
Chrysotile-treated	2.18	2.14	12.48

a The cells were treated with asbestos (1.0 mg/ml) for 2 hours at 37°C and then incubated with 20 µM [^{3}H]-benzo[a]pyrene for 30 minutes.

the covalent binding of the metabolites to the macromolecular fraction of the cells was increased 2-fold in the presence of chrysotile (Table 9). In another experiment, 3T6 cells were grown in the presence of [^{3}H]-benzo[a]pyrene for 72 hours with and without the fibers. When the cells were treated with Canadian chrysotile for 24 hours before adding benzo[a]pyrene, there was a reduction in benzo[a]pyrene metabolism. When the cells were concurrently exposed to the hydrocarbon and asbestos

TABLE 9

BINDING OF BENZO[A]PYRENE METABOLITES TO THE MACROMOLECULAR FRACTION OF MOUSE EMBRYO FIBROBLAST CELLS (3T3)[a]

Condition	Product bound (nmols/mg of protein)
Control	11.8
Chrysotile-treated	25.0

a The cells were treated with asbestos (1.0 mg/ml) for 2 hours at 37°C and then incubated with 20 µM [^{3}H]-benzo[a]pyrene for 30 minutes.

and maintained for 72 hours, there was no change in the metabolism. Eneanya et al. (18) also noticed a decrease in the overall metabolism of benzo[a]pyrene in Syrian hamster embryo cells treated with NIEHS intermediate chrysotile.

Benzo[a]pyrene Metabolism in
Pulmonary Alveolar Macrophages

It is known that pulmonary alveolar macrophages metabolize benzo[a]pyrene to proximate and ultimate carcinogens (19). It was of interest, therefore, to examine the effect of asbestos on the metabolic activation of benzo[a]pyrene in these cells. Preliminary experiments with guinea pig alveolar macrophages showed the stimulation of irreversible protein binding by benzo[a]pyrene metabolites when the cells were preincubated with chrysotile for 1 hour at 37°C. Further work is necessary to ascertain whether macrophage membrane oxidases, capable of activating benzo[a]pyrene, are stimulated by asbestos exposure.

Effect of Asbestos on the In Vitro Metabolic
Activation of Benzo[a]pyrene in Guinea Pig Lung Microsomes

Guinea pig lung microsomes have been shown to possess high prostaglandin synthetase activity and very little mixed function oxidase (16). The metabolic activation of benzo[a]pyrene in this system appears to be mediated by prostaglandin synthetase (16). Our preliminary experiments indicate a

considerable increase in the covalent binding of benzo[a]-
pyrene metabolites to protein in guinea pig microsomes, when
arachidonic acid is used as a cofactor. This result suggests
that asbestos stimulates prostaglandin synthetase which could
mediate the activation of benzo[a]pyrene, resulting in
increased macromolecular-binding metabolites.

DISCUSSION

The modifying effect of asbestos on aryl hydrocarbon
hydroxylase activity could have important consequences in
vivo. Interference of asbestos with the detoxification of
benzo[a]pyrene may result in a slower pulmonary clearance of
the carcinogen, thereby increasing its retention time. This
would explain the greater induction by polycyclic aromatic
hydrocarbons of benzo[a]pyrene hydroxylase in lymphocytes,
isolated from asbestos workers than from controls, reported
by Naseem et al. (20), and confirmed by us (P.J. O'Brien,
unpublished results). With asbestos, large amounts of
material are retained in the tissues, unlike other carcinogenic agents which are excreted or metabolized (21). If the
mineral were to retard the early metabolism of benzo[a]pyrene,
the presence of residual retained asbestos may ensure a continuing exposure of the tissue to the carcinogen. Thus,
benzo[a]pyrene may remain biologically active in the lung for
a prolonged duration. The retention of unmetabolized benzo-
[a]pyrene in the lungs would accentuate the carcinogenic
stimulus insofar as the risk of bronchogenic carcinoma
correlates with the duration of exposure to the carcinogen
(9).

It is known that heavy metals, such as nickel, chromium,
and cobalt, are found on the surface of asbestos in trace
amounts (1,17). Dixon et al. (22) have postulated that these
metals can inhibit benzo[a]pyrene hydroxylase and thereby
could indirectly contribute to a carcinogenic stimulus. Our
preliminary studies, however, indicate that the in vitro
effect of particles on aryl hydrocarbon hydroxylase may not
be due to associated heavy metals. Canadian chrysotile
appears to increase the microsome-catalyzed irreversible
binding of benzo[a]pyrene to microsomal protein. Of the
asbestos-associated heavy metals, only Cr^{2+} increased this
binding. Hart et al. (23) observed an enhanced binding of
benzo[a]pyrene metabolites to DNA in NIEHS intermediate
chrysotile-pretreated human fibroblasts. Since this form

of chrysotile undergoes least elemental change during intracellular uptake, the authors do not ascribe this modifying effect to metal ions.

The observed modification of the metabolism of benzo[a]pyrene in fibroblasts by asbestos fibers further confirms our results obtained with the in vitro studies on the microsomal metabolism of benzo[a]pyrene. Prior treatment of cells or microsomes with asbestos fibers seems to be necessary for the observed effects. It is likely that the fibers stimulate membrane oxidases which may be involved in the bioactivation of benzo[a]pyrene. It is evident that mechanisms other than particle-enhanced carcinogen transport (7) are likely to be operative in carcinogenesis.

While decreasing the overall in vitro metabolism of benzo[a]pyrene, asbestos seems to enhance the binding of benzo[a]pyrene metabolites to macromolecules. It is likely that asbestos redirects the metabolism of benzo[a]pyrene resulting in the formation of more reactive intermediates. In the presence of asbestos, polycyclic aromatic hydrocarbons could be increasingly activated by another enzyme system, such as the prostaglandin synthetase (24). Sirois et al. (25) have recently shown that asbestos causes a high increase in alveolar macrophage phospholipase A activity and prostaglandin synthesis. Therefore, a tumor promotor role for asbestos is conceivable. Very recently, tumor promotion by chrysotile asbestos has been unequivocally demonstrated by Topping and Nettesheim (26). Their studies also reveal that asbestos enhances tumor induction by a mechanism not involving the adsorptive properties of the fibers.

SUMMARY

In vitro studies indicate that asbestos modifies aryl hydrocarbon hydroxylase activity of microsomes and fibroblasts. Asbestos fibers could also alter the activities of epoxide-metabolizing enzymes. These effect of the fibers may have important consequences in vivo inasmuch as aryl hydrocarbon hydroxylase and epoxide-metabolizing enzymes are involved in the metabolic disposition of polycyclic aromatic hydrocarbons. By retarding the rapid metabolism of the carcinogenic hydrocarbons, asbestos fibers could increase the residence time of the carcinogens in the target tissue, thus accentuating the carcinogenic stimulus. While decreasing the metabolism of benzo[a]pyrene, the particles appear to enhance the binding of benzo[a]pyrene metabolites to macromolecules. In the presence of the fibers, benzo[a]pyrene might be increasingly activated by enzyme systems other than mixed function oxidase. Prior exposure of microsomes and whole cells to asbestos seems to be necessary for manifestation of the observed effects. It is likely that the fibers stimulate membrane oxidases which are capable of activating benzo[a]pyrene.

ACKNOWLEDGMENTS

The financial support of the National Cancer Institute of Canada is gratefully acknowledged. We are very thankful to Mrs. Linda Ross for her generous help in cell culturing, and Dr. R. Green for making his facilities available. Our thanks are also due to Dr. A. Liepins for supplying fibroblasts. The secretarial assistance of Miss Michelle Connolly is very much appreciated.

REFERENCES

1. Selikoff, I.J. and Lee, D.H.K. (1978): Asbestos and disease, Academic Press, Inc., New York, pp. 307-336.
2. Harris, C.C. (1977): Respiratory carcinogenesis. In: Lung Cancer: Clinical diagnosis and treatment, edited by M.J. Straus, pp. 1-17, Grune & Stratton, New York.
3. Nettesheim, P., Topping, D.C. and Jamasbi, R. (1981): Host and environmental factors enhancing carcinogenesis in the respiratory trace, Ann. Rev. Pharmacol. Toxicol., 21:133-163.
4. Selikoff, I.J., Hammond, E.C. and Churg, J. (1968): Asbestos exposure, smoking and neoplasia, J. Am. Med. Assoc., 204:106-112.
5. Lunden, R.E., Jr., Lloyd, J.W., Smith, .EM., Archer, V.E., and Holaday, D.A. (1969): Mortality of uranium miners in relation to radiation exposure, hardrock mining and cigarette smoking-1950 through September 1967, Health Phys., 16:571-578.
6. Lloyd, J.W. (1971): Long-term mortality study of steel-workers, V. Respiratory cancer in coke plant workers, J. Occup. Med., 13:53-68.
7. Lakowicz, J.R., Englund, F. and Hidmark, A. (1978): Particle-enhanced membrane uptake of a polynuclear aromatic hydrocarbon: A possible role in cocarcinogenesis, J. Natl. Cancer Inst., 61:1155-1159.

8. Pylev, L.M., Roe, F.J.C., and Warwick, G.P. (1969): Elimination of radioactivity after intratracheal instillation of tritiated 3,4-benzo[a]pyrene in hamsters, Br. J. Cancer, 23:103-115.
9. Shabad, L.M., Pylev, L.N., Krivosheevo, T., Kulagina, F. and Nemenko, B.A. (1974): Experimental studies on asbestos carcinogenicity, J. Natl. Cancer Inst., 52:1175-1180.
10. Kandaswami, C. and O'Brien, P.J. (1980): Effects of asbestos on membrane transport and metabolism of benzo[a]pyrene, Biochem. Biophys. Res. Commun., 97:794-801.
11. Kandaswami, C. and O'Brien, P.J. (1981): Pulmonary metabolism of benzo[a]pyrene: Effect of asbestos, Biochem. Pharmacol., 30:811-814.
12. Maxwell, K.W., Dietz, T., and Marcus, S. (1964): An in situ method for harvesting quinea pig alveolar macrophages, Am. Rev. Resp. Dis., 89:579-580.
13. Grover, P.L., Hewer, A. and Sims, P. (1974): Metabolism of polycyclic hydrocarbons by rat-lung preparations, Biochem. Pharmacol., 23:323-332.
14. Sims, P. (1970): Qualitative and quantitative studies on the metabolism of a series of aromatic hydrocarbons by rat-liver preparations, Biochem. Pharmacol., 19:795-818.
15. Hill, D.L. and Shih, T.W. (1975): Inhibition of benzo[a]pyrene metabolism catalyzed by mouse and hamster lung microsomes, Cancer Res., 35:2717-2723.
16. Sivarajah, K., Anderson, M.W. and Eling, T.E. (1978): Metabolism of benzo[a]pyrene to reactive intermediate(s) via prostaglandin biosynthesis, Life Sci., 23:2571-2578.
17. Roy-Chowdhuri, A.K., Mooney, T.R., and Reeves, A.L., (1973): Trace metals in asbestos carcinogenesis, Arch. Environ. Health, 26:253-255.
18. Eneanya, D.I., Bernard, F.B., and Hart, R.W. (1979): Asbestos (chrysotile intermediate) alters the metabolism of benzo[a]pyrene in Syrian hamster embryo cells, Proceedings of the 11th International Congress of Biochemistry, Toronto, p. 687, Abstract 13-5-255, (Poster 13).
19. Harris, C.C., Hsu, I.C., Stoner, G.D., Trump. B.F., and Selkirk, J.K. (1978): Human pulmonary alveolar macrophages metabolize benzo[a]pyrene to proximate and ultimate mutagens, Nature, 272:633-634.
20. Naseem, S.M., Tishler, P.V., Anderson, H.A., and Selikoff, I.J. (1978): Aryl hydrocarbon hydroxylase in asbestos workers, Am. Rev. Respir. Dis., 118:693-700.

21. Seidman, H., Selikoff, I.J., and Hammond, E.C. (1979): Short term asbestos work exposure and long-term observation. In: Health Hazards of Asbestos Exposure, edited by I.J. Selikoff and E.C. Hammond, pp. 61-89. The New York Academy of Sciences, New York.
22. Dixon, J.R., Lowe, D.B., Richards, D.E., Cralley, L.J. and Stokinger, H.E. (1970): The role of trace metals in chemical carcinogenesis: Asbestos cancers, Cancer Res., 30:1068.
23. Hart, R.W., Daniel, F.B., Kindig, O.R., Beach, C.A., Joseph, L.B., and Wells, R.C. (1980): Elemental modifications and polycyclic aromatic hydrocarbon metabolism in human fibroblasts, Environ. Health Perspect., 34:59-68.
24. O'Brien, P.J. (1981): Peroxide mediated metabolic activation of carcinogens. In: Lipid Peroxides in Biology and Medicine, edited by K. Hagi, Academic Press (in press).
25. Sirois, P., Rola, Pleszczynski, M. and Bégin, R. (1980): Phospholipase A activity and prostaglandin release from alveolar macrophages exposed to asbestos, Prostaglandins Med., 5:31-37.
26. Topping, D.C. and Nettesheim, P. (1980): Two-stage carcinogenesis studies with asbestos in Fischer 344 rats, J. Natl. Cancer Inst., 65:627-630.

MOLECULAR SPECTRA OF POLYCYCLIC AROMATIC HYDROCARBONS

W. KARCHER, R.J. FORDHAM, A. NELEN, R. DEPAUS, J. DUBOIS,
and Ph. GLAUDE
Commission of the EC, JRC, Petten Establishment, P.O. Box 2,
1755 ZG Petten, The Netherlands.

INTRODUCTION

In industrialised countries, appreciable quantities of polycyclic aromatic hydrocarbons (PAH) are continually released into the environment (so far more than 150 PAH have been positively identified in car exhausts (1), 108 in the waste gas from oil-fired heaters (2), 280 in tobacco smoke (3), 106 in coal tar (4) and 150 in petroleum and related products (5)) and a systematic control of environmental and occupational samples presents a rather difficult task.

Ideally, pure samples of each individual PAH needing identification and quantification should be available for comparison and calibration. However, due to the enormous number of isomers and derivatives which are encountered, this situation is unlikely to be attained in the foreseeable future.

Instead, the availability of high quality spectra and related data may be seen as a more realistic alternative. A systematic collection and evaluation of the UV and fluorescence spectra would facilitate determination by HPLC and low temperature fluorescence spectra may sometimes be used for the identification and determination of certain PAH without prior separation (6). Also, NMR spectra can serve for the unambiguous identification of isomers which cannot be distinguished by other techniques.

MATERIALS AND METHODS

Materials

Samples were prepared with a minimum purity of 99% by either separation, enrichment and purification from natural or synthetic sources or purification of commercial samples or synthesis following published routes.

Measurement of Spectra

As far as possible, the spectra were recorded under

standardised conditions. Thus, for the UV and fluorescence spectra the same solutions in cyclohexane at a concentration of approximately 5 mg.l^{-1} were used in a Perkin-Elmer model 555 UV and a Perkin-Elmer model MPF-44 with a DCSU-2 spectrum correcto, respectively.

Mass spectra were recorded using a Ribermag model R10-10C quadrupole MS at 70 eV using GCMS as the mode of introduction.

NMR spectra were determined using a Bruker WP80 DS FT instrument for both proton (in C_6D_6 or $CDCl_3$ at 80 MHz) and carbon-13 (in deutero-acetone or chloroform at 20 MHz). Concentrations in the range 10-50 g.l^{-1} were used.

<u>Spectral Atlas</u>. An atlas of the spectra recorded and evaluated in this study is being prepared and will be published separately.

For presentation, the spectra of each compound are arranged in the following sequence:
- UV and fluorescence spectra
- Mass spectra recorded by both quadrupolar and magnetic instruments
- Proton and carbon-13 NMR spectra.

The spectra are accompanied by tables giving the molar absorption coefficients (extinctions) and wavelengths, relative fluorescence intensities and wavelengths, relative mass fragment abundances, chemical shifts and coupling constants <u>etc</u>. together with some details of experimental conditions (see Fig. 1).

RESULTS AND DISCUSSION

As it is not feasible to present here a large number of PAH spectra in detail, the discussion is limited to a few representative spectra. Selected mass, UV and fluorescence and NMR spectra are shown in Figs. 2-4 & 6. Particular attention is directed towards two benzonaphthothiophene isomers and their nitrogen-containing homologues (benzocarbazoles). One of these isomers, benzo[b]naphtho[2,1-d]thiophene (BN21Th) is of some environmental importance in view of its occurrence in industrial effluents and its 'fingerprinting' role for distinguishing emissions from diesel and petrol engines (7,8).

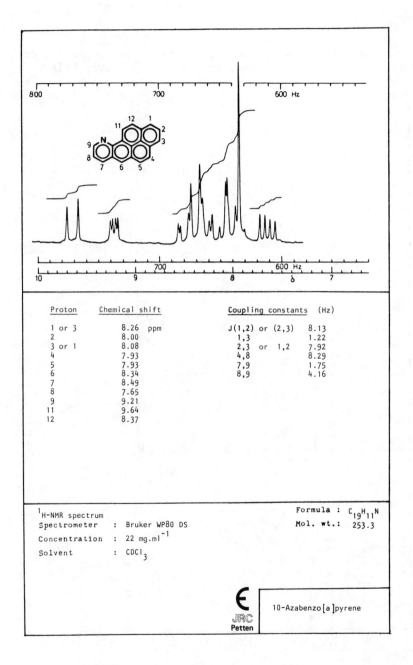

FIGURE 1 Example of Spectral Atlas entry for 10-Azabenzo-[a]pyrene.

Mass Spectra

As has been noted previously (9) with the isomeric thiophene derivatives of mass 258, MS gives very little help

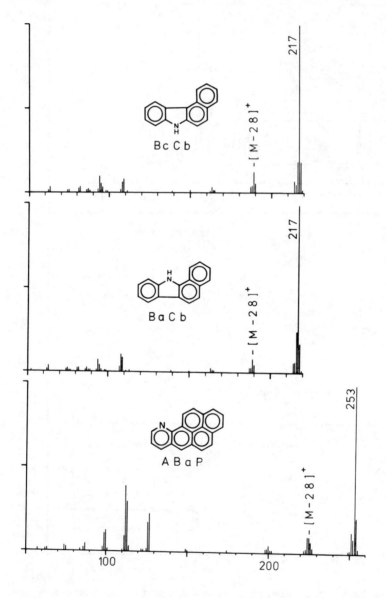

FIGURE 2   Quadrupole mass spectra of N-PAH

in differentiating isomers. The spectra of benzo[b]naphtho-[1,2-d]thiophene (BN12Th) and BN21Th are practically identical. A similar observation may be made for the structurally similar benzo[a]carbazole (BaCb) and benzo[c]carbazole (BcCb) (see Fig. 2).

In general, it is found that the M-1 fragment is less abundant than the M-2 except for the benzocarbazoles where presumably the proton attached to the N is easily lost.

A second characteristic feature of the benzocarbazoles is the presence of a fragment at m/z 189 (M-28) which may arise from an equivalent fragmentation to that producing an m/z 189 (M-45) fragment from the benzonaphthothiophenes and other S-PAH. This M-28 fragment is found also with 10-azabenzo-[a]pyrene (ABaP) and, in addition to the odd mass number of the parent ion, may be considered confirmatory of the presence of a nitrogen in either a ⟩NH or a ⟩N structure (see Fig. 2).

The conclusion remains that the most important use of MS is the information given regarding the mass of the parent ion and consequently the molecular formula. Structural identification requires other techniques.

UV and Fluorescence Spectra

The picture that emerges from a close examination of the UV absorption spectra is one which appears to show no coherent pattern of similarities or differences between the isomeric pairs or the homologous pairs. Nor is a close match possible with the corresponding carbocyclic PAH, the benzofluorenes or benz[a]anthracene. In contrast, greater similarities exist between chrysene and BN12Th as shown by the wavelengths of the absorption maxima:

Chrysene	257	<u>267</u>	280	293	305	318	342	353	360 nm
BN12Th	255	<u>263</u>	280	293	307	319	333	343	350 nm

The UV spectra of isomeric pairs show several significantly different, but non-predictable, features (see Figs. 3 & 4). These features aid identification but may hinder quantification using UV detectors in HPLC. For example, at 254 nm the approximate molar absorption coefficient of BN12Th is 42000 while that of BN21Th is 35000 and of benzo[b]naphtho[2,3-d]thiophene (not shown) is 32000. However, BN21Th has its λ_{max} at 252 nm (ε = 52000), and errors in setting wavelengths with detectors using monochromators or thermal effects (band broadening) could lead to significant quantification errors or to significantly non-linear calibration curves.

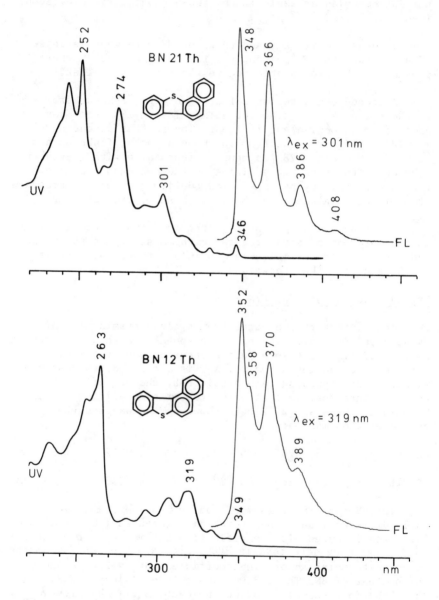

FIGURE 3 UV/fluorescence spectra of benzonaphthothiophenes in cyclohexane at RT.

It may be noticed that all four materials illustrated exhibit a similar pattern of weak absorption peaks in the a-region (long wavelength) and this pattern is repeated in the

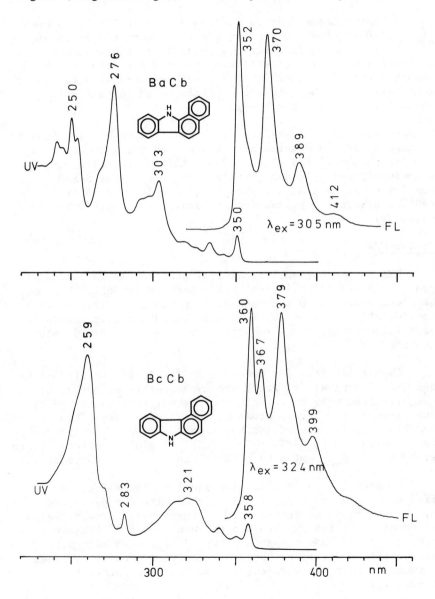

FIGURE 4   UV/fluorescence spectra of benzocarbazoles in cyclohexane at RT.

room temperature fluorescence spectra also shown. Here again, however, there are sufficient differences in pattern and wavelengths to differentiate isomers from each other.

An interesting feature was reported recently by Colmsjö et al. (10) in the low temperature fluorescence spectra of S-PAH. Whereas no quasilinear low temperature fluorescence emission (Shpol'skii effect) was observed for dibenzo-, benzonaphtho- and dinaphthothiophenes, such spectra were found for some S-PAH with the thiophene ring fused on all sides (e.g. phenanthro[4,5-bcd]thiophene).

As was reported earlier (9), these compounds also tend to be more active in mutagenicity tests than the thiophenes showing no Shpol'skii effect. Thus, this correlation may conceivably be applied to predict the more active S-PAH in complex environmental and industrial samples and so limit the tedious separation procedures necessary to identify important mutagens.

Proton NMR

NMR is the one analytical tool practically guaranteed to provide 'a priori' differentiation between isomers. This may be demonstrated with the BaCb/BcCb and BN12Th/BN21Th pairs by a consideration of some chemical shifts and coupling constants associated with protons subject to what may be termed "steric effects".

Chemical shifts for bay-protons 1 & 11 for BN12Th & BcCb and 1 & 12(equivalent) for benzo[c]phenanthrene (BcPh) reflect the differing proximities of other aromatic rings caused by the sizes of the hetero-atoms or equivalent groups (see Table 1 and Fig. 5). The closer another aromatic ring, the higher is the chemical shift (-NH- $<$-S- $<$-C=C-). A similar effect occurs with protons 6 & 7 of BaCb & BN21Th and 4 & 5 of chrysene (C).

Considering the "heteroatom side" of the molecules, we find some mesomeric effects associated with the p-electrons of the heteroatoms. For example, a relatively large difference in chemical shifts is observed between BaCb & BN21Th for proton 1 or 10 when compared to the differences seen with steric effect described above. In addition, the differences between isomers are particularly marked (see shifts for proton 1).

Less variability is found with coupling constants which are nearly identical for carbocyclic and heterocyclic

TABLE I

CHEMICAL SHIFTS IN PPM (IN $CDCl_3$ FROM TMS FOR PROTON (n)

PAH	steric effects				heteroatom effects				$|J_{1,2}-J_{3,4}|$ (Hz)
BcCb	(1)	8.79	(11)	8.58	(6)	7.62	(8)	7.4+0.1	0.23
BN12Th	(1)	9.03	(11)	8.87	(6)	7.91	(8)	8.0$\overline{0}$	0.41
BcPh[a]	(1)=(12)	9.14							0.68
BaCb	(6)	8.07	(7)	8.06	(1)	7.62	(10)	7.24	0.26
BN21Th	(6)	8.18	(7)	8.22	(1)	8.15	(10)	7.97	0.01
C	(4)	8.79	(5)	8.73					0.40

a in C_6D_6

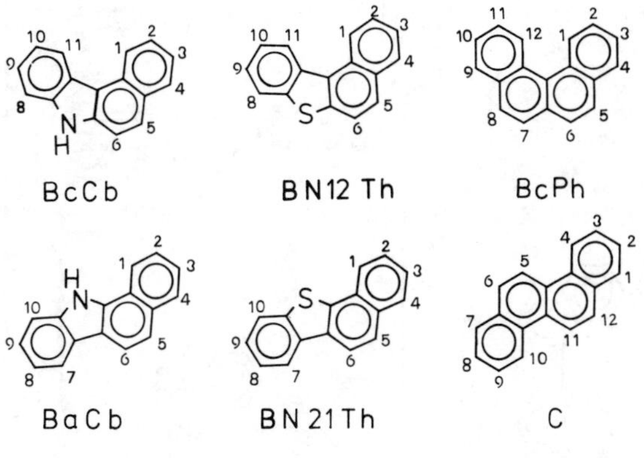

FIGURE 5 Structures of PAH

TABLE 2

COUPLING CONSTANTS OF PROTON-4 IN Hz (± 0.03 Hz) of DIFFERENT PAH

PAH	$J_{1,4}$	$J_{2,4}$	$J_{3,4}$	$J_{4,5}$	$J_{4,6}$
BaCb	+ 0.74	+ 1.32	+ 8.18	- 0.45	+ 0.28
BcCb	+ 0.72	+ 1.34	+ 8.15	- 0.46	+ 0.25
BN12Th	+ 0.79	+ 1.26	+ 8.33	-	-
BN21Th	+ 0.71	+ 1.22	+ 8.25	- 0.43	+ 0.27
Naphthalene	+ 0.74	+ 1.24	+ 8.28	- 0.45	+ 0.23

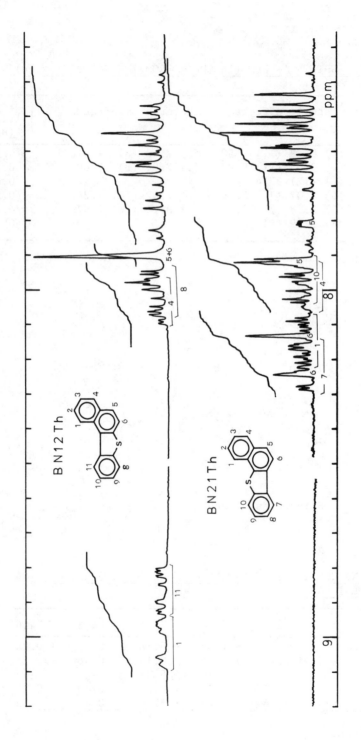

FIGURE 6   80 MHz proton spectra of benzonaphthothiophenes in CDCl$_3$

analogues. Table 2 shows various couplings of H4 for the carbazoles and thiophenes and a comparison with values for the primary residue (naphthalene) (11). This similarity of spectral fine structure, however, can serve to assist in the assignment of chemical shifts to protons. Despite the similarities, the difference between J(1,2) and J(3,4) from the same naphthalene moiety provides a clear distinction between isomers (Table 1). This is another manifestation of the steric effects described above. The difference is largest for BcPh (0.68 Hz) decreasing to 0.23 Hz for BcCb and again from 0.4 Hz (chrysene) to 0 Hz for BN21Th. Here, the reversal (BN21Th <BaCb) is caused by the proton of the ⟩NH group.

Conclusions

Although the wide availability of calibration materials of known purity is highly desirable for the analysis of PAH in complex environmental matrices, a more practical approach may be represented by a comprehensive collection of the molecular spectra. Thus, availability of their UV and fluorescence spectra will facilitate quantification by HPLC, and in turn, mass spectra indicate the presence of heteroatoms (peaks at M-45 and M-28 for S-PAH and N-PAH respectively). In the case of structurally similar isomers, however, these spectra rarely permit direct identification. Unambiguous structural identification can only be obtained by an interpretation of the NMR spectra.

ACKNOWLEDGEMENTS

A. Nelen is supported by a grant from the Commission of the EC, Directorate General for Research, Science and Education.

REFERENCES

1. Grimmer, G., Böhnke, H., and Glaser, A. (1977): Investigation on the carcinogenic burden by air pollution in man, XV - Polycyclic aromatic hydrocarbons in automobile exhaust gas - An inventory, Zbl. Bakt. Hyg., I. Abt. Orig. B 164: 218-234.
2. Herlan, A. (1977): Kanzerogene polyzyklische Aromaten und Metaboliten als mögliche Bestandteile von Emissionen, Zbl. Bakt. Hyg., I. Abt. Orig. B 165:174-191.

3. Snook, M.E., Severson, R.F., Higman, H.C., Arrendale, R.F., and Chortyk, O.T. (1976): Polynuclear aromatic hydrocarbons of tobacco smoke: Isolation and identification, Beitr. zur Tabakforschung 8:250-272.
4. Lang, K.F., and Eigen, I. (1967): Im Steinkohlenteer nachgewiesene organische Verbindungen. Fortschritte der chemischen Forschung Vol. 8, Springer Verlag, Berlin/New York, pp. 91-170.
5. Grimmer, G., and Böhnke, H. (1978): Polyzyklische aromatische Kohlenwasserstoffe und Heterozyklen - Beziehung zum Reifegrad von Erdölen des Gifhorner Troges, Erdöl u. Kohle 31:272-277.
6. Yang, Y., D'Silva, A.P. and Fassel, V.A. (1981): Laser-excited Shpol'skii spectroscopy for the selective excitation and determination of polynuclear aromatic hydrocarbons, Anal. Chem., 53:894-899.
7. Grimmer, G., Böhnke, H. and Glaser, A. (1977): Polycyclische aromatische Kohlenwasserstoffe im Abgas von Kraftfahrzeugen, Erdöl u. Kohle - Erdgas - Petrochemie, vereinigt mit Brennstoffchemie, 30:411-417.
8. Pott, F. and Grimmer, G. (1979): Polycyclische aromatische Kohlenwasserstoffe, VDI-Kolloquium-Umwelt, Nr. 2, pp. 122-125.
9. Karcher, W., Nelen, A., Depaus, R., van Eijk, J., Glaude, P. and Jacob, J. (1981): New results in the detection, identification and mutagenic testing of heterocyclic aromatic hydrocarbons. In: Polynuclear Aromatic Hydrocarbons - Chemical analysis and Biological Fate, edited by M. Cooke and A.J. Dennis, pp. 317-27, Battelle Press, Columbus (Ohio).
10. Colmsjö, A.L., and Ostman, C. (1981): The utility of the Shpol'skii Effect for the detection of POM. Paper presented at 6th Internat. Symp. on PAH, Columbus, 1981 (this Vol.).
11. Crecely, R.W. and Goldstein, J.H. (1970): Complete PMR analysis of naphthalene as an eight spin system, Org. Magn. Res. 3:405-16.

POLYCYCLIC AROMATIC HYDROCARBONS IN LETTUCE. INFLUENCE OF A
HIGHWAY AND AN ALUMINIUM SMELTER.

BONNY LARSSON AND GREGER SAHLBERG
Food Laboratory, National Food Administration, Box 622,
S-751 26 Uppsala, Sweden

INTRODUCTION

The Swedish National Food Administration runs a monitoring program for polycyclic aromatic hydrocarbons (PAH) in various foods, including vegetables that have been contaminated via deposition from polluted air. A study in the German Democratic Republic (1) has shown that vegetables are a major source of benzo[a]pyrene (BaP) in the human diet, especially when they are grown in industrial areas. From a food hygiene point of view it is important to study how local PAH sources, such as automobile traffic, residential heating systems and various industrial activities, influence the levels in plant foodstuffs. In this study our interest was focused on a highway and an aluminium smelter.

Studies on the deposition of contaminants near roadways have previously mainly concerned lead. It is well documented that lead accumulates in vegetables along roadways in concentrations varying with traffic flow, distance from the road, exposure time and external plant characteristics (2,3). Less is known about the roadside distribution of PAH. In the present study the occurrence of PAH in lettuce grown in an allotment area close to a highway was studied. Lettuce was chosen because it is a commonly grown crop with a large surface exposed to air pollution. The aims were to monitor the levels of PAH in relation to distance from the driving lane and, secondly, to examine whether accumulated PAH could be removed by washing.

The emission of BaP from an aluminium smelter in Sundsvall, a Swedish industrial town of 60 000 inhabitants, has previously been estimated to be 0.26 kg/h (4). In order to study the influence of this source on locally grown food crops, samples were collected during summer 1980. Lettuce was taken from private gardens in Sundsvall at various distances from the aluminium smelter. The results of the analysis of PAH in these lettuce samples are discussed in the present report.

The use of glass capillary gas chromatography enables simultaneous determination of a number of PAH compounds. The relative abundance of individual compounds in the sample, i.e. the

PAH profile, gives some indication of the source(s) of contamination.

MATERIALS AND METHODS

Sampling

Lettuce plants were grown from seed in the laboratory and planted out in an area of allotments close to a major highway (traffic flow 44 000 motor vehicles/24 h) in a non-industrial area outside the city of Stockholm. The lettuce was planted at a distance of 12-50 m from the driving lane, where space could be obtained. The samples were collected at the end of July 1979, after an exposure time of 6 weeks. Where possible, plants not shielded from the highway by natural obstacles (bushes etc.) were selected. Part of the sample material was washed as follows: the leaves were separated and immersed in tap water at $20^{\circ}C$ for 2 min, drained and air-dried. All samples were chopped and stored deep-frozen until analyzed.

Samples of lettuce were taken from private gardens in Sundsvall at a distance of 0.5-6.5 km from the aluminium smelter in June and July 1980. The majority of the samples were taken north-west of the smelter as south-eastly winds are predominant during the summer. The samples were chopped and stored deep-frozen until analyzed.

Method

After addition of an internal standard (400 ng of ββ-binaphthyl) the sample (60-80 g) was digested in 80 ml of boiling methanolic 2M KOH for 3-4 hours. The further extractions and silica column clean-up were performed according to the method of Grimmer and Böhnke (5), modified in that the solvent volumes used in the initial extraction steps were reduced to 1/5. The purified sample solution was concentrated under a stream of nitrogen to ca 50 µl. The gas chromatographic analysis was performed on a Pye Unicam GCD gas chromatograph with a 50 m x 0.30 mm SE-54 glass capillary column. The flame ionization detector was connected to an electronic integrator. Sample volumes of 1-4 µl were injected by the falling needle technique. Hydrogen was used as carrier gas at a flow of 3 ml/min. The temperature program was $165^{\circ}C$ for 6 min, $4^{\circ}C$/min to $255^{\circ}C$, held at $255^{\circ}C$ for 30 min. The peaks were identified by comparing the retention times with those of standards. The identified compounds were quantified by comparing the integrated peak areas with that of the internal standard.

RESULTS AND DISCUSSION

Lettuce grown close to a highway

The average PAH levels found in lettuce grown close to a major highway are listed in Table 1. A total of 15 PAH compounds

TABLE 1

PAH LEVELS (µG/KG FRESH WEIGHT) IN LETTUCE GROWN AT VARIOUS DISTANCES FROM A HIGHWAY[a]

Compound[b]	Distance from driving lane (m)					
	12	17	18	32	38	50
Phe	7.5	2.9	1.8	6.1	2.9	5.0
Ant	0.3	0.2	<0.1	0.2	0.1	0.2
2-MPh	1.6	0.7	0.5	1.0	0.5	0.6
1-MPh	1.6	0.7	0.6	0.8	0.7	0.7
Flu	9.1	2.9	4.3	4.6	2.8	5.3
Py	10.4	4.0	5.2	5.0	3.4	5.8
BaA	4.6	0.7	1.5	0.9	0.7	0.9
Ch/Tri	7.1	1.4	2.7	2.7	1.8	3.3
BbF	7.3	0.5	1.5	0.7	1.0	0.6
Bj,kF	6.1	0.5	1.4	0.5	0.9	0.4
BeP	6.7	0.7	1.6	0.5	0.9	0.8
BaP	6.2	0.5	1.2	0.3	0.6	0.5
Per	1.7	0.2	0.3	0.2	0.2	<0.1
IPy	8.3	0.4	0.9	0.3	0.5	0.6
BghiP	10.8	0.6	1.3	0.5	0.7	0.5
ΣPAH	89.3	16.9	24.9	24.3	17.7	25.2

a Average of 2 samples at each distance
b Phe = Phenanthrene   BbF = Benzo[b]fluoranthene
   Ant = Anthracene   Bj,kF = Benzo[j]fluoranthene &
   2-MPh = 2-Methylphenanthrene   Benzo[k]fluoranthene
   1-MPh = 1-Methylphenanthrene BeP = Benzo[e]pyrene
   Flu = Fluoranthene   BaP = Benzo[a]pyrene
   Py = Pyrene   Per = Perylene
   BaA = Benz[a]anthracene   IPy = Indeno[1,2,3-cd]pyrene
   Ch/Tri = Chrysene &   BghiP = Benzo[ghi]perylene
            Triphenylene   ΣPAH = Sum of 15 PAHs

were identified. High PAH concentrations were found in the samples taken closest to the highway, i.e. 12 m from the driving lane. The maximum levels of BaP and ΣPAH were 7.1 and 102 µg/kg, respectively. Samples taken further away, at a distance of 17-50 m, showed a decreased and fairly constant level, with concentrations of BaP and ΣPAH varying between 0.2-1.6 µg/kg and 15-27 µg/kg, respectively. Lead analyses performed on the same sample material showed that lead was distributed along the highway in a similar way to PAH (Figure 1). The BaP levels found in

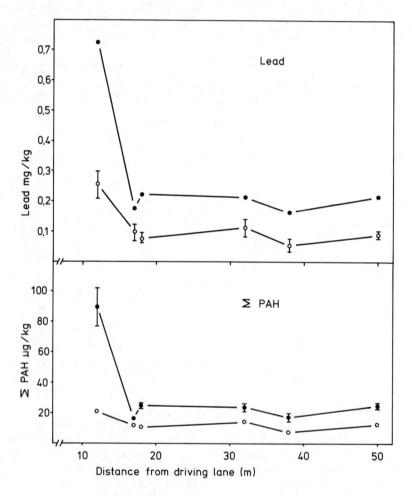

FIGURE 1. Distribution of PAH and lead in washed (o) and unwashed (●) samples of lettuce, grown at various distances from a major highway (44 000 motor vehicles/24 h).

the present study are higher than those reported from Norway for samples of lettuce grown within 10 m of a highway (0.2-0.3 µg/kg) (6).

Washing the lettuce had little effect on phenanthrene (Phe) levels but it considerably reduced the levels of high molecular PAH compounds (Table 2). Only 5-33 % of the BaP remained after washing. On average 50 % of the total PAH was removed. An explanation of this selective removal of PAHs by washing could be that the low molecular compounds in the vapour phase penetrate the waxy layer of the plant surface, whereas the particle-bound high molecular material remains on the surface in a more exposed position. Previously reported results indicate a maximum reduction of 10-20 % in total PAH levels in vegetables (1,7). The marked reduction in the PAH content of lettuce obtained by washing found in the present study is of importance when considering the potential risk to the consumers.

TABLE 2

EFFECT OF WASHING ON PAH LEVELS IN LETTUCE

ΣPAH before washing (µg/kg)	Distance from the driving lane (m)	% of various PAHs left after washing[a]						
		Phe	Flu	Py	BaA	Bap	BghiP	ΣPAH
102	12	96	33	41	8	6	3	21
16	17	100	79	103	57	5	10	76
23	18	118	49	47	25	12	27	45
26	32	44	60	71	37	33	10	54
15	38	95	52	54	33	12	40	52
23	50	86	57	53	33	12	10	55
Mean		90	55	61	32	13	17	50

[a] One lettuce taken from each zone was divided in two, one part was washed before analysis, the other was analysed directly.

The PAH profiles in the lettuce samples varied with the distance from the highway. In Figure 2 the average profiles in lettuce grown 12-50 m (12 samples) and 12 m (2 samples) from the

PAH IN LETTUCE

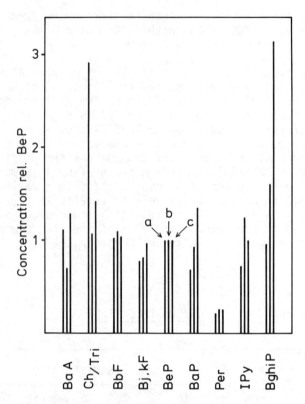

FIGURE 2. PAH profile in (a) lettuce grown 12-50 m (aver. of 12 samples) and (b) 12 m (aver. of 2 samples) from a major highway, compared to (c) a typical PAH profile from automobile exhaust, ECE-reglement 15 (Grimmer et al (8)).

driving lane are compared to a typical PAH profile from automobile exhaust (8). The comparison includes only compounds measured in both types of sample, the concentrations relative to benzo[e]pyrene (BeP) are shown. The PAH profile in the lettuce samples from the 12 m zone is fairly similar to the automobile exhaust profile. The lower concentration of benzo[ghi]perylene (BghiP) in the lettuce samples could be due to selective washoff by rainfall.

Lettuce grown near an aluminium smelter

The results of the analysis of PAH in lettuce from private gardens in Sundsvall are schematically shown on the map in Figure 3. The concentration of BaP and ΣPAH in samples taken at a

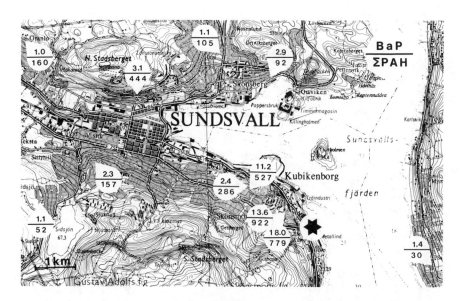

FIGURE 3. Levels of BaP and ΣPAH in lettuce (µg/kg fresh weight) from sampling sites in Sundsvall. The location of the aluminium smelter is indicated by a star.

distance of 0.5-6.5 km from the aluminium smelter varied between 1.0-18.0 µg/kg and 30-922 µg/kg, respectively. The highest levels were found in samples grown 0.5-1.5 km downwind of the smelter. Two samples from villages outside Sundsvall, 13 and 20 km from the smelter, contained <30 µg/kg of ΣPAH. The PAH levels in the lettuce samples from Sundsvall are generally higher than those previously reported for Nordic leafy vegetables (6, 9,10). It should also be noted that the BaP concentrations in all but one sample exceeded 1 µg/kg, the present tolerance for smoked meat products in the Federal Republic of Germany.

Figure 4 shows a comparison between average PAH profiles in lettuce samples grown at distances of 0.5-1.5 km from the aluminium smelter and 12-18 m from the highway, respectively. A significant difference can be observed in the content of fluoranthene (Flu), which is the dominant component (approx. 30 % of ΣPAH) in all samples from the Sundsvall area. Furthermore, the comparatively high concentrations of high molecular PAH, particularly BghiP, observed in the lettuce samples collected close to the highway and also in automobile exhaust (8), are not observed in the lettuce grown near the aluminium smelter.

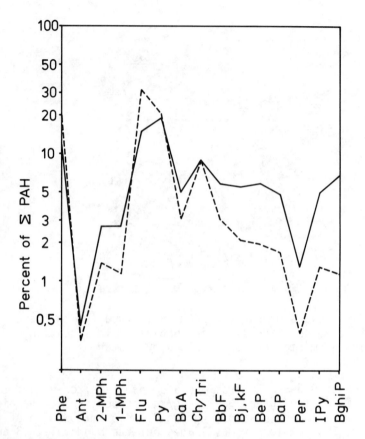

FIGURE 4. Average PAH profile in samples of lettuce grown 0.5-1.5 km from an aluminium smelter (---) and 12-18 m from a major highway (—).

The high relative concentration of Flu observed in the lettuce from Sundsvall is also found in emission samples from the aluminium smelter (4). The gradient in the concentrations of PAH observed in the lettuce samples collected downwind of the smelter indicates that it is probably the major source of PAH.

Comparison of Swedish results

The PAH level in the lettuce samples collected around the aluminium smelter in Sundsvall is the highest that has been observed in lettuce in Sweden. Comparable levels are found only in occasional samples taken within 15 m from major highways. In

Table 3 the PAH levels in lettuce from various sites in Sweden are summarized. It appears from the table that the contamination of lettuce is primarily due to the influence of local, recognized PAH emitters and that the PAH levels in the plants are highly dependent on the distance from the source. However, when comparing the impacts of the various sources it should be borne in mind that the PAH levels obtained are influenced also by meteorological conditions (wind, rainfall etc.) during the cultivation period.

TABLE 3

PAH IN LETTUCE FROM DIFFERENT SITES IN SWEDEN[a]

Site, distance from source	Number of samples	BaP (μg/kg) Median	BaP (μg/kg) Range	ΣPAH (μg/kg) Median
Highway,				
8-15 m	13	1.0	<0.1-31.7	50
15-50 m	16	0.4	0.2-1.6	26
Airport,				
150-800 m	7	0.5	0.1-1.0	24
Alum. smelter,				
0.5-1.5 km	4	12.4	2.4-18.0	654
2.0-6.5 km	8	1.3	1.0-3.1	128
Industrial areas	13	0.1	<0.1-7.3	13
Residential,				
urban	8	0.2	<0.1-1.0	13
country	8	0.1	<0.1-0.4	12

a   Results reported by Bjørseth et al. (9) are included.

ACKNOWLEDGEMENTS

Anders Eriksson is thanked for skillful technical assistance.

REFERENCES

1. Fritz, W. (1971): Umfang und Quellen der Kontamination unserer Lebensmittel mit krebserzeugenden Kohlenwasserstoffen. Ernährungsforschung, 16:547-557.
2. Havre, G. and Underdal, B. (1976): Lead contamination of vegetation grown close to roads. Acta Agric. Scand., 26: 18-24.
3. Fowles, G.W.A. (1976): Lead content of roadside fruit and berries. Food Chem., 1:33-39.
4. Gränges Aluminium, Sundsvall, Sweden: Personal communication.
5. Grimmer, G. and Böhnke, H. (1975): Polycyclic aromatic hydrocarbon profile analysis of high-protein foods, oils and fats by gas chromatography. J. Assoc. Off. Anal. Chem., 58:725-733.
6. Kveseth, K., Sortland, B. and Støbet, M.-B. (1981): Polycyclic aromatic hydrocarbons in leafy vegetables, a comparison of the Nordic results. Nordic PAH Project, No 8, Central Institute for Industrial Research, Oslo.
7. Grimmer, G. and Hildebrand, A. (1965): Der Gehalt polycyclischer Kohlenwasserstoffe in verschiedenen Gemüsesorten und Salaten. Deut. Lebensm.-Rundschau, 61:237-239.
8. Grimmer, G., Naujack, K.-W. and Schneider, D. (1980): Changes in PAH-profiles in different areas of a city during the year. In: Polynuclear Aromatic Hydrocarbons: Chemistry and Biological Effects, Fourth International Symposium, A. Bjørseth and A.J. Dennis, Eds., Batelle Press, Columbus, Ohio.
9. Bjørseth, A., Støbet, M.-B. and Lunde, G. (1978): Analyses of polycyclic aromatic hydrocarbons in lettuce from Sweden. Nordic PAH Project, No 2, Central Institute for Industrial Research, Oslo (in Norwegian).
10. Pyysalo, H. (1979): Analyses of polycyclic aromatic hydrocarbons (PAH) in Finnish leafy vegetables. Nordic PAH Project, No 3, Central Institute for Industrial Research, Oslo.

IDENTIFICATION AND MUTAGENICITY OF NITROGEN-CONTAINING POLYCYCLIC AROMATIC COMPOUNDS IN SYNTHETIC FUELS

DOUGLAS W. LATER*, MILTON L. LEE*, RICHARD A. PELROY**
AND BARY W. WILSON**
*Department of Chemistry, Brigham Young University, Provo, Utah 84602; **Pacific Northwest Laboratory, Operated by Battelle Memorial Institute, Richland, Washington 99352

INTRODUCTION

Nitrogen-containing polycyclic aromatic compounds (N-PAC) constitute a major fraction of coal liquids and shale oils (1) and are of particular concern for several reasons. Nitrogen-containing compounds adversely affect catalytic refining processes (2), contribute to the instability of stored fuels (3), and are responsible for noxious NO_x emissions upon combustion (4). Exposure of plant workers to synfuel process streams and products that contain N-PAC, and formation of N-PAC during combustion of nitrogen-rich fuels pose both occupational and environmental concerns because many N-PAC are known mutagens and/or carcinogens (5). On the other hand, coal-derived products could provide a valuable source of N-PAC, e.g. quinolines, for the manufacture of pharmaceuticals, pesticides, dyes, and herbicides (1,6).

The detailed characterization of synfuel products and by-products can provide invaluable information relating to the environmental implications of synfuel production and the potential end uses of the products. This paper describes an effective analytical methodology for the separation and identification of N-PAC in these materials, and the results obtained from mutagenicity assays of isolated fractions.

MATERIALS AND METHODS

Samples

Two solvent refined coal (SRC) liquids were used for this study. An SRC II heavy distillate (HD) from the Fort Lewis pilot plant, and an SRC I process solvent (PS) from the Wilsonville pilot plant were examined in this study. The SRC II HD had a boiling point range of 260°C to 450°C. Both materials were of pilot plant origin and hence should not be considered representative of future commercial scale products.

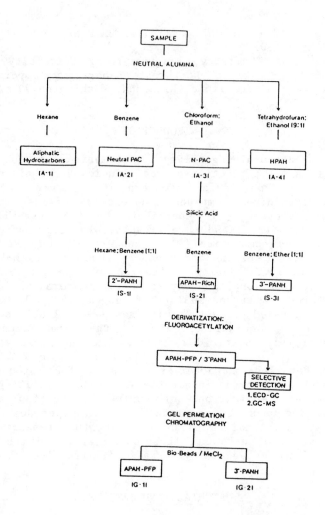

FIGURE 1. Chemical class fractionation scheme.

Methods

Separation Methods. A schematic diagram of the total separation procedure is shown in Figure 1. An initial separation on neutral alumina was used to obtain an N-PAC fraction. Subsequently, silicic acid adsorption chromatography was used to provide separation of the N-PAC into three additional groups according to the functionality of the nitrogen heteroatom in the aromatic ring system: (a) fraction S-1, secondary polycyclic aromatic nitrogen heterocycles (2°-PANH, e.g. carbazole), (b) fraction S-2, amino polycyclic aromatic hydrocarbons (APAH, e.g. aminoanthracene), (c) fraction S-3, tertiary polycyclic aromatic nitrogen heterocycles (3°-PANH, e.g. acridine). Details of these separation methods are outlined and discussed elsewhere (7). Next, the APAH-rich fraction was derivatized with pentafluoropropionic anhydride for further separation and selective detection (8).

TABLE 1

POSSIBLE CLASSES OF POLYCYCLIC AROMATIC NITROGEN HETEROCYCLES

NAME	STRUCTURE	CHEMICAL CLASS	NUMBER OF ISOMERS
Phenanthridine		3°-PANH	8
Carbazole		2°-PANH	6
9-Aminophenanthrene		APAH	8
9-Cyanophenanthrene		CPAH	8
9-Nitrophenanthrene		NPAH	8
		Total	38

A complete separation of the APAH-fluoroamide derivatives from the 3°-PANH was achieved using gel permeation chromatography (GPC). A GPC column (1.5m x 11mm i.d.; Ace Glass Incorporated) was packed with approximately 70g of Bio-Beads SX-12 (200-400 mesh; BIO-RAD) and eluted with methylene chloride (CH_2Cl_2) at a gravity flow rate of approximately 0.5 ml/min. The column effluent was monitored with an Altex Model 330 analytical optical unit (UV absorption; 254 nm filter) equipped with a gravity flow cell. Standard compounds were used to determine optimum elution parameters for complete resolution of the pentafluoropropylamide derivatives (APAH-PFP) and 3°-PANH on this GPC system.

Gas Chromatography and Gas Chromatography-Mass Spectrometry. High resolution capillary column gas chromatography (GC) was performed on a Hewlett-Packard (HP) 5880 gas chromatograph equipped with a 25m x 0.20mm i.d. fused silica capillary column statically coated with SE-52 (0.17 μm film thickness). Samples were introduced with the capillary vaporization injection system operated in the splitless mode. The oven temperature was held for 2 min at 40°C, then temperature programmed at a rate of 4°C/min to 265°C. Both flame ionization (FID) and ^{63}Ni electron capture (ECD) detectors were used.

The gas chromatograph-mass spectrometer system (GC-MS) was a HP 5982A quadrupole mass spectrometer operated in the electron impact mode at 70 eV electron energy. GC conditions were similar to those already described. Spectra were acquired and processed with a HP 5934A data system.

Mutagenicity Assays. Agar plate mutagenicity assays were performed essentially as described by Ames et al. (9). Salmonella typhimurium TA98, specific for frame shift mutations, and dimethyl sulfoxide solvent were used. The assay plates were inverted and incubated for 24-36 h. Revertant colonies were automatically counted with a colony counter.

RESULTS AND DISCUSSION

Extensive investigation and characterization of N-PAC in synthetic fuel products is currently receiving considerable attention. Perhaps the most rapid and complete separation of N-PAC from synfuel products was reported by Schiller and Mathiason (10,11). A modified version of their alumina adsorption column chromatographic method was used to obtain N-PAC fractions in this work. Although the total N-PAC can be rapidly isolated from synfuel materials on neutral alumina, several subclasses of different nitrogen functionality occur together in this fraction. Table 1 shows five different subclasses of N-PAC that may be present in synthetic fuel products. If only the parent compounds and alkylated homologues for the three-ring N-PAC are considered, the number of possible compounds easily exceeds one hundred. The resolution and identification of components in such a complex mixture represents a formidable task even with the use of high resolu-

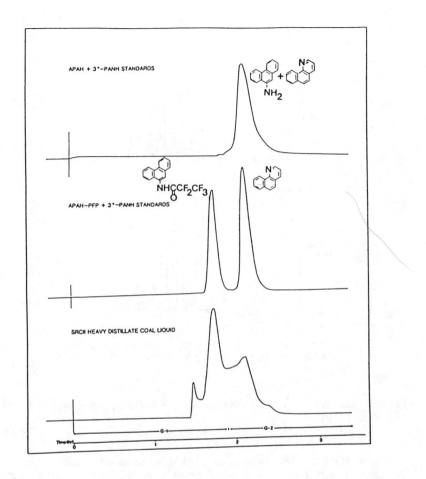

FIGURE 2. Separation of APAH-PFP and 3°-PANH by gel permeation chromatography.

tion capillary GC technology. Another complication is that differences in concentration levels of N-PAC obscures minor classes of nitrogen aromatic compounds. For these reasons and others discussed by Later et al. (7), the N-PAC isolate was subfractionated on silicic acid to give 2°-PANH, APAH-rich, and 3°-PANH fractions.

Fluoroacetylation of the APAH-rich fraction (S-2, see figure 1) not only enables the selective detection of these compounds as described by Later et al. (8) but also confers sufficiently different properties on the APAH so that they can

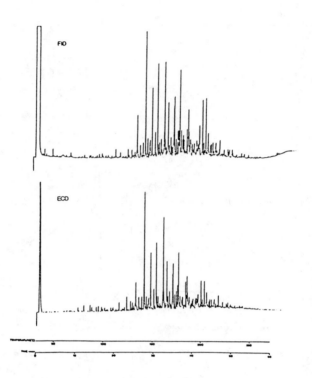

FIGURE 3. Dual gas chromatograms, FID/ECD, of fraction G-1 of the SRC II HD.

be separated by GPC from the 3°-PANH, some of which also elute in this fraction. Figure 2 shows the separation of the APAH-PFP and 3°-PANH in Fraction S-2 of the SRC II HD. GPC on Bio-Beads (a porous styrene-divinylbenzene copolymer) with a CH_2Cl_2 eluent is principally size exclusion chromatography, and the PFP-derivatized APAH elute first due to their larger molecular size as compared to the 3°-PANH. The dual FID/ECD capillary gas chromatograms shown in Figure 3 demonstrates the quality of this separation in that virtually every compound present produced an ECD response. The response ratio for benzo[f]quinoline as compared to 9-aminophenanthrene-PFP derivative is approximately 1:50,000. Therefore, only PFP-derivatized APAH produced an ECD response.

Using the separation scheme outlined in Figure 1, a rapid and detailed characterization of N-PAC in synthetic fuel products is possible. Discrete chemical classes are obtained and identification of individual compounds by retention indices

TABLE 2

WEIGHT DISTRIBUTION AND MUTAGENIC ACTIVITY OF CHEMICAL CLASSES IN SOLVENT REFINED COAL LIQUIDS.

FRACTION	COMPOUND CLASS	WEIGHT PERCENT DISTRIBUTION[a]		SPECIFIC MUTAGENIC ACTIVITY[b]	
		SRC I PS	SRC II HD	SRC I PS	SRC II HD
A-1	ALIPHATIC HYDROCARBONS	19.4%	5.5%	-	-
A-2	NEUTRAL PAC	36.4%	56.5%	-	-
A-3	N-PAC	15.4%	20.0%	64.87 Rev/µg	198.00 rev/µg
A-4	HPAC	16.0%	14.0%	-	-
S-1	2°-PANH	4.5%	11.9%	-	-
S-2	APAH	1.0%	1.7%	267.00 Rev/µg	917.94 Rev/µg
S-3	3°-PANH	2.8%	3.0%	34.04 Rev/µg	162.00 Rev/µg

[a] Average of three determinations.
[b] (1) Specific mutagenic activity in revertants per microgram using TA98. (2) Linear regression methods used to calculate specific mutagenic activity from dose response data. (3) All correlation coefficients for values reported are greater than 0.87. (4) Controls: 1 µg of 1-aminoanthracene = 197 ± 3 rev/plate, 10 µg of benzo[a]pyrene = 214 ± 33 rev/plate.

(12) and capillary GC-MS is straightforward. Another advantage of this method is that the fractions from the separation procedure can be assayed for biological activity, providing better correlations between chemical class type and genotoxicity.

The weight distribution and results from Ames assay using TA98 are presented in Table 2 and Figure 4. For the alumina fractions, activity was essentially confined to the N-PAC fractions for both coal liquids. Furthermore, testing of the silicic acid chromatographic fractions revealed that greater than 70% of the normalized activity for both the SRC I PS and SRC II HD is attributed to the APAH. Lower levels of activity

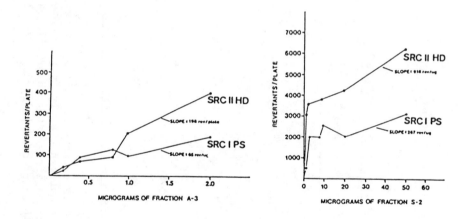

FIGURE 4. TA98 dose response curves for fraction A-3 and fraction S-2 of the SRC I PS and the SRC II HD.

were observed for the 3°-PANH. These results are supported by findings in other studies (13-15) that the APAH are the determinant mutagens in synthetic fuel products. A significant difference in the levels of mutagenic activity between the SRC I PS and the SRC II HD was observed. Possible reasons for this observation can be obtained by comparison of data given in Figure 5 and Table 3. First, the APAH are in lower concentration in the SRC I PS as compared to the SRC II HD. Secondly, the APAH composition in the SRC I PS is principally composed of one- and two-ring compounds, whereas two-, three-, and four-ring APAH were identified in the SRC II HD. Thus, it

NITROGEN-CONTAINING PAH

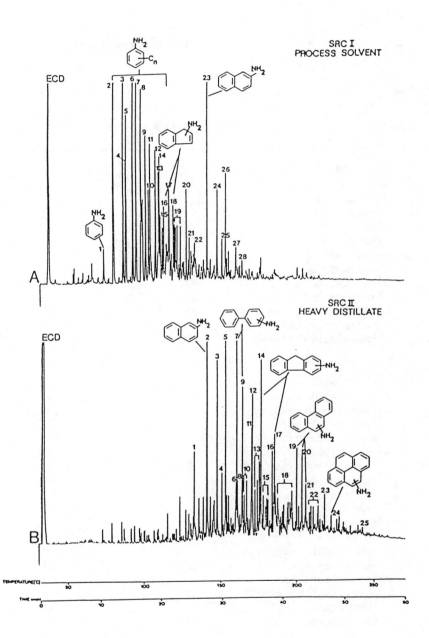

FIGURE 5. ECD gas chromatograms of fraction G-1 from (A) the SRC I PS and (B) the SRC II HD. Peak numbers refer to the compounds listed in Table 3.

TABLE 3

IDENTIFICATION AND SEMI-QUANTITATION OF APAH IN AN SRC I PS AND AN SRC II HD

COMPOUND	PEAK NUMBER	APPROXIMATE[a] CONCENTRATION ($\mu g/g$)
A. SRC I PS		
aniline	1	4.3
C_1-anilines	2,3,4	115.3
C_2-anilines	5-10	119.2
C_3-anilines	11-15	59.9
C_4-aniline	17	18.0
aminoindans	16,18	16.0
C_1-aminoindan	19,20	32.8
C_2-aminoindans	21	5.6
1-aminonaphthalene	22	2.8
2-aminonaphthalene	23	29.9
C_1-aminonaphthalenes	24,25,26	39.1
C_2-aminonaphthalenes	27,28	17.3
B. SRC II HD		
1-aminonaphthalene	1	42.9
2-aminonaphthalene	2	161.4
C_1-aminonaphthalenes	3,4,5	226.9
C_2-aminonaphthalenes	6,8	86.3
3-aminobiphenyl	7	141.5
4-aminobiphenyl	9	95.8
C_1-aminobiphenyls	10-13	327.1
C_2-aminobiphenyls	15	96.5
aminofluorenes	14,16,17	121.2
C_1-aminofluorenes	18	125.3
aminoanthracenes/phen-anthrenes	19,20,21	153.2
C_1-aminoanthracenes/phen-anthrenes	22	104.1
aminofluoranthene/amino-phenylnaphthalene	23	31.0
aminopyrene	24	18.0
aminochrysene	25	18.7

[a] Semi-quantitation was accomplished using 1-aminonaphthalene as an internal standard. The results reported are the average of two determinations.

appears that there is a trend toward higher mutagenic activity with increasing number of rings within the APAH.

ACKNOWLEDGMENT

This work was supported by the Department of Energy, Contract number V-B84843AV, with Pacific Northwest Laboratory. The authors thank Robert M. Campbell for his technical assistance.

REFERENCES

1. Dong, M., Schmeltz, I., LaVoie, E., and Hoffman, D. (1978): Aza-arenes in the respiratory environment: Analysis and assays for mutagenicity. In: Carcinogenesis, Vol. 3: Polynuclear Aromatic Hydrocarbons, edited by P.W. Jones and R.I. Freudenthal, pp. 97-108, Raven Press, New York.
2. Hartung, G.K., and Jewell, D.M. (1962): Carbazoles, phenazines and dibenzofuran in petroleum products, methods of isolation, separation and determination, Anal. Chim. Acta, 26:514-527.
3. Drushel, H.V., and Sommers, A.L. (1966): Isolation and identification of nitrogen compounds in petroleum, Anal. Chem., 38:19-28.
4. Axworthy, A.E., Schnelder, G.R., Shuman, M.D., and Dahan, V.H.; Pb 250373, U.S. NTIS PB Rep., 1976, 35 pp.
5. Lee, M.L., Novotny, M.V., and Bartle, K.D. (1981): Analytical Chemistry of Polycyclic Aromatic Compounds. Academic Press, New York, 446 pp.
6. White, C.M., Schweighardt, F.K., and Schultz, J.L. (1978): Combined gas chromatographic-mass spectrometric analyses of nitrogen bases in light oil from a coal liquefaction product, Fuel Process. Technol., 1:209-215.
7. Later, D.W., Lee, M.L., Bartle, K.D., Kong, R.C., and Vassilaros, D.L. (1981): Chemical class separation and characterization of organic compounds in synthetic fuels, Anal. Chem., 53:1612-1620.
8. Later, D.W., Lee, M.L., and Wilson, B.W. (1982): Selective detection of amino polycyclic aromatic compounds in solvent refined coal. Anal. Chem., 54:(in press).
9. Ames, B.N., McCann, J.M., and Yamasaki, E. (1975): Methods for detecting carcinogens and mutagens with salmonella/ mammalian-microsome mutagenicity test, Mutat. Res., 31:347-364.

10. Schiller, J.E., and Mathiason, D.R. (1977): Separation method for coal-derived solids and heavy liquids, Anal. Chem., 49:1225-1228.
11. Schiller, J.E. (1977): Nitrogen compounds in coal-derived liquids, Anal. Chem., 49:2292-2294.
12. Lee, M.L., Vassilaros, D.L., White, C.M., and Novotny, M. (1979): Retention indices for programmed-temperature capillary-column gas chromatography of polycyclic aromatic hydrocarbons, Anal. Chem., 51:768-774.
13. Wilson, B.W., Petersen, M.R., Pelroy, R.A., and Cresto, J.T. (1981): In-vitro assay for mutagenic activity and gas chromatographic-mass spectral analysis of coal liquefaction material and the products resulting from its hydrogenation, Fuel, 60:289-294.
14. Ho, C.-h., Clark, B.R., Guerin, M.R., Ma, C.Y., and Rao, T.K. (1979): Aromatic nitrogen compounds in fossil fuels - A potential hazard?, ACS, Div. Fuel Chem., Preprints, 24(1):281-291.
15. Later, D.W., Lee, M.L., Pelroy, R.A., and Wilson, B.W. (1981): Mutagenicity of chemical class fractions from a coal liquid, (submitted for publication).

HIGH PERFORMANCE SEMI-PREPARATIVE LIQUID CHROMATOGRAPHY AND LIQUID CHROMATOGRAPHY-MASS SPECTROMETRY OF DIESEL ENGINE EMISSION PARTICULATE EXTRACTS.

S. P. LEVINE*, L. M. SKEWES**, L. D. ABRAMS*, AND
A.G. PALMER, III*
* O.H. Materials Company, Scientific Services Division, Box 551, Findlay, Ohio 45840; ** Ford Motor Company, Scientific Research Staff, Analytical Sciences Dept., Dearborn, Michigan 48121

INTRODUCTION

A variety of techniques have been utilized to identify the constituents and assess the biological activity of the soluble organic fraction (SOF) of diesel engine exhaust particulate material.[1,2] Preparative liquid chromatography (PLC) has been used to fractionate and concentrate mixtures containing polycyclic aromatic hydrocarbons (PAH) prior to analysis.[1-7,9] PLC techniques such as open-column LC[4-7], analytical (Type 1) high performance liquid chromatography (HPLC)[3], and automated coupled-column HPLC[9] have been used for the fractionation of SOF and of fuel feedstocks. Recent trends in HPLC column packings have emphasized the use of microparticles which provide 10,000 - 20,000 theoretical plates/meter for certain analytical and preparative applications.[8,10,11] In the first phase of our study, the use of a silica microparticulate packing in conjunction with a large diameter (8 mm) separation column and commercially available HPLC instrumentation resulted in a maximum allowable sample size of 15-25 mg, with little or no reduction of resolution when compared to analytical HPLC separations. This method fits into the semi-preparative, or Type 2 PLC classification (as defined by Verzele[8]), which is best suited for the fractionation of compounds present in complex mixtures. The solvent program used was designed to provide separation of paraffins from PAH, of nitro-PAH from PAH, and of the oxygenated-PAH sub-fractions.

In the second phase of our study, an analytical column was utilized to investigate the use of HPLC directly interfaced to MS for the one-step analysis of diesel particulate extracts. The use of HPLC-MS for the separation, identification and quantification of complex mixtures has been well-documented in recent years.[12,13] The use of the moving-belt HPLC-MS interface has gained wide acceptance, and significant work has been performed on improving the performance of this interface by design modifications.[14] The application of this improved system to the analysis of diesel particulate extracts was the objective of this study.

MATERIALS AND METHODS

All PAH standards were obtained from Aldrich Chemical Company, except for 1-nitropyrene and dinitropyrene which were obtained from C. King (Michigan Cancer Center) and R. Mermelstein (Xerox Corp.), respectively. All solvents were distilled-in-glass and quality-controlled to ensure less than 5 ppb phthalate-type impurities (Burdick and Jackson).[15,16] Chromatography was done during the PLC phase of the study with a Varian 5060 HPLC using Perkin-Elmer 75 AC UV (254 nm) and Schoeffel FS 970 fluorescence (254 nm excitation, 320 nm cut off) detectors. A Spectrum 101 signal amplifier/noise filter was used to condition the output of the fluorescence detector. Detector monitoring was performed with strip chart recorder and/or with a Varian 401 data system. The chromatographic column was 7.9 mm x 25 cm packed with 10 micron Microporasil (Waters Assoc.).[10] The solvent gradient was as follows: Hold 100% hexane for 5 min after injection, linear change for 5 min to 5% methylene chloride, linear change for 25 min to 100% methylene chloride, hold with 100% methylene chloride for 10 min, linear change for 10 min to 100% acetonitrile, hold with acetonitrile for 5 min, step change to tetrahydrofuran (THF) and hold for 10 min, step change back to acetonitrile and hold for 5 min (end of chromatographic run), step change from acetonitrile to methylene chloride back to hexane at 10 min intervals, re-equilibrate for 15 min before the next injection. The solvent flow rate was 4.5 cc/min. All preparative injections were made using a Waters U6K injector and 200 ul of methylene chloride solutions containing 50-100 mg of SOF/ml.

During the HPLC-MS phase of the study, chromatography was done with an Altex 110A-332 HPLC using an Altex AOU detector at 254 nm. The chromatographic column was an Altex Ultra sphere-Si packed with 5 micron silica. The solvent gradient was as follows: Hold 100% hexane for 5 min after injection, linear change for 15 min to 100% methylene chloride, hold with 100% methylene chloride for 10 min, step change to acetonitrile and hold for 15 min, step change to tetrahydrofuran and hold for 10 min (end of chromatographic run), return to initial conditions as in the PLC runs. The solvent flow rate was 0.7 cc/min. All injections were made using a Rheodyne 7120 valve.

The LC was interfaced to a Finnigan 4000 quadrupole mass spectrometer using a modified Finnigan moving belt interface.[14] The operating conditions for the interface were as follows: drive chamber infrared heater at maximum, vaporization heater 325°C, cleanup heater 350°-375°C, vacuum chamber belt heaters held at temperature sufficient to keep chambers warm.

The mass spectrometer was operated in the positive ion electron impact mode. Analyses were performed using both full scanning and multiple ion detection (MID) capabilities of the mass spectrometer. In the full scanning mode, the mass range 100-500 AMU was scanned in 1.95 seconds with a settle time of 0.05 seconds at the bottom of the range. In the MID mode, up to eight mass ions were scanned with a dwell time of 0.200 seconds per ion. A 0.5 AMU window, which was centered about the exact mass of the selected ion, was utilized to optimize the signal to noise ratio.

The spectrometer was tuned for maximum sensitivity and resolution using perfluorotributylamine. The electron energy was 70.0 eV, the filament emission current was 0.50 amp and the source block was maintained at a temperature of 330°C.

RESULTS AND DISCUSSION

Semi-Preparative Liquid Chromatography (PLC)

The PLC analytical procedure is illustrated in Figure 1 along with the UV and fluorescence profiles obtained from the PLC of a typical Diesel SOF sample. The aliphatic compounds and alkylbenzenes, which make up 40-80% of a typical SOF sample,[1,3] are separated into the HC fraction with 90-95% efficiency (as determined mass spectrometrically). (All fraction designations follow the nomenclature given in references 1 and 3, and are illustrated in Figure 1.) An injection in which there is column overloading, usually by the HC fraction, is characterized by decreased column efficiency, which results in reduced detail in the alpha-fraction envelope, and by the pyrene and benzo[a]pyrene (BaP) components eluting at shorter retention times. The efficient separation of the HC and alpha fractions, which is achieved by the use of the initial hexane hold period, permits an approximately 4 to 5 fold improvement in maximum allowable sample size up to the 20 mg level. Work is underway to increase the maximum allowable sample size to the 50-100 mg range through the use of 20 mm diameter columns.

HPPLC AND HPLC-MS OF DIESEL ENGINE EMISSION PARTICULATE EXTRACTS

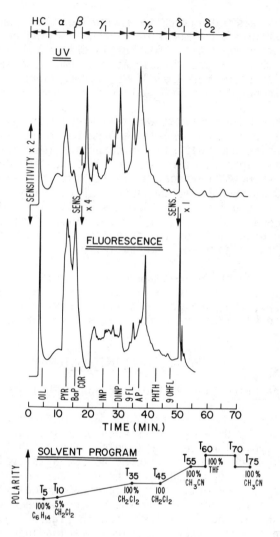

FIGURE 1 Typical PLC Profile of SOF Sample:

Top: Fraction designations (HC through delta)[1,3]
Middle: UV and Fluorescence detector output tracings with <u>approximate</u> retention times of standards noted (Pyr=pyrene, BaP=benzo[a]pyrene, Cor=coronene, 1NP=1-nitropyrene, 9FL=9-fluorenone, DINP=dinitropyrene, AP=aminopyrene, PHTH=bis(2-ethylhexyl) phthalate, 9OHFL=9-hydroxy-flourene).
Bottom: Solvent program and time scale.

The gamma regions of the fractionation scheme are potentially the most important due to the presence of 1-nitropyrene and related nitro-PAH compounds. These substances characteristically have high mutagenic activity on the Ames bioassay test,[1,17] and must therefore be separated cleanly from both the alpha+beta and from the delta fractions, as well as from bis (2-ethylhexyl)phthalate which emerges late in the gamma-2 region. (Phthalates are ubiquitous contaminants of solvents and glassware and will therefore be present under even the most carefully controlled conditions.) The objective of resolving the Ames-active constituents of the mixture from each other and from interfering compounds is accomplished using this PLC technique.

The retention times of components in the gamma-2 and delta regions vary according to the immediate past history of the column. Specifically, the degree of activity of the silica and the silica-solvent interface layers can greatly affect the retentivity of the silica towards polar compounds[18], e.g., 9-hydroxyfluorene. For this reason, a mixture of retention time standards must be chromatographed daily, and the column activity adjusted with dry hexane, tetrahydrofuran, or a solution of 1% water in acetonitrile. (The use of water should be kept to an absolute minimum to avoid column degradation.)

The recovery of the sample after PLC fractionation is not readily determined because of the wide variation in sample type, ranging from 75% by weight unoxidized hydrocarbons (oil and un-burned fuel) in diesel emission samples to 65% by weight highly oxygenated PAH compounds in ambient air samples. Furthermore, three separate but related criteria for recovery should be used: recovery as Ames activity[17], recovery of injected mass, and recovery of individual components. Typically, mass recoveries in the range of 60-80% have been achieved without the final THF elution step. THF elution results in mass recoveries of up to 100% probably due to increased recovery of highly polar species[19]. The use of methanol has been discussed with respect to increased recovery of polar species[1]. This solvent must be used with care to avoid significant changes in retention times of polar components in subsequent runs, and to avoid slow degradation of column efficiency. The recoveries of nitropyrene, 9-fluorenone, and 9-hydroxy fluorene were determined through the injection either of a standard or of a standard mixed with engine oil (a simulated SOF matrix), collection of the appropriate peak fraction, and reinjection of that fraction. In all cases, comparison of the peak areas indicated a 100% recovery (within experimental error) for these PAH derivatives. The questions of recovery of other constituents of diesel SOF, and of the Ames activity, are presently being investigated and will be reported at a later date.

The use of simultaneous UV and fluorescence monitoring of HPLC separations of PAH-containing mixtures has been advocated.[20] This procedure is used in these PLC separations, but interpretation is difficult because of the large number of unidentified constituents in the SOF. However, it was observed that typical compounds that elute in the gamma and delta region have strong UV absorbance characteristics but little or no fluorescence emissions. The fluorescence trace shows more individual peaks in the delta region than does the UV trace. Nitro-PAH standards can be monitored using a UV detector, but there is insufficient selectivity and sensitivity to monitor these compounds during an actual analysis.

High Performance Liquid Chromatography-Mass Spectrometry (HPLC-MS)

The objective of investigating the direct HPLC-MS analysis of diesel particulate extracts was to see if the multi-step PLC-concentration of fractions-MS/GC-MS procedure could be reduced to one step. Less than satisfactory results (detailed in Table 1) were obtained. However, useable mass chromatograms were obtained which, together with retention time information, could serve to allow the confirmation of the presence of certain previously identified components of the extract. Figure 2 illustrates the region of the chromatogram in which 1-nitropyrene elutes. This chromatogram illustrates the 5-10 ng lower limit of detection of HPLC-MS for 1-nitropyrene.

CONCLUSIONS

Semi-preparative (Type 2) high performance liquid chromatography has been applied successfully to the fractionation of the SOF of diesel engine exhaust particulate material. A capacity of 15-25 mg has been achieved along with a clean separation of HC, PAH, mildly oxygenated PAH, and highly polar oxygenated fractions. The use of directly coupled HPLC-MS for this application has been investigated and has been found to be less than satisfactory.

TABLE I

PERFORMANCE OF COUPLED HPLC-MS SYSTEM

Function	Results
Resolution	Degraded somewhat from HPLC-UV results. For many components GC^2 high resolution separation is a requirement.
Sensitivity	Limited by HPLC column capacity (~100 µg) which is limited by maximum flowrate to belt interface (~1 cc/min). Oil fraction of SOF overloads column and degrades separation. Limit of detection of nitropyrene ~7 ng.
HPLC-MS Interface	Requires optimization for each type of compound. Difficult to optimize for PAH-acids.
Other	All HPLC-separated fractions are deposited in source requiring frequent source cleanup. Prefractionation of extract would improve performance for all factors listed above, but would defeat objective of one-step analysis.

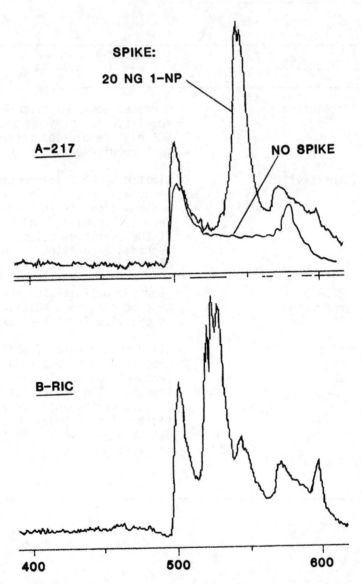

FIGURE 2. HPLC-MS trace of region in which 1-nitropyrene (1-NP) elutes:
A- Mass chromatogram of m/z 217 with and without 1-NP spike, plus diesel particulate extract
B- Reconstructed total ion chromatogram of 1-NP spiked extract

ACKNOWLEDGMENTS

We thank R. Gorse for the samples, M. Paputa and A. Durisin for technical assistance, W. Pierson, D. Schuetzle, R. Gorse, J. Dahlgran, and P. Dymerski for helpful suggestions, and T. Riley for MS analyses.

REFERENCES

1. Schuetzle, D. (1981): Int. J. Environ. Anal. Chem., 9:93.
2. Uden, P.C., Siggia, S., and Jensen, H.B., editors, (1978): Advances in Chemistry Series 170: Anal. Chemistry of Liquid Fuel Sources. Washington D.C.
3. Schuetzle, D. (1981): Proceedings 74th Air Poll. Control Assoc. Conf., Paper 81-56.4.
4. Sorell, R.K. and Reding, R. (1979): Analysis of polynuclear aromatic hydrocarbons in environmental waters by high-pressure liquid chromatography, J. Chromatogr., 185:655-670.
5. Szepsey, L., et al. (1981): Rapid method for the determination of polycyclic aromatic hydrocarbons in environmental samples by combined liquid and gas chromatography, J. Chromatogr., 206:611-616.
6. Funkenbush, E.F., et al. (1979): Soc. Auto. Engin. Congress, Paper 790418.
7. Disango, F.P., Uden, P.C., and Siggia, S. (1981): Isolation of one-ring aromatics in shale oil by Sephadex LH-20 gel, Anal. Chem., 53:721-722.
8. Verzele, M. and Geeraert, E. (1980): Preparative liquid chromatography, J. Chromatogr. Sci., 18:559-579.
9. Ogan, K. and Katz, E. (1980): Perkin-Elmer Analytical Study 78, Perkin-Elmer Corp., Norwalk, CT.
10. Majors, R.E. (1980): Recent advances in HPLC packings and coatings, J. Chromatogr. Sci., 18:488-511.
11. Strack, D.L. and Abbot, M.L. (1979): Waters Bulletin B24, Waters Assoc., Milford, MA.
12. Symposium on LC/MS (1981): In: Proc. of the 29th Ann. Conf. on MS and Allied Topics, pp. 467-486.
13. Symposium on LC/MS (1980): In: Proc. of the 28th Ann. Conf. on MS and Allied Topics, TPMOC8-TPMOC12, RPMP12-RPMP15.
14. Dymerski, P.P. (1980): ibid., RPMP 21.
15. Bowers, W.D., et al. (1981): Trace impurities in solvents commonly used for gas chromatographic analysis of environmental samples, J. Chromatogr., 206:279-288.
16. Huber, J., personal communication (1981): Burdick and Jackson Laboratories, Musekgon, MI.

17. Salmeen, I., et al., personal communication (1981): Ford Motor Co., Dearborn, MI.
18. Scott, R.P.W. (1980): J. Chromatogr. Sci., 18:297.
19. Reference 2, op. cit., Chapter 8.
20. Marsh, S. and Grandjean, C. (1978): Combined ultraviolet absorbance and fluorescence monitoring, J. Chromatogr., 147:411-414.

THE RELATIVE CONTRIBUTION OF PNAs TO THE MICROBIAL

MUTAGENICITY OF RESPIRABLE PARTICLES FROM URBAN AIR

J. LEWTAS, A. AUSTIN, L. CLAXTON, R. BURTON, and R. JUNGERS
U.S. Environmental Protection Agency, Research Triangle
Park, North Carolina

INTRODUCTION

For many years, researchers have considered the respirable particles from urban air as potential contributors to the incidence of lung cancer in urban areas (1-3). Organic extracts of urban particulate matter have been shown to be carcinogenic in animals (4,5) and capable of transforming cells in culture (6,7). Although carcinogenic polynuclear aromatic hydrocarbons such as benzo(a)pyrene were found in these organic extracts, they did not account for all of the carcinogenic activity observed. The Ames Salmonella typhimurium bioassay has recently been used by many investigators to examine the mutagenicity of air particles (8) . These studies have generally utilized the standard high volume air sampler, which uses a glass fiber filter, to collect a broad spectrum of particles in the air, including both the respirable (<5μm) and the larger non-respirable particles. The development of the massive air volume sampler (MAVS) (see Figure 1) by Henry and Mitchell (9) and recent improvements on the sampler now permit the collection of gram-quantities of size-fractionated ambient air particles. The MAVS employs two impactors that collect the 20-to 3.5-μm (Stage I) and 3.5-to 1.7-μm (Stage II) particles, followed by an electrostatic precipitator (ESP) that collects the particles <1.7 μm (Stage III) without filtration (See Figure 1). Recent studies have been conducted to evaluate the use of this sampler for the collection of air particles for analysis of mutagenic activity and polycyclic aromatic hydrocarbon content of the particulate organic matter less than 1.7 μm (10).

The objective of this study was to quantitatively determine the relative contribution of the polynuclear aromatic (PNA) and other chemical class fractions to the microbial mutagenicity of the respirable particles from urban air. The MAVS was utilized to collect size-fractionated ambient air particles at two urban sites. Since the greatest mass and mutagenicity is observed in the organics from the smallest particles (<1.7 μm), these organics were chemically fractionated by class and bioassayed.

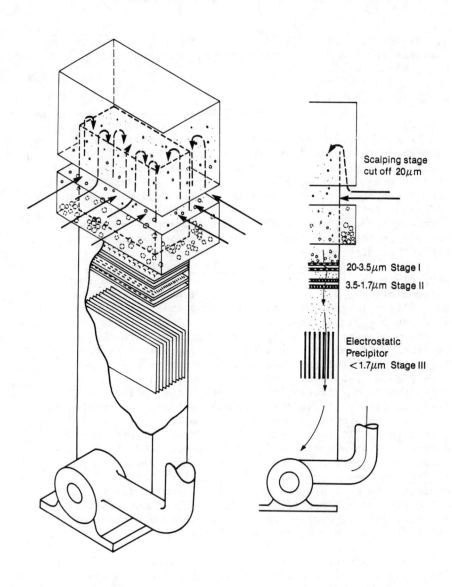

FIGURE 1.   Massive air volume sampler (MAVS).

MATERIALS AND METHODS

Sample Collection, Extraction and Fractionation

MAVS were located in midtown Manhattan, New York City, NY and in an industrial site in Philadelphia, PA. The air particles from each stage of the MAVS were separately scrapped from the collector plates and soxhlet-extracted with dichloromethane (DCM) for 24 h. The DCM-extracted organics from the <1.7-μm particles (Stage III) were fractionated by a solvent partitioning scheme developed for use with air particles and described in detail elsewhere (11). This solvent partitioning fractionation procedure, outlined in Figure 2, results in the following fractions:

1. Acid fraction, which contains both weak and strong organic acids such as phenols and carboxylic acids.
2. Base fraction, which contains organic Bronsted bases such as amines.
3. Non-polar neutral (N-1), cyclohexane soluble fraction, which contains neutral organics less polar than napthalene.
4. Intermediate polarity, PNA fraction (N-2) from a nitromethane wash, which preferentially solubilizes condensed ring aromatics.
5. Polar neutral (N-3), methanol soluble fraction, which contains materials more polar than the PNA hydrocarbons.

The N-2 fraction from the New York sample was further fractionated by preparative high pressure liquid chromatographic separation on silical gel using gradient elution (C. Sparacino, Research Triangle Institute, personal communication). The initial fraction obtained with elution with 2% DCM in hexane contained the PNA hydrocarbons. The other fractions eluted with 50% DCM in hexane, DCM, and 10% methanol in DCM contained more polar neutral compounds.

All fractions were taken to dryness under nitrogen, weighed, and dissolved in dimethylsulfoxide (DMSO) for bioassay.

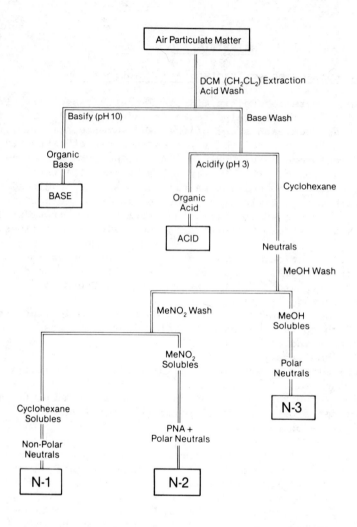

FIGURE 2. Solvent partitioning fractionation scheme.

Microbial Mutagenicity Methods

The Salmonella typhimurium/microsome plate incorporation assay was performed as described by Ames et al. (12) with the following modifications: the standard minimal concentration of histidine was added directly to the plate instead of the overlay, and the plates were incubated for 72 rather than 48 h. The fractions compared at one site were bioassayed simultaneously in the same experiment for each tester strain. Tester strains TA98 and TA100 were employed in these studies, although only the data from TA98 is reported here. The studies were performed with triplicate plates at five doses with and without metabolic activation. Activation was provided by a 9000 x g supernatant of liver from Aroclor 1254-induced CD rats. The data was analyzed using a non-linear model (13) to determine the slope of the dose-response curve. Weighted mutagenicities were determined for each fraction based on the mutagenicity model slope (rev/µg) and the percent of the total mass recovered from each fraction. The weighted mutagenicities were then used to determine the percent of mutagenicity attributed to each chemical fraction.

RESULTS

The mutagenicity of the total extractable organics from Stages I, II, and III, with and without metabolic activation, from New York and Philadelphia were compared to a Los Angeles freeway site previously reported (8) (See Figure 3). At all three sites, the extract from the smaller respirable particles collected at Stage III were significantly more mutagenic than from the larger particles collected at Stage I. At the Philadelphia and Los Angeles sites, the organic matter from the particles <1.7 µm (Stage III) were the most mutagenic. At the New York site, there was no significant difference between the mutagenicities of the total organics from Stages II and III, either with or without activation. Although the organics from the <1.7 µm particles were mutagenic in the absence of the S9 activation system, the addition of S9 increased the mutagenicity of the organics from all three sites.

The percent mass, mutagenicity slope (rev/µg) and percent of mutagenicity contributed by each fraction (Acid, Base, N-1, N-2, N-3) for Philadelphia and New York are shown in Table 1. The acid fraction from both sites represented about 5% of the mass; however, at the Philadelphia site, this fraction was significantly more mutagenic both in the absence and presence of activation, contributing 41% of the direct-acting activity. This contrasts with the results from the New York site, where the acid fraction contributed 23% of the activity without metabolic activation (direct-acting activity). The base fraction contributed less than 1% of the mass in the Philadelphia sample and 4% of the mass in the New York sample. At both sites, the mutagenicity of this fraction was significantly increased with the addition of S9 activation. The Philadelphia base fraction (in the presence of S9) was the most mutagenic fraction (11.5 rev/μg) and contributed 19% of the mutagenic activity. The New York base fraction contributed more to the mass, was less mutagenic (0.1 rev/μg), and contributed less than 1% of the mutagenic activity in the presence of S9 activation.

TABLE 1

RELATIVE PERCENT MICROBIAL MUTAGENICITY IN EACH CHEMICAL FRACTION USING SALMONELLA TYPHIMURIUM, TA98

	Philadelphia			New York		
	% Mass[a]	Slope[b]	% Mutag[c]	% Mass[a]	Slope[b]	% Mutag[c]
Without (-S9) Metabolic Activation						
ACID	4.6	0.8	41.3	5.3	0.2	23.4
BASE	0.7	0.2	1.4	4.2	0.0	0.0
N-1[d]	53.2	0.0	0.0	69.0	0.0	0.0
N-2[e]	9.0	0.05	4.7	6.9	0.3	53.9
N-3[f]	32.4	0.1	52.6	14.6	0.07	22.7
With (+S9) Metabolic Activation						
ACID	4.6	2.2	23.8	5.3	0.2	1.8
BASE	0.7	11.5	19.3	4.2	0.1	0.6
N-1[d]	53.2	0.0	0.0	69.0	0.0	0.0
N-2[e]	9.0	0.06	12.4	6.9	8.1	79.8
N-3[f]	32.4	0.6	44.5	14.6	0.9	17.8

[a] % Mass = % total organic weight.
[b] Slope = measured as rev/μg organics/plate.
[c] % Mutag = % total mutagenic activity.
[d] N-1 = CH Solubles (NPN).
[e] N-2 = MeNO$_2$ Solubles (PNA's & PN).
[f] N-3 = MeOH Solubles (PN).

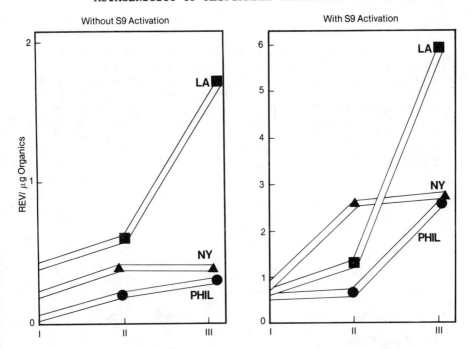

FIGURE 3. Mutagenic activity of size-selected ambient air particles from Los Angeles, Philadelphia, and New York (slope values for TA98 S. typhimurium.)

The non-polar neutral fraction (N-1) at both sites contained most of the mass (53% and 69%). This fraction was not mutagenic at either site in the absence or presence of S9 activation. This is consistent with previous studies that have detected primarily non-mutagenic aliphatic hydrocarbons in this fraction (11). The N-2 fraction, in which the PNA hydrocarbons are found, was significantly more mutagenic in the New York sample than in the Philadelphia sample and contributed nearly 80% of the mutagenic activity. Further fractionation of the N-2 fraction from New York to isolate the PNA hydrocarbons from more polar constituents resulted in only 7% of the mass and 2 to 5% of the mutagenicity isolated in the PNA hydrocarbon fraction (2% DCM in hexane), as shown in Figure 4. The largest sub-fraction, the DCM fraction, contained 89 to 97% of the mutagenicity.

The polar neutral fractions from both sites contained from 15 to 32% of the mass. In the absence of metabolic activation, 53% of the mutagenicity of the Philadelphia sample and 23% of the New York sample was found in the N-3, polar neutral fraction. The addition of metabolic activation increased the mutagenic (slope) activity of both samples. In the presence of metabolic activation, 45% of the mutagenicity of the Philadelphia sample and 18% of the New York sample was found in this fraction. The polarity of the compounds in this fraction make identification by conventional gas chromatography/mass spectrometry difficult.

	2% DCM in Hexane	50% DCM in Hexane	DCM	10% MeOH in DCM
% Mass:	7.0%	0	77.2%	15.8%
Mutagenicity: rev/μg – S9	0.16	—	0.28	0.10
+ S9	1.28	—	5.02	0.14
% Mutagenicity: – S9	4.6%	—	88.9%	6.6%
+ S9	2.2%	—	97.1%	0.6%

(Branching from N-2 into the four fractions above)

FIGURE 4. HPLC fractionation of N-2 New York sample.

DISCUSSION

The MAVS permitted the collection of size-fractionated ambient air particles. Microbial mutagenicity bioassay of the organics associated with each particle size fraction showed that the <1.7 μm in the Philadelphia sample and those <3.5 μm in the New York sample were significantly more mutagenic than the larger particles. This finding is consistent with that previously reported by Teranashi et al. (14) and Pitts et al. (15), both of whom employed cascade impactors attached to a high volume sampler that collects the particles <1.1 μm by filtration.

Chemical fractionation and bioassay of the organics extracted from the particles <1.7 μm in Philadelphia and

New York showed that although there were significant
differences between the two sites, over 80% of the
mutagenicity recovered was accounted for in the acid, N-2
(nitromethane soluble) and N-3 (methanol soluble) fractions
at both sites. Only in the Philadelphia sample did the
base fraction make a substantial (19%) contribution to the
mutagenicity in the presence of metabolic activation. The
least polar neutral fraction, N-1, contained over 50% of
the mass at both sites and was non-mutagenic both without
and with metabolic activation.

Although the two neutral fractions, N-2 and N-3,
containing the PNA hydrocarbons and more polar organics,
accounted for over 50% of the mutagenicity at both sites,
the two sites differed significantly in the relative
mutagenicity observed in these two fractions. In the New
York sample, 80% of the mutagenic activity was in the N-2
fraction, which contains the PNA hydrocarbons. This
fraction showed a significant (27X) increase in activity
with the addition of metabolic activation. Further HPLC
fractionation and bioassay of this fraction showed
significantly more mutagenic activity in the moderately
polar DCM fraction than in the PNA hydrocarbons. In the
Philadelphia sample, more mutagenic activity was observed
in the most polar, N-3, fraction.

These findings show that although the PNA fraction
(N-2) accounts for over 50% of the mutagenicity at both
sites, significant mutagenicity was observed in both the
organic acid and polar neutral (N-3) fractions. Although
significant differences were observed between the two
sites, over 80% of the mutagenicity at both sites can be
accounted for in the same three fractions (acids, N-2 and
N-3). These studies point to the need for identification
of compounds present in the acidic and polar neutral
fractions that are mutagenic and potentially carcinogenic.

ACKNOWLEDGMENTS

The authors acknowledge the assistance of Barbara
Elkins, Northrop Services Inc., for assisting in the
editing and preparation of this manuscript.

REFERENCES

1. Kotin, P., and Falk, H.L. (1963): Atmospheric factors in pathogenesis of lung cancer, Adv. Cancer Res., 7:475-514.
2. Carnon, B.W. and Meier, P. (1973): Air pollution and pulmonary cancer, Arch. Environ. Health, 22:207.
3. Hoffman, D. and Wynder, E.L. (1976): Environmental respiratory carcinogenesis. In: Chem. Carcino., ed. by C.E. Searle, pp. 324-365, American Chemical Soc., Washington, D.C.
4. Leiter, J., Shimkin, M.D. and Shear, M.J. (1942): Production of subcutaneous sarcomas in mice with tars extracted from atmospheric dusts, J. Natl. Cancer Inst., 3:155-165.
5. Hueper, W.C., Kotin, P., Tabor, E.C., Payne, W.W., Falk, H. and Sawicki, E. (1962): Carcinogenic bioassays on air pollutants, Arch. Pathol., 74:89-116.
6. Freeman, A.E., Price, P.J., Bryan, R.J., Gordon, R.J., Gilden, R.V., Kelloff, G.J. and Heubek, R.J. (1971): Transformation of rat and hamster embryo cells by extracts of city smog. Proc. Natl. Acad. Sci. USA, 68:445-449.
7. Gordon, R.J., Bryan, R.J., Rhim, J.S., Demoise, C., Wolford, R.G., Freeman, A.E., and Huebner, R.J. (1973): Transformation of rat and mouse embryo cells by a new class of carcinogenic compounds isolated from city air. Int. J. Cancer, 12: 223-227.
8. Lewtas Huisingh, J. (1981): Bioassay of particulate organic matter from ambient air. In: Short-Term Bioassays in the Analysis of Complex Environmental Mixtures 1980, edited by M. Waters, S. Sandhu, J. Lewtas Huisingh, L. Claxton and S. Nesnow, pp. 9-20, Plenum Press, New York.
9. Henry, W.M. and R.I. Mitchell. Development of a large sample collector of respirable particulate matter. EPA-600/4-78-009. U.S. Environmental Protection Agency, Research Triangle Park, North Carolina, 1978, pp. 1-42.
10. Jungers, R., Burton, R., Claxton, L. and Lewtas Huisingh, J. (1980): Evaluation of collection and extraction methods for mutagenesis studies on ambient air particulate. In: Short Term Bioassays in the Analysis of Complex Environmental Mixtures 1980, edited by M. Waters, S. Sandhu, J. Lewtas Huisingh, L. Claxton and S. Nesnow, pp. 45-66, Plenum Press, New York.

11. Kolber, A., Wolff, T., Hughes, T., Pellizzari, E., Sparacino. C., Waters, M., Lewtas Huisingh, J. and Claxton, L. (1980): Collection, chemical fractionation, and mutagenicity bioassay of ambient air particualte. In: Short Term Bioassays in the Analysis of Complex Environmental Mixtures 1980, edited by M. Waters, S. Sandhu, J. Lewtas Huisingh, L. Claxton and S. Nesnow, pp. 21-44, Plenum Press, New York.
12. Ames, B.N., McCann, J. and Yamaski, E. (1975): Methods for detecting carcinogens and mutagens with the Salmonella/mammalian microsome mutagenicity test, Mutat. Res., 31: 347-364.
13. Stead, A.G., Hasselblad, V., Creason, J.P., and Claxton, L. (1981): Modeling the Ames test, Mutat. Res., 85: 13-27.
14. Teranashi, K., Hamada, K., Tekeda, N., and Watanabe, H. (1977): Mutagenicity of the tar in air pollutants. Proc. 4th Int. Clean Air Congress, Tokyo, pp. 33-36.
15. Pitts, J.N., Van Cauwenberghe, K.A., Grosjean, D., Schmid, J.P., Fitz, D.R., Belser, W.L., Knudson, G.B. and Hynds, P.M. (1978b): Chemical and microbiological studies of mutagenic pollutants in real and simulated atmospheres. In: Application of Short-term Bioassays in the Fractionation and Analysis of Complex Environmental Mixtures. Edited by M.D. Waters, S. Nesnow, J.L. Huisingh, S.S. Sandhu, and L. Claxton, pp. 353-379, Plenum Press, New York.

EFFECT OF CATALYST ON PAH CONTENT OF AUTOMOTIVE DIESEL EXHAUST PARTICULATE

IVAN E. LICHTENSTEIN
Johnson Matthey, Inc.
Malvern, PA 19355

INTRODUCTION

The increasing popularity of the diesel vehicle as a fuel-efficient means of transportation has focused attention on an undesirable feature of its operation: the generation of large amounts of exhaust particulate. Diesel particulate has been found to contain more PAH, specifically benzo(a)pyrene (BaP), than does particulate from the corresponding catalyst-equipped gasoline-powered automobile (1), and is therefore of considerable environmental concern. Legislation to sharply reduce the permissible level of diesel particulate emissions is currently being debated.

Johnson Matthey is developing a device to control diesel particulate emissions. An integral part of the exhaust manifold, it traps particulate, and, by the action of its platinum-group metal catalyst, enables the combustion of trapped hydrocarbons (as well as of gaseous HC and CO) to proceed at the relatively low temperatures prevailing in diesel exhaust.

The JM catalyst/trap (JMCT) concept has been tested on a diesel vehicle at Southwest Research Institute (SWRI). Comparisons of total particulate emissions using the original (baseline) and JMCT-equipped vehicle were carried out using various test cycles and collection filter types. Of great interest was whether, in addition to greatly reducing total diesel particulate relative to baseline (2), JMCT would specifically remove PAH. Consequently, a set of particulate laden filters from the SWRI tests were analyzed in our laboratory. The results of this study are detailed below.

EXPERIMENTAL

Apparatus and Chemicals

An HPLC system consisting of a dual pump chromatograph (equipped with a solvent switching device), a fluorescence detector, and a data collector, similar to that described by Ogan et al. (3), was used. Silica and ODS columns were set up for normal and reverse phase chromatography, respectively. HPLC-grade solvents were used throughout. PAH standards were obtained locally.

HPLC Parameters

Normal Phase Chromatography. A solvent program at constant flow of 1.5 ml/min was chosen. Parameters were as follows: T equilibrium, 10 minutes at 90% hexane, 10% dichloromethane (DCM); T_1, 5 minutes at 10% DCM; T_2, 5 minutes linearly to 100% DCM; T_3, 15 minutes at 100% DCM or methanol (MeOH): T_{3D}, 6 minutes-DCM, T_{3M}, 4 minutes-MeOH, T_{3D1}, 5 minutes-DCM; T_4, 5 minutes linearly to 10% DCM.

DCM was chilled before use to avoid formation of gas bubbles in the pump and detector while using this volatile solvent. A 500 μl sample loop was used to provide sufficient sample extract for fractionation and subsequent reverse phase separation and quantitation of individual PAHs. The fluorescence detector was set at low sensitivity in light of this. Wavelength settings of λex = 300 nm, λem = 400 nm at 10 nm slit widths were used to monitor PAH groups eluted by this procedure.

Reverse Phase Chromatography. A flow program, with solvent composition 97% MeOH, 3% H_2O was chosen for this work. Parameters were as follows: T equilibrium, 5 minutes at 0.3 ml/min; T_1, 10 minutes at 0.3 ml/min; T_2, 5 minutes linearly to 1.0 ml/min; T_3, 10 minutes at 1.0 ml/min. A 20 μl sample loop was used for extracts and standards. Fluorescence settings of λex = 305 nm, λem = 430 nm at "normal" sensitivity were used to obtain chromatograms representative of the broad spectrum of PAH's found in the non-polar fractions separated by normal phase chromatography. Settings of λex = 383 nm, λem = 430 nm were used for the selective determination of BaP in these fractions (4).

Procedure for Sample Analysis

The 47 mm diameter filters (and appropriate filter blanks) were Soxhlet-extracted for 6 hours with DCM. Extracts were evaporated at very low heat, re-dissolved in a few ml of DCM, and transferred in several portions, via a syringe containing a 0.5 micron filter, to 10 ml volumetric flasks. After dilution to volume with filtered DCM, the volumetrics were stored in a freezer pending analysis.

Normal phase chromatography was carried out on 500 μl aliquots of sample extracts. The fractions eluting between 1 and 4 minutes were collected in vials, and evaporated to dryness under nitrogen. Residues were reconstituted to 500 μl with acetonitrile. Aliquots of 20 μl or less were then put through the reverse phase program.

RESULTS AND DISCUSSION

Normal Phase Chromatography

Representative chromatograms are shown in Figure 1. The mix used for 1A contained six of the PAHs cited in the EPA's priority list: naphthalene, phenanthrene, pyrene, benzo(b) fluoranthene (BbF), BaP, and benzo (g,h,i) perylene (ghi). All of these non-polars eluted well within the 1-4 minute period used for collection of sample fractions taken for later reverse phase analysis. The JMCT-treated extract (1B) shows a small to moderate peak at 1.4 minutes, and a very small one at 8.8 minutes. There is also a small peak at 26 minutes. The baseline extract (1C) has a large fluorescence peak at 1.4 minutes, a small peak at 9.1 minutes, and shallow, broader emission in the 11 minute region. A sharp, moderate-sized peak at 26 minutes followed by broader emission centering at 27 minutes completes the baseline spectrum.

For purposes of comparison, fluorescence in the 0-13 minute region will be considered due to non-polar and "transitional" PAH compounds, and will be designated the "N" region, to avoid confusion with the "α" and "β" divisions described by other workers (5). Species eluting after 13 minutes by this program, the polar and highly polar PAH derivatives, will be designated "P" ("γ" and "δ" have been used to describe sub-fractions in this region).

The integrated areas corresponding to the N and P regions for JMCT and baseline samples can be compared to obtain an approximate measure of the ability of the catalyst trap to reduce the PAH content of diesel particulate. These integrated fluorescence intensities, corrected for blank emissions and for differing exhaust tunnel dilution factors to bring the test results to a common basis for comparison, are shown in Table 1. Per cent reductions of N and P region PAHs by JMCT relative to baseline, calculated from the fluorescence data, are likewise shown. Results for Pallflex-collected particulate indicate that JMCT has eliminated better than 95% of the non-polar PAHs and 90% of their polar derivatives, on average, from the corresponding baseline samples. Clean-up for the various portions of the FTP cycle was best for the hot transient (HT) part, slightly lower for the cold transient (CT) and cold stabilized (CS) parts. JMCT likewise gave excellent PAH abatement for the steady-state (85 kph) and fuel economy (HFET) cycles. Results on glass and fluoropore media for the composite four-bag FTP test are consistent in terms of per cent PAH reduction with the FTP portions run separately on Pallflex. Overall "N" contents were significantly lower than on Pallflex, but this was not true for the polars, and this may reflect higher reactivity of PAHs trapped on glass and fluoropore. Lee et al. (6) have called attention to the problem of substrate-induced PAH degradation.

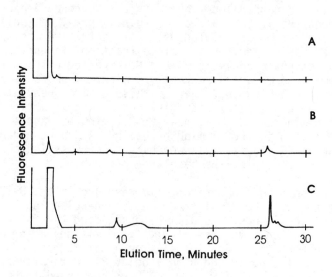

FIGURE 1. Representative Normal Phase Chromatograms

A. Synthetic PAH Mix - Naphthalene, Phenanthrene, Pyrene Benzo (B) Fluoranthene (BBF), Benzo (A) Pyrene (BAP), Benzo (G, H, I) - Perylene (GHI).

B. JMCT - Treated Exhaust, Hot Transient Portion of Federal Test Procedure Cycle (FTPht), Pallflex Filter.

C. Baseline Exhaust, Same Test Cycle and Filter as B.

TABLE 1

NORMAL PHASE CHROMATOGRAPHY OF DIESEL PARTICULATE: REDUCTION OF NON-POLAR (N) AND POLAR (P) PAHs[a] BY JOHNSON MATTHEY CATALYST/TRAP

Filter Type	Test	[b]Fluorescence Peak Areas				Percent Reduction JMCT vs. Baseline		
		Baseline		JMCT				
		N	P	N	P	N	P	[c]ΣN + P
Pallflex	FTPct	98.1	12.1	4.4	1.8	96	85	94
Pallflex	FTPcs	92.8	8.4	7.0	1.0	92	88	92
Pallflex	FTPht	119.9	13.4	4.0	<0.2	97	>98	97
Pallflex	HFET	47.7	11.6	0.7	<0.3	99	>97	98
Pallflex	85 kph	40.2	5.0	1.1	0.1	97	98	97
Glass	FTP4b	31.6	10.4	1.5	0.6	95	94	95
Fluoropore	FTP4b	30.5	23.5	1.5	1.2	95	95	95

[a] Non-Polars: 0-13 Minute Region of Chromatograms
Polars: >13 Minute Region of Chromatograms

[b] Net vs. Appropriate Filter Blanks and Corrected for Particulate Dilution Factors
(Ratio of Total Particulate Collected, in MG/KM, to Particulate on Test Filter, in MG)

[c] Assumes Equal Weight to N and P Areas

Reverse Phase Chromatography

Chromatograms for the PAH synthetic, and for the 1-4 minute fractions of the same JMCT/baseline samples whose normal phase chromatograms are shown in Figure 1, are given in Figures 2 and 3. Data is summarized in Table 2.

Total "N" Emissions. Figure 2A shows that representative PAHs give satisfactorily resolved peaks using the flow program chosen for this work. Known amounts of these PAHs taken through the normal phase, evaporation, and reverse phase steps gave a similar spectrum and calculated recoveries of better than 90% each. Elution times were: pyrene - 7.8, BbF - 10.3, BaP - 11.9, ghi - 13.3 minutes under these conditions. Naphthalene does not appear due to its weak fluorescence at the λex, em settings chosen, and phenanthrene is barely observed, just before the pyrene peak. A run at λex = 280 nm, λem = 340 nm gave a moderate-intensity, broadly tapering peak at 6.7 minutes, corresponding to the naphthalene-phenanthrene elution region. The flow program serves to delay the elution sequence in order to obtain a better separation of this emission region, which shows up strongly in the samples (Figs 2B and 2C), from peaks in the pyrene region.

The JMCT sample, Fig 2B, shows small broad emissions peaking at 6.6 and 7.8 minutes. The 7.8 minute peak corresponds to the pyrene peak for the standard mix, and it is reasonable to assume that pyrene is its main component - though with contributions from species such as fluoranthene or chrysene resulting in broadening. Of the later, very small peaks, only those at 13.4 minutes (in the ghi region) and another at 17.5 minutes were recognized by the data processor for integration.

A far richer spectrum is observed for the baseline sample, Fig 2C. Major peaks at 6.6 and 7.9 minutes are followed by more or less distinctive peaks at 10.3, 11.2, 11.9, 13.3, 15.7, and 17.4 minutes. The 7.9 peak can again be associated with pyrene, though the later appearing shoulder shows that there are other PAHs in this region. The small peak at 10.3 coincides with the BbF region. The larger peak at 11.2 minutes may be benzo(k)fluoranthene or other PAH found to elute between BbF and BaP in programs of this kind (3), (4), (7). The smaller peak at 11.9 minutes correlates with the BaP peak of the synthetic mix. Its identity was confirmed by spiking this sample with BaP and repeating the run; as the dotted line in Fig 2C shows, the spike peak appears where the presumptive BaP peak in the original was found. Other substantial PAH peaks are noted in the ghi region and even further out (less polar, possibly higher-order ring PAHs).

BaP-Selective Scans. Figure 3A shows the dramatic effect on the chromatogram for the PAH synthetic of changing λex from 305 to 383 nm. The pyrene and BbF peaks are greatly reduced, while BaP and ghi are markedly increased, BaP by more than 2-fold. Little effect is noted on the JMCT sample's spectrum, Figure 3B, except to damp down the peaks formerly observed at 6.6 and 7.8 minutes. Again, peaks further out are not strong enough to be recognized for integration by the data collector. A much clearer picture emerges in the baseline sample's spectrum, Fig 3C. Activity at 6.6 minutes is somewhat reduced, that in the 8 minute region greatly reduced as compared with Fig 2C. The major peak is now at 11.9 minutes, as expected for BaP from the synthetic run. The peak at 13.3 minutes, likely associated with ghi, is also enhanced.

PAH Reduction by JMCT. Reverse phase chromatography thus confirms the initial normal phase finding that JMCT significantly reduces the amount of PAH in diesel particulate (Fig 2), and further suggests that this PAH abatement applies to compounds of particular concern with respect to mutagenic potential-specifically, BaP (Fig 3). It is therefore of

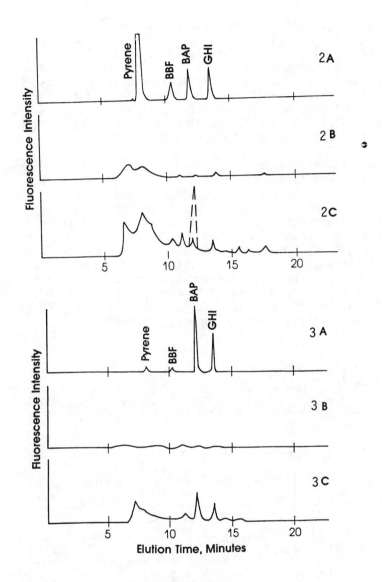

FIGURES 2 AND 3. Representative reverse phase chromatograms of 1-4 minute normal phase fractions.
Figure 2: λex = 305 nm, λem = 430 nm
Figure 3: λex = 383 nm, λem = 430 nm
A. Synthetic PAH mix.
B. JMCT-treated exhaust, FTPht cycle, Pallflex filter
C. Baseline exhaust, same test cycle and filter as B

interest to compare the integrated fluorescence signals for BaP and for total "N" fraction emissions of the JMCT and baseline samples. These calculations, as total fluorescence per test km, are shown in Table 2. As the comparative fluorescence data indicates, JMCT cleaned up more than 90%, on average, of the engine-out BaP from the cold transient, cold stabilized, and hot transient portions of the FTP cycle. Overall "N" fraction PAH reduction was better than 95%, consistent with the normal phase results (Table 1).

From the fluorescence of known amounts of BaP, the sample fluorescence data can be translated directly into ug BaP per test km - probably the most significant measure of PAH abatement by the JMCT-equipped vehicle. On a time-weighted basis, BaP generation using JMCT is well under 1 µg/km, whereas baseline averages more than 5 µg/km for the FTP cycle. Finally, BaP as a percentage of the original particulate can be calculated. As shown in the table, BaP in ppm for the JMCT samples is less than or equal to those for the corresponding baseline samples. And this means that reduction of PAH by JMCT at least keeps pace with its overall reduction of diesel particulate - a very encouraging result.

TABLE 2

REVERSE PHASE CHROMATOGRAPHY OF DIESEL PARTICULATE. REDUCTION OF BAP AND TOTAL "N" FRACTION EMISSIONS BY JOHNSON MATTHEY CATALYST TRAP

[a]Test	Type	[b]Fluorescence Peak Areas $\times 10^{-5}$		% Reduction JMCT vs. Baseline		BAP Generated	
		BAP	[c]N	BAP	[c]N	µG Per Test KM	As PPM of Particulate
FTPct	Base	0.36	4.22	—	—	12	36
FTPct	JMCT	0.05	0.15	86	96	1.7	35
FTPcs	Base	0.025	3.26	—	—	0.8	3.5
FTPcs	JMCT	<0.002	0.01	>90	99	<0.1	<2.7
FTPht	Base	0.20	5.57	—	—	6.7	22
FTPht	JMCT	<0.002	<0.005	>99	>99	<0.1	<2.9

[a] All on Pallflex 47 mm Diameter Filters
[b] Corrected for Blank and for Analytical and Particulate Tunnel Dilutions, to Give Total Calculated Fluorescence Per Test KM
[c] Non-Polars From 1-4 Minute Fraction, Normal Phase Chromatography

Further Work Planned

A vehicle study of 80,000 km is under way to assess the durability of JMCT's diesel particulate abatement. Experiments similar to those described above will be carried out to determine PAH reduction by JMCT as a function of catalyst aging. This work will be extended to characterization of the polar PAH fractions, specifically the nitro PAHs which have been shown to constitute a significant part of the direct-acting mutagens in diesel particulate (5), (8). And preparative HPLC, culminating in bio-assay tests of mutagenicity, will be carried out on pertinent samples.

REFERENCES

1. Springer, K. J., and Baines, T. M. (1979): Emissions from diesel versions of production passenger cars. In: The Measurement and Control of Diesel Particulate Emissions, Progress in Technology Series Number 17, pp. 207-231, Society of Automotive Engineers (SAE), Warrendale, PA.
2. Enga, B. E., Buchman, M. F., and Lichtenstein, I.E.: Catalytic control of diesel exhaust emissions. Manuscript in preparation for presentation at SAE's International Congress and Exposition, Detroil, 1982.
3. Ogan, K., Katz, E., and Slavin, W. (1979): Determination of polycyclic aromatic hydrocarbons in aqueous samples by reversed-phase liquid chromatography, Anal. Chem., 51: 1315-1320.
4. Swarin, S. J., and Williams, R. L. (1980): Liquid chromatographic determination of benzo(a)pyrene in diesel exhaust particulate: Verification of the collection and analytical methods. In: Polynuclear Aromatic Hydrocarbons: Chemistry and Biological Effects, edited by A. Bjørseth and A. J. Dennis, pp. 771-790, Battelle Press, Columbus.
5. Schuetzle, D., Lee, F.S.-C., Prater, T. J., and Tejada, S. B. (1981): The identification of polynuclear aromatic hydrocarbon (PAH) derivatives in mutagenic fractions of diesel particulate exhaust, Intern. J. Environ. Anal. Chem., 9: 93-144.

6. Lee, F. S.-C., Pierson, W. R., and Ezike, J. (1980): The problem of PAH degradation during filter collection of airborne particulates - An evaluation of several commonly used filter media. In: Ref. 4, pp. 543-563.

7. Choudhury, D. R. (1981): Applications of on-line high performance liquid chromatography/rapid scanning ultraviolet spectroscopy to characterization of polynuclear aromatic hydrocarbons in complex mixtures. In: <u>Chemical Analysis and Biological Fate: Polynuclear Aromatic Hydrocarbons,</u> edited by M. Cooke and A. J. Dennis, pp. 265-276, Battelle Press, Columbus.

8. Schuetzle, D., Prater, T., Riley, T., Durisin, A., and Salmeen, I. (1980): Analysis of nitrated derivatives of PAH and determination of their contribution to Ames assay mutagenicity for diesel particulate extracts. Presented at Fifth International Symposium on Polynuclear Aromatic Hydrocarbons, Columbus, October 28-30, 1980.

MEASUREMENT OF POTENTIALLY HAZARDOUS POLYNUCLEAR AROMATIC HYDROCARBONS FROM OCCUPATIONAL EXPOSURE DURING ROOFING AND PAVING OPERATIONS (1)

MURUGAN MALAIYANDI*†, A. BENEDEK**, A.P. HOLKO, AND J.J. BANCSI**, *Bureau of Chemical Hazards, Environmental Health Directorate, Health Protection Branch, Health and Welfare Canada, Tunney's Pasture, Ottawa, Canada K1A 0L2; **Zenon Environmental Enterprises Ltd., Hamilton, Ontario, Canada, L8P 3V3; † To whom correspondence should be addressed.

INTRODUCTION

Since the enactment of the 1958 Amendment (2) to the Food, Drug and Cosmetic Act of the United States, with particular reference to carcinogenic compounds, considerable interest has been evinced toward the analysis and survey of such compounds in various consumer products and in the environment. Particularly, attention has been focused on the polynuclear aromatic hydrocarbons (PAHs), since many members of this class exhibit carcinogenic properties (3) and because of their ubiquity as a consequence of wide-spread utilization of coal, petroleum and other fossil energy sources.

The mortality rate from lung cancer has increased about 30 times since 1900 (4). The etiological studies on lung cancer incidence have indicated air pollution as one of the causes and implicated PAHs as air pollutants capable of causing lung cancer (4). These PAHs are emitted into the atmosphere and the compounds having molecular weight (between 120 and 300 Daltons) are preferentially bound to airborne particulates (< 1 um) that can be inhaled and deposited in lungs of humans (5,6). Further, materials like coal tar pitch and asphalt used in roofing and paving operations are known to contain numerous PAHs and these materials could pose a potential hazard to the personal health of individuals working with such materials.

Asphalt is a component of crude oil which on fractional distillation leaves a heavy residue called straight run asphalt and has a low PAH content. However, straight run asphalt is subjected to a softening process by air blowing at $200°$-$280°C$ to attain higher penetration power. By this process, components in the residue might undergo oxidation and polymerization which might increase the PAH content in asphalt.

Coal tar pitch is a complex mixture of PAHs which includes a myriad of toxic substances listed as priority pollutants by the Environmental Protection Agency (7).
Coal tar pitch is shown to contain ca. 2000 ppm of the carcinogen, benzo(a)pyrene (BaP) (8,9) whereas asphalt contains typically 1 ppm of the above carcinogen (9). The enhanced levels of PAHs in coal tar pitch can be explained on the basis of two competing processes, namely pyrolysis and pyrosynthesis occurring during coking process of coal leading to coke and coal tar pitch, although bituminous coal itself is known to contain a number of PAHs (10). The bituminous matter at high temperatures is partially cracked to smaller unstable molecules which recombine to form thermodynamically more stable larger molecules of PAH (11). These PAHs during asphalting and roofing operations find themselves in the form of particulates or in gaseous state in ambient air.

Most air sampling techniques for PAHs make use of glass fiber filters and silver frits in Hi-Volume samplers to pass large volumes of air at a constant flow rate. Depending on the experiment protocol, the duration of air sampling may vary from several hours to several months. Van Vaeck and co-workers (12) have demonstrated that the PAH profile by this technique will not be a representative sample as to the actual PAH content at the time of sampling and inhaled by human beings. It has also been shown that the equilibrium vapor concentrations of some PAHs are significantly high (13,14) and that they are not retained by glass fiber filters. Therefore, it is conceivable that, even though these PAHs are adsorbed on particulates and in spite of their high boiling points, these PAHs evaporate even at room temperatures and remain in the gaseous phase. Hence, during long term sampling, even after deposition on the filters, the PAHs may be "blown off" to some degree depending upon the ambient temperature. Further, Pitts and co-workers (15) observed that, during large volume air sampling using glass fiber filters, BaP has been converted to its K-region epoxide which has been identified as a DNA-binding metabolite of BaP in several biological systems (16).

Worker exposure to PAHs may occur during the preparation of coal tar pitch melt and at the time of application of hot pitch or hot asphalt during roofing and paving activities. Exposure of workers to the fumes from coal tar pitch and asphalt in open atmosphere paving and roofing operations has not been thoroughly investigated and documented. The intent of this investigation is to evaluate exposure of roofing and paving crew members to PAH-laden fumes derived from coal tar pitch and asphalt using personal sampling techniques (17). In order to trap the PAHs "blown off" from the filtering train, a Tenax GC packing (18) was incorporated in the sampling cartridge along with fiberglass filters and backstop membranes. This system provides quantitative data on PAH content in particulates and in vapor phase during a normal 8-hour work period.

MATERIALS AND METHODS

PAH Compounds - Pyrene (P), Perylene (Per), Benzo (g,h,i,) perylene [B(ghi)Per] and Indeno (1,2,3,-cd) pyrene (Ip) from Aldrich Chemical Co. Ltd., Milwaukee, Wis.; Benzo (a) pyrene (BaP), 7,12-Dimethylbenz (a) anthracene (DMBA), and Dibenz (a,h) anthracene [DB (ah) A] from Eastman Kodak Co., Rochester, N.Y.; Chrysene (Chry), and Fluoranthene (F) from J.T. Baker Co. Ltd. Phillipsburg, N.J.; Benz (a) anthracene (BaA) from General Biochemicals Inc. Chagrin Falls, OH.; and Benzo (k) fluoranthene (BkF) from Air Pollution Directorate, Environment Canada, Ottawa, Ontario, were obtained.

Florisil - Mandel Scientific Co. Ltd., Rockwood, Ontario.

Tenax GC - (60-80 mesh) Chromatographic Specialties, Ltd., Brockville, Ontario.

Fiber Glass Filter - (37 mm dia), binderless type Millipore AP 4000 - 37 - 305 and Cellulose Back Stop Support - Millipore A 100 - 3700, Millipore Corporation, Missisauga, Ontario.

Silver-frit membrane - 0.8 um, porosity, Selas Corporation, Flotronic Division, Huntington Valley, PA.

Cyclohexane - Glass-distilled, Caledon Laboratories Ltd., Georgetown, Ontario.

Acetonitrile - Spectrograde, Analy. Chem. Ltd. Markham, Ontario.

Sampling Pumps - DuPont Model P-2500, DuPont of Canada Instrument Division, Toronto, Ontario or Bendix Model BDX-44, Levitt Safety Ltd., Toronto, Ontario.

High Pressure Liquid Chromatograph - Varian Model 8500 with Vydac 201 TP reverse phase column and Turner Model 111 Fluorometer.

L.C. Conditions - Eluent - acetonitrile-water (75:25 v/v); Pressure - 105 kg/cm^2; Flow-rate 1 ml/min; Sample Size - 50 ul; Injection Loop - 50 ul; Detector - λ ex 250 nm, λ em 370 nm for F, P, BaA, Chry, DMBA, Per, BkF, BaP and λ ex 285 nm, λ em 418 nm for B(ghi)Per, DB(ah)A and IP.

Sampling of Raw Materials - About 70 g of hot mix paving asphalt or about 0.5 g of roofing asphalt or coal tar pitch was accurately weighed, Soxhleted with cyclohexane for 48 hrs. The extracts were concentrated under subdued light and chromatographed on a florisil column (recovery levels have been pre-determined) using cyclohexane eluent according to the method of Smillie et al. (18). After careful concentration of the eluate to about 2 ml, the residue was taken twice in 2 ml of acetonitrile and evaporated to near dryness. The concentrate was redissolved in 1 ml of acetonitrile and this solution was used for analysis.

Air Sampling Train Assembly:

Two types of sampling train assemblies were used. The type I, shown in Figure 1 consists of an inlet containing a binderless fiber glass filter followed by cellulose back stop and a 10 cm deep Tenax GC packing (19) backed by a cellulose backstop to retain Tenax packing. The type II sampler differs from type I by that a 0.8 um silver frit membrane is placed immediately after the front cellulose back stop and before the Tenax GC packing followed by the back stop.

The sampling trains were connected to DuPont P-2500 or Bendix BDX-44 personal air sampling pumps by means of a 30 cm thick-walled tygon tubing. Prior to air sampling, the filter assemblies were Soxhleted with cyclohexane for 48 hrs, heated to about 100°C for fifteen minutes and cooled in a vacuum desiccator.

The sampling train type II was used at work sites V & VIII and type I was used at the remaining sampling sites. The sampling trains of type I were extracted as a unit to obtain a composite value for PAHs in particulates and gaseous PAHs. The sampling train of the type II was dismantled to separately Soxhlet the Tenax GC packing and the remaining train to obtain PAH content in the gaseous phase and in particulates respectively. The extracts were concentrated as described for concentrating cyclohexane eluate from florisil column under "Sampling of Raw Materials".

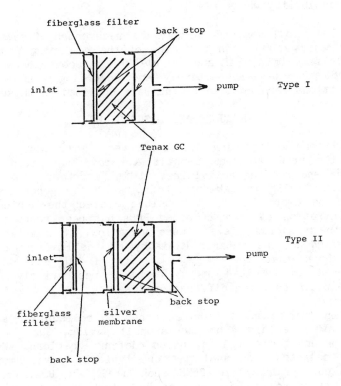

FIGURE 1. Sampling Train Assembly

Analysis:

The analysis of PAHs was carried out by high pressure liquid chromatographic techniques according to the method of Smillie et al. (18). The separation of synthetic mixtures of standard PAHs, the air sample and raw material extract concentrates were performed under the same conditions. The baselines were adjusted when necessary after changing detector wavelengths. The separated PAHs in the environmental samples were identified with retention volumes and their specific fluorescence responses as compared with those of the standards. Quantitative data were obtained by computing the HPLC peak height with concentration from the standard calibration curve constructed for each individual PAH. The detection limit with this system was 0.2 - 0.5 ug/ml of the extract depending upon individual PAH and the reproducibility was ± 10%.

In a collaborative study, the recoveries of seven PAHs were nearly 100%. In the case of the remaining four PAHs namely BaA, DMBA, BaP and [B(ghi)Per], recoveries were 80, 80, 80 and 64% respectively and accordingly appropriate corrections were applied to the data obtained on these PAHs.

RESULTS AND DISCUSSIONS

Prior to studying the exposure of the work crew to volatile products from asphalt and coal tar pitch during paving and roofing activities, it is important to know the levels of PAHs in ambient air to which the general public would be exposed. In the United States the ambient air concentration of some selected PAHs has been reported (20) and their average levels were very low. A similar study conducted by the Ontario Ministry of the Environment (20) has shown that the average levels of BaP and BkF were observed to be 1.44 and 3.31 ng/m^3 respectively in the ambient air from ten different cities.

In this study, we have chosen to collect air samples and raw materials from two asphalt paving, three asphalt roofing and three coal tar pitch roofing locations. We have also selected to analyze for typical, environmentally prevalent quadricyclic PAHs such as F, P, BaA, Chry and DMBA, pentacyclic PAHs such as BkF, Per, DB(ah) A and BaP, and hexacyclic PAHs such as B(ghi) Per and IP. Initially we have analyzed raw asphalt mix, roofing asphalt melt and coal tar pitch melt used at the eight job sites for the content of the above eleven individual PAHs and this is given in

TABLE 1

PAH CONCENTRATION IN RAW MATERIAL SAMPLES FROM EACH SITE

Site	Site Description	Percent Asphalt in Hot Mix	PAH Concentration - ug/g											Total PAH Concentration (mg/g)
			F	P	B(a)A	Chry	DMBA	Per	B(k)F	B(a)P	B(ghi)Per	DB(ah)A	IP	
I	Asphalt Paving	5.06	3.34	10.3	101	84.5	4.75	1.67	1.98	1.28	1.59	3.73	--	0.218
II	Asphalt Paving	4.81	2.58	8.97	79.8	75.0	3.88	1.32	1.72	1.24	0.87	5.55	--	0.183
III	Asphalt Roofing	N/A	13.1	INT	382	384	8.96	10.4	2.78	8.06	5.49	6.74	--	0.826
IV	Asphalt Roofing	N/A	14.8	31.6	167	121	10.8	3.49	6.14	5.40	4.23	ND	--	0.364
V	Asphalt Roofing	N/A	5.16 --	5.49* 6.75	107 5.02*	45.0 6.65*	0.71 --	2.26 6.66*	0.29 --	1.80 4.34*	4.28 3.46*	0.39 --	-- --	0.174 --
VI	Coal Tar Pitch Roofing	N/A	28300	17500	241000	66600	2160	4300	3680	7350	2060	1470	--	376
VII	Coal Tar Pitch Roofing	N/A	32500	23600	324000	88000	2950	5840	4500	11000	3980	1680	2460	500
VIII	Coal Tar Pitch Roofing	N/A	19800 17750†	22600 9225†	169000 6610†	42500 2495†	1070 --	2090 7301†	1670 1080†	4290 1920†	754 1560†	317 265†	616 --	265 --

N/A - Not applicable
INT - Interference

* Average values from Wallcave et al (1971)
† Average values from Lijinsky et al (1963)

Table 1.

As reported earlier (8,9), the data indicate that the concentrations of PAHs and their total levels in coal tar pitch are at least two to three orders of magnitude higher than those found in asphalt. The average values obtained for Per, BaP and, B(ghi)Per are fairly in agreement with the results obtained by Wallcave and coworkers (9), although the concentration of the quadricyclic PAHs are found in higher concentrations in the asphalt surveyed in this study than in those reported by Wallcave et al. (9). Further, levels of these eleven compounds in coal tar pitch reported in this study are also generally higher than the values reported by Lijinsky et al.(8). This is ascribed to the fact that these researchers have found varying amounts of these PAHs even in a single sample using different aliquots and that the analytical technique used in our study is much more precise than the method used by Lijinsky and coworkers (8).

Table 2, 3 and 4 show the concentrations of the eleven PAHs and their total levels present in a cubic meter of ambient air and air samples collected in the breathing zone of the workers. During asphalt paving operations, the average concentrations of some of the quadricyclic PAHs in ambient air sampled upwind far away from paving location seem to be in excess of 100 ng/m^3. However, the levels of BaP and BkF are about 2 to 4 times higher than the levels reported earlier (20).

The data indicate that, at the paving sites, the air sampled in the breathing zone of work crews apparently contains higher levels of all the PAHs than those found in ambient air. Moreover, the levels of penta- and hexacyclic aromatic hydrocarbons are about 3 to 8 times more than their levels found in ambient air. Furthermore, on the basis of these results, comparatively the machine drivers could be exposed to higher levels of PAHs than the level wheel operators (spreading and compacting operations).

The concentration of PAHs per cubic meter of ambient air at the upwind locations of asphalt roofing sites are also higher than the values reported earlier (20) and the range is found to be between 0.45 and 0.95 ug/m^3 (Table 3). The cartridges used to sample air in the breathing zone of the applicators and the kettlemen contain approximately the same level of total PAHs (14-18 ug/m^3) except those used at

TABLE 2

PAH CONCENTRATIONS FOUND IN THE WORKING ATMOSPHERE DURING PAVING OPERATIONS

Site Description	Source of Sample	Volume of Air Sample (Litres)	PAH Concentration - µg/m³											Total PAH Concentration (µg/m³)
			F	P	B(a)A	Chry	DMBA	Per	B(k)F	B(a)P	B(ghi)P per	DB(ah)A	IP	
Asphalt Paving Site I	Ambient Air	740	1.190	0.51	0.23	0.19	0.01	0.01	0.01	Trace	0.04	0.01	0.01	2.21
	Machine Driver	580	INT*	0.85	6.59	2.49	0.02	0.06	0.03	0.02	0.10	Trace	0.04	10.20
	Level Wheel Operator	580	0.92	0.51	3.79	0.19	0.07	Trace	0.01	0.01	0.03	0.01	0.03	5.57
Asphalt Paving Site II	Ambient Air	870	0.48	0.04	0.03	0.10	ND*	0.01	0.01	Trace	0.03	ND	0.01	0.71
	Machine Driver	870	0.78	2.14	8.78	1.05	0.14	Trace	0.05	Trace	0.03	0.01	0.01	12.99
	Level Wheel Operator	870	0.91	0.33	2.25	0.71	0.03	0.01	0.01	0.01	0.03	0.01	0.01	4.32

*INT - Interference
*ND - Not Determined

TABLE 3

PAH CONCENTRATIONS FOUND IN THE WORK - ENVIRONMENT DURING ASPHALT ROOFING OPERATIONS

Site Description	Source of Samples	Volume of Air Sample (Litres)	PAH-Concentration - $\mu g/m^3$											Total PAH Concentration ($\mu g/m^3$)
			F	P	B(a)A	Chry	DMBA	Per	B(k)F	B(a)P	B(ghi)Per	DB(ah)A	IP	
Asphalt Roofing Site III	Ambient Air	720	0.53	0.10	0.11	0.10	0.01	Trace	0.01	0.01	0.04	ND	ND	0.91
	Applicator	440	1.09	1.50	11.70	2.09	0.64	0.05	0.14	0.08	0.32	ND	ND	17.6
	Kettleman	540	10.30	5.30	68.00	26.50	0.26	0.59	0.48	0.43	0.44	0.15	ND	112.45
Asphalt Roofing Site IV	Ambient Air	860	0.22	0.21	0.17	0.19	0.04	0.01	0.03	0.01	0.10	ND	ND	0.98
	Applicator	820	1.87	0.63	9.32	5.25	0.07	0.03	0.07	0.04	0.02	ND	ND	17.27
	Kettleman	840	1.41	0.46	8.53	3.92	0.05	0.06	0.05	0.04	0.05	ND	ND	14.53
Asphalt Roofing Site V	Ambient Air	644	0.09	0.05	0.14	0.10	Trace	0.02	0.01	Trace	0.03	0.01	ND	0.45
	Applicator	620	0.72	0.19	9.63	5.34	0.14	0.09	0.08	0.05	0.11	ND	ND	16.35
	Kettleman	680	1.97	1.25	11.40	5.10	0.05	0.02	0.18	0.08	0.27	0.03	ND	20.35

ND - Not Determined

TABLE 4

PAH CONCENTRATIONS FOUND IN THE WORK ENVIRONMENT DURING COAL TAR PITCH ROOFING OPERATIONS

Site Description	Source of Samples	Volume of Air Samples (Litres)	PAH Concentration - $\mu g/m^3$										Total PAH Concentration ($\mu g/m^3$)
			F	P	B(a)A	Chry	DMBA	Per	B(k)F	B(a)P	B(ghi)P	DB(ah)A	IP	
Coal Tar Pitch Roofing Site VI	Ambient Air	810	0.33	0.35	0.59	0.40	0.07	0.01	0.06	0.01	0.07	0.04	ND	1.93
	Applicator	590	97.40	36.70	64.80	26.40	0.81	0.62	0.96	0.40	ND	ND	ND	228.09
	Kettleman	780	144.00	110.00	257.00	798.00	1.37	3.49	2.78	4.22	3.24	0.35	0.98	1325.43
Coal Tar Pitch Roofing Site VII	Ambient Air	660	0.33	0.34	1.59	0.79	0.05	0.08	0.08	0.08	0.11	0.04	0.05	3.54
	Applicator	510	154.00	162.00	145.00	77.60	1.22	1.93	2.03	1.22	1.48	0.80	0.38	547.66
	Kettleman	540	51.40	188.00	61.90	32.00	0.88	0.83	1.00	0.62	0.28	0.11	0.11	337.13
Coal Tar Pitch Roofing Site VIII	Ambient Air	690	0.70	0.28	3.78	0.74	0.02	0.01	Trace	0.02	0.01	ND	Trace	5.59
	Applicator	340	53.00	44.80	109.00	28.10	0.28	0.57	0.60	0.93	0.28	ND	0.08	237.36
	Kettleman	440	87.30	87.30	523.00	152.00	16.30	6.97	5.19	11.30	4.16	1.49	1.81	896.82

ND - Not Determined

site III where the cartridge used by the kettleman had unusually high levels of the PAHs compared to the ones used by the applicator and other kettlemen at sites IV and V.

At the Coal tar pitch roofing sites VI to VIII (Table 4) the cartridges used for sampling ambient air at upwind locations are found to contain higher levels of some of the individual PAHs and the total PAHs (1.9 to 5.7 ug/m^3) than the ambient air cartridges used at asphalt roofing sites (Table 3). Moreover, the cartridges carried by kettlemen have shown higher levels of PAHs than the cartridges used by applicators except at site VII, where the reverse was observed. From the data, it is obvious that the cartridges used by coal tar pitch workers contained about 10 to 25 times more of the penta- and hexacyclic PAHs than the cartridges used by the crew employed in asphalt roofing.

Considering the data on total PAHs and the carcinogen BaP, higher levels were found in the cartridges used by applicators and kettlemen of the roofing operations than in those used by work crew involved in paving operations. It is also apparent that the crew employed in coal tar pitch roofing activity are likely to be exposed to higher amounts of PAHs than those occupied with asphalt roofing. Furthermore, it is noteworthy that in coal tar pitch roofing operations the cartridges used in the breathing zone of kettlemen contained more PAHs and BaP than those used by applicators. This would indicate that kettlemen would be exposed to higher concentrations of these pollutants than applicators.

An attempt was made to correlate exposure levels in roofing operations with total PAH concentration and representatives of a quadricyclic and pentacyclic aromatic hydrocarbon. In Table 5 are given the levels of total PAHs, BaA and BaP in coal tar pitch and in asphalt.

Although the concentration of total PAHs in raw coal tar is about 800 times more than in asphalt, the crew working with coal tar pitch are exposed only to about 15 times more to PAHs than those working with asphalt. If the BaA levels are considered, there are about 1100 times more BaA in coal tar than in asphalt; however, the exposure of coal tar pitch workers is about 10 times more to this aromatic hydrocarbon than that of the asphalt workers. In spite of the fact that the BaP content of coal tar pitch is about 1500 times more than that found in asphalt, the occupational exposure to BaP is only about 20 times greater

TABLE 5

THE RELATIONSHIP BETWEEN EXPOSURE LEVELS IN ROOFING OPERATIONS AND PAH CONTENT OF THE RAW MATERIAL SAMPLES

	Coal Tar Pitch		Asphalt	
	$\bar{x}$	S.D.	$\bar{x}$	S.D.
Total PAH in Raw Material	381,000 ug/g	118,000	454 ug/g	335
Total PAH in Air	476.000 ug/m^3	260.000	33.100 ug/m^3	38.700
B(a)A in Raw Material	244 000 ug/g	77,600	219 ug/g	145
B(a)A in Air	193.000 ug/m^3	177.000	19.800 ug/m^3	23.700
B(a)P in Raw Material	7,500 ug/g	3,360	5.09 ug/g	3.14
B(a)P in Air	3.110 ug/m^3	4.250	0.119 ug/m^3	0.152

from coal tar pitch operations than from asphalt roofing activities. Further, although the temperature of asphalt melt was about 30°C higher than coal tar pitch, it appears that extremely high concentrations of PAHs in coal tar pitch could be responsible for greater occupational exposure of workers employed in coal tar pitch roofing than in asphalt roofing activities.

As mentioned earlier, in long term air sampling extending from several days to several months using Hi-volume samplers, the PAHs originally adsorbed on to filter elements could be "blown off" from filtering trains of the cartridges. Also it has been reported (21,22) that organic pollutants with high vapour pressure could volatilize from particulate matter embedded in the filter while using large volume sampling devices. Van Vaeck et al. (12) have observed several high molecular weight pollutants (below 250 Daltons) were desorbed in varying degrees from the filter into gaseous phase while no component above 250 Daltons were lost by the Hi-Volume Sampling technique.

In order to verify the "carry over" effect, the air samples were collected in the breathing zone of the workers using type II sampling train for 4 to 6 hours at sites V and VIII. After sampling, the cartridges were dismantled and the filtering elements and the Tenax GC packings were separately extracted, concentrated and analyzed.

The results of PAH analysis of filter elements and Tenax GC packing are given in Table 6. As reported by Cautreels et al. (23) and Handa and coworkers (24), the lower molecular weight PAHs such as F, P, BaA and Chry are found in higher amounts in net Tenax GC packing than in the filter elements used to collect the ambient air upwind of site V. The ambient air sampled upwind from site V is however, found to contain less of low molecular weight PAHs in the Tenax GC packing than in the packing used to sample the ambient air upwind of site VIII. This is in correspondence with their concentrations in raw materials.

When individual PAH contents of the filters used in sampling ambient air upwind of sites V and VIII are compared, about 8 to 50 times more of the low molecular weight PAHs are observed in the filter of the ambient air upwind location of site VIII than in the one used at the upwind location of site V. This would indicate that the low molecular weight PAHs are predominant in gaseous state owing to their high volatility (13) or they are "blown off" from

TABLE 6

ANALYSIS OF FILTER ELEMENTS AND TENAX RESIN FROM SITES V AND VIII - PAH CONCENTRATION IN μg/m³

Site Description	Source	F	Phe	B(a)A	Chry	DMBA	Per	B(k)F	B(a)P	B(ghi)Per	DB(ah)A	IP
V Asphalt Roofing	Ambient Air Filters	0.0103	0.004	0.049	0.049	ND	0.004	0.003	0.002	0.004	0.003	ND
	Ambient Air Tenax	0.077	0.046	0.084	0.049	0.003	0.010	0.007	0.002	0.028	0.003	ND
VIII Coal Tar Roofing	Ambient Air Filters	0.472	0.138	2.350	0.439	0.006	0.007	0.004	0.010	0.003	ND	ND
	Ambient Air Tenax	0.228	0.138	1.520	0.304	0.015	0.006	0.004	0.006	0.010	ND	0.004
V Asphalt Roofing	Applicator Filters	0.644	0.135	9.950	5.230	0.144	0.018	0.019	0.034	ND	ND	ND
	Applicator Tenax	0.725	0.565	0.077	0.108	ND	0.076	0.058	0.017	0.106	ND	ND
VIII Coal Tar Roofing	Applicator Filters	49.900	43.900	104.000	27.100	0.262	0.535	0.579	0.891	0.212	ND	0.079
	Applicator Tenax	3.120	0.953	5.060	1.020	0.018	0.035	0.021	0.038	0.068	ND	ND
V Asphalt Roofing	Kettleman Filters	1.950	1.270	12.300	5.526	0.048	0.179	0.097	0.066	0.106	0.016	ND
	Kettleman Tenax	0.211	0.105	0.153	0.074	0.007	0.047	0.102	0.024	0.187	0.019	ND
VIII Coal Tar Roofing	Kettleman Filters	86.700	87.000	521.000	151.000	14.200	6.940	5.160	11.200	4.140	1.490	1.800
	Kettleman Tenax	0.598	0.270	2.380	1.130	2.100	0.030	0.095	0.036	0.024	ND	0.011

the filters and are captured by Tenax GC packing. In terms of penta- and hexacyclic aromatic hydrocarbon content in ambient air from upwind location of site V, the Tenax GC packing is found to contain about 1 to 8 times more of these compounds than the filter. Similarly, at the upwind location of site VIII the levels of high molecular weight aromatics in Tenax GC packing are about 1 to 3 times more than those found in filters.

Comparison of levels of PAHs in filter and Tenax GC packing of the cartridge used by applicator at site V indicate that these compounds are found in higher concentrations in Tenax GC packing than in filter elements except for BaA, Chry and BaP levels. In the case of PAH contents present in filter and Tenax GC packing used by applicator at site VIII, only less than about 5% of all PAHs are found in Tenax GC packing compared to their levels in filter. An analogous observation is made in the case of the filter and Tenax GC packing used by kettlemen at both sites V and VIII although the levels of PAHs in the Tenax GC packing of kettleman at site VIII are much higher than in the packing used at site V.

It would appear that, in the case of asphalt roofing activity, most of the PAHs are in the vapour phase whereas the PAHs emanating from coal tar pitch are in the form of condensed particulate matter or aerosols. It is noteworthy that, even in such a short sampling period significant amounts of PAHs escape from filter elements. The fact that considerable amount of the PAHs are found in the Tenax GC packing is somewhat disconcerting since most of the Hi-Volume air sampling for PAH analysis is carried out using filter elements alone. It is apparent that inclusion of Tenax GC resin in the sampling train would give more accurate measure of PAHs present in air samples than the results obtained with the use of filter elements alone. However, filter elements are necessary to screen off the aerosols and particulates.

Conclusion:

The data indicate the levels of PAHs found in coal tar pitch are two to three orders of magnitude higher than those found in asphalt. It would appear that during asphalt paving activity, the machine drivers could be exposed to comparatively higher concentrations of PAHs than the applicators. During asphalt roofing operations the exposure levels to PAHs of applicators and kettlemen would be nearly the same. However, during coal tar roofing activities, the kettlemen would be exposed to higher levels of PAHs than the applicators. It was observed that higher exposure levels to PAHs could result from coal tar roofing than from asphalt roofing activities. Since the quadricyclic and pentacyclic PAHs could escape even during Low Volume air sampling, the use of Tenax GC packing along with filters and silver frits is recommended to obtain accurate measurements of PAHs during air sampling.

ACKNOWLEDGEMENT

Thanks are due to D.T. Wang, R.D. Smillie and O. Meresz of the Ontario Ministry of the Environment for suggestions and help, P. Toft and D.T. Williams for their interest in this research and P. Blais and G. LeBel for reviewing the manuscript.

REFERENCES

1. This work was carried out under the contract # 682/80-81 of Health and Welfare Canada, Government of Canada.
2. U.S. Laws, Statutes, Etc., "Food Additives Amendment of 1958", (Sept. 6, 1958, Public Law 85-929, 72 Stat. 1784, April 7, 1961, P.L. 87-19, 75 Stat. 42). In: Food Drug Cosmetic Law Reports, pp 4301-4302, Chicago: CCH 1975.
3. Suess, M.J. (1976): The environmental load and cycle of polycyclic aromatic hydrocarbons, Sci, Tot. Environ., 6:239-250.
4. Maugh, T.H. (1974): Chem. Carcinogenosis: A long neglected field blossoms, Science, 183:940-944.
5. Kertesz-Saringer, M., Meszaros, E., and Varkonyi, T. (1971): On the size distribution of benzo[a]pyrene containing particles in urban air, Atmos. Environ., 5:429-431.
6. Miguel, A.H. and Friedlander, S.K. (1978): Distribution of benzo[a]pyrene and coronene with respect to particle size in Pasadena aerosols in submicron range, Atmos. Environ., 12:2407-2413.
7. U.S. Environmental Protection Agency (April 1977): Sampling and Analysis Procedures for Screening Industrial Effluents for Priority Pollutants, Environmental Monitoring Support Laboratory, Cincinnati, Ohio.

8. Lijinsky, W., Domsky, I., Mason, G., Ramahi, H.Y., and Safovi, T. (1963): The chromatographic determination of trace amounts of polynuclear hydrocarbons in petroleum mineral oil and coal tar, Anal. Chem., 35:952-956.
9. Wallcave, L., Garcia, H., Feldman, R., Lijinsky, W., and Shubik, P. (1971): Skin tumorigenesis in mice by petroleum asphalts and coal tar pitches of known polynuclear aromatic content, Toxicol. Appl. Pharmacol., 18:41-52.
10. Tye, R., Horton, A.W., and Rapien, I. (1966): Benzo[a]-pyrene and other aromatic hydrocarbons extractable from bituminous coal, Amer. Ind. Hyg. Assn. J., 27:25-28.
11. Hoffmann, D. and Wynder, E.L. (1977): Organic Particulate Pollutants-Chemical Analysis and Bioassay for Carcinogenicity. Air Pollution, Vol. 2, Ed. A.C. Stern, p. 361, Academic Press, New York.
12. Van Vaeck, L., Broddin, G., and Van Cauwenberge, K., (1980): On the relevance of air pollution measurements of aliphatic and polyaromatic hydrocarbons in ambient particulate matter, Bio. Med. Mass. Spect., 7:473-483.
13. Murray, J.J., Pottie, R.F., and Pupp, C. (1974): The vapor pressures and enthalpies of sublimation of five polycyclic aromatic hydrocarbons, Can. J. Chem., 52:557-563.
14. Pupp, C., Lao, R.C., Murray, J.J., and Pottie, R.F. (1974); Equilibrium vapour concentrations of some polycyclic aromatic hydrocarbons, As_4O_6 and SeO_2 and the collection efficiencies of these air pollutants, Atmos. Environ., 8:915-925.
15. Pitts, J.N., Jr., Lokensgard, D.M., Ripley, P.S., Van Chauwenberghe, K.A., Van Vaeck, L., Shaffer, S.D., Thill, A.J., and Belser, W.L., Jr. (1980): "Atmospheric" epoxidation of benzo[a]pyrene by ozone: Formation of the metabolite benzo[a]pyrene-4,5-oxide, Science, 216:1347-1349.
16. Grover, P.L. and Sims, P.C. (1970): Interactions of the K-region epoxides of phenanthrene and dibenz(ah)anthracene with nucleic acids and histone, Biochem. Pharmacol., 19:2251-2259.
17. NIOSH Manual of Analytical Methods, (April 1977) 2nd Edition, Vol. 1, Part 1, pp. 217-222.
18. Smillie, R.D., Wang, D.T. and Meresz, O. (1978): The use of a combination of ultraviolet and fluorescent detectors for the selective detection and quantitation of polynuclear aromatic hydrocarbons by high pressure liquid chromatography, J. Environ. Sci. Health, A13:47-59.
19. Gallant, F.F., King. J.W., Levins, P.L., and Piecewicz, J. (1978): Characterization of sorbent resins for use in environmental sampling, EPA 600/7-78-054.

20. Anonymous - Ontario Ministry of the Environment (1979): Polynuclear Aromatic Hydrocarbons - ARB-TD-Report No. 58-79.
21. Rondia, D. (1965): Sur la volatilite des hydrocarbures polycycliques, *Air Water Pollut.*, 9:113-121.
22. Della Fiorentina, H., De Weist, F., and De Graeve, J. (1975): Determination par spectrometrie infra rouge de la matiere organique non-volatiles associee aux particules en suspension dans l'air. II. Facteurs influencant l;indice aliphatique, *Atmos. Environ.*, 9:517-522.
23. Cautreels, W. and Van Cauwenberghe, K. (1978): Experiments on the distribution of organic pollutants between airborne particulate matter and the corresponding gas phase, *Atmos. Environ.*, 12:1133-1141.
24. Handa, T. Kato, Y., Takaki, Y., Tadahiro, I., and Hidetsuru, M. (1980): In situ emission levels of polynuclear aromatic hydrocarbons from gasoline and diesel engine vehicles on an expressway, *J. Environ. Sci. Health,* A15:573-599.

THE POLYNUCLEAR AROMATIC HYDROCARBON CONTENT OF SMOKED FOODS IN THE UNITED KINGDOM

A.S. McGILL, P.R. MACKIE, E. PARSONS, C. BRUCE and R. HARDY
Torry Research Station, 135 Abbey Road, Aberdeen AB9 8DG, Scotland.

INTRODUCTION

Smoked fish, bacon and cheese are popular commodities in the UK and because of the possibility of introducing undesirable components from the smoking process an investigation into the possible toxicity of these foods is underway at our laboratory. Part of this study has been to establish the concentrations of polynuclear aromatic hydrocarbons (PNAH) present in smoked produce on sale in Britain and to assess the influence of the smoking technique adopted so that improvements can be introduced if necessary.

In this preliminary communication the results of the quantitative analysis of major PNAH carcinogens are reported. A more complete account of the work will be published elsewhere.

EXPERIMENTAL

All the samples of fish, sausages and all but two of the bacon samples were obtained from producers situated throughout the UK. Only one cheese sample was obtained from a producer; the remaining cheese and two bacon samples were obtained from local retailers as were the vegetables.

The PNAH from 20 g of the foodstuff were extracted and isolated in the manner described previously (1). Bacon samples have high fat contents which can interfere with the subsequent work-up procedure. To minimise this, the extract was made up to 100 ml in a volumetric flask using chloroform and an aliquot equivalent to 10 g of sample used for the analysis. To remove chlorophyll and carotenoid pigments, vegetable isolates were further purified by the solvent system used by Grimmer (2). Some of the fish and all bacon, cheese and vegetable samples were spiked with deuterated anthracene (4.82 ng) and pyrene (11.1 ng) to confirm recoveries.

The PNAH were quantitatively analyzed by gas chromatography (GC) on a methyl silicone quartz capillary column

(25 m, 100°C—260°C @ 3°/min) coupled to a V.G. micromass 16F mass spectrometer equipped with a V.G. 2250 data system. The mass spectrometer was operated in the electron impact and selective ion recording modes.

RESULTS

The recoveries of 6 individual carcinogens, namely benz(a)anthracene, indeno(1,2,3-cd)pyrene, dibenzo(a,h) anthracene, benzo(a)pyrene, benzo(ghi)perylene, 7,12-dimethylbenz(a)anthracene, two non-carcinogens, namely benzo(k)fluoranthene and 9,10-dimethylanthracene and one carcinogen mixture (chrysene + triphenylene) are given in Table 1.

TABLE 1

RECOVERIES OF POLYNUCLEAR AROMATIC HYDROCARBONS ADDED TO FISH AT A CONCENTRATION OF 2.5 ppb

Compound	Recovery
Benz(a)anthracene	65%
Indeno(1,2,3-cd)pyrene	106%
Dibenz(a,h)anthracene	103%
Benzo(a)pyrene	94%
Benzo(ghi)perylene	108%
7,12-Dimethylbenz(a)anthracene	97%
Benzo(k)fluoranthene	89%
9,10-Dimethylanthracene	96%
Chrysene + Triphenylene	96%
d-10 Anthracene	104%
d-10 Pyrene	96%

In all, 48 individual or groups of isomeric PNAH components such as dimethyl and ethyl naphthalenes were assessed. The total PNAH and BaP content of the foods are presented in Figures 1 and 2 respectively and the carcinogen concentrations in Table 2.

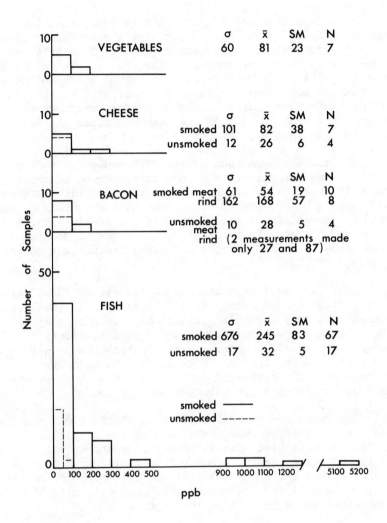

FIGURE 1. Total PNAH distribution for vegetables, smoked and unsmoked fish, bacon and cheese.
σ standard deviation; x̄ mean; SM standard error of the mean; N No. of samples.

DISCUSSION

It can be seen from Figure 1 that the smoking of foods on the whole increased the total PNAH content. The increase is not uniform and this is particularly evident in the analysis of the smoked fish which have a mean PNAH content of 245 ppb with a standard deviation of 676. This is illustrated more clearly in the histogram in the figure. It is quite possible that if as many samples of smoked bacon and cheese as fish had been analyzed, a wider range of distribution of PNAH in these commodities would also have been observed. It is apparent that in most of the unsmoked foods the PNAH content lies in the 0-100 ppb range.

A detailed comparison of the PNAH contents of fish smoked in modern mechanical kilns with those produced in traditional smoke 'houses' (i.e. exposed directly to smouldering sawdust) show that a higher proportion of the latter contained total PNAH in excess of 100 ppb. This was observable also in the bacon samples and, in particular, the rinds of most of the traditionally smoked samples contained the highest concentrations of PNAH.

Some care must be exercised in the use of these values. In the first instance they do not represent the total PNAH present but only the 48 individual or groups of components that the mass spectrometer was programmed to identify and quantify. Secondly, standards of all these compounds were not available and the values given for the majority of the hydrocarbons are estimates derived from the observations made on those components for which we had a standard.

The basis for estimates where no standard was available was the observation that the molecular ion production for most of the unsubstituted PNAH standards is comparatively constant as is the molecular ion product from similarly substituted PNAH. It was therefore assumed that this response relationship would hold for the unknown PNAH and estimates of their concentrations were made on this basis.

As far as is known, no comparable results have been quoted in the literature although fluorescence values for hydrocarbons have been quoted. However, because of the inherent variation in fluorescence associated with PNAH compounds, it is not clear at this stage how such values compare with those obtained in the present study.

Table 1 shows that the recovery of all but one of the standard PNAH was very good considering the extremely small amounts used. Furthermore, it was noted that the mass spectrometer response in all instances was linear over the 0.01-100 ng range on the column.

A summary of the concentrations of the recognized carcinogenic PNAH together with previously reported values for these substances in smoked foods is presented in Table 2.

In the past when analyzing for PNAH components, large quantities of material had to be used to obtain sufficient sample for analysis, and because of the methods of analyses available complex fractionation techniques had to be used to isolate individual components prior to their analysis. Such isolation procedures were often difficult to carry out and often not wholly successful. In addition, the possibility of extraneous contamination introduced in the work-up procedures was always high. For these reasons early results on concentrations of PNAH, especially individual components, must be treated with some reserve. Now with modern analytical methods of fractionation such as capillary gas liquid chromatography and very sensitive specific detectors such as the mass spectrometer many of these problems have been overcome and greater reliance can now be placed on the results so obtained especially when calibration with known standards is possible.

The values found in this present work are in reasonably close agreement with those reported by other workers (3,4). In the system used at Torry Research Station, chrysene is measured with triphenylene, as these compounds are not separated on the column used and they also give rise to the same parent ion. The b, k and j benzofluoranthene isomers are separated but, because only one isomer was available as a standard, they have been grouped as a single result. For the same reason, the dibenzanthracenes and C2 isomers of benzanthracene/phenanthrene have been reported as a group. This accounts for the fact that in some instances higher concentrations of these PNAH groups were obtained when compared with other individual PNAH compounds.

As with the total PNAH measured, the concentration of various carcinogens in smoked fish is dependent on the method of smoking and on the raw material. Although most foodstuffs examined here are cold smoked, that is their temperature does not rise above $30^\circ C$, some are hot smoked, that is the temperature in the kiln is allowed to rise to $80^\circ C$. Our results show that the quantities of individual and total PNAH are similar in cold and hot smoked foods.

It is noteworthy that over 88% of all smoked food samples (Figure 2) and 86% of fish samples had BaP contents of less than 1 ppb. With minor variations, the other carcinogens

PNAH CONTENT OF SMOKED FOODS IN UK

TABLE 2

CARCINOGEN CONCENTRATIONS FOUND FOR FOODS EXAMINED IN THIS STUDY COMPARED WITH THOSE REPORTED IN THE LITERATURE (ppb)

		BaP	BaA	BcPA	CH	CH/TP	IP	BPR	BFL	DBA	DBP	C2 BA/PA
FISH												
SMOKED:	REPORTED	0-63 (22)*	0-189 (9)*		0-173 (6)*		0- 9 (3)*	0-10 (9)*	0-72 (4)*	0- 2.4 (3)*		
SMOKED:	UK (67)	0-18	0- 86	0- 15		0-143	0-37	0-25	0-55	0-77	0	0- 55
UNSMOKED:	UK (17)	Tr- 0.35	Tr- 0.09	0.01- 0.09		0.30- 1.73	0- 0.33	Tr- 0.39	0- 1.03	0- 0.46	0	0- 4.87
BACON												
SMOKED:	REPORTED	0-30 (10)*	0- 18 (5)*		0- 21 (3)*		0- 2.5 (2)*	0-11 (6)*	0.35-15 (2)*	0- 9.3 (3)*	1.0 (1)*	
SMOKED:	UK MEAT (10)	0.01- 0.14	Tr- 0.33	Tr- 0.18		0.20- 1.15	Tr- 0.11	Tr- 0.12	0.06- 0.40	0- 0.13	0	0- 2.54
	RIND (10)	0.02- 0.45	0.02- 0.64	0.03- 0.36		0.30- 2.49	0.04- 0.38	0.03- 0.31	0.10- 1.04	0- 0.38		Tr-11.47
UNSMOKED:	UK MEAT (4)	0.01- 0.04	0.02- 0.03	0.03- 0.04		0.30- 0.57	0.01- 0.03	0.03- 0.04	0.06- 0.17	0,Tr		0- 1.04
	RIND (2)	0.02, 0.02	0.02 , 0.03	0.05, 0.06		0.20 , 0.68	0.02, 0.03	0.03 , 0.03	0 , 0.12	Tr, Tr		0 , 0.08
SAUSAGES												
SMOKED:	REPORTED	0- 2.9 (12)*	0- 18 (7)*		0- 9.2 (3)*		0, 1.0 (1)*	0- 0.5 (5)*	0- 3.5 (1)*	0- 0.60 (4)*		
SMOKED:	UK (4)	0.04- 0.26	0.04- 0.38	0.05- 0.21		0.45- 1.4	0.04- 1.40	0.06- 0.27	0.19- 0.75	0- 0.10		1.92- 3.28
UNSMOKED:	UK (2)	0.03, 0.26	0.04 , 0.13	0.05, 0.10		0.47, 1.4	0.05, 0.18	0.05, 0.19	0.20, 0.51	0.05, 0.14		1.22, 2.86
CHEESE												
SMOKED:	REPORTED	0- 0.9 (4)*	0 (2)*				0,<0.02 (2)*	0- 0.60 (3)*	0.45 (1)*	0 (1)*	0 (1)*	
SMOKED:	UK (7)	0.01- 1.62	0.01- 5.59	0.02- 1.32		0.08- 16	0- 2.45	0- 1.57	0.02- 3.87	0- 0.84		0.08- 2.16
UNSMOKED:	UK (4)	Tr- 0.03	0.01- 0.02	0.01- 0.05		0.06- 0.26	0- 0.03	Tr- 0.06	0.01- 0.09	0		Tr- 1.54
VEGETABLES												
	REPORTED LETTUCE	1.2 -13 (2)*	0.30- 15 (1)*					2.98-10 (1)*		0.56- 1.01 (1)*		
	UK (7) LETTUCE AND CABBAGE	0- 1.42	0.05- 3.17	0.05- 1.50		0.12- 15	0- 1.92	0- 1.39	0- 5.20	0- 0.39	0	0.24- 9.4

BaP = Benzo(a)pyrene; BaA = Benz(a)anthracene; BcPA = Benzo(c)phenanthrene; CH = Chrysene; TP = Triphenylene; IP = Indeno(1,2,3-cd)pyrene; BPR = Benzo(ghi)perylene
BFL = Benzofluoranthenes; DBA = Dibenzanthracenes; DBP = Dibenzopyrenes; C2 BA/PA = C2 isomers of Benzanthracene/phenanthrene
( )* = No. of literature reports; ( ) = No. of samples analysed in this study; Tr = < .004 ppb

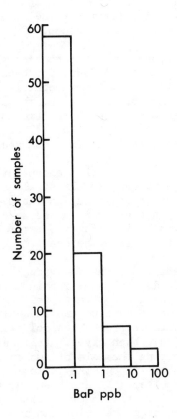

FIGURE 2. BaP distribution in all smoked foods.

present had a similar distribution.

Generally, the smoking of food increases the concentration of all carcinogens although in some instances the increase was probably less than the experimental error of the method.

Almost without exception, smoked bacon contained lower levels of carcinogenic PNAH than smoked fish. Thus, only one sample of bacon meat contained more than 50 ppt of BaP; the rinds contained slightly more but only marginally so. On the whole, the amounts found in the smoked product were only slightly greater than those found in the unsmoked, suggesting that compared with fish only low concentrations of smoke constituents are deposited on bacon.

In the 4 samples of sausages, the concentrations found were marginally higher than those found for bacon. In the UK smoked cheeses are imported from a few countries. There is only one major UK producer so, once again, the number of analyses carried out was small. With the exception of one German cheese, concentrations tended to be less than those in fish and more comparable with those in bacon and sausages.

It is interesting to compare the concentrations in smoked foods with those in vegetables. In almost all cases the BaP content of the lettuce and cabbage lay in the 0 – 1.5 ppb range which, on the whole, was higher than in the majority of smoked foods. A few analyses indicated that the outer leaves contained higher concentrations than the heart.

With only a few exceptions the individual PNAH carcinogen concentrations of all foodstuffs analyzed in this study have rarely exceeded 1 ppb. We are not in a position to judge whether the concentration of carcinogens found in the smoked foods analysed are acceptable from a toxicological viewpoint as no clear guidelines have been given. Legislation has been passed in Germany preventing the sale of smoked foods containing more than 1 ppb of BaP but the toxicological grounds for this are not wholly clear. It has been suggested that the European Economic Community will enact similar legislation with respect to the sale of smoke solutions used in the preparation of smoked foods. If this occurs then, taking into account the very small amounts of smoke solution used in the preparation of smoked food products, the concentration of carcinogens introduced will be extremely small, probably less than 10 ppt.

It is estimated that the annual per capita consumption of smoked fish, bacon and green vegetables (cabbage, sprouts, lettuce) in the UK amounts to approximately 600 g, 1280 g and 8845 g respectively. The total ingested content of BaP from these sources would therefore amount on average to 1 µg per capita per annum for fish, 0.51 µg per capita per annum for bacon and 3.3 µg per capita per annum for vegetables.

REFERENCES

1. Mackie, P.R., Hardy, R., Whittle, K.J., Bruce, C., and McGill, A.S. (1980): The tissue hydrocarbon burden of mussels from various sites around the Scottish coast. In: <u>Polynuclear Aromatic Hydrocarbons: Chemistry and Biological Effects</u>: 4th International Symposium, edited by A. Bjørseth and A.J. Dennis, pp. 379-393, Battelle Press, Ohio.
2. Grimmer, H., Böhnke, H., and Naujack, K.W. (1978): Simultaneous gas chromatographic profile analysis of carcinogenic polycyclic aromatic compounds: Polycyclic aromatic hydrocarbons, carbazoles and acridines/aromatic amines, <u>Z. analyt. Chem.</u>, 290: 147.
3. Lo, Mei-Tein., and Sandi, E. (1978): Polynuclear aromatic hydrocarbons (polynuclears) in foods, In: <u>Residue Rev.</u>, 69: 34-86.
4. Howard, J.W., and Fazio, T. (1980): Review of polycyclic aromatic hydrocarbons in foods: Analytical methodology and reported findings of polycyclic aromatic hydrocarbons in foods, <u>J. Assoc. Off. Anal. Chem.</u>, 63(5):1077-1104.

MAGNETIC CIRCULAR DICHROISM AS AN ANALYTICAL TOOL FOR SUBSTITUTED POLYNUCLEAR AROMATIC HYDROCARBONS.

JOSEF MICHL and GEORGE H. WEEKS
Department of Chemistry, University of Utah, Salt Lake City, Utah 84112

INTRODUCTION

The identification and quantitative determination of polynuclear aromatic hydrocarbons, their substituted derivatives, and heterocyclic analogues is of great current interest. Several spectroscopic methods are in use, such as fluorescence spectroscopy and gas chromatography-mass spectrometry. The differentiation of positional isomers remains a difficult task, particularly since only small samples are often available and authentic specimens are not always readily accessible for comparison. At the same time, it is an important task, given the frequently vastly different biological properties which such isomers exhibit.

The purpose of this paper is to point out the potential analytical usefulness of magnetic circular dichroism (MCD), a relatively new spectroscopic technique which possesses some attractive features suitable for the qualitative and quantitative analysis of polynuclear aromatic hydrocarbons, their substituted derivatives and heterocyclic analogues. In this method, the circular dichroism (optical activity) of a solution of the sample is measured in the presence of magnetic field parallel to the light propagation direction. Under these conditions, all substances are optically active. The most interesting features of the method in the present context are the sensitivity of MCD spectra to positional isomerism and the predictability of the observed peak signs from first principles.

MATERIALS AND METHODS

A sample of 1-fluorobenzanthracene was obtained from Professor R.S. Becker (University of Houston). The absorption spectrum was measured in a Cary 17 spectrophotometer and MCD on a JASCO 500 C spectropolarimeter equipped with a 15-kG electromagnet supplied by JASCO. Molar magnetic ellipticity $[\Theta]_M$ per unit magnetic field in units of $deg.L.m^{-1}.mol^{-1}G^{-1}$ is shown in the spectrum. The measurements were performed on $6 \times 10^{-4} - 2 \times 10^{-5}$ M solutions in spectral grade cyclohexane using a 1 cm pathlength.

RESULTS AND DISCUSSION

Experimental Aspects

MCD signal intensity is proportional to the strength of the magnetic field. The homogeneity of the field is unimportant and it is quite easy to obtain field strengths of the order of 50-60 kG using superconducting magnets. The increase in sensitivity is paid for by the inconvenience and expense of working with liquid helium, and may well be unnecessary for work with cyclic π-electron systems, which yield relatively intense MCD signals.

Sample preparation is similar as in UV absorption work; concentration of the sample is best chosen so as to yield an optical density 0.5-1.5 in the wavelength range of interest.

Theory

Background. A quantitative calculation of MCD spectra from first principles represents a formidable task. Fortunately, however, it has been found that for certain classes of molecules already very simple theoretical procedures suffice to produce quite reliable predictions of the qualitative features of the low-energy region of the spectra, *i.e.* the absolute signs and relative magnitudes of peaks corresponding to the first few transitions in UV absorption. One of the structural types which can be handled in this way are aromatic molecules, defined here as all those derivable formally from a fully conjugated perimeter containing n atoms and 4N+2 electrons. Among these are benzenoid hydrocarbons, derived from an uncharged perimeter (n = 4N+2), and non-alternant hydrocarbons containing an odd-membered ring and derived from an uncharged (*e.g.*, azulene) or a charged (*e.g.*, fluoranthene) perimeter, as well as all of their substituted derivatives and heterocyclic analogues.

The theoretical tool which provides a qualitative understanding of the MCD spectra of aromatics is known as the perimeter model. The theoretical treatment has been described in detail in ref. 1; a qualitative summary can be found in ref. 2. Results important for the present purposes are outlined in the following.

MCD of L and B Bands. The UV and MCD spectra of aromatics begin with two relatively weak $\pi\pi^*$ transitions labeled L_1 and L_2 in the order of increasing energy, followed at higher energies by two strong $\pi\pi^*$ transitions labeled B_1 and B_2 (in symmetrical molecules, they are usually referred to as L_b, L_a, B_b, B_a, where the subscripts refer to the direction of the transition moment). The MCD intensity of each of these four bands contains two contributions. One of these is proportional to a perimeter magnetic moment μ^- which is relatively small, the other to a perimeter magnetic moment μ^+ which is an order of magnitude larger. The proportionality constants in the μ^- contributions are nearly independent of molecular structure. They provide weakly negative or vanishing increments to the MCD intensity of the L_1 and L_2 transitions, a more strongly negative increment to the MCD intensity of the B_1 transition, and a comparable but positive increment to the MCD intensity of the B_2 transition. In contrast, the proportionality constants contained in the μ^+ contributions depend on the molecular structure quite sensitively. The molecular property which determines their sign is the relative size of the energy splitting of the highest two occupied π molecular orbitals derived from the perimeter, ΔHOMO, and the energy splitting of the lowest two free π molecular orbitals derived from the perimeter, ΔLUMO (Figure 1). If they are equal (ΔHOMO=ΔLUMO),

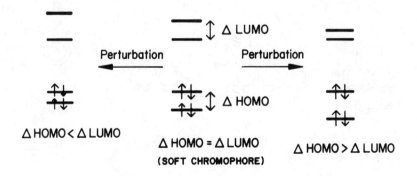

FIGURE 1. Illustration of the sensitivity of a soft chromophore to perturbations: energies of the highest two occupied and lowest two unoccupied molecular orbitals.

the proportionality constants vanish and so do the μ^+ contributions. In such a case, the μ^- contributions alone are observed in the MCD spectrum. Then, the MCD intensity of the two long-wavelength transitions L_1 and L_2 is weak and negative or zero (the latter usually takes the form of some positive and some negative vibrational fine structure peaks within the same electronic band). The B_1 band is more strongly negative and the B_2 band more strongly positive. Unsubstituted polynuclear benzenoid hydrocarbons such as I - V are prime examples of this group of compounds. In this case, the theory predicts and experiment confirms a nearly vanishing UV and MCD intensity for the L_b band and a stronger UV intensity with a negative MCD peak for the L_a band [in some larger polycyclic benzenoid hydrocarbons, more significant deviations from the exact equality of ΔHOMO and ΔLUMO apparently occur (3)].

If ΔHOMO is larger than ΔLUMO, the signs of the μ^+ contributions to MCD intensities are -,+,-,+, for L_1, L_2, B_1, and B_2, respectively. If ΔHOMO is smaller than ΔLUMO, the signs of the contributions are just the opposite: +,-,+,-, respectively. The magnitude of the contributions is dictated by the magnitude of the difference ΔHOMO-ΔLUMO; the larger the latter, the larger the former. Because of the relative size of the moments μ^- and μ^+, the μ^+ contributions tend to dominate the spectra unless ΔHOMO and ΔLUMO are nearly equal. This is particularly true of the L_1 and L_2 bands which have relatively small μ^- contributions. Hydrocarbons which are formally derived from a charged (4N+2)-electron perimeter, such as VI and VII, are typical representatives of this class of compounds.

Hard and Soft MCD Chromophores. The sensitivity of the MCD signs of an aromatic chromophore to minor perturbations such as introduction of substituents or aza heteroatoms varies widely. The MCD spectra of those chromophores for which ΔHOMO is quite different from ΔLUMO to start with can be expected to be quite insensitive to such perturbations. They are referred to as hard chromophores. On the other hand, the signs in the MCD spectra of chromophores for which ΔHOMO is nearly equal to ΔLUMO are likely to be totally dominated by the effects of such weak perturbations which can make either ΔHOMO or ΔLUMO larger. They are referred to as soft chromophores. The MCD signs which are most susceptible to change are those for which the μ^- contributions are the smallest, i.e., those of the two L bands. These are generally also the easiest ones to measure.

From the above, it is clear that all benzenoid hydrocarbons are soft MCD chromophores and that their L_b band and, to a lesser degree, L_a band, will be particularly sensitive to minor structural perturbations. This is the basis for the analytical utility of MCD spectroscopy for the derivatives and heterocyclic analogues of polynuclear benzenoid hydrocarbons.

MCD Signs and Molecular Structure of Perturbed Polynuclear Benzenoid Hydrocarbons. In order to predict the change of the MCD signs of the two L bands of a polynuclear benzenoid hydrocarbon upon perturbation such as substitution or introduction of a heteroatom, it is only necessary to estimate the direction of the change which the perturbation causes in the quantity ΔHOMO-ΔLUMO. This can be done at various levels of sophistication. The simplest procedure, quite adequate in almost all instances, is the use of first order perturbation theory. It leads to the conclusion that the change in ΔHOMO-ΔLUMO is dictated jointly by the electronic nature of the substituent and by its location on the molecular perimeter; effects of multiple weak perturbations are additive (4).

The most important property of the substituent is its π-electron conjugative (mesomeric) effect. This can be π-electron withdrawing (+E: CN, CHO, COOH, NO_2, etc.) or π-electron donating (-E: alkyl, halogen, OCH_3, NH_2, etc.). The σ-electron inductive effect I of the perturbation plays a subordinate role and it appears that it only needs to be considered when the E effect is absent. The most common example is the aza replacement (-CH = → -N=), which has a +I effect (increased electronegativity at the perturbed center).

The important property of the position of substitution in the benzenoid hydrocarbon is the relative magnitude of the squares of the coefficients in the top two occupied and bottom two unoccupied molecular orbitals (MOs) of π symmetry, derived from the molecular perimeter. This determines the relative sensitivity of the energies of the two orbitals to perturbation at this position. In alternant hydrocarbons ("odd-soft chromophores"), the MO coefficients of the occupied orbitals are related in a particular way to those of unoccupied orbitals, so that in the last analysis, all necessary information can be obtained by inspection of the squares of the MO coefficients of the highest occupied molecular orbital and the second highest occupied molecular orbital. Those positions in which the former is larger are called dominant (D), those in which the latter is larger are called subdominant (S), those in which they are approximately equal are called neutral (N).

Standard programs permit a rapid computation of the coefficients of SCF MOs of any benzenoid hydrocarbon of interest. Even more simply, one can look up the coefficients of Hückel MOs in published tables (5), nearly always with the same result. The labels of positions of substitution in a selected group of benzenoid hydrocarbons based on inspection of Hückel MOs are collected in Figure 3.

The effect of E and I perturbations in D, N, and S positions on the MCD of the lower of the L bands, as predicted by first-order perturbation theory, is listed in Table 1. A plus sign means a more positive or less negative MCD peak relative to the unperturbed parent hydrocarbon, a negative sign means a less positive or more negative MCD peak. The effects on the MCD of the upper of the two L bands are just the opposite. In the absence of nπ* transitions, the two L transitions correspond to the lowest two observed bands in the spectra.

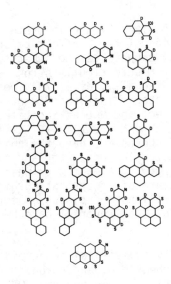

FIGURE 3. Classification of positions in selected polynuclear benzenoid hydrocarbons.

The effect on the first vibrational component (the 0-0 peak) of the longest-wavelength transition is particularly useful to monitor in practice.

TABLE 1

EFFECT OF PERTURBATIONS ON MCD INTENSITY OF THE L_1 BAND IN BENZENOID HYDROCARBONS

Perturbation	Type of Position[a]		
	D	N	S
+E	+	0, then +	-, then 0, then +
-E	-	0, then -	+, then 0, then -
+I	+	0	-
-I	-	0	+

a   Changes which occur as the E effect gradually becomes stronger are noted.

Three comments on Table 1 are in place. First, it is seen that the direction of the E effect of substitution in subdominant positions changes as the substituent becomes stronger, so that for a sufficiently strong substituent, all positions respond equally, as if they were of type D. In practice, weakly interacting substituents such as alkyls and halogens affect the MCD as indicated on the left hand side of the column under S (+E: -, -E: +). Only very strong substituents such as NH_2 have the ability to cause the inversions indicated in the center and on the right hand side of the column. The ease with which this happens depends on the structure of the parent chromophore in a predictable way but will not be discussed further here. Only a few examples of such reversal have been observed so far.

Second, in highly symmetrical benzenoid hydrocarbons such as benzene, triphenylene, and coronene, the highest two occupied MO's have the same energy (HOMO is degenerate). These are referred to as double-soft chromophores and the expected substituent effects are +E: +, -E: -, +I: 0, -I: 0.

Third, the magnitude of the effect of a given substituent in a position of a given type, say D, will depend on the actual size of the squares of the MO coefficients. This information is not given in Figure 3 but can be obtained readily from published tables (5).

<u>Consequences for Analytical Chemistry</u>. The theoretical results described above have immediate consequences of interest in analytical chemistry. If the nature of a weakly interacting substituent on a benzenoid hydrocarbon is known, a comparison of the MCD of its L bands with that of the parent hydrocarbon will show whether the position of substitution is of the D, N, or S type. In most cases this will not lead to a unique identification of the position of substitution, since more than one position of each kind is present, but it will narrow down the possibilities considerably. If some of the positional isomers in question are available for comparison, it may well be possible to limit the choices further by consideration of the magnitude of the effect. Conversely, if the position of substitution is known, information on the nature of the substituent can be obtained. The analysis is somewhat more complicated for very strongly interacting E substituents such as NH_2 and CHO.

In addition to such "a priori" use of the understanding of the MCD signs, it is also useful to note that the various positional isomers of a substituted benzenoid hydrocarbon are much more likely to differ widely in their MCD spectra than in their ordinary UV or fluorescence spectra. This suggests that even a purely empirical use of MCD, based on comparison with spectra of authentic samples, is likely to be of interest. Indeed, it has been demonstrated on the example of the three isomeric methylpyrenes that a quantitative analysis of their mixtures at the 1-μg level is much more reliably performed by MCD than by UV, simply because the MCD spectra are much less alike (6).

Comparison with Experiment

Previous Results. Table 2 summarizes the MCD signs of the first $\pi\pi^*$ transition (0-0 peak) observed for a variety of derivatives and hetero analogues of polynuclear hydrocarbons, some soft chromophores (I - IV) and some hard chromophores (VI, VII). This provides a nice illustration of the sensitivity of the former and insensitivity of the latter to weak perturbations. The Table also gives the signs predicted by the simple theory from Hückel MO coefficients (in the case of the hard chromophores, from orbital energies). It is seen that already this simplest approximation yields an excellent agreement with experiment.

1-Fluorobenzanthracene. The previous results were obtained on derivatives of hydrocarbons which were highly symmetrical and thus likely to obey the simple theory better. They were also relatively small and of limited biological activity. We have therefore selected a derivative of benzanthracene (V) for further testing of the theory. 1-Fluorobenzanthracene contains a -E substituent in a subdominant position and would be expected to exhibit a positive first MCD peak by the simple rules. Figure 4 shows its UV and MCD spectra and demonstrates that the expectation is indeed fulfilled.

Conclusions and Future Outlook

It appears that the perimeter model treatment of the MCD of aromatics is not only a nice demonstration of the power of simple MO models but also has some consequences of possible practical interest. The remarkable yet predictable sensitivity of MCD to positional isomerism in derivatives and hetero analogues of polynuclear benzenoid

TABLE 2

EFFECT OF SUBSTITUTION ON THE MCD SIGN OF THE L$_1$ BAND IN POLYNUCLEAR AROMATIC HYDROCARBONS (6,7)

Hydrocarbon	Substituent	Position	Observed Sign	Expected Sign
SOFT CHROMOPHORES[a]				
Naphthalene (I)				
	CH$_3$, OH, NH$_2$, F	1(D)	−	−
	CH$_3$, F	2(S)	+	+
	OH, NH$_2$	2(S)	−	b
	CN, CHO	1(D)	+	+
	CN, CHO	2(S)	+	b
	(N)	1(D)	+	+
	(N)	2(S)	−	−
Anthracene (II)				
	F, NH$_2$	1(D)	−	−
	F, CH$_3$	2(S)	+	+
	NH$_2$	2(S)	0	b
	CH$_3$, F, Cℓ, Br	9(D)	−	−
	(N)	9(D)	+	+
Phenanthrene (III)				
	(N)	1(D)	+	+
	(N)	4(S)	+	− (weak)
	(N)	9(D)	+	+
Pyrene (IV)				
	F, CH$_3$, OH, NH$_2$	1(D)	−	−
	F, CH$_3$, OH, NH$_2$	2(S)	+	+
	F, CH$_3$, NH$_2$	4(D)	−	−
	COOC$_2$H$_5$	1(D)	+	+
	COOC$_2$H$_5$	2(S)	−	−
	COOC$_2$H$_5$	4(D)	+	+
HARD CHROMOPHORES[c]				
Acenaphthylene (VI)				
	CN, Br	1	+	+
	F	2	+	+
	F	3	+	+
	F	4	+	+

TABLE 2 (CONTINUED)

Fluoranthene (VII)

F, NH$_2$, COOH	1	+	+
F, NH$_2$	2	+	+
F, NH$_2$, (N), COOH	3	+	+
F, NH$_2$, COOH	7	+	+
F, NH$_2$	8	+	+

[a] A positive sign indicates that the substitution or aza replacement (N) made the MCD signal of the 0-0 component more positive or less negative relative to the parent hydrocarbon. In almost all cases, this means that the signal is positive. A negative signs means the opposite, and in almost all cases this means that the signal is negative.

[b] The expected sign agrees with that observed after it is recognized that the substituent is of the "strong" type - cf. Table 1. This complication can only occur for E substituents in positions of type S.

[c] The MCD sign of the 0-0 component is given.

hydrocarbons can be utilized in analytical applications. The sample requirements are comparable to those in UV spectroscopy, the instrumentation is commercially available, and the interpretations are straightforward.

A considerable further increase in sensitivity and selectivity could be achieved by using fluorescence to detect the MCD effect (8).

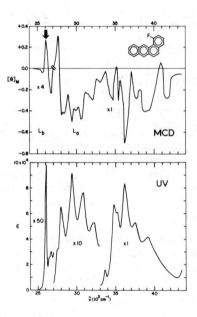

FIGURE 4. MCD and UV spectra of 1-fluorobenzanthracene in cyclohexane. The arrow indicates the 0-0 band of the L_b transition.

ACKNOWLEDGEMENT

This work was supported by NIH grant GM-21153.

REFERENCES

1. Michl, J. (1978): MCD of cyclic π-electron systems. I. Algebraic solution of the perimeter model for the A and B terms of high-symmetry systems with a (4N+2)-electron [n]annulene perimeter, J. Am. Chem. Soc., 100:6801-6811; Michl, J. (1978): MCD of cyclic π-electron systems. II. Algebraic solution of the perimeter model for the B terms of systems with a (4N+2)-electron [n]annulene perimeter, J. Am. Chem. Soc., 100:6812-6818.

2. Michl, J. (1980): Electronic structure of aromatic π-electron systems as reflected in their MCD spectra, Pure Appl. Chem., 52:1549-1563.

3. Pedersen, P.B., Thulstrup, E.W., and Michl, J.: Magnetic circular dichroism of cyclic π-electron systems. 24. Polarizations and assignments in some pentacyclic and hexacyclic benzenoid hydrocarbons, Chem. Phys., In press.

4. Michl, J. (1978): MCD of cyclic π-electron systems. III. Classification of cyclic β-chromophores with a (4+2)-electron [n]annulene perimeter and general rules for substituent effects on the MCD spectra of soft chromophores, J. Am. Chem. Soc., 100:6819-6823.

5. Streitwieser, A, and Brauman, J.I. (1965): "Supplemental tables of molecular orbital calculations", Pergamon: Oxford.

6. Waluk, J.W., and Michl, J. (1981): Determination of positional isomers of methylpyrenes and other polycyclic aromatic hydrocarbons by magnetic circular dichroism, Anal. Chem., 53:236-239.

7. Kolc, J., Thulstrup, E.W., and Michl, J. (1974): Excited singlet states of fluoranthene. I. Absorption, linear and magnetic circular dichroism and polarized fluorescence excitation of the fluorofluoranthenes, J. Am. Chem. Soc., 96:7188-7202.; Thulstrup, E.W., and Michl, J. (1976): Electronic states of acenaphthylene. Linear dichroism in stretched polyethylene and magnetic circular dichroism, J. Am. Chem. Soc., 98:4533-4540; Thulstrup, E.W., Downing, J.W., and Michl, J. (1977): Excited singlet states of pyrene. Polarization directions and magnetic circular dichroism of azapyrenes, Chem. Phys., 23:307-319; Kenney, J.W., III, Herold, D.A., and Michl, J. (1978): MCD spectroscopy of cyclic π-electron systems. XVI. Derivatives of acenaphthylene, J. Am. Chem. Soc., 100:6884-6887; Vašák, M., Whipple, M.R., and Michl, J. (1978): MCD spectroscopy of cyclic π-electron systems. VII. Aza analogs of naphthalene, J. Am. Chem. Soc., 100:6838-6843; Whipple, M.R., Vašák, M., and Michl, J. (1978): MCD spectroscopy of cyclic π-electron systems. VIII. Derivatives of naphthalene, J. Am. Chem. Soc., 100:6844-6852; Steiner, R.P., and Michl, J. (1978): MCD spectroscopy of cyclic π-electron systems. XI. Derivatives and aza analogs of anthracene, J. Am. Chem. Soc., 100:6861-6867;

Vašák, M., Whipple, M.R., and Michl, J. (1978): MCD spectroscopy of cyclic π-electron systems. XII. Aza analogs of phenanthrene, J. Am. Chem. Soc., 100:6867-6871; Vašák, M., Whipple, M.R., Berg, A., and Michl, J. (1978): MCD spectroscopy of cyclic π-electron systems. XIII. Derivatives of pyrene, J. Am. Chem. Soc., 100:6872-6877; Dalgaard, G.P. and Michl, J. (1978): MCD spectroscopy of cyclic π-electron systems. XVII. Derivatives of fluoranthene, J. Am. Chem. Soc., 100:6887-6892.

8. Sutherland, J.C., and Low, H.H. (1976): Proc. Natl. Acad. Sci., 73:276-280.

BINDING OF BENZO[a]PYRENE-DIOL EPOXIDES AND -TRIOL CARBONIUM IONS TO DNA

KENNETH J. MILLER, JONATHAN J. BURBAUM, and JOSEF DOMMEN
Department of Chemistry, Rensselaer Polytechnic Institute, Troy, New York 12181.

INTRODUCTION

Steric effects are important in the binding of the ultimate carcinogens, benzo[a]pyrene diol epoxides (BPDEs) and the benzo[a]pyrene triol carbonium ions (BPTCs) to DNA. The ability of diaxial (da) and diequatorial (de) conformers of four stereoisomers, (±) 7β,8α-dihydroxy-9α,10α-epoxy-7,8,9,10-tetrahydrobenzo[a]pyrene, denoted by BPDE I(±), and (±) 7β,8α-dihydroxy-9β,10β-epoxy-7,8,9,10-tetrahydrobenzo[a]pyrene, denoted by BPDE II(±), to bind to DNA is examined theoretically. These compounds are illustrated in Figure 1. It is first shown that intercalation of BPDEs into DNA cannot physically permit binding to the amino groups on guanine, adenine or cytosine for an acceptable covalent bond length and proper hybrid configurations about the reacting atoms even if the DNA deforms and the base pairs separate to the maximum extent of 8.25 Å to expose these groups. However, BPDE II(-) is in a favorable position for phosphorylation. Then it is shown that outside binding to N2 on guanine cannot occur unless the B-DNA structure deforms by unwinding which also forces a separation of adjacent base pairs to permit access to N2 on guanine by C10 on BPDE. Finally, it is suggested that the stereoselectivity results from two sources: the appropriate sequences in DNA and the steric hindrance of the BPDEs. The first occurs because pyrimidine (p) purine sequences unwind selectively. The latter occurs because the benzo[a]pyrene moiety has a preferred orientation in the minor groove of the DNA in an open form. The correlation between distance of closest approach and stereoselectivity exists only with respect to the orientation of the benzo[a]pyrene moiety.

The binding of BPDE I(+) to N2 on guanine, as well as N4 on cytosine and N6 on adenine is well known (1-5). There is also evidence for the formation of a bond to oxygen on the phosphate group (6-9); however, hydrolysis conditions have limited analyses to the covalently bound adduct involving the amino groups. Although the BPDEs appear to intercalate (10-11), the covalently bound adduct is bound to the outside of the DNA (12-17), and this binding is accompanied by unwinding of the DNA (18). The purpose of this theoretical investigation is to suggest a mechanism in which both the DNA and the BPDEs play a role in the steroselectivity. It is shown that the addition products should be trans for BPDE I(+) and II(+) and cis for BPDE I(-) and II(-). As this paper deals with the steric effects, the selective metabolism of a particular enantiomer of the two stereoisomers and their chemical reactivity toward N2 on guanine is beyond the scope of this investigation.

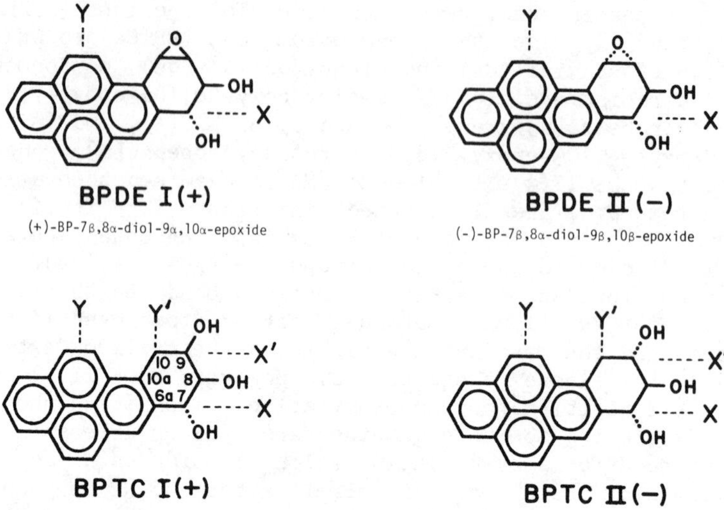

BPDE I(+)
(+)-BP-7β,8α-diol-9α,10α-epoxide

BPDE II(-)
(-)-BP-7β,8α-diol-9β,10β-epoxide

BPTC I(+)

BPTC II(-)

FIGURE 1. BPDE I(+) and II(-), and their corresponding triol carbonium ions, BPTC I(+) and II(-).

BINDING SITES IN DNA AND BINDING ENERGIES

An algorithm to generate nucleic acid structures has been developed to define possible conformations of DNA (19). The B-DNA structure is known to be the most favored for base pairs separated by 3.38 Å and perpendicular to the helical axis. In addition, for base pairs separated by 6.76 Å to accept an intercalant, three families of intercalation sites, denoted I, II and III, have been obtained for a tetramer duplex extracted from a section of the B-DNA double helix for which the helical angles are $26°$, $20°$ and $8°$ (20,21). The DNA is unwound by $10°$, $16°$ and $28°$, respectively. At a helical angle of $35.5°$, and with base pairs separated by 3.38 Å, B-DNA is most unwound. Further unwinding requires separation of the base pairs, otherwise, the backbone would be compressed (19). Hence, the use of these conformations both as intercalation sites, as well as examples of unwound forms of DNA in this investigation. These unwound forms reduce the steric hindrance for the approach of an adduct in the case of outside binding. In addition, a conformation with the base pairs fully separated to 8.25 Å is used to present the case of the most unhindered approach of the BPDEs to N2 on guanine. The calculations are performed with dimer duplex units in the notation of previous work (20,21).

Binding energies, relative to infinite separation of the BPDEs and BPTCs and DNA, and binding positions are calculated for the interaction of a molecule with DNA in each of the conformations of DNA. The interfragment interactions between the molecule and DNA are added to the intrafragment interactions consisting of the change in the conformational energy of the molecule as it conforms to the binding site. They consist of Coulomb, van der Waals attractive and repulsive (6-14 potential), and torsional energy terms. The net charges on the BPDEs and the resulting BPTCs are obtained with the MINDO/3 molecular orbital theory (22). It should be noted that the choice of a carbonium ion as the attacking agent is not essential in this investigation. Rather, the ability of the adduct to fit onto the DNA without steric interference during covalent bond formation is being examined.

INTERCALATION OF BPDE INTO DNA AS A STEP IN COVALENT BINDING

Intercalation of the diaxial and diequatorial conformers of BPDE I(+) and BPDE II(-) is shown in Figure 2. These isomers are constrained between planar base pairs in the site primarily through steric considerations. The results for the binding orientations are assumed to be nearly identical for the other two enantiomers BPDE I(-) and BPDE II(+). Only site I was considered because previous experience with intercalation of planar chromophores favored this site by at least 10 kcal. Sample calculations with sites II and III also support this decision. Minimum energy conformations calculated for insertion into the major and minor grooves revealed a global minimum for each conformer of the two isomers. The distances between C10 and the amino nitrogens on guanine, cytosine and adenine range from 3.9 Å to 7.4 Å. In these orientations, covalent binding is not possible. If the C10 and amino nitrogen atoms were directly over each other, then the minimum distance would be 3.38 Å, which is the minimum separation arising from stacking of planar molecules, and the bond angles would be 90°. Therefore, if intercalation occurs, covalent binding to an amino group is impossible

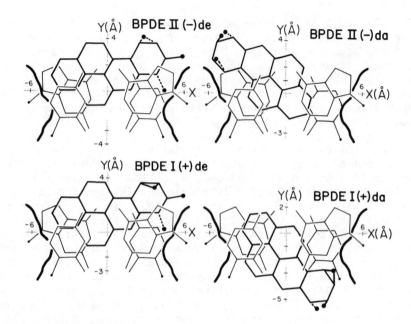

FIGURE 2. Optimum orientation of the diaxial and diequatorial forms of BPDE I(+) and II(-) in intercalation site I with base pairs G·C viewed along the helical axis. The solid curve, or steric contour, defines the inaccessible region of the backbone.

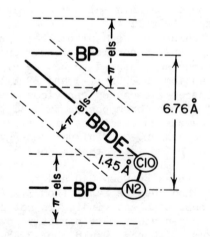

FIGURE 3. Spatial representation of the π-systems of each base pair (BP) and BPDE in an orientation required for covalent bond formation.

without a further separation of the base pairs and/or a withdrawal of the pyrene moiety from the site to allow the C10 of the BPDEs to approach the amino nitrogens more closely for the formation of a covalent bond. In Figure 3, the spatial requirements for proper hybridization are shown to scale. The π-system of the BPDE intersects the base pairs. The severe steric constraint with the upper base pair can be relieved by opening the DNA from this classical intercalation structure, and this leads to outside binding either through a forced separation of the upper base pair or through a reorientation of the BPDE.

OUTSIDE BINDING

Results for the calculation of the energy of interaction for the approach of BPTC to N2 on guanine of B-DNA and of DNA unwound to several conformations are presented in subsequent sections. Two representative conformers, da and de, of each of the four carbonium ions, BPTC I($\pm$) and BPTC II($\pm$), are chosen for study with both cis and trans addition. If the distance of closest approach, R, is defined as that which occurs when the energy increases by approximately 10% from the minimum, then the possibility of bond formation between C10 on the BPTCs and N2 on guanine can be assessed. The reaction coordinates, shown in Figure 4, consist of the distance between C10 on BPTC and N2 on guanine, the bond angle, α, the dihedral angle, β, and three Euler angles. Each BPTC is translated in the space-fixed (X, Y, Z) coordinate system of the DNA so that C10 becomes the origin of the new (X', Y', Z') system shown in Figures 1 and 4. Then it is rotated through Euler angles in the prime system. For a given distance, R, the energy of interaction between the BPTC adduct and DNA, and the conformational energy of BPTC is minimized with respect to α, β and the Euler angles. With this procedure, the bond angles are not restrained to be those which are chemically acceptable; however, they are examined afterwards to determine whether proper hybrid configurations can exist on the reacted atoms. This approach greatly enhances the minimization procedure, and it permits the use of several rigid DNA structures as representative examples of possible conformations in the dynamics of the mechanism. The conformation of the benzo-ring of BP is chosen to be the more stable half-chair form (23). Formation of the (+) and (-) forms of BPTC I and BPTC II results in a carbonium ion for which the bond between C10 and the amino group is axial with diequatorial orientations of the 7,8-diol groups and vice versa. For the special case of outside binding, the van der

Waals radius of C10 on BPTC is set equal to zero, and one hydrogen atom on N2 is removed. These adjustments insure that a large repulsion between N2 and C10 cannot occur as this interatomic distance decreases for possible formation of a covalent bond between BPTC and guanine.

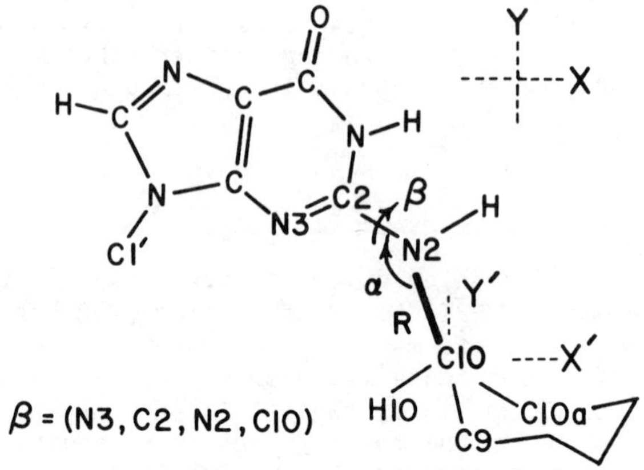

FIGURE 4. Reaction coordinates for the approach of C10 on BPTC to N2 on guanine. Three Euler angles define the orientation about C10.

UNWINDING AS A NECESSARY STEP IN COVALENT BINDING

The inability of B-DNA to accommodate the BPTCs derived from the BPDEs, and the need to unwind the DNA to expose the N2 on guanine is illustrated best by steric contours of the binding site. First, the minimum energy orientation of the BPDE adduct on guanine is taken as an assumed conformation. Then a steric contour about the DNA is drawn in the plane of the pyrene moiety. These steric contours define the inaccessible region of the DNA. They are calibrated so that a chemical structure can be superimposed on the contour. Unfavorable contacts occur when atomic positions on the molecule penetrate the contour. This representation is discussed in detail elsewhere(24) and in this volume (25). In Figure 5, trans addition of BPDE I(+)da is represented in several conformations of DNA for several orientations about N2 on guanine. The view is into the minor groove. The pyrene moiety is oriented approximately 30° relative to the helical axis as found experimentally (14). As the DNA unwinds, the base pairs separate from the normal stacking distance of

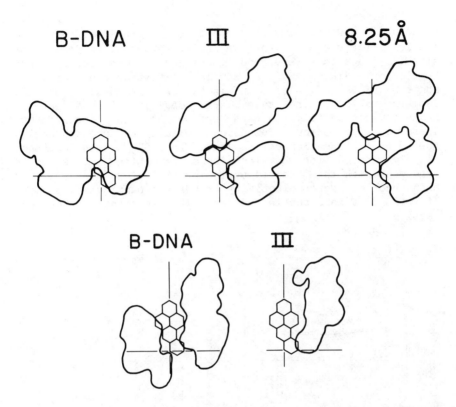

FIGURE 5. Steric contours in the plane of (C7,C6a,C10a,C10) (———) and 1.2 Å below (- - -) the plane for the trans-BPDE I(+)da-guanine adduct in B-DNA and unwound forms are shown for the BPDE I(+)-guanine adduct. The optimum conformation is taken as the one about a free guanine (upper), and in the presence of each form of DNA (lower).

3.38 Å (B-DNA) through 6.76 Å (site III) to 8.25 Å in the fully open form. Thus the space available for the pyrene increases as the DNA unwinds. Based on an examination of the steric contours, it is clear that the DNA must unwind from the B-form. That the fit is very much improved in the fully extended form is obvious. However, of all the possible conformations of DNA, these have been chosen to illustrate a necessary step in the mechanism leading to covalent binding. Certainly, it is clear from Figure 5 that a few contacts are not ideal if the optimum orientation of the BPTC about N2 is assumed. For example, the hydrogen atoms in the 2, 3 and 11 positions are unfavorable. This is in sharp contrast to all positions exhibiting unfavorable contacts in the B-DNA structure. However, a slight adjustment of the BPTC due to steric interference of the DNA in the unwound form denoted by site III leads to a very good fit as seen in the lower right portion of Figure 5. A similar situation occurs for cis addition of BPDE II(-)de. In fact, an examination of all isomers exhibits the same large repulsive effects for both trans and cis addition to B-DNA but not to the open forms. These steric contours demonstrate clearly the need to unwind the B-DNA conformation before any of the BPDEs can approach closely enough for covalent bond formation. This is in agreement with the observed unwinding (18) of $30°$ to $35°$. The phenomenon of stereoselectivity by DNA for particular isomers of BPDE is a more subtle problem. It is examined in the next section.

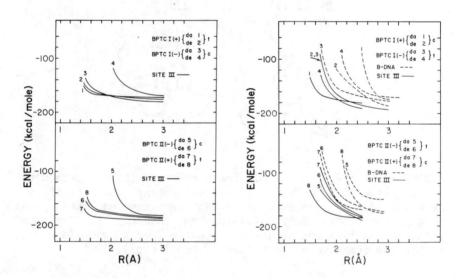

FIGURE 6. Accessibility of the da and de conformers of BPTC I(±), II(±) to ↑C·G, G·C↓ (left) and ↑G·C, C·G↓ (right) of B-DNA and of an unwound form denoted by site III.

STEREOSELECTIVITY OF BPDEs BY DNA

In Figure 6 the interaction energy between each of the BPTCs is presented for the approach to B-DNA and to one unwound conformation denoted by site III. In all cases, B-DNA exhibits a very large increase in energy well before a covalent bond can form. Although the conformers of the various isomers corresponding to curves 1, 4, 5 and 8, and 1,

TABLE 1

CLOSEST APPROACH OF BPTCs (DERIVED FROM BPDEs) TO UNWOUND DNA*

BPDE	R Å	α deg	γ deg	δ deg	ε deg	Adduct Type**
I(+)da	2.5	-	-	-	-	t
I(+)de	1.7	134	103	126	109	t
I(−)da	1.7	128	95	139	100	c
I(−)de	2.5	-	-	-	-	c
II(−)da	2.5	-	-	-	-	c
II(−)de	1.7	143	115	110	118	c
II(+)da	1.7	148	109	115	119	t
II(+)de	1.7	150	98	112	100	t
I(+)da	1.7	125	105	101	137	c
I(+)de	2.5	-	-	-	-	c
I(−)da	2.0	106	91	87	127	t
I(−)de	2.0	102	119	83	142	t
II(−)da	2.5	-	-	-	-	t
II(−)de	2.0	111	91	91	124	t
II(+)da	2.0	108	92	91	122	c
II(+)de	2.0	111	109	97	138	c

* For the distance of closest approach, (R), the angles about N2, α , and C10, γ = (N2, C10, C9), δ = (N2, C10, C10a) and ϵ = (N2, C10, H10) are given for trans and cis addition. Data in the upper (lower) portion of this table correspond to curves in the left (right) hand portion of Figure 6.
** A view into the minor groove for the two different orientations of the benzo[a]pyrene moiety is indicated.

2, 3, 6, 7 and 8 of the left and right portions of the Figure 6 approach site III most closely, an examination of hybrid configurations about C10 provides a further selection. The two orientations of the benz[a]pyrene moiety of the BPTCs in the minor groove of DNA are termed adduct types in Table 1. Each type presents the epoxide to N2 on guanine for trans or cis addition according to the particular enantiomer of each stereoisomer. In the upper part of the table, this adduct type results in trans addition of the I(+) isomer, cis addition of the I(−) isomer, etc., which are related by mirror symmetry. By reorienting the benz[a]pyrene moiety, the adduct type in the lower part of the table results in an interchange of the trans and cis addition assignments. As viewed into the minor groove of DNA, the position of guanine in the lower base pair must be C·G and G·C as shown for the two types, respectively, to permit an optimum approach to DNA.

The first step in selectivity occurs when the DNA opens. Of the two possibilities, the purine(p)pyrimidine sequences open most easily, and these sequences result in products listed in the upper half of Table 1. Thus BP can be A·T or G·C. The energy required to open DNA to yield guanine in the optimum position for cis addition of BPDE I(+), etc. shown in the lower half of Table 1 requires an additional 6 kcal if BP is C·G and up to 14 kcal if BP is C·G. The energy values required in this step are listed in Table V of reference 21. The second step occurs when the benzo[a]pyrene moiety lies in the minor groove so that it can bind to N2 on guanine. Thus the combination of the opening of DNA with the minimum energy and the manner in which the BPDEs lie in the minor groove yield the adduct types listed in the upper part of Table 1. As the DNA conformations have not been fully relaxed, ideal bond lengths and bond angles about N2 and C10, the atom about which BPTCs are orientated, have not been realized. Thus, in the upper part of Table 1, the BPDEs which sterically are most favourable for binding are I(+)de and I(-)da for trans and cis addition. For several remaining cases, one may argue that an adjustment in and to slightly different values would only relocate C10 with little change in the contact with the DNA. Then, II(+)da, II(+)de and II(-)de for trans, trans and cis addition should also be included. In fact, by choosing α to be near an experimental of 123°, the configuration about C10 approaches the ideal tetrahedral configuration at the cost of only a few kcal. Thus, the principal criterion for adduct formation appears to lie in an acceptable bond length without a large steric repulsion by DNA after opening to pyrimidine (p) purine sequences. Thus trans addition of BPDE I(+)de occurs rather than cis addition, and this is observed experimentally (26).

The mechanism for the binding of the BPDEs to DNA may be summarized as follows: The DNA opens selectively to sequences shown in the upper part of Table 1. If the BPDEs intercalate, no covalent bond occurs. Rather hydrolysis to a tetraol may occur which is also found to intercalate (17). However, if the BPDEs bind to the outside of DNA in the minor groove, then covalent bond formation occurs with the preference for trans addition of BPDE I(+) and II(+) and cis addition for BPDE I(-) and II(-). That two binding sites are observed experimentally, but the covalently bound adduct is not intercalated (17), and that BPDE I(+) binds to N2 on guanine for trans addition after the DNA is unwound by the covalently bound BPDE (18) has been rationalized by this theoretical analysis. However, the further selection of BPDE I(+) as carcinoigenic and BPDE II(-) as mutagenic (27-30) requires an analysis of the metabolites produced and the chemical reactivity of the various BPDEs toward N2 on guanine.

ACKNOWLEDGEMENTS

This work was supported by the National Institutes of Health under Grant CA-28924, and by a grant of computer time from Rensselaer Polytechnic Institute.

REFERENCES

1. Jeffrey, A. M., Weinstein, I. B., Jennette, K. W., Grzeskowiak, K., Nakanishi, K., Harvey, R. G., Autrup, H., and Harris, C. (1977): Structures of benzo(a)pyrene-nucleic acid adducts formed in human and bovine bronchial explants, Nature (London), 269: 348-350.
2. Jennette, K. W., Jeffrey, A. M., Blobstein, S. H., Beland, F. A., Harvey, R. G., and Weinstein, I. B. (1977): Nucleoside adducts from the in vitro reaction of benzo[a]pyrene-7,8 dihydrodiol 9,10-oxide or benzo[a]pyrene 4,5-oxide with nucleic acids, Biochemistry, 16: 932-938.
3. Meehan, T., Straub, K., and Calvin, M. (1977): Benzo(a)pyrene diol epoxide covalently binds to deoxyguanosine and deoxyadenosine in DNA, Nature, 269: 725-727.
4. Jeffrey, A. M., Grzeskowiak, K., Weinstein, I. B., Nakanishi, K., Roller, P., and Harvey, R. G. (1979): Benzo(a)pyrene-7,8-dihydrodiol 9,10-oxide adenosine and deoxyadenosine adducts: Structure and stereochemistry, Science, 206: 1309-1311.
5. Ivanovic, V., Geacintov, N. E., Yamasaki, H., and Weinstein, I. B. (1978): DNA and RNA adducts formed in hamster embryo cell cultures exposed to benzo[a]pyrene, Biochemistry, 17: 1597-1603.
6. Mager, R., Huberman, E., Yang, S. K., Gelboin, H. V., and Sachs, L. (1977): Transformation of normal hamster cells by benzo(a)pyrene diol-epoxide, Int. J. Cancer, 19: 814-817.
7. Koreeda, M., Moore, P. D., Wilson, P. G., Levin, W., Conney, A. H., Yagi, H., and Jeffrey, A. M. (1978):

Binding of benzo[a]pyrene 7,8-diol-9,10-epoxides to DNA, RNA, and protein of mouse skin occurs with high stereoselectivity, Science, 199: 778-781.

8. Jeffrey, A. M., Jennette, K. W., Blobstein, S. H., Weinstein, I. B., Beland, F.A., Harvey, R. G., Kasai, H., Miura, I., and Nakanishi, K. (1976): Benzo[a]pyrene nucleic acid-derivative found invivo structure of a benzo[a]pyrene tetrahydrodiol epoxide-guanosine adduct, J. Amer. Chem. Soc., 98: 5714-5715.

9. Moore, P. D., Koreeda, M., Wisloki, P. G., Levin, W., Conney, A. H., Yagi, H., and Jerina, D. M. (1977): Drug Metabolism Concepts, edited by D. M. Jerina, Amer. Chem. Soc., Wash. D. C., pp 127-154.

10. Drinkwater, N. R., Miller, J. A., Miller, E. C., and Yang, N.-C. (1978): Covalent intercalative binding to DNA in relation to the mutagenicity and hydrocarbon epoxides and N-acetoxy-2-acetylaminofluorene, Cancer Res., 38: 3247-3255.

11. Yang, N.-C., Ng, L.-K., Neoh, S. B., and Leonov, D. (1978): A spectrofluorimetric investigation of calf thymus DNA modified by BP diolepoxide and 1-pyrenyloxirane, Biochem. Biophys. Res. Commun., 82: 929-934.

12. Prusik, T., Geacintov, N. C., Tobiasz, C., Ivanovic, V., and Weinstein, I. B. (1979): Fluorescence study of the physico-chemical properties of a benzo(a)pyrene 7,8-dihydrodiol 9,10-oxide derivative bound covalently to DNA, Photochem. Photobiol., 29: 223-232.

13. Prusik, T., and Geacintov, N. E. (1979): Fluorescence properities of a benzo(a)pyrene 7,8 dihydrodiol 9,10-oxide-DNA adduct. Conformation and effects of intermolecular DNA interactions, Biochem. Biophys. Res. Commun., 88: 782-790.

14. Geacintov, N. E., Gagliano, A., Ivanovic, V., and Weinstein, I. B. (1978): Benzo[a]pyrene-7,8-dihydrodiol 9,10-oxide covalently bound to DNA, Biochemistry, 17: 5256-5262.

15. Frenkel, K., Grunberger, D., Boublik, M., and Weinstein, I. B.(1978): Conformation of dinucleoside monophosphates modified with benzo[a]pyrene-7,8-dihydrodiol 9,10-oxide as measured by circular dichroism, Biochemistry, 17: 1278-1282.

16. Lefkowitz, S. M., Brenner, H. C., Astorian, O. G., and Clarke, R. H. (1979) Optically detected magnetic resonance study of benzo[a]pyrene-7,8-dihydrodiol 9,10-oxide covalently bound to DNA, FEBS Letters, 105: 77-80.

17. Ibanez, V., Geacintov, N. E., Gagliano, A. G., Brandimarte, S., and Harvey, R. G. (1980): Physical

binding of tetraols derived from 7,8-dihydroxy-9,10-epoxybenzo[a]pyrene to DNA, J. Amer. Chem. Soc., 102: 5661-5666.
18. Gamper, H. B., Straub, K., Calvin, M., and Bartholomew, J. C. (1980): DNA alkylation and unwinding induced by benzo[a]pyrene diol epoxide: Modulation by ionic strength and superhelicity, Proc. Natl. Acad. Sci. (USA), 77: 2000-2004.
19. Miller, K. J. (1979): Interactions of molecules with nucleic acids. I. An algorithm to generate nucleic acid structures with and application to the B-DNA structure and a counterclockwise helix, Biopolymers, 18: 959-980.
20. Miller, K. J., and Pycior, J. F. (1979): Interaction of molecules with nucleic acids. II. Two pairs of families of intercalation sites, unwinding angles and the neighbor-exclusion principle, Biopolymers, 18: 2683-2719.
21. Miller, K. J. (1981): Three families of intercalation sites for parallel base pairs: A theoretical model, In: Proceedings of the Second SUNYA Conversation in the Discipline Biomolecular Stereodynamics, Vol. II, edited by R. H. Sarma, pp 469-486, Adenine Press, New York.
22. Bingham, R. C., Dewar, M. J. S., and Lo, D. H. (1975): Ground states of molecules. XXV. MINDO/3. An improved version of the MINDO semiempirical SCF-MO method, J. Amer. Chem. Soc., 97: 1285-1293.
23. Lavery, R., and Pullman, B. (1979): Theoretical model study of the reactivity of benzo(a)pyrene diol epoxide with the amino groups of the nucleic acid bases, Int. J. Quant. Chem., XVI: 175-188.
24. Miller, K. J., Macrea, J., and Pycior, J. F.(1980): Interactions of molecules with nucleic acids. III. Steric and electrostatic energy contours for the principal intercalation sites, prerequsites for binding, and the exclusion of essential metabolites from intercalation, Biopolymers, 19: 2067-2089.
25. Miller, K. J., Kowalczyk, P., and Segmuller, W.: Representations of binding sites for PAHs in DNA with computer graphics techniques, (this volume).
26. Jeffrey, A. M., Weinstein, I. B., Jennette, K. W., Grezeskowiak, K., Nakanishi, K., Harvey, R. G., Autrup, H., and Harris, C. (1977): Structures of benzo(a)pyrene-nucleic acid adducts formed in human and bovine bronchial explants, Nature, 269: 348-350.
27. Huberman, E., Sachs, L., Yang, S. K., and Gelboin, H. V.(1976): Identification of mutagenic metabolites of benzo[a]pyrene in mammalian cells, Proc. Natl. Acad. Sci. (USA), 73: 607-611.
28. Wood, A. W., Chang, R. L., Levin, W., Yagi, H., Thakker,

D. R., Jerina, D. M., and Conney, A. H. (1977): Differences in mutagenicity of the optical enantiomers of the diastereomeric benzo[a]pyrene 7,8-diol-9,10-epoxides, Biochem. Biophys. Res. Commun., 77: 1389-1396.

29. Wood, A. W., Wislocki, P. G., Chang, R. L., Levin, W., Lu, A. Y. H., Yagi, H., Herandez, O., Jerina, D. M., and Conney, A. H. (1977): Mutagenicity and cytotoxicity of benzo(a)pyrene benzo-ring epoxides, Cancer Res., 37: 3358-3366.

30. Buening, M. K., Wislocki, P. G., Levin, W., Yagi, H., Thakker, D. R., Akagi, H., Koreeda, M., Jerina, D. M., and Conney, A. H. (1978): Tumorigenicity of the optical enantiomers of the diastereomeric benzo[a]pyrene 7,8-diol-9,10-epoxides in newborn mice: Exceptional activity of (+)-7β,8α-dihydroxy-9α,10α-epoxy-7,8,9,10-tetrahydrobenzo[a]pyrene, Proc. Natl. Acad. Sci. (USA), 75: 5358-5361.

REPRESENTATIONS OF BINDING SITES FOR PAHs IN DNA WITH COMPUTER GRAPHICS TECHNIQUES

KENNETH J. MILLER, PAUL J. KOWALCZYK and WOLFGANG SEGMULLER
Department of Chemistry, Rensselaer Polytechnic Institute, Troy, New York 12181.

INTRODUCTION

The representation of molecules embedded in a receptor site requires clear pictures which provide conformational information about the species involved and permit qualitative analyses of the process. Applications include receptor sites which may be nucleic acids, and molecules which bind to the receptor such as antitumor and carcinogenic agents. On a molecular level, the interactions between receptor and molecules involve the enhancement of conformational and electrostatic interactions, and even covalent bond formation. A new method of representing a space filling receptor site is presented. It incorporates the details of the atoms in a steric surface in a manner which permits one to test whether or not a chemical structure will fit into the site. The method is a unification of many existing techniques. This paper deals with a description of the mathematical algorithm used, and in conjunction with a paper in this volume (1), the inability of BPDEs to bind to N2 on guanine while DNA is in the B-form. The need to unwind DNA to accommodate the BPDE, already discussed (1), is shown with steric contours by the very much improved fit of the BPDEs as DNA is unwound.

GENERATION OF STERIC COMPOUNDS

To generate the steric contours, a test atom, t, is used to probe the surface of a receptor. First the atomic positions of the DNA are defined for one of its conformations. Then the DNA is oriented in a coordinate system so that the contours can be drawn in a series of equally spaced planes along a z-axis. These contours represent the distance of closest approach defined when a test atom makes contact with the surface. The transition from the usual space filling models (2,3) to this method is shown in Figure 1. The path of hydrogen (H) represents the steric contour calculated when the 6-14 potential is zero, namely, when it begins to become repulsive. A test atom of hydrogen is used because the periphery of the PAHs contain hydrogen. This method of a soft sphere analog is described in detail elsewhere (4).

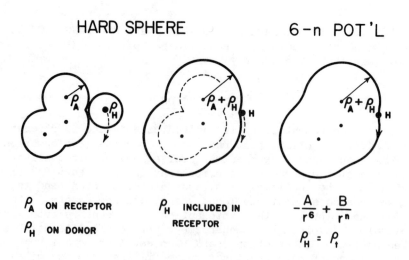

FIGURE 1. Representation of the path of H where the van der Waals radius ρ_H is first on the donor and then included in the receptor in a hard sphere model, and finally in the present method with a 6-14 potential.

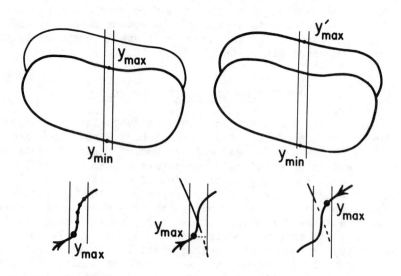

FIGURE 2. Steric contours with hidden regions defined by pairs of representative points in each strip, and the method of drawing subsequent contours with revised pairs of points which define new hidden regions.

The principal feature of this method lies in the incorporation of the van der Waal radius of the test atom into that of the receptor. Thus, chemical structures of molecules, rather than their space filling models, can be superimposed on the surface. If the atomic positions of the molecule do not touch the surface, the contacts are favorable, whereas, penetration of the surface results in poor contacts. Thus, visually one can determine the cause of a poor fit. The contours are generated in order so that the first is closest to the viewer. They are drawn as shown in Figure 2. In the xy plane of the screen, the x-direction is divided into strips. As the first contour is drawn, a pair of points, y(min) and y(max) for each strip is saved. They define the hidden region for the next contour. Points of the second contour are drawn only if they do not fall within the pair [y(min), y(max)] of the corresponding strip of the first contour. Thus, only the visible portion of the second contour is drawn, and a new hidden region, [y(min), y'(max)], is defined for each strip. Because the region is divided into strips the problem of resolution arises. Upon entering a strip, the first point is saved. As the second contour is drawn, it either intersects or falls short of the first contour depending on the direction in which the first contour was drawn. This problem is illustrated at the bottom of Figure 2. The contours are generated in the xy direction in a plane of constant z by extrapolation according to a procedure developed by Wahl (5). First the 6-14 potential

$$U(x,y) = C,$$

is recast as $y = f(x)$, differentiated at point P_0 to yield

$$(dy/dx)_0 = - (\partial U/\partial x)_0 / (\partial U/\partial y)_0,$$

and, with the second derivative (not given), the increment

$$\Delta y = (dy/dx)_0 \Delta x + (d^2y/dx^2)_0 \Delta x^2$$

is calculated for a given Δx in the extrapolation procedure. This increment is obtained by choosing a constant step Δs along the arc length

$$\Delta s = \sqrt{1 + (dy/dx)_0} \, \Delta x.$$

In Figure 3, extrapolation from point P_0 to $P_1^{(2)}$ is shown. If $P_1^{(2)}$ lies on a new contour $C^{(2)}$ which is too far from C, then a step perpendicular to the tangent at $P_1^{(2)}$ is taken to move closer to C, i.e. to $P_1^{(1)}$.

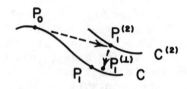

FIGURE 3. Generation of a contour by extrapolation from P_0 to $P_i^{(2)}$ with a possible correction from $P_i^{(2)}$ to $P_i^{(\perp)}$ to get back to the contour C.

STERIC CONTOURS OF BPDE I(+) IN DNA

To demonstrate the space available in the minor groove for the binding of the BPDEs to N2 on guanine, steric contours are presented for the addition of the corresponding triol carbonium ions, BPTCs, to DNA. Examples of cis and trans addition of BPDE I(+) to B-DNA, and to an unwound form of DNA, represented by site III, are presented in Figures 4 and 5. The preference for trans over cis addition through opening of the DNA to selected sequences is discussed in detail elsewhere (1). The steric fit of both adducts shown in Figure 4 demonstrates the severe steric hindrance to BPDE I(+) by B-DNA. A substantial portion of the pyrene moiety is embedded in the DNA. Shown is the optimum orientation about the N2 on guanine for an N2-C10 distance of 1.7 Å. Pivoting the BPTC about N2 will relieve the repulsion of the pyrene moiety at the expense of the benzo end. However, unwinding the DNA separates the adjacent base pairs, and rotates the base pair containing guanine away from the pyrene moiety. The addition of BPDE I(+) to this form is shown in Figure 5. The upper base pair and associated backbone is well removed from a substantial portion of the pyrene moiety and the fit about N2 on guanine is very much improved. The tight fit in B-DNA has been previously reported (6,7), but the unwinding of and selectivity by the DNA is suggested by these studies. That all isomers do not fit well in B-DNA, whereas, all can be accommodated by an unwound form can be understood through structural considerations. Because C10 is coplanar with the pyrene moiety, one may expect that the steric fit in DNA of all isomers will depend on the conformation of the benzo-ring and the orientation of the triol substituents. Unwinding the DNA to accomodate the pyrene moiety suggests that all of the isomers, are equally accessible to DNA. This is rationalized by noting that the pyrene moiety can fit into the minor groove about N2 on guanine in the same manner for each isomer because of the same hybridization about C10. Then one only needs to

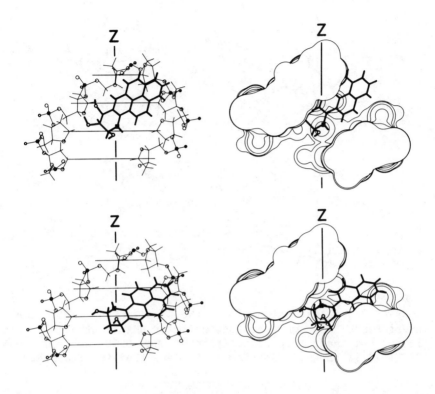

FIGURE 4. Illustration of the steric interference for the BPDEs to B-DNA. Cis (lower) and trans (upper) addition of the corresponding BPDE I(+) is shown. View is along an axis perpendicular to the helical axis (Z) of DNA and an axis containing the glycosidic carbon atoms C1' of base pair 2. The contours are spaced at 0.48 Å.

choose a conformation of the benzo moiety and the triol in the appropriate axial or equatorial form to achieve a fit. In Table 1 of reference (1), the study of the accessibility of the BPDEs to DNA (1) lists examples of each isomer which can approach N2 on guanine for possible covalent bond formation. Therefore, the analysis of the problem of steric fit of BPDEs results in the proposal that the DNA must unwind to accomodate BPDE I(+) in agreement with experimental observation (8), and that the problem of stereoselectivity requires the incorporation of other factors as the chemical reactivity of the various isomers with guanine, and the metabolism to selected isomers.

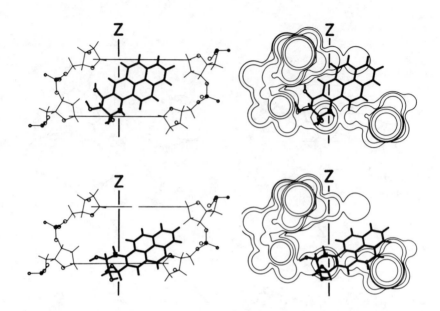

FIGURE 5. Illustration of the reduction of steric interference through the unwinding of DNA to site III. Cis (lower) and trans (upper) addition of the BPTC derived from BPDE I(+) is shown. The view is as described in Figure 4.

ACKNOWLEDGEMENTS

This work was supported by the National Institutes of Health under Grants CA-27241 and CA-28924 and by a generous grant of computer time by Rensselaer Polytechnic Institute.

REFERENCES

1. Miller, K. J., Burbaum, J., and Dommen, J.: Binding of benzo[a]pyrene-diol epoxides and -triol carbonium ions to DNA, (this volume).
2. Lee, B., and Richards, F. M. (1971): The interpretation of protein structures: Estimation of static accessibility, J. Mol. Biol., 55: 379-400.
3. Alden, C. J., and Kim, S.-H. (1979): Accessible surface areas of nucleic acids and their relation to folding, conformational transition and protein recognition, In: Stereodynamics of Molecular Systems, Pergamon, Elmsford, N.Y., pp 331-350.
4. Miller, K. J., Macrea, J., and Pycior, J. F., (1980): Interactions of molecules with nucleic acids. III. Steric and electrostatic energy contours for the principal

intercalation sites, prerequsites for binding, and the exclusion of essential metabolites from intercalation, Biopolymers, 19: 2067-2089.
5. Wahl, A. C. (1966): Molecular orbital densities: Pictorial studies, Science, 151: 961-967.
6. Beland, F. A. (1978): Computer-generated graphic models of the N^2-substituted deoxyguanosine adducts of 2-acetylaminofluorene and benzo[a]pyrene and the O^6-substituted deoxyguanosine adduct of 1-naphthylamine in the DNA double helix, Chem.-Biol. Interactions, 22: 329-339.
7. Lavery, R., and Pullman, B. (1979): Theoretical model study of the reactivity of benzo(a)pyrene diol epoxide with the amino groups of the nucleic acid bases, Int. J. Quant. Chem., XVI: 175-188.
8. Gamper, H. B., Straub, K., Calvin, M., and Bartholomew, J. C. (1980): DNA alkylation and unwinding induced by benzo[a]pyrene diol epoxide: Modulation by ionic strength and superhelicity, Proc. Natl. Acad. Sci. (USA), 77: 2000-2004.

NECESSITY FOR EXTRANUCLEAR METABOLISM? AN ALTERNATE SCENARIO FOR BaP AND/OR REDUCED A-RING 7, 12-DMBA-INDUCED TRANSFORMATION OF HUMAN FIBROBLASTS, IN VITRO

GEORGE E. MILO[2,4], THOMAS O. MASON[2], DONALD T. WITIAK[1,4], M. INBASEKARAN[1,4], FREDERICK D. CAZER[1,4], and WILLIAM R. GOWER[3]
[1]Division of Medicinal Chemistry and Pharmacognosy, College of Pharmacy; [2]Department of Physiological Chemistry, College of Medicine; [3]Department of Surgery, College of Medicine; [4]Comprehensive Cancer Center, 410 West 12th Avenue, The Ohio State University, Columbus, Ohio 43210.

SUMMARY

The polynuclear aromatic hydrocarbon (PAH) 1,2,3,4,-tetrahydro-7,12-dimethylbenz(a)anthracene, (TH-DMBA) was found to be a carcinogen when evaluated on human foreskin fibroblasts as measured by anchorage-independent growth and cellular neoplasia. Under conditions for transformation, radiolabelled [^{3}H-G] – TH-DMBA was examined for its biotransformation to oxygenated metabolites and intracellular distribution under conditions for transformation. Under these conditions we recovered 95% of the [^{3}H-G]-TH-DMBA as the non-oxygenated PAH. It also was interesting to note that [^{3}H-G] - TH-DMBA localized in the nucleus, 1.4 x 10^7 molecules/nuclear residue compared to 1.6 x 10^9 molecules per nuclear residue for [^{3}H-G] -benzo(a)pyrene (BaP) and non-detectable amounts per nuclear residue for [^{3}H-G] -7,12-dimethyl benzo(a)anthracene (DMBA). The transformation indices as measured by anchorage-independent growth were found to be for, BaP treated cells, 16-26; 7,12 – DMBA, 0; TH-DMBA, 84-123; 2,3-dihydro-5,11-dimethyl-1H-cyclopent(a)anthracene (CP-DMA), CP-DMA 18-21; BPDE-(I)anti 26-33. These data imply a mechanism for PAH-induced carcinogenesis in human cells that does not involve extranuclear activation as a prerequisite for biotransformation of the PAH to an ultimate carcinogenic form. If covalent binding is involved in PAH- induced carcinogenesis then less than 1% of the intracellular TH-DMBA associated with nuclear macromolecules is responsible for the alteration in nuclear function resulting in carcinogenesis. Therefore, we propose that there are a small number of critical sites in the human fibroblast genome whose perturbation is essential to BaP genotypic modification which results in the expression of human cell carcinogenesis.

INTRODUCTION

Historically, AHH activation and subsequent binding of oxygenated PAH to DNA (adduct formation) has been correlated with the critical nuclear event in carcinogenesis. Recently, Diamond et al (1) have reported that human cells (Hep G2 liver parenchymal cell line) activate BaP to oxygenated metabolites that bind to DNA. We, (2), however, have demonstrated that it is not necessary to activate BaP at the P450 microsomal level to induce a carcinogenic event in human cells. When BaP was added to proliferating human foreskin fibroblast cells in vitro, hydrocarbon-DNA-adducts were found, while no ethyl acetate extractable oxygenated metabolites were detected (3). In addition, we demonstrated that unmetabolized BaP is transported to and localized in the nucleus by a lipoprotein complex (4). These observations indicated that in the case of BaP the requisite metabolic activation takes place within the nucleus.

DMBA, a noncarcinogen in this system (2), does not localize in the nucleus, but was found randomly dispersed throughout the cytoplasm, (2). By contrast, the A-ring reduced form of this PAH, TH-DMBA is a potent carcinogen in this cell system. Recognizing that the "bay region" diol epoxide would not be a likely metabolite of TH-DMBA we investigated the metabolism and cellular compartmental distribution of TH-DMBA under conditions for human cell carcinogenesis, (5).

MATERIALS AND METHODS

Chemical Synthesis of Analogues

Benzo(a)pyrene diol epoxide (I) racemic mixture of the ± form, BaP and DMBA were obtained from the NCI repository (sponsored in part by NCI-DCCP, Bethesda, MD.). TH-DMBA(16) and CP-DMA were organically synthesized by the Comprehensive Cancer Center at Ohio State University. TH-DMBA was radio-labeled [G^3-H]-TH-DMBA by Midwest Research Institute, Kansas City, Missouri, 64110. The radiolabeled compound was purified by hplc at the Comprehensive Cancer Center, The Ohio State University.

The CP-DMA analogue, a new compound, was prepared from its known quinone precursor, (15) using methodology essentially identical to that employed in the preparation of TH-DMBA (16).

Carcinogen Treatment of Cell Cultures

Logarithimically growing human foreskin cultures were treated with each radiolabelled PAH prior to the sixth population doubling (PDL), (6). Human foreskin cultures were treated under conditions that would result in the expression of a transformation event, (6). In brief, cell populations at PDL 4 were seeded in T-150 plates at a cell density of 25,000 cells per cm^2. Twenty-two hours following seeding, plates were refed with 20 ml of Dulbecco's modified Eagles Minimum Essential Medium minus arginine and glutamine (7) to reduce the labelling index of 26% to less than 0.1%. When there was less than 1 radiolabelled nucleus per 10^4 cells, the Dulbecco's Modified MEM was removed and the cell cultures were refed with 20 ml of complete growth medium, (CM) containing arginine and glutamine. At this time the radiolabelled PAH compounds were individually added to the cultures (5). After the treated population had entered S phase of the cell cycle, (10 hrs. later), and traversed through S, (8.2 hrs later), the CM containing each radiolabelled PAH was removed and the cultures washed with fresh CM 3 times. The extracellular medium, and cell fraction (8) were extracted with 3 volumes of spectrar grade ethyl acetate.

Preparation of Subcellular Fraction(s)

Cells were harvested by scraping the T-150 plates with a rubber policeman collected by centrifugation at 600 x g for 10 min and washed twice by resuspension in 5 ml volumes of cold salt supplemented Dulbecco's phosphate buffered saline (9) at pH 7.4. The cytoplasmic preparation was prepared as described previously, (4), and the nuclei prepared by resuspending the cells in Buffer A (10mM Tris-maleate buffer, pH 7.5 containing 1mM DTT, 3mM calcium acetate and 2mM magnesium acetate) at 1.0 x 10^7 cells/ml, then disrupted by homogenization in a Teflon-steel homogenizer. When the cells were 85% disrupted as determined by Trypan blue dye exclusion, (10), the lysate was brought to 1.7M sucrose, and layered over Buffer A containing 2.0M sucrose, and centrifuged in an SW-41 rotor at 40,000xg for 60 min. The crude nuclear pellet was resuspended in Buffer A containing 1.0M sucrose and centrifuged at 10,000xg for 10 min. This procedure provided nuclei free from cytoplasmic contamination, as determined by contrast-interference

Nomarski microscopy, with yields of 30-40%. The nuclear residue was identified as that part of the radioactive fraction remaining after 3 extractions with ethyl acetate. For analysis the residue was solubilized with NCS (Amersham), titrated to pH 7.0, mixed with scintillation cocktail (Beckman, TM) and radioactivity measured in a Beckman LS 9000.

High Pressure Liquid Chromatography (hplc)

Whole cells, isolated nuclei and cytoplasmic fractions were extracted with 3 vol of ethyl acetate. The organic and aqueous phases were separated, the organic phase passed over 7-gms of anhydrous sodium sulfate, filtered through a fine sintered glass funnel, then dried under argon. The resulting residue was stored at -20°C until analyzed by hplc.

hplc Analysis of PAH Metabolites

Profiles of BaP and DMBA metabolites extracted from human foreskin fibroblasts treated under transformation conditions have been previously reported, (4,8). Twenty-five µl aliquots of ethyl acetate extracts were analyzed on a Beckman Model 332 MP liquid chromatograph using a Zorbax ODS column, (250mm x 6.2mm), (4).

Extracts were analyzed for metabolites using a linear gradient of 60-100% methanol for 90 minutes and then holding the 100% methanol for an additional 10 minutes. One hundred 1 minute fractions were collected at a rate of 0.8ml/minute. Two ml of Instagel scintillation cocktail was added to each fraction which were counted in a Beckman LS 9000, (38% efficiency).

Transformation Sequence

PAH treated cells were evaluated for their ability to exhibit anchorage-independent growth in soft (0.33%) agar and cellular neoplasia. Two days following the initiation of PAH treatment the treated cell cultures were passaged at a 1:2 split ratio into CM supplemented as described elsewhere (11). These treated populations were serially passaged at a 1:10 split ratio for 16 PDL (1:10 split ratio = 3.3 PDL). At this time the treated cells were seeded into 0.33% agar, (7). The soft agar overlay containing 50,000 cells/25cm^2 well was incubated at 37°C - 4% CO_2-enriched air atmosphere, (7). After 4 weeks the colonies containing 50 cells or more were removed and seeded into CM. As the cells grew they were passaged once at a 1:4 split ratio and then evaluated for cellular neoplasia. Organ cultures of chick embryonic skin, (CES) were prepared from 9 day old fertilized eggs, (12).

The CES were layered onto an enriched agar base containing 10 parts 1% Agar in Earle's balance salt solution, 4 parts chick embryo extract and 4 parts of FBS. One hundred thousand cells from populations exhibiting anchorage independent growth and untreated populations were seeded onto the CES. Three days later the organ cultures were removed fixed in Bouin's solutions and 5μ transverse sections stained with hematoxylin and eosin and examined for invasiveness.

RESULTS

Subcellular Compartmentalization of ^3H-B a P, ^3H-DMBA and ^3H-TH-DMBA

As reported previously by Ekelman and Milo, (2) BaP localized in the nucleus in higher concentrations than that adsorbed to the charcoal-treated cytoplasmic preparations. This association of radiolabelled PAH with the cytoplasmic preparation was subsequently identified as a specific lipoprotein complex. Chromatographic analysis of the adsorbed BaP was shown to be principally BaP and not oxygenated metabolites, (4). Data presented in Table 1 indicated that for [^3H-G]-BaP, the nucleus retains approximately 34% of residual counts (BaP) after exhaustive extraction of the 2.5 x 10^6 nuclei. However, hplc evaluation of the ethyl acetate extractable material indicated that <2% of the extractable radiolabelled BaP located in the nucleus was in the form of oxygenated metabolites.

Examination of the intracellular distribution of 7,12-DMBA indicated that a substantial amount entered the cell and localized in the cytoplasmic fraction, (1.3 x 10^6 molecules/cell). However, when the cytoplasm was partitioned with activated charcoal no residual radioactivity could be detected, (Table 1). When nuclei were examined for the presence of radiolabel we found 1.00 x 10^6 molecules in the nuclear compartment. However, extraction of 2.8 x 10^6 nuclei with 3 volumes of ethyl acetate removed over 95% of the radioactivity. When the nuclear residue was examined following solubilization no radioactivity was detected above background.

TABLE 1

LOCALIZATION OF RADIOLABEL PAH IN TRANSFORMABLE
HUMAN FORESKIN FIBROBLASTS

Cell Compartment	Treatment	Maximum PAH Molecules Per Compartment
Whole Cell Charcoal treated	BaP 13 μM	3.2×10^{10}
Cytoplasm		1.12×10^{9}**
Nucleus		4.7×10^{9}
Non-extractable Nuclear Residue		1.61×10^{9}
Whole Cell Charcoal treated	DMBA 16 μM	4.17×10^{7}
Cytoplasm		N.D.
Nucleus		1.00×10^{6}
Non-extractable Nuclear Residue		N.D.*
Whole Cell Charcoal treated	TH-DMBA 24 μM	1.4×10^{10}
Cytoplasm		3.8×10^{9}**
Nucleus		4.2×10^{8}
Non-extractable Nuclear Residue		1.4×10^{7}

Eight million logarithmically growing cells at 70% confluent density were blocked in G_1 release and exposed to PAH for 12 hrs. The optimum localization of the PAH in the nucleus occurred 16 hrs following initiation of treatment of the cells. The specific activity of PAH added to the population was either BaP 2.5 Ci/mmol (13 μM); or DMBA 0.06 Ci/mmol (16 μM); or TH-DMBA 0.06 Ci/mmol (24 μM). We obtained a yield of 31% nuclei per isolation.
*N.D. = Non-detectable
**The counts presented here do not add up to 100% recovery. The total cts associated with the dextran coated charcoal treated cytoplasm is lower than the untreated fraction prior to charcoal treatment. When those values are included in the total recovery of the radiolabel from the cellular compartments was in excess of 95%.

Examination of the localization of TH-DMBA (24 µM) indicated that the pattern of localization of this compound in the cells was similar to the localization of BaP and quite unlike DMBA. We found in excess of 3 times the amount of radiolabel associated with the cytoplasmic fraction compared to BaP. When we examined the distribution of 32 µM TH-DMBA (data not shown in Table 2) throughout the cell we found over 6 times the radioactivity associated with the charcoal treated cytoplasmic preparation compared with BaP (13µM)-treated cells. Also, we observed a 125% increase in the number of PAH molecules/nuclear compartment of 32µM TH-DMBA-treated cells compared to 24µM TH-DMBA-treatment, (1.4 x 10^7 molecules/compartment). Like the nuclear residue preparation of BaP treated cells, we did note a progressive increase in the association of the radiolabel with the nuclear residue compared to the charcoal treated cytoplasm in 24-32 µM treated cells. At the higher concentration of TH-DMBA (32 µM) we found 4.70 x 10^7 molecules per compartment compared to 1.4 x 10^7 molecules per nuclear residue compartment for the lower concentration of TH-DMBA.

A Typical hplc elution profile of [^{3}H-G-] TH-DMBA is illustrated in Fig. 1. An identical profile was obtained from the ethyl acetate extract of cytoplasmic nuclear and whole cell preparations, (Fig. 1). When we evaluated the functional capacity of treated cells to exhibit anchorage - independent growth, TH-DMBA treated cells formed 84 colonies per 100,000 seeded cells in experiment one and 123 colonies per 100,000 cells in a second experiment. BaP, BPDE(I) and CP-DMA transformed the cells at about the same frequency i.e. 16-33 colonies per 100,000 seeded cells. DMBA-treated cells formed no colonies in soft agar, (Table 2). All of the PAH-transformed populations formed colonies in soft agar when evaluated on CES exhibited cellular invasiveness, (Table 2). Cellular tumors were evaluated as simulated fibrosarcomas.

DISCUSSION

Many reports exist describing the metabolism of BaP to specific oxygenated metabolites by human fibroblast cultures. Oxygenation of the BaP also has been reported to take place in the microsomal-plasma membrane P450 aryl hydrocarbon

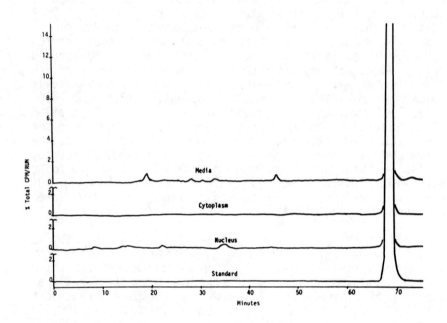

FIGURE 1. The profile illustrated here represents a metabolic profile of TH-DMBA by human foreskin cells under conditions of transformation (11). The oxidinate is expressed as the percent of the CPM plotted as a function of time of elution. The flow rate was 0.8 ml/min. The recovery of the radiolabel [^{3}H-G]-TH-DMBA was 98.6% of that administered to the column. The designate TH-DMBA indicates where the marker TH-DMBA eluted from the column.

TABLE 2

GROWTH OF PAH TRANSFORMED CELLS IN 0.33% AGAR OVERLAY-
2.0% AGAR BOTTOM LAYER AND INCIDENCE OF CELLULAR NEOPLASIA ON CES

Treatment[a]	Concentration ug/ml	No. of Colonies/10^5 Cells Seeded in S.A.[b]	No. of Neoplasms No. of CES Seeded[c]
B a P	1	26	6/6
B a P	1	26	
BPDE(I)	0.1	26	6/6
BPDE(I)	0.1	33	
TH-DMBA	1	84	6/6
TH-DMBA	1	123	
CP-DMA	1	21	6/6
CP-DMA	1	18	
7,12-DMBA	1	0	ND*
7,12-DMBA	1	0	

a) Concentrations of parent polynuclear hydrocarbons were described in Milo et al. (11). Benzo(a)pyrene diol epoxide (1) was used at its effective cytotoxic dose of 50%, (E.D. 50) and the cells were treated for 30 minutes in S instead of 16 hrs as for the unmetabolized PAH compounds (5). Each compound was evaluated twice for its ability to transform cells. The duplication of each treatment represents 2 different experiments for each PAH.
b) Eight 25 cm^2 wells were seeded with 50,000 treated cells per well into 2 ml of 0.33% agar supplemented with growth medium over a 5 ml 2% growth medium supplemented agar base (11). Values reported here represent the mean values per 8 different wells.
c) Values cited here are for the number of CES organs exhibiting evidence for cellular neoplasia 3 days after the CES were seeded with 10^5 treated and untreated cells. Untreated cells exhibited no cellular invasiveness (5,12).
* N.D. = Not evaluated on CES. However, these populations were evaluated in nude mice and found to be negative (Ekelman and Milo 1978 Cancer Res 38, 3031. See note added in proof.)

hydroxylase complex (13). Extracellular release of these metabolites has been identified with the carcinogenesis process without directly measuring a transformation event.

Isolation of a cytoplasmic complex from human foreskin fibroblasts that bound unmodified BaP was the first indication that unmetabolized BaP-induced carcinogenesis in these cells and that such an induction was not comparable to microsomal P450 metabolized BaP-induced carcinogenesis in animal systems. Moreover, in non-proliferating confluent dense populations, transformable by BaP, AHH activity was optimal, (14). Subsequent fractionation of logarithmically growing cells (4) into cytoplasmic and nuclear compartments revealed that BaP was localized in the nucleus, (Table 1). There appears to be a progressive uptake of radiolabel over the 12 hr treatment period, (Table 1).

Using TH-DMBA wherein a bay region dihydrodiol epoxide metabolite would be unlikely, we nonetheless observed (9) that this compound could induce transformation of human cells. Possibly another portion of the molecule, perhaps ring D, represents the critical site of metabolism in this hydrocarbon, but additional work is necessary to establish whether metabolism per se is needed for induction of a carcinogenic event.

It is also interesting to note (Table 1) that we can now detect non-extractable radioactivity in the nuclear residue. When the concentration of TH-DMBA was increased from 24 µM to 32µM a 33% increase in parent PAH, we observed a 125% increase in localization of PAH molecules/per nuclear compartment. Again, we found by hplc analysis that TH-DMBA like BaP is not metabolized to any detectable amount by the plasma-membrane associated oxygenase enzymes, (Fig. 1). Moreover, when we evaluated the 5 membered A-ring analogue of 7,12 DMBA (CP-DMA) we could still induce a carcinogenic event. Moreover, the incidence of growth of colonies in soft agar of the CP-DMA transformed cells was comparable to BaP transformed cells and not the TH-DMBA transformed cell populations. In the case of CP-DMA it would be impossible to obtain a classical dihydrodiol 1,2-epoxide in the bay region. Benzo(a)pyrene-7,8 enediol-9,10-epoxide (I) anti form, a known presumed ultimate carcinogen, induces a carcinogenic event in human cells and this is compatible with the proposed bay region epoxide theory of carcinogenesis. However, this proposal may not be compatible with the process for TH-DMBA or its less lipophilic CP-DMA analogue-induced transformation of human cells.

Data presented here imply that there is another mechanism for PAH-induced carcinogenesis in human cells that does not involve extranuclear oxygenation of PAH to bay region epoxides by extranuclear enzymes. From data published elsewhere (15) we know that oxygenated BaP interacts with DNA. This oxygenation appears to occur at the nuclear level. The resultant reactive adduct species is 7R-dG- BPDE-I. Treatment of these cells with exogenous BPDE-I also produces the same adducts (3). If the requirement for dG-BPDE-I adduct is the critical nuclear event then it is possible that BPDE-I, the critical oxygenated species, is site specific and only forms dG adducts at sensitive sites in the DNA created by modulation with either insulin or anthralin or TPA (7) of the nuclear protein surrounding the carcinogenic site. However, we know that BPDE-I- interacts with all available exposed dG sites and that modulation can augment the carcinogenic process (7). In the case of TH-DMBA, reaction with DNA must be either highly site specific or metabolism followed by covalent binding to the nuclear macromolecules in the classical sense is not required.

However, we propose the following; expression of carcinogenesis, a response to the nuclear carcinogenic insult, requires specific modulation of nuclear proteins protecting the critical site in order for the initial stages of carcinogenesis to be expressed.

ACKNOWLEDGEMENT: We wish to acknowledge the contribution(s) by Dr.'s Ramon Tejwani and Karen Ekelman. The data presented here from Dr. Ekelman's effort was extracted from her Ph.D. thesis presented to the Ohio State University in partial requirement for her Ph.D. The thesis is entitled, "Uptake, binding to cytoplasmic protein, translocation to the nucleus and morphological transformation of human cells by benzo(a)pyrene and 7,12-dimethyl-benzanthracene". This work reported here was supported in part by an award from AF 49620-80-C-00 85 and National Cancer Institute 25907.

REFERENCES

1. Diamond, L., Kruszewski, F., Aden, D., Knowles, B., and Baird, W. (1980): Metabolic activation of benzo a pyrene by a human hepatoma cell line Carcinogenesis, 1:871-876.
2. Ekelman, K., and Milo, G. (1978): Cellular uptake, transport and macro-molecular binding of benzo(a)pyrene and 7,12 dimethylbenz(a)anthracene by human cells in vitro, Cancer Res., 38:3026-3032.
3. Tejwani, R., Jeffrey, A., and Milo, G. (1982): Benzo(a)pyrene diol epoxide DNA adduct formation in transformable and non-transformable human foreskin fibroblast cells in vitro. Carcinogenesis (submitted).
4. Tejwani, R., Nesnow, S., and Milo, G., (1980): Analysis of intracellular distribution and binding of benzo(a)pyrene in human diploid fibroblasts, Cancer Letters, 10:57-65.
5. Tejwani, R., Witiak, D., Inbasekaran, N., Cazer F., and Milo, G. (1981): Characteristics of benzo(a)pyrene and A-ring reduced 7,12-DMBA induced neoplastic transformation of human cells in vitro, Cancer Letters, 13:119-127.
6. Milo, G. and DiPaolo, J. (1978): Neoplastic transformation of human diploid cells in vitro after chemical carcinogen treatment, Nature (London), 275:130-132.
7. Milo, G., and DiPaolo, J. (1980): Presensitization of human cells with extrinsic signals to induce chemical carcinogenesis. Int. J. of Cancer, 26:805-812.
8. Milo, G., Trewyn, R., Tejwani, R., and Oldham, J. Intertissue variation in benzo(a)pyrene metabolism by human skin, lung and liver, in vitro. In: advisory group for aerospace research and development. North Atlantic Treaty Organization, AGARG CP-309: B7-1 to B7-9.
9. Cazer, F., Inbasekaran, M., Loper, J., Tejwani, R., Witiak, D., and Milo, G., Human cell neoplastic transformation with benzo(a)pyrene and A-region reduced analogue of 7,12-dimethyl a anthracene. Symposium on Polynuclear Aromatic Hydrocarbons. Battelle Memorial Inst. 5:499-506.
10. Noyes, I., Milo, G., and Cunningham, C., (1980): Establishment of proliferating human epithelial cells in vitro from cell suspensions of neonatal foreskin. Tissue culture Assoc. Lab. Manual (Rockville) 5:1173-1176.

11. Milo, G., Oldham, J., Zimmerman, R., Hatch, G., and Weisbrode, S. (1981): Characterization of human cells transformed by chemical and physical carcinogens in vitro, 17:719-730.
12. Donahoe, J., Noyes, I., Milo, G., and Weisbrode S. (1981): A comparison of expression of anchorage independent growth with neoplasia of carcinogen transformed human fibroblast in nude mouse and chick embryonic skin in vitro. In Vitro (Rockville) In Press.
13. Huberman, E., and Sachs, L. (1973): Metabolism of the carcinogenic hydrocarbon benzo(a)pyrene in human fibroblast and epithelial cells. Int. J. Cancer, 11:412-418.
14. Milo, G., Blakeslee, J., Yohn, D., and Di Paolo, J. (1978): Biochemical activation of AHH activity, cellular distribution of polynuclear hydrocarbon metabolites and DNA damage by PNH products in human cells in vitro. Cancer Res., 38:1638-1644.
15. Crawford, R., Levine, S., Elofson, R., and Sandin, R. (1957): Observations on some alkyl substituted anthracenes and anthraquinones. J. Amer. Chem. Soc., 79:3154-3157.
16. Inbasekaran, M., Witiak, D., Barone, K., and Loper, J.(1980): Synthesis and mutagenicity of A-ring reduced analogues of 7,12-dimethylbenz(a)anthracene. J. Medicinal Chem., 23:278-281.

PROPERTIES OF ^{3}H-LABELED 1-NITROPYRENE DEPOSITED ONTO COAL FLY ASH

A. MOSBERG, G. FISHER, D. MAYS, R. RIGGIN, M. SCHURE*,
J. MUMFORD**
Battelle Columbus Laboratories, Columbus, Ohio 43201,
*Department of Chemistry, Colorado State University, Fort
Collins, Colorado, **U.S. Environmental Protection Agency,
Genetic Bioassay Branch, Health Effects Research Laboratory,
Research Triangle Park, North Carolina.

ABSTRACT

Nitropyrene was synthesized by nitration of pyrene and purified by liquid chromatography. Vapor phase adsorption was performed using a specially designed apparatus. Chemical analysis of solvent extractable nitropyrene coated onto the fly ash showed that adsorption was erratic with concentrations ranging from 0 to 250 ppm. Later studies with ^{3}H-nitropyrene indicated that the irreproducibility of the coating procedure was probably due to the inefficiency of the extraction technique. Most of the radiolabeled nitropyrene was not extracted from the fly ash surface. Isotopic exchange studies and tissue culture media extraction studies indicated minimal isotopic exchange and approximately 10 percent extraction of nitropyrene from fly ash.

INTRODUCTION

Recent studies indicate that polycyclic organic matter (POM) is deposited on the surface of coal fly ash. Natusch (1) has indicated that POM adsorption takes place after cooling of the flue stream from a power plant. Christ et al. (2) reported that organic mutagens are present on the surface of stack-collected coal fly ash. Fisher et al. (3) extended these observations to describe a process of adsorption and surface stabilization of chemical mutagens on fly ash. Most recently, Hansen et al. (4) reported that nitroaromatic hydrocarbons may be present on the surface of the fly ash studied by Chrisp et al. and Fisher et al.

Due to physical and biological importance of surface interactions and the relatively high mutagenic activity of nitropyrene, vapor phase adsorption studies were performed to produce nitropyrene-coated coal fly ash for subsequent biological experimentation.

MATERIALS AND METHODS
SYNTHESIS AND PURIFICATION OF NITROPYRENE

The nitration of pyrene to yield nitropyrene was carried out using purchased pyrene (unlabeled) (6). The product consisted chiefly of 1-nitropyrene but was contaminated with dinitropyrenes, trinitropyrenes, and pyrene. An attempt to prepare this material with little or no dinitropyrene contamination by using mild conditions and only the stoichiometric amount of nitric acid resulted in a large amount of unreacted pyrene.

The unlabeled nitropyrene was purified by elution through a silica gel column. Aldrich 923 grade, 100 to 200 mesh, silica gel was conditioned by drying at 110°C for 3 hours. About 120 grams of the conditioned silica gel was packed into 500 mm long, 25 mm diameter glass columns fitted with 500 ml solvent reservoirs. Two column volumes of methylene chloride were passed through the column followed by one column volume of hexane. About 2 mm of hexane was left above the silica gel.

The nitropyrene to be purified was weighed (no more than 2 mg per gram of silica gel) and placed in a 125 ml Erlenmeyer flask. The nitropyrene was dissolved in a minimum amount of methylene chloride and 10 grams of silica gel was added to the solution. The methylene chloride was removed using a nitrogen purge. When dry, the silica gel-nitropyrene was added to the top of the column. About 25 ml of hexane was used to rinse the flask and then added to the top of the column. The hexane was eluted to the top of the column. This was followed by a mixture of 60 percent methylene chloride to 40 percent hexane. The clear eluent eluting before the yellow nitropyrene was discarded.

Fractions were collected at 100 ml intervals and analyzed by gas chromatography to determine the purity of the nitropyrene. Fractions were collected until the presence of dinitropyrene was detected or nitropyrene was no longer present. The nitropyrene obtained was better than 99 percent pure by high performance liquid chromatography (HPLC) analysis. From 750 milligrams of crude nitropyrene, 227 milligrams of the purified nitropyrene was obtained.

Because the first attempts to prepare pyrene by a route suitable for ^{14}C label incorporation was not successful, the possibility of ^3H labeling by isotope exchange was examined. The use of ^3H has the disadvantage of possible ^3H loss by isotope exchange with ^1H in the intended application. It has two advantages, however; (1) the label may be inserted at the last synthetic step, thus minimizing label losses in a multi-step process, and (2) the label, as tritiated water, is relatively inexpensive.

Procedures have been reported in the literature for the exchange of protons with the ^{3}H of tritiated water (7,8). Pyrene is one of the compounds reported to exchange well, but no nitroaromatics have been mentioned.

New England Nuclear Company (NEN) (Boston, MA) was selected for preparation of the tritiated material. Two 50 mg aliquots of purified 1-NP were sent to NEN for tritiation specified by the following procedure. Fifty mg of precursor was dissolved in 0.3 ml of very dry dimethylformamide, then 150 mg of platinum black catalyst and 25 Curies of tritiated water was added. The reaction mixture was stirred to 18 hours at 80°C. Labile tritium was removed in vacuo using benzene/ethanol (1:1) as the solvent. After filtration from the catalyst, the product was again dried in vacuo and then dissolved into 20.0 ml of benzene.

The tritiation resulted in preparation of two lots of ^{3}H-NP. Table 1 presents the radiochemical and chemical specifications. The specific activity of Batch 1 (6.3 mCi/mg) was less than that of Batch 2 (15.1 mCi/mg).

TABLE 1

RADIOCHEMICAL AND CHEMICAL SPECIFICATIONS OF ^{3}H-NITROPYRENE (NP)

	Batch #1		Batch #2
Radioactivity[a] (mCi)	316.0		756.0
Specific Activity[a] (MCi/mg)	6.3		15.1
Total Mass NP (mg)[b]	38.0		47.0
Impurities[b] (%)	8.3		13.2
Combined Mass Recovery (%)	---	38.00	---
Combined Radioactivity Recovery (%)	---	0.45	---
Specific Activity After Purification (mCi/mgNP)	---	0.13	---

[a]Data provided by New England Nuclear
[b]Analyzed by Battelle

The ^{3}H-nitropyrene was received dissolved in benzene and subsequently purified as previously described. Before purification, the solution was analyzed by gas chromatography to determine the nitropyrene concentration (Table 1). The solvent was removed from the purified ^{3}H-nitropyrene using a slow nitrogen purge and the ^{3}H-nitropyrene was stored in the darkness at -20°C. The overall yield of radioactive nitropyrene (NP) after tritiation of 100 mg at NEN and purification at Battelle was 38 percent by mass and 0.45 percent by activity (Table 1).

CHEMICAL ANALYSIS

Chemical analysis of NP was performed by HPLC and gas chromatography (GC). The following protocol was used to analyze solutions containing nitropyrene by liquid chromatography:

Instrument:	Altex 312 MP with 100A pumps
Column:	Spherisorb-ODS, 5 micron, 250 x 4.6 mm
Mobile Phase:	70 percent acetonitrite/ 30 percent water
Flow:	1 ml/min
Detector:	LDC model 1203 at 254 μm
Injection Volume:	5 μl

Under these conditions, nitropyrene eluted at 10.8 minutes and could be detected at levels above 1 μg injected onto the column. Quantification was by area integration of the unattenuated detector output using a Hewlett-Packard 1000 computer with Computer Inquiry System software.

The following protocol was used to analyze solutions containing nitropyrene solutions by GC:

Instrument:	Hewlett-Packard 5830A
Column:	3 x 2 mm I.D., glass, packed with SP2100 on 100/200 mesh Supelcoport
Detector:	Flame ionization detector
Detector Temperature:	240°C
Injector Temperature:	200°C
Column Temperature:	200°C for 2 minutes, programmed to rise at 10°C/min to 280°C
Injection Volume:	2 μl

Under these conditions, nitropyrene eluted at 5.3 minutes and could be detected at levels above 10 ng injected onto the column. Quantification was by area integration or by peak height when very low levels of nitropyrene were found.

Fly Ash Extraction Procedure

About 50 mg of fly ash was weighed accurately and placed in a six-gram vial with Teflon-lined screw cap. Ten milliliters of benzene were added and the vial was capped and placed in an ultrasonic bath for 30 minutes. The vial was then removed and the fly ash allowed to settle. The liquid was transferred to a Kuderna-Danish tube and concentrated to about 1 ml.

Media Extraction and Analysis Procedure

Extraction of tissue culture medium (M199 with Hanks salts, 10 percent fetal calf serum, and 1 percent antibiotic/antimycotic) was performed on 25 ml samples. Media samples were spiked with 125 μl of nitropyrene in methylene chloride or DMSO depending upon the conditions of the experiment. Spiking amounts were 200 to 250 μg and 2.0 to 2.5 μg into the 25 ml samples. Incubated samples were incubated for 20 hours at 37°C.

The samples were extracted by adding 25 ml of distilled water and 5 ml of acetonitrile followed by four extractions with 50 ml of methylene chloride. The methylene chloride extracts were combined and concentrated to 2 ml using a Kuderna-Danish tube or a nitrogen purge for the tritiated material.

Fly Ash Spiked Media Extraction Procedure

About 50 mg of nitropyrene-coated fly ash was added to 100 ml of media in medium bottles. The solution was well mixed and then incubated in darkness for 20 hours. After incubation, the solution was centrifuged (about 200 RPM) to remove fly ash particles and filtered through a 0.22 micron Nuclepore filter to obtain 25 ml of sample for extraction and analysis. A water-acetonitrile soultion was extracted four times with 50 ml of methylene chloride. The methylene chloride extracts were combined and concentrated using nitrogen purge to a volume of 2 ml. The samples were diluted 1:1 with methanol for analysis by HPLC.

Fly Ash Properties

On initiation of the project, a theoretical calculation was made to determine the limit of pyrene adsorption by fly ash. This calculation assumes the fly ash to be composed of 3 μm spheres of density 2 g/cm^3.

The results indicate that in theory, the limit of molecular monolayer concentration on the fly ash is 675 ppm. Fly ash was provided by EPA.

Surface area determinations were made by 5 point gas adsorption techniques at Particle Data Laboratories (Elmhurst, IL) for both the U.S. EPA fly ash and a fly ash sample chosen by Battelle to simulate the EPA sample. The Battelle fly ash sample was used in trial runs to develop laboratory adsorption techniques and to refine material handling procedures before using the limited EPA fly ash sample. The results indicated that the surface areas of the EPA and Battelle fly ash samples were 3.42 m^2/gram and 4.92 m^2/gram, respectively. The size distributions of the U.S. EPA and Battelle fly ash samples were VMD 2.33 μm (σg = 1.4) and 3.65 μm (σg = 1.95), respectively, based on particle volume measurements by the electrical sensing zone method.

Before use in coating experiments, the fly ash samples were heated for 4 hours at 350°C to remove previously adsorbed organic matter.

Vapor Phase Coating Apparatus

For this study, Battelle used a vapor phase adsorption system (9) for deposition of polycyclic aromatic hydrocarbons onto the surface of fly ash. This system is designed as an open loop flow device in which nitrogen gas is heated in a stainless steel heat exchanger to a desired temperature at which vapor phase nitropyrene can evolve at a controlled concentration (Figure 1). The heated nitrogen gas is then passed through a diffusion cell to transport the nitropyrene vapor to another heated box in which the temperature is elevated by approximately 5 degrees. Here, a second stainless steel heat exchanger elevates the temperature of the nitropyrene-nitrogen gas mixture to avoid condensation. The gas phase is passed through an expanded fly ash bed supported by a coarse glass frit and contained in a glass cylindrical exchange chamber. As the gas passes through the fly ash, the fly ash is mixed and the exchange of vapor phase nitropyrene with the surface of the fly ash is accomplished. Upstream a

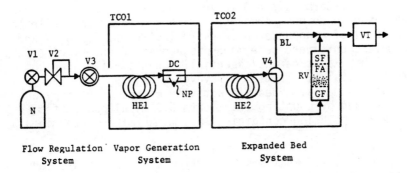

FIGURE 1.

N - high purity nitrogen
    supply
V1,V2 - 2-stage pressure
    regulator
V3 - flow controller (2000
                        frit
TCO1, TCO2 - temperature
    controlled oven
HE1, HE2 - heat exchanger
DC - diffusion cell
NP - nitropyrene source

V4 - 3-way valve
B1 - bypass loop
GF - glass frit support
FA - expanded fly ash bed
SF - stainless steel, 2 m   ml/min)
RV - expanded bed reactor
    vessel
VT - vapor trap

stainless steel frit, with a 2 micron diameter pore size, reduces the possibility of fly ash moving out of the expanded fly ash reactor bed. Nitrogen passed through the fly ash bed then flows out of the heated atmosphere and is either condensed or collected in a filter bed before release into the atmosphere. The three major components of this flow system are described in Figure 1 and include the gas flow regulation system, the polycyclic aromatic hydrocarbon vapor generator system, and the expanded adsorbent bed system.

Tritium Determination

Tritium was counted in a Searle Mark III scintillation counter (Model Number 6880). Samples counted directly were dissolved in 25 ml of Aquasol (New England Nuclear) contained in low potassium, borosilicate glass scintillation vials (Scientific Products). Count times were typically 2 minutes. Sample disintegrations per minute (DPM) were calculated by the instrument from sample counts per minute by using the counter efficiency versus a standard curve derived from the ratio of reference to standard channel counts.

A Tri-Carb Sample Oxidizer (Packard Instrument Company) was used to combust samples. Tritiated water was collected in 15 ml of monophase-40 (Packard Instrument Company) and sample counts were determined as above. Recovery of tritium from ^{3}H-hexadecane was essentially quantitative when combusted in this system. Tritium carryover was less than 0.05 percent between samples. Specific activity of the ^{3}H-nitropyrene was determined by counting triplicate 5.0 ml aliquots of a ^{3}H-NP standard solution containing 16.0 mg NP/ml as determined by HPLC.

Results of analysis of tritium standards indicated approximately 100 percent recovery of radioactivity combusted to tritiated water. Overall counting efficiency of the scintillation counter was approximately 90 percent of the theoretical activity.

RESULTS AND DISCUSSIONS

Extraction Efficiency

Preliminary studies were performed with nitropyrene added in solution to previously heated fly ash to evaluate solvent extraction efficiency. Benzene was found to be a more efficient extractant of spiked fly ash than extraction with methylene chloride (Table 2). Benzene was therefore used for subsequent fly ash analyses. However, because of the difficulty of removing benzene from the tissue culture medium and because extraction studies of nitropyrene in media indicated high efficiency for methylene chloride extraction, methylene chloride was used for this phase of the program (Table 2).

Initial Coating Experiments

Results of initial evaluations demonstrated complete adsorption of pyrene by the fly ash bed for 12 hours of a vapor phase coating experiment. Pyrene was initially used because of the ease of handling compared to the mutagenic nitropyrene. Agitation of the adsorption bed did not produce channeling of pyrene through the bed during this evaluation. When pyrene began to pass the adsorption bed, the concentration of the vapor phase pyrene available upstream of the reactor bed rapidly increased from 0 percent to 100 percent in about 30 minutes.

TABLE 2

EFFICIENCY OF EXTRACTION OF NITROPYRENE

COAL FLY ASH SPIKED WITH NITROPYRENE SOLUTION

Spiked Sample (ppm)	Extraction Solvent	Recovery (%)
100	Benzene	77
200	Benzene	64
1000	Benzene	90
100	Methylene Chloride	31
200	Methylene Chloride	49
1000	Methylene Chloride	57

MEDIA SPIKED WITH NITROPYRENE SOLUTION[a]

Medium	Extraction Trial	Recovery (%)	Mean ± S.D.
Phosphate-Buffered Saline	1	82.7	84.9 ± 1.9
	2	86.2	
	3	86.7	
	4	84.0	
Medium 199	1	82.4	75.7 ± 11.8
	2	84.4	
	3	77.3	
	4	58.6	
Medium 199 + 10% FCS	1	74.9	74.6 ± 2.1
	2	72.0	
	3	77.2	
	4	74.3	

[a]Final concentration of 10 μg nitropyrene/ml medium; all media were extracted with methylene chloride.

The findings showed that vapor phase coating was sufficiently efficient for the open loop method and that the vapor phase would not pass through the adsorbent bed without saturating the majority of adsorption sites.

Nitropyrene Coating Experiments

The diffusion cell of the adsorption apparatus was maintained at 156-160°C and the adsorption bed at 162-164°C. The results of 14 adsorption studies using non-tritiated and tritiated NP are presented in Table 3.

Although the duration of coating was altered in an attempt to enhance concentrations, no relationship was observed between time and concentration (Table 3). The ^{3}H-nitropyrene was coated on four separate occasions (Experiments 12-15). Although analysis of fly ash from experiment 11 indicated 5 ppm extractable nitropyrene, the pooled sample from experiments 12, 14 and 15 contained 200 ppm nitropyrene. Material from experiment 13 was not included due to contamination in sample processing.

In addition to chemical/radiochemical analysis of the coated material, light microscopy (LM) and scanning electron microscopy (SEM) were also used to evaluate the efficacy of the coating process employed. These techniques are very sensitive indicators for a number of physical and morphological characteristics, due to their ability to detect particles or structures in the picogram and femtogram range. In this study, light and electron microscopy were used to evaluate the homogenity and evenness of coating, the amount of free nitropyrene present as crystalline nitropyrene, and the gross morphological/physical nature of the coating. In addition, SEM was used to evaluate surface texture and crystal habit of any free or surface crystals visible.

In all phases of microscopic evaluation, uncoated parent material was directly compared with treated fly ash. Although surface texture was observed microscopically, it did not appear qualitatively or quantitatively different in the nitropyrene-coated fly ash than in the uncoated parent material. Inhomogeneity or unevenness of coating was not observed since the coating itself was not directly observable microscopically, either by SEM or LM. No morphologic changes were apparent in shape, size, color, or surface structures and texture. No free crystals of nitropyrene were observed by any of the techniques employed, even after scanning numerous fields of hundreds of particles each. In summary, there was no difference between coated and uncoated fly ash that could be detected microscopically, either by SEM at magnifications up to 20,000X or by LM using transmitted or reflected, brightfield or polarized illumination at magnifications up to 900X.

TABLE 3

SUMMARY OF FLY ASH COATING EXPERIMENTS

Experiment No.	Mass (g) of Fly Ash Coated	Duration (hr) of Coating	of Nitropyrene ($\mu g/g$)
1	1.5	15	<4
2	1.5	18	220
3	1.4	45	45
4	2.1	45	59
5	2.1	45	59
6	1.2	30	91
7	2.1	19	35
8	1.1	15	15
9	1.1	15	17
10	1.3	39	110
11	2.3	62.5	98
12a	2.3	21.5	<5, b200
13a	2.4	21	contaminated
14a	2.6	32	b200
15a	2.6	38.5	b200

a3H-Nitropyrene
bConcentration of NP in pooled sample of 12, 14 and 15

^{3}H-Nitropyrene Coated Fly Ash

Analyses of the ^{3}H-NP coated fly ash from the four experimental runs indicated activity varying from approximately 50,000 to 100,000 DPM/mg (Table 4). Material from runs 12, 14, and 15 were combined and the resulting activity of 110,000 DPM/mg was observed. The apparent discrepancy between combined activity and individual sample activity is most likely due to the relatively high coefficient of variation observed for runs 12, 14, and 15.

TABLE 4

RADIOACTIVITY STUDIES OF ^{3}H-NITROPYRENE COATED FLY ASH

	Run 1	Run 2	Run 3	Run 4	Combined (1,3,4)
Sample - Total Activity (DPM/mg FA)					
1	88,438	39,380	96,944	68,717	110,872
2	113,216	50,544	62,864	57,107	105,503
3	89,544	50,573	90,635	63,372	100,600
4	78,335	64,231	129,386	65,319	124,485
$\bar{X}$	92,383	51,182	94,957	63,065	110,365
S.D.	± 14,776	± 10,171	± 27,313	± 5,811	± 10,056

Sample - ^{3}H Remaining on Fly Ash Extracted with Benzene (DPM/mg FA)[a]

1	73,649
2	68,189
3	75,325
4	67,974
$\bar{X}$	73,534
S.D.	± 3,813

Sample - ^{3}H Extracted from Fly Ash (DPM/mg FA)[a]

	Extraction 1	Extraction 2	$\bar{X}$ + S.D.
1	39,011	42,577	41,134
2	39,664	43,287	± 2,112

$$\% \ ^3H \ \text{Recovery} = \frac{73,534 + 41,134}{110,365} = 104\%$$

[a] Combined runs 1, 3 and 4

Analysis of benzene extracts of the fly ash indicated that approximately 37 percent of the total activity was extracted. Similarly, analysis of the residual activity remaining on the fly ash after extraction indicated 68 percent of the total activity was unextracted (Table 4). Excellent radioactivity closure was obtained; the activity of the extracted and residual samples was 104 percent of the un-extracted, untreated fly ash sample. As noted in Table 1, the specific activity of the purified ^{3}H-nitropyrene was 0.13 mCi/mg.

^{3}H-NP and ^{3}H-NP Coated Fly Ash in Tissue Culture Medium

In order to evaluate isotopic exchange and extraction efficiency for nitropyrene alone and nitropyrene-coated fly ash, ^{3}H-NP (8.0 and 0.080 g/ml) and one ^{3}H-NP-coated fly ash (480 μg/ml) were used. The fly ash concentration was chosen to provide extractable ^{3}H-NP used, i.e., 0.080 vs 0.096 μg ^{3}H-NP/ml. It was found that 74.0 and 75.7 percent of the total tritium recovered was extractable from the media containing the direct additions of ^{3}H-NP at 8.0 and 0.08 μg/ml, respectively (Table 5). This compares favorably with the chemical determination of NP in the methylene chloride extracts of aqueous media which showed an average recovery of 74.6 percent of added NP (Table 2). Based on a specific activity of 279,000 DPM/g ^{3}H-NP, the NP extracted from the incubation media was 5.73 ± 0.15 and 0.0569 ± 0.024 μg NP/ml, respectively, for high and low spiked incubations according to radioactivity determinations. Chemical analysis yielded similar results with extraction of 6.01 ± 0.57 and 0.0559 ± 0.0003 μgNP/ml, respectively, for the high and low spiked incubations. These results demonstrate that within experimental error the extractable tritium follows the NP in the directly spiked media. Therefore, isotopic exchange is considered to be minimal under these conditions. Aliquots of the medium counted for total tritium showed that 96.9 ± 5.1 percent and 93.9 ± 3.9 percent, respectively, of the total tritium added was recovered from the high and low directly spiked media.

Results of incubation of the ^{3}H-NP coated fly ash in tissue culture medium are summarized in Table 5. Radioactivity analysis indicated complete (106 percent) recovery of added ^{3}H-NP. Filtration studies demonstrated that 13 percent of the radioactivity and 11 percent of the mass of ^{3}H-NP (based on radioactivity analysis) dissolved in the medium. Extraction studies demonstrated that 8.6 percent of the radioactivity and 7.9 percent of the mass of the ^{3}H-NP dissolved in the media

TABLE 5

SUMMARY OF ^{3}H-NITROPYRENE STUDIES IN TISSUE CULTURE MEDIUM

I. ^{3}H-NP (8.0 µg/ml)

	Mass of NP based on Radioactivity (µg NP/ml)[a]			Mass of NP based on Chemical Analysis (µg NP/ml)		
	$\bar{X} \pm$ S.D.	n	(%)[b]	$\bar{X} \pm$ S.D.	n	(%)[b]
Total Recovery	7.75 ± 0.41	4	(96.9)	8.01 ± 0.76	4	(100)
Extractable	5.73 ± 0.15	4	(71.6)	6.01 ± 0.57	4	(75.1)
Residual	0.23 ± 0.01	4	(2.9)	na[c]		

II. ^{3}H-NP (0.080 µg/ml)

	Radioactivity (µg NP/ml)[a]			Mass (µg NP/ml)		
	$\bar{X} \pm$ S.D.	n	(%)[b]	$\bar{X} \pm$ S.D.	n	(%)[b]
Total Recovery	0.0751 ± 0.0031	4	(93.9)	0.0745 ± 0.0004	4	(93.1)
Extractable	0.0569 ± 0.0024	4	(71.1)	0.0559 ± 0.0003	4	(69.9)
Residual	0.0026 ± 0.001	4	(3.2)	na[c]		

III. ^{3}H-NP-Coated Fly Ash (480 µg Fly Ash/ml)[a]

	Radioactivity (µg NP/ml)[a]			Mass (µg NP/ml)		
	$\bar{X} \pm$ S.D.	n	(%)[b]	$\bar{X} \pm$ S.D.	n	(%)[b]
Total Recovery	0.2017 ± 0.0101	6	(106)	--		
Filterable	0.0255 ± 0.0043	3	(13.2)	0.0200 ± 0.0049	4	(10.5)
Extractable	0.0164 ± 0.0047	4	(8.6)	0.0150 ± 0.0037	4	(7.9)
Residual	0.0044 ± 0.0062	4	(2.2)	na[c]		

a. Mass determination from previous specific activity measurement of 279,000 DPM/µg ^{3}H-NP recovery
b. Percent recovery of total added
c. Residual samples could not be chemically analyzed

could be extracted with methylene chloride. Residual activity of the extracted medium was 2.2 percent based on the direct scintillation analysis of an aliquot and 0.9 percent based on analysis after sample oxidation. The specific activity of fly ash recovered after incubation in medium was 122,313 DPM/mg which was similar to that of the starting material (110,365 ± 10,056 DPM/mg). Although some discrepancies were observed between direct analysis of soluble extracts and analysis after oxidation, these data generally are in agreement. These studies show that ^{3}H-^{1}H exchange in tissue culture medium from either ^{3}H-nitropyrene or ^{3}H-nitropyrene coated fly ash occurs at a minimal rate under the conditions of this experiment and should not influence other studies using these materials.

SUMMARY AND CONCLUSIONS

Initially, ^{14}C-labeled nitropyrene was to be used. Due to difficulties during synthesis, tritiated nitropyrene was substituted for further experiments. Nitropyrene was synthesized by nitration of pyrene and purified by liquid chromatography. Vapor phase adsorption was performed using the apparatus described by Miguel et al. (9). Chemical analysis of solvent extractable nitropyrene coated onto the fly ash showed that adsorption was erratic with concentrations ranging from 0 to 250 ppm. Later studies with ^{3}H-nitropyrene indicated that the irreproducibility of the coating procedure was probably due to the inefficiency of the extraction technique. In agreement with Griest and Guerin (10) most of the radiolabeled nitropyrene was not extracted from the fly ash surface. Isotopic exchange studies and tissue culture media extraction studies indicated minimal isotopic exchange and approximately 10 percent extraction of nitropyrene from fly ash.

These studies demonstrate the feasibility of coating coal fly ash with vapor phase, radiolabeled polynuclear aromatic hydrocarbons. A comparison of chemical analysis of radioactivity indicates that vapor phase-deposited nitropyrene is not readily extracted from fly ash surfaces. In particular, benzene extraction of fly ash indicate that 32 percent of the total nitropyrene is extractable. Incubation of ^{3}H-nitropyrene with tissue culture media resulted in solubilization of 13 percent of the radioactivity and 11 percent of the mass. The ^{3}H-^{1}H exchange in tissue culture medium from either ^{3}H-nitropyrene or ^{3}H-nitropyrene-coated fly ash occurs at a minimal rate and should not influence subsequent studies with these materials.

More work is necessary to improve the reproducibility of the vapor phase-coating procedures. Based on the tissue culture studies, protein binding of nitropyrene with serum proteins should be evaluated to determine the potential biological significance.

ACKNOWLEDGEMENTS

The authors gratefully acknowledge the financial support of the U.S. Environmental Protection Agency (Contract No.

68-02-3169). Dr. D.F.S. Natusch was a great aid in the planning of these studies. We also acknowledge the synthesis of nitropyrene by Dr. H. M. Grotta and the excellent technical assistance and advice of B. Prentice, J. Browning, P. Mondren and S. Summer.

REFERENCES

1. Natusch, D.F.S. (1978): Potentially carcinogenic species emitted to the atmosphere by fossil fueled power plants, Environ. Health Perspect., 2:79-90.
2. Chrisp, E.E., Fisher, G.L., and Lammert, N. (1978): Mutagenicity of respirable coal fly ash, Science, 199:73-75.
3. Fisher, G.L., Chrisp, C.E., and Raabe, O.G. (1979): Physical factors affecting the mutegenicity of fly ash from a coal-fired power plant, Science, 205:879-881.
4. Hansen, L.D., Fisher, G.L., Chrisp, C.E., and Eatough, D.J. (1981): Chemical properties of bacterial mutagens in stack collected coal fly ash. In: Proceedings of the Fifth International Symposium on Polynuclear Aromatic Hydrocarbons, edited by M. Cooke and A.J. Dennis, pp. 507-517, Battelle Press, Columbus, Ohio.
5. Fruend, M. Fleischer, K. (1974): Uben eine syntbretische bildungweise des pyrens, Justus Liebigs Annalen der Chemie, 402:77-82.
6. Vollmann, H. Becker, H. Corell, M., and Streeck, H. (1937): Beitrage zur kentniss des pyrens und seiner deviate, Justus Liebigs Annalen der Chemie, 531:1-159.
7. Long, M.A., Garnett, J.L. Vining, R.F.W. (1975): A new simple method for rapid tritium labeling of organics using organoaluminum dihalide catalysts, Journal of the American Chemical Society, 94:8632-8633.
8. Long, M.A., Garnett, J.L., and Vining, R.F.W. (1975): Rapid deterioration and tritiation of organic compounds using organometallic and elemental halides as catalysts, Journal of the Chemical Society, Perkin Transactions II, 1298-1303.
9. Miguel, A.H., Korfmacher, W.A., Wehrg, L., Mamaton, G., and Natusch D.F.S. (1979): Apparatus for vapor-phase adsorption of polycyclic organic matter onto particulate surfaces, Environ. Sci, and Tech., 13:1229-1232.
10. Griest, W.C. and Guerin, M.R. (1979): EA-1092 Electric Power Research Institute Report, Palo Alto, CA.

AMBIENT PARTICULATE AND BENZO (α) PYRENE CONCENTRATIONS FROM RESIDENTIAL WOOD COMBUSTION, IN A MOUNTAIN COMMUNITY

DENNIS J. MURPHY (1), ROY M. BUCHAN*, DOUGLAS G. FOX**
*Occupational Health and Safety Section, 110 Veterinary Science, Colorado State University, Fort Collins, CO 80523;
**U.S. Forest Service, Rocky Mountain Forest and Range Experiment Station, 240 W. Prospect St., Fort Collins, CO 80526.

INTRODUCTION

Consumption of wood for space heating in the United States appears to be steadily increasing, judging from the dramatic increase in sales of wood burning stoves and homes with fireplaces during the past 8 years (1, 2, 3).

Telluride, Colorado, is a small, resort community in the San Juan Mountains, at an elevation of 2,682 meters above sea level, in a steep-walled, narrow mountain valley, with relatively poor meteorological ventilation. The low wind speeds and frequent temperature inversions inhibit dispersion of particulates. This results in high ambient particulate concentrations, even though the area has relatively low source strength.

Residential wood combustion (RWC) provides approximately 25% of the annual heating demand, and is the major source of particulates and polycyclic organic matter compounds.

METHODS AND MATERIALS

Four sites were selected, and each location was equipped with two high volume air samplers and meteorological monitoring equipment. Air samples were collected for a 13-day period, during January 1980.

---

(1) Dennis Murphy is currently a consultant with Occusafe, Inc. 1040 S. Milwaukee, Wheeling, IL 60090

Site 1 (Figure 1) was located at ground level, in the municipal park, 1 km east of the town. No sources were in the immediate area. Site 2 was on the roof of a building in town, at an elevation of 5 m above ground. This site was chosen for its location in the center of town, representative of the area with the highest emission. Site 3 was at ground level, in a field, 2.4 km west of town. Source strength in the area was very low, and the site was expected to receive particulates and POMs generated in town and carried to the site by down-valley winds. Site 4, 2.8 km southwest of town at an elevation 220 m above the valley floor, served as a control because it was not directly influenced by sources in the valley. This site was on a lift building, on the mountain of the Telluride Ski Company. The high volume air sampler was located 6 m above ground, on the roof of the ski lift.

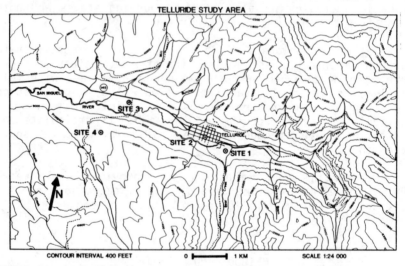

<u>FIGURE 1</u>. Topographic map of Telluride study area including sampling site locations.

All sites were also equipped with MRI (Meteorological Research Inc.) mechanical weather stations (6). Temperature, wind speed, and wind direction data were collected continuously and reduced to 1-hour averages.

Air Sampling Protocol

TSP and POM samples were collected on Gelman type AE, 8-inch by 10-inch, glass fiber filters, using high volume (hi-vol) air samplers. Hi-vols were operated for 24 hours (0000-2400), and two hi-vols were located at each site so that continuous sampling could be maintained. All air sampling equipment was calibrated at altitude in Telluride. Before and after the field study, all glass fiber filters were numbered, dried for 24 hours, and weighed to the nearest 10^{-5} gram using a Model HL-52 Mettler Electrobalance. Filters were then placed in an air-tight, plastic container to prevent contamination.

Laboratory Procedures

All samples were returned to the laboratory at Colorado State University. Total suspended particulate (TSP) was then determined in micrograms per cubic meter ($\mu g/m^3$). The filters were cut in half, one half was placed in a freezer; the other half was used for analysis of POMs.

POMs were extracted from the filters using high pressure liquid chromatography (HPLC) grade, "distilled in glass," benzene in Soxhlet extractors. Extraction time for the filters was 8 hours, using 250 ml of benzene. Following extraction, the samples were concentrated to approximately 100 ml using a nitrogen blanket. The samples were transferred into 250-ml, amber glass bottles. Samples were then stored in a refrigerator.

Analysis was performed using a high pressure liquid chromatograph (HPLC). Twenty ml of sample was placed in a graduated centrifuge tube and concentrated to 0.1 ml using a nitrogen blanket to remove the benzene. Twenty ml of HPLC grade, "distilled in glass," methylene chloride was then placed in the centrifuge tube as the solvent in place of benzene. This procedure was repeated three times to remove as much benzene as possible, as benzene was not compatible with the HPLC column.

Before injection into the HPLC, the samples were concentrated to a volume of 0.1 ml using a nitrogen blanket; 25 µl of sample was injected into a 25 cm µBondpak C_{18} column. Methanol and water mobile phase was used 80:20 for 35 minutes and 90:10 for 15 minutes. A pressure of 3,300 psi was used with a flow rate of 1 ml/min. The reservoir, column, and detectors were all maintained at room temperature. BaP was detected by ultraviolet (UV) and fluorescence detectors. Data for TSP and BaP were corrected to standard temperature and pressure.

RESULTS AND DISCUSSION

Concentration data for total suspended particulates (TSP) and benzo(a)pyrene (BaP) at the four sampling sites are listed in Table 1. Site 1 had the lowest concentrations of both TSP and BaP of all sites in the valley floor. TSP ranged from 10 to 45 µg/m^3, with a mean concentration of 20 µg/m^3 and a standard deviation of 9.5. BaP ranged from 0.6 to 3.4 ng/m^3, with a mean of 1.3 ng/m^3 and standard deviation of 0.9.

The highest concentrations of TSP and BaP were at Site 2. TSP ranged from 36 to 89 µg/m^3. The mean was 61 µg/m^3 and standard deviation of 16.4. BaP concentrations were also highest at Site 2, ranging from 3.7 to 14.8 ng/m^3 with a mean concentration of 7.4 ng/m^3 and standard deviation of 3.2.

At Site 3 pollutants were generally found in concentrations which were lower than Site 2 but higher than Site 1. TSP ranged from 11 to 62 µg/m^3 with a mean of 36 µg/m^3 and standard deviation of 15.8. BaP ranged from 0.04 to 7.4 ng/m^3 with a mean of 2.4 ng/m^3 and standard deviation of 2.0.

Considerable problems were experienced with the electrical system at Site 4, the control site. Five samples were not collected and two were terminated before completing the 24-hour sample. The highest concentration found, 26 µg/m^3, was only a 12-hour sample. Other samples at this site ranged from 3 to 9 µg/m^3. As the 12-hour sample was included in the statistical analysis the mean concentration was 7.5 µg/m^3 and the standard deviation was 7.7. BaP was not detected in any of the samples from the control site.

The meteorological data from Sites 1, 2, and 3 were typical of mountain valley wind patterns. A down-valley wind (1-3 mph) blew from the east and southeast generally after sundown; a stronger, up-valley wind blew during the day, once the south-facing slopes had warmed up. The mountains and valley winds were responsible for dispersion of the pollutants away from Site 2.

Site 2 had the highest concentrations of both TSP and BaP, because it was located in the area of greatest source strength. The correlation coefficient between TSP and BaP was 0.61 ($p = 0.02$). This relatively low correlation probably resulted from the fact that a portion of the TSP was from various fugitive dust sources.

TABLE 1

CONCENTRATIONS OF TSP AND BaP BY SITE, TELLURIDE, COLORADO, JANUARY 7-19, 1980. (CORRECTED TO STANDARD TEMPERATURE AND PRESSURE)

	Site 1		Site 2		Site 3		Site 4	
	TSP	BaP	TSP	BaP	TSP	BaP	TSP	BaP
7	21	0.8	NS	NS	53	3.1	NS	NS
8	26	1.0	88	7.5	40	1.8	9	ND
9	14	1.1	65	8.1	38	2.9	4*	ND
10	45	1.3	68	4.7	21	0.04	26ψ	ND
11	18	0.9	62	11.3	26	1.6	5	ND
12	21	3.4	89	14.8	43	7.4	NS	NS
13	14	1.1	59	7.6	30	2.0	NS	NS
14	11	0.6	36	6.3	17	0.2	NS	NS
15	10	0.6	45	4.6	35	1.5	NS	NS
16	27	3.0	61	5.5	62	3.0	4	ND
17	26	1.8	58	7.3	61	5.1	5	ND
18	13	0.8	NS	NS	35	2.3	3	ND
19	14	0.9	46	3.7	11	0.5	4	ND
$\bar{X}$	20	1.3	61	7.4	36	2.4	7.5	–
SD	9.5	0.9	16.4	3.2	15.8	2.0	7.7	–

TSP in $\mu g/m^3$
BaP in ng/m^3

NS – No Sample
ND – Not Detectable
* 15 hour sample
ψ 12 hour sample

RESIDENTIAL WOOD COMBUSTION

At Site 3, the correlation coefficient between TSP and BaP was 0.69 (p = 0.005). The TSP and BaP were carried to Site 3 by down-valley winds at night. This was also the time of the most wood burning in town.

When data for TSP and BaP from the three valley sites were combined, the correlation was 0.82 (p = 0.001).

TSP concentrations at Site 4 represent background levels, except for the 12-hour sample of January 10, which was 26 µg/m^3. During that day, there were strong, up-valley winds, and it is hypothesized that pollutants which had drained down valley the night before or particulates from fugitive dust sources were carried back to Site 4 by the strong up-valley winds.

This hypothesis is supported by the data presented in Figure 2, which is the ratio of TSP:B(a)P by site and sample day. As mentioned above, during January 10, 1980 strong, up-valley winds prevailed most of the entire 24-hour period. An elevated ratio of TSP:B(a)P was observed at all three valley sites, especially at Site 3, on this day, suggesting that the major contributor to TSP was fugitive dust from some unknown down-valley source.

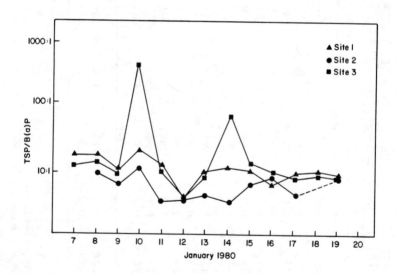

FIGURE 2. Ratio of TSP:B(a)P observed during Jaunary 7-19, 1980, at sites 1, 2, and 3, Telluride, Colorado.

On January 12, 1980, however, down-valley winds of low velocity (1-3 mph) dominated the 24-hour sampling period. The ratio at all three valley sites was very similar, suggesting a common source was responsible for this observation.

CONCLUSIONS

The mean concentration of BaP at Site 2, 7.4 ng/m^3, was relatively high for a mountain community. The value was considerably higher than the mean concentration of 2.6 ng/m^3 found in Denver, Colo., during the Winter Aerosol Study of 1978 (5). The authors of the Denver Study note "even though this average is biased by choosing the filters from high pollution days, it is still within the normal urban BaP range of 2-7 ng/m^3." Based on the BaP concentrations found in Denver and concentrations found at 26 National Air Sampling Network Stations reviewed recently by Faoro and Manning (4), the levels of BaP found in Telluride appear to be high, especially for an isolated community with low source strength.

The concentrations of BaP found in the center of Telluride appears to be several times higher than what is usually found in United States metropolitan areas. Levels were considerably lower outside of town and BaP was not detected at the control site. Based on other indirect data available for Telluride, residential wood combustion is suspected as the primary source of BaP during the study period.

REFERENCES

1. Statistical Abstracts of the United States. U.S. Department of Commerce, Washington, D.C., 1975, 1055 pp.
2. Preliminary Characterization of Emissions from Wood-fired Residential Combustion Equipment. U.S. Environmental Protection Agency, Research Triangle Park, North Carolina, March 1980.
3. Budiansky, S. (1980): Bioenergy: The lesson of wood burning, Environ. Sci. and Tech., 14(7):769-771.
4. Faoro, R. B., and Manning, J. A. (1981): Trends in Benzo(a) Pyrene, APCA J., 31(1):62-71.

5. Countess, R. J., Wolff, G. T., and Cadle, S. H. (1980): The Denver winter aerosol: A comprehensive chemical characterization, APCA J., 30(11):1194-1200.
6. The use of trade and company names is for the benefit of the reader; such use does not constitute an official endorsement or approval of any service or product by the U.S. Department of Agriculture to the exclusion of others that may be suitable.

OXIDATION OF PHENANTHRENE WITH A CARBONYL OXIDE

ROBERT W. MURRAY, SHAILENDRA KUMAR,
Department of Chemistry, University of Missouri-St. Louis, St. Louis, Missouri 63121

INTRODUCTION

It is now well established that many polycyclic aromatic hydrocarbons (PAHs) are converted by metabolic activation to actual or ultimate carcinogens (1-13). A major chemical characteristic of the active metabolites is that they are electrophiles which can become covalently attached to biological macromolecules, including DNA, and, in some cases, initiate the carcinogenic process. In the case of PAHs the most potent metabolites are epoxy diols (6). The epoxy (or oxirane) functionality provides the required electrophilic character. Metabolites or synthetic compounds containing only oxirane functional groups are frequently mutagenic (14), presumably because of their DNA bonding capacity.

The oxidative metabolic processes referred to are catalyzed by monooxygenase enzymes (MOX). Efforts to understand the basic chemical processes brought about through the intervention of these complex molecules frequently involves the use of chemical model systems. In some cases these model systems are devised on the basis of general mechanistic guidelines, such as the oxene theory proposed by Hamilton (15). A number of chemical model systems for the MOX have been proposed and studied. Since the discovery of the NIH shift (16) however, the number of viable model systems has been considerably reduced because of the failure of many systems to meet this more demanding mechanistic criterion. One such surviving system is that using a carbonyl oxide as the oxygen transfer species as first suggested by Hamilton (17), and also studied by Jerina et al. (18) and ourselves. Some time ago we developed (19) a more convenient source of carbonyl oxides which source has permitted us to study the chemistry of these interesting oxidants in more detail. In particular we have been engaged in detailed studies of carbonyl oxides as MOX models.

These studies are of particular importance to the subject of environmental health since carbonyl oxides can be expected to be formed in polluted atmospheres via the reaction between two well established components of such atmospheres, namely ozone and olefins. Since such atmospheres also frequently contain a variety of PAH, including a number of pre-carcinogens, we have speculated (20) that some PAH may become partially or completely activated in such atmospheres without the intervention of metabolic processes, that is via oxidations brought about by carbonyl oxides. Pertinent to this speculation are the growing number of reports (20-26) that extracts of atmospheric particulate matter have a higher carcinogenic activity than that associated with their PAH content. Carbonyl oxides thus are of interest both as models for the MOX enzymes and as potential contributors to PAH activation in the environment.

In order for carbonyl oxides to be considered viable models for the monooxygenase enzymes they must meet several tests relative to their chemical reactivity. One such test is the ability to epoxidize olefinic materials. Hamilton and Keay (27) have shown that an α-carbonyl carbonyl oxide, produced by ozonization of an alkyne, is capable of carrying out such epoxidations. Using the non-ozone method of producing a carbonyl oxide we have also been able to demonstrate the epoxidation capacity of these species (28). A second test for an MOX model is that it be able to oxidize aromatic substrates. Furthermore such oxidations should follow the now well-known details of the enzymatic oxidations, namely that they proceed via an arene oxide and, in suitably-substituted substrates, that they display the NIH shift. We earlier (29) described the oxidation of naphthalene by a carbonyl oxide. We have since described several cases (30) in which carbonyl oxides have been observed to carry out intramolecular oxidations of aromatic rings. Jerina et al. (18) have reported a case in which a carbonyl oxide oxidized anisole-4-D with accompanying NIH shift. We now have carried out carbonyl oxide oxidations of aromatic compounds in which NIH shifts of chlorine and methyl groups have been observed (31). Observance of the NIH shift argues strongly for the intermediacy of arene oxides in these oxidations. Likewise the observed distribution (29) (85/15 = α/β) of naphthols in the carbonyl oxide oxidation of naphthalene is very close to that known to be produced (32) in the isomerization of naphthalene-1,2-oxide. Nevertheless we felt that isolation of an arene oxide in a carbonyl oxide oxidation was a desirable goal which would provide unequivocal evidence for the intermediacy of such species in these oxidations. We describe here the results of such a study.

MATERIALS AND METHODS

Materials

Absolute alcohol and 95% ethanol were purchased from U.S.

Industrial Chemical Company, New York, N.Y. The other solvents were obtained from Fisher Scientific Company, Fairlawn, N.J. Methylene chloride was purified by stirring it overnight with conc. sulfuric acid followed by washing with water, 5% sodium carbonate, and again with water. It was then dried over calcium chloride and distilled under nitrogen over calcium hydride. Benzophenone, m-chloroperbenzoic acid, and 9-phenanthrol were purchased from Aldrich Chemical Company, Milwaukee, Wisconsin. Phenanthrene and 95% hydrazine hydrate were obtained from Eastman Kodak Company, Rochester, N.Y. ⓟ-Rose Bengal (33) was purchased from Hydron Laboratories, Inc., New Brunswick, N.J. Active manganese dioxide was prepared as described in the literature (34). Alumina-activity 1 (80-200 mesh), purchased from Fischer Scientific Co., Fairlawn, N.J., was converted to alumina-activity 3 by shaking it with 6% water by weight. TLC glass plates (20 x 20cm) precoated with 1.0 mm thickness silica gel with fluorescent indicator UV-254 were purchased from Brinkmann Instruments, Inc., Des Plaines, Illinois. HPLC cyano-silica column (8 mm x 60 cm), cyano-10, was purchased from Varian Aerograph, Palo Alto, Calif.

Instrumentation

High Performance liquid chromatography was performed on a Varian model 5020 Liquid chromatograph using a variable wavelength Vari-Chrom detector set at 254 nm. Mass spectra were recorded at 70 eV ionizing voltage on an Associated Electronic Industries MS 1201-B Spectrometer. Infrared Spectra were recorded on a Perkin Elmer Model 337 Grating Infrared Spectrophotometer using matched 0.1 mm sodium chloride solution cells purchased from Wilmad Glass Company, Buena, N.J. Melting points were determined on a Thomas-Hoover capillary melting point apparatus and on a Dynamics Optics hot stage microscope.

Methods

Benzophenone was converted to its hydrazone by refluxing it in 95% hydrazine hydrate (10 g; 0.2 mol) containing 15 ml of absolute ethanol. The hydrazone was recrystallized from 95% ethanol and had mp 97-98°C; lit (35) mp 98°. A solution of 5.8 g (30 mmol) of benzophenone hydrazone, 3.0 g of anhydrous magnesium sulfate, and 60 ml of methylene chloride was stirred in an ice water bath. To this rapidly stirred solution was added, in one portion, 9.2 g of active manganese dioxide (34). Stirring was continued for 2 hours at 0° and for 1 hour at room temperature. Solid material was removed by filtration and washed with methylene chloride.

Removal of solvent by rotary evaporation gave a dark maroon oil which gave 5.2 g (88%) of dark red needles of diphenyldiazomethane when cooled, mp 30°C (lit mp 35°C) (36). The photolysis apparatus consisted of a pyrex vessel equipped with a jacketed immersion well, a fritted gas inlet at the bottom, a gas outlet fitted with a Dewar condenser, and a side neck for the introduction of reactants. Irradiation was by a General Electric DWY 600 Watt lamp contained in the immersion well. During photooxygenation the entire apparatus was immersed in an ice water bath and the reaction mixture was maintained at ca. 5°C by circulating ice water through the jacket of the immersion well. A solution of 1.0 g (5.0 mmol) of phenanthrene in 75 ml of methylene chloride was placed in the photolysis apparatus and to it was added 200 mg of ⓟ-Rose Bengal (33) as sensitizer. Irradiation and oxygen bubbling were begun and a solution of 1.95 g (10 mmol) of diphenyldiazomethane in 50 ml of methylene chloride was added in 3 ml aliquots at 4 minutes intervals during the photooxidation. Irradiation was discontinued after the red color of the diazo compound had been discharged (1 hour). The reaction mixture was worked up using two different methods. In the first method (37) the reaction mixture was concentrated to about 20 ml and 3 g of alumina (activity 3, 80-200 mesh) was added as adsorbent and residual solvent removed in vacuo. The coated alumina was transferred to a chromatography column (2 cm diam.) filled with alumina (activity 3, 10 cm) and the column capped with anhydrous sodium sulfate (10 cm). The column was eluted with 200 ml of methylene chloride, the eluate concentrated to 20 ml and the residue again deposited on 3 g of alumina (activity 3) using rotary evaporation. The coated alumina was transferred to an alumina column (activity 3, 100 g). The column was eluted with hexanes until most of the phenanthrene had been removed. The column was then developed with successive mixtures of diethyl ether (0-10%) in hexanes. The presence of phenanthrene-9,10-oxide was monitored by comparing its TLC R_f value (benzene as developing solvent) with that of an authentic sample. In the second workup method methylene chloride was removed from the reaction mixture by rotary evaporation, the residue dissolved in ether and analyzed by TLC using 20 x 20 cm, 1.0 mm thickness silica gel plates and a mixture of hexanes and diethyl ether (6:1) as developing solvent. The TLC fractions having the same R_f values as those of authentic phenanthrene-9,10-oxide and 9-phenanthrol were removed from the TLC plates using diethyl ether as extracting solvent. Phenanthrene-9,10-oxide was further purified using high performance liquid chromatography (HPLC) and a cyanosilica column (8 mm x 50 cm) with elution (4 ml/minute) at room temperature using 8% methylene chloride in hexane as solvent. Under these conditions the oxide has a retention time of 19.3 minutes. The phenanthrene-9,10-oxide was identified by comparing its TLC R_f values (2 solvent systems), HPLC retention time, mass spectral, IR, and melting point data with those of the authentic material. The oxide had mp 103-104°C (lit mp 104-105°C) (38) and mixed mp 103-104°C. The yield of phenanthrene-9,10-oxide was determined by calibrating HPLC peak areas using known quantities of authentic material and the cut and weigh method. The yield was 8.4%. The 9-phenanthrol was identified by comparing its TLC R_f value, mass spectral, and IR data with those of the authentic material. The yield of 9-phenanthrol was 2.9% as determined by weighing TLC-isolated material. A number of control experiments were run. The photooxidation was run 1) with no added diazo compound and 2) with no irradiation. The reaction was also carried out with 1.9 g (10.4 mmol) of benzophenone added instead of diazo compound. A similar reaction was carried out with benzophenone substituting for diazo compound but with no sensitizer present. A summary of product yield data for all of the photooxidation reactions is given in Table 1. In a separate experiment 30 mg (0.15 mmol) of phenanthrene-9,10-oxide was exposed to the conditions of the sensitized photooxidation reaction. Analysis of the reaction mixture indicated that the oxide was converted into 9-phenanthrol in 27% yield. Authentic phenanthrene-9,10-oxide was prepared according to the method of Ishikawa et al. (37).

RESULTS AND DISCUSSION

Oxidation of phenanthrene with benzophenone carbonyl oxide gives phenanthrene-9,10-oxide and 9-phenanthrol. A control experiment indicated that the oxide is converted to 9-phenanthrol under the reaction conditions so that it is likely that the phenanthrol is not a primary reaction product of the oxidation. Griffin et al. (39, 40) had earlier shown that the oxide is converted to 9-phenanthrol, among other products, under the conditions of rose bengal sensitized photooxidation. These same workers have also suggested (39, 40) that the photosensitized oxidation of phenanthrene could also produce phenanthrene-9,10-oxide. We have run a number of control reactions to test for the possibility of direct production of the oxide via singlet oxygen oxidation of phenanthrene as well as by a variety of other reaction conditions (Table 1).

In experiment 1 (Table 1) singlet oxygen oxidation of diphenyldiazomethane is used to produce the carbonyl oxide, benzophenone oxide, as previously described (19). Under these

TABLE 1

OXIDATION OF PHENANTHRENE

Expt. No	Reaction Conditions	% yield (oxide)	% yield (ol)
1[a]	$(C_6H_5)_2CN_2$ + $^1O_2$ + Phenanthrene	8.4	2.9
2[a]	$^1O_2$ + Phenanthrene	0.6	1.0
3	$(C_6H_5)_2CN_2$ + $^3O_2$ + Phenanthrene	0	0
4[a]	$(C_6H_5)_2C=O$ + $^1O_2$ + Phenanthrene	1.2	1.3
5[b]	$(C_6H_5)_2C=O$ + $^3O_2$ + Phenanthrene	0.4	0.1

a Experiments in which Rose Bengal sensitization used.
b With irradiation.

conditions the highest yield of phenanthrene-9,10-oxide is obtained. As shown in experiment 2 direct singlet oxygen oxidation of phenanthrene gives a small yield of the oxide as suggested earlier by Griffin et al. (39, 40). Experiment 3 indicates that no ground state oxygen oxidation of phenanthrene is occurring under the reaction conditions. Control experiments 4 and 5 were prompted by earlier observations of Shimizu and Bartlett (41). These workers had shown that photooxidation of olefins in the presence of benzophenone leads to epoxidation of the olefins. The authors have suggested that the epoxidizing species is an adduct between benzophenone and oxygen. Since our reaction conditions involving diphenyldiazomethane lead to the production of benzophenone it was necessary to determine whether such an adduct could also directly oxidize phenanthrene. As shown in experiment 5 some oxidation of phenanthrene does occur when benzophenone is present during photooxidation. In experiment 4 Rose Bengal sensitization is used as well as having benzophenone present. The increased yield of oxide in experiment 4 over experiment 5 is presumably due to the combined contributions of direct singlet oxygen oxidation as well as oxidation via a mechanism involving benzophenone, possibly the mechanism suggested by Shimizu and Bartlett (41). Whatever the details of the oxidation mechanism the observation that some oxidation of phenanthrene occurs under the conditions of experiment 5 is itself interesting and has implications for the chemistry of polluted atmospheres. The data of Table 1 indicate that carbonyl oxide oxidation is the major source of phenanthrene-9,10-oxide in the phenanthrene oxidation.

By following the production of phenanthrol and phenanthrene-9,10-oxide with time under the conditions of experiment 1 we have been able to show that the phenanthrol is not a primary product of carbonyl oxide oxidation, but results entirely from isomerization of the oxide under the reaction conditions. We have also been able to show that this isomerization is a photochemical and not a thermal process under our reaction conditions. We have not been able to obtain any evidence for the formation of 2,3,4,5-dibenzoxepin under our reaction conditions. Griffin et al. had earlier (39) indicated that the phenanthrene-9,10-oxide is converted to this oxepin under the conditions of singlet oxygen oxidation.

These experiments add support to the use of carbonyl oxides as MOX models since it has now been shown that carbonyl oxide oxidation of a PAH gives an arene oxide, specifically a K-region oxide. At the same time the results strengthen our earlier speculation (26) that PAH may become partially activated in polluted atmospheres via a mechanism involving ozone-olefin derived carbonyl oxides.

ACKNOWLEDGEMENTS

We gratefully acknowledge support of this work by the National Institutes of Health through grant no. 1 R01 ES01984.

REFERENCES

1. Jerina, D.M. and Daly, J.W. (1974): Arene oxides: A new aspect of drug metabolism, Science, 185: 573-582.
2. Boyland, E. (1950): The biological significance of metabolism of polycyclic compounds, Biochem. Soc. Symp., 5: 40-54.
3. Boyland, E. and Sims, P. (1965): The metabolism of benz[a]anthracene and dibenz[a,h]anthracene and their 5,6-epoxy-5,6-dihydro derivatives by rat-liver homogenates, Biochem. J., 97: 7-16.

4. DePierre, J.W. and Ernster, L. (1978): The metabolism of polycyclic hydrocarbons and its relationship to cancer, Biochem. Biophys. Acta,, 473: 149-186.
5. Selkirk, J.K. (1977): Benzo(a)pyrene carcinogenesis: a biochemical selection mechanism, J. Tox. Environ. Health, 2: 1245-1258.
6. Sims, P., Grover, P.L., Swaisland, A., Pal, K., and Hewer, A. (1974): Metabolic activation of benzo(a)pyrene proceeds by a diolepoxide, Nature, 252: 326-328.
7. Heidelberger, C. (1975): Chemical carcinogenesis, Ann. Rev. Biochem., 44: 79-121.
8. Sims, P. and Grover, P.L. (1974): Epoxides in polycyclic aromatic hydrocarbon metabolism and carcinogenesis, Advan. Cancer Res., 20: 165-274.
9. Boyland, E. and Sims, P. (1962): The metabolism of phenanthrene in rabbits and rats: dihydrodihydroxy compounds and related glucosiduronic acids, Biochem. J., 84: 571-582.
10. Miller, E.C. and Miller, J.A. (1974): Biochemical mechanisms of chemical carcinogenesis. In: The Molecular Biology of Cancer, edited by H. Busch, pp. 377-402. Academic Press, New York.
11. Gelboin, H.V. (1969): Microsome dependent binding of benzo[a]pyrene to DNA, Cancer Res., 29: 1272-1276.
12. Grover, P.L. and Sims, P. (1968): Enzyme-catalyzed reactions of polycyclic hydrocarbons with deoxyribonucleic acid and proteins in vitro, Biochem. J., 110: 159-160.
13. Nebert, D.W. and Gelboin, H.V. (1968): Substrate-inducible microsomal aryl hydroxylases in mammalian cell culture, J. Biol. Chem., 243: 6242-6249.
14. Ames, B.N., Sims, P. and Grover, P.L. (1972), Epoxides of carcinogenic polycyclic hydrocarbons are frameshift mutagens, Science, 1976: 47-49.
15. Hamilton, G.A. (1964): Oxidation by molecular oxygen. II The oxygen atom transfer mechanism for mixed-function oxidases and the model for mixed-function oxidases, J. Am. Chem. Soc., 86: 3391-3392.
16. Guroff, G., Reifsnyder, C.A. and Daly, J. (1966): Retention of deuterium in p-tyrosine formed enzymatically from p-deuteriophenylalanine, Biochem. Biophys. Res. Comm., 24: 720-724.
17. Hamilton, G.A. and Giacin, J.R. (1966): Oxidation by molecular oxygen. III. Oxidation of saturated hydrocarbons by an intermediate in the reaction of some carbenes with oxygen, J. Am. Chem. Soc., 88: 1584-1585.

18. Jerina, D.M., Boyd, D.R. and Daly, J.W. (1970): Photolysis of pyridine N-oxide: An oxygen atom transfer model for enzymatic oxygenation, arene oxide formation, and the NIH shift, Tetrahedron Lett., 457-460.
19. Higley, D.P. and Murray, R.W. (1974): Oxidation of diazo compounds with singlet oxygen. Formation of ozonides, J. Am. Chem. Soc., 96: 3330-3332.
20. Murray, R.W., Kumar S., and Agarwal, S.K. (1981): Carbonyl oxide chemistry and chemical carcinogenesis, Abstracts of Papers, 64th Canadian Chemical Conference, Halifax, Nova Scotia, p. 71
21. Moller, M. and Alfheim, I. (1980): Mutagenicity and PAH analysis of airborne particulate matter, Atmos. Environ., 14: 83-88.
22. Kotin, P., Falk, H.L., Mader, P., and Thomas, M. (1954): Aromatic hydrocarbons. I. Presence in the Los Angeles atmosphere and the carcinogenicity of atmospheric extracts, Arch. Ind. Hyg. Occup. Med., 9: 153-163.
23. Hueper, W.C., Kotin, P., Tabor, E.C., Payne, W.W., Falk, H. and Sawicki, E. (1962): Carcinogenic bioassays on air pollutants, Arch. Pathol., 74: 89-116.
24. Epstein, S.S., Joshi, S., Andrea, J., Mantel, N., Sawicki, E., Stanley, T. and Tabor, E.C. (1966): Carcinogenicity of organic particulate pollutants in urban air after administration of trace quantities to neonatal mice, Nature, 212: 1305-1307.
25. Rigdon, R.H. and Neal, J. (1971): Tumors in mice induced by air particulate matter from a petrochemical industrial area, Tex. Rep. Biol. Med., 29: 109-123.
26. Freeman, A.E., Price, P.J., Bryan, R.J., Gordon, R.J., Gilder, R.V., Kelloff, G.J., and Huebner, R.J. (1971): Transformation of rat and hamster embryo cells by extracts of city smog, Proc. Natl. Acad. Sci. U.S.A., 68: 445-449.
27. Keay, R.E. and Hamilton, G.A. (1976): Alkene epoxidation by intermediates formed during the ozonation of alkynes, J. Am. Chem. Soc., 98: 6578-6582.
28. Hinrichs, T.A., Ramachandran, V., and Murray, R.W. (1979) Epoxidation of olefins with carbonyl oxides, J. Am. Chem. Soc., 101: 1282-1284.
29. Chaudhary, S.K., Hoyt, R.A., and Murray, R.W. (1976): Oxidation of naphthalene by an intermediate in the singlet oxygen oxidation of diphenyldiazomethane: A chemical model for the monooxygenase enzymes, Tetrahedron Lett., 4235-4236.
30. Kumar, S. and Murray, R.W. (1980): An intramolecular oxenoid oxidation, Tetrahedron Lett., 4781-4782.

31. Murray, R.W. and Kumar S., unpublished results.
32. Jerina, D.M., Daly, J.W. Witkop, B.,Zaltzman-Nirenberg, P., and Udenfriend, S. (1970); 1,2-Naphthalene oxide as an intermediate in the microsomal hydroxylation of naphthalene, Biochemistry, 9: 147-155.
33. Schaap, A.P. Thayer, A.L., Blossey, E.C., and Neckers, D.C. (1975): Polymer-based sensitizers for photooxidations, II., J. Am. Chem. Soc., 97: 3741-3745.
34. Attenburrow, J. Cameron, A.F.B., Chapman, J.H., Evans, R.M.,Hems,B.A., Jansen, A.B.A., and Walker, T. (1952): A synthesis of vitamin A from cyclohexanone, J. Chem. Soc., 1094-1111.
35. Miller, J.B. (1959): Preparation of crystalline diphenyldiazomethane, J. Org. Chem., 24: 560-561.
36. Reimlinger, H. (1954): Darstellung und eigenschaften einiger diaryl-diazomethane, Ber., 97: 3493-3502.
37. Ishikawa, K., Charles, H.C. and Griffin, G.W. (1977): Direct peracid oxidation of polynuclear hydrocarbons to arene oxides, Tetrahedron Lett., 427-430.
38. Newman, M.S. and Blum, S. (1964): A new cyclization reaction leading to epoxides of aromatic hydrocarbons, J. Am. Chem. Soc., 86: 5598-5600.
39. Dowty, B.J.,Brightwell, N.E., Laseter, J.L., and Griffin, G.W. (1974): Dye-sensitized photooxidation of phenanthrene, Biochem. Biophys. Res. Comm., 57: 452-456.
40. Patel, J.R., Politzer, I.R., Griffin, G.W., and Laseter, J.L. (1978): Mass spectra of the oxygenated products generated from phenanthrene under simulated environmental conditions, Biomed. Mass Spec., 5: 664-670.
41. Shimizu, N. and Bartlett, P.D. (1976): Photooxidation of olefins sensitized by α-diketones and by benzophenone. A practical epoxidation method with biacetyl, J. Am. Chem. Soc. 98: 4193-4200.

COMPARISON OF THE SKIN TUMOR INITIATING ACTIVITIES OF EMISSION EXTRACTS IN THE SENCAR MOUSE.

S. NESNOW*, L. L. TRIPLETT**, T. J. SLAGA**
*Carcinogenesis and Metabolism Branch, U.S. Environmental Protection Agency, Research Triangle Park, North Carolina;
**Biology Division, Oak Ridge National Laboratory, Oak Ridge, Tennessee.

INTRODUCTION

The incomplete combustion of fossil fuels results in the emission of particulate and organic vapor-phase components to the atmosphere. The particulate phase of these emissions contains organic materials adsorbed onto the particulate matrix. These organic materials have been subjected to intense chemical analysis, fractionation and characterization (1-3). Emissions from gasoline engines, collected as condensates, were tumorigenic when applied dermally to mice (4-7). Extracts from particles collected from a gasoline engine were also tumorigenic on mouse skin (8). Kotin et al. (9) reported that extracts of particles collected from diesel engines were active in producing tumors on strain A mice, while Mittler and Nicholson (7) reported little tumorigenic activity from diesel exhaust condensates.

We have previously reported that extracts from particulate emissions from coke oven, roofing tar, and several diesel and gasoline engines produced papillomas when applied to the skin of SENCAR mice (10,11), a mouse strain that has been shown to be highly sensitive to chemical carcinogens (12,13). This paper describes the results of a systematic study of the ability of extracts of particulate emissions from various sources to induce benign and malignant tumors in SENCAR mice.

MATERIALS AND METHODS

Sample Generation and Isolation

The details of sample generation and isolation have been reported elsewhere (2). Particulate emissions were collected from a 1973 preproduction Nissan Datsun and a 1977 Mustang vehicle (Table 1), each which was mounted on a chassis dynamometer with a repeated highway fuel economy cycle of 10.24 mi, an average speed of 48 mph, and a running time of 12.75 min. Particle samples were collected with a dilution tunnel in which

TABLE 1

PARTICULATE EMISSION SAMPLES BIOASSAYED ON SENCAR MICE

Source	Emission Type
Nissan-Datsun 220 C Vehicle[a] (preproduction model, 1973)	Combustion-Diesel
Mustang II - 302 Vehicle[a] (with cataylst, 1977)	Combustion-Unleaded Gasoline
Topside Coke Oven Emissions[b]	Pyrolytic process
Roofing Tar Emissions[c]	Particles emitted by heating pine-based pitch at 182-193°C

a Particulate emissions collected on Teflon-coated fiberglass filters after cooling and dilution of the exhausts, which were generated using a standard highway fuel economy transit cycle.
b Particulate emissions collected atop a coke oven battery by electrostatic precipitation.
c Particulate emission collected from a tar pot with Teflon socks in a baghouse.

the hot exhaust was diluted, cooled and filtered through Pallflex Teflon-coated fiberglass filters. Topside coke oven samples were collected from the top of a coke oven battery at Republic Steel, Gadston, AL, by use of a Massive Air Volume Sampler. Because of the topside ambient location and local wind conditions, an unknown portion of this emission sample contains particles from the local urban environment. The roofing tar emission sample was collected from a conventional tar pot with external propane burner. Pine-based tar was heated to 182° to 193°C and emissions were collected with a 1.8 m stack extension and Teflon socks in a baghouse. It should be noted that only one source was used for each sample, therefore each sample may not be representative of a particular technology. All samples were Soxhlet extracted with dichloromethane, which was then removed by evaporation under dry nitrogen gas.

Tumor Experiments

Studies involved 80 7- to 9-week-old SENCAR mice per treatment group (40 of each sex) (14). Animals were housed in plastic cages (10 per cage) with hardwood chip bedding, under yellow light, fed Purina chow and water ad libitum, and maintained at 22° to 23°C with 10 changes of air per hour. All mice were shaved with surgical clippers two days before the initial treatment and only those mice in the resting phase of the hair cycle were used. All samples at all doses were applied as a single topical treatment in 0.2 ml of acetone, except for the 10 mg dose, which was administered in five daily doses of 2 mg. One week after treatment, 2.0 µg of the tumor promoter 12-O-tetradecanoyl phorbol-13-acetate (TPA) (in 0.2 ml acetone) was administered topically twice weekly. Skin tumor formation was recorded weekly, and papillomas greater than 2 mm in diameter and carcinomas were included in the cumulative total if they persisted for one week or longer. At six months the numbers of papillomas per surviving animal were recorded for statistical purposes.

Statistical Analysis Methods for Papilloma Data

Analyses were carried out on papilloma scorings performed 24 to 26 weeks after initiation. Two types of statistical analyses were performed. Tumor incidence data were analyzed by a probit model according to Finney (15), which takes into account the numbers of spontaneous tumors occuring in the TPA control groups (16). The tumor multiplicity data were analyzed by a nonlinear Poisson model with correction for background according to Hasselblad (17).

RESULTS

The results of the tumor initiation experiments on SENCAR mice for benzo(a)pyrene (BaP) and for topside coke oven, Nissan, roofing tar and Mustang extracts are shown in Tables 2 to 4. Animals were scored at six months for papilloma formation and at one year for carcinomas. The carcinoma data represent the cumulative number of carcinomas found in each treatment group and therefore include tumors on both living and dead animals. The BaP, topside coke oven, Nissan, and roofing tar samples produced an 89% or greater papilloma incidence at the highest dose level applied. Tumor multiplicity ranged from 5 to 6 papillomas per mouse in the roofing tar and Nissan samples to greater than 7 in the BaP and topside coke oven samples. These groups of animals also produced a significant number of squamous cell carcinomas, ranging from 13 to 48% of the mice bearing carcinomas at the highest dose evaluated (Tables 2,4).

The Mustang sample (Table 3) was tested at doses from 0.1 to 3 mg/mouse due to sample limitations. The papilloma per mouse response was maximal in the female animals at 3 mg/mouse and activity plateaued at 2 to 3 mg/mouse in the male animals. Twenty percent of the female mice produced carcinomas at the highest dose tested (Table 4). The lack of a monotonic dose response across the complete dose range tested in the Mustang sample may indicate a toxic response. Damage to the skin epidermal cells will result in lower expression of the tumorigenic response.

TABLE 2

SENCAR MOUSE SKIN TUMORIGENESIS, BENZO[A]PYRENE - TUMOR INITIATION

Dose (µg/mouse)	No. Mice Surviving	Mice with Papillomas[a] (%)	Papillomas per Mouse[a]	Mice with Carcinomas (%)	Carcinomas per Mouse[b]
0 (M)	37	8	0.08	5	0.05
0 (F)	39	5	0.05	0	0
2.52 (M)	40	45	0.50	5	0.07
2.52 (F)	39	31	0.44	5	0.05
12.6 (M)	40	73	1.8	20	0.20
12.6 (F)	37	57	1.1	23	0.23
50.5 (M)	39	100	5.8	25	0.25
50.5 (F)	40	75	2.8	20	0.20
101 (M)	38	95	10.2	30	0.33
101 (F)	38	97	7.9	25	0.25

a Scored at 6 months.
b Cumulative score after one year.

DISCUSSION

Quantitative methods for the analysis of tumor data are many and employ tumor incidence, tumor multiplicity, and tumor latency data. Statistical methods have been employed using Poisson and other distribution assumptions, as well as both uni- and multivariate analytical approaches (18,19). We have chosen to apply a nonlinear Poisson model to the papilloma incidence data in order to compare the tumorigenic activities

TABLE 3

SENCAR MOUSE SKIN TUMORIGENESIS – TUMOR INITIATION RESULTS FOR FOUR SAMPLES SCORED AT SIX MONTHS FOR PAPILLOMAS

Dose (mg/mouse)	No. Mice Surviving				Mice with Papillomas (%)				Papillomas per Mouse			
	Nissan	Mustang	Coke Oven	Roofing Tar	Nissan	Mustang	Coke Oven	Roofing Tar	Nissan	Mustang	Coke Oven	Roofing Tar
0.1 (M)	37	39	40	40	0	5	13	10	0	0.05	0.13	0.13
0.1 (F)	39	39	40	39	3	13	10	15	0.03	0.23	0.20	0.21
0.5 (M)	38	39	40	40	26	13	73	28	0.34	0.15	1.6	0.35
0.5 (F)	39	38	40	39	23	18	70	13	0.39	0.24	1.8	0.15
1.0 (M)	40	40	37	39	33	18	95	38	0.38	0.18	2.6	0.41
1.0 (F)	38	37	39	40	39	10	72	45	0.53	0.13	2.0	0.80
2.0 (M)	35	39	38	39	66	22	95	36	1.1	0.24	4.0	0.62
2.0 (F)	40	39	39	38	58	21	90	37	1.6	0.23	3.5	0.45
3a/10b (M)	38	34	39	39	89	18	100	100	5.5	0.24	7.1	6.4
3a/10b (F)	38	40	40	40	97	23	100	95	5.6	0.28	7.7	5.7

a Mustang sample applied at 3 mg/mouse.
b Nissan, topside coke oven and roofing tar samples applied at 10 mg/mouse.

TABLE 4

SENCAR MOUSE SKIN TUMORIGENESIS - TUMOR INITIATION RESULTS FOR FOUR SAMPLES: CUMULATIVE CARCINOMA SCORE AFTER ONE YEAR

Dose (mg/mouse)		Mice with Carcinomas (%)				Carcinomas per Mouse			
		Nissan	Mustang	Coke Oven	Roofing Tar	Nissan	Mustang	Coke Oven	Roofing Tar
0.1	(M)	0	5	0	5	0	0.03	0	0.05
0.1	(F)	5	13	8	10	0.05	0.13	0.08	0.10
0.5	(M)	13	0	5	10	0.13	0	0.05	0.10
0.5	(F)	10	10	15	18	0.10	0.10	0.15	0.18
1.0	(M)	20	5	15	5	0.20	0.05	0.15	0.05
1.0	(F)	13	10	3	15	0.13	0.10	0.03	0.15
2.0	(M)	13	15	13	13	0.13	0.15	0.13	0.13
2.0	(F)	15	13	10	15	0.15	0.13	0.10	0.15
$3^a/10^b$	(M)	36	5	13	23	0.36	0.05	0.15	0.25
$3^a/10^b$	(F)	31	20	20	48	0.31	0.20	0.23	0.48

a Mustang sample applied at 3 mg/mouse.
b Nissan, topside coke oven, and roofing tar samples applied at 10 mg/mouse.

of complex mixtures from emission extracts. The model calculates the number of papillomas per mouse for a 1 mg treatment dose. This model assumes a Poisson distribution of tumors, that tumor multiplicity is related to dose, that the response may be nonlinear, and that there is a background response. Estimates from the models are presented only if they are in the range from which the data were obtained and if the observed data adequately fit the calculated model (Table 5).

Results from the nonlinear Poisson model suggest the following ranking: topside coke oven > Nissan > roofing tar > Mustang. The values calculated are only estimates and in some cases all the assumptions made to derive the estimates are only partially fulfilled.

A probit model has been chosen to evaluate the tumor incidence data. The probit model examines animals with tumors (regardless of multiplicity) and animals without tumors and calculates the dose that produces tumors in 50% (over background) of the animals. Results from the probit analysis suggest the ranking: BaP > topside coke oven > Nissan = roofing tar. These are not the only models which can be applied to these data, and although they appear effective in this case, more effort is being placed in improving statistical

TABLE 5

SENCAR MOUSE SKIN TUMOR INITIATION
ESTIMATES FROM TWO MODELS BASED ON PAPILLOMA DATA AT SIX MONTHS[a]

		Nonlinear Poisson		Probit	
		Papillomas/Mouse at 1 mg	95% Confidence Intervals	Dose for 50% Papilloma Incidence (TID$_{50}$), mg	95% Confidence Intervals
Benzo(a)pyrene	M	b		0.0036	0.0021 - 0.0062
	F			0.0091	0.0057 - 0.015
Topside Coke	M	2.2[c]	2.00 - 2.40	0.30	0.22 - 0.40
Oven	F	2.0[c]	1.90 - 2.20	0.42	0.31 - 0.58
Nissan	M	0.49[c]	0.38 - 0.63	1.60	1.2 - 2.2
	F	0.68[c]	0.57 - 0.79	1.50	1.1 - 1.9
Roofing Tar	M	0.38[c]	0.30 - 0.49	1.8	1.2 - 2.7
	F	0.44[c]	0.35 - 0.55	2.1	1.5 - 2.8
Mustang	M	0.17	0.12 - 0.24	d	

a Estimates calculated from model according to Material and Methods.
b Not calculated since data were obtained at a lower dose range.
c The distribution of tumors at all dose levels was not Poisson as the variances exceeded the means.
d Not calculated since tumor incidence was <50%.

and modeling techniques and in statistical analysis of the carcinoma data.

It is compelling to postulate that the BaP in these complex mixtures could account for their tumorigenic activity, since mouse skin is exquisitely sensitive to this agent. The results presented here reveal that a single application of less than 4 µg of BaP will yield a 50% tumor incidence as a tumor initiator. However, the relationship between BaP content in each mixture, and papilloma response for each mixture, is not linear (Figure 1). Probably none of the activity of the topside coke oven sample can be explained by BaP content, as the BaP-induced tumor response at the BaP level in the topside coke oven sample is quite small. Even the BaP level in the Nissan sample (11 µg/10 mg extract) can only account for 20 to 30% of the papilloma response elicited by the Nissan sample. Other components of the mixtures may play an important role in their tumorigenic activities.

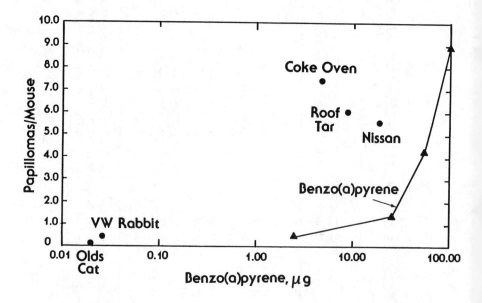

FIGURE 1. Relationship of skin tumor initiating activities of complex mixtures to their BaP levels: comparison to skin tumor initiating activity of pure BaP. The VW Rabbit, Caterpillar and Oldsmobile diesel sample values were calculated from data previously reported (11).

In addition to the tumorigenesis studies described above, detailed gross and histopathological analyses of selected animals have been undertaken. Further results from these detailed pathological studies on the formation of internal tumors and the appearance of tumors with longer latency periods will be presented at a later date.

ACKNOWLEDGEMENTS

The authors wish to thank R. L. Bradow, R. H. Jungers, B. D. Harris, T. O. Vaughan, R. B. Zweidinger, K. M. Cushing, J. Bumgarner, and B. E. Gill for the sample collection, preparation, and characterization, C. Evans, A. Stead, J. Creason, and V. Hasselblad for the ADP and statistical assistance, and C. J. Alden and L. J. Jones for assistance in preparation of this manuscript. The research was sponsored by the U.S. Environmental Protection Agency, contract no. 79D-X0526, under the Interagency Agreement, U.S. Department of Energy no. 40-728-78, and the Office of Health and Environmental Research, U.S. Department of Energy, under contract no. 7405 eng-26 with the Union Carbide Corporation.

REFERENCES

1. Huisingh, J., Bradow, R., Jungers, R., Claxton, L., Zweidinger, R., Tejada, A., Bumgarner, J., Duffield, F., Waters, M., Simmon, V., Hare, C., Rodriguez, C. and Snow, L. (1979) Application of bioassay to the characterization of diesel particle emissions. In: _Application of Short-term Bioassays in the Fractionation and Analysis of Complex Environmental Mixtures._ Waters, M. D., Nesnow, S., Huisingh, J. L., Sandhu, S. S. and Claxton, L., eds. Plenum Press: New York, pp. 383-418.
2. Huisingh, J. L., Bradow, R. L., Jungers, R. H., Harris, B. D., Zweidinger, R. B., Cushing, K. M., Gill, B. E., and Albert, R. E. (1980) Mutagenic and carcinogenic potency of extracts of diesel and related environmental emissions: study design, sample generation, collection, and preparation. In: _Health Effects of Diesel Engine Emissions. Proceedings of an International Symposium._ Pepelko, W. E., Danner, R. M. and Clarke, N. A., eds. Vol 2, Washington, D.C., U.S. Govt. Printing Office (EPA Publication No. EPA-600/9-80-057b), pp. 788-800.

3. Waters, M. D., Sandhu, S. S., Huisingh, J. L., Claxton, L. and Nesnow, S. eds. (1981): Short-term Bioassays in the Analysis of Complex Environmental Mixtures, II. Plenum Press, New York.
4. Hoffmann, D. and Wynder, E. L. (1963): Studies on gasoline engine exhaust. J. Air Pollut. Contr. Assoc., 13:322-327.
5. Hoffmann, D., Theisz, E. and Wynder, E. L. (1965) Studies on the carcinogenicity of gasoline exhaust. J. Air Pollut. Contr. Assoc., 15:162-165.
6. Misfeld, J. and Timm, J. (1978): The tumor-producing effect of automobile exhaust condensate and fractions thereof. Part III: Mathematical-statistical evaluation of the test results, J. Environ. Path. and Toxicol., 1:747-772.
7. Kotin, P., Falk, H. L. and Thomas, M. (1954): Aromatic hydrocarbons. II. Presence in the particulate phase of gasoline-engine exhausts and the carcinogenicity of exhaust extracts, AMA Arch. Ind. Hyg. Occup. Med., 9:164-177.
8. Mittler, S. and Nicholson, S. (1957): Carcinogenicity of atmospheric pollutants, Ind. Med. and Surg., 26:135-138.
9. Kotin, P., Falk, H. L., and Thomas, M. (1955): Aromatic hydrocarbons. III. Presence in the particulate phase of diesel-engine exhausts and the carcinogenicity of exhaust extracts, AMA Arch. Ind. Health, 11:113-120.
10. Slaga, T. J., Triplett, L. L. and Nesnow, S. (1980): Mutagenic and carcinogenic potency of extracts of diesel and related environmental emissions: two-stage carcinogenesis in skin tumor sensitive mice (SENCAR). In: Health Effects of Diesel Engine Emissions. Proceedings of an International Symposium. Pepelko, W. E., Danner, R. M., Clarke, N. A. eds. Vol 2. Washington, D.C. U.S. Govt. Printing Office (EPA publication No. EPA-600/9-80-057b), pp. 874-897.
11. Nesnow, S., Triplett, L. L., and Slaga, T. J. (1981): Tumorigenesis of diesel exhaust, gasoline exhaust, and related emission extracts on SENCAR mouse skin. In: Short-term Bioassays in the Analysis of Complex Environmental Mixtures, II. Waters, M., Sandhu, S. S., Lewtas, J., Claxton, L. and Nesnow, S., eds. New York: Plenum Press, pp. 277-297.
12. DiGiovanni, J., Slaga, T. J. and Boutwell, R. K. (1980): Comparison of the tumor-initiating activity of 7,12-dimethylbenz(a)anthracene and benzo(a)pyrene in female SENCAR and CD-1 mice, Carcinogenesis, 1:381-389.

13. Hennings, H., Devor, D., Wenk, M. L., Slaga, T. J., Former, B., Colburn, N. H., Bowen, G. T., Elgjo, K., and Yuspa, S. H. (1981): Comparison of two stage epidermal carcinogenesis initiated by 7,12-dimethylbenz(a)anthracene or N-methyl-N'-nitro-N-nitrosoguanidine in newborn and adult SENCAR and Balb/c mice, Cancer Res., 41:773-779.
14. Boutwell, R. K. (1964): Some biological aspects of skin carcinogenesis, Progr. Exptl. Tumor Res., 4:207-250.
15. Finney, P. J. (1971): Probit Analysis, Cambridge University Press, Cambridge.
16. Hasselblad, V., Stead, A. G., and Creason, J. P. (1980): Multiple probit analysis with a nonzero background, Biometrics, 36:659-663.
17. Stead, A. G., Hasselblad, V., Creason, J. P. and Claxton, L. (1981): Modeling the Ames test, Mutation Res., 85:13-27.
18. Drinkwater, N. R. and Klotz, J. H. (1981): Statistical methods for the analysis of tumor multiplicity data. Cancer Res., 41:113-119.
19. Gart, J. J., Chu, K. C., and Tarone, R. E. (1979): Statistical issues in interpretation of chronic bioassay tests for carcinogenicity, J. Nat. Cancer Institute, 62:957-974.

AN INVESTIGATION OF FACTORS INFLUENCING THE SEPARATION OF POLYNUCLEAR AROMATIC HYDROCARBONS BY LIQUID CHROMATOGRAPHY

U. D. NEUE, B. P. MURPHY, J. CROOKS
Waters Associates, Inc., Milford, Massachusetts 01757

ABSTRACT

A new packing material has been developed for the chromatographic determination of the Polynuclear Aromatic Hydrocarbons under scrutiny in the United States and internationally. The chromatographic parameters influencing retention were investigated with special emphasis on the influence of temperature. Enthalpy effects are shown to be related to the structure of the compounds under investigation.

We chromatographed isomers at various temperatures and made a Van't Hoff plot of the capacity factors. The slope of the line from this plot is related to enthalpy. It is clear from the slope of this plot that the more linear the molecule, the more it will be affected by changes in temperature. As a result, temperature can be used as a variable to improve the separation.

INTRODUCTION

The application of high pressure liquid chromatography (HPLC) to determine polynuclear aromatic hydrocarbons (PAH) in a variety of matrices has been widespread (1-3). In the literature there has been much work reporting the effects of altering the conditions of analysis such as mobile phase composition and stationary phase (5-8). In this work we held all the conditions of analysis constant and only allowed temperature variation.

Experimental

Instrumentation - A Waters Associates Model 204 Liquid Chromatograph equipped with a Model 440 UV Absorbance Detector at 254nm wavelength and a Houston Instruments OmniScribe Strip Chart Recorder were used. Constant temperature was maintained using a prototype oven from a Model 150C liquid chromatograph equipped with a proportional controller. Subambient temperatures were achieved using an ice bath with constant stirring.

Column - A Radial-PAK™ PAH Cartridge (10cm X 5mm i.d.) from Waters Associates was used. The cartridge is packed with 10μm spherical silica particles bonded with an octadecylsilane. The cartridge was compressed in an RCM-100® Radial Compression Module.

　　Reagents - LC grade acetonitrile from Waters Assocaties was used. Chrysene, benz(a)anthracene, pyrene, phenathrene, anthracene and benz(a)pyrene were obtained from Chem Service (West Chester, PA). Dibenz(a,h)anthracene, benzo(g,h,i) perylene and indeno(1,2,3-cd)pyrene were obtained from Analabs, Inc. (North Haven, Connecticut).

　　Eluents and Samples - The mobile phase was 70% acetonitrile and 30% water (V/V). The mobile phase was mixed, degassed and allowed to warm to room temperature before using. The samples were individually dissolved in 100% acetonitrile and injected separately.

Analysis

　　The Radial-PAK™ PAH was compressed in the RCM-100® Radial Compression Module. The RCM-100® was placed in the prototype oven and connected to the instrument. A temperature was dialed in on the proportional controller and when the oven reached the selected temperatures, fifteen minutes were allowed to lapse so that the oven and cartridge would reach equilibrium. The cartridge was then equilibrated for fifteen minutes in 70/30 acetonitrile/water (V/V) at 4ml/minute flow rate. The samples were injected separately along with acetone as an indicator for void volume. This process was repeated five times at elevated temperatures and once at subambient temperature in an ice bath. The temperature range over which we worked was 3-53°C.

Evaluation of the Capacity Factor

　　The retention time of acetone was taken as t_o and the capacity factor, k, was calculated from the retention time of the sample, t_R, by the relationship $k = (t_R-t_o)/t_o$. Enthalpies were calculated by linear-regression analysis of the capacity factor data.

RESULTS

　　Figure 1 shows the structure of the isomers chromatographed (or similar sized compounds in the case of B(BHI)P and I(1,2,3-cd)P) along with the relationship between the capacity factors and temperature. For any given temperature the change in enthalpy and entropy changed by:

$$\Delta G = \Delta H - T\Delta S \qquad (1)$$

Rearranging Equation 1 and differentiating, we get:

$$\frac{d(\ln K)}{dT} = \frac{\Delta H^0}{RT^2} \qquad (2)$$

From Equation 2 we see that variation in the equilibrium constant with temperature is dependent on the standard heat of the reaction[9].

If we integrate assuming that ΔH is essentially temperate-independent, we get:

$$\ln K = \frac{-\Delta H}{R} \cdot \frac{1}{T} + \text{Constant} \qquad (3)$$

By plotting $\ln K$ versus $1/T$ (^{0}K), we get a straight line with the slope equal to $-\Delta H/R$. Figures 2 and 3 are Van't Hoff plots for the three-ringed isomers and similar sized six-ringed compounds. Examining Figure 2, we find that as we lower the temperature of the chromatographic analysis, the compounds, phenanthrene and anthracene are retained longer on the column. Also note that the capacity factor (k') for anthracene is increasing at a faster rate than the capacity factor for phenanthracene . From Figure 1, the structure of anthracene is more linear than the structure of phenanthrene.

Figure 3 is a Van't Hoff plot for indeno(1,2,3-cd)pyrene and benzo(g,h,i)perylene. Note here that as we lower the temperature of analysis, the capacity factors increase, but at differing rates such that the compounds reverse the order of elution at approximately 38^{0}C. Furthermore, in Figure 4, the six-ringed compound, benzo(g,h,i)perylene, and the five-ringed compound, dibenz(a,h)anthracene, reverse order of elution at approximately 23^{0}C. Referring again to Figure 1, we find that dibenz(a,h)anthracene and indeno(1,2,3-cd)pyrene are more linear than benzo(g,h,i)perylene. Examining Figure 1 more closely, we find that the more linear the molecule (for similar sized compounds such as isomers), the more it will be affected

HPLC SEPARATION FACTORS

by changes in temperature (affect is taken as the steepness of the slope of the line). To state this differently, if two compounds have similar capacity factors but are different in structure, changes in temperature will most likely result in dissimilar capacity factors[10].

COMPOUND	STRUCTURE	RELATIONSHIP	CORRELATION
Phenanthrene	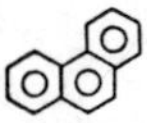	Y = 4835X − 14.09	.9532
Anthracene	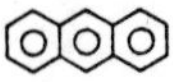	Y = 6698X − 19.59	.9833
Pyrene		Y = 2451X − 6.80	.9986
Benz(a)anthracene	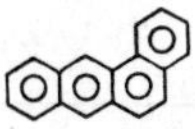	Y = 3479X − 9.66	.9965
Chrysene	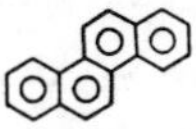	Y = 3779X − 10.59	.9965
Benz(a)pyrene		Y = 5036X − 13.93	.9992
Dibenz(a,h)anthracene	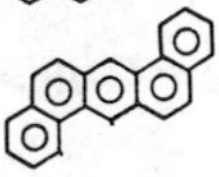	Y = 5666X − 15.53	
Benzo(g,h,i)perylene	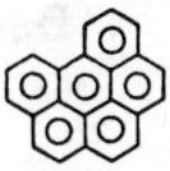	Y = 5136X − 13.74	.9905
Indeno(1,2,3-cd)pyrene	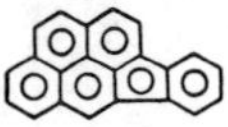	Y = 5998X − 16.50	.9982

Relationship: $Y = \ln k$; $x = 1/T$ (°K); slope $= -\frac{H}{R}$; Y intercept is related to entropy.

Correlation shows the strength of the relationship between $\ln k$ and $1/T$ (1.00 is a perfect correlation).

Figure 1

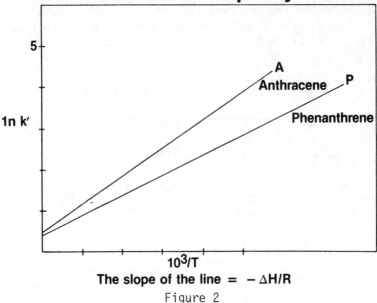

Figure 2

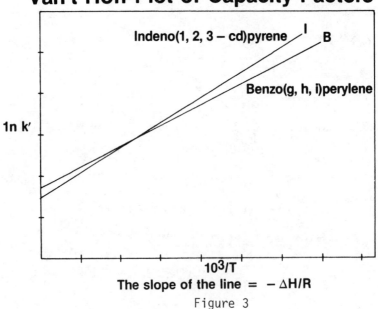

Figure 3

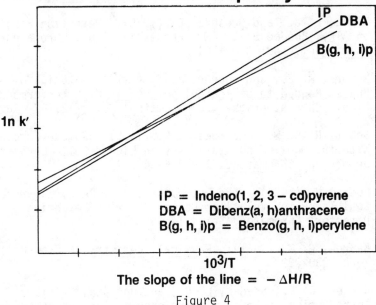

Figure 4

CONCLUSION

Based on our findings, we feel that it is important that you monitor the temperature at which you run your chromatographic analysis. Variations in your chromatography may be explained by variations in temperature in the lab. You could use temperature as a variable to improve your separation.

References

1. Ogan, K., Katz, E., and Slavin, W. (1979): Determination of Polycyclic Aromatic Hydrocarbons in Aqueous Samples by Reversed-Phase Liquid Chromatography, Anal. Chem., 51:1315-1320.

2. Colmsjo, A. L., and MacDonald, J. C. (1980): Column-Induced Selectivity in Separation of Polynuclear Aromatic Hydrocarbons by Reversed-Phase, High Performance Liquid Chromatography, Chromatographia, 13:350-352.

3. Euston, C. B., and Baker, D. R. (1979): Trace Analysis of Water Pollutants by Automated HPLC, Am. Lab., 11:91-100.

5. Kikta, E. J., and Grushka, E. (1976): Retention Bahavior on Alkyl Bonded Stationary Phases in Liquid Chromatography, Anal. Chem., 48:1098-1104.

6. Majors, R. E., and Hopper, M. J. (1974): Studies of Siloxane Phases Bonded to Silica Gel for Use in High Performance Liquid Chromatography, J. Chromatog. Sci., 12:767-778.

7. Scott, R. P. W., and Kucera, P. (1975): Solute Interaction With the Mobile and Stationary in Liquid-Solid Chromatography, J. Chromatog., 112:425-442.

8. Bodin, H. (1976): The Determination of Benzo(a)Pyrene in Coal Tar Pitch Volatiles Using HPLC with Selective UV Detector, Chromatog. Sci., 14:391-395.

9. Barrow, G. M. (1961): Physical Chemistry, McGraw-Hill, New York, 230-232.

10. Karger, B. L., Snyder, L. R., and Horvath, C. (1973): An Introduction to Separation Science, Wiley-Interscience, New York, 58-61.

COMPARISON OF NITRO-AROMATIC CONTENT AND DIRECT-ACTING MUTAGENICITY OF DIESEL EMISSIONS

MARCIA G. NISHIOKA*, BRUCE A. PETERSEN*, AND JOELLEN LEWTAS**
*Battelle-Columbus Laboratories, 505 King Avenue, Columbus, Ohio 43201; **Office of Research and Development, Health Effects Research Laboratory, U.S. Environmental Protection Agency, Research Triangle Park, North Carolina 27711.

INTRODUCTION

The increasing number of automobiles and light-duty trucks powered by diesel engines has generated concern over the emissions associated with these engines. Diesel engines have a higher particulate emission rate than that of gasoline catalyst engines. The extractable organics from both diesel and gasoline particle emissions have been found to be mutagenic and carcinogenic (1). Recently, nitro-substituted polynuclear aromatic hydrocarbons (nitro-PNA, nitro-aromatics) have been identified in diesel particle extracts (2). Several of these nitro-aromatics are very potent bacterial mutagens (3).

A study was carried out to identify and quantify nitro-aromatic compounds in the extract of particulate material from three different diesel engines (Datsun Nissan diesel 220C, Oldsmobile diesel 350, and VW Rabbit diesel) and one gasoline engine (Ford Mustang II). The operating and sampling conditions have been described elsewhere (4). Mutagenic assay data were also collected on these extracts using the Salmonella typhimurium TA98 bioassay, without S9 metabolic activation. The results of these two studies were compared to determine whether the concentration of nitro-aromatics detected can fully account for the direct-acting mutagenic activity indicated by the bioassay data.

TECHNICAL APPROACH

Chemical

Prior to analysis by mass spectrometry, each particulate extract was separated into four chemical compound classes using open-bed silica gel column chromatography: (1) aliphatic hydrocarbons--hexane elution, (2) polycyclic aromatic hydrocarbons--hexane:benzene (1:1 v/v) elution, (3) moderately polar neutrals--methylene chloride elution, and (4) highly polar neutrals, primarily oxygenated compounds--methanol elution. Fraction #2 was expected to contain the mononitro-aromatics, fraction #3 was expected to contain the di- and trinitro-aromatics, and fraction #4 was expected to contain the oxygenated nitro-aromatics. The quantity of silica gel and the volume of elution solvent used was adjusted for the amount of total organic material available for fractionation.

A high resolution gas chromatography/negative chemical ionization mass spectrometry (HRGC/NCIMS) method using on-column injection was developed to analyze the particulate extracts. The instrumentation consisted of a Finnigan Model 4000 GC/MS interfaced to a Finnigan/INCOS Model 2300 Data System. Methane was selected as the reagent gas. A J&W DB-5 bonded, fused silica capillary column was used. The on-column injection was made at 40°C, and the GC was temperature programmed from 40-320°C at 10°C/minute. Standard solutions and fractions #2, #3, and #4 of each extract were analyzed by this method.

Standard solutions containing the following nitro-aromatic compounds

1-nitronaphthalene	2,7-dinitrofluorene
2-nitrofluorene	2,7-dinitrofluorenone
9-nitroanthracene	1,6-dinitropyrene
1-nitropyrene	1,3,6-trinitropyrene

were prepared over the concentration range 0.1-10 ng/μl for the calibration curve. D_7-nitronaphthalene was chosen as the internal standard and it was spiked into all standards and samples at the level of 2.5 ng/μl. The response factor for each compound relative to the internal standard was calculated over the calibration range. Nitro-aromatic compounds not present in the standard which were detected in samples were quantified using the response factor of the most similar compound from the standard. In addition, separate solutions of 1,3- and 1,8-dinitropyrene were analyzed to determine the elution order of the three dinitropyrene isomers.

The relative retention time (relative to d_7-nitronaphthalene) and mass spectrum of each standard compound was used to assign specific isomer identification to a few of the nitro-aromatics detected in the extracts. The other nitro-aromatic isomers detected were identified by appropriate mass spectra and retention times. The availability of other specific nitro-aromatic isomers may allow assignments to be made of those compounds.

The methanol fractions (#4) of the Nissan and VW extracts were also analyzed by HRGC/MS and high resolution mass spectrometry (HRMS) to determine the formulas of the most polar compounds. From these data, structures and compound classes were tentatively assigned.

Biological

An aliquot of the total extract from each sample was tested in several short-term bioassays as reported previously (1). The slope of the dose response curve (rev/µg) in the Salmonella typhimurium TA98 plate incorporation assay (5) without metabolic activation (-S9) was utilized in this study to compare the nitro-aromatic content and direct-acting mutagenicity of each sample extract.

RESULTS

Twenty-one different nitro-aromatics were tentatively identified in the diesel engine extracts. However, only 1-nitropyrene was detected in the gasoline engine extract. In all cases, the 1-nitropyrene was the nitro-aromatic compound detected in greatest quantity, followed by the nitrophenanthrene/anthracene isomers. Quantities of the detected nitroaromatics are given in Table 1 and vary over the range of 0.1 ppm to 589 ppm in the extracts.

Mono-nitro derivatives of the PAHs at molecular weight 228 (such as benz[a]anthracene and chrysene) and molecular weight 252 (such as benzo[a]pyrene, perylene and benzofluoranthenes) were also detected. Two nitropyrenone isomers were tentatively identified in the Nissan and Oldsmobile samples and the three dinitropyrene isomers were identified in the VW sample.

The chromatograms from the NCI HRGC/MS analysis of the hexane:benzene and methylene chloride fractions of the Nissan diesel extract are shown in Figures 1 and 2, respectively. In spite of the relatively rapid temperature program rate, 10°C/minute, the excellent resolving power of the capillary column was maintained. As shown in Figure 1, analyses were complete in 30 minutes, achieving baseline separation for every significant peak. This resolving power enhanced the accuracy for quantifying closely eluting structural isomers.

In Figure 1, the peaks with intensity greater than 25% of the largest peak were identified as PAHs, with the exception of the peak at scan #1275. This peak was identified as

TABLE 1.
QUANTIFICATION OF NITRO-AROMATICS DETECTED IN EXTRACTS OF ENGINE PARTICULATE

Compound	Concentration in Extract, wt ppm				Relative Retention Time (a)
	Nissan Diesel	VW Rabbit Diesel	Olds-mobile Diesel	Mustang II Gasoline	
1-nitronaphthalene	0.5	0.7	0.3	ND	1.003
Nitronaphthalene/azulene isomer	0.1	0.6	0.2	ND	1.031
Nitronaphthalene/azulene isomer	0.9	1.2	0.3	ND	1.042
Nitrofluorene isomer	ND (b)	1.0	0.5	ND	1.426
2-nitrofluorene	ND	ND	0.4	ND	1.478
Nitrofluorene isomer	1.9	1.4	1.1	ND	1.503
Nitrophenanthrene/anthracene isomer	97.8	285.0	63.1	ND	1.494
Nitrophenanthrene/anthracene isomer	2.7	33.0	25.4	ND	1.545
1-nitropyrene	406.8	589.3	107.2	2.5	1.842
Nitromethylpyrene isomer	23.0	45.6	22.0	ND	1.843
Nitromethylpyrene isomer	1.9	3.3	1.9	ND	1.874
Nitro 228 (c)	3.0	1.1	1.4	ND	1.881
Nitro 228	0.8	1.3	0.2	ND	1.938
Nitro 228	3.9	5.6	1.2	ND	1.977
Nitromethyl 228	2.6	ND	ND	ND	2.132
Nitro 252 (d)	28.5	10.0	1.1	ND	2.232
Nitropyrenone isomer	4.8 (T) (e)	ND	2.1 (T)	ND	1.901
Nitropyrenone isomer	11.5 (T)	ND	3.9 (T)	ND	1.916
1,3-dinitropyrene	ND	0.6	ND	ND	2.090
1,6-dinitropyrene	ND	0.6	ND	ND	2.126
1,8-dinitropyrene	ND	0.4	ND	ND	2.149

(a) Relative to internal standard, d_7-nitronaphthalene.
(b) Not detected.
(c) PAH isomers of MW 228-benz(a)anthracene, chrysene, benz(c)phenanthrene, etc.
(d) PAH isomers of MW 252-B(e)P, B(a)P, perylene, benzofluoranthenes, etc.
(e) Tentative identification.

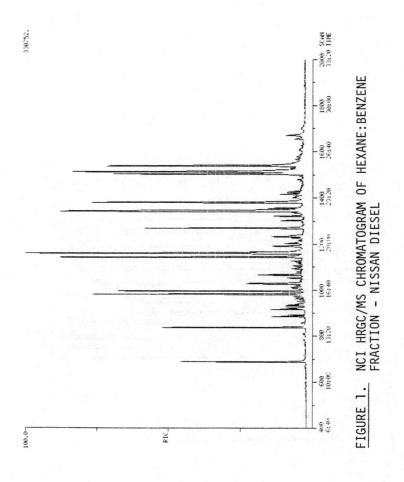

FIGURE 1. NCI HRGC/MS CHROMATOGRAM OF HEXANE:BENZENE FRACTION - NISSAN DIESEL

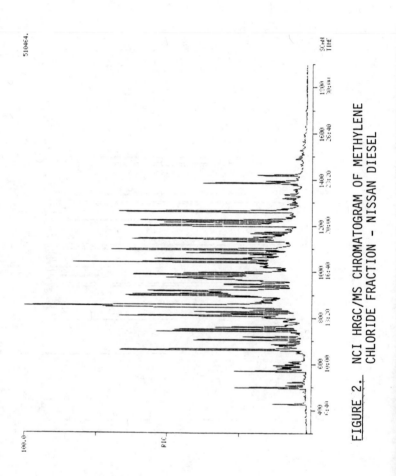

FIGURE 2. NCI HRGC/MS CHROMATOGRAM OF METHYLENE CHLORIDE FRACTION - NISSAN DIESEL

1-nitropyrene. The methlyene chloride fraction was considerably more complex and included primarily carboxaldehyde and semi-quinone derivatives of PAH. The largest peak (scan #850) was identified as 9-fluorenone.

The mutagenic activity (rev/µg) by linear regression analysis for the engine emission extracts and selected nitro-aromatic compounds is given in Table 2. Note that the commercially available 1-nitropyrene appears to be contaminated with di-nitropyrenes.

Analyses by high-resolution mass spectrometry and electron impact HRGC/MS were used to determine that most of the compounds in the methanol fraction of the Nissan and VW Rabbit extracts were quinones, ketones, and dicarboxylic acid anhydride derivatives of PAHs. Representative compounds tentatively identified in the methanol fractions are given in Table 3.

TABLE 2

MUTAGENIC ACTIVITY OF ENGINE EXTRACTS AND PURE NITRO-AROMATICS

Sample	Mutagenic Activity, rev/µg TA98,-S9	Reference
Nissan Diesel	12.0	
VW Rabbit Diesel	3.7	
Oldsmobile Diesel	5.8	
Mustang II Gasoline	1.1	
1-nitronaphthalene	0.3	3
2-nitronaphthalene	1.2	3
2-nitrofluorene	65.9	3
1-nitropyrene (commercially available)	2,111	6
1-nitropyrene (99.99% pure)	918	6
1,3-dinitropyrene	98,000	3
1,6-dinitropyrene	124,500	3
1,8-dinitropyrene	250,000	3
nitrophenanthrene (nitration mix)	224	7
nitroanthracene (nitration mix)	1.8	7
nitrochrysene (nitration mix)	190	7
nitroperylene (nitration mix)	654	7

TABLE 3

REPRESENTATIVE COMPOUNDS TENTATIVELY IDENTIFIED IN METHANOL FRACTIONS

Tentatively Identified Compound	Accurate Mass	Measured Mass Assignment Accuracy, ppm	Formula	Structure
Fluorenone	180.0574	-2.7	C13 H8 O	
Acenaphthylenedione	182.0367	-3.4	C12 H6 O2	
Anthracenone	194.0730	0.5	C14 H10 O	
Naphtho-pyrandione	198.0315	-0.2	C12 H6 O3	
Anthracendione	208.0522	-3.7	C14 H8 O2	
Pyrenedione	232.0523	0.4	C16 H8 O2	
Phenanthrenediquinone	238.0264	6.3	C14 H6 O4	
Cyclopenta(def)-chrysenone	254.0731	-1.1	C19 H10 O	
Benzo(ghi)fluor-anthenedione	256.0522	6.7	C18 H8 O2	
Pyrene-pyrandione	272.0471	3.8	C18 H8 O3	

NITRO-AROMATIC CONTENT AND BIOASSAY OF DIESEL EMISSIONS

CONCLUSIONS

Negative chemical ionization high-resolution gas chromatography/mass spectrometry with on-column injection (NCI HRGC/MS) provides sensitivity and selectivity for the detection of nitro-aromatics. The limit of detection is approximately 50 pg for the mono-nitro-aromatics and di-nitro-pyrenes in the full mass scan data acquisition mode.

The NCI on-column injection HRGC/MS method provides significant benefits for the analysis of nitro-aromatics over the conventional EI GC/MS methods. It provides for injection into the chromatographic system at a lower temperature in order to avoid degradation of the thermally labile nitro-aromatics. The chemical ionization technique is more gentle than electron impact and thus enhances sensitivity for detection of a molecular ion by decreasing the amount of fragmentation. The negative chemical ionization technique is especially sensitive and selective for the detection of nitro-aromatics. This is due to the electro-negative nature of the nitro substitution which is highly susceptible to attachment of a thermal electron from the reagent gas plasma. Theoretically, the greater the nitro substitution the lower the limit of detection. This enhanced sensitivity has been observed only when an extremely well deactivated GC column was used. Column active sites presumably cause the irreversible adsorption of nitro-aromatics and the magnitude of the effect is greater for the di- and tri-nitro compounds than for the mono-nitro compounds. In practice, sensitivity for mono- and di-nitro compounds is approximately the same. For the reasons listed above, the NCI on-column injection HRGC/MS method is the most sensitive and selective method identified to date for the analysis of nitro-aromatics.

For each extract, the percentage of measured mutagenic activity due to 1-nitropyrene was calculated and is given in Table 4. The value of 918 rev/µg was used as the mutagenic activity of 1-nitropyrene for these calculations. In addition, the percentage of measured mutagenic activity due to those nitro-aromatics detected by mass spectrometry whose activity is known (Table 2) was calculated for each extract. These values are included in Table 4. Two assumptions were made for these calculations: (1) the nitropyrenone isomers were assumed to be the oxidation products of dinitropyrene isomers originally present in the extract. The mutagenic activities of 1,8- and 1,6-dinitropyrene were applied to the nitropyrenone isomer concentrations for the calculations of a "worst case",

TABLE 4

PERCENTAGE CONTRIBUTION OF NITRO-AROMATICS TO DIRECT-ACTING MUTAGENIC ACTIVITY OF EXTRACTS

Sample	Measured Mutagenic Activity[a] rev/µg	Calculated Percentage Contribution to Mutagenic Activity due to	
		1-Nitro-Pyrene	Identified Nitro-Aromatics
Nissan Diesel	12.0	3.1	25.4
Oldsmobile Diesel	5.8	1.7	19.4
VW Rabbit Diesel	3.7	14.6	22.9
Mustang II Gasoline	1.1	0.2	0.2

a = TA98,-S9 assay, Reference 5.

and (2) the nitration mixes produced synthetically were assumed to be similar in isomer distribution to that found in the diesel extracts. The mutagenic activity of each nitration mix was applied to the appropriate isomers detected in the extracts.

Since the calculated contribution to mutagenic activity from identified nitro-aromatics is well below 100 percent, several conclusions should be considered: (1) nitro-aromatics whose activities are unknown to date contribute significantly to the direct acting activity, (2) compounds other than nitro-aromatics, present in the extracts, are potent mutagens, or (3) nitro-aromatics originally present in the extract oxidized or decomposed with time.

The identification of compounds present in the methanol fractions and the detection of the two nitropyrenone isomers tend to support the hypothesis of oxidation occurring in the diesel engine extracts during storage. This hypothesis is further supported by data from bioassays of the Nissan and Oldsmobile extracts which had been stored for two years. These recent bioassays indicated much reduced mutagenic activity and higher toxicity for the stored extract than measured originally in the fresh extract. Additionally, quinones and keto-aromatics are generally toxic but not mutagenic. It is possible that the nitro-aromatics may have originally accounted for a larger percentage of the activity, but with storage, many of the nitro-aromatics may have been oxidized to quinones and keto-aromatics.

These observations may indicate that a greater concentration of nitro-aromatics might have been originally present in the extracts. This possibility is currently under investigation.

REFERENCES

1. Nesnow, S. and Huisingh, J.L. (1980): Mutagenic and carcinogenic potency of extracts of diesel and related environmental emissions: Summary and discussion of the results. In: Health Effects of Diesel Engine Emissions, Vol. II, edited by W.E. Pepelko, R.M. Danner, and N.A. Clarke, EPA-600/9-80-057b, U.S. Environmental Protection Agency, Cincinnati, Ohio, pp 898-912.
2. Petersen, B.A., Chuang, C., Margard, W., and Trayser, D. (1981): Identification of mutagenic compounds in extracts of diesel exhaust particulates. In: Proceedings of the 74th Annual APCA Meeting, Philadelphia, Pennsylvania.
3. Rosenkranz, H.S., McCoy, E.C., Sanders, D.R., Butler, M., Kiriazides, D.K., and Mermelstein, R. (1980): Nitropyrenes: Isolation, identification and reduction of mutagenic impurities in carbon black and toners, Science, 209:1039-1043.
4. Huisingh, J.L. Bradow, R.L., Jungers, R.H., Harris, R.D., Zweidinger, R.B., Cushing, K.M., Gill, B.E., and Albert, R.E. (1980): Mutagenic and carcinogenic potency of extracts of diesel and related environmental emissions: Study design, sample generation, collection and preparation. In: Health Effects of Diesel Engine Emissions, Vol. II, edited by W.E. Pepelko, R.M. Danner, and N.A. Clarke, EPA-600/9-80-057b, U.S. Environmental Protection Agency, Cincinnati, Ohio, pp. 788-800.
5. Claxton, L.D. (1980): Mutagenic and carcinogenic potency of diesel related environmental emissions: Salmonella bioassay. In: Health Effects of Diesel Engine Emissions, Vol. II, edited by W.E. Pepelko, R.M. Danner, and N.A. Clarke, EPA-600/9-80-057b, U.S. Environmental Protection Agency, Cincinnati, Ohio, pp. 801-807.
6. Claxton, L. and Kohan, M., Personal communication.
7. Rosenkrantz, H.S., Personal communication.

AN EVALUATION OF PAH CONTENT, MUTAGENICITY

AND CYTOTOXICITY OF RICE STRAW SMOKE

H. OLSEN[1], J. YEE[1,2], T. MAST[1], J. WOODROW[1],
G. FISHER[1,3], J. SEIBER[1], D. HSIEH[1]
[1]Department of Environmental Toxicology, University of California, Davis, CA 95616; [2]California Air Resources Board, P.O. Box 2815, Sacramento, CA 95812; [3]Battelle Columbus Laboratories, 505 King Avenue, Columbus, OH 43201.

INTRODUCTION

The half-million acres of rice fields in Northern California produce more than one million tons of rice annually with about the same quantity of rice straw, of which 90% is generally disposed of by burning (1). In the Sacramento Valley, rice straw burning is of particular concern because it generates almost 10% of the Valley's total yearly output of carbon monoxide and hydrocarbons and 5% of its atmospheric particulate matter. The wind patterns in the Sacramento Valley coupled with the topography of the valley and the short period twice yearly when rice straw is burned make burning of this agricultural waste a significant source of air pollution.

Three studies were conducted to investigate the chemical composition and biological activities of rice straw smoke particulate matter (SPM): chemical analyses for elemental composition and organic constituents, mutagenicity assay using the Ames Salmonella/ microsome mutagencity test (2), and cytotoxicity assay using an _in vitro_ pulmonary alveolar macrophage (PAM) test (3).

MATERIALS AND METHODS

Sampling

Rice straw smoke particulate matter (SPM) was collected from two different sources: burning fields and the UC Riverside Burn Tower (a controlled combustion facility designed to simulate conditions of field burning). Samples for chemical analysis or the Ames test were collected on glass fiber filters with high volume air samplers. The UC Riverside Burn Tower samples used for chemical analysis for organic constituents and Ames testing were extracted by sonication with benzene:methanol (1+1). Samples for the PAM cytotoxicity test were collected on acrylic plastic filters, followed by a multistep extraction process during which the ethanol soluble organic matter and solvent insoluble particles were recovered for testing.

AN EVALUATION OF RICE STRAW SMOKE

Chemical Analysis

Elemental analysis was completed by particle induced x-ray emission spectrometry at Crocker Nuclear Laboratory, UC Davis. Identification of organic constituents in rice straw SPM extract was accomplished by gel permeation chromatography (Sephadex LH-20, tetrahydrofuran and methanol as eluting solvents) (4,5) followed by computerized gas chromatography-mass spectroscopic analysis. A 30 m X 0.25 mm i.d. SE-54 coated fused silica capillary column with a column temperature program of 80-260°C at 8° per minute was used.

Ames Mutagen Assay

The Ames assay was performed as described by Ames et al. (2) with refinements to technique as described by Pitts (6).

Pulmonary Alveolar Macrophage Cytotoxicity Assay

The PAM cytotoxicity assay was performed as described by Fisher et al. (3) with modifications which include the use of bovine PAM (7).

RESULTS AND DISCUSSION

Chemistry

Some compounds identified in Fraction 5 of the gel permeation fractionation procedure included substituted aromatics and PAHs, specifically cresol isomers, xylenol isomers, methyl benzaldehyde, styrene, cresol acetate, coumaran, acenaphtene, biphenyl, methylbenzothione isomer, dimethylnaphthalene isomer, trimethylindene isomer, fluorene, carbazole, phenanthrene, anthracene, methylphenanthrene, cyclopentanophenanthrene, fluoranthene, pyrene, 1,2-benzanthracene, chrysene, triphenylene, 3Hbenzo[e]indole-2-carboxylic acid, and thiazole[5,4d]pyrimidine-5-ethylamino. Compound identification was done in part by comparing mass spectral fragmentation patterns and relative GC retention times with those of standards. Where standards were not available spectral data are compared with published spectra (8) or those stored in the computerized NIH-EPA Chemical Information System. Elemental analysis of particulate samples collected near burning rice fields confirmed that samples were from burning plant material. This confirmation is based on the enrichment of K (from burning plant material) relative to Ca (a typical soil marker), i.e., an increase in the K/Ca ratio.

Ames Mutagen Assay

Fractionated rice straw SPM extracts chemically analyzed by GC-MS were also tested in the Ames mutagen assay (Figure 1). Figure 1 compares percent, by weight and mutagenic response, of each gel permeation fraction with the unfractionated extract. Eighty-one percent of the mutagenic activity of unfractionated material was recovered in Fractions 1-6 with Fraction 5 containing most of the recovered mutagenic activity (71% of unfractionated extract). Specific mutagenic activity of unfractionated material was 2169 revertants/mg while that of Fraction 5 was 2311 revertants/mg.

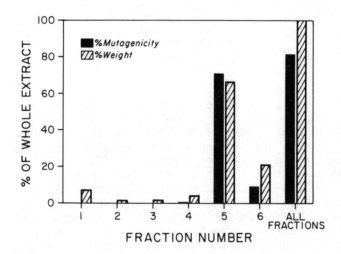

FIGURE 1. Mutagenicity of fractionated rice straw smoke samples from UC Riverside Burn Tower were evaluated with <u>Salmonella typhimurium</u> TA 98 with S-9 (Arochlor induced) activation. Percent mutagenic activity and weight of Fractions 1-6 and all fractions combined are based on unfractionated extract being 100%. See reference (5) for further details.

Table 1 lists some of the compounds identified in Fraction 5, and notes whether they are mutagenic in the Ames assay and what combustion source they have previously been identified in. Results indicate that PAHs are responsible for an unquantified portion of the mutagenic activity.

AN EVALUATION OF RICE STRAW SMOKE

TABLE 1

SOME PAHS IDENTIFIED IN FRACTION 5 BY GC-MS SYSTEM (COMPARISON WITH OTHER AEROSOLS AND BIOLOGICAL EFFECTS)

Compounds Identified in Fraction 5[A]	Aerosol Previously Identified As Containing this Compound	Mutagenic Response In Salmonella Typhimurium
Phenanthrene	Smoke from Leaf Burning[9]	Negative[12]
Anthracene	Smoke from Leaf Burning[9]	Negative[12]
Pyrene	Smoke from Leaf Burning[9]	Negative[12]
Fluorene	Soot[10]	Negative[12]
Acenaphthene	Soot[10]	Positive[10]
Carbazole	Soot[10]	Negative[10]
Fluoranthene	Urban Aerosol[11]	Negative[11]
1,2-Benzanthracene	Urban Aerosol[11]	Positive[11]
Chrysene	Urban Aerosol[11]	Positive[11]
Triphenylene	Urban Aerosol[11]	Positive[11]

A - See Chemistry Section

All rice straw SPM extracts tested were mutagenic in this assay. All samples tested required S-9 enzyme activation to exhibit maximal mutagenic activity, indicating the presence of promutagens in the extracts. Rice straw SPM was found to be mutagenic to <u>Salmonella typhimurium</u> TA 98, TA 100, TA 1537, and TA 1538 but not TA 1535, indicating the presence of a large portion of frame-shift mutagens. Specific mutagenic activity of the field samples and the UC Riverside Burn Tower samples overlapped in the region of 1,000-2,000 revertants/mg. Comparison of rice straw SPM with a sample of urban particulate matter (from downtown Los Angeles, CA) based on the number of revertants per m^3 of air sample showed both aerosols to have similar potency (52-94 revertants/m^3 versus 164 revertants/m^3).

Pulmonary Alveolar Macrophage Cytotoxicity Assay

The recognized significance of the pulmonary alveolar macrophage in the biological defense against invading foreign particles has prompted development of assays for evaluating the cytotoxicity of environmental toxicants to PAM. A bovine PAM <u>in vitro</u> test allowed simultaneous measurement of the effect of toxicants on PAM phagocytic ability, attachment, adherence, and viability. The PAM were exposed to rice straw SPM in the respirable size range (<3.8μm aerodynamic diameter) at various dose levels for twenty-one hours incubation time. At the 0.03 mg/ml level, significant decreases in phagocytic ability were observed but influences on adherence and viability were not evident. In general phagocytic ability appears to be the most sensitive function in response to the toxic action of rice straw SPM. Using this test, the toxicity was compared for silica, rice straw SPM, coal fly ash, and glass beads (Figure 2). Glass beads had no toxic effect on PAM function. Silica particles caused extensive cell lysis at the higher dose levels tested and phagocytic ability could not be measured. The types of particulate matter decrease in relative toxicity in the order silica > rice straw SPM ≥ coal fly ash.

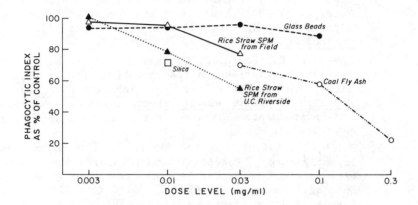

FIGURE 2. PAM phagocytic response to various types of particles: silica (quartz, by weight 98% < 5 μm diameter), rice straw smoke SPM from the field and the UC Riverside Burn Tower (50% < 3.8 um aerodynamic diameter), coal fly ash from a power plant (2.2 um volume median diameter), and a no effect reference particle glass beads (3-5 um volume median diameter). Phagocytic index is defined as the percent of 200 PAM ingesting test spheres.

The potential health implications of rice straw SPM can be evaluated by comparing the relative toxicity of rice straw SPM to two evironmentally important particles, silica and coal fly ash. The PAM, in addition to its role as a scavenger in the lung, is capable of activating promutagens (13). Therefore the possibility exists that PAM may play a role in activating potentially mutagenic substances encountered by PAM in the pulmonary region of the lung.

ACKNOWLEDGEMENTS

This paper is based on research completed in partial fulfillment of Agreement Number A-8-093-31 with the California Air Resources Board.

REFERENCES

1. Miller, M. (1978): Personal Communication. California Rice Research Advisory Board.
2. Ames, B. N., McCann J., and Yamasaki, E. (1975): Methods for detecting carcinogens and mutagens with Salmonella/microsome mutagenicity test, Mutation Research, 31:347-364.
3. Fisher, G. L., McNeill, K. L., Whaley, C. B., and Fong, J. (1978): Attachment and phagocytosis studies with murine pulmonary alveolar macrophages. J. Reticuloendothel Society, 24:243-252.
4. Royer, R. E., Sturm, J. C., and Walter, R. A. (1979): Extraction and fractionation of organics from diesel exhaust particles. In: Inhalation Toxicology Research Institute Annual Report, Lovelace Biomedical and Environmental Research Institute, Lovelace, N. M., pp. 207-210.
5. Hsieh, D., Seiber, J., and Fisher, G. (Principal Investigators). Potential Health Hazards Associated with Particulate Matter Released from Rice Straw Burning. Contract A-8-093-31, California Air Resources Board, Sacramento, CA, 1981, 230 pp.
6. Pitts, J. (Principal Investigator). Geographical and Temporal Distribution of Atmospheric Mutagens in California. Contract A-7-138-30. California Air Resources Board, Sacramento, CA, 1980, 92 pp.
7. Valentine, R., Goettlich-Riemann, W., Fisher, G.L., and Rucker, A.B. (1981): An elastase inhibitor from isolated bovine pulmonary macrophages, Proc. Soc. Exp. Biol. Med., 68:238-244.
8. Cornu, A. and Massot, R. (1966): Compilation of Mass Spectral Data. Hayden and Farr Ltd.
9. Friedman, L. and Calabrese, E.J. (1977): The health implications of open leaf burning, Reviews on Environmental Health, 2(4):257-277.
10. Kaden, D.A., Hites, R.A., and Thilly, W.G. (1979): Mutagenicity of soot and associated polycylic aromatic hydrocarbons to Salmonella typhimurium, Cancer Res., 39:4152-4159.
11. Tokiwa, H., Morita, K., Takeyoshi, H., Takahashi, K., and Ohnishi, Y. (1977): Detection of mutagenic activity in particulate air pollutants, Mutation Research, 48:237-248.

12. McCann, J., Choi, E., Yamasaki, E., and Ames, B.N. (1975): Detection of carcinogens as mutagens in the Salmonella/microsome test: assay of 300 chemicals, Proc. Nat. Acad. Sci. USA, 72(12):5135-5139.

13. Harris, C.C., Hsu, I.C., Stoner, G.D., Trump, B.F., and Selkirk, J.K. (1978): Human pulmonary alveolar macrophages metabolized benzo[a]pyrene to proximate and ultimate mutagens, Nature, 272:633-634.

MUTAGENIC ACTIVATION AND INACTIVATION OF NITRO-PAH COMPOUNDS BY MAMMALIAN ENZYMES

T. C. PEDERSON and J-S. SIAK
Biomedical Science Department, General Motors Research Laboratories, Warren, MI 48090-9055

INTRODUCTION

The genotoxic properties of polycyclic aromatic compounds found in combustion products can be significantly altered by formation of nitro-substituted derivatives. The direct-acting mutagenic activity of diesel exhaust particle extracts in the Salmonella mutation assay may be mostly due to nitroaromatic compounds activated by bacterial nitroreductase enzymes (1,2). 1-Nitropyrene and other monosubstituted nitro-PAH compounds are present in these extracts (2-14), but chromatographic separations demonstrated that much of the mutagenic activity is associated with more polar components (2). Our more recent studies employing TLC and HPLC separation techniques indicate that the particle extract contains multinitro-substituted pyrenes (15). The predominant compounds appear to be 1,6-dinitro- and 1,8-dinitropyrene. Mutagenic activity attributable to these compounds was identified using Salmonella strain TA98NR and the recently-developed dinitropyrene-resistant strains TA98/1,8-DNP$_6$ and TA98NR/1,8-DNP$_2$ (16,17).

The nitroreductase-deficient strain TA98NR was also used in the Salmonella/S9 assay to examine the comparative ability of mammalian enzymes to activate the mutagenicity of diesel particle extracts. Under appropriate assay conditions, an activation of mutagens in diesel particle extract was demonstrated as the difference between assays with and without NADPH (18). The activity of 1-nitropyrene was similarly increased in the TA98NR/S9 mutation assay. The activation of 1-nitropyrene involved only the microsomal fraction of the S9 preparation, but both cytosol and microsomal enzymes contributed to the activation of the particle extract.

The mutagenic activities of the multinitropyrenes, which exceed the activity of 1-nitropyrene by more than two orders of magnitude, are greater than that of any other known compound in the Salmonella assay (19). The relationship between the mutagenicity of the multinitropyrenes in bacteria and their possible geotoxic effects in mammalian systems depends in part on the comparative ability of mammalian enzymes to activate these compounds. In this paper, we describe the contrasting effects of rat liver S9 enzyme activities on the mutagenic action of mono- and di-substituted nitropyrenes.

MATERIALS AND METHODS

Bacterial Mutagenicity Assays

Mutagenic activity was measured in histidine-requiring strains of <u>Salmonella typhimurim</u> (20) as previously described (2) using strain TA98 from Dr. Bruce Ames at the University of California, Berkeley, and strains TA98NR, TA98/1,8-DNP$_6$ and TA98NR/1,8-DNP$_2$ from Dr. Herbert Rosenkranz at Case Western Reserve University School of Medicine, Cleveland, Ohio. Mutagenic activity is expressed as the mean number of revertants on triplicate plates plus or minus standard deviations.

Preparation of Rat Liver S9 Enzymes

The S9 enzyme preparation was obtained using published procedures (20) from the liver of male Sprague-Dawley rats (200-250 g) pretreated with Arochlor 1254 (IP injection of 500 mg/kg) 4 days prior to sacrifice. The preparation was stored at -80°C and used in the mutation assay as previously described (18). The cytosol and microsomal fractions from this S9 preparation were obtained by centrifugation at 110,000 times g. The microsomal pellet was resuspended in a volume of 1.15% KCl equal to that of the decanted supernatant (protein concentrations were 5 and 20 mg/mL respectively).

Source of Nitropyrenes and Diesel Particle Extract

1-Nitropyrene was purchased from Aldrich Chemical Company and purified by thin layer chromatography (2). The dinitropyrenes, 1,3-dinitro-, 1,6-dinitro- and 1,8-dinitro-, were obtained from Dr. R. Mermelstein of Xerox Corporation, Rochester, NY. Diesel exhaust particles were collected from the undiluted exhaust of a 1978 GM 5.7 L diesel engine (constant speed = 1600 rpm, load = 110 N.m; equivalent to an automotive cruise speed of 80 km/h) by electrostatic precipation (2) and extracted with dichloromethane in a Soxhlet apparatus (22). The extracted material, accounting for 6.4% of the particulate mass, was redissolved in dichloromethane (20 mg/mL) and stored in the dark at -20°C.

RESULTS AND DISCUSSION

Bacterial Activation of Nitro-PAH by Multiple Nitroreductases

The identification of multisubstituted nitropyrenes in certain photocopy toners (23,24) also served to demonstrate the multiplicity and specificity of bacterial nitroreductase activities which activate these compounds. Salmonella strain TA98NR, resistant to the mutagenic action of niridazole and other nitroaromatic compounds, retained an unaltered sensitivity to 1,6- and 1,8-dinitropyrene. A dinitropyrene-resistant strain, TA98/1,8-DNP$_6$, and a double mutant, TA98NR/1,8-DNP$_2$, were subsequently derived confirming the existence of multiple nitroreductase activities (16,17).

The mutagenicity of the nitropyrene reference compounds and a diesel particle extract towards strains TA98, TA98NR and TA98/1,8-DNP$_6$ are compared in Figure 1. The mutagenicity of each compound is normalized by its activity in strain TA98 to illustrate the proportionate activity with the niridazole- and dinitropyrene-resistant strains. All three dinitropyrenes exhibited less activity with strain TA98/1,8-DNP$_6$ but only 1,3-dinitropyrene was less active with strain TA98NR. The mutagenicity of the diesel particle extract was reduced by 40% with TA98NR and by more than 70% with TA98/1,8-DNP$_6$.

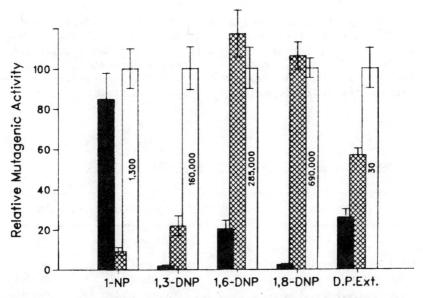

FIGURE 1. Proportionate mutagenic activities of the nitropyrene reference compounds and a diesel particle extract (D.P. Ext.) in two nitroreductase-deficient derivatives of strain TA98. The mutagenicities in strains TA98NR (hatched columns) and TA98/1,8-DNP$_6$ (solid columns) are shown as a percentage of the activity in TA98 (open columns). Inscribed on each column for strain TA98 is the specific mutagenic activity expressed as revertants per µg per plate.

The 1-nitropyrene used in these experiments was purified by thin layer chromatography to remove contaminating multi-nitropyrenes. This sample and a purified preparation from another laboratory (19) have considerably less mutagenic activity than has been reported elsewhere (24-26). Moreover, it exhibits a 90% reduction of activity in TA98NR but no significant change in TA98/1,8-DNP$_6$.

Activation and Inactivation by Rat Liver S9 Enzymes

The NADPH-dependent enzyme activities in rat liver S9 preparations increased the mutagenic activity of both diesel particle extract and 1-nitropyrene when assayed using strain TA98NR and appropriate assay conditions (18). The activations required low concentrations (2.5-10 μL/plate) of the S9 preparation. In addition, the S9 components interferred with the direct-acting mutagenicity of the particle extract and the S9-dependent increase in activity was only evident as the difference between assays with and without NADPH.

It was thought probable that the mutagenicity of the dinitropyrenes, like that of 1-nitropyrene, would be increased by S9 enzymes, but as shown in Table 1, the NADPH-dependent activity of the S9 preparation markedly decreased the mutagenicity of all the dinitropyrenes. Table 1 also shows for comparison the NADPH-dependent increase in activity of

TABLE 1

MUTAGENIC ACTIVITIES IN THE PRESENCE OF RAT LIVER S9 ENZYMES*

	Net TA98NR Revertants/Plates	
	-NADPH	+NADPH
1,3-Dinitropyrene, 10 ng/plate	55 ± 10	16 ± 4
1,6-Dinitropyrene, 2 ng/plate	238 ± 14	40 ± 13
1,8-Dinitropyrene, 1 ng/plate	280 ± 14	38 ± 11
Diesel Particle Extract, 30 μg/plate	133 ± 9	199 ± 14
1-Nitropyrene, 0.5 μg/plate	12 ± 2	363 ± 30

*All assays contain 2.5 μL of S9 per plate

1-nitropyrene and diesel particle extract under the same assay conditions. Although the decreased mutagenicity of the dinitropyrenes in the Salmonella/S9 enzyme assay indicates these compounds may be metabolized in mammalian systems without genotoxic effects, subsequent experiments with separated microsomal and cytosolic S9 enzymes revealed there are both activating and inactivating enzymes in the S9 preparation.

The Role of S9 Microsome and Cytosol Enzyme Activities

The NADPH-dependent activation of 1-nitropyrene was catalyzed by S9 microsomal enzymes, while both cytosol and microsomal enzymes contributed to the activation of diesel particle extract (18). The role of cytosol and microsomal enzymes in the inactivation of dinitropyrenes was similarly examined. The results of these experiments are shown in Table 2. The NADPH-dependent effects are expressed as the percent change in the activity of the control assay lacking NADPH. The inactivation of the dinitropyrenes was catalyzed by the microsomal fraction. However, NADPH added alone or in the presence of cytosol enzymes increased the mutagenic activity of the

TABLE 2

CONTRASTING EFFECTS OF MICROSOME AND CYTOSOL ENZYMES*

		Net TA98NR Revertants/Plate		
		No Enzymes	Plus Microsomes	Plus Cytosol
1,3-Dinitropyrene	−NADPH	209 ± 77	50 ± 12	179 ± 32
10 ng/plate	+NADPH	398 ± 31	7 ± 5	763 ± 36
	Change	+90%	−85%	+326%
1,6-Dinitropyrene	−NADPH	302 ± 19	282 ± 17	347 ± 16
2 ng/plate	+NADPH	606 ± 24	19 ± 5	878 ± 98
	Change	+101%	−93%	+153%
1,8-Dinitropyrene	−NADPH	374 ± 51	196 ± 13	392 ± 21
1 ng/plate	+NADPH	633 ± 49	7 ± 6	1129 ± 51
	Change	+69%	−95%	+188%
Diesel Particle Extract	−NADPH	407 ± 30	332 ± 37	313 ± 10
30 µg/plate	+NADPH	464 ± 22	486 ± 11	578 ± 32
	Change	+12%	+45%	+85%

*Concentrations of both microsomes and cytosol were the equivalent of 5 µL S9/plate as described under Methods.

dinitropyrenes. The NADPH-dependent increase in activity was greater in the presence of the cytosol enzymes which suggests the increase includes both cytosol-dependent and cytosol-independent reactions. Activation of diesel particle extract was again found to be catalyzed by both microsomal and cytosol enzymes. In the previous study, a small NADPH-dependent increase in activity in the absence of any S9 enzyme was judged statistically insignificant (18), but as shown in Table 2, a similar increase was again observed and is probably attributable, at least in part, to the presence of dinitropyrenes in the particle extract.

The experiments described in Tables 1 and 2 were conducted using strain TA98NR for the purpose of comparing the results with those of our earlier study. Since TA98NR retains a normal sensitivity to the mutagenicity of 1,6-dinitro- and 1,8-dinitropyrene, the effects of the S9 enzymes were also examined using strain TA98NR/1,8-DNP$_2$ and a twenty fold greater concentration of 1,8-dinitropyrene. These experiments also compare specificities for NADH and NADPH. The results in Figure 2 demonstrate that the mutagenic activity was again

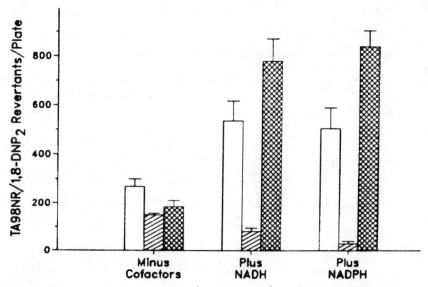

FIGURE 2. Activation and inactivation of 1,8-dinitropyrene (20 ng/plate) by S9 enzymes with strain TA98NR/1,8-DNP$_2$. NADH and NADPH were present as indicated (2.5 µmole/plate) in assays containing either no mammalian enzymes (open columns), and S9 microsome fraction (slashed columns), or cytosol fraction (hatched columns) at concentrations of 5 µL per plate.

decreased by microsomal activity and increased by NADH and NADPH in the absence or presence of cytosol activity. The cytosol-dependent increase, as well as the cytosol-independent increase, exhibited no specificity for NADH or NADPH, but the inactivation by microsomal enzymes was greater in the presence of NADPH. Both the microsomal inactivation and cytosol-dependent activation appear to be enzyme catalyzed reactions since heat denaturations of the enzyme preparation eliminated these activities.

In summary, these studies have examined the effect of mammalian enzyme activity on the mutagenicity of mononitro- and dinitropyrenes in the Salmonella mutation assay and support the following conclusions.
(1) The bacterial mutagenicity of 1-nitropyrene is increased by the NADPH-dependent enzyme activity of the rat liver S9 preparation, but the dinitropyrenes are inactivated.
(2) The microsomal fraction of the S9 preparation catalyzes both the activation of 1-nitropyrene and inactivation of the dinitropyrenes. However, the cytosol fraction in absence of microsomal enzymes increased the mutagenicity of the dinitropyrenes. The NAD(P)H-dependent activation of dinitropyrenes is partly attributable to a cytosol-independent reaction.
(3) The NADPH-dependent increase in mutagenicity of diesel particle extract in the Salmonella/S9 assay involves both multiple extract components and multiple S9 enzymes. The dinitropyrenes presumably contribute to the cytosol-catalyzed activation. The much smaller effect of microsomal enzymes on the mutagenicity of the particle extract must reflect competing activation and inactivation reactions as evidenced by the difference between 1-nitropyrene and the dinitropyrenes.

REFERENCES

1. Pederson, T. C. and Siak, J-S. (1980): Characterization of direct-acting mutagens in diesel exhaust particulates by thin layer chromatography, American Chemical Society, Division of Environmental Chemistry, Extended Abstracts, 20(2):533-535.
2. Pederson, T. C. and Siak, J-S. (1980): The role of nitroaromatic compounds in the direct-acting mutagenicity of diesel particle extracts, J. Appl. Toxicology, 1:54-60.
3. Rodriguez, C. F., Fisher, J. B., and Johnson, D. E. (1980): Characterization of organic constituents in diesel exhaust particulates. In: Health Effects of Diesel Engine Emissions: Proceedings of an International Symposium, EPA-600/9-80-075a, 1:34-48 edited by W. E. Pepelko, R. M. Danner and N. A. Clark.

4. Schuetzle, D., Lee, F. S. C., Prater, T. J., and Tejada, S. B. (1981): The identification of polynuclear aromatic hydrocarbon derivatives in mutagenic fractions of diesel particulate extracts, Int. J. Environ. Anal. Chem., 9:93-144.
5. Gibson, T. L., Ricci, I., and Williams, R. L. (1981): Measurement of polynuclear aromatic hydrocarbons, their derivatives, and their reactivity in diesel automobile exhaust. In: Chemical Analysis and Biological Fate: Polynuclear Aromatic Hydrocarbons, edited by M. Cook and A. J. Dennis, pp. 707-717, Battelle Press, Columbus.
6. Zweidinger, R. B. (1981): Emission factors from diesel and gasoline powered vehicles, EPA 1981 Diesel Emissions Symposium, Oct. 5-7, Raleigh, NC, Extended Abstracts.
7. Yergey, J. A., Risby, T. H., and Lestz, S. S. (1981): The chemical characterization of diesel particulate matter, EPA 1981 Diesel Emissions Symposium, Oct. 5-7, Raleigh, NC, Extended Abstracts.
8. Riley, T., Prater, T., Schuetzle, D., Harvey, T. M., and Hunt, D. (1981): The analysis of nitrated polynuclear aromatic hydrocarbons in diesel exhaust particulates by MS/MS techniques, EPA 1981 Diesel Emissions Symposium, Oct. 5-7, Raleigh, NC, Extended Abstracts.
9. Salmeen, I., Durisin, A. M., Prater, T. J., Riley, T., and Schuetzle, D. (1981): Contribution of 1-nitropyrene to direct acting Ames assay mutagenicities of diesel particulate extracts, EPA 1981 Diesel Emissions Symposium, Oct. 5-7, Raleigh, NC, Extended Abstracts.
10. Braddock, J. N. (1981): Emission of diesel particulate and particulate mutagens at low ambient temperature, EPA 1981 Diesel Emissions Symposium, Oct. 5-7, Raleigh, NC, Extended Abstracts.
11. Henderson, T. R., Sun, J. D., Royer, R. E., Clark, C. R., Harvey, T. M., Hunt, D. F., Fulford, J. E., Lovett, A. M., and Davidson, W. R. (1981): GC/MS and MS/MS studies of direct-acting mutagens in diesel emissions, EPA 1981 Diesel Emissions Symposium, Oct. 5-7, Raleigh, NC, Extended Abstracts.
12. Lang, J. M., Carlson, R. A., Snow, L., Black, F. M., Zweidinger, R., and Tejada, S. (1981): Characterization of particulate emissions from in-use gasoline fuels motor vehicles, EPA 1981 Diesel Emissions Symposium, Oct. 5-7, Raleigh, NC, Extended Abstracts.

13. Xu, X. B., Nachtman, J. P., Jin, Z. L., Wei, E. T., Rappaport, S., and Burlingame, A. L. (1981) Isolation and identification of mutagenic nitroarenes in diesel-exhaust particulates, EPA 1981 Diesel Emissions Symposium, Oct. 5-7, Raleigh, NC, Extended Abstracts.
14. Nishioka, M. G., Petersen, B. A., and Lewtas, J. (1981): Comparison of nitro-PHA content and mutagenicity of diesel emissions, EPA 1981 Diesel Emissions Symposium, Oct. 5-7, Raleigh, NC, Extended Abstracts.
15. Pederson, T. C., and Siak, J-S. (1981): Dinitropyrenes: Their probable presence in diesel particle extracts and consequent effect on mutagenic activations by NADPH-dependent S9 enzymes, EPA 1981 Diesel Emissions Symposium, Oct. 5-7, Raleigh, NC, Extended Abstracts.
16. Rosenkranz, H. S., McCoy, E., Mermelstein, R., and Speck, W. T. (1981): A cautionary note on the use of nitro-reductase-deficient strains of Salmonella typhimurium for the detection of nitroarenes as mutagens in complex mixtures including diesel exhaust, Mutat. Res., 91:103-105.
17. Mermelstein, R., Rosenkranz, H. S., and McCoy, E. C. (1981): Mutagenicity in bacteria of nitroarenes. In: Symposium on the Genotoxic Effects of Airborne Agents, Brookhaven National Laboratory, Feb. 9-11.
18. Pederson, T. C., and Siak, J-S. (1981): The activation of mutagens in diesel particle extract with rat liver S9 enzymes, J. Appl. Toxicol., 1:61-66.
19. Mermelstein, R., Kiriazides, D. K., Butler, M., McCoy, E. C., and Rosenkranz, H. S. (1981): The extraordinary mutagenicity of nitroarenes in bacteria, Mutation Research, 89:187-196.
20. Ames, B. N., McCann, J., and Yamasaki, E. (1975): Methods for detecting carcinogens and mutagens with the Salmonella/mammalian microsome mutagenicity test, Mutat. Res., 31:347-364.
21. Chan, T. A., Lee, P. S., and Siak, J-S. (1981): Diesel particulate collection for biological testing. Comparison of electrostatic precipitation and filtration, Env. Sci. Technol., 15:89-93.
22. Siak, J-S., Chan, T. L., and Lee, P. S. (1980): Diesel particulate extracts in bacterial test systems. In: Health Effects of Diesel Engine Emissions: Proceedings of an International Symposium, EPA-600/9-80-075a, 1:245-262, edited by W. E. Pepelko, R. M. Danner and N. A. Clark.

23. Lofroth, G., Hefner, E., Alfheim, I., and Moller, M. (1980): Mutagenic activity in photocopies, Science, 209:1037-1039.
24. Rosenkranz, H. S., McCoy, E. C., Sanders, D. R., Butler, M., Kiriazides, D. K., and Mermelstein, R. (1980): Nitropyrenes: Isolation, identification, and reduction of mutagenic impurities in carbon black and toners, Science, 209:1039-1043.
25. Wang, C. Y., Lee, M-S., King, C. M., and Warner, P. O. (1980): Evidence for nitroaromatics as direct-acting mutagens of airborne particulates, Chemosphere, 9:83-87.
26. Tokiwa, H., Nakagawa, R., Morita, K., and Ohnishi, Y. (1981): Mutagenicity of nitro derivatives induced by exposure of aromatic compounds to nitrogen dioxide, Mutat. Res., 85:195-205.

COMPARISON OF BENZO(A)PYRENE AND DIESEL PARTICULATE EXTRACT IN THE SISTER CHROMATID EXCHANGE ASSAY IN VIVO AND IN UTERO AND THE MICRONUCLEUS ASSAY

MICHAEL A. PEREIRA[a], LOFTON McMILLAN[a], P. KAUR[b], AND P.S. SABHARAWAL[b]
[a]U.S. Environmental Protection Agency, Health Effects Research Laboratory, Cincinnati, Ohio 45268; [b]Environmental Health Research and Testing, Inc., 3217 Whitfield Avenue, Suite 11, Cincinnati, Ohio 45220

INTRODUCTION

The micronucleus and sister chromatid exchange (SCE) assays are two in vivo assays proposed for screening chemicals and complex mixtures for genotoxic activity. It has been predicted that the number of diesel powered passenger automobiles in the U.S.A. will increase and by 1985 will occupy 25% of the new car market (1). The particulate matter present in diesel exhaust emissions has been demonstrated to contain polycyclic aromatic hydrocarbons including benzo(a)pyrene (BaP) and mutagenic activity (2). As part of the evaluation of the possible genotoxic activity of diesel exhaust emissions in mammals, we determined the effect of an extract of diesel particulate in these two assays.

The in utero exposure of the fetus to chemical mutagens can result in fetotoxicity, neoplasia, congenital anomalies, and heritable abnormalities. The in vivo SCE has been extended to inlcude the detection of chemical induced SCE in rodent fetus exposed in vitro (3-7). In this communication, we also examined the induction of SCE in Syrian hamster fetal liver by in utero exposure to benzo(a)pyrene and an extract of diesel particulate matter.

MATERIAL AND METHODS

Cyclophosphamide, 5-bromo-2'-deoxyuridine (BrdU) and trioctanoin were purchased from Sigma Chemical Co., St. Louis, MO; benzo(a)pyrene (BaP; Gold label; 99+% purity) from Aldrich Chemical Co., Milwaukee, WI; dimethylsulfoxide (DMSO) from Burdick and Jackson Lab., Inc., Muskegon, MI, and Hoechst 33258 from Calbiochem.-Behring Corp., La Jolla, CA.

Golden Syrian hamsters weighing 80-100 gm were purchased from Engle's Laboratory Animals (Farmersburg, IN), Chinese hamsters (80-90 gm) from Northeastern University (Boston, MA) and B6C3F1 mice (2-3 months of age) from Harlan Industries, Indianapolis, IN. The animals were maintained in accordance with the standards set forth in the "Guide for the Care and Use of Laboratory Animals" of the Institute of Laboratory Animal Resources, National Research Council. They received Purina Laboratory Chow (Ralston Purina Co.) and water ad libitum.

Diesel emission was produced as previously described by Hinners et al. (8) by one of two Nissan-CN-6 diesel 6 cylinder engines coupled to a Chrysler torque-flite automatic transmission Model A-727 and mounted on an Eaton-Dynamometer Model 753-DG. The particles were collected on telfon coated T60A20 type fitlers (Pallflex Products Corp.) and extracted for 24 hrs. in a Soxhlet extraction apparatus using methylene chloride as the eluant as previously described (9). The mass of the extract represented 23% of the particulate matter. The extract was made up to the desired concentration in dimethyl sulfoxide (DMSO) by solvent exchange.

The animals were sacrificed by cervical dislocation and the bone marrow cells harvested by the method of Schmidt (10) as previously described (9). The slides of the cell suspensions were strained with May Grunwald solution followed by Giemsa. For each animal, the number of micronucleated polychromatic erythrocytes in 1000 such cells was determined as recommended by von Ledebur and Schmid (11).

A slight modification as previously described (9) of the BrdU pellet implanation procedure of Allen et al. (12) for in vivo sister chromatid exchange was used. Briefly, twenty-four hours prior to sacrifice a 60 mg pellet of BrdU was implanted under the skin. Two hours prior to sacrifice the animals were injected intraperitoneally with colchicine (10 mg/Kg bd. wt.). The bone marrow cells were flushed from the canal with 0.075 M KCl and the contents from both femurs combined. The slides of the cell suspension were strained by the Hoechst-Giemsa blacklight method of Goto et al. (13). Twenty-five metaphases from each animal were evaluated for the number of SCE. The mitotic index was determined by counting the number of dividing cells per 1,000 cells.

For determination of in utero SCE, pregnant Syrian hamsters were treated on day 12 of gestation with BrdU by implanation of a pellet under the skin. The diesel particulates extracts or benzo(a)pyrene were administered two hrs. later by intraperitoneal injection. Eighteen hours after the BrdU was given, the animals were sacrificed. Fetal livers from each litter (10-14 fetuses) were pooled in 10 ml of calcium and magnesium free HBSS. After 20 minutes of incubation at 37°C, the livers were gently crushed with a spatula against the side of the flask to release the cells from the connective tissue. The suspension was treated with colcemid for 2 hrs. at 37°C in a CO_2 incubator. The cells were collected by centrifugation and resuspended in 5-6 ml of 0.075 M KCl. After incubation for 10 minutes at 37°C, the cells were collected by centrifugation. The collected cells were fixed with ice-cold Carnoy's solution and stained by the Hoechst-Giemsa black-light method (13).

RESULTS AND DISCUSSION

TABLE 1

EFFECT OF BENZO(A)PYRENE ON SCE

Treatment mg/Kg	Bone Marrow Cells[a] Mice	Fetal Liver Cells[b] Hamsters
0	5.67 ± 0.45	6.5 ± 0.34
50	-	9.5 ± 0.50
100	13.49 ± 1.01	11.3 ± 0.45
125	-	12.5 ± 0.65

a   The BaP was administered i.p. 48 hrs. prior to sacrifice. A 60 mg pellet of BrdU was implanted under the skin 24 hrs. prior to sacrifice. Twenty-five metaphases/mouse were scored for SCE. Results are the mean of the number of SCE per metaphase from 6 mice ± SE.

b   The 300 mg BrdU pellet was implanted under the skin of pregnant hamsters on day 12 of gestation. Two hrs. later, the BaP was administered i.p. and the hamsters sacrificed 18 hrs. later. The fetal livers of a litter were pooled and 25 metaphases/litter scored for SCE. Results are the mean of the number of SCE/metaphases from 6 litters ± SE.

TABLE 2

EFFECT OF AN EXTRACT OF DIESEL PARTICULATE MATTER ON MICRONUCLEI AND SCE

TREATMENT mg/kg		BONE MARROW CELLS				FETAL LIVER CELLS HAMSTERS
		MICE		HAMSTERS		
		MICRONUCLEI[a]	SCE[b]	MICRONUCLEI[a]		SCE[c]
0	(0)[d]	0.37 + 0.06	5.67 + 0.45	0.12 + 0.03		5.4 + 0.52
80	-	-	-	0.32 + 0.10		8.3 + 0.59
160	(200)	0.32 + 0.04	-	0.13 + 0.05		10.0 + 0.63
320	(400)	0.55 + 0.13	9.35 + 0.60	0.10 + 0.06		10.3 + 0.56
640	-	-	-	0.15 + 0.04		11.7 + 0.77
800	(800)	0.80 + 0.20	-	0.13 + 0.05		13.6 + 1.00
	(1000)	0.80 + 0.20	-	-		-

[a]The extract was administered i.p. 30 hrs. prior to sacrifice. The number of micronucleated cells was determined per 1000 polychromatic erythrocytes. Results are the mean of the percentage of micronucleated cells from 6 animals + SE.

[b]The extract was administered i.p. 48 hrs. prior to sacrifice. A 60 mg pellet of BrdU was implanted under the skin 24 hrs. prior to sacrifice. Twenty-five metaphases/mouse were scored for SCE. Results are the mean per metaphase from 6 mice + SE.

[c]The 300 mg BrdU pellet was implanted under the skin of pregnant hamsters on day 12 of gestation. Two hrs. later, the extract was administered and the hamsters sacrificed 18 hrs. later. The fetal livers of a litter were pooled and 25 metaphases/litter scored for SCE. Results are the mean of the number of SCE per metaphase from 6 litters + SE.

[d]The concentrations enclosed by parentheses were administered to the mice and the other concentrations to hamsters.

The effect of BaP on the frequency of SCE in bone marrow cells of mice and in fetal hamster liver cells on day 13 of gestation is presented in Table 1. BaP caused a dose-related increase in the frequency of SCE in fetal liver cells and at a dose of 125 mg/kg resulted in almost a doubling of SCE. A doubling in the frequency of SCE was obtained in mice treated with 100 mg/kg. A higher dose of BaP (500 mg/kg) has been reported by Heddle and Bruce (14) not to increase the incidence of micronuclei in polychromatic erythrocytes of mice. Therefore, the SCE assay would appear to be more sensitive to BaP than the micronucleus assay.

The effect of a diesel particulate extract in the micronucleus and SCE assays is presented in Table 2. The extract more than doubled the frequency of SCE in bone marrow cells of mice and in hamsters fetal liver cells. The increase frequency of SCE in fetal liver cells was dose-dependent. The micronucleus assay in bone marrow cells of mice and hamsters was insensitive to the genotoxic material present in the extract. As with BaP, the SCE assay compared to the micronucleus assay was more sensitive to the genotoxic material presented in a complex mixture of an extract of diesel particulate matter.

The in utero SCE assay has been proposed for inclusion in programs design to screen chemicals and complex mixtures for their ability to cause congenital anomalies, teratogenesis and heritable abnormalties (3-7). We have demonstrated that the in utero SCE assay can detect possible hazard to the fetus as the result of an exposure to BaP or to a complex mixture containing genotoxic material. Therefore, the in utero SCE assay appears to warrant further evaluation for inclusion in a program to screen chemicals and complex mixtures for genotoxic hazard to the fetus.

ACKNOWLEDGEMENTS

This work was supported in part by contracts No. 68-03-3004 and 68-03-1527 with the U.S. Environmental Protection Agency, Health Effects Research Laboratory, Cincinnati, Ohio.

REFERENCES

1. U.S. Environmental Protection Agency (1978): Health effects associated with diesel emisssions. U.S. EPA-600/1/78-063.

2. Huisingh, J., Bradow, R., Jungers, R., Claxton, L., Zweidinger, R., Tejada, S., Bumgarner, J., Duffield, F., Waters, M., Simmon, V.F., Hare, C., Rodriquez, C. and Snow, L. (1978): Application of bioassays to the characterization of diesel particle emissions. In: Application of Short-Term Bioassays in the Fractionation and Analysis of Complex Environmental Mixtures, edited by M. Waters, S. Nesnow, J. Huisingh, S. Sandhu, and L. Claxton, pp. 383-418. New York: Plenum Press.
3. Basler, A. (1979): Sister chromatid exchanges in vivo in Chinese hamster embryonic liver cells exposed transplacentally to BrdU, Cytogenet. Cell Genet., 24:193-196.
4. Kram, D., Bynum, G.D., Senula, G.C. and Schneider, E.L. (1979): In utero sister chromatid exchange analysis for detection of transplacental mutagens. Nature (London). 279:531.
5. Kram, D., Bynum, G.D., Senula, G.C., Bickings, C.K. and Schneider (1980): In utero analysis of sister chromatid exchange: Alterations in susceptibility to mutagenic damage as a function of fetal cell type and gestational age. Proc. Natl. Acad. Sci. (USA) 77:4784-4787.
6. Knuutila, S., Harkki, A., Rossi, L., Westermarck, T., Lappalinen, L. and Rantanen, P. (1979): In vivo method for sister chromatid exchanges in Chinese hamster foetal and bone marrow cells. Hereditas. 91:23-26.
7. Allen, J.W., El-Nahas, E., Sanyal, M.K., Dunn, R.L., Gladen, B. and Dixon, R.L. (1981): Sister-chromatid exchange analysis in rodent material, embryonic and extraembryomic tissues:Transplacental and direct mutagen exposure. Mutation Res. 80:297-311.
8. Hinners, R.G., Burkart, J.K., Malanchuk, M. and Wagner, W.D. (1980): Facilities for diesel exhaust studies. In: "Health Effects of Diesel Engine Emissions, edited by W.E. Pepelko, R.M. Danner, and N.A. Clarke, EPA-600/9-80-0516, pp. 681-697.
9. Pereira, M.A. (1981): Genotoxicity of diesel exhaust emissions in laboratory animals. In: Toxicological Effects of Emissions from Diesel Engines, edited by J. Lewtas, New York: Elsevier North Holand (In Press).
10. Schmid, W. (1977): The micronucleus test, In: Handbook of Mutagenicity Test Procedure, edited by B.J. Kilbey, M. Legator, W. Nichols, and C. Ramel, pp. 235-242, Elsevier, Amsterdam.

11. Von Ledebur, M. and Schmid, W. (1973): The micronucleus test, Methodological aspects. Mutation Res. 19:109-117.
12. Allen, J.W., Shuler, C.F. and Lat, S.A. (1978): Bromodeoxyiridine tablet methodology for in vivo studies of DNA synthesis. Somatic Cell Genet., 4:393-405.
13. Goto, K., Maeda, S., Kano, Y., and Sugiyama, T. (1978): Factors involved in differential Giemsa-staining of sister chromatids. Chromosoma, 66:351-359.
14. Heddle, J.A. and Bruce, W.R. (1977): Comparison of tests for mutagenicity or carciogenicity using assays for sperm abnormalities, formation of micrponuclei, and mutation in Salmonella, In: Origins of Human Cancer, edited by H.H. Hiatt, J.D. Watson and J.A. Winsten pp. 1549-1557, Cold Spr. Harb. Lab. Press.

ANALYSIS OF PAH IN DIESEL EXHAUST PARTICULATE BY HIGH RESOLUTION CAPILLARY COLUMN GAS CHROMATOGRAPHY/MASS SPECTROMETRY

BRUCE A. PETERSEN, CHENG C. CHUANG, TIMOTHY L. HAYES, DAVID A. TRAYSER
Battelle Columbus Laboratories, 505 King Avenue, Columbus, Ohio 43201

INTRODUCTION

The combustion of fuel in diesel engines is known to emit polynuclear aromatic hydrocarbons (PAH), some of which, such as benzo(a)pyrene, are known or suspected carcinogens. The form and magnitude in which these PAH appear in the exhaust of diesel engines are matters of increasing interest and importance because of the mounting use of diesel engines. PAH measurement in diesel exhaust, extremely complex in itself, is further complicated because PAH emission is influenced by the engine speed and load, cycle of operation, fuel-to-air ratio, operating temperature, oil consumption, and fuel composition. Reliable quantitative analytical techniques need to be established before potential effects of PAH on the environment can be assessed. The CAPE-24 Committee of the Coordinating Research Council is funding a multi-year study at Battelle. The objective is to develop methodology for determining the PAH content in diesel engine exhaust that is emitted to the atmosphere. (1)

This study is being conducted in two phases. Phase I is being directed towards (1) construction and activation of a heavy-duty diesel engine test facility and (2) development of methods for the accurate measurement of the PAH emitted to the atmosphere from this system. Phase II will use these methods to determine the PAH emissions as a function of fuel consumption, engine design and engine operation.

One aspect of the Phase I program has been to develop an analytical method for the extraction and measurement of PAH in diesel exhaust particulate. The technical approach for this study was based on the use of stable isotopically labeled PAH as tracer compounds. The isotope selected was the perdeutero derivative where each proton (H^1) on the PAH nucleus was replaced with a deuteron (H^2). The perdeutero compounds are chemically similar to the native PAH, and differ only in molecular weight. Therefore, compound loss due to reaction, adsorption and inefficient separation should be the same for both.

The objective of this study was to assess the influence of the analytical procedure on the recovery of PAH. Three experiments were carried out. In each experiment, known quantities of the perdeuterated PAH were introduced at various steps in the analytical preparation of diesel exhaust particulate filter samples. Recovery of the perdeuterated PAH was used to determine if a significant loss of PAH occurs during the preparation of particulate filter samples for analysis. This paper discusses the experimental design and results of this study.

MATERIALS AND METHODS

Five PAHs have been used to represent the PAH class of compounds emitted from diesel engines. These are:

>   Pyrene
>   Benz(a)anthracene
>   Chrysene
>   Benzo(a)pyrene
>   Perylene

Recently, benzo(e)pyrene has been added to this list. It was however, not used in this study. Perdeuterated derivatives of these compounds are commercially available from Merck and Company, St. Louis, Missouri.

The procedures used for the recovery of PAH from diesel particulate were as follows:

- preparation of filters for particulate collection
- collection of diesel exhaust particulate filter samples
- extraction of particulate filter samples
- fractionation of particulate extracts
- GC/MS analysis of PAH

Each of these procedures are discussed in this section.

Preparation of Filters for Particulate Collection

The particulate samples were collected on Pallflex T60A20 Teflon-impregnated glass fiber filters. The filters were circular, 4-inches in diameter, and cut from a 12" x 100' roll. Filters were conditioned for at least 12 hours in a temperature

(68°F) and humidity (50 percent relative humidity) controlled room before weighing on a Mettler MA-5 analytical balance. After weighing the filters were stored between two Pyrex watch glasses and sealed with Teflon tape. Storage time between weighing and particulate collection was typically less than 24 hours.

Filters were loaded into the Pyrex filter holders at the test engine facility, and mounted on the end of the particulate sampling probe. After collection of particulate, the filters were removed from the holder, then stored between the watch glasses and returned to the balance room. They were equilibrated for 12-16 hours in the balance room and weighed.

Extraction of Particulate Filter Samples

Soluble organic material was removed from the particulate filter by Soxhlet extraction with 350 ml of toluene. The filter particulate samples were folded in half and placed in the extractor. Extraction was carried out for 16 hours at a 10 minute cycle rate, after which the solvent was concentrated to 5 ml by rotary evaporation and an aliquot (50-100 µl) removed and evaporated to dryness for gravimetric analysis. The residue was accurately weighed (constant weight) to the nearest microgram on a Mettler ME30 microbalance and this aliquot weight was used to determine the total organic extractable mass in the particulate.

Fractionation of the Particulate Extract

The particulate extract was partitioned by means of silica gel chromatography to isolate the PAH fraction. A schematic representation of the fractionating procedure is shown in Figure 1. The solvent extract of the particulate samples (350 ml) was first concentrated to 5 ml using rotary evaporation before fractionation. The extract was partitioned into aliphatic, PAH, and polar fractions by open column liquid chromatography on 5 percent H_2O-deactivated silica gel. The silica gel columns (0.55 cm i.d. x 7 cm long in a 15 cm disposable Pasteur pipette) were packed in a hexane slurry and the gel retained with a silanized glass wool plug. Columns were prepared for each organic solution to be partitioned. An additional column was also prepared to check the accuracy of the silica gel deactivation before partitioning the neutral organic solution. This was done by measuring the volume of hexane required to elute 100 µg of anthracene. The migration of the anthracene was monitored by a 366 nm UV lamp, and the volume of hexane was measured during the migration. When the

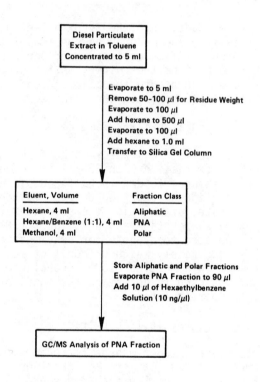

FIGURE 1. Diagram of the procedure for the fractionation and analysis of PAH in extracts of diesel particulate.

silica gel was deactivated 5 percent, anthracene would start to elute from the column after the addition of 7 ± 0.5 ml of hexane. For 3 percent and 7 percent deactivation, the volumes of hexane required were 11 ± 0.5 ml and 5 ± 0.5 ml, respectively.

Upon assurance that the silica gel was 5 percent H_2O-deactivated, the neutral organic compounds were further fractionated. Three elution solvents were used. They were applied to the column, and the eluent collected in the following sequence: 4 ml of hexane, 4 ml of hexane/benzene (1:1 v/v), 4 ml of methanol. The collected fractions correspond to the aliphatic, PAH and polar fractions. The PAH fraction was evaporated to 90 µl and 100 ng of hexaethylbenzene (10 µl of a 10 ng/µl solution) added. The mixture was then analyzed by GC/MS to quantify the PAH.

Gas Chromatographic/Mass Spectrometric Analysis

Capillary column gas chromatography/mass spectrometry (GC/MS) provides the primary basis for identification and quantification of PAH in the extracts of diesel particulate. The advantages of this technique are (1) high sensitivity for trace level detection, (2) specificity for unequivocal identification and (3) versatility for the separation of large numbers of compounds. For quantification of PAH, the mass spectrometer was operated using Selected Ion Monitoring (SIM). In this mode, the mass spectrometer concurrently monitors one or more ions characteristic of a specific compound during its expected elution time from the gas chromatographic column. Because the ions which are monitored are unique to the compound being analyzed, other compounds will not contribute to the ion current signal and therefore will not interfere with the analysis.

Electron impact (20 eV) capillary column GC/MS/SIM was used to obtain data on the deuterated tracer and engine PAH used in this study. The ions monitored correspond to the molecular ion of both the tracer and engine PAH. Since the gas chromatographic properties of the deuterated and engine PAH are virtually identical, they co-elute into the mass spectrometer ion source. By monitoring these ions at their elution time, the engine PAH can be unequivocally identified. Quantification of both deuterated and engine PAH is accomplished by comparison of their integrated ion current response to that of the hexaethylbenzene internal standard and interpolation of standard curves.

Five point standard response curves for each PAH were established and span a mass range of 0.5 to 50 ng of each PAH injected on column. Each point on the curve is the average of three measurements. Standard curves were prepared for every experiment involving PAH measurement and when quality control check samples indicated the curves were no longer satisfactory.

In order to ensure that all GC/MS data acquired on this project were of high quality, the total GC/MS system performance was evaluated each day before undertaking analysis of particulate samples. The performance evaluation included (1) a mass calibration, (2) evaluation of mass spectrometric resolution and (3) GC/MS analysis of a PAH test mixture.

Recovery of PAH From Diesel Particulate

Three experiments were carried out in this study. In each experiment, known quantities of the perdeuterated PAH were introduced, or spiked, at various steps in the analytical workup of particulate filter samples. Recovery of the perdeuterated PAH was used to determine if a significant loss of PAH occurs during the work-up. The three spiking experiments were the following:

(1) Solvent spike: injecting a solution of the perdeuterated PAH into the solvent reservoir before Soxhlet extraction of particulate filter samples

(2) Filter spike: introducing a solution of the perdeuterated PAH directly onto particulate filter samples before Soxhlet extraction

(3) Vapor spike: vaporizing the perdeuterated PAH into an air stream and sampling the air stream through particulate filter samples before Soxhlet extraction.

Two levels of perdeuterated PAH were used in each experiment. These levels were approximately one-half the quantity of engine PAH (low level) and twice the quantity of engine PAH (high level) in a specific particulate filter sample. The levels used are presented in Table 1. Each experiment was carried out in triplicate for a total of 18 measurements (3 methods x 2 levels x triplicates).

TABLE 1
QUANTITY OF DEUTERATED PAH USED IN SPIKE STUDY

	LOW LEVEL SPIKE, ng	HIGH LEVEL SPIKE, ng
PYRENE	1000	4000
BaA	100	400
CHRYSENE	1000	4000
BaP	10	40
PERYLENE	10	40

Collection of Particulate Filter Samples

Twenty-one particulate filter samples were collected in one day of test engine operation for this study. Eighteen filter samples were used for the spike study. PAH were measured on the remaining three filter samples to determine the low and high levels for spiking. The test engine condition was the rated speed/half-load mode. Particulate sampling time was 45 minutes at an average flow rate of 1.62 scfm. The temperature of the tunnel was maintained at 125°F using a dilution ratio of 15.3 to 1. At these dilution and sampling conditions, approximately 8-9 mg of particulate was collected on each filter. Average engine, exhaust, dilution and sampling conditions of the test are listed in Table 2.

RESULTS AND DISCUSSIONS

Table 3 presents the results of the spiking study. In general, the recoveries are high, repeatable, and are considered satisfactory from an analytical viewpoint. There is relatively minor difference between the results of the solvent spike and filter spike methods. For the vapor spike method, there is an improvement in the recovery for pyrene

TABLE 2

AVERAGE ENGINE, EXHAUST AND SAMPLING CONDITIONS IN PAH RECOVERY STUDY

Engine Conditions:	
Engine speed, rpm	2098
Percent rated load	51.6
Fuel rate, lb/hr	70
Exhaust Conditions:	
HC concentration, ppm C	116
NO_x concentration, ppm	780
CO concentration, ppm	271
CO_2 concentration, percent	6.6
O_2 concentration, percent	12.3
Exhaust particulate concentration, mg/m^3	60.4
Exhaust sample flow rate, scfm	12.4
Time exhaust sample in transfer pipe, sec.	0.65
Exhaust sample inlet velocity, ft/sec.	10.8
Exhaust sample temperature, °F	
Transfer pipe inlet	836
Transfer pipe outlet	632
Particulate Sampling Conditions:	
Particulate sampling time, min.	45
Dilution tunnel flow rate, scfm	202
Dilution air inlet velocity, ft/sec.	10.8
Dilution air to exhaust sample velocity ratio	1.0
Residence time of diluted exhaust in tunnel, sec.	3.19
Dilution ratio	15.3
Dilution tunnel temperature, °F	125
Sampling face velocity, cm/sec.	8.5
Filter temperature, °F	114

TABLE 3

RECOVERY OF PERDEUTERATED PAH FROM PARTICULATE FILTER SAMPLES

AVERAGE PERCENT RECOVERED ± S.D.

PAH	SPIKE LEVEL	SOLVENT SPIKE	FILTER SPIKE	VAPOR SPIKE
PYRENE	(L)	83±5	87±3	101±4
	(H)	94±6	99±3	96±7
BaA	(L)	75±7	87±2	86±5
	(H)	90±6	99±5	89±5
CHRYSENE	(L)	87±2	85±4	93±7
	(H)	94±4	91±2	100±1
BaP	(L)	72±5	77±2	62±4
	(H)	83±3	86±2	72±16
PERYLENE	(L)	79±2	76±4	53±1
	(H)	88±3	87±4	74±4

and chrysene. No change is observed for BaA. A decrease in recovery is observed for BaP and perylene. Specific discussions of each experiment are presented in the following paragraphs.

Solvent Spike Method

The objective of this experiment was to determine whether the organics extracted from diesel particulate adversely affect PAH recovery. Six particulate filter samples were used in this experiment. Each was placed in a separate Soxhlet extractor; three for the low level spike and three for the high level spike. Toluene (350 ml) was placed in the solvent reservoir of each extractor and spiked with the appropriate level of perdeuterated PAH.

Recovery results for the perdeuterated PAH at the low level spike ranged from an average of 72 ± 5 percent for BaP to 87 ± 2 percent for chrysene. Slightly higher recoveries were obtained at the high level spike and the averages ranged 83 ± 3 percent for BaP to 94 ± 4 percent for chrysene. The recoveries at both levels are considered acceptable from an analytical viewpoint. Recovery of perdeuterated PAH did not appear to be influenced by organic material which was extracted from particulate.

Filter Spike Method

The objective of this experiment was to determine whether the particulate material on filters adversely affects the recovery of PAH. The procedure for spiking was as follows. Particulate filters were placed on watch glasses, and a methylene chloride solution (100 μl) containing all the perdeuterated PAH was slowly introduced onto the center of the filter. The rate of introducing the solution allowed the particulate to adsorb the solvent without the solvent passing through. Three filters were spiked at the low tracer level and three at the high tracer level.

Recovery of the perdeuterated PAH ranged from 76 ± 4 percent (perylene) to 87 ± 2 percent (BaA) for the low level spike and 86 ± 2 percent (BaP) to 99 ± 3 percent (pyrene) for the high level spike. Again, the recoveries are slightly higher at the high spike level. These results indicate that a small portion may be inherently lost through the analytical procedures.

Vapor Spike Method

In this experiment, the perdeuterated PAH were spiked as vapors onto particulate filter samples. The objective was to introduce the PAH to the particulate without the use of a solvent. The apparatus used in these experiments is illustrated in Figure 2. It consists of a temperature controlled probe, a glass "T", a capillary sample tube, a filter holder and a XAD-2 adsorbent trap. The trap was used as a back-up to collect any PAH which may have passed through the filter.

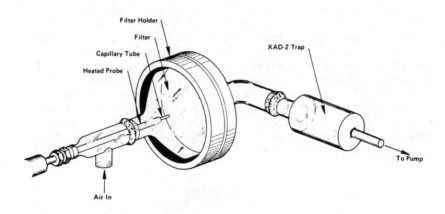

FIGURE 2. Apparatus used for vapor spiking of PAH onto particulate filter samples.

The procedure for spiking was as follows. A solution of perdeuterated PAH was first placed into the capillary tube using a microliter syringe and the solvent allowed to evaporate at room temperature. The capillary was then inserted into the probe and the probe positioned 5 cm above the surface of a particulate filter. An air flow of 2 liter/min. was then established by pulling a stream of air in at the glass "T", over the probe, and through the filter and trap. After 10 minutes, the probe was rapidly heated to 280°C (in $\sim$2 minutes) which vaporized the PAH in the capillary. The vapors were then carried by the airstream to the particulate filter sample.

The probe was kept at 280°C for 10 minutes to assure all the PAH were vaporized. Results from bench tests using blank filters confirmed that all the PAH were vaporized from the capillary under these conditions.

Three filters were spiked with the low level of tracers and three were spiked with the high level. The filters were removed from the holders and analyzed for PAH. The traps, capillary and filter holder were also analyzed for PAH. In all vapor spike tests, all the perdeuterated PAH was found on the particulate filters. No perdeuterated PAH was found in the traps.

Recoveries ranged from 86 ± 5 percent (BaA) to 101 ± 4 percent (pyrene) for the low level and 89 ± 5 percent (BaA) to 100 ± 1 percent (chrysene) for the high level. BaP and perylene recovery was lower than the recovery measured by the other two spiking experiments. Their range was 53 ± 1 percent (perylene) and 62 ± 4 percent (BaP) at the low level spike and 72 ± 16 percent (BaP) and 74 ± 4 percent (perylene) at the high level. The lower recoveries of BaP and perylene is not clearly understood. Possible reasons may be either oxidation at the high probe temperatures or a strong adsorption with the particulate.

CONCLUSIONS

The conclusions which can be made on the basis of this study are as follows:

(1) Soluble organic material extracted from particulate filter samples does not influence the recovery of PAH through the analytical procedure used for PAH analysis

(2) Particulate matter on filters does not influence the recovery of PAH through the analytical procedure used for PAH analysis

(3) Recovery of perdeuterated PAH in all experiments is high and repeatable

(4) Solvent spiking or filter spiking techniques can be used to evaluate the influence of the analytical procedures on the measurement of PAH

(5) Quantities of the perdeuterated PAH should approximate the quantity of engine PAH if their recovery values will be used to correct for loss through the analytical procedure

ACKNOWLEDGEMENT

We respectfully acknowledge support of this work by the Coordinating Research Council under the CAPE-24-72 program.

REFERENCES

(1) Petersen, B. A., Chuang, C. C., Kinzer, G. W., Meehan, P. W., Riggin, R. M., and Trayser, D. A. "Diesel Engine Emissions of Particulates and Associated Organic Matter" First Annual Report, NTIS Publication # PB 80 221963, July, 1980.

SIMPLIFIED REAL-TIME PAH MEASUREMENT TECHNIQUES

TIMOTHY L. PIERCE, DENNIS R. JAASMA
Department of Mechanical Engineering, Virginia Polytechnic
Institute and State University, Blacksburg, Virginia 24061

INTRODUCTION

The development of combustion systems which emit less polynuclear aromatic hydrocarbons (PAH) is hampered by the high cost and time-consuming nature of current PAH measurement techniques. Typical practice to quantify PAH emissions is to accumulate samples on filters or organics traps such as XAD-2, extract the samples one or more times, and analyze the extracts using various types of chromatography and mass spectrometry. The end results are typically expressed as emission factors (mass of pollutant emitted per mass of fuel consumed) for specific compounds or families of compounds. The accuracy of the emission factors thus obtained is subject to grave uncertainty, as illustrated by the work of Sonnichen et al. (1). They sampled the stack of a coal-fired power plant while simultaneously metering known amounts of deuterated PAH into their sample probe. The recovery of these compounds ranged from two orders of magnitude less than the amount spiked to several times the amount spiked into the system, showing that there is great difficulty in quantifying PAH emissions from combustion systems which emit particulate matter. Problems with extraction efficiencies and handling losses can account for underestimation by at least a factor of three (2), while catalytic effects due to the filter material can cause problems due to degradation of the sample (3). The realization that it is very difficult to accurately quantify PAH emissions has given new impetus to the search for simplified measurement techniques. Smith and Levins (4) have reported a simple, inexpensive, and rapid test which monitors the sensitized fluorescence of PAH in contact with filter paper and which can be used to determine relative amounts of PAH in complex samples. Their test has been shown to be capable of giving reasonable order of magnitude estimates of PAH emissions from fireplaces and airtight stoves (5). The airtight stove is believed to be the greatest national source of PAH emissions, accounting for perhaps 38% of all PAH emissions (6).

It is our desire to develop a real-time technique for measuring PAH emission rates from hand-fired wood or coal stoves. Real-time measurements would be valuable because they could be used to rapidly quantify the effects of controllable variables such as air flows and could also be used to quantify the amount of PAH emitted during different parts of the burn cycle, thereby identifying the places where design improvements can be most effective at reducing PAH emissions. The measurement technique must also be simple and relatively inexpensive if it is to be useful to the stove industry, which is composed of many small and technologically unsophisticated manufacturers.

Two approaches are under consideration. The first is an automated version of the Smith and Levins spot test and the second is the use of a proxy which correlates with PAH emissions. High precision and accuracy are not expected to be achieved, but it is believed that both approaches have the potential to become useful tools for the development of improved combustion systems.

MATERIALS AND METHODS

Apparatus

The real-time version of the spot test is called the "moving tape sampler" (MTS) and is designed to give direct measurements of PAH. It is now being evaluated for use in determining woodstove emissions. A schematic of the apparatus is shown in Fig. 1. The MTS is supported near the top of a woodstove stack, where product gas temperatures are ca. 150°C. A variable speed motor is used to draw a 2.5 cm wide filter tape through a 15 cm diameter exhaust stack. Tape speeds on the order of 1 cm/s are currently in use. The tape is drawn through the stack so the flow impinges directly against the flat surface of the tape. This allows PAH-containing particulate matter to collect on the tape as it passes through the stack. After leaving the stack the tape is twisted by a set of guides so that the bottom side (the side originally facing upstream) is facing up. The guides are arranged so the bottom side does not come in contact with the guides. A syringe pump then adds sensitizer fluid (naphthalene in methylene chloride, 60 g/liter), to the "up" side of the tape, the side which was previously facing into the flow stream. The pump is set to a flow that just wets the entire width of tape, and the methylene chloride is evaporated as the tape passes through a dryer. As it enters the detector section, the tape is illuminated with a 254 nm ultraviolet light source and a photomultiplier tube is used to detect the amount of visible fluorescence from the sample and sensitizer. The light from the tape passes through a 450 nm long-pass optical filter before reaching the photomultiplier. The filter is used

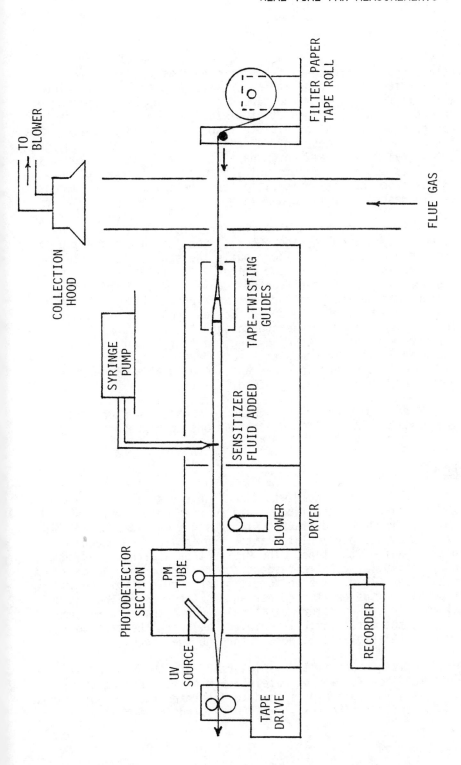

FIGURE 1. MTS schematic.

to improve the signal to noise ratio. Improvement occurs because the fluorescence from the tape/sample/sensitizer combination tends to be at longer wave lengths than the light which is received from the tape/sensitizer combination in the absence of sample. The output of the photomultiplier tube indicates the amount of visible fluorescence and thus represents, in some way, the amount of PAH collected on the tape. The output signal is recorded on a chart recorder. The tape speed in use gives about a 1 minute delay between the sample collecting on the tape and the photomultiplier tube detecting the amount of fluorescence from the sample.

A 20 x 25 cm glass fiber filter (Gelman 61635) is contained within a collection hood located just above the top of the 6 m high stack. A blower pulls the stack exhaust plus some ambient air through the filter. The filter holder is designed so that filters can be readily changed, and thus filters can be inserted for brief periods of time. The dilution of the exhaust by room air causes the effluent temperature to drop prior to passage through the filter. This temperature was not monitored during the tests reported here. The glass filters capture the "total" PAH emissions during the time the filters are in place and are used as a standard to which the MTS signal is compared.

A gas sampling train is used to sample the stack, allowing real-time CO_2 and CO concentrations to be recorded.

Procedures

The stove is loaded with kindling and split wood, after which a fire is established. A 3 cm wide metal strip is inserted through the stack just beneath the filter tape. This strip prevents particulate matter from collecting on the bottom half of the tape. The MTS is then started, causing the photomultiplier tube to respond to the light from the tape and sensitizer fluid. The photomultiplier's output voltage is recorded on the chart recorder and is taken as an indication of zero PAH (zero reference signal). Once the zero reference is set, the metal strip is removed, allowing the tape to accumulate sample as it passes through the stack.

A clean glass filter is inserted into the hood and is left in place for approximately 1.5 minutes. Concomitantly, the gas sampling train is used to determine the concentrations of CO_2 and CO. The real-time PAH results for the interval during which the filter is in place can be located on the chart recorder, taking into account the 1 minute time for the tape to reach the photomultiplier.

The PAH data reported herein are for two firings of the stove. During the first firing, three samples were obtained on the glass filters, while four samples were obtained during the second firing.

The filters are analyzed in the following manner. The entire filter is placed in a Soxhlet apparatus and extracted for 24 hours using methylene chloride. The resulting solution is evaporated on a flash evaporator with a Snyder column to remove the methylene chloride, and the remaining sample is dissolved in benzene and reduced to 4 ml. One milliliter of this sample is transferred to Sep-Pak. A first fraction is eluted by passing hexane through the Sep-Pak, and a second fraction is obtained using benzene with 1% methanol. Both fractions are run on an HP 5830 A gas chromatograph using a 6 m, 3% SP 2100 column. A flame ionization detector is used. The PAH emission factors presented herein represent emissions of the following compounds.

Naphthalene
Acenaphthene
Acenaphthylene
Fluorene
Phenanthrene
Anthracene
Fluoranthene

Pyrene
Benz[a]anthracene
Chrysene
Benzo[a]pyrene
Indeno[1,2,3-cd]pyrene
Benzo[k]fluoranthene
Benzo[ghi]perylene

The PAH measured by the extraction and subsequent GC workup is presumed to be the actual PAH emission during the time the filter was in place, and is compared to the MTS signal.

RESULTS

Since there is no known way to generate real-time PAH emission data to compare with the instantaneous MTS signal, the MTS signal has been integrated with respect to time to give a "signal area". This area is hoped to be proportional to the amount of PAH emitted during the time the glass filter is in place. An example of the MTS output showing the signal area is given in Fig. 2. Figure 3 shows the correlation between signal area and the PAH emissions determined from the filter

REAL-TIME PAH MEASUREMENTS

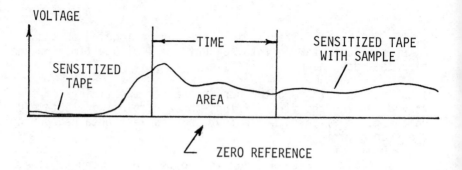

FIGURE 2. MTS output

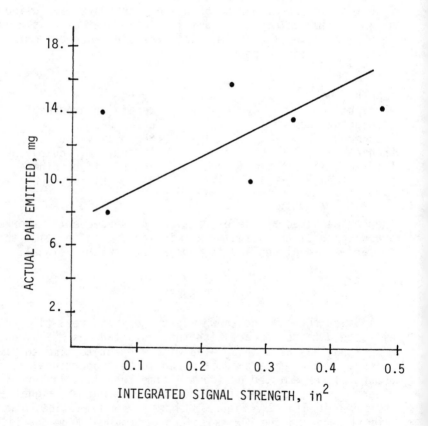

FIGURE 3. Real-time results compared to conventional results.

catch. The PAH emission factors were computed using the sampling time and an estimated stack gas flow of 0.6 Nm^3/minute.

The CO and PAH emission factors for each test are plotted in Fig. 4, along with emission factors taken from the literature. No NO_x measurements were obtained during the present work, but literature values of PAH and NO_x emission factors are also presented in Fig. 4 to illustrate the possible utility of NO_x as an "inverse proxy."

DISCUSSION

The correlation between the integrated MTS signal and the glass filter PAH measurements (Fig. 3) is believed to be relatively good in light of the variables which are uncontrolled and probably affect the amount of PAH which adheres to the tape. The PAH accumulation on the glass filter during the 90 seconds it is in place is expected to be linear with respect to stack flow (NM^3/s) and the concentration of particulate PAH in the flow (ng/Nm^3). For this preliminary work, questions concerning vapor-phase PAH passing through the glass fiber filter are not being considered. However, the collection rate of PAH on the tape is expected to depend on factors such as the size of the particles in the flow and the temperature of the stack flow. For example, consider a stack flow of a given rate and particulate PAH concentration. If the particles are small, they will do a better job of following the stack gas flow, and less PAH will accumulate on the tape than if the particles were larger. Also, since the stack temperature is higher than the temperature of the diluted mixture as it passes through the glass filter, it is possible that higher stack temperatures may allow vapor-phase PAH to bypass the tape while the dilution is still adequate to condense most of the PAH before the mixture reaches the glass filter. Another factor is that the particulate onto which the PAH is adsorbed or condensed may have different fluorescence-quenching efficiencies, depending on the chemical make-up of the soot particles. It is not clear at this time whether the fluorescence of the tape/sensitizer/sample combination is primarily due to fluorescence of PAH on particles, or whether the sensitizer solution performs an in-situ extraction which moves PAH from particles onto regions of the tape where particles have not covered it.

The above mentioned difficulties suggest that the MTS might be more effective if it was located in a flow stream of relatively constant temperature and if the particle accumu-

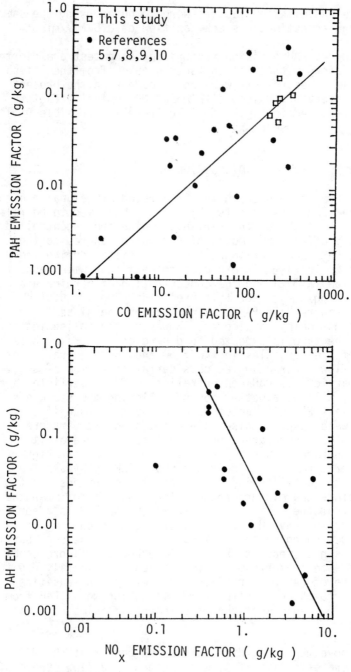

FIGURE 4. Possible proxy compounds for PAH.

lation on the tape could be freed of dependence on aerodynamic factors. These goals can perhaps be reached by locating the MTS in a dilution tunnel and using an "active" sampling system which relies on a pump to draw a known flow rate of gas through the tape.

Although the MTS has only been tested for its utility in determining woodstove PAH emissions, it is possible that it will be useful for workplace PAH monitoring when used with an active sampling system. Admittedly there can be problems such as low particle collection efficiencies, but it may be possible to calibrate the system to compensate for this.

Proxy Compounds

The first compound examined as a possible proxy compound is CO. Figure 4 shows CO and PAH emission factors obtained from our work, and also shows literature values of emission factors for the combustion of coal, wood, gas, and oil (5, 7, 8, 9, 10). The line drawn through the data shows that all of the PAH emission factors plotted, with exception of one, can be predicted to within better than an order of magnitude by a straight line which follows the general trend of the data. Given the difficulties of measuring actual PAH emissions and the wide variety of fuels and combustors represented, this degree of correlation is about all one could hope for. Regarding other possible proxies, it is of interest to note that BaP appears to be a particularly poor proxy for woodstove PAH emissions. Allen and Cooke's data (7) for woodstoves show the PAH/BaP ratio ranging from 13 to 1300.

No NO_x measurements were obtained during this study, but literature values of emission factors for residential combustors allow Fig. 4 to be constructed. Because the conditions which favor NO_x formation (high temperature and high oxygen concentration) are ones which favor destruction of PAH, it is expected that high NO_x emissions will correspond to low PAH emissions, as shown by the figure. The straight line is a curve fit which correlates 90% of the data points to within a factor of six, a correlation which is good considering the variety of fuels, combustors, and measurement techniques which are represented. It is also possible that the correlations for both NO_x and CO will improve as better PAH data are accumulated.

In conclusion, it has been shown that the MTS can be used to obtain real-time measurements of the fluorescence due to PAH which is being emitted from a woodstove. Adequate data have not yet been obtained to determine what relationship, if any, exists between the MTS photomultiplier signal and the PAH emission rate of the stove. The MTS may function better in a dilution tunnel setting rather than in a stack where the gases have time-varying temperature and velocity. The MTS concept may also be useful for workplace monitoring of PAH. Both NO_x and CO have been investigated as proxies to imply the PAH content of exhaust from residential combustors, and the available data indicate that slightly better than order of magnitude PAH estimates can be obtained using either proxy. On the basis of the data available, NO_x seems to be the better choice of proxy.

ACKNOWLEDGEMENTS

The authors are indebted to the National Institute of Health for a Biomedical Research Support Grant which has made this effort possible. We also acknowledge the help of Jean Dickinson and Roderick Young, of the VPI&SU Biochemistry Department, who performed the PAH analysis.

REFERENCES

1. Sonnichsen, T. W., McElroy, M. W., and Bjorseth, A. (1979): Use of PAH Tracers During Sampling of Coal Fired Boilers. In: <u>Polynuclear Aromatic Hydrocarbons: Chemistry and Biological Effects</u>, edited by A. Bjorseth and A. J. Dennis, pp. 617-632, Battelle Press, Columbus, Ohio.
2. Griest, W. H., Caton, J. E., Guerin, M. R., Yeatts, Jr., L. D., and Higgins, C. E. (1979): Extraction and Recovery of Polycyclic Aromatic Hydrocarbons from Highly Sorptive Matrices Such as Fly Ash. In: <u>Polynuclear Aromatic Hydrocarbons: Chemistry and Biological Effects</u>, edited by A. Bjorseth and A. J. Dennis, pp. 543-563, Battelle Press, Columbus, Ohio.
3. Lee, F. S. C., Pierson, W. R., and Ezike, J. (1979): The Problem of PAH Degradation During Filter Collection of Airborne Particulates - An Evaluation of Several Commonly Used Filter Media. In: <u>Polynuclear Aromatic Hydrocarbons: Chemistry and Biological Effects</u>, edited by A. Bjorseth and A. J. Dennis, pp. 543-563, Battelle Press, Columbus, Ohio.

4. Smith, E. M. and Levins, P. L. (1979): Sensitized Fluorescence Detection of PAH. In: <u>Polynuclear Aromatic Hydrocarbons: Chemistry and Biological Effects</u>, edited by A. Bjorseth and A. J. Dennis, pp. 973-982, Battelle Press, Columbus, Ohio.
5. DeAngelis, D. G., Ruffin, D. S., and R. B. Reznik. Preliminary Characterization of Emissions from Wood-fired Residential Combustion Equipment. EPA-600/7-80-040, U.S. Environmental Protection Agency, Research Triangle Park, North Carolina, 1980, 146 pp.
6. Peters, J. A., DeAngelis, D. G., and Hughes, T. W. (1980): An Environmental Assessment of POM Emissions from Residential Wood-Fired Stoves and Fireplaces. In: <u>Chemical Analysis and Biological Fate: Polynuclear Aromatic Hydrocarbons</u>, edited by M. Cooke and A. J. Dennis, pp. 571-581, Battelle Press, Columbus, Ohio.
7. Allen, J. M., and W. M. Cooke. Control of Emissions from Residential Wood Burning by Combustion Modification. EPA-600/7-81-091, U. S. Environmental Protection Agency, Research Triangle Park, North Carolina, 1981, 92 pp.
8. DeAngelis, D. G., and R. B. Reznik. Source Assessment: Coal-fired Residential Combustion Equipment Field Test, June 1977. EPA-600/2-78-004o, U. S. Environmental Protection Agency, Research Triangle Park, North Carolina, 1978, 83 pp.
9. Hangebrauk, R. P., Von Lehmden, D. J., and Meeker, J. E. (1964): Emissions of Polynuclear Hydrocarbons and Other Pollutants from Heat-Generation and Incineration Processes, <u>Journal of the Air Pollution Control Association</u>, 14(7): 267-278.
10. Hubble, B. R., Stetter, J. R., Gebert, E., Harkness, J. B. L., and Flotared, R. D. (1981): Experimental Measurements of Emissions from Residential Wood-Burning Stoves. In: <u>Residential Solid Fuels: Environmental Impacts and Solutions</u>, pp. 29.

THE METABOLISM OF NITRO-SUBSTITUTED POLYCYCLIC AROMATIC HYDROCARBONS IN SALMONELLA TYPHIMURIUM

MICHAEL A. QUILLIAM[*], F. MESSIER[*], C. LU[**], PAUL A. ANDREWS[*], BRIAN E. McCARRY[*], AND DENNIS R. McCALLA[**]
Departments of Chemistry[*] and Biochemistry[**], McMaster University, Hamilton, Ontario, Canada, L8S 4M1

INTRODUCTION

Since the work of Pitts (1) there has been increasing interest in the possible environmental hazards posed by nitro derivatives of polycyclic aromatic hydrocarbons (nitro-PAHs). More recently, several nitropyrenes were found as trace impurities in carbon black used in older batches of xerographic photocopier toners (2,3) and it has been reported that nitro-PAHs probably account for much of the mutagenic activity of particulate material from diesel exhaust (4). Nitro-PAHs as a class are powerful direct acting mutagens in the Ames Salmonella assay (1,3). From the limited results available, it appears that these compounds are also mutagenic to mammalian cells but that they are more potent in Salmonella (5,6).

In the present work, we have investigated (a) the metabolism of 1-nitropyrene and 1,8-dinitropyrene, (b) the formation of DNA adducts of 1-nitropyrene, and (c) the fractionation of nitroreductase enzymes of Salmonella strains using a new assay for 1-nitropyrene reduction. Some of this work has already been reported (7) and will only be summarized here.

MATERIALS

1-Nitropyrene and 1,8-dinitropyrene were prepared by treating pyrene with nitric acid in acetic anhydride. Preparative HPLC was used to ensure a high degree of purity. 1-Aminopyrene was synthesized by hydrogenating a sample of purified 1-nitropyrene at atmospheric pressure over Adam's catalyst in methanol. The acetylation of 1-aminopyrene in dichloromethane with acetic anhydride yields N-acetyl-1-aminopyrene. Tritiated 1-nitropyrene, synthesized by the nitration of tritiated pyrene, was purified on an analytical reverse phase HPLC column to afford products with specific activities of 12.6 and 16.4 Ci/mmole in two preparations. The tritiated pyrene was prepared by the hydrogenolysis of 1-bromopyrene with tritium gas (reduction performed by Amersham Corp., Oakville, Ontario).

Four strains of Salmonella typhimurium kindly provided by Dr. B. N. Ames, Berkeley, CA, were used: TA98 (his D3052, Δ uvr B rfa'plus pKM 101 plasmid; TA100 his G46, Δ uvr B rfa plus pKM 101 plasmid: TA1538 his D3052, Δ uvr B rfa and TA1978 the uvr+ analogue of TA1538. TA100 F50, a nitrofuran resistant mutant which is partially deficient in nitroreductase activity was obtained from Dr. H. S. Rosenkranz, New York, N.Y. Cultures were maintained on nutrient agar and grown in Davis-Mingioli medium supplemented with glucose, histidine and biotin.

RESULTS AND DISCUSSION

Analysis of Metabolites

Reverse phase HPLC analysis of extracts of Salmonella typhimurium TA98 incubated with [^{3}H]-1-nitropyrene diluted with cold carrier revealed a number of metabolites (see Figure 1). The principal metabolites were identified by mass spectrometry, and confirmed by retention times of synthesized reference compounds, to be 1-aminopyrene and N-acetyl-1-aminopyrene.

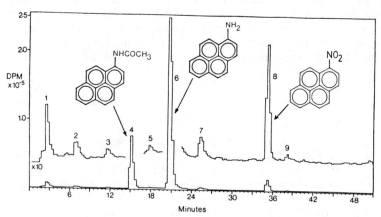

Figure 1: HPLC radiochromatogram of 1-nitropyrene and its metabolites

However, the rate of metabolism of 1-nitropyrene by Salmonella typhimurium TA98 was found to be very much slower than that of the mutagen nitrofurazone, a nitrofuran derivative, under the same conditions. A time course study showed the exponential loss of 1-nitropyrene with a half-life of about 7 hours and the concomitant appearance of the two major metabolites.

A similar study using 1,8-dinitropyrene revealed several metabolites, the principal one being established by mass spectrometry (see Figure 2) as 1-amino-8-nitropyrene. Preliminary rate studies have indicated that the rate of reduction of 1,8-dinitropyrene is very similar to that of 1-nitropyrene.

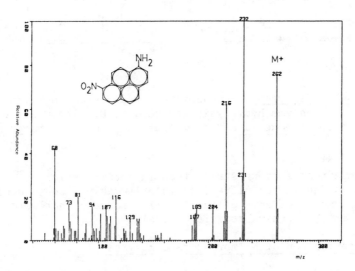

Figure 2: Mass spectrum of principal metabolite of 1,8-dinitropyrene

Formation of DNA Adducts

When [^{3}H]-1-nitropyrene was incubated with <u>Salmonella typhimurium</u>, radioactivity became tightly associated with DNA. After the isolated DNA was enzymatically hydrolyzed to its constituent nucleosides, this radioactivity appeared in low molecular weight components (see Figure 3).

The extent of binding of radioactivity from nitropyrene to DNA in strain TA98 increased with the time of incubation and was dependent on the amount of mutagen available per cell. The maximum levels of binding were of the order of one molecule of nitropyrene per 10^6 nucleotides in DNA. The amount of tritium associated with DNA was markedly reduced in a mutant strain known to have reduced nitroreductase activity and also,

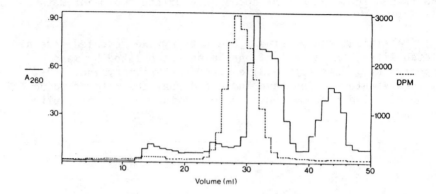

Figure 3: Sephadex G-25 radiochromatogram of enzyme hydrolysis products of DNA isolated from Salmonella after exposure to [^{3}H]-1-nitropyrene

in a repair-proficient strain than in repair-deficient strains. The latter difference is presumably a consequence of the removal of some of the labelled adducts that were formed in the repair-proficient strain.

Characterization of Nitroreductase Enzymes

In order to investigate the enzymes responsible for the reduction of 1-nitropyrene, crude enzyme extracts from different strains of S. typhimurium were fractionated by ammonium sulfate precipitation and chromatographed on DEAE-52 cellulose (see Figure 4). The chromatographic fractions were then assayed for nitroreductase activity using both nitrofurazone and 1-nitropyrene. While the reduction of nitrofurazone could be monitored spectrophotometrically (8), a new spectrofluorimetric assay was developed to examine 1-nitropyrene reduction. The emission of the principal reduction product, 1-aminopyrene, was easily quantified at 445 nm using a Perkin-Elmer MPF-44 Spectrofluorimeter with excitation at 390 nm. Under these conditions there was low background fluorescence, insignificant quenching by the bacterial extracts or by the assay mixture and a linear relationship between the measured fluorescence and the concentration of 1-aminopyrene. The increasing fluorescence corresponded to the appearance of 1-aminopyrene as determined by HPLC. Since the reductase enzymes utilized NADH and NADPH equally well, a NADH (NADPH) regeneration system using glucose-6-phosphate and glu-

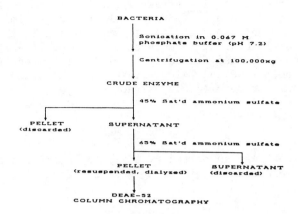

Figure 4: Isolation of nitroreductase enzymes

cose-6-phosphate dihydrogenase was incorporated into the assay mixture.

Figure 5 shows the nitroreductase activities of the DEAE-52 cellulose column fractions from the TA 100 strain toward nitrofurazone and 1-nitropyrene. Clearly, only certain enzymes that reduce nitrofurazone can also reduce 1-nitropyrene.

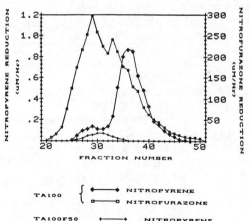

Figure 5: Nitroreductase activities of fractions eluted from DEAE-52 cellulose column toward 1-nitropyrene and nitrofurazone

Since the Salmonella strain TA100 F50 selected for its resistance to nitrofurazone shows substantially diminished 1-nitropyrene reductase activity, it appears likely that one of the major nitrofurazone reductases also reduces 1-nitropyrene. In addition, the specific activity of 1-nitropyrene reduction is low in both TA98 and TA100 strains, on the order of 5 nmoles/min/mg protein (crude extract).

ACKNOWLEDGEMENT

This work was supported by a Strategic Grant from the Natural Sciences and Engineering Research Council of Canada (NSERC).

REFERENCES

1. Pitts, J.N. Jr., et al., (1979): Atmospheric reaction of polycyclic aromatic hydrocarbons: Facile formation of mutagenic nitro derivatives, Science, 202: 515-518.
2. Lofroth, G., et al., (1980): Mutagenic activity of photocopies, Science, 209: 1037-1039.
3. Rosenkranz, H.S., et al., (1980): Nitropyrenes: Isolation, identification and reduction of mutagenic impurities in carbon balck and toners, Science, 209: 1039-1043.
4. Schuetzle, D., Lee., F.S.-C. and Prater, T.J. (1981): The identification of polynuclear aromatic hydrocarbons (PAH) derivatives in mutagenic fractions of diesel particulate extracts. Intern. J. Environ. Anal. Chem., 9: 93-144.
5. Burrell, A.D., et al., (1981): Genetic toxicity of 2,4,7-trinitrofluorene-9-one in the Salmonella assay, L5178 TK+/- mouse lymphoma cell mutagensis assay and sister chromatid exchange assay, Abstracts of the Environmental Mutagen Society, 121-122.
6. Arlett, C.F., et al., (1981): Mutagenic effects in human and mouse cells by a nitropyrene. AUI/BNL Symposium on "The Genotoxic Effects of Airborne Agents."
7. F. Messier, et al., (1981): Metabolism of 1-nitropyrene and formation of DNA adducts in Salmonella typhimurium. Carcinogensis, (in press).
8. D. W. Bryant, et al., (1981): Type 1 nitroreductases of Escherichia coli, Can. J. Microbiol., 81-86.

RATIONAL SELECTION OF SAMPLE PREPARATION TECHNIQUES FOR THE MEASUREMENT OF POLYNUCLEAR AROMATICS

W. K. ROBBINS AND F. C. McELROY
Analytical and Information Division, Exxon Research and Engineering, Linden, NJ 07036

INTRODUCTION

PNA-Enrichment Techniques

Polynuclear aromatic hydrocarbons (PAH) and their heterocyclic analogs (collectively PNAs) are often found as mixtures of closely related isomers in complex matrices. Although some PNAs may be measured directly by sophisticated instrumentation, most individual species cannot be measured reliably until PNA-rich fractions have been isolated. Because PNAs can influence process design, product quality, and biological activity, numerous separation schemes have been developed for the fractionation of petroleum or alternate fuels. This paper discusses the measurement of distribution coefficients (K_Ds) for the evaluation of PNA-enrichment techniques and describes the application of K_D data to the development of a separation scheme for isolating parent PAHs and N-PNAs from shale oil samples.

Among the published procedures, a few afford the isolation of heterocyclics (S-PNA, N-PNA, O-PNA) as well as PAH from alternate fuel sources (1,2,3). Some separations emphasize an in-depth multi-step isolation of a single subclass such as parent PAH (4) while others emphasize a broad, multi-dimensional functional group fractionation (5). Although both approaches have been widely accepted, a multitude of alternatives have been suggested. Although empirical correlations have been developed between the chromatographic separations and the physical properties of compounds, reagents, and solvents (6), these are not generally applied to non-chromatographic techniques.

Distribution Characteristics

The distribution coefficients (K_D) of model compounds can be used to compare different separation techniques because they are intrinsic, intensive properties of the separation equilibria, i.e., they are independent of the quantities used. The distribution coefficient (K_D) is defined as the ratio of analyte concentration in the reagent phase to that in the solvent phase when the two immiscible phases are under dynamic equilibrium. Where both phases are liquids, extraction efficiencies can be calculated from K_D, the relative phase volumes, and the number of equilibrations with fresh reagent (7). Where the reagent phase is a solid, the specific elution volume (V_R, ml/g) is the sum of K_D and a constant k_i where k_i is the specific interstitial volume fraction (i.e., the ml/g of solid which may be occupied by the solvent).

The distribution characteristics of PNAs have been reported for some PNA separations. The K_D fpr PAH in some extraction systems have appeared (7). In some chromatographic studies, the "capacity factor" k' (=K_D/k_i) have been reported instead of V_R. If the constant k_i is known, then V_R may be calculated. However, the chromatographic data have generally been reported for one set of conditions. To evaluate parameter effects, more K_Ds for model compounds are needed.

For many of these techniques, such data can be collected with the shake vial tests which have been described previously for both liquid-liquid extraction (7) and for liquid-solid adsorption (8). When available, ^{14}C-PNAs were used and the analysis was performed by radio assay. Otherwise, the analyses were performed by measurement of the UV absorbance at an appropriate wavelength.

RESULTS AND DISCUSSION

Acid Extraction

In some procedures, 1N HCl is used for the isolation of "basic N" compounds (9,10). The K_Ds for N-heterocyclic and NH_2-PNAs in the 1N HCl/methyl-t-butyl ether extraction pair (Table 1) show that 1N HCl is only efficient for the extraction of the heterocyclics and amino compounds with three or fewer rings. If all the "basic-N" compounds are to be isolated in a single fraction, alternate separation techniques will be required.

TABLE 1

DISTRIBUTION COEFFICIENTS FOR N-HETEROCYCLICS
IN 1N HCl/METHYL-t-BUTYL ETHER

Compound	K_D	Percent Extracted* Into Acid
Acridine	0.0	100
Dibenz(a,h)acridine	2.6	98
9 Aminofluorene	15.6	100
9 Aminophenanthrene	4.2	99
1 Aminoanthracene	2.7	98
1 Aminopyrene	0.48	69
3 Aminofluoranthene	0.22	45
6 Aminochrysene	0.15	35

*Calculated for 3 equal volume extractions.

NMP Extractions

Previous work (7) has shown that PAH may be enriched by solvent extraction with N-methyl pyrrolidone, (NMP), dimethyl sulfoxide (DMSO), and dimethylformamide (DMF). Of the three, NMP exhibited the largest K_Ds under comparable conditions. Consequently, the shake vial technique has been used to measure K_Ds for PAH, neutral N-PNAs, and basic N-PNAs in a cyclohexane/NMP:H$_2$O (4:1) extraction pair (Table 4). Three characteristics were observed for this system: (1) For a given number of rings, the K_Ds follow the order neutral N > Basic N $\geq$ PAH; (2) K_Ds increase with number of rings within a class; and (3) K_Ds are sharply reduced by alkyl substitution. Although NMP extractions have been able to reduce the alkyl interferences in the measurement of PAH in shale oils by roughly an order of magnitude (8), the parent PAH in general are still masked by substantial quantities of alkyl derivatives.

TABLE 2

DISTRIBUTION COEFFICIENTS FOR PNAs IN THE CYCLOHEXANE/NMP:H_2O (4:1) SYSTEM

PAH	K_D	Neutral N-PNA	K_D	Basic N-PNA	K_D
PHEN*	2.0	CAR	23	ACR	10.6
PYR	3.0				
--1Me	2.2				
BaA	5.6	BACAR	33	BCACR	5.0
--7Me	2.8				
--7,12DiMe	1.5				
BaP	6.1				
DB(a,h)A	10.2	DB(c,g)CAR	50	DB(a,h)ACR	8.4

*Abbreviations: PHEN=phenanthrene; BaA=benz(a)anthracene; h)A=dibenz(a,h)anthracene, BaP=Benzo(a)pyrene; CAR=carbazole; B(A)CAR. Benz(a)carbazole DB(c,g)CAR=dibenz(c,g)carbazole; Q=Quinoline; ACR=acridine; B(c)ACR=benzo(c)acridine; DB(a,h)ACR=dibenz(a,h)acridine

Adsorption Chromatography

Adsorption chromatography is one of the most widely used techniques for obtaining functional group fractions from oils (5, 6). A number of characteristic differences between adsorbents are generally recognized: (1) the elution of PAH from alumina is more sensitive to the size of the π aromaticity than silicic acid or Florisil; (2) the elution volume of neutral N-PNA increase and basic N-PNA decrease as the pH of alumina increases; (3) the separations obtained on silicic acid or Florisil at the same levels of activation are similar; and (4) silicic acid provides a better functional group separation than alumina. However to design a separation scheme, one must know the properties of the specific adsorbent(s) to be used.

To quantify the differences between adsorbents, the shake vial technique has been used to determine K_Ds for a wide range of PAH and N-PNAs at several eluent strengths on a number of adsorbents at different levels of water deactivation. These K_D values have been combined with specific interstitial volume fraction (ki) to calculate the specific elution volumes ($V_R = K_D + k_i$) (Table 3). Only a few of the most essential points have been included.

TABLE 3

SPECIFIC ELUTION VOLUMES V_R (ml/g) FOR SELECTED ADSORBENTS

Adsorbent	Biosil A	Florisil-PR	Aluminas Neutral
%H_2O Deactivation	10 %	0	2 %
Specific Interstitial k_i (ml/g)	1.5	1.3	0.6
A. PAH (%CH_2Cl_2 in cyclohexane)	(3)	(25)	(10)
Phen*	2.7	2.3	1.4
BaA	3.3	2.8	4.3
DB(a, h)A	4.2	4.7	21.5
BaP	4.4	6.0	7.0
B. Neutral N-PAH % (CH_2Cl_2 in Cyclohexane)	(50)	(50)	(50)
CAR	2.8	5.0	3.5
DB(c, g)CAR	4.1	6.0	11.0
C. Basic N-PAH (% CH_2Cl_2 in Cyclohexane)	(100)	(**)	(50)
Q	23.5	5.3	5.3
ACR+	43	5.2	5.5
B(C)ACR	3.1	5.2	1.0
DB(a, b)ACR	2.5	--	--

*Abbreviations given in Table 2
**3% CH_3OH in CH_2Cl_2

SAMPLE PREPARATION TECHNIQUES

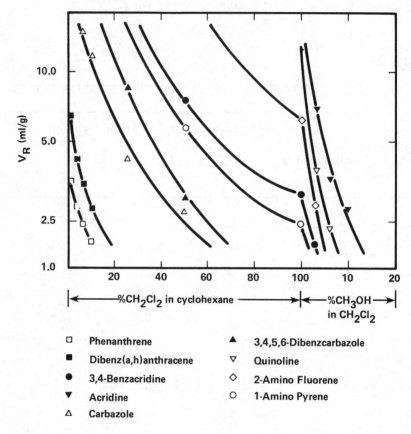

- □ Phenanthrene
- ■ Dibenz(a,h)anthracene
- ● 3,4-Benzacridine
- ▼ Acridine
- △ Carbazole
- ▲ 3,4,5,6-Dibenzcarbazole
- ▽ Quinoline
- ◇ 2-Amino Fluorene
- ○ 1-Amino Pyrene

FIGURE 1. Specific elution volume (V_R) for PNAs on 10% H_2O deactivated Biosil A

From this study, 10% H_2O-deactivated silicic acid was considered to be the most generally useful adsorbent for the preliminary isolation of aromatics and their N-heterocyclic analogs into recognizable functional group classes. More V_R data is summarized in Figure 1. From these curves a column procedure has been designed to yield five fractions: saturates, aromatics, neutral N-heterocyclics, basic N-heterocyclics and polar-N compounds. The use of this column for the initial fractionation of 1 g shale oil samples has been reported previously (8).

LH-20/IPA Gel Filtration

The elution of compounds from Sephadex LH-20 by isopropyl alcohol (IPA) has been reported to follow a unique pattern (11). Aliphatics elute by gel permeation order followed by aromatics which are retained by a combination of π and H-bonding properties. This has led to the use of the LH-20/IPA system for isolating a 3-5 ring PAH-rich subfraction from aromatics isolated from lube oil samples by silicic acid chromatography (4).

To define the behavior of PAH and N-PNA more generally, the concentration of model compounds was determined in 2 ml aliquots taken during the elution of a 10 g LH-20 column with IPA. (The shake vial technique cannot be applied because the LH-20 swells.) These results have been combined with those reported by Oelert (11) for 1-3 ring compounds to calculate the V_R presented in Table 4.

TABLE 4

V_R FOR PNAs IN SEPHADEX LH-20/IPA GEL FILTRATION

Compound	(ml/g)	Compound	(ml/g)
Aromatics		Basic N-Heterocyclics	
Benzene	(2.7)*	Pyridine	2.6 (2.6)
Naphthalene	(3.3)	Quinoline	3.0
Anthracene	(4.3)	Acridine	3.6 (3.6)
Chrysene	(6.2)	1,2 Benzacridine	4.8
Neutral N-Heterocyclics		Amino - PAH	
Indole	4.8 (4.6)	Aniline	(4.7)
Carbazole	5.8 (5.5)	Dimethylaniline	(3.0)
--2,3 Dimethyl	4.2 (4.4)	1-Naphthyl Amine	(7.1)
--1,2,3,4 Tetrahydro	4.6 (5.7)	2-Amino fluorene	8.5
1,2 Benzocarbazole	7.2	3 Amino fluoranthene	15.0
		6-Amino chrysene	17.5

*Values in parentheses calculated from Oelert (11)

SAMPLE PREPARATION TECHNIQUES

The V_{RS} increase with the amount of π aromaticity within each class with little effect of alkyl substitution. For example, tetrahydrocarbazole (a C_4 indole) has a V_R close to indole itself. Furthermore, the V_R for the PAH and the three N-PNA types reveal that H-bonding plays a role in the separation. The LH-20/IPA system, therefore, may be adopted to obtain PNA-rich subfractions on the basis of ring size from each of the aromatic or heterocyclic fractions generated by the polarity separation on the silicic acid.

When applied to the neutral N silicic acid fraction of a shale oil, for example, a 10 g LH-20/IPA column concentrated 77% ^{14}C-carbazole into the expected fraction (50-100 ml) which contained only 5% of the original fraction mass. The bulk (93%) of that mass, which eluted ahead of the N-PNAs, was found to consist of a mixture of alkyl indoles and aliphatic ketones when analyzed by GC/MS and IR techniques. However, the "enriched" N-PNA subfraction still contains too many components to permit GC measurement of the parent carbazole or its benzologs (Figure 2). The similar FID and NPD chromatograms for that fraction demonstrate that most of the isolated components contain N. Independent MS data have shown that shale oils contain numerous C_1-C_5 substituted isomers which apparently co-elute with the parent compounds (10).

Sephadex LH-20/NMP Column Partition Chromatography

The successful isolation of parent PAH from derivative alkyl aromatics in lube oils has been previously reported by Grimmer (4) who used specified volumes of hexane to elute the parent PAH from an LH-20 column loaded with DMF:H_2O (85:15). The volumes used can be related to the K_Ds previously established for the PAH in a cyclohexane/DMF:H_2O (9:1) extraction system (4). Studies of extraction systems have shown that the cyclohexane/NMP pair can be used over a wider range of conditions. Consequently, a LH-20/NMP partition procedure was investigated.

Column partition chromatography is, in effect, an extract in which the solvent phase is mobile and the reagent phase is stationary on a support. If the support is inert, the elution volume for a compound is controlled by its K_D for the two phases. The distribution coefficients determined for extractions may be used to estimate the partition column behavior for NMP:H_2O (4:1) supported on LH-20. Because LH-20 swells in this matrix, the phase

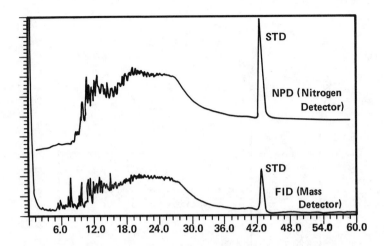

FIGURE 2. Chromatograms of a shale oil neutral nitrogen fraction. On a 1/8 x 6' 5% SP-2100/ supelcoport column programmed from 80 to 300°C at 8°/min. with He flow 50 ml/min.

ratios can only be estimated. The volume of cyclohexane required to elute ^{14}C-carbazole thru a 5 g LH-20/NMP column (Table 5) can be used with the K_Ds to estimate the elution volumes of the other compounds. The K_D in Table 2 show that compounds with more than C_2 alkyl substitution can be well resolved from their parent compounds by column partition chromatography on LH-20/NMP.

The actual elution volumes for ^{14}C-carbazole have also been determined for a neutral-N fraction of a shale oil (Table 5). The results show that this column retains the carbazole (by virtue of its high K_D) while allowing the bulk of the sample (alkyl isomers) to be eluted. This results in a hundred-fold enrichment with 66% of the activity being recovered in 0.6% of the mass. The LH-20/NMP procedure gives similar enrichments for the neutral N- PNAs in shale oils obtained after both silicic acid and LH-20/IPA separations (Table 6). The chromatograms demonstrate that the ~1000 fold enrichment by the three chromatographic procedures is sufficient to allow carbazole to be observed as a discrete species.

TABLE 5

DISTRIBUTION OF ^{14}C-CARBAZOLE
ACTIVITY IN 4:1 NMP:H$_2$O PARTITION CHROMATOGRAPHY

*mL Pre-Equilibrated Cyclohexane	Solvent Blank Percent Total ^{14}C Activity	Raw Shale Percent Total ^{14}C Activity	Percent of SiO$_2$ Neutral N Fraction Weight
0-50	0.2	4.6**	93.6
50-100	0.4	0.3	2.7
100-150	<0.1	<0.1	1.2
150-200	2.2	0.8	0.9
200-250	19.1	8.4	0.4
250-300	28.6	35.0	0.3
300-350	26.6	31.0	0.3
350-400	16.7	7.5	0.1
400-450	4.9	1.2	0.2
450-500	1.1	0.5	0.2
Total Recovered	99.8	89.3	

*Volume used to elute a 5.0 g LH-20 column.
**Counts variable, light and temperature sensitive. These are attributed to chemiluminescence and are not included in the total.

TABLE 6

ISOLATION OF ^{14}C-CARBAZOLE ADDED TO SHALE OIL

	% OF ORIGINAL MASS	% ^{14}C-CARBAZOLE	ENRICHMENT (%^{14}C/% MASS)
Non-Volatile Base-Neutrals	75.3	100	1.32
Neutral N-Fraction	15.2	96.3	6.3
Neutral N-PNA-Subfraction	1.7	87.6	51.5
Carbazole Rich Subfraction	<0.05	60.7	>1214

SAMPLE PREPARATION TECHNIQUES

Integrated Separation Schemes

From the K_D obtained in this study, three of the techniques studied have been incorporated into a separation scheme for the isolation of parent PNAs from shale oils. The silicic acid column is used to separate the sample by polarity into aromatics, neutral N, and basic N fractions which are generated then separately sub-fractionated by gel filtration with LH-20/IPA. This gel filtration concentrates the PAH and N-PNA into sub-fractions while eliminating those components with ≤ 2 ring aromaticity. Finally, isomeric alkyl derivatives are eliminated by partition chromatography on LH-20/NMP.

The other techniques studied were omitted from this scheme, but their K_D data can still be used. Acid extraction, which did not extract all the basic N with equal efficiency, could be used to separate NH_2-PNA from heterocyclic N-PNA if the need were to arise. The sensitivity of PNA K_D to alkyl substitution in the NMP extraction was utilized for development of the partition column for the isolation of the parent PNAs. To summarize, the measurement of K_Ds for model PNA has provided the basis for selecting the most effective separation scheme for isolating selected PNAs from real samples.

REFERENCES

1. Willey, C.; Iwao, M.; Castle, R. N.; Lee, M. L. (1981): Anal. Chem., 53, pp. 400-407.
2. Jewell, D. M.; Albaugh, E. W.; Davis, B. E.; Ruberto, R. (1974): Ind. Eng. Chem., Fundanm., 13, pp. 278-282.
3. Farcasiu, M. (1977): Fuel, 50, pp. 9-14.
4. Grimmer, G.; Bohnke, H. (1976): Chromatographics, 9, pp. 30-39.
5. Snyder, L. R.; Buell, B. E. (1968): Anal. Chem., 40, pp. 1295-1302.
6. Snyder, L. R. (1968): Principles of Adsorption Chromatography, Dekker, N.Y.
7. Robbins, W. K. (1979): Solvent Extraction of Polynuclear Aromatic Hydrocarbons, in Polynuclear Aromatic Hydrocarbons, A. Bjorseth and A. J. Dennis, ed., Buttelle Press., pp. 841-861.
8. Robbins, W. K.; Blum, S. C. (1980): Analysis of Shale Oil Liquids for Polynuclear Aromatic Hydrocarbons and Their N-Heterocyclic Analogs, in Health Effects Investigation of Oil Shale Development, Ann Arbov Science, pp. 45-71.
9. Hertz, H. S., et al., (1980): Anal. Chem., 52, pp. 1650-1657.
10. Jones, A. R.; Guerin, M. R.; Clark, B. R. (1977): Anal. Chem., pp. 1766-1771.
11. Oelert, H. N. (1969): Z. Anal. Chem., 244, pp. 91-101.

THE EFFECT OF 5,6-BENZOFLAVONE ON THE MUTAGENICITY OF PAH

MICHAEL F SALAMONE, PATRICIA BELTZ AND MORRIS KATZ*
Microbiology Section, Ontario Ministry of the Environment,
P.O. Box 213, Resources Road, Rexdale, Ontario M9W 5L1 Canada
*Centre for Research on Environmental Quality, York University,
Downsview, Ontario M3J 1P3 Canada

INTRODUCTION

Polycyclic aromatic hydrocarbons (PAH) are products of the combustion of fossil fuels and organic matter, and their origin can be from anthropogenic as well as natural sources. PAH have long been a public health concern and certain PAH, in particular 7,12-dimethylbenzanthracene (DMBA) and benzo(a)-pyrene (BaP) have been linked to tumorigenesis and carcinogenesis. However, recent studies with agents such as 5,6- and 7,8-benzoflavone have shown that animals can be protected from some of the deleterious health aspects of PAH. Both pulmonary and skin tumors in mice, as well as mammary tumors in rats, resulting from exposure to DMBA, were shown to be inhibited by prior treatment with 5,6-benzoflavone (beta-napthoflavone, BNF) (1,2,3). Although the precise mechanism of this protection is unclear, it is thought to be related to the ability of BNF to influence the induction of specific enzymes, such as aryl hydrocarbon hydroxylase (AHH), necessary for the metabolism of the PAH. Normally, in vivo, these enzymes are responsible for the metabolism of the inert PAH into metabolites which can be easily eliminated from the body. Paradoxically, however, during this detoxification process, tumorigenically or carcinogenically active intermediate metabolites can be formed. These intermediates in turn are able to react with nucleophilic macromolecules which may initiate carcinogenesis. Under some circumstances, BNF is thought to protect by inhibiting the induction of these enzymes, while in other situations, BNF may induce significantly large quantities of these mixed function oxidases so as to bring about a rapid metabolism of both the parent PAH and its subsequent metabolic intermediates into polar, water soluble substances which can be safely excreted.

Since DMBA and BaP produce mutagenic endpoints as well as tumorigenic and carcinogenic endpoints, it is of interest to determine whether BNF also protects against the mutagenic aspects of these genotoxic agents. We report here the results of such studies.

MATERIALS AND METHODS

Animals

Two strains of mice were used in this study, the B5C3F1 hybrid mouse produced from parent strains C57Bl/B6 females and C3H/HeB males (Charles River Danada Inc.) and the inbred strain DBA-2J (Jackson Labs.). The strains differ in their genetic capability to induce AHH. The B6C3F1 mice are classified as homozygous responsive relative to their ability to induce the AHH enzyme system, while the DBA-2J mice are considered homozygous non-responsive for inducing AHH.

All experiments were carried out using female mice approximately 8-10 weeks old.

Chemicals

The test chemicals included 7,12-dimethylbenzanthracene (DMBA), benzo(a)pyrene (BaP), cyclophosphamide (CP) and ethyl methane sulfonate (EMS). These were used at doses of 25mg/kg (DMBA), 186 mg/kg (BaP), 43 mg/kg (CP) and 150 mg/kg (EMS) with mice pretreated with 5,6-benzoflavone (BNF) at 80 mg/kg/ treatment or corn oil at 0.1 ml. Chemicals were dissolved in dimethylsulfoxide (DMBA, BaP), corn oil (5,6-BNF) or saline (ethyl methane sulfonate, EMS; cyclophosphamide, CP) and administered by intraperitoneal injection.

Experimental Protocol

Animals were given two treatments of either 5,6-BNF or corn oil at 48 and 24 h prior to exposure to a single injection of the mutagenic agent. Animals were sampled at various intervals up to 120 h after treatment with the mutagen. Bone marrow slides were then prepared, stained and scored as described elsewhere (4,5). The data at each sample point represents the average of 3-6 mice. The variability in the number of animals per data point reflects losses due to mortality or bone marrow toxicity.

RESULTS AND DISCUSSION

The effect of BNF on the mutagenicity of DMBA and BaP in B6C3F1 mice is presented in Figures 1 and 2. Both DMBA and BaP produced fairly strong mutagenic responses, as indicated by micronuclei formation in polychromatic erythrocytes, in those animals pretreated only with the corn oil solvent. In case, a mutagenic response was first notices between 20 and

30 h after treatment. The maximum mutagenic response with DMBA occurred at about 60 h while that for BaP occurred at about 65 h. For those mice pretreated with BNF, the % micronucleated PCE value did not exceed that of the non-treated mice during the entire 78 h sampling span with BaP or the 96 h sample period for DMBA.

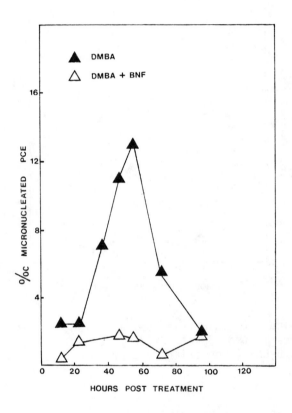

FIGURE 1. The mutagenic response, as a function of time, of a single treatment of DMBA in B6C3F1 mice, given a pretreatment of either 5,6-BNF (open triangles) or corn oil (closed triangles).

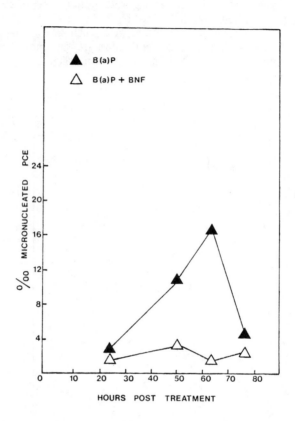

FIGURE 2. The mutagenic response, as a function of time, of a single treatment of BaP in B6C3F1 mice, given a pretreatment of either 5,6-BNF (open triangles) or corn oil (closed triangles).

These data clearly indicate that BNF is able to protect the hemopoietic centres in these mice from the clastogenic action of DMBA and BaP. Whether other tissues or organs can also be protected from PAH genotoxic damage by pretreatment with BNF needs to be investigated. That BNF protects against both mutagenic and carcinogenic endpoints is of interest as this may reflect a correlation between mutagenesis and carcinogenesis.

In order to check whether this protective nature of BNF is restricted to PAH, the effect of BNF on two non-PAH mutagens, cyclophosphamide (CP) and EMS, was also investigated. These two clastogens were selected as CP is a promutagen which is metabolized differently from that of PAH while EMS does not require metabolic activation to express its mutagenic effect.

Figure 3 depicts the clastogenic influence of CP on BNF pretreated and non-pretreated mice. A comparison of the time response curves suggests that no significant protection is provided by BNF pretreatment as both curves are fairly identical. Similarly, with EMS, (Figure 4), BNF did not protect the animals from the EMS-mediated genotoxic damage. These data imply that the mechanism of protection afforded by BNF against clastogenic agents may be similar to that hypothesized for its protective role with tumorigenic and carcinogenic compounds.

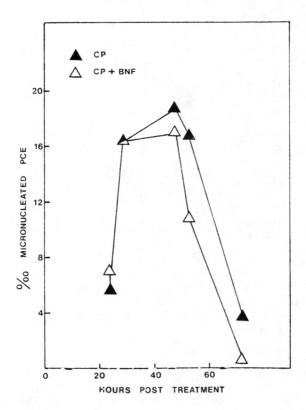

FIGURE 3. The mutagenic response, as a function of time, of a single treatment of CP in B6C3F1 mice, given a pretreatment of either 5,6-BNF (open triangles) or corn oil (closed triangles).

As stated, the mechanism of this protection is unclear, since BNF may act by inhibiting or inducing specific enzymes or it may function as a competitive inhibitor of the mutagenic agent.

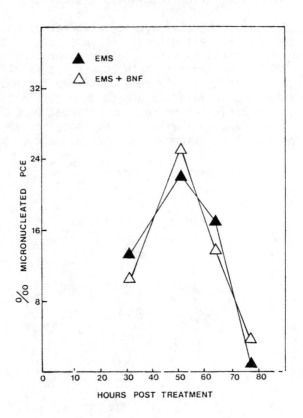

FIGURE 4. The mutagenic response, as a function of time, of a single treatment of EMS in B6C3F1 mice, given a pretreatment of either 5,6-BNF (open triangles) or corn oil (closed triangles).

If the protection mechanism involves the modulation of AHH enzymes, then a strain of mice which is non-inducible for AHH should not be protected by BNF. However, preliminary data with DBA-2J mice, which are unable to induce this enzyme (6), shows a degree of protection by BNF similar to that observed above with B6C3F1 mice (Figure 5). This suggests that the method of protection may not involve the modulation of AHH levels.

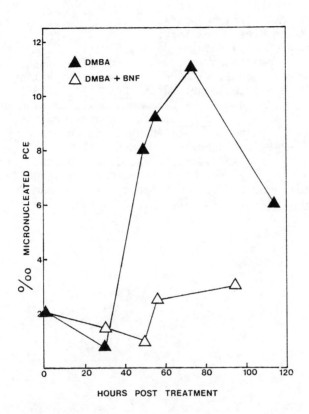

FIGURE 5. The mutagenic response, as a function of time, of a single treatment of DMBA in DBA-2J mice, given a pretreatment of either 5,6-BNF (open triangles) or corn oil (closed triangles).

REFERENCES

1. Wattenberg, L. W. and Leong, J. L. (1967): Inhibition of 9,10-dimethylbenzanthracene (DMBA) induced mammary tumorigenesis, Fed. Proc., 26:692.
2. Wattenberg, L. W. and Leong, J. L. (1968): Inhibition of the carcinogenic action of 7,12-dimethylbenzanthracene by beta-napthoflavone, Proc. Soc. Exp. Biol. Med., 128:940-943.
3. DiGiovanni, J., Slaga, T. J., Berry, D. L., and Juchau, M. R. (1980): Inhibitory effects of environmental chemicals on polycyclic aromatic hydrocarbon

carcinogenesis. In: *Carcinogenesis, Vol. 5: Modifiers of Chemical Carcinogenesis*, edited by T. J. Slaga, pp. 145-168, Raven Press, New York.
4. Salamone, M. F., Heddle, J. A., Stuart, E., and M. Katz. (1980): Towards an improved micronucleus test: Studies on 3 model agents, mitomycin C, cyclophosphamide and dimethylbenzanthracene, *Mutat. Res.*, 74:347-356.
5. Heddle, J. A. and M. F. Salamone. (1981): The micronucleus assay I: *in vivo*. In: *Short-Term Tests for Chemical Carcinogens*, edited by H. Stich and R. H. C. Sans, pp. 243-249, Springer-Verlag, New York.
6. Thomas, P. E., Kouri, R. E. and J. J. Hutton. (1972): The genetics of aryl hydrocarbon hydroxylase induction in mice: A single gene difference between C57Bl/6J and DBA-2J, *Biochem. Genet.* 6:157-162.

SYNTHESIS OF PAH CONTAINING CYCLOPENTA-FUSED RINGS

R. SANGAIAH*, A. GOLD*, G.E. TONEY*, S.H. TONEY**,
R. EASTERLING***, L.D. CLAXTON*** AND S. NESNOW***
*Environmental Sciences and Engineering 201 H, University of
North Carolina-Chapel Hill, N.C. 27514; **Northrop Services,
Inc.-Environmental Sciences, Research Triangle Park, N.C. 27709;
***Genetic Toxicology Division, U.S. Environmental Protection
Agency, Research Triangle Park, N.C. 27711.

INTRODUCTION

The mutagenic and carcinogenic activity of cyclopenta(cd)
pyrene (CPP;I) (ref. 1,2) has evoked considerable interest in
polycyclic aromatic hydrocarbons (PAH) containing cyclopenta-
fused rings (ref. 3-7). These PAH's are a potential human
health hazard because of their presence in pyrogenic airborne
particles. The cyclopenta-PAH are nonalternant PAH, a subset
of biologically active PAH whose metabolism has not been
extensively studied. Analysis of metabolic profiles of these
compounds may contribute insights into mechanism of oxidative
bioactivation and provide additional molecular descriptors for
structure-activity correlations currently under development.

Studies on the metabolism of CPP have indicated (ref. 2,8)
that the five-membered ring, which is highly olefinic and bears
large electron density in the highest occupied molecular orbital (HOMO), may be a preferred site of metabolism via epoxidation. Molecular orbital (M.O.) calculations on reactivity of
oxides on the 5-membered rings (ref. 9) show that they can open
to carbonium ions with large resonance stabilization energies
($\Delta E_{deloc}/\beta$), that are directly correlated with biological
activity (ref. 10), and therefore may be ultimate mutagenic/
carcinogenic metabolites. Among the largest stabilization
energies are those of carbonium ions generated from epoxides on
the cyclopenta ring of the benzaceanthrylene isomers (ref. 9).
In order to investigate the metabolism of these compounds and
assess their biological activity, we have embarked on the
synthesis of this series of PAH's. We report here the synthesis and preliminary bioassay results on benz(j)aceanthrylene
(cholanthrylene [CA]; II).

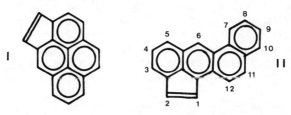

SYNTHESIS OF PAH CONTAINING CYCLOPENTA-FUSED RINGS

MATERIALS AND METHODS

Synthesis of benz(j)aceanthrylene

4(3-phenanthryl)butyric acid. This compound was synthesized from phenanthrene by a previously reported method (ref. 11,12).

8-oxo-8,9,10,11-tetrahydrobenzo(a)anthracene. The previously synthesized acid (10 g) was heated in polyphosphoric acid (350 g) with mechanical stirring for 3 h. at 80°. The reaction mixture was poured into a large excess of ice water and extracted with ether. The ether extract was washed sequentially with bicarbonate and water, dried over Na_2SO_4 and evaporated to yield the desired ketone which was recrystallized from ethanol (yield, 6.5 g, 70%).

Ethyl 2(8[8-hydroxy-8,9,10,11-tetrahydrobenzo(a)anthryl] acetate. This compound was synthesized by a Reformatski reaction as previously reported (ref. 12), employing activated zinc dust (ref. 13). (2.7 g, 87%).

8-benzanthrylacetic acid. A solution of the previously obtained ester (2.7 g) and p-toluene sulfonic acid (100 mg) in benzene (300 ml) was refluxed for 2 h. After washing 2X with brine, the benzene was evaporated to give a mixture of exo- and endocyclic dehydration products utilized directly in the following step.

The mixture of esters (2.4 g) and 2,3-dichloro-5,6-dicyanobenzoquinone (2.0 g) was refluxed in dry benzene for 4 h., the reaction mixture placed directly onto an alumina column and eluted with benzene: 30-60° petroleum ether (1:1) to yield the aromatized ester (2.1 g, 88%). Hydrolysis to the acid was accomplished by heating the ester (2 g.) in methanolic KOH (22 g. KOH, 10 ml H_2O, 40 ml NaOH) for 1 h. on a steam bath (1.6 g, 88%).

1-oxo-1,2-dihydrobenzo(j)aceanthrylene. The acid (1.6 g) was stirred in anhydrous HF for 15 h. at room temperature. After evaporation of the HF under a stream of nitrogen, the solid residue was dissolved in benzene:ether (1:1, 300 ml), washed with aqueous NaOH and water and dried. Evaporation of the solvent yielded crude ketone which was purified by chromatography on silica with benzene eluent (1.25 g, 88%).

1-hydroxy-1,2-dehydrobenz(j)aceanthrylene. Sodium borohydride (≈ 1 g.) was added to a solution of the ketone (1 g.)

in DMF (30 ml) and stirred at ambient temperature for 6 h. Addition of the reaction to 500 ml saturated brine followed by extraction of ether, 2X washing with water, drying over Na_2CO_3 and evaporation of the ether yielded the alcohol (0.95 g, 95%).

Benz(j)aceanthrylene (II). A solution of the alcohol (0.95 g.) and p-toluene sulfonic acid (10 mg) in benzene (400 ml) was refluxed for 45 min., washed 2X with water, dried and evaporated to yield the crude PAH. Some polymerization, as evidenced by dark red insoluble material, appeared to have occurred, even under the mild conditions of dehydration. Purification was accomplished by column chromatography on silica with hexane eluant yielding orange plates (0.55 g., 62%), m.p. 170-171°. Material for bioassay was further purified by thin layer chromatography on silica and by reverse phase HPLC (80:20:methanol:water to 100% methanol over 10 min.). By accurate mass measurement the elemental composition of the molecular ion corresponded to $C_{20}H_{12}$:expected, 252.0939; found; 252.0947. Fragmentation pattern of the mass spectrum obtained by electron impact $M^{+}_{\cdot}$, 252; major fragments at m/e 250$(M-H_2)^+$, 126 M^{++}, 125 $(M-H_2)^{++}$. UV-Vis spectrum (hexane), λ_{max} (ε x 10^{-4}):414 (0.92), 392(0.86), 379(0.71), 360(0.53), 313(2.38), 302(2.42), 278(2.70), 261(3.64), 259(3.62), 221(3.66).

The mutagenicity data on CA was obtained by methods similar to those of Ames et al. 1975 (ref. 14). CA was dissolved in dimethylsulfoxide and bioassayed at fourteen concentrations ranging from 0.03 µg to 100 µg per plate. Each dose was tested in triplicate with and without metabolic activation using strains TA98, TA100, TA1535, TA1537, and TA1538. The appropriate dose of CA was added to a 7% agar overlay tube containing 100 µl of Salmonella typhimurium. When the metabolic activation system was used, 500 µl was added directly before vortexing and pouring the tube. The metabolic activation system consisted of an Aroclor 1254 induced liver S-9 from Charles River CD rats. CA was bioassayed at twenty-four concentrations of S-9 to determine the optimal dose. The inverted plates were incubated in the dark at 37° for three days. At the end of this time plates were scored using an automatic colony counter. The assay was performed under yellow light to protect light sensitive compounds.

Metabolism of CA by Aroclor-1254 Induced Rat Liver Microsomes

The incubation mixture consisted of 60 µM CA, an NADPH generating system, 0.5 mg/ml microsomal protein and phosphate buffer pH 7.4. After 15 min., the incubation mixture was extracted with ethyl acetate-acetone (2:1), and the organic extracts evaporated and chromatographed by HPLC on a reverse-phase C-18 column. The chromatographic conditions were: Flow rate, 1.5 mℓ/min., solvent A, water; solvent B, methanol; initial concentration 80% B; final concentration 100%B; change, 2%/min. Cholanthrylene elutes at 20 min. under these conditions.

RESULTS AND DISCUSSION

The synthesis (Scheme I) was designed to take advantage of Friedel-Crafts acylations at the preferred C3 position of the phenanthrene starting material and C7 of the intermediate 8-benzanthryl acetic acid (III). A yield of 12% overall was achieved. The physico-chemical properties of the product are in accord with the CA structure. The correct elemental composition was established by an accurate mass measurement performed on the molecular ion of material. The mass spectrometric fragmentation pattern (with an electron impact source) was typical of PAH, having the molecular ion ($M^+_.$) as base peak and additional peaks corresponding to $(M-H_2)^+$, M^{++} and $(M-H_2)^{++}$ (ref. 15).

In the ^{1}H NMR (Fig. 1), the two bay region protons (H6 and H7) are shifted to very low field by ring current deshielding (ref. 16,17). Absence of interring proton coupling, characteristic of PAH, results in the appearance of H6 as a low field singlet. H7 is a doublet through coupling to H8 which appears as a doublet of doublets collapsing to a doublet, as expected, on decoupling from H7. The AX quartet arising from H1 and H2 on the etheno-bridge is readily apparent from its upfield position in accord with the high degree of localization of the C1-C2 double bond and the small coupling characteristic of cisoid olefinic protons. The assignment of the AX quartet was verified by collapse of H1 to a singlet on decoupling from H2. Similar quartets have been reported for the protons on the etheno bridges of acephenanthrylene derivatives (ref. 18) and CPP (ref. 3).

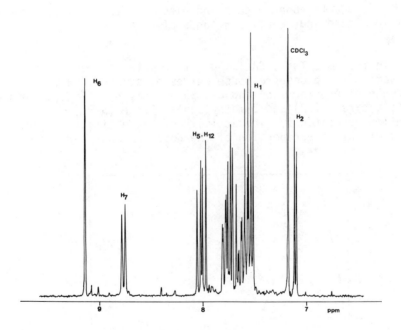

FIGURE 1. ^{1}H NMR (250 MHz, methylene chloride-d_2) of benz(j)aceanthrylene.

Like other known cyclopenta-PAH, CA is highly colored and does not fluoresce under black light. The extreme acid lability of CA (see Materials and Methods) is consistent with high, asymmetrically distributed electron density on the etheno bridge (Fig. 2) leading to facile protonation with formation of a benzylic carbonium ion.

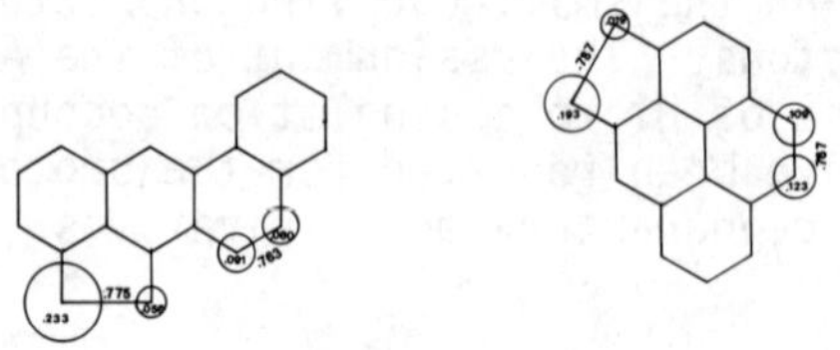

FIGURE 2. Huckel bond orders and electron densities for the most localized and reactive π-bonds of CA and CPP.

In the Ames test, CA is an active mutagen requiring S-9 activation. The dose-response curves of strain TA98 in the presence of Aroclor 1254-induced S9 are given in Fig. 3 for CA, benzo(a)pyrene (BP) and 3-methylcholanthrene (3-MC) under identical conditions. The activity of CA, estimated by the slope of the linear portion of the dose response curve, is

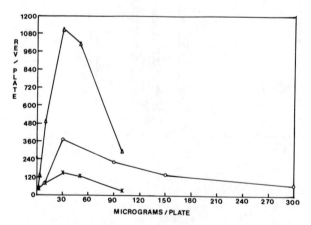

FIGURE 3. Comparison of dose-response curves of CA, Δ; BP, o; 3MC, x, in strain TA98 in the presence of 0.79 mg/plate S9 protein.

comparable to BP and greater than 3-MC. CA is less toxic than BP, resulting in greater activity at high concentrations. Activity of CA over the five tester strains (Table 1) is similar to that of BP and unlike that of CPP which is highly active in TA1538 but does not exhibit any enhancement of activity in the plasmid-bearing derivative strain TA98. CA is inactive in strain TA1535, which is sensitive to base pair modification, suggesting that CA, like other PAH, is a frame-shift mutagen.

TABLE 1.

COMPARISON OF CA ACTIVITY IN AMES STRAINS

	TA98	TA100	TA1537	TA1538	TA1535
Revertants/μg[a]	70	173	6	8	0

[a] Average slope of the dose response curve corrected for toxicity (ref. 19).

Activity in TA98 as a function of S-9 concentration for CA is maximal at a 5-fold lower concentration than for BP or 3MC (Fig. 4). The S-9 dependence of CA is similar to that observed for

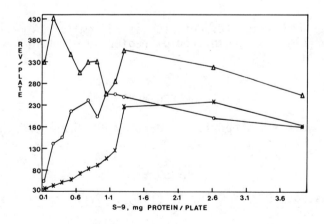

FIGURE 4. S-9 dependence of mutagenesis of CA, Δ; BP, o; and 3MC, x, at 10 μg/plate in strain TA98.

CPP relative to BP and may reflect one-step activation via epoxidation of the 5-membered ring.

Work on CPP has demonstrated that metabolism occurs primarily at the etheno bridge and to a lesser extent at the K-region (ref. 8,18). The electron densities and bond orders at the etheno bridge and K-region of CA are very similar to the corresponding sites of CPP (Fig. 2), suggesting that CA may also be metabolized at these two locations. The HPLC trace at 254 nm of the metabolite mixture from CA metabolism by Aroclor-1254 induced rat liver microsomes (Fig 5) shows that the two major metabolites elute in the dihydrodiol region. This result is consistent with the predicted metabolism at the etheno bridge and K-region. Future investigations will be concerned with identification of these metabolites and determination of the relative extent to which bay region and non-bay region metabolites are responsible for biological activity.

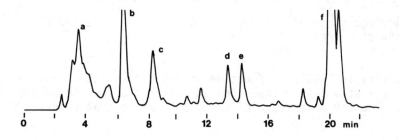

FIGURE 5. HPLC trace (UV detector at 254 nm) of CA metabolites: a, highly polar; b,c, dihydrodiol region; d,e, phenol region; f, unmetabolized CA.

ACKNOWLEDGEMENT

This work was supported in part by PHS Grant CA 28622.

REFERENCES

1. Eisenstadt, E. and Gold, A. (1978): Proc. Nat'l Acad. Sci. USA, 75:1667.
2. Gold, A., Newnow, S., Moore, M., Garland, H., Curtis, G., Howard, B., Graham, D. and Eisenstadt, E. (1980): Cancer Res. 40:4482.

3. Gold, A., Eisenstadt, E. and Schultz, J. (1978): Tetrahedron Lett., p. 4491.
4. Ittah, Y. and Jerina, D.M. (1978): Tetrahedron Lett., p. 4495.
5. Konieczny, M. and Harvey, R.G. (1979): J. Org. Chem. 41: 2158.
6. Ruehle, P.H., Fischer, D.L. and Wiley, J.C. (1979): J. Chem. Soc., Chem. Comm., p. 302.
7. Krishnan, S. and Hites, R. (1981): Anal. Chem. 53:342.
8. Gold, A. and Eisenstadt, E. (1980): Cancer Res. 40:3940.
9. Fu, P.P., Beland, F.A. and Yang, S.K. (1980): Carcinogenesis 1:725.
10. Jerina, D.M., Lehr, R.E., Yagi, H., Hernandez, O., Dansette. P.M., Wislocki, P.G., Wood, A.H., Chang, R.L., Levin, W. and Conney, A.H. (1976) in: In Vitro Metabolic Activation and Mutagenesis Testing, edited by F.J. DeSerres, J.R. Bend and R.M. Philpott, p. 159 ff, Elsevier/North Holland Biomedical Press, New York.
11. Backmann, W.E. and Bradbury, J.T. (1937): J. Org. Chem. 2: 175.
12. Bachmann, W.E. (1938): J. Org. Chem. 3:434.
13. Frankenfeld, J.W. and Werner, J.J. (1981): J. Org. Chem. 34:3689.
14. Ames, B.N., McCann, J. and Yamasaki, E. (1975): Mutat. Res. 31:347.
15. Benyon, J.H. (1968): The Mass Spectra of Organic Molecules. Elsevier, New York, 129 pp.
16. Poole, J.A., Schneider, W.G. and Bernstein, H.J. (1959): High Resolution Nuclear Magnetic Resonance. McGraw-Hill, New York, 180 pp.
17. Bartle, K.D., Jones, D.W. and Pearson, J.E. (1967): J. Mol. Spectry. 24:330.
18. Eisenstadt, E., Shpizner, B. and Gold, A. (1981): Biochem Biophys. Res. Commun. 100:965.
19. Stead, A.G., Hasselblad, V., Creason, J.P. and Claxton, L. (1981): Mutat. Res. 85:13.

MUTAGENESIS TESTING OF GASOLINE ENGINE OILS

C. A. SCHREINER AND C. R. MACKERER
Mobile Environmental and Health Sciences Laboratory,
P. O. Box 1029, Princeton, New Jersey 08540

INTRODUCTION

Used gasoline engine crankcase oil has been reported to contain significant amounts of polycyclic aromatic hydrocarbons (4). Many of these compounds are known mutagens and, therefore, it is not surprising that used motor oils are mutagenic in the Ames Salmonella/mammalian microsome bioassay (3,4). The mutagenic compounds present in used oils are believed to be generated during the combustion process and to accumulate with time.

In the published proceedings of the 1979 Battelle symposium, Peake and Parker (3) reported that not only used motor oils but also several commercially available unused oils were mutagenic in the Ames assay in the absence of metabolic activation. This activity was lost with the addition of a rat liver metabolizing mixture. Among these oils was the 100% synthetic based engine lubricant, Mobil 1.

After the Peake and Parker report appeared, we tested Mobil 1 in the Ames system, together with composite samples of both new and used oils provided by the American Petroleum Institute (API). In addition, testing was conducted on a retained sample of the original unused Mobil 1 5W-20 oil evaluated by Drs. Peake and Parker. The mutagenicity results for the API composite oil samples have been compared with data from mouse skin painting studies currently in progress under the sponsorship of API.

The strong mutagenic activity of used motor oil reported by Peake and Parker and Payne et al. (4) was confirmed. However, no evidence of mutagenicity with extracts of Mobil 1 or the composite new oil sample with or without metabolic activation was observed, a finding also corroborative of Payne et al. who observed no mutagenic activity in five commercial unused oils.

MATERIALS AND METHODS

Oil Samples

The Mobil 1 5W-20 sample first tested (Code # RN 1669AAB-2) in this study was representative of the commercial product available in 1979, at the time of the previously reported study (3). The second Mobil 1 5W-20 sample was that tested by Peake and Parker in 1979 and 60 ml were generously provided by K. Parker; this sample is referred to as the Parker sample. The unused and used oils provided by the American Petroleum Institute were composites of 15 commercially available 10W-40 motor oils. The composite used oil was produced by operating gasoline engines with each of the 15 component oils for 5000 miles and then blending equally to form a composite sample.

Extraction of Oil Samples

Samples (20 ml) of each of the three oils tested were extracted with 10 ml of dimethyl sulfoxide (DMSO) (2). The oil-DMSO mixtures were vigorously shaken, allowed to stand at room temperature for 30 min., and then centrifuged at 350 x g for 1 minute to separate the phases. The lower DMSO phase was drawn off with a pipet and either used immediately in the mutagenesis assay or after storage in the dark at 4^oC for periods not exceeding two weeks.

Ames Mutagenesis Assay

The method of Ames et al. (1) was strictly adhered to in all assays. The five standard Salmonella strains, TA98, TA100, TA1535, TA1537 and TA1538, were used in testing three of the DMSO extracts both with and without metabolic activation. Aroclor 1254-induced rat liver 9000 x g supernatant (S9) was purchased from Litton Bionetics, Kensington, MD. The Mobil 1 extract was evaluated at eight doses in strains TA98 and TA100 in the preliminary toxicity phase of the assay, and at four doses in the full mutagenesis test. The Parker sample was tested in strains TA98 and TA100 at the doses employed by Peake and Parker (2). The two composite oils were tested directly in the mutagenesis assay at eight doses in all five strains. Doses were selected to span the complete range of toxicity.

Spontaneous reversion rates and negative solvent control rates were monitored in each test. Positive control compounds not requiring metabolic activation and their concentrations were 2-nitrofluorene (2-NF; 20 µg/plate) for TA98 and TA1538, 9-aminoacridine (9-AA; 2 µg/plate) for TA1537 and either 4-nitroquinoline-$\underline{N}$-oxide (4-NQNO; 10 µg/plate) or $\underline{N}$-methyl-$\underline{N}$'-

nitro-N-nitrosoguanidine (MNNG; 2 μg/plate) for strains TA100 and TA 1535. 2-Aminoanthracene (2-AA; 2 μg/plate) was the positive control with activation for all strains except TA1537.

In Vivo Carcinogenesis Studies

The ongoing API-sponsored mouse skin painting studies are comprised of 50 male C_3H/HEJ mice per treatment group. Animals are being treated with 50 mg of undiluted used or unused composite motor oil twice per week for 104 weeks or until the appearance of the first tumor. All mice will be observed for the remainder of their lifetime. The used oil study was in the 104th week and the unused oil study was in the 69th week as of October 1, 1981. Negative control groups include two concurrent groups of shaved mice and one group of mice treated with 50 μl toluene on the same schedule as the experimental groups. Positive control animals are painted with 0.05 % or 0.15% benzo(a)pyrene in a toluene vehicle twice weekly.

RESULTS AND DISCUSSION

Of the three oil samples tested, only the used composite had detectable mutagenic activity. Table 1 shows the revertant values obtained with the used composite oil in the five standard Ames strains with and without metabolic activation.

TABLE 1

AN AMES SALMONELLA/MAMMALIAN MICROSOME MUTAGENESIS ASSAY OF API USED MOTOR OIL COMPOSITE FROM GASOLINE ENGINES

Dosage Groups	Sal. Strains	TA98		TA100		TA1535		TA1537		TA1538	
	Activation	-S9	+S9	-S9	+S9	-S9	+S9	-S9	+S9	-S9	+S9
	Cells Seeded	1.3×10^8		1.0×10^8		1.0×10^8		1.1×10^8		1.2×10^8	
0.0 μl (50.0 μl DMSO)		22	36	150	120	24	14	6	7	15	30
0.22 μl DMSO Extract of Used Motor Oil Composite		35	86	181	205	24	16	10	17	30	76
1.10 μl DMSO Extract of Used Motor Oil Composite		62	249	207	387	30	23	18	76	42	279
2.20 μl DMSO Extract of Used Motor Oil Composite		71	346	222	468	24	38	26	114	64	395
11.03 μl DMSO Extract of Used Motor Oil Composite		180	751	406	839	30	45	73	196	165	535
22.05 μl DMSO Extract of Used Motor Oil Composite		255	968	541	1103	20	29	88	272	259	627
50.0 μl DMSO Extract of Used Motor Oil Composite		358	1170	718	1322	14	25	105	268	315	755
100.0 μl DMSO Extract of Used Motor Oil Composite		359	1543	924	1400	T	14	T	203	270	712
200.0 μl DMSO Extract of Used Motor Oil Composite		381	1490	T	1380	0	10	0	65	T	161
2.0 μg 2-aminoanthracene		32	3184	172	3197	24	302	N.A.	N.A.	13	1100
10.0 μg 4-Nitroquinoline-N-oxide		N.A.	N.A.	2839	N.A.	N.A.	N.A.	N.A.	N.A.	N.A.	N.A.
20.0 μg 2-nitrofluorene		987	N.A.	N.A.	N.A.	N.A.	N.A.	N.A.	N.A.	1576	N.A.
9-aminoacridine		N.A.	N.A.	N.A.	N.A.	N.A.	N.A.	114	N.A.	N.A.	N.A.
2.0 μg N-methyl-N'-nitro-N-nitrosoguanidine		N.A.	N.A.	N.A.	N.A.	137	N.A.	N.A.	N.A.	N.A.	N.A.

Comments: N.A. - Not applicable due to strain specificity; T - Microcolonies

Considering that the DMSO extraction procedure concentrated the mutagens in the oil by at most two-fold, the mutagenic response, particularly with metabolic activation, is fairly strong. Toxicity was observed as either a reduction in revertants or formation of microcolonies at the highest dose (200 µl/plate). The observed mutagenic activity of used oil is entirely consistent with previous reports (3,4), and demonstrates the appropriateness of the solvent extraction procedure and dosing schedule used.

Tables 2 and 3 present the analogous data obtained in the assays of Mobil 1 and the unused composite oil sample, respectively. Table 4 presents the results of testing with the Parker sample in strains TA98 and TA100. Figures 1 and 2 graphically compare the mutagenic activity of all four motor oils in strains TA98 and TA100. No increase in revertant frequency compared to concurrent negative controls was observed for the unused composite motor oil or for Mobil 1 5W-20 with or without metabolic activation. Unused Mobil 1 5W-30, a more current formulation of the synthetic-based lubricant was also tested and no mutagenic activity was observed.

TABLE 2

AN AMES SALMONELLA/MAMMALIAN MICROSOME MUTAGENESIS ASSAY OF MOBIL 1 5W-20 MOTOR OIL

Dosage Groups	Sal. Strains	\multicolumn{10}{c}{Average No. of Revertant Colonies per Plate}
		TA98		TA100		TA1535		TA1537		TA1538	
	Activation	-S9	+S9	-S9	+S9	-S9	+S9	-S9	+S9	-S9	+S9
	Cells Seeded	1.3×10^8		1.3×10^8		1.6×10^8		1.4×10^8		1.2×10^8	
0.0 µl (DMSO)		18	30	115	126	17	14	6	7	8	24
0.22 µl Mobil 1 extract		17[b]	32[b]	126[b]	124[b]	-	-	-	-	-	-
1.10 µl Mobil 1 extract		20[b]	32[b]	126[b]	105[b]	-	-	-	-	-	-
2.20 µl Mobil 1 extract		19	29	124	124	17	14	6	6	7	25
11.03 µl Mobil 1 extract		16	32	129	153	15	10	7	6	6	11
22.05 µl Mobil 1 extract		24	30	148	151	17	10	6	9	9	10
40.0 µl Mobil 1 extract		27	34	112	166	18	12	7	7	7	11
50.0 µl Mobil 1 extract		14[b]	29[b]	142[b]	130[b]	-	-	-	-	..	-
100.0 µl Mobil 1 extract		21[b]	25[b]	83[b]	109[b]	-	-	-	-	-	-
200.0 µl Mobil 1 extract		7[b]	22[b]	T[b]	T[b]	-	-	-	-	-	-
2.0 µg 2-aminoanthracene		25	2158	138	2203	16	306	N.A.	N.A.	8	1988
2.0 µg N-methyl-N'-nitro-N-nitrosoguanidine		N.A.	N.A.	1231	N.A.	472	N.A.	N.A.	N.A.	N.A.	N.A.
20.0 µg 2-nitrofluorene		1305	N.A.	N.A.	N.A.	N.A.	N.A.	N.A.	N.A.	570	N.A.
25.0 µg 9-aminoacridine		N.A.	N.A.	N.A.	N.A.	N.A.	N.A.	21	N.A.	N.A.	N.A.

Comments: N.A. - Not applicable due to strain specificity; b - Data from preliminary toxicity/mutagenicity test in TA98, TA100 only; T - Microcolonies

The lack of mutagenicity in the unused oil extracts is in agreement with the findings of Payne et al., who tested used oil

MUTAGENESIS OF ENGINE OILS

and five commercial brands of unused oil in Ames strain TA98 and found activity only in the used oil extract. The Parker sample was toxic at doses of 20 μl and 100 μl per plate in TA98 and at 100 μl per plate in TA100 in the absence of S-9; however, no increase in revertants such as previously reported was observed in either strain.

TABLE 3

AN AMES SALMONELLA/MAMMALIAN MICROSOME MUTAGENESIS ASSAY OF API UNUSED OIL COMPOSITE

Dosage Groups	Sal. Strains	\multicolumn{10}{c	}{Average No. of Revertant Colonies per Plate}
		TA98		TA100		TA1535		TA1537		TA1538	
	Activation	-S9	+S9	-S9	+S9	-S9	+S9	-S9	+S9	-S9	+S9
	Cells Seeded	1.4×10^8		1.2×10^8		1.0×10^8		1.1×10^8		1.3×10^8	
0.0 μl (DMSO)		23	37	149	107	18	14	5	7	14	29
0.22 μl DMSO Extract of Unused Motor Oil Composite		14	31	152	121	32	11	6	5	12	28
1.10 μl DMSO Extract of Unused Motor Oil Composite		25	37	150	116	29	10	7	5	11	18
2.20 μl DMSO Extract of Unused Motor Oil Composite		24	32	142	113	29	8	4	7	13	20
11.03 μl DMSO Extract of Unused Motor Oil Composite		20	27	149	96	29	8	9	8	11	17
22.05 μl DMSO Extract of Unused Motor Oil Composite		20	24	140	87	25	12	5	7	9	17
50.0 μl DMSO Extract of Unused Motor Oil Composite		19	19	113	79	22	15	5	5	8	7
100.0 μl DMSO Extract of Unused Motor Oil Composite		15	17	T	73	23	12	3	4	T	T
200.0 μl DMSO Extract of Unused Motor Oil Composite		15	19	0	T	T	T	T	T	T	T
2.0 μg 2-aminoanthracene		32	2531	157	2676	24	302	N.A.	N.A.	21	397
10.0 μg 4-Nitroquinoline-N-oxide		N.A.	N.A.	2676	N.A.	N.A.	N.A.	N.A.	N.A.	N.A.	N.A.
20.0 μg 2-nitrofluorene		1758	N.A.	N.A.	N.A.	N.A.	N.A.	N.A.	N.A.	2474	N.A.
9-aminoacridine		N.A.	N.A.	N.A.	N.A.	N.A.	N.A.	114	N.A.	N.A.	N.A.

Comments: N.A. - Not applicable due to strain specificity; T - Microcolonies

TABLE 4

AN AMES SALMONELLA/MAMMALIAN MICROSOME MUTAGENESIS ASSAY OF PARKER SAMPLE OF MOBIL 1

Dosage Groups	Sal. Strains	\multicolumn{4}{c	}{Average Number Revertant Colonies per plate}
		TA98		TA100	
	Activation	-S9	+S9	-S9	+S9
	Cells Seeded	1.2×10^8		1.3×10^8	
0.0 μl (DMSO)		21	26	129	121
5.0 μl Parker Sample Extract		21	24	124	133
7.5 μl Parker Sample Extract		25	26	125	131
10.0 μl Parker Sample Extract		30	27	123	131
20.0 μl Parker Sample Extract		T	27	134	134
100.0 μl Parker Sample Extract		T	26	109	132
2.0 μg 2-aminoanthracene		26	1602	184	430
2.0 μg N-methyl-N'-nitro-N-nitrosoguanidine		N.A.	N.A.	597	N.A.
20.0 μg 2-nitrofluorene		1545	N.A.	N.A.	N.A.

Comments: T - Microcolonies
N.A. - Not applicable due to strain specificity

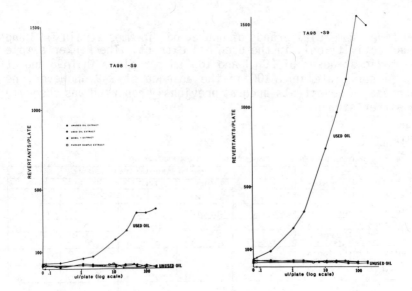

FIGURE 1. TA98: ■Unused oil extract; ◆Used oil extract; ●Mobil 1 extract; □Parker sample extract

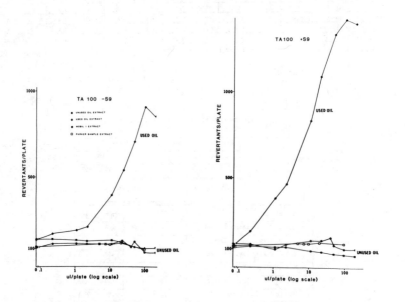

FIGURE 2. TA100: ■Unused oil extract; ◆Used oil extract; ●Mobil 1 extract; □Parker sample extract

Toxicity was produced at the highest doses with all of the extracts providing evidence of interaction of the test material with the bacteria; also, that bacteriotoxic components were present to approximately the same extent in both the unused and the used oils. The latter observation is particularly important since it confirms the validity of the dosing schedule used in testing the unused oils.

These findings obtained in the Ames test correlate well with in vivo data obtained from an ongoing API mouse skin painting study (Table 5). When samples of the same two used and unused composite oil samples were administered to mice in controlled studies, only the used oil composite induced tumors with the first appearing after 29 weeks of treatment. Of the sixteen tumor-bearing mice observed after 106 weeks, seven bear highly malignant tumors. In contrast, after 69 weeks of administration, the unused oil has produced no skin tumors.

TABLE 5

API USED AND UNUSED COMPOSITE MOTOR OILS - MUTAGENIC AND CARCINOGENIC DATA

Dose	Ames Mutagenesis Test 0.22-200.0 µl/plate	Mouse Skin Painting Test 50 mg/male mouse; 2x/wk
Unused Motor Oil	Negative ± S9	Negative: No tumors in 69 weeks
Used Motor Oil	Positive ± S9	Positive: 16 tumor bearing mice/50 mice in 106 weeks

Taken together, these data strongly indicate that the mutagenic and carcinogenic components known to be present in used engine oils are produced during engine operation and probably as a consequence of combustion. Although the unused refined composite sample is a complex mixture of aliphatic and aromatic hydrocarbons, apparently none of these compounds are activated either in the in vitro Salmonella/mammalian microsome system or in the more metabolically complex mouse skin carcinogenesis assay. It is even less likely that the totally synthetic-based and therefore chemically defined product, Mobil 1, would possess inherent mutagenic activity.

Further studies are currently underway to compare the chemical constituents of the mutagenic and carcinogenic used motor oil with the inactive unused oil and to explore the chemical processes involved in the formation of mutagenic components. Although it is apparent that the mutagenic polycyclic aromatic hydrocarbons contribute significantly to the mutagenicity of the used oils, it is equally clear that other mutagens, not requiring metabolic activation, are also present. Identifying and characterizing the biological activities of these direct-acting mutagens will also constitute a principal part of these continuing studies.

ACKNOWLEDGEMENTS

The authors express their appreciation to the American Petroleum Institute for its cooperation in providing composite oil samples and data from the ongoing mouse skin painting studies.

REFERENCES

1. Ames, B. N., Mc Cann, J. and Yamasaki, E. (1975): Methods for detecting carcinogens and mutagens with the Salmonella/mammalian-microsome mutagenicity test. Mutat. Res. 31:347-363.
2. Parker, K., personal communication.
3. Peake, E. and Parker, K. (1980): Polynuclear aromatic hydrocarbons and the mutagenicity of used crankcase oils. In: Polynuclear Aromatic Hydrocarbons Chemistry and Biological Effects, A. Bjørseth and A. J. Dennis, Eds., pp. 1025-1039, Batelle Press, Columbus, Ohio.
4. Payne, J. F., Martins, I. and Rabimtula, A. (1978): Crankcase oils: are they a major mutagenic burden in the aquatic environment? Science 200: 329-330.

THE EFFECT OF TEMPERATURE ON THE ASSOCIATION OF POM WITH AIRBORNE PARTICLES

M.R. SCHURE* AND D.F.S. NATUSCH**
Department of Chemistry, Colorado State University, Fort Collins, Colorado 80523. *present address: Department of Chemistry, University of Utah, Salt Lake City, Utah 84112.
**Director, Liquid Fuels Trust Board, P.O. Box 17, Lambton Quay, Wellington, NEW ZEALAND.

INTRODUCTION

It is now well established that polycyclic aromatic hydrocarbons (PAHs) are formed during the combustion of carbonaceous fuels (1). These compounds are formed initially as vapors (1,2), yet, with the possible exception of anthracene and phenanthrene, are invariably found to be particle associated in ambient air.

Several possibilities exist for the gas-to-particle conversion of PAH but it is the postulate of this paper that the process of primary importance involves adsorption onto coentrained particulate matter during or following emission. Such adsorption is, of course, highly dependent on temperature and in the following sections we present both theoretical and experimental evidence for temperature dependent adsorption. The theoretical basis employed is quite general but is evaluated for the specific case of selected PAHs adsorbing onto fly ash in the stack system of a coal-fired power plant. On the basis of the theoretical treatment employed, together with experimentally determined parameters, a model is developed which is capable of predicting the rate and extent of PAH adsorption onto coal fly ash continuously throughout the stack system of a coal-fired power plant.

THEORETICAL

Consider a vapor with concentration [P] and a surface with concentration of surface sites [S]. The vapor can adsorb on the surface forming a surface bound state [P·S]. This can be written as:

$$[P] + [S] \underset{\text{desorption}}{\overset{\text{adsorption}}{\rightleftharpoons}} [P \cdot S] \qquad \{1\}$$

If the adsorption rate is associated with a rate constant k_1, and the desorption rate is associated with a rate constant k_{-1}, a rate equation can be written for the case where the vapor concentration is constant and forms a monolayer on the surface:

$$\frac{d[P \cdot S]}{dt} = k_1[P][S](1 - \theta) - k_{-1}[P \cdot S], \quad \theta = \frac{[P \cdot S]}{[S]}. \quad \{2\}$$

Under steady state conditions, Equation {2} can be integrated easily to give:

$$\theta = \frac{[P \cdot S]}{[S]} = \frac{b[P]}{1 + b[P]}. \quad \{3\}$$

Equation {3} is commonly known as Langmuir's isotherm, which relates the concentration of vapor with the fraction of the surface (θ) that is covered by adsorbed molecules at equilibrium. The constant b is equal to k_1/k_{-1}. This formulation is inadequate for studying the fate of a vapor in the environment or in a power plant since the vapor is not at constant concentration but is depleted upon adsorption.

Equation {2} can be modified to include the depletion of vapor. The new equation is referred to as the vapor-depletion Langmuir equation and can be written in differential form as:

$$\frac{d[P \cdot S]}{dt} = k_1([P] - [P \cdot S])[S](1 - \theta) - k_{-1}[P \cdot S]. \quad \{4\}$$

The corresponding equilibrium solution of Equation {4} is obtainable in closed form.

To evaluate the temperature dependence of adsorption, the rate constants must be explicitly defined in terms of physical constants. Using absolute rate theory (4,5) it can be shown (6) that:

$$k_1 = \sqrt{\frac{RT}{2\pi M}} \; \sigma \; \exp[-\varepsilon_a/RT]$$

and $$k_{-1} = \frac{8\pi^3(kT)^2}{h^3} \sqrt{I_x I_y} \; \frac{\sigma_{s,2D}}{\sigma_{s,3D}} \; \exp[-\varepsilon_d/RT] \quad ,$$

where R, k, T, and h are the gas constant, Boltzmann's constant, the absolute temperature, and Planck's constant, respectively. The quantities ε_a and ε_d are the activation energies of adsorption and desorption. The quantities σ, $\sigma_{s,2D}$, $\sigma_{s,3D}$, I_x, and I_y are the surface area of an adsorbed molecule, the two dimensional and three dimensional symmetry numbers (see ref. 4) and the x and y axes principal moments of inertia of the molecule. The PAHs are assumed to sit flat on the surface and the z axis is perpendicular to the surface.

Other adsorption models of higher complexity can be modified to include the vapor depletion effect. These include the Temkin and Freundlich adsorption models which postulate a distribution of surface site energies (the Langmuir model assumes all surface sites have equal energy), and the BET and Halsey models which postulate multilayer adsorption. These models are reviewed in references 6, 7, and 8.

In order to evaluate the sorption behavior of a vapor phase molecule it is necessary to determine the temperature dependence of two quantities. These are the mole fraction adsorbed, χ, and the half time, $t_{\frac{1}{2}}$, to establish equilibrium. The quantity χ is given as:

$$\chi = \frac{[P \cdot S]}{[P]} \quad , \quad \{5\}$$

where [P] is the initial concentration of PAH in the vapor phase. The quantity $t_{\frac{1}{2}}$ can be expressed in closed form solution and is given in reference 6. In the next section the various quantities which are used for the evaluation of Equation {5} will be discussed.

PARAMETERS

To evaluate Equation {5} as a function of temperature, the quantities k_1, k_{-1}, [P], and [S] must be chosen.

The physical constants needed to evaluate the rate constants, with the exception of ε_a and ε_d, are available in the literature. The quantity ε_a, the activation energy for adsorption, is set to zero for physical adsorption involving van der Waal's forces (9). This is not the case when chemical bonds are broken, as for chemisorption, so for the purpose of illustration the activation energies examined here are 0, 5, and 10 kcals per mole.

The activation energy for desorption is usually higher than the thermodynamic heat of vaporization of a pure substance (9). Pyrene, for which all calculations herein are performed, has a heat of vaporization of 24 kcals per mole; the activation energies for desorption chosen here for illustration, therefore, are 25, 30, and 35 kcals per mole.

The amounts of vapor and surface sites present per unit volume in a coal-fired power plant can be estimated from measurements of PAH concentrations measured in several coal-fired power plants as published by Hangebrauk et al. (10). For the purpose of illustration the pyrene vapor phase concentrations considered are 10^{-2}, 10^{-1}, 1, and 10 ppb by volume at 25°C. The number of surface sites per unit volume is obtainable from

$$[S] = S_m w_p \cdot 10^8 / \sigma$$

where [S] is the number of surface sites per cc, S_m is the mass specific surface area, w_p is the particle loading (μg per cubic meter) and σ is the surface area of an adsorbed molecule. S_m is chosen to be two square meters per gram on the basis of our own measurements. w_p is chosen for illustration to be 10^5, 10^6, and 10^7 μg per cubic meter. For pyrene σ is 90.95 Å.

RESULTS

The dependence of χ on temperature is illustrated in Figures 1a thru 1d. It can be seen from these plots that the

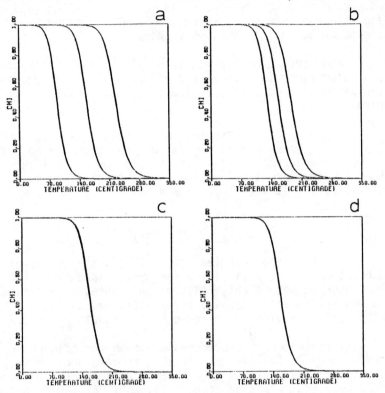

FIGURE 1a-d. Evaluation of χ, the fraction of vapor adsorbed, for the vapor-depletive Langmuir equation. The numbers in brackets are used for all calculations except when one quantity is varied.

a. ε_d = 25, [30], 35 kcals per mole
b. w_p = 10^5, [10^6], 10^7 µg per m^3
c. [P] = 10^{-2}, 10^{-1}, [1], 10 ppb
d. ε_a = [0], ε_d = 30 kcals per mole
   ε_a = 5, ε_d = 35 kcals per mole
   ε_a = 10, ε_a = 40 kcals per mole

energy of desorption is the most sensitive parameter in determining the temperature range where adsorption becomes significant. Thus, a 15 percent change in ε_d shifts the temperature at which adsorption becomes important by 70°C. This is contrasted to a 10-fold change in particle loading (and surface site concentration) which shifts the temperature at which adsorption becomes important by about 25°C.

As determined from Figure 1c, varying the adsorbate concentration seems to have little effect on the temperature dependence of adsorption, at least for the conditions illustrated here.

By varying ε_a but keeping the quantity $(\varepsilon_d - \varepsilon_a)$ constant it can be seen from Figure 1d that there is no effect on the temperature dependence of adsorption at equilibrium. There is, however, a very large effect on the adsorption half-time, $t_{\frac{1}{2}}$, as seen from Figure 2d.

ADSORPTION ENERGETICS

The adsorption energies (specifically the energy of desorption) are very sensitive parameters in determining the temperature range at which adsorption becomes significant. The adsorption energies have been determined by curve fitting experimental results to the theoretical equation for χ.

The experimental procedures employed for adsorption studies were as follows. A variable temperature spectrophotometric cell was filled with either a ferromagnetic, mineral, or carbonaceous fraction of coal fly ash. At 400°C, and after outgassing the ash under vacuum, known amounts of pyrene, benzo(a) pyrene, or fluoranthene dissolved in methylene chloride were injected into the cell. The temperature was then lowered successively and the vapor phase concentration determined by UV absorption spectrophotometry. The fraction of the vapor adsorbed is thus determined as a function of temperature.

Figure 3 shows the result of the adsorption of pyrene onto the various fractions of coal fly ash. The Temkin model is used to curve fit the data as it gives the best least-squares fit.

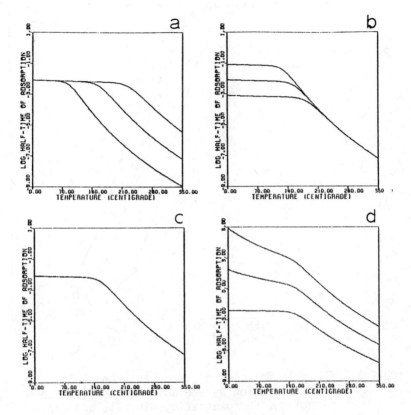

FIGURE 2a-d. Evaluation of $t_{1/2}$, the half-time of adsorption, for the vapor-depletive Langmuir equation. The quantity log $t_{1/2}$ is shown. The numbers in brackets are used for all calculations except when one quantity is varied.

a. ε_d = 25, [30], 35 kcals per mole

b. w_p = 10^5, [10^6], 10^7 µg per m^3

c. [P] = 10^{-2}, 10^{-1}, [1], 10 ppb

d. ε_a = 10, ε_d = 40 kcal per mole
ε_a = 5, ε_d = 35 kcal per mole
ε_a = [0], ε_d = 30 kcal per mole

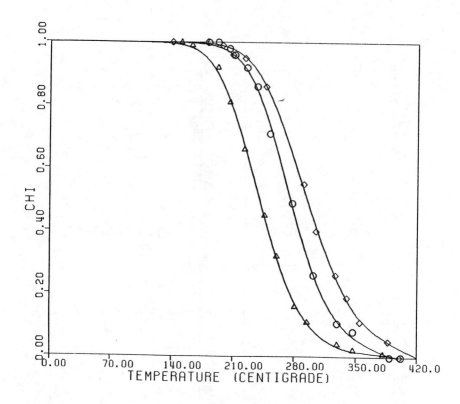

FIGURE 3. The mole fraction of vapor adsorbed versus temperature for pyrene adsorbed on (left to right) the ferromagnetic, mineral, and carbonaceous fractions of coal fly ash. The lines are the least-squares fit of the vapor-depletion Temkin equation.

ONE-DIMENSIONAL MODEL OF POM ADSORPTION

IN A COAL-FIRED POWER PLANT

Availability of sorption energies makes it possible to model the sorption behavior of PAH in a coal-fired power plant. For this purpose a one-dimensional model based on adsorption theory has been developed to predict the location at which the vapor-surface association becomes important.

The time-temperature profile is used as the integration path of ixj differential equations which are representative of the adsorption of adsorbate i on surface j. Such calculations have been carried out for pyrene, fluoranthene, and benzo(a)pyrene adsorbing in the stack system of a power plant onto mineral, ferromagnetic, and carbonaceous components of coal fly ash. The results are illustrated in Figure 4 for pyrene but are similar for fluoranthene and benzo(a)pyrene. The input parameters for this calculation are summarized in Table 1.

TABLE 1

Parameters for Figure 4.

$[P]_{PYRENE}$ = 1 ppb

$[P]_{BAP}$ = 0.1 ppb

$[P]_{FLUORANTHENE}$ = 0.1 ppb

w_p = 10^6 µg per cubic meter

fraction of mineral ash = 0.96

fraction of ferromagnetic ash = 0.03

surface area of mineral ash = 2.09 m^2 per gram

surface area of magnetic ash = 0.25 m^2 per gram

surface area of carbonaceous ash = 63 m^2 per gram

fraction of carbonaceous ash = 0.01

FIGURE 4 (next page). One dimensional model for pyrene adsorption in a coal-fired power plant. The chi plot (top to bottom) is the total fraction of pyrene vapor adsorbed, the fraction of vapor adsorbed on carbonaceous ash, and fraction of vapor adsorbed on mineral ash. The fraction of pyrene vapor adsorbed on ferromagnetic particles is too small to be observed. The theta plot has the same ordering as chi, with respect to surface type. The term P/P_0 is the fractional saturation pressure ratio.

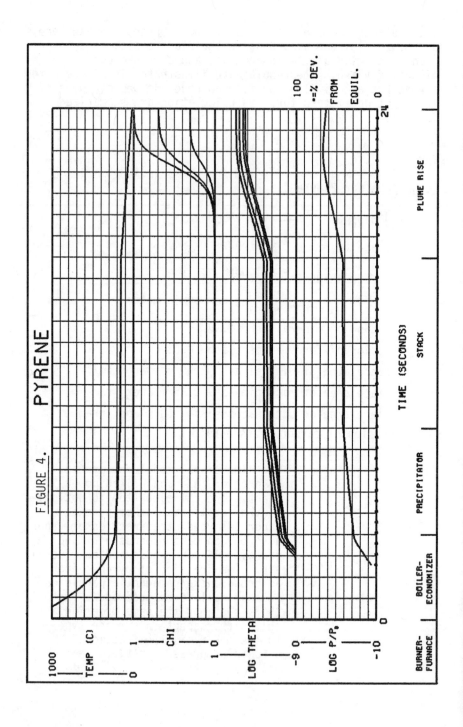
FIGURE 4. PYRENE

DISCUSSION

Figure 4 predicts that in the higher temperatures encountered in the stack system of a coal-fired power plant the fraction of vapor adsorbed is very low but vapor adsorption occurs rapidly as the vapor particle mixture cools. This is consistent with the findings of Tomkins (11) who collected coal fly ash samples concurrently both inside the stack (temp ~290°C) and from the emitted plume (~5°C) of a coal-fired power plant. As measured by total fluorescence the benzene extract of the emitted fly ash was approximately 1000 times greater than that of the fly ash that was collected inside the stack.

Another prediction which can be made on the basis of Figure 3 is that most of the PAH will be adsorbed by the carbonaceous fraction of ash which provides much of the surface area even though the percentage mass of this fraction is usually less than one-percent of the total ash. As demonstrated by Soltys (11), the carbonaceous ash can not be solvent extracted efficiently since PAHs are tenaciously held by this substrate. This suggests that many of the PAH emission factors that have been determined to date may be erroneous.

ACKNOWLEDGEMENTS

The work described herein was supported by a research grant no. DE-AC02-78EV04960.A003 from the U.S. Department of Energy.

REFERENCES

1. Committee on Biological Effects of Atmospheric Pollutants: Particulate Polycyclic Organic Matter, National Academy of Sciences, Washington, D.C., 1972.
2. Badger, G.M.: Mode of formation of carcinogens in human environment, Nat. Cancer Inst. Monograph, 9:1-16.
3. Scientific and Technical Assessment Report on Particulate Polycyclic Organic Matter, Publication EPA-600/6-75-001, Office of Research and Development, U.S. Environmental Protection Agency, Washington, D.C., 1975.
4. Glasstone, S., Laidler, K.J., and Eyring, H. (1941): The Theory of Rate Processes. McGraw-Hill Book Company.

5. Laidler, K.J., Glasstone, S., and Eyring, H. (1940): J. Chem. Phys., 8:659.
6. Schure, M.R. (1981): The effect of temperature upon the transformation of polycyclic organic matter, Ph.D. Dissertation, Colorado State University, Fort Collins, Colorado.
7. Young, D.M. and Crowell, A.D. (1962): Physical Adsorption of Gasses. Butterworths.
8. Adamson, A.W. (1976): Physical Chemistry of Surfaces. John Wiley and Sons.
9. Brunauer, S. (1945): The Adsorption of Gasses and Vapors, I. Princeton Unviersity Press.
10. Hangebrauk, R.P., Von Lehmden, D.J., and Meeker, J.E. (1964): J. Air Pollution Control Assoc., 14:267.
11. Soltys, P.A. (1980): The extraction behavior of PAH from coal fly ash, Masters Thesis, Colorado State University, Fort Collins, Colorado.

ENHANCED BENZO[A]PYRENE METABOLISM IN HAMSTER EMBRYONIC CELLS EXPOSED IN CULTURE TO FOSSIL SYNFUEL PRODUCTS

D. D. SCHURESKO*, G. D. GRIFFIN*, M. C. MacLEOD**, AND J. K. SELKIRK**
*Health and Safety Research Division, **Biology Division, Oak Ridge National Laboratory, Oak Ridge, Tennessee 37830

INTRODUCTION

Polynuclear aromatic (PNA) compounds, including numerous potent carcinogens and mutagens, are major constituents of the products, process solvents, and wastes produced during the conversion of coal and oil shale to liquid and gaseous fuels. Human exposure to tars and oils in fossil fuel conversion and utilization facilities, and to synthetic fuel-derived end products constitutes one of the major health hazards associated with synthetic fossil fuels (1). Biochemical indicators of PNA exposure and body burden would be of great utility to the tasks of estimating and controlling health risks associated with PNA toxicity.

The enzyme aryl hydrocarbon mono-oxygenase (AHM), which is involved in the initial metabolism of PNAs to excretable polar derivatives, is present in many tissues including liver, lung, skin, and kidney, as well as in blood lymphocytes (2). AHM converts PNAs to reactive epoxide intermediates which can be subjected to further detoxification metabolism, or which can bind to cellular macromolecules, thereby causing toxic effects. In vitro or in vivo exposure to PNA compounds and to cigarette smoke has been shown to cause induction or increased AHM activity in many tissues and cell types, including human peripheral blood lymphocytes and pulmonary alveolar macrophages (4). Given these facts, it seems appropriate to investigate the potential for monitoring human PNA "dose" by measuring AHM activity directly in accessible tissues, such as lymphocytes and pulmonary macrophages, and indirectly through the levels of PNA metabolites and conjugates in physiologic fluids.

The variability in AHM activity measurements with human lymphocytes is well established (3, 4). This variability may result from the intrinsic heterogeneity of lymphocyte populations or from the requirement for stimulation from the resting state for expression of AHM induction. Thus in carrying out preliminary investigations of AHM induction by fossil-synfuel products, we have chosen to work with the

well-characterized hamster embryo fibroblast AHM system
(5, 6). We have demonstrated that the AHM system enzymes of
these cells are stimulated by much lower concentrations of
some synfuel materials compared to individual PNA compounds
(9), a fact which may be related to the comparatively lower
carcinogenic potential (10) of the synfuel materials.

MATERIALS AND METHODS

Electrostatic precipitator tar from a low-Btu coal
gasifier, process solvent from a direct coal liquefaction
facility, and an intermediate distillate from a catalytic
coal liquefaction process were obtained from the ORNL
Analytical Chemistry Division repository. Benzo[a]anthracene
(BaA) was obtained from Eastman Organic Chemicals, Inc.
Tertiary cultured Syrian hamster embryo fibroblasts (HEFs)
were prepared as described (6) and exposed for 24 hours to
culture medium containing pure PNA compounds or fossil synfuel products. Control and exposed cells were harvested with
trypsin, washed, and finally resuspended in phosphate-
buffered saline. Viable (impermeable to trypan blue) cell
counts were obtained by hemocytometry. Incubation mixtures,
consisting of approximately 10^6 cells, 3 micromoles $MgCl_2$,
5 nanomoles NADPH*, and 0.24 nanomoles of tritium-labelled
benzo[a]pyrene (BaP, Amersham, 0.2 and 6.0 Ci/mmol) in 1.5
mℓ of phosphate-buffered saline, were analyzed spectrofluoro-
metrically before and after incubation for 2 hours at 37°C.
This analysis closely parallels the so-called direct assay
for total AHM activity (7), a variant of which has recently
been used in flow cytofluorometric AHM activity determinations (8). For reference, the alkali-extractable, non-
conjugated phenol metabolites were assayed by extracting each
incubation mixture with 3 volumes of hexane (following the
addition of 1.5 mℓ of acetone to stop the reaction), then
subsequently extracting 1.0 mℓ of the hexane phase with
3.0 mℓ of 1.0 N NaOH. This latter analysis closely parallels
the assay developed by Nebert and Gelboin for hydroxylase
activity determinations (5). Finally, the ethyl-acetate
extractable metabolites from several incubation mixtures were
pooled and analyzed by high performance liquid chromatography, according to the procedure of Selkirk and co-workers
(13), and quantified by liquid scintillation counting.

---

*Reduced nicotinamide adenine dinucleotide phosphate

RESULTS AND DISCUSSION

The fluorescence emission spectra of one such incubation mixture are shown in Fig. 1; fluorescence was excited at the

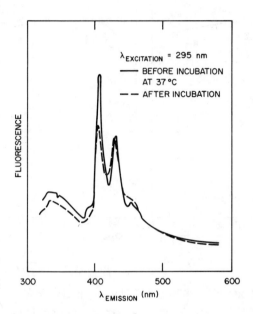

FIGURE 1. Fluorescence spectra of HEF incubation mixture.

295 nm BaP absorption maximum. In the solid curve (pre-incubation), one can see the BaP emission spectrum superimposed upon the background fluorescence spectrum of the cells. The dashed curve, obtained after incubation, shows a decreased BaP component plus an unresolved emission component at longer wavelengths. This latter component is indicative of the formation of phenol metabolites and their conjugates; dihydrodiol BaP metabolites, on the other hand, fluoresce at shorter wavelengths than does BaP due to the reduced size of the aromatic moiety in these derivatives. The extent of metabolism in a given mixture can be determined as the fractional change in the height of the principal BaP emission maximum at 404 nm.

Fig. 2a depicts AHM activity vs. exposure data for HEF cells exposed to coal gasification by-product tar. Bearing in mind that AHM activity is underestimated in the plateau

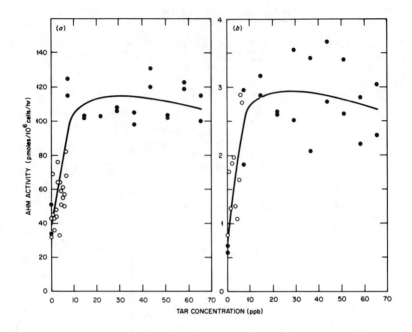

FIGURE 2. (a) Specific AHM activity in p-moles BaP metabolized per 10^6 cells per hour vs. tar concentration. (b) Specific AHM activity in p-moles extractable phenol metabolites per 10^6 cells per hour vs. tar concentration. The open and closed circles denote HEF cells from the same tertiary culture exposed and analyzed independently.

region of this curve, due to high percentage substrate metabolism, we can see that at least a 2.6-fold induction of AHM activity has resulted from exposure to medium containing sub-parts-per-million amounts of tar. For comparison, with pure PNA compounds such as benzo[a]anthracene, the same level of induction occurs at much higher exposure levels, typically in the parts-per-million range (see Fig. 4 and Ref. 5). The AHM-associated hydroxylase activity of these tar-exposed HEF cells, assayed by quantitation of the alkali extractable phenol metabolites (Fig. 2b), correlates well with the total AHM activity.

These observations are borne out by chromatographic analysis of the non-polar BaP metabolites (Fig. 3 and Table I). The solid and dashed curves of Fig. 3 are, respectively, the

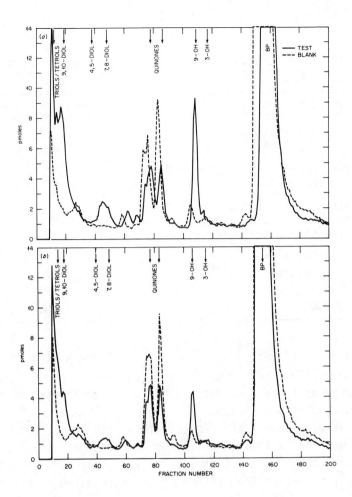

FIGURE 3. (a) BaP metabolism of gasifier tar induced hamster embryo fibroblasts. (b) BaP metabolism of unexposed hamster embryo fibroblasts.

metabolite profiles obtained from test and blank incubation mixtures of HEFs exposed to 21.8 ppb gasifier tar (Fig. 3a) and from unexposed HEFs (Fig. 3b). Non-polar metabolites were extracted twice with 2.5 volumes of ethyl acetate (added to the blanks prior to incubation for 2 hours at 37°C), redissolved in methanol, and analyzed on a Spectra-Physics 3500-B liquid chromatograph fitted with a 1-m Zorbax ODS column (DuPont Inst., Wilmington, DE) by elution with a

TABLE 1

AHM INDUCTION IN HEFs EXPOSED TO 22 PPB COAL GASIFICATION TAR

Indicator (Method)	Induction Ratio
Loss of Substrate BaP (Fluorometric)	2.6[a]
Production of Alkali-Extractables (Radiometric)	3.8[b]
Substrate Loss/Production of Individual Metabolites (Radiochromatographic)	
Substrate	2.2[c]
9,10-diol	2.6[c]
7,8-diol	2.6[c]
9-OH	3.1[c]

[a] Computed as $(X_I^{t_1} - X_I^{t_0}) / (X_C^{t_1} - X_C^{t_0})$, where X denotes the 404 nm BP fluorescence of control (C) and induced (I) HEFs measured at times t_0 and t_1 at the start and finish of the 2-hour incubation period.

[b] Same as a, except that X denotes total alkali-extractable tritium, and t_0 and t_1 denote the time at which the reaction in the blank and test samples was stopped.

[c] Same as b, except that X denotes the integrated peak areas of the corresponding metabolite profiles (Fig. 3).

30-70% methanol and water gradient. The arrows indicate the peaks of a ^{14}C-labelled standard mixture of metabolites which was co-injected with the incubation mixture extracts. The major BaP metabolites produced by the gasifier tar induced cells are the 7,8 and 9,10-diols, and the 9-OH. The 3-OH, which is a major metabolite of rat liver cells and lymphocytes (11), is evident, but not well resolved in Fig. 3. These results agree well with the composite intra- and extra-cellular metabolism of BaP auto-induced HEFs (6). The lower level of BaP quinones in the test vs. blank profiles may be due to further metabolism of quinone impurities present in the BaP substrate material. Quinones are among the recognized decomposition products of BaP (12).

Induction ratios determined from total BaP metabolism and from the production of specific metabolites are listed in Table I. The agreement between the extent of metabolism and the induction as measured by the fluorometric "direct" assay and as measured by the decrease in ^{3}H-labelled substrate is quite good. Induction ratios as determined by the production of individual metabolites and by the production of alkali-extractables are somewhat comparable. Note that 3-OH, 9-OH, and 9,10-diol metabolites contribute significantly to the hydroxylase activity as determined by the Nebert method (14).

Induction levels comparable to those reported here have been observed for HEFs exposed to ppb-ppm levels of coal liquefaction process liquids (Fig. 4). The fact that significant AHM induction occurs at much lower exposures to some

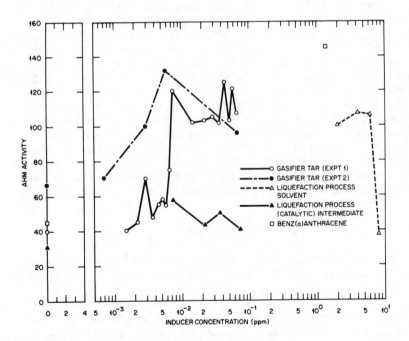

FIGURE 4. Specific AHM activity (p-moles BaP metabolized per 10^6 cells per hour) vs. inducer concentration (ppm) as determined by the fluorometric total AHM assay.

synfuel materials than to pure PNAs may be due to synergistic induction effects of the synfuel PNA constituents. It may also be caused by as yet unidentified highly potent AHM inducers in the synfuel materials, or perhaps by synfuel constituents which themselves are not AHM inducers, but which greatly enhance the susceptibility of HEF cells to induction. It is worth noting at this point that the higher levels of individual PNA exposure required for AHM induction are not much less than exposure levels which are markedly cytotoxic, and which apparently saturate the cellular uptake capacity for PNAs[+]. Thus, the presence of compounds in the synfuel mixtures which enhance the transport and binding of synfuel PNAs to cellular sites involved in AHM induction (e.g., receptor proteins, chromatin) may result in the increased potency of these complex mixtures as AHM inducers.

Finally, we have observed that BaP metabolism in control, hydrocarbon, and synfuel induced cells is equally sensitive to exogenous NADPH[*]. This indicates that although trypsin-harvested HEF cells are somewhat leaky to large polar molecules, the enhanced BaP metabolism in the synfuel exposed cells is not due to increased cellular permeability to NADPH. We may not at present rule out the possibility that the observed AHM catalysis induction is due to enhanced substrate transport to AHM sites, or to AHM alterants present in the synfuel materials. We have not observed discrepancies between the BaP fluorescence emission spectra of control and induced cells, however, indicating that the BaP-binding matrices in control and induced cells are at least of similar polarity. Measurements of the rate of substrate uptake, and measurements of the cytochrome P-450 content of control and induced cells are clearly required.

CONCLUSIONS

The utilization of AHM induction as a biochemical indicator of exposure to PNAs is supported by these observations of

---

[+]We have, for example, observed benz[a]anthracene (BaA) excimer fluorescence in HEFs exposed to growth media containing 2×10^{-10} moles BaA per 10^6 cells (or $\sim 10^8$ molecules per cell). Exposure levels 2-3 fold less have been reported as optimal for AHM induction (5, 9).

[*]Despite the permeability to exogenous NADPH, trypsin harvested HEFs are viable in that they are impermeable to trypan blue. Futhermore, trypsin-harvested secondary and tertiary HEF cultures achieve confluence rapidly, clearly demonstrating viability.

AHM induction in cultured mammalian cells exposed to fossil synfuel materials. However, the interaction of synfuel materials with the AHM system of these hamster embryonic cells is not simply dependent upon the total PNA content of each material. The possibility that the variability in carcinogenic potential of synfuel materials is linked to AHM induction capability should clearly be explored. Thus the relationship between AHM induction potential and carcinogenesis potential is of generic interest to the problem of synfuel toxicity, and of importance to the use of AHM activity as a biological indicator of toxic exposures.

ACKNOWLEDGEMENTS

This research is sponsored by the Office of Health and Environmental Research, US Department of Energy, under contract W-7405-eng-26 with the Union Carbide Corporation. The technical assistance of Kris Dearstone, Betty Mansfield, and Stan Dinsmore has been invaluable to this effort.

REFERENCES

1. National Institute for Occupational Safety and Health: Criteria for a Recommended Standard... Occupational Exposure to Coal Tar Products. Rockville, MD, US Department of Health, Education, and Welfare, Public Health Service, Center for Disease Control, NIOSH, 1978.
2. Gelboin, H.V. (1967): Carcinogens, enzyme induction, and gene action, Advances in Cancer Research,10:1-81.
3. Atlas, S. A., Vesell, E. S., and Nebert, D. W. (1976): Genetic control of interindividual variations in the inducibility of aryl hydrocarbon hydroxylase in cultured human lymphocytes, Cancer Research, 36:4619-4630.
4. Griffin, G. D., and Schuresko, D. D. (1981): Aryl hydrocarbon mono-oxygenase activity in human lymphocytes. ORNL/TM-7721, Oak Ridge National Laboratory, Oak Ridge, TN 37830.
5. Nebert, D. W., and Gelboin, H. V. (1968): Substrate-inducible microsomal aryl hydroxylase in mammalian cell culture, J. Biol. Chem., 243: 6242-6249.

6. MacLeod, M. C., Cohen, G. M., and Selkirk, J. K. (1979): Metabolism and macromolecular binding of the carcinogen benzo[a]pyrene and its relatively inert isomer benzo[e]pyrene by hamster embryo cells, Cancer Research, 39:3463-3470.
7. Yang, C. S., and Kicha, L. P. (1978): A direct fluorometric assay of benzo[a]pyrene hydroxylase, Analytical Biochemistry, 89:154-163.
8. Miller, A. G., and Whitlock, J. P. (1981): Novel variants in benzo[a]pyrene metabolism, J. Biol. Chem., 256:2433-2437.
9. Schuresko, D. D., Bostick, W. D., Dinsmore, S. R., and Mrochek, J. E. (1980): Induction of cytochrome-P450 mixed function oxidase (MFO) activity in hamster embryo culture by exposure to coal-derived tars, 7th Annual Meeting of the Federation of Analytical Chemistry and Spectroscopy Societies, Philadelphia, PA.
10. Holland, J. M. (1980): H-coal pilot plant project: quarterly progress report for the period ending June 30, 1980. ORNL/TM-7469, pp. 10-16.
11. Selkirk, J. K., Croy, R. G., Whitlock, J. P., and Gelboin, H. V. (1980): In vitro metabolism of benzo[a]pyrene by human liver microsomes and lymphocytes, Cancer Research, 35;3651-3655.
12. Katz, M., Chan, C., Tosine, H., and Sakuma, T. (1979): Relative rates of photochemical and biological oxidation in vitro of polynuclear aromatic hydrocarbons, In: Polynuclear Aromatic Hydrocarbons, 3rd Int. Symp. on Chemistry and Biology, edited by P.W. Jones and P. Leber, pp. 171-189, Ann Arbor Science Publishers.
13. Selkirk, J. K., Croy, R. G. and Gelboin, H. V. (1974): Benzo[a]pyrene metabolites: Efficient and rapid separation by high-pressure liquid chromatography, Science, 184:169-171.
14. Selkirk, J. K., Yang, S. H., and Gelboin, H. V. (1976): Analysis of benzo[a]pyrene metabolism in human liver and lymphocytes and kinetic analysis of benzo[a]pyrene in rat liver microsomes, In: Carcinogenesis 1, Polynuclear Aromatic Hydrocarbons: Chemistry, Metabolism, and Carcinogenesis, edited by R.I. Freudenthal and P.W. Jones, Raven Press, New York.

ANALYSIS OF PAH METABOLITES BY NOVEL TRIPLE QUADRUPOLE MASS SPECTROMETRY

B.I. SHUSHAN and T. SAKUMA
SCIEX INC., 55 Glencameron Rd., #202, Thornhill, Ontario
L3T 1P2, Canada.

D.A. ROKOSH and M.F. SALAMONE
Microbiology Section, Laboratory Services Branch, Ontario
Ministry of the Environment, P.O. Box 213, Rexdale, Ontario,
M9W 5L1, Canada.

INTRODUCTION

The technique of triple quadrupole mass spectrometry (1), a form of tandem mass spectrometry (MS/MS), has recently been shown to be a powerful methodology for the rapid screening of chemical species of interest in complex matrices. The determination of 2,3,7,8-tetrachlorodibenzo-p-dioxin in extracts of fish has been reported by SCIEX (2). The same technique has also been used to determine the presence of trinitrofluorenone in carbon black, chlorinated dibenzofurans in polychlorinated biphenyls and nitropyrenes in crude extracts of diesel engine exhaust (3).

The MS/MS system is also a potentially useful tool for the measurement of small quantities of organic compounds produced by enzymatically mediated biological reactions. This system can be applied directly to crude enzyme extracts, with minimal alteration of the sample integrity and with minimal storage and handling time delays. These advantages would circumvent the inherent problems associated with extraction, separation and derivatization of metabolites which would be required in conventional gas chromatography-mass spectrometry (GC/MS) techniques.

An example of such an enzymatic reaction, which results in the formation of small quantities of complex and perhaps transient metabolites, is the activation of chemical promutagens by mammalian microsomal enzymes. In this study we provide documentation of the use of the MS/MS system in the measurement of metabolites produced by the activation of benzo(a)pyrene (BaP) by Aroclor induced rat-liver microsomal enzymes.

The Instrument

The TAGA™ 6000 is a triple quadrupole mass spectrometer system, and was described at the last PAH Symposium (2).

Microlitre quantities of the sample were deposited on a stainless steel mesh probe and the probe was introduced into the ionization chamber. The sample was thermally desorbed by a voltage pulse and this vaporized sample was ionized through carefully selected ion/molecule reactions. The ionic species of interest were selected by the first quadrupole analyzer (Q1). This selected ion was then subjected to collision induced fragmentation (CID) in the second quadrupole region (Q2). The analysis of fragment ions occurred in Q3 to obtain the CID spectra of these compounds.

In the present study, Q1 and Q3 were operated at unit mass resolution while Q2 was operated in radio frequency (rf) only (ion confinement). The CID energy was 25 V and the gas pressure was 1.054×10^{-3} mmHg, argon. The quantitation of metabolites was carried out by selected ion monitoring of the parent and several intense daughter (or fragment) ions.

Chemicals

PAH and metabolite standards were obtained from Aldrich Chemicals, Milwaukee, Wisconsin and used without further purification. Distilled-in-glass quality solvents were obtained from Caledon Chemicals, Caledon, Ont. Analar grade dimethyl sulfoxide was obtained from BDH Chemicals, Toronto, Ont.

Metabolism By Microsomal Extracts

Metabolic activation of BaP was achieved using a 10% suspension of Aroclor 1254 induced rat-liver microsomal enzymes (Litton Bionetics, Lot number DG074) in a buffered reaction mixture whose final volume of 2.0 mL contained 9 μmoles NADP (Sigma) and 10 μmoles glucose-6-phosphate (Sigma). This mixture was warmed to 37°C prior to the addition of 50 μg BaP in 50 μl dimethyl sulfoxide. This solution was incubated at 37°C with gentle shaking. Samples were removed at times of 0, 10, 20 and 30 minutes, immediately mixed with 1.0 mL n-hexane, then stirred by vortex for 20 seconds. To enhance recovery of metabolites, 3 mL n-hexane and 1 mL diethyl ether were added to the mixture.

This resultant solution was mixed with a vortex stirrer, then sonicated. The organic phase was recovered and diluted 5 fold with n-hexane from which a 3 μl sample was taken for MS/MS analysis. An additional flask containing enzyme reaction mixture but lacking BaP was incubated at 37°C for identical time intervals. These flasks were extracted by an identical procedure and served as reaction mixture controls.

RESULTS AND DISCUSSION

CID Spectra of Some Selected Substituted and Unsubstituted PAH

Figure 1a is a CID spectrum of benzo(a)pyrene, a model compound for unsubstituted PAH. Only the parent ion was observed. Fragmentation did not occur even at higher CID gas pressures or CID energy up to 100 V. Other unsubstituted PAH were also analyzed, but very little or no fragmentation was observed. However, if the MS/MS technique is carried out on oxygenated PAH metabolites, some fragmentation does take place in the CID process. This is demonstrated in Figure 1b, the CID spectrum of 6-hydroxybenzo(a)pyrene. The fragmentation of the molecular ion of an oxygenated PAH can be used to separate the metabolite from a potentially interfering unsubstituted PAH having the same molecular weight.

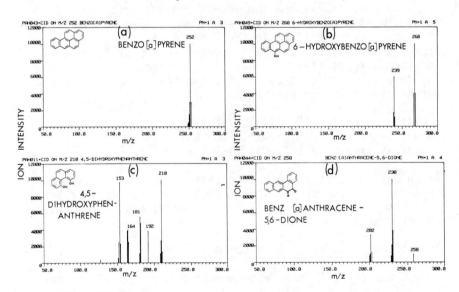

FIGURE 1. CID spectra of some selected substituted and unsubstituted PAH.

Figure 1c shows the result of the CID method carried out on the molecular ion of 4,5-dihydroxyphenanthrene (m/z=210).

The salient fragmentations are the losses of H_2O (m/z=192), HCO (m/z=181), HCO+OH (m/z=164) and HCO+CO (m/z=153). McClusky et al. (4) have reported the use of a mass-analyzed ion kinetic energy spectrometry (MIKES) with a collision gas cell in the second field-free region for the identification of similar metabolites. In their study, the salient fragmentations are the loss of 29 and 58, with much smaller losses of 17 and 34 for the high mass fragment ions. These fragmentations are most likely the sequential loss of CHO twice, and also HO twice. In contrast to their finding, the TAGA™ 6000 produces intense fragment ions due to the loss of H_2O and to the sequential losses of HCO and CO. The peak intensity due to the loss of 2HCO is very small when compared to the results obtained by McClusky et al. (4).

Figure 1d is a CID spectrum of benz(a)anthracene-5,6-dione. The major fragmentations are due to the loss of 28 (CO) and 56 (2CO). This fragmentation pattern agrees with trends observed on electron impact spectra of a few benzo(a)pyrenediones (5).

In order to examine the isomeric specificity by this MS/CID/MS method, 3 hydroxypyrene isomers were analyzed under the conditions specified in the Experimental section. These are pyrene substituted with a hydroxyl group at 1, 2 and 4 positions, respectively. These isomers produced the following results, when a number of CID spectra were averaged and normalized to the molecular ion.

1-hydroxypyrene: 218 (100%), 189 (89%);
2-hydroxypyrene: 218 (100%), 200 (8%), 189 (46%);
4-hydroxypyrene: 218 (100%), 189 (56%).

By comparing relative peak intensities, one can differentiate these 3 hydroxypyrene isomers. Because of the difficulty in obtaining standards, a similar study could not be carried out on BaP metabolite isomers.

Traditional Spectra of Reaction Mixtures

Crude extracts were examined at first by applying only radio frequency (rf) fields to Q1 and Q2 for ion collimation and by scanning Q3 (rf + direct current (dc) fields) for mass analysis. This mode of the TAGA™ 6000 operation produces results similar to those obtainable with a single mass analyzer. Figure 2 is a control-corrected mass spectrum of 25 μL of BaP reaction mixture after 20 minutes incubation. As one can see, it is a complex mixture and no positive

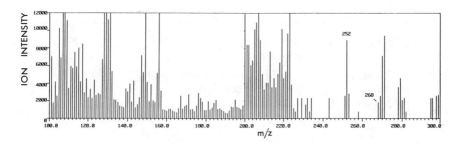

FIGURE 2. C_6H_6 APCI Mass Spectrum of S-9 BaP Extract.

identification can be made from this spectrum. However, the peaks at m/z = 252 and 253 are most likely due to the starting material BaP, and the peak at m/z=268 may be attributed to monohydroxy-BaP.

CID Spectra of Selected Molecular Ions

In order to positively identify these peaks, Q1 was set to select only the m/z=252 and 268 parent ion species by applying rf and dc fields. The MS/CID/MS mode carried out on these ions produced spectra very similar to those shown in Figure 1.

Calibration Curve of $[M-HCO]^+$ Fragment Ion From Hydroxy-BaP

Calibration curves were prepared by flash-vaporizing different amounts of a 6-hydroxybenzo(a)pyrene standard. The parent or molecular ion m/z=268 was selected by Q1 and Q3 was set to monitor the m/z=239 $[M-HCO]^+$ fragment or daughter ion. Figure 3 is an example of such calibration curves. In this instance, a calibration curve was prepared for 0 - 100 nanograms of the standard and no interference peak was observed. The desorption curve (response vs time) in the inset is a typical system response to hydroxy-BaP present in crude extracts from the incubation mixtures.

The absence of interference and the linear response in the usual workup range show that this method is a useful tool in the rapid screening of metabolites and reaction monitoring of complex metabolic reactions.

Response to Molecular Ion of BaP and S-9 Incubation Time

The disappearance of the starting material, BaP, was monitored by the MS/CID/MS method. Both Q1 and Q3 were set

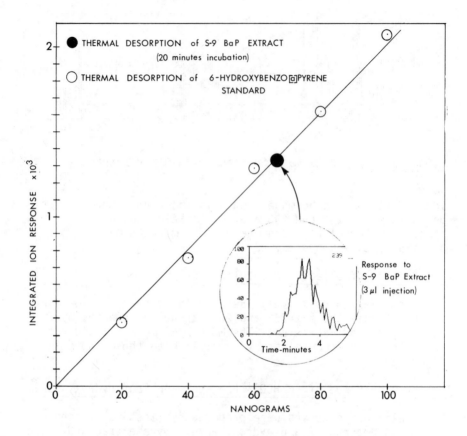

FIGURE 3. Response to [M-HCO]+ Fragment Ions Formed by CID of Hydroxybenzo[a]pyrene Molecular Ions.

at m/z=252 and the CID gas was applied to obtain the time dependant loss of unsubstituted BaP from the microsomal enzyme reaction mixture.

Figure 4 shows the system's response to the BaP molecular ion present in the crude extracts versus incubation time. Approximately 32% of the original BaP was lost over a half hour. The desorption curve shown in the inset was obtained by spiking the S-9 preparation (control) with approximately 530 pg of BaP as an internal standard. The system's response to BaP is linear within the usual working range, and can be determined within a minute, in comparison to the lengthy procedure required to complete similar analysis by other methods.

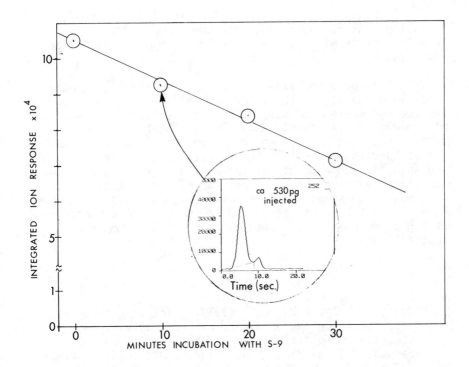

FIGURE 4. Response to Molecular Ion of Benzo[a]pyrene vs S-9 Incubation Time.

CONCLUSION

Because of the space limitation, only a portion of our work is presented. This method demonstrates a number of advantages over the traditional methods for the analysis of metabolites of PAH and other similar derivatives.

(1) Analytical Speed. A fraction of time is required to complete the analysis of metabolites, when compared to traditional methods, i.e., extraction, sample clean-up, derivatization and confirmation by GC/MS. Rapid screening is possible while conducting an incubation run.

(2) High Degree of Analytical Confidence. The MS/CID/MS method is very selective to chemical species of interest. Unambiguous identification can often be made with ease. Because of the high speed and minimal sample treatment, the sample integrity can be maintained.

(3) Minimal Interference. Because of a number of filtering systems associated with the TAGA™ 6000, interference is minimal and direct mixture analysis can be realized.

It must be emphasized that this is only a preliminary study into the use of the TAGA™ 6000 MS/MS system as a tool for the measurement of substrates and products of enzymatically mediated reactions. However, the capabilities of the system, enabling measurement of metabolites with a minimum of sample handling as well as the ability of the system to isolate specific segments of the metabolite matrix, make it an excellent tool for such studies. It is our intention to continue these studies with specific emphasis placed on the resolution of products generated in the biological activation of PAH promutagens by induced mammalian microsomal enzyme systems.

REFERENCES

1. Yost, R.A. and Enke, C.G. (1979): Triple quadrupole mass spectrometry for direct mixture analysis and structure elucidation, Anal. Chem., 51: 1251A-1264A.
2. Sakuma, T., Davidson, W.R., Lane, D.A., Thomson, B.A., Fulford, J.E. and Quan, E.S.K. (1981): The rapid analysis of gaseous PAH and other combustion related compounds in hot gas streams by APCI/MS and APCI/MS/MS. In: Chemical Analysis and Biological Fate: Polynuclear Aromatic Hydrocarbons, edited by M. Cooke and A.J. Dennis. pp. 179-188. Battelle Press, Columbus, OH.
3. Fulford, J.E., Sakuma, T. and Lane, D.A. (1981): Real-time analysis of exhaust gases using triple quadrupole mass spectrometry (MS/MS). The Sixth International Symposium on Polynuclear Aromatic Hydrocarbons, Columbus, OH.
4. McClusky, G.A., Huang, S.K.S., Moore, C.J. and Selkirk, J.K. (1981): Application of MS/MS techniques to the identification of polycyclic aromatic hydrocarbon metabolites. The 28th Annual Conference on Mass Spectrometry and Allied Topics, New York, N.Y.
5. Katz, M., Chan, C., Tosine, H. and Sakuma, T. (1979): Relative rates of photochemical and biological oxidation (in vitro) of polynuclear aromatic hydrocarbons. In: Polynuclear Aromatic Hydrocarbons, edited by P.W. Jones and P. Leber, pp. 171 - 189, Ann Arbor Science, Ann Arbor, MI.

DIOL-EPOXIDE REACTIVITY OF METHYLATED POLYCYCLIC AROMATIC HYDROCARBONS (PAH): RANKING THE REACTIVITY OF THE POSITIONAL MONOMETHYL ISOMERS.

B. D. SILVERMAN*, JOHN P. LOWE** *IBM Thomas J. Watson Research Center, Yorktown Heights, New York 10598; **Department of Chemistry, Pennsylvania State University, University Park, Pennsylvania 16802.

INTRODUCTION

Methylation at the different positions of polycyclic aromatic hydrocarbons (PAH) has been used in the past as a probe to determine which positions on the molecule are actively engaged in the carcinogenic process (1). More recently, methylation on the ring of the parent hydrocarbon that requires metabolic activation to yield the highly reactive bay-region diol-epoxide (2) has been used as a means of identifying this ultimate carcinogenic metabolite (3,4). There has also been recent discussion concerning certain methyl group substitutions on terminal rings that apparently do not completely block formation of the diol precursors of vicinal diol-epoxides(5-7).

Methyl substitution at positions adjacent to the ring that require metabolic activation to yield a bay-region diol-epoxide has also been a subject of considerable interest. The presence of a methyl group is known to induce a diaxial conformation of an adjacent diol (8,9). Methyl substitution at the peri-position should therefore induce a diaxial conformation of the hydroxyl groups of the diol precursor of the bay-region diol-epoxide (10,11). Since one expects poor conversion of such diol to the bay-region diol-epoxide and consequently an inhibition of carcinogenicity, this has been a subject of interest (10,11).

Methyl group substitution at the other position adjacent to the angular benzo-ring of a bay-region diol-epoxide metabolite, namely "bay-region methyl group substitution," is known to yield a carcinogenically active monomethyl isomer relative to the other positional isomers (12-14).

Methyl group substitutions that are neither on nor adjacent to the angular benzo-ring have received much less attention and discussion in the recent experimental literature. This is apparently not due to a lack of interest in the effects of such substitutions (11), but may well be due to difficulties in understanding how such remote† methyl substitutions might alter relevant metabolic pathways.

When metabolic factors do not preclude formation of a bay-region diol-epoxide and stereospecific factors, e.g., diaxial diol conformation, do not inhibit diol-epoxide reactivity, the *predicted chemical reactivity for positional isomers of* benzo-ring diol-epoxides of individual hydrocarbons has been paralleled by their respective mutagenicity and/or tumorigenicity (10).

In other words, of all possible isomeric vicinal diol-epoxides on the terminal rings of PAH, it is the bay-region diol-epoxide that has been calculated to be the most reactive and for several PAH investigated found to be the most mutagenic and/or carcinogenic (10). Since bay-region diol-epoxide reactivity appears therefore to be an important ingredient in determining relative mutagenicity and/or carcinogenicity, it is therefore reasonable to calculate how the bay-region diol-epoxide reactivity depends upon the placement of methyl groups at the different positions of a given PAH.

The purpose of the present paper is to point out and emphasize the close correspondence between the observed carcinogenicity of remotely methylated isomers of certain PAH and their calculated bay-region diol-epoxide reactivities (15-17). One therefore expects that the effect of methylation on molecular reactivity as well as on the direction of metabolic pathways should be an important ingredient in atttempts at ranking the relative carcinogenicity of methylated PAH.

METHODS

The original 'bay-region theory' calculation (18) made use of Dewar reactivity numbers (19) which are essentially a measure of carbocation stabilization energies. It was shown that of the isomeric diol-epoxides of a particular PAH, one should expect the bay-region diol-epoxide to be the most reactive. Subsequent calculations of the effect of methylation on bay-region diol-epoxide reactivity have utilized simple Hückel (15), Indo (16) and Gaussian 70 (17) molecular orbital programs. If consideration is restricted to remote[†] substitutional sites, the ranking of diol-epoxide reactivities of methylated PAH determined by the different methods is essentially the same. There is a simple reason for this (20). If the methyl group is remote from the carbocation center, stabilization of the carbocation occurs primarily by the "through bond" mechanism for methyl stabilization of a cation. The degree of carbocation stabilization then depends primarily upon the magnitude of the lowest unoccupied molecular orbital (LUMO) at the site of substitution. One can, therefore, rank the relative degree of carbocation stabilization for different monomethyl isomers by simply examining the magnitude of the LUMO at each of the potential substitutional sites. It is almost a trivial exercise to obtain the simple Hückel magnitudes (19,20). Another demonstration utilizing simple Hückel MO amplitudes has recently shown why the bay-region diol-epoxide is calculated to be the most reactive of all possible isomeric vicinal diol-epoxides of a particular PAH (21). Therefore, in the present paper the ranking of diol-epoxide reactivity (carbocation stabilization) of methylated PAH will be made by examination of the simple Hückel LUMO magnitudes at the different positions.

RESULTS

A two-stage system of tumorigenesis has been recently used to evaluate the tumor initiating activity of the monomethylbenzo(a)pyrenes(MBP)(14). This procedure reveals differences in carcinogenic activity of the different monomethyl isomers not previously observed (22). Methyl substitutions at the peri position 6 and at the remote positions 2 and 5 yield molecules significantly less active than do the other remote substitutions.

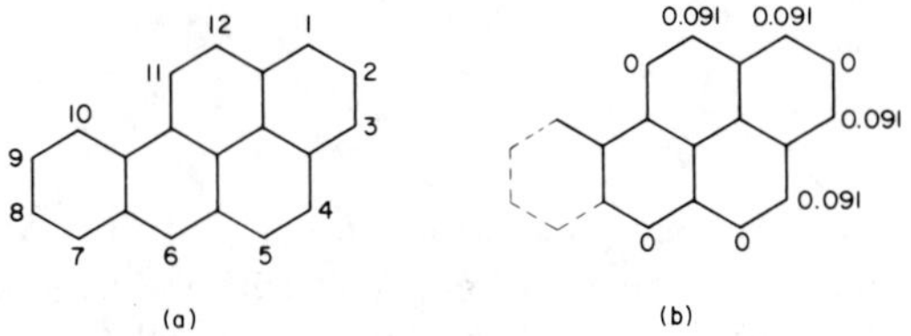

FIGURE 1. (a) benzo(a)pyrene. (b) the planar conjugated portion of the benzo(a)pyrene carbocation.

Fig. 1b shows the square of the simple Hückel LUMO amplitudes at the different substitutional sites. It should be noted that there are nodes at the peri (position 6) and the bay-region (position 11) positions. This is a general feature for all odd-alternant bay-region carbocations. Among the remote positions there are nodes at positions 2 and 5. It is therefore reasonable that 2,5 and 6-methyl benzo(a)pyrene are observed to be the least active group of monomethyl isomers. Since the degree to which a methyl group stabilizes a carbocation is greater, the greater the LUMO amplitude at the site of substitution, methyl substitution at position 2,5 or 6 is expected to yield little, if any carbocation stabilization. For the peri methyl substituted isomer one therefore expects not only poor conversion of the diol to a vicinal diol-epoxide but also little carbocation stabilization associated with the vicinal diol-epoxide that has been converted.

5-methylchrysene(5-MeC) is the most carcinogenic of the chrysene monomethyl isomers (23). It is of interest that not only the 1,2 diol has recently been found to be active, but the 7,8 diol as well (24). The former is the diol precursor of the bay-region diol-epoxide of 5-MeC, while the latter is the diol precursor of the bay-region diol epoxide of 11 MeC, in our notation (Fig. 2).

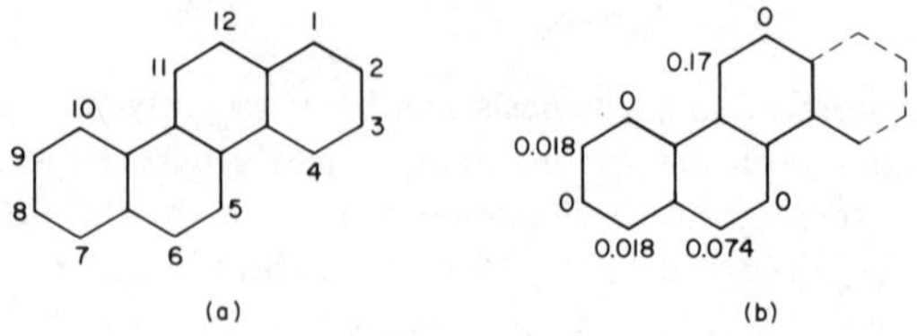

FIGURE 2. (a) chrysene (b) the planar conjugated portion of the chrysene carbocation

The carcinogenic activity associated with the 7,8 diol of 5-MeC (1,2 diol of 11-MeC in our notation) is therefore consistent with the large orbital amplitude found at position 11 prior to methyl substitution. The relative magnitudes of the other orbital amplitudes also suggest that 6-MeC should exhibit somewhat greater activity than isomers obtained by methyl substitution on the terminal ring. This has apparently not been observed.

There have been studies of the mutagenicity and carcinogenicity of the methylated dihydrocyclopentaphenanthrenes(12) as well as further extensive studies of the highly carcinogenic bay-region methyl substituted cyclopentaphenanthrene, 15,16-dihydro-11-methylcyclopenta(a)phenanthren-17-one(25). The only other methyl substituted isomer exhibiting significant carcinogenic and mutagenic activity is 7-methylcyclopenta(a)phenanthrene (12). The carbocation of 15,16-dihydrocyclopenta(a)phenanthren-17-one has the same Hückel LUMO square amplitudes (Fig. 3b) regardless of whether the oxygen is treated as a heteroatom, as a methylenic carbon, or even if the entire 17,18 branch is omitted. This means that at this simple level, this molecular cation is indistinguishable from phenanthrene (Fig. 5b). More refined molecular orbital calculations (26) show that the effect of the oxygen remains negligible within the context of the present discussion.

Figure 3b. then shows that the only remote substitutional site of the two available for methyl substitution (positions 7 and 12) that has significant orbital amplitude associated with it is position 7. The moderate carcinogenic and mutagenic activity observed (12) for 7-methylcyclopentaphenanthrene is then consistent with the relatively large orbital amplitude at this position.

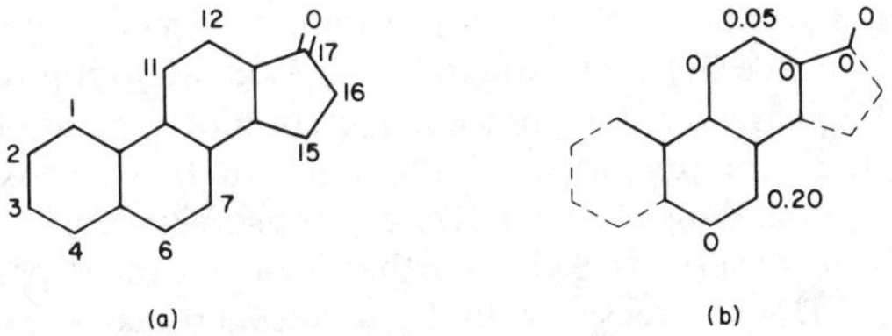

FIGURE 3. (a) 15,16-dihydrocyclopenta(a)phenanthren-17-one. (b) the planar conjugated portion of the cyclopentaphenanthrene carbocation.

A number of different carcinogenicity studies (1) on the methylated benzo(a)anthracenes (MBA) have shown that substitutions at the remote positions 6,7 and 8 yield carcinogenically active molecules whereas substitutions at positions 9, 10 and 11 yield relatively inactive molecules.

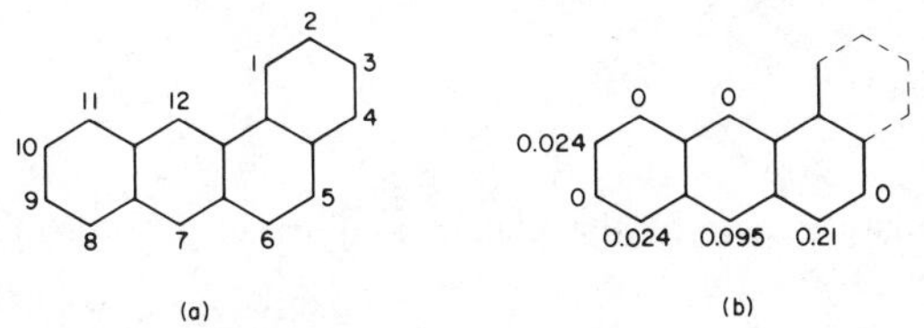

FIGURE 4. (a) benzo(a)anthracene. (b) the planar conjugated portion of the benzo(a)anthracene carbocation.

Examination of the squares of the LUMO amplitudes (Fig. 4b) for benzo(a)anthracene(BA) shows the results of carcinogenicity studies to be consistent with all of these values except for position 8. Detailed molecular orbital calculations for more representative geometries (16,26) in fact yield a somewhat greater bay-region carbocation stabilization energy for methyl substitution at position 7 than at position 6. Such calculations do, however, yield a relatively small carbocation stabilization energy for methyl group substitution at position 8. This is consistent with the small square of the LUMO amplitude at this position (Fig. 4b). The correspondence between carcinogenicity and methyl group induced bay-region carbocation stabilization must be treated with extra caution in the case of BA. Even though the diol precursor of the bay-region diol-epoxide as well as the bay-region diol-epoxide have been shown to exhibit enhanced activity (27-29), very little BA is metabolized by rat liver microsomes to the bay-region diol precursor (30). Substitution of a methyl group at position 8 is also known to enhance 3,4 diol formation (5, 31). Methyl group substitution at position 8 of BA apparently does not completely block metabolism to the 8,9 diol (5). This is the diol precursor of the 8,9-diol-10,11-epoxide for which significant amounts of binding to the DNA of mouse skin and hamster embryo cells has recently been observed (32). Furthermore, recent experiments on the inhibition of $\Phi X174$ DNA infectivity in E. coli spheroplasts by diol-epoxides show significant inhibition for one diastereomer of the 8,9-diol-10,11-epoxide of BA (33).

Recent tumor initiation studies of the mono and certain dimethyl phenanthrenes (34,35) have shown 1,4-dimethylphenanthrene(1,4-DMP) and 4,10-DMP to exhibit tumorigenic activity. It was suggested (34) that this could be due to inhibition of 9,10 diol formation and direction of metabolism to the diol precursor of the bay-region diol-epoxide.

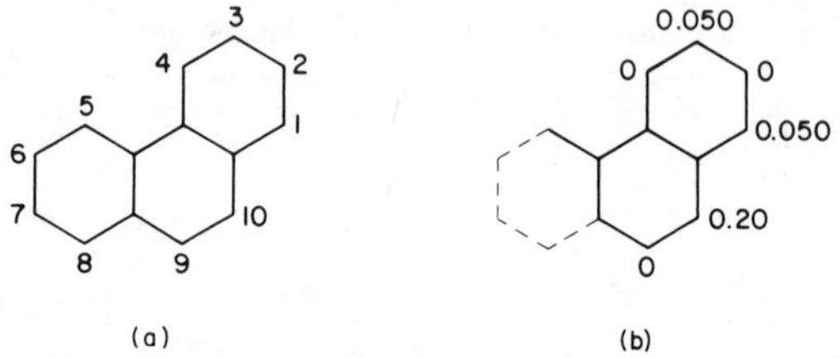

FIGURE 5. (a) phenanthrene. (b) the planar conjugated portion of the phenanthrene carbocation.

Fig. 5b shows the square of the orbital amplitudes for the bay-region carbocation of phenanthrene. Note that this carbocation Hückel substrate is the same as the substrate previously examined in connection with the cyclopentaphenanthrenes (Fig. 3b). Consistent with the observed activity of 1,4-DMP and 4,10-DMP, it is seen that positions 1 and 10 both are associated with nonvanishing LUMO amplitude. 4,10-DMP has, however, been found to be less active than 1,4-DMP (35). More detailed calculations (26) with inclusion of steric crowding arising from the presence of the methyl group in a bay region do not change the qualitative inferences we have drawn from simple Hückel theory.

As one last example, we will make some predictions concerning bay-region carbocation stabilization and carcinogenicity of methylated dibenzo(a,h)anthracene (DBA). To the best of our knowledge, the methylated isomers of this material have not yet been examined for their relative tumor initiating activity. DBA should be an appropriate PAH to study since the diol precursor of the bay-region diol-epoxide has been found to be a major metabolite induced by a rat liver microsomal system (36).

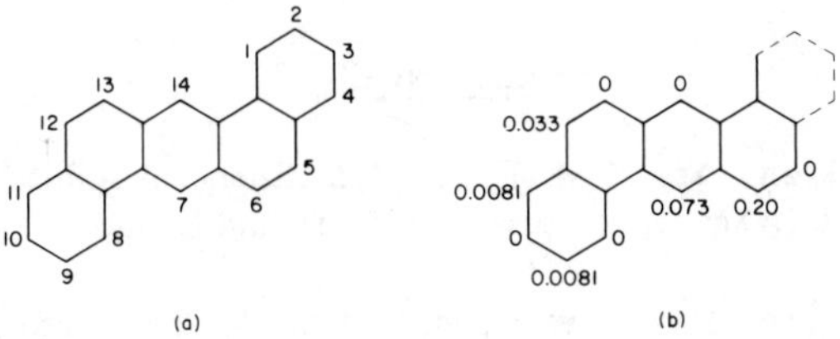

FIGURE 6. (a) dibenzo(a,h)anthracene. (b) the planar conjugated portion of the dibenzo(a,h)anthracene carbocation

Fig. 6b shows the square of the orbital amplitudes for the bay-region carbocation of DBA. Examination of these values leads one to suspect that aside from bay-region methyl substitution at position 14, methyl substitution at positions 6 and 7 should yield the most highly tumorigenic substituted monomethyl isomers.

CONCLUSIONS

The present paper has examined the effect of methylation on the bay-region carbocation stability of several PAH. It was shown that a correspondence exists between methyl group induced carbocation stability and carcinogenicity for what we have defined to be *remote methyl substitutions*. This is in no way meant to suggest that other factors arising from methyl group substitution, e.g., the direction of metabolism, are unimportant. The present paper does, however, emphasize that the effect of methylation on carbocation stability is one other important ingredient that should be taken into account in attempts at interpreting carcinogenicity and other related data.

ACKNOWLEDGEMENTS

We would like to thank A. Jeffrey and E. LaVoie for preprints of their work. We are also indebted to S. K. Yang for keeping us informed of relevant experimental literature.

† a remote methyl substitution is a substitution only at a position that is not on the ring requiring metabolic activation to yield a bay-region diol-epoxide and not at either of the two positions adjacent to this ring.

REFERENCES

1. Arcos, J. C. and Argus, M. F. (1974): Chemical Induction of Cancer, section 5.1.1.2, Academic Press, New York and London.

2. Sims, P., Grover, P. L., Swaisland, A., Pal, K., and Hewer, A. (1974): Metabolic activation of benzo(a)pyrene proceeds by a diol-epoxide, Nature, 252:326-328.

3. Jerina, D. M. and Daly, J. W. (1977): Oxidation at Carbon. In: Drug Metabolism, edited by D. V. Parke and R. L. Smith, pp. 13-32, Taylor and Francis, London.

4. Jerina, D. M., Yagi, H., Levin, W., and Conney, A. H. (1977): Carcinogenicity of benzo(a)pyrene. In: Proceedings of the Alfred Benzon Symposium, No. 10. Drug design and adverse reactions, pp. 261-271. Academic Press, New York.

5. Yang, S. K., Chou, M. W., Weems, H. B., and Fu, P. P., (1979): Enzymatic formation of an 8,9-diol from 8-methylbenz(a)anthracene, Biochem. Biophys. Res. Commun., 90:1136-1141.

6. Jeffrey, A. M., Kinoshita, T., Santella, R. M., Grundberger, D., Katz, L., and Weinstein, I. B. (1980): The Chemistry of Polycyclic Aromatic Hydrocarbon-DNA Adducts. In: Carcinogenesis: Fundamental Mechanisms and Environmental Effects, edited by B. Pullman, P.O.P.Ts'o and H. Gelboin, pp. 565-579, D. Reidel Pub. Co., Dordrecht.

7. Wong, T. K., Chiu, P., Fu, P. P., and Yang, S. K. (1981): Metabolic study of 7-methylbenzo(a)pyrene with rat liver microsomes: Separation by reversed-phase and normal-phase high performance liquid chromatography and characterication of metabolites, Chem-Biol Interactions, 36:153-166.

8. Tierney, B., Burden, P., Hewer, A., Ribeiro, O. and Walsh, C. (1979): High-performance liquid chromatography of isomeric dihydrodiols of polycyclic hydrocarbons, J. Chromatography, 176:329-335.

9. Zacharias, D. E., Glusker, J. P., Harvey, R. G. and Fu, P. P. (1977): Molecular Structure of the K-region cis-dihydrodiol of 7,12-dimethylbenz(a)anthracene, Cancer Res., 37:775-782.

10. Jerina, D. M., Sayer, J. M., Thakker, D. R., Yagi, H., Levin, W., Wood, A. W. and Conney, A. H. (1980): Carcinogenicity of Polycyclic Aromatic hydrocarbons: The bay-region theory. In: Carcinogenesis: Fundamental Mechanisms and Environmental Effects, edited by B. Pullman, P.O.P. Ts'o and H. Gelboin, pp. 1-12, D. Reidel Pub. Co., Dordrecht.

11. Yang, S. K., Chou, M. W., and Fu, P. P. (1980): Metabolic and structural requirements for the carcinogenic potencies of unsubstituted and methyl-substituted polycyclic aromatic hydrocarbons. In: Carcinogenesis: Fundamental Mechanisms and Environmental Effects, edited by B. Pullman, P.O.P.Ts'o and H. Gelboin, pp. 143-156, D. Reidel Pub. Co., Dordrecht.

12. Coombs, M. M., Dixon, C., and Kissonerghis, A. (1976): Evaluation of the mutagenicity of compounds of known carcinogenicity, belonging to the benzo(a)anthracene, chrysene, and cyclopenta(a)phenanthrene series, using Ames' test, Cancer Res., 36:4525-4529.

13. Hecht, S. S., Amin, S., Rivenson, A., and Hoffmann, D. (1979): Tumor initiating activity of 5,11-dimethylchrysene and the structural requirement favoring carcinogenicity of methylated polynuclear aromatic hydrocarbons, Cancer Lett., 8:65-70.

14. Iyer, R. P., Lyga, J. W., Secrist III, J. A., Daub, G. H. and Slaga, T. J. (1980): Comparative tumor-initiating activity of methylated benzo(a)pyrene derivatives in mouse skin, Cancer Res.,40:1073-1076.

15. Smith, I. A., and Seybold, P. G. (1978): Methylbenz(a)anthracenes: Correlations between theoretical reactivity indices and carcinogenicity. In: International Journal of Quantum Chemistry: Quantum Biology Symposium,5:311-320.

16. Poulsen, M. T., and Loew, G. H. (1981): Quantum chemical studies of methyl and fluoro analogs of chrysene: Metabolic activation and correlation with carcinogenic activity, Cancer Biochem. Biophys., 5:81-90.

17. Silverman, B. D. (1981): Carcinogenicity of methylated benzo(a)pyrene: Calculated ease of formation of the bay-region carbonium ion, Cancer Biochem. Biophys., 5:207-212.

18. Jerina, D. M., Lehr, R. E., Yagi, H., Hernandez, O., Dansette, P. M., Wislocki, P. G., Wood, A. W., Chang, R. L., Levin, W., and Conney, A. H., (1976): Mutagenicity of benzo(a)pyrene derivatives and the description of a quantum mechanical model which predicts the ease of carbonium ion formation from diol-epoxides. In: In Vitro Metabolic Activation and Mutagenesis Testing, edited by F. de Serres, J. R. Fouts, J. R. Bend and R. M. Philpot, pp. 159-177, Elsevier Press, Amsterdam.

19. Dewar, M. J. S. (1969): The Molecular Orbital Theory of Organic Chemistry, McGraw-Hill Book Co., New York. Chapter 8.

20. Silverman, B. D. and Lowe, J. P.: Carcinogenicity of methylated hydrocarbons: Effect of methylation on the calculated diol-epoxide reactivity, Cancer Biochem. Biophys.,(in press).

21. Lowe, J. P. and Silverman, B. D. (1981): Simple molecular orbital explanation for "bay-region" carcinogenic reactivity, J. Am. Chem. Soc. 103:2852-2855.

22. Lacassagne, A., Zajdela, F., Buu-Hoi, N. P. Chalvet, O. and Daub, G. H. (1968): Activité Cancérogène elévée des mono-di-, et trimethylbenzo(a)pyrènes, Int. J. Cancer, 3:238-243.

23. Hecht, S. S., Bondinell, W. E. and Hoffmann, D. (1974): Chrysene and methylchrysenes: Presence in tobacco smoke and carcinogenicity, J. Nat. Cancer Inst., 53:1121-1133.

24. Hecht, S. S., Rivenson, A and Hoffmann, D. (1980): Tumor-initiating activity of dihydrodiols formed metabolically from 5-methylchrysene, Cancer Res., 40:1396-1399.

25. Coombs, M. M., Bhatt, T. S., Kissonerghis, A. and Vose, C. W. (1980): Mutagenic and Carcinogenic metabolites of the carcinogen 15,16-dihydro-11-methylcyclopenta(a)phenanthren-17-one, Cancer Res., 40:882-886.

26. B. D. Silverman, unpublished results.

27. Levin, W., Thakker, D. R., Wood, A. W., Chang, R. L., Lehr, R. E., Jerina, D. M. and Conney, A. H. (1978): Evidence that benzo(a)anthracene 3,4-diol-1,2-epoxide is an ultimate carcinogen on mouse skin, Cancer Res., 38:1705-1710.

28. Slaga, T. J., Huberman, E., Selkirk, J.K., Harvey, R. G. and Bracken, W. M. (1977): Carcinogenicity and mutagenicity of benzo(a)anthracene diols and diol-epoxides, Cancer Res., 38: 1699-1704.

29. Hemminki, K., Cooper, C. S., Ribeiro, O., Grover, P. L. and Sims, P. (1980): Reactions of 'bay-region' and non-'bay-region' diol-epoxides of benzo(a)anthracene with DNA: evidence indicating that the major products are hydrocarbon $-N^2-$ guanine adducts, Carcinogenesis, 1:277-286.

30. Thakker, D. R., Nordqvist, M., Yagi, H., Levin, W., Ryan, D., Thomas, P., Conney, A. H. and Jerina, D. M. (1979): Comparative metabolism of a series of polycyclic aromatic hydrocarbons by rat liver microsomes and purified cytochrome P-450. In: Polynuclear Aromatic Hydrocarbons,

edited by P. W. Jones and P. Leber, pp. 455-472, Ann Arbor Science Pub., Inc., Ann Arbor.

31. Wislocki, P. G., Fiorentini, K. M., Fu, P. P., Chou, M. W., Yang, S. K., and Lu, A. Y. H. (1981): Tumor-initiating activity of the dihydrodiols of 8-methylbenz(a)anthracene and 8-hydroxymethylbenz(a)anthracene, Carcinogenesis, 2:507-509.

32. Cooper, C. S., Ribeiro, O., Hewer, A., Walsh, C., Pal, K., Grover, P. L. and Sims, P. (1980): The involvement of a 'bay-region' and a non-'bay-region' diol-epoxide in the metabolic activation of benzo(a)anthracene in mouse skin and in hamster embryo cells, Carcinogenesis,1:233-243.

33. Harvey, R. G. (1981): Activated metabolites of carcinogenic hydrocarbons, Acc. Chem. Res., 14:218-226.

34. LaVoie, E. J., Tulley-Freiler, L., Bedenko, V. and Hoffman, D (1981): Mutagenicity, tumor initiating activity and metabolism of methylphenanthrenes, Cancer Res., 41:3441-3447.

35. LaVoie, E. J., Tulley-Freiler, L., Bedenko, V. and Hoffmann, D. (1981): On the metabolic activation of methylated phenanthrenes, In: Proceedings of the Seventy-Second Annual Meeting of the American Association for Cancer Research, pp. 94.

36. Nordqvist, M., Thakker, D. R., Levin, W., Yagi, H., Ryan, D. E., Thomas, P. E., Conney, A. H. and Jerina, D. M. (1979): The highly tumorigenic 3,4-dihydrodiol is a principal metabolite formed from dibenzo(a,h)anthracene by liver enzymes, Mol. Pharmacol.,16:643-655.

ON-LINE MULTIDIMENSIONAL LIQUID CHROMATOGRAPHIC DETERMINATION

OF POLYCYCLIC ORGANIC MATERIAL IN COMPLEX SAMPLES

W. J. SONNEFELD AND W. H. ZOLLER*: W. E. MAY AND S. A. WISE**
*Department of Chemistry, University of Maryland, College
Park, Maryland 20742; **Organic Analytical Research Division,
National Bureau of Standards, Washington, D.C. 20234

INTRODUCTION

The successful analysis of individual components in complex mixtures requires an analytical system with sufficient resolving power. The complexity of the sample matrix is often the determining factor in deciding which analytical technique should be used. Chromatographic separations are commonly employed for the analysis of complex mixtures to reduce the complexity of the sample as seen by the detection system. In many cases the sample components can be sufficiently separated using a single chromatographic column. However, more complex samples generally require the use of certain forms of sample fractionation prior to the final determination. The approach in which a selective fraction from one chromatographic column is transferred to one or more secondary columns for further separation, is called multidimensional chromatography (1).

Samples for determination of polycyclic organic material (POM) are classically fractionated into basic, neutral, and acidic fractions by liquid-liquid partitioning [see, for example (2)]. More recently Wise et al. (3) have recommended a high performance liquid chromatographic fractionation of polycyclic aromatic hydrocarbons (PAH) on a chemically-bonded amine phase. The affinity for PAH is based on the number of aromatic carbons with alkylation of the parent ring having little effect on the retention of the compounds. Thus fractions can be collected which include a given parent set of PAH isomers and their alkyl substituted homologs. Similar fractionation schemes have been proposed using a bonded nitro phase (4), a bonded diamino phase (5), and a mixed bonded cyano/amino phase (6). Each of these fractionations is believed to involve a π-π interaction between the non-polar PAH and the lone pair of electrons in the nitrogen containing stationary phase. Although these packing materials exhibit vastly different affinities toward PAH (7), the relative affinities (retention indices) for individual PAH on each of these materials have not been shown to be significantly different. One such comparison (8) has shown that the diamine and μBondapak™¹ amine phases have similar relative affinities for PAH.

Although the weakly-attracted PAH behave similarly in a variety of the nitrogen containing LC packing materials under normal phase solvent conditions, the more polar POM [e.g., N-heterocyclic (N-PAH), nitro and amine substituted PAH] have different relative affinities towards the different packing materials (8). Therefore a sample can be fractionated for the isolation of the polar POM on various stationary phases to obtain distinctly different fractions, all of which contain the analyte. The optimum packing material to be used in a given fractionation scheme will then depend upon several factors. These factors include: the complexity of the sample, the selectivity of the analytical column, and the selectivity of the detection system.

The aforementioned multidimensional techniques (3-6) are off-line sequential methods. Each involves the collection of a desired fraction containing the analyte from a normal-phase system, evaporation of the solvent to near dryness and redissolution of the fraction in an appropriate solvent for a final reversed-phase determination.

Multidimensional chromatographic techniques using both normal- and reversed-phase systems can be used in on-line methods either by 'heart' cutting (1) or on-column concentration (1,7). In the 'heart' cutting method, very small volumes of the fraction must be injected into the reversed-phase system due to solvent miscibility and strength requirements. This leads to severe deterioration of the system sensitivity. The precision and accuracy of 'heart' cutting techniques are also affected adversely by any change in the primary column efficiency or retention time. This method is currently limited to qualitative determinations where quantitation of the analyte is of secondary importance.

The on-column concentration utilizes a third 'concentrator' column chosen such that three criteria are realized (7): the affinity of the concentrator column must be sufficient to retain the analyte through the trapping and solvent

---

¹Certain commercial equipment, instruments or materials are identified in this paper in order to adequately specify the experimental procedure. Such identification does not imply recommendation or endorsement by the National Bureau of Standards, nor does it imply that the materials or equipment identified are necessarily the best available for the purpose.

evaporation steps; the affinity for the analyte species using reversed-phase solvent conditions must be sufficiently low to allow focusing onto the analytical column; and the analyte must not degrade under the conditions used. The desired fraction is then concentrated on this third column and the normal-phase solvent evaporated using an inert gas purge. The 'concentrator' column is then inserted into the reversed-phase analytical system and the fraction transferred onto the analytical column using gradient elution focusing.

This on-line method has been used to date only for PAH determinations (7). A bonded amine packing material was used for fractionation and a bonded diamine material was used as the concentrator column. The analytical column consisted of a bonded octadecylsilane material. The materials used for fractionation and concentration have similar relative affinities (i.e., selectivities) towards PAH. This relationship allows the isolation of fractions equivalent to those obtained in the off-line method used previously by this laboratory. This on-line system yields an improvement in the system detection limits compared to both the off-line method and on-line heart cutting approaches. This method also permits both accurate and precise qualitative and quantitative determination of individual PAH in complex samples.

This paper contains a description of the on-line column affinity sequencing method for the determination of more polar substituted PAH.

MATERIALS AND METHODS

Figure 1 is the schematic diagram of the on-line multidimensional system. A semipreparative (8 mm x 30 cm) bonded aminosilane column (μBondapak NH_2™) is used for the normal-phase fractionation. An analytical (4.6 mm x 25 cm) octadecylsilane column (Vydac™ 201TP 5 μm or Zorbax™ ODS) is used for the reversed phase column. The concentrator columns used were: Chromegabond™ diamine (30 cm x 4.6 mm), Nucleosil™ NH_2 and NO_2 (25 cm x 40 mm), and Zorbax™ cyano (25 cm x 4.6 mm).

A detailed description of the operating sequence has been previously published (7). All solvents were commercially available HPLC-grade and were used without further purification. The normal-phase solvents consisted of methylene chloride/n-pentane or ethyl ether/n-pentane solutions. The reversed-phase solvents were water and acetonitrile. Nitrogen was employed as the purging gas after passing through a

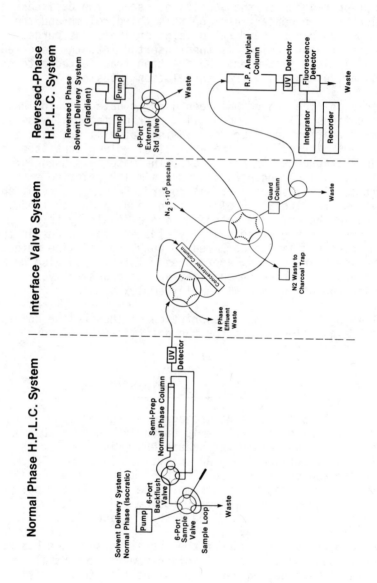

FIGURE 1. Schematic diagram of the on-line multidimensional system.

molecular sieve and magnesium perchlorate trap. All POM were purchased from commercial sources with the exception of 3-nitroperylene (obtained from the NCI Chemical Repository at IIT Research Institute, Chicago, Illinois) and 6-aminobenzo-[a]pyrene (obtained through the courtesy of M. Lee, Brigham Young University, Provo, Utah). All POM were used as received.

The primary purpose of this work was to determine if the selected compounds could be qualitatively and quantitatively determined by this on-line technique. The chromatographic behavior of selected compounds on each column was first determined under a variety of mobile phase conditions in order to verify the applicability of the column affinity focusing technique for these classes of compounds. Standard solutions of the analytes dissolved in the appropriate solvents were then used to test for the percent recovery of the compounds. Between 100 — 1500 ng of analyte material was used in each case.

RESULTS AND DISCUSSION

The on-line multidimensional LC method has been previously applied to three different matrices for the determination of individual PAH. An example of the results obtained are contained in Table 1. We have demonstrated that the PAH can be efficiently collected, the solvent evaporated, and the PAH quantitatively focused onto the reversed-phase analytical column. The more polar polycyclic organic materials (e.g., nitro and amine substituted PAH) investigated can also be fractionated on the μBondapak NH_2 column. The diamine column which is used as the 'concentrator' column has a higher affinity for the POM as well as the PAH previously studied. Therefore the primary criterion that the analyte species be efficiently collected and concentrated on the diamine concentrator column is valid for these classes of compounds.

The POM which were studied can be transferred onto the reversed-phase analytical column using gradient elution focusing without observable deterioration of the chromatographic integrity (peak shape, column efficiency, retention time). This verifies that the second criterion for the on-line multidimensional system is valid for these polar compounds. This also demonstrates that the retention times can be used for qualitative identification of the individual species.

TABLE 1

COMPARISON OF ON-LINE MULTIDIMENSIONAL RESULTS WITH OFF-LINE HPLC RESULTS

Compound	Multidimensional HPLC (on-line)[a]	Multidimensional HPLC (off-line)[a]	Certified
	SRM 1580 (Shale Oil) (Concentration in µg/g)		
Fluoranthene	53.8 ±3.5 (3)[b]	53 ± 2 (9)	54 ±10
Pyrene	109.7 ±7.1 (3)	107 ± 8 (10)	104 ±18
Perylene	2.32 ± .15(5)	3.9 ± .6(11)	3.4± 2.2
Benzo[a]pyrene	20.7 ±2.2 (8)	23 ± 1 (8)	21 ± 6
	SRC II Coal Liquid (Concentration in mg/g)		
Fluoranthene	3.15 ±.04 (3)	3.3 ±.16 (3)	
Pyrene	6.65 ±.05 (3)	6.0 ±.2 (3)	
Perylene	0.024±.002(3)		
Benzo[a]pyrene	0.133±.008(5)	0.134±.007(3)	
Benzo[e]pyrene	0.153±.006(3)	0.143±.005(3)	
Benzo[k]fluoranthene	0.062±.001(5)		
	Wilmington Crude Oil (Concentration in µg/g)		
Fluoranthene	2.4 ± .3(3)	3 ± 1(3)	
Pyrene	14.8 ±2.2(3)	14 ± 2(3)	

a Uncertainty is 1 standard deviation.
b (n) indicates the number of replicate measurements.

The third and final criterion for the success of this approach is that quantitative recoveries of the analyte be achieved under the collection/purging conditions employed. We have found that the more polar POM studied are not recovered from the diamine concentrator column quantitatively in all cases. The recovery is compound specific but in general is inversely proportional to the capacity factor using normal-phase solvents.

The solvent evaporation step can be accomplished by purging either at ambient temperature ($\sim$22 °C) or at elevated temperatures (>50 °C). We have noted an increase in the recovery for most of the compounds studied when the system is purged at ambient temperatures without heating the concentrator column (Table 2). This necessitates a longer purging time (60 minutes) than was recommended previously for PAH. As evidenced in Table 2 the recoveries are not very reproducible when using the bonded diamine column.

Studies using bonded amine, nitro, and cyano functional groups have shown similar losses from the interface column. The losses exhibited by this system are consistent with an 'irreversible' adsorption phenomenon when the analyte is concentrated onto the silica-based packing materials. We believe that this occurs to a greater extent at the higher temperature due to more efficient removal of the mobile phase.

An attempt to improve the recovery from the interface by modifying the diamine concentrator column with glycol prior to collection of the analyte has proven to be partially successful for the nitro substituted PAH. The reproducibility of recovery has also improved (Table 3) for most compounds studied. This modification involves the injection during the concentrator column re-equilibration step (7) of the 100 μL aliquots of pure glycol directly onto the concentrator column followed by a 20 mL purge using the normal-phase solvent. However, this reduces the affinity of the diamine column for these compounds and thus requires that the collected volume be reduced accordingly (7). This modification diminishes the 'irreversible' adsorption effect of the analyte observed during concentration and solvent evaporation onto a bonded-silica surface. The analyte is instead concentrated onto a less active glycol coated bonded-silica surface. Other methods for modifying the concentrator column are currently under investigation. The recoveries of amine substituted PAH are presently being studied using this glycol modification.

TABLE 2

PERCENT RECOVERY OF ANALYTE POM FROM DIAMINE CONCENTRATOR COLUMN IN MULTIDIMENSIONAL SYSTEM

Compound	Purged with Heat (~70 °C)	Purged at Ambient Temperature (~22 °C)
2-nitronaphthalene	78 ± 5 (5)	87 ± 7 (6)
2-nitrofluorene	86 ± 6 (5)	95 ± 4 (5)
9-nitroanthracene	78 ± 3 (4)	89 ± 6 (5)
1-nitropyrene	85 ± 5 (5)	92 ± 3 (4)
1-aminonaphthalene	49 ± 3 (3)	80 ± 12 (3)
2-aminofluorene	30 ± 8 (3)	52 ± 14 (3)
2-aminoanthracene	8 ± 6 (3)	35 ± 20 (3)
6-aminochrysene	47 ± 8 (3)	77 ± 4 (3)
6-aminobenzo[a]pyrene	40 ± 9 (3)	80 ± 22 (2)

Uncertainty is 1 standard deviation; (n) indicates the number of replicate measurements.

TABLE 3

PERCENT RECOVERY OF NITRO-SUBSTITUTED PAH FROM GLYCOL MODIFIED DIAMINE CONCENTRATOR COLUMN PURGED WITH HEAT (~70 °C)

2-nitronaphthalene	98.5 ± 1.4 (3)
2-nitrofluorene	99.2 ± 2.2 (3)
9-nitroanthracene	87.2 ± 2.6 (3)
1-nitropyrene	101.5 ± 9.7 (4)
3-nitroperylene	94.6 ± 3.5 (3)

Uncertainty is 1 standard deviation; (n) indicates the number of replicate measurements.

SUMMARY

The on-line multidimensional system has been demonstrated to yield accurate and precise determinations of PAH in complex mixtures. The more polar polycyclic organic compounds can also be analyzed but quantitative information must take into account analyte losses in the system when using a diamine or modified diamine packing material as the concentrator column. These recovery factors must be ascertained for each analyte.

REFERENCES

1. Majors, R. E. (1980): Multidimensional High Performance Liquid Chromatography, J. Chrom. Sci., 18:571-579.

2. Schmeltz, I. (1967): The Nitromethane soluble, neutral fraction of cigarette smoke, Phytochemistry, 6:33-38.

3. Wise, S. A., Chesler, S. N., Hertz, H. S., Hilpert, L. R., and May, W. E. (1977): Chemically-Bonded Aminosilane Stationary Phase for the High-Performance Liquid Chromatographic Separation of Polynuclear Aromatic Compounds, Anal. Chem., 49:2306-2310.

4. Blumer, G. P., Zander, M. (1977): Hochdruck-flüssigkeitschromatographisches Verhalten von Biarylen, Kondensierten Aromaten und Aza-aromaten an mit Nitrophenyl-Gruppen modifiziertem Kieselgel, Fresenius Z. Anal. Chem., 228:277-280.

5. Chmielowiec, J., George, A. E. (1980): Polar Bonded-Phase Sorbents for High Performance Liquid Chromatographic Separations of Polycyclic Aromatic Hydrocarbons, Anal. Chem., 52: 1154-1157.

6. Tomkins, B. A., Reagan, R. R., Caton, J. E., Griest, W. H. (1981): Liquid Chromatographic Determination of Benzo[a]pyrene in Natural, Synthetic, and Refined Crudes, Anal. Chem., 53:1213-1217.

7. Sonnefeld, W. J., Zoller, W. H., Wise, S. A., May, W. E. On-Line Multidimensional Liquid Chromatographic Determination of Polynuclear Aromatic Hydrocarbons in Complex Samples, Anal. Chem. (in press).

8. Sonnefeld, W. J. (unpublished data).

MULTIDIMENSIONAL LIQUID CHROMATOGRAPHY

CREDIT

The authors are grateful to the Department of Energy's Office of Environment for support of this work under contract 82EV72015-001 and .001A.

This work is taken in part from a dissertation to be submitted to the Graduate School, University of Maryland, by W. J. Sonnefeld in partial fulfillment of the requirement for the PhD degree in Chemistry.

ENRICHMENT OF PAH AND PAH DERIVATIVES FROM AUTOMOBILE EXHAUSTS, BY MEANS OF A CRYO-GRADIENT SAMPLING SYSTEM.

ULF STENBERG+, ROGER WESTERHOLM+, TOMAS ALSBERG+, ULF RANNUG++, ANNICA SUNDVALL++
+ Department of Analytical Chemistry, Arrhenius Laboratory,
++Department of Toxicology Genetics, Wallenberg Laboratory,
University of Stockholm, S-106 91 Stockholm, Sweden

INTRODUCTION

The demand for reliable risk assessments concerning air pollution from motor vehicles has created an especial interest in the technology of sampling polynuclear aromatic hydrocarbons, PAH, and related compounds. Earlier sampling methods were based on enrichment of undiluted gases, (1,2) i.e., the emission was trapped prior to any reaction with ambient air. However, to simulate the fate of the exhausts immediately after the emission into air, a dilution tunnel can be employed (3). With this technique it is anticipated that adsorption/condensation on particles and/or certain reactions take place before sampling. Introduction of the dilution technique had the result that most of the chemical analyses and mutagenicity tests have been carried out only on particulate samples.

This paper describes a trapping device which permits sampling of both volatiles and particles, from either undiluted or diluted exhausts. The utility of the equipment is demonstrated by means of PAH analyses and mutagenicity tests.

MATERIALS AND METHODS

Sampling System

We have selected a particulate trapping system followed by condensation at cryo temperatures, i.e., "gas phase" enrichment. However, it is necessary to emphasize that "gas phase" is not an unambiguous definition since the composition will vary, among other factors, with exhaust temperature or dilution ratio.

The sampling system consists of a filter holder and cooling equipment, FIGURE 1, mounted on a mobile rack of approximately 2 m height. The pump unit can be used either as a proportional sampler, using a reference signal from the air intake of the carburettor, or as a constant flow sampler compensating for the drop in pressure.

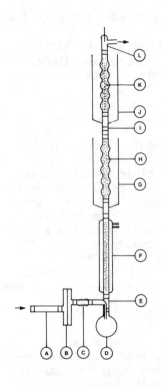

FIGURE 1. Sampling equipment.
A. Flexible metal tubing, 250 mm
B. Filter holder, stainless steel, ⌀ 127 mm
C. Teflon bellow
D. Round bottom flask, 1000 mL
E. Normal glass joint
F. Ice/water condenser, filled with glass beads ⌀ 6 mm
G. Cryo vessel, dry ice/ethanol
H. Condenser cartridge, filled with glass rings 8x8 mm, 6x6 mm and glass beads ⌀ 5 mm
I. Ball glass joint
J. Cryo vessel, liquid nitrogen
K. Condenser cartridge, filled with glass rings 6x6 mm, and glass beads ⌀ 5 mm, ⌀ 2 mm
L. Teflon/glass joint, temp. sensor

Vehicle Test Procedure

The test vehicles, gasoline 2.1 L and diesel 2.3 L, were driven on a chassis dynamometer according to FTP-72. Undiluted gas was sampled at the tailpipe, and diluted gas in a dilution tunnel (0.25 x 4.6 m) connected to a Horiba CVS. The approximate dilution ratios in these tests were 1:14, gasoline sample 1 and 2 and 1:10 sample 3. The diesel sample was diluted approximately 1:8.

Extraction and Analysis

Before sampling, the filters (Gelman A/E) were washed in ethanol and heated for 1 h at 120 °C. After use they were extracted with acetone for 18 hrs. The condensers were extracted with acetone in a specially designed Soxhlet apparatus. Of the extracts 75 % (approx. 4 mL) were used for mutagenicity tests.

PAH analyses, described elsewhere (4), were performed by capillary gas chromatography with FID and internal standards, and by means of mass spectrometry, (JEOL D-300 connected to an INCOS 2000 computor system, EI 70 eV, scanning rate 35-350 amu/sec, resolution 700.)

Mutagenicity Testing

Mutagenicity tests were performed as described by Ames et al. (5), using the Salmonella typhimurium strains TA98 and TA100. All samples were tested with and without the addition of a metabolizing system (S9), and the results were determined by regression analysis.

RESULTS AND DISCUSSION

Sampling System

Before sampling authentic automobile exhausts, the utility of the system was evaluated by different laboratory tests. For this purpose a small dilution tunnel was built in order to generate different artificial sample streams. First, the physical arrangements of the filling material in the condensers were optimized to obtain as large a sample amount as possible before clogging, (ice and dry ice formation). By using different filling materials (glass rings and beads of different sizes) about 1.5 m^3 sample volume could be sampled at a maintained flow of 75 L/min, (air saturated with water at 25 °C and 7% CO_2).

This type of cryogenic trapping device will inevitably create aerosols in the sample stream. However, ice and dry ice formation in the two last condensers enhances the trapping surface, and thus prevents the aerosol from penetrating the system. This was confirmed the following way: cyclohexane, C-6, was mixed with dry air to 52 ppm in the laboratory dilution tunnel. 32 L/min of this mixture was sampled and with a filter (held at -70 °C) placed at top of the sampling system, i.e. after the condensers. The concentration of C-6 found in the sample stream present between the condensers and the filter was 6 ppm, while after passing through the filter it contained less than 0.5 ppm. This shows that C-6 formed fine particles and, in spite of the filling material, this aerosol could not be completely retained. When introducing water and CO_2 in the sample stream, as described previously, a rapid formation of ice and dry ice crystals takes place and the efficiency increases. However, this entails that, before this larger surface area has been built up, less efficiency should be encountered during the first seconds at real sampling.

Furthermore, other trace components of different volatility were generated in the sample stream and the trapping efficiency was calculated, TABLE I. These break-through studies were performed be means of gas chromatography-FID with a heated loop injector.

TABLE 1

TRAPPING EFFICIENCY FOR ARTIFICIAL SAMPLE STREAMS

Compound	Flow(L/min)	C_i(ppm)[a]	Trapping %[b]
n-pentane	71.7	52.3	93 ± 4
	54.5	78.2	98 ± 1
cyclohexane	70.6	45.0	97 ± 1
	34.5	50.0	98 ± 3
chlorobenzene	74.4	17.1	100 ± 1
isobuturic aldehyde	34.3	46.6	100 ± 1

a=inlet concentration, detection limit 0,5ppm
b=$(1-C/C_i)100$, C=outlet concentration, 3-5 determinations.

All these experiments were performed with an entrance temperature of 25 °C which gave exit condenser temperatures of +8 °C, -55 °C and -155 °C respectively. Thus, the temperature in the liquid nitrogen condenser indicated the possibility of trapping NO. This gas is prevalent in automobile exhausts and has a freezing point of -164 °C. In order to investigate if NO was enriched, a 200 ppm sample stream was generated. The trapping of NO was not quantified, but after disassembling of the liquid nitrogen condenser, NO was observed as a bright blue solid.

Gas Phase PAH

Chemical analyses and mutagenicity tests performed on samples taken simultaineously in diluted and undiluted exhausts have, to our knowledge, not been reported previously. Our results show that the dilution of gasoline exhausts entailes an increased adsorption of PAH onto particles, FIGURE 2. However, this effect is only significant for components such as benzo(a)anthracene, chrysene and cyclopenteno(cd)pyrene which will then be particle associated to more than 50%, compared to approximately 30% for undiluted gases. For components such as phenanthrene and fluoranthene the dilution effect is negligiable. Lee et al. (6) have reported somewhat different results, using another technique. Our results also show that components such as the benzopyrene isomers can still be found in the gas phase after dilution. However, for components with molecular weight ≥ 252, dilution is not necessary since most of these are adsorbed onto the particles already in the tailpipe.

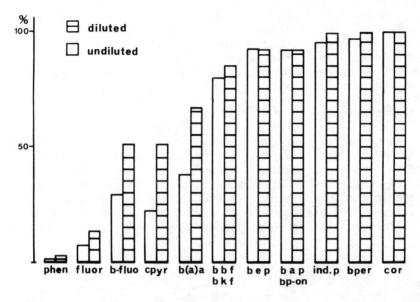

FIGURE 2. Percent particle associated PAH in gasoline exhaust, FTP, cold start. phen=phenanthrene, fluor=fluoranthene, b-fluo= benzo(ghi)fluoranthene, cpyr=cyclopenteno(cd)pyrene, b(a)a= benzo(a)anthracene, bbf/bkf=benzo(b&k)fluoranthenes, bep=benzo-(e)pyrene, bap=benzo(a)pyrene, bp-on=benzo(cd)pyrenone, ind.p= indeno(1,2,3-cd)pyrene, bper=benzo(ghi)perylene, cor=coronene

Nitro PAH

The difference in sampling of undiluted and diluted exhausts could also be reflected in the amount of reactive gases which passes the filter and condensers. Reactions with nitrogen oxides and PAH have been suggested, and it has been shown that nitro-derivatives of certain PAH can be found in diesel exhausts (7). However, it has also been reflected upon the occurrence of e.g. 1-nitropyrene in diesel exhausts as an artifact of sampling, and to investigate this the following experiments were performed. One sampling stream from the dilution tunnel was diverted to three parallel filter and ice/water cooled condensers. Immediately in front of two of the filters, NO_2 was added to give total levels of 5 and 20 ppm respectively. The experiment was performed twice and with both a diesel and a gasoline vehicle. With MS/Multiple Ion Detection mono-nitro-phenanthrene and -pyrene were detected in the particulate diesel emission. However, there was no difference between the three filters, i.e. during sampling no nitration occured which

increased the amount found. In the particulate emission from the gasoline vehicle no nitrated phenanthrene or pyrene could be detected. The detection limit was approx 0.05 µg/m³ exhaust. The amount found in the diesel emission was 0.13 µg/m³, (nitropyrene in diluted exhausts).
However, unstable components like b(a)p and cyclopenteno(cd)-pyrene were degraded when adding NO_2, at both levels.

Mutagenic effects

Mutagenicity tests were performed both on the particulate phase and the gas phase from diluted as well as undiluted gasoline and diesel exhausts. The results from samples of diluted exhausts, FIGURE 3, show that all the samples were mutagenic to S. typhimurium TA 100. Mutations were also induced in strain TA 98 although the effect of the gas phase was lower. Thus biologically active components are present also in the gas

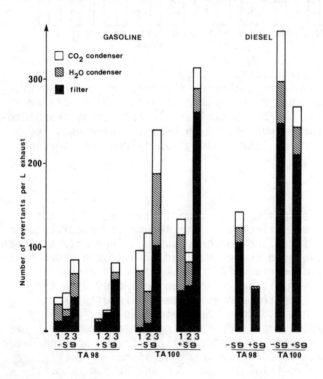

FIGURE 3. The mutagenic effect of samples from diluted gasoline and diesel exhausts on Salmonella typhimurium TA 98 and TA 100 in the presence and absence of a metabolizing system (S9).

phase. The contribution of the gas phase to the total mutagenicity seems to be more important in the absence of S9, than in the presence of S9, both with gasoline and diesel exhausts. With gasoline exhausts the particulate phase normally gives a higher mutagenic effect with S9 than without. This cannot be seen with the gas phase. The distribution of high and low molecular weight PAH between the particulate and the gas phase, FIGURE 2, may explain this difference in mutagenicity. The results also (8) indicate that the relative contribution of mutagenicity from PAH is smaller in the gas phase than in the particulate phase.

ACKNOWLEDGEMENT

Special thanks are due to Karl Erik Egebäck and Gunnar Tejle at Air Pollution Research Laboratory, Motor Vehicle Section at Studsvik. This work has been sponsored by the National Swedish Environment Protection Board.

REFERENCES

1. Gross, G.P. (1972): Gasoline composition and vehicle exhaust gas polynuclear aromatic content. NTIS PB 209 955
2. Grimmer, G., Hildebrandt, A., and Böhnke, H. (1973): Sampling and analytics of polycyclic aromatic hydrocarbons in automobile exhaust gas: 1. Optimization of the collecting arrangements, Zbl. Bakt. Hyg., I. Abt. Orig., 158:22-34.
3. Hare, C.T., Springer, K.J., and Bradow, R.L. (1976): Fuel and additive effects on diesel particulate-development and demonstration of methodology, SAE, 760130:527-555.
4. Stenberg, U., Alsberg, T., and Bertilsson, B.M. (1981): A comparison of the emission of polynuclear aromatic hydrocarbons from automobiles using gasoline or a methanol/gasoline blend, SAE, 810441.
5. Ames, B.N., McCann, J., and Yamasaki, E. (1975): Methods for detecting carcinogens and mutagens with the Salmonella/mammalian-microsome mutagenicity test. Mutation Res., 31:347-364.
6. Lee, F.S-C., Prater, T.J., and Ferris, F. (1979): PAH emission from a stratified-charge vehicle with and without oxidation catalyst: sampling and analysis evaluation. In: Polynuclear Aromatic Hydrocarbons, edited by P.W. Jones and P. Leber, pp. 83-110, Ann Arbor Science Press, Ann Arbor.
7. Schuetzle, D., Riley, T., Prater, T.J., Harvey, T.M., and Hunt, D.F. (1981): The identification of nitrated derivatives of PAH in diesel particulates, Anal. Chem., (in press).
8. Rannug, U., Sundvall, A., Westerholm, R., Alsberg, T., and Stenberg, U. (1982): Proceedings from EPA Symposium-Application of short term bioassays in the analysis of complex environmental mixtures, January 25-27, 1982, Chapel Hill, NC.

SUNLIGHT ACTIVATION OF SHALE OIL BYPRODUCTS AS MEASURED BY GENOTOXIC EFFECTS IN CULTURED CHINESE HAMSTER CELLS

GARY F. STRNISTE, DAVID J. CHEN, AND RICHARD T. OKINAKA
Genetics Group, Life Sciences Division, Los Alamos National Laboratory, Los Alamos, New Mexico 87545

INTRODUCTION

Activation of certain classes of promutagens/procarcinogens can be accomplished by exposure to various radiation sources. We have been studying the process of photoactivation with near ultraviolet light (NUV) of polycyclic aromatic hydrocarbons, both at the molecular and cellular levels (2,3,5,6), and have extended these studies toward assessing potential health and environmental problems resulting from photoactivation of complex waste streams generated by the developing oil shale industry (1,4,7,8,9). Retort processes currently in use in the production of shale oil generate significant quantities of "process waters" which contain a wide spectrum of UV-absorbing, organic material. Photoactivation of these waters with an artificial source of NUV results in genotoxic events in cultured mammalian cells (1,4,7,8,9). Since significant amounts (2-4%) of solar radiation reaching the earth's surface is NUV, we were concerned about potential biological effects resulting from solar-irradiated waste streams. This paper summarizes new and previously published data (8) concerning the induction of both cytotoxicity and mutagenicity in cultured Chinese hamster cells (line CHO) after their exposure to a particular oil shale retort process water and natural sunlight.

MATERIALS AND METHODS

Retort Process Water

The water used throughout this study was obtained from a holding tank for shale oil produced in an above-ground retort utilizing shale deposits from the Green River Formation located in the western United States. The water was filtered through Whatman #42 paper and a Millipore 0.2 μ Swinnex unit before use.

Tissue Culture

CHO cells (AA8-4) obtained from Dr. L. Thompson (Lawrence Livermore National Laboratory) were cultured under conditions described elsewhere (7). The protocols for determining cytotoxicity (colony formation assay) and mutagenicity at the hypoxanthine-guanine phosphoribosyl transferase (HGPRT) locus have been previously reported (7,8). The plating efficiencies for non-treated cells were regularly between 90 and 100%.

Irradiation

The source of artificial NUV (300-400 nm) was two parallel 15 W black lights (General Electric #15T8BLB). The incident fluence was 5.6 J per m^2 per sec. For experiments utilizing natural sunlight, petri dishes with attached, retort water treated cells were exposed for various intervals of time on the rooftop of the Health Research Laboratory at the Los Alamos National Laboratory, Los Alamos, New Mexico. The incident fluence of sunlight (200-2500 nm) was between 850-950 J per m^2 per sec. Details of these irradiation procedures have been published elsewhere (6,7,8).

RESULTS AND DISCUSSION

Our previous observations in which shale oil byproducts were transformed into genotoxic agents upon exposure to artificial NUV (1,4,7,9) suggested that natural sunlight may have photoactivating potential. To test this, CHO cells were pretreated with a surface retort process water (1:270 dilution for 1 h in the dark) and subsequently exposed for various times to natural sunlight. The survival data (in terms of colony forming ability) are shown in Figure 1 (right panel). For comparison, the results of an experiment in which artificial NUV was used as the radiation source are also shown (left panel). The exposure time of sunlight necessary to reduce cell survival to 37% (D_{37}) in the treated population is about 50 sec, whereas non-treated CHO cells are inactivated to 37% surviving fraction only after 6 min exposure to sunlight. This represents a sevenfold increase in the photosensitivity of CHO cells to sunlight when pretreated with the process water as compared to untreated cells. The measured fluence for sunlight (200-2500 nm wavelength) in Los Alamos, NM, during January 1981 (when

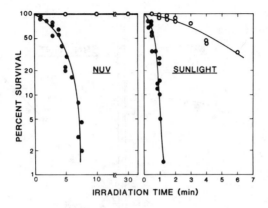

FIGURE 1. Cytotoxicity induced in CHO cells by oil shale retort process water and artificial NUV or sunlight. Open symbols represent cells exposed to light only whereas closed symbols represent cells exposed to process water (1:270 dilution) 1 h prior to irradiation.

the experiments were performed) averaged 900 J per m^2 per sec of which we estimate ~ 3% is NUV (300-400 nm). Thus, the dose of NUV from sunlight necessary to reduce survival to 37% is about 1400 J per m^2, strikingly similar to the dose of artificial NUV (5 min or 1680 J per m^2) necessary to elicit a similar cytotoxic response.

In Figure 2 we show the number of HGPRT deficient mutants induced per 10^5 viable cells as a function of cell survival for CHO cells pretreated with oil shale process water and subsequently exposed to either artificial NUV (left panel) or sunlight (right panel). At 37% survival, there is about a sevenfold increase in the number of mutants induced by sunlight for cells pretreated with process water compared to untreated cells after subtraction of the background mutation frequency (~ 1 per 10^5 cells). It should be noted that no genotoxic effects were seen in CHO cells exposed to artificial NUV only (up to 30 min exposure).

To examine the role of the NUV component of natural sunlight in this photoactivation process, an experiment was performed in which special "cut-off" light filters were placed between the source (sun) and the target (plated

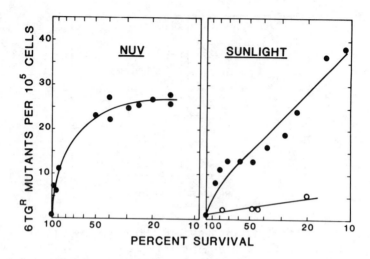

FIGURE 2. Mutagenicity induced at the HGPRT locus in CHO cells by oil shale retort process water and artificial NUV or sunlight. The numbers of mutants induced at each dose point have been corrected for plating efficiencies determined at time of selection. Open symbols represent cells exposed to light only whereas closed symbols represent cells exposed to the process water (1:270 dilution) 1 h prior to irradiation.

cells). The results are presented in Table 1. Using the colorless glass, sharp-cut UV filter (< 0.1% transmission for light of 320 nm or less), similar cytotoxic and mutagenic effects were observed compared to treated and non-filtered, sun exposed cells. However, when a sharp-cut yellow glass filter was employed, which transmits < 0.1% light at 400 nm or less, cell survival increased and cell mutagenicity decreased substantially. These data suggest that the component of sunlight responsible for photoactivation is primarily NUV although a measurable portion of this photoactivating potential resides at wavelengths > 400 nm.

Experiments currently in progress are employing other "cut-off" filters to determine the potential of violet-blue visible light (400-470 nm) in photoactivating these complex mixtures. Preliminary results using a daylight fluorescent lamp (primarily > 400 nm wavelength radiation) indicate photo-induced genotoxicity in retort process water treated CHO cells.

TABLE 1

WAVELENGTH DEPENDENCE OF GENOTOXICITY INDUCED IN CHO CELLS BY SUNLIGHT-ACTIVATED RETORT PROCESS WATER

Treatment	Filter	Cell Survival[b]	Mutagenicity[c]
PW[a] (dark control)	None	100%	2
PW plus sun(60 sec)	None	15%	36
PW plus sun(65 sec)[d]	WG 360[e]	22%	31
PW plus sun(66 sec)[d]	GG 420[f]	82%	12

a   PW: above-ground oil shale retort process water at 1:270 dilution.
b   Survival measured by colony formation assay.
c   Mutagenicity measured at the HGPRT locus and reported as the number of mutants induced per 10^5 viable cells.
d   Exposure time adjusted for transmittance correction factor for particular filter.
e   Schott Optical Glass Inc. colorless glass, sharp-cut UV filter.
f   Schott Optical Glass Inc. sharp-cut yellow glass filter.

ACKNOWLEDGMENTS

We thank Ms. J. Bingham and Ms. E. Wilmoth for their technical assistance. Also, we wish to thank Ms. V. Strniste for editing and typing this manuscript. This work was funded by the Department of Energy.

REFERENCES

1. Chen, D.J. and Strniste, G.F. (1981): Induction of mutations at five genetic loci in CHO cells by near ultraviolet light-activated shale oil process water, Environ. Mutagenesis, 3:308.
2. Hoard, D.E., Ratliff, R.L., Bingham, J.M., and Strniste, G.F. (1981): Reactions induced in vitro between model DNA and benzo(a)pyrene by near ultraviolet light, Chem.-Biol. Inter., 33:179-194.

3. Strniste, G.F. and Brake, R.J. (1980): Photoactivity of near ultraviolet light-irradiated benzo(a)pyrene, Fed. Proc., 39:2011.
4. Strniste, G.F. and Brake, R.J. (1980): Toxicity and mutagenicity of shale oil retort product waters photoactivated by near ultraviolet light, Environ. Mutagenesis, 2:268.
5. Strniste, G.F., Martinez, E., Martinez, A.M., and Brake, R.J. (1980): Photo-induced reactions of benzo(a)pyrene with DNA in vitro, Cancer Res., 40:245-252.
6. Strniste, G.F. and Brake, R.J. (1981): Cytotoxicity in human skin fibroblasts induced by photoactivated polycyclic aromatic hydrocarbons, In: Polycyclic Aromatic Hydrocarbons, edited by M. Cooke and A.J. Dennis, pp. 109-118, Battelle Press, Columbus, Ohio.
7. Strinste, G.F. and Chen, D.J. (1981): Cytotoxic and mutagenic properties of shale oil byproducts. I. Activation of retort process waters with near ultraviolet light, Environ. Mutagenesis, 3:221-231.
8. Strniste, G.F., Chen, D.J., and Okinaka, R.T. (1981): Genotoxic effects of sunlight-activated waste waters, J. Natl. Cancer Inst., (in press).
9. Strniste, G.F., Martinez, E., and Chen, D.J. (1981): Light-activation of shale oil byproducts, In: The Los Alamos Integrated Oil Shale Health and Environmental Program. A Status Report. LA-8665-SR, Los Alamos National Laboratory, Los Alamos, NM, pp. 61-64.

CHEMICAL AND ENZYMATIC OXIDATIONS OF SUBSTITUTED BENZO(A)PYRENES

PAUL D. SULLIVAN, LARRY E. ELLIS, LUZ M. CALLE AND IGNACIO J. OCASIO
Department of Chemistry, Ohio University, Athens, Ohio 45701.

INTRODUCTION

Several recent studies have suggested that multiple pathways may exist for the activation of polycyclic aromatic hydrocarbons to ultimate carcinogenic forms (1,2). However, there is convincing evidence that the primary ultimate carcinogenic form of benzo(a)pyrene (BP) is a 7,8-diol-9,10-epoxide derivative (3,4). However substitution, particularly at positions 7,8,9 or 10, might block the formation of the diol-epoxide and other activation pathways may be important for such derivatives. The mutagenicity and carcinogenicity of a collection of substituted BPs are summarized in Table I. While several of these compounds are more mutagenic than the parent compound, none have been reported to be more carcinogenic. Although it should be stressed that several 7,8,9 or 10-substituted BPs exhibit weak carcinogenicity. In view of this it is certainly worthwhile to investigate the oxidative behavior of such BP derivatives to further understand possible activation pathways. This paper reports a summary of our investigations into chemical and enzymatic oxidations of 6-, 7- and 10-methyl BPs, 6,10- and 7,10-dimethyl BPs, 7-, 8-, 9- and 10-fluoro BPs, and the parent compound itself using electron paramagnetic resonance (EPR) spectroscopy to detect radical intermediates and high performance liquid chromatography (HPLC) to separate stable reaction products.

MATERIALS AND METHODS

Samples of 6-methyl, 6-ethyl, 6-methoxy and 6-acetoxy-BP were supplied by Dr. E. Cavalieri; 10-methyl, 7,10-dimethyl, and 7-, 8-, 9- and 10-fluoro-BP were obtained from Dr. M. S. Newman, 7-methyl and 4,5,7-trimethyl-BP from Dr. R. G. Harvey and the other dimethyl derivatives (except the 6,10-) from Dr. G. H. Daub. 6-Methyl, 10-methyl and 6-hydroxymethyl-BP were also synthesized by literature methods (13-14) and 6,10-dimethyl and 6-hydroxymethyl-10-methyl-BP are new compounds, the synthesis of which will be reported elsewhere.

TABLE 1

MUTAGENICITY AND CARCINOGENICITY OF SUBSTITUTED BENZO(A)PYRENES

BP Derivative	Mutagenicity[a] (Revs/nmole)	Carcinogenicity[b]
6-Hydroxymethyl-	268	++
6-Methyl-	199	+++
6-Hydroxymethyl-10-methyl-	163	N.D.
8-Fluoro-	160	++
1,6-Dimethyl-	71	++
10-Methyl-	64	+
6,10-Dimethyl-	60	N.D.
7-Fluoro-	54	+
Benzo(a)pyrene	41	+++
10-Fluoro-	29	++
9-Fluoro-	26	++
3,6-Dimethyl-	20	-
7-Methyl-	14	++
6-Ethyl-	9	-
6-Acetoxy-	8	-
6-Methoxy-	6	+
2,3-Dimethyl-	0	+
4,5-Dimethyl-	0	+
7,10-Dimethyl-	0	-
4,5,7-Trimethyl-	0	-
1,3-Dimethyl-	0	++

a. Towards strain TA98 of Salmonella typhimurium, results from our laboratory (12 and unpublished), revertants ±10%.
b. Ref 2,5-11; results for fluorinated BP's communicated by Drs. E. C. and J. A. Miller.

Experimental Methods

In all cases except with H_2SO_4 1-2 mg of the hydrocarbon were dissolved in acetone or benzene for the reaction.

Microsomes. The microsomes from male Sprague-Dawley rats induced with β-napthoflavone were prepared according to standard procedures (15). Incubations consisted of 1 ml of microsomes in tris-chloride buffer, 6-8 mg of NADPH and the hydrocarbon in 100 μl of acetone.

Horseradish Peroxidase-Hydrogen Peroxide (HRP/H_2O_2). 10 Mg of HRP (type II from Sigma) in 500 µl water and 10 µl 30% H_2O_2 were added to the hydrocarbon in acetone and the mixture incubated for 30 minutes at 37°C.

Fenton's Reagent. 5 µl of 30% H_2O_2 and 10 µl of 1 M $FeSO_4$ were added to the hydrocarbon and the mixture vortexed for 5 min.

Trifluoroacetic Acid-Hydrogen Peroxide (TFA/H_2O_2). 10 µl of TFA and 10 µl of 30% H_2O_2 were added to the hydrocarbon in benzene and the mixture vortexed for 5 min.

Thallium Tristrifluoroacetate-Trifluoroacetic Acid (TTFA/TFA). 1-3 µl of 0.27 M TTFA in TFA were added to the hydrocarbon in benzene and the solution was vortexed for 5 min.

Sulfuric Acid (H_2SO_4). The hydrocarbon was dissolved in 0.1 ml of conc H_2SO_4 and allowed to stand. For HPLC analysis H_2O was added before $CHCl_3$ extraction, whereas for EPR measurements the sample was placed directly in a capillary tube.

HPLC. The reaction mixtures containing acetone were extracted with $CHCl_3$, the solvent evaporated and the residue dissolved in 30 µl of MeOH. Those run in C_6H_6 were filtered, evaporated and redissolved in MeOH. Using a Tracor liquid chromatograph with solvent programmer and 254 nm detector the products were separated on a µ-Bondapack C_{18} column. The composition of the solvent reservoirs was (A) 60% MeOH, 39.9% H_2O, 0.1% H_3PO_4 and (B) 85% MeOH, 14.9% H_2O, 0.1% H_3PO_4. A 20% nonlinear concave gradient program was used at a rate of 1.25%/min with initial conditions 10% B and final conditions 100% B. Benzo(a)pyrene products were identified by comparison with genuine samples of BP metabolites. Retention time windows for dihydrodiols (diols), quinones and phenols of substituted BP's were calculated by taking the retention times of the known BP metabolites and multiplying by the ratio of the retention time of the substituted hydrocarbon over the retention time of BP itself.

EPR. Spectra were recorded on a Varian E-12 spectrometer using a dual sample cavity; g values were obtained using a perylene radical anion as a secondary standard.

RESULTS AND DISCUSSION

Oxidation Products of BP and Alkylated Derivatives

    Previous studies with the parent compound have established

that a common pathway of BP oxidation which occurs under a variety of chemical and enzymatic conditions is oxidation to 6-hydroxy-BP followed by autoxidation to a 6-oxy-BP radical intermediate leading to BP quinones as stable end products (16,17). Thus, those oxidizing conditions, i.e. TFA/H_2O_2, Fenton's, HRP/H_2O_2 and microsomes, which give rise to an EPR signal of the 6-oxy-BP radical also give BP quinones (1,6-, 3,6- and 6,12-) on HPLC separation. Microsomal oxidation of BP also takes place via epoxidation leading to dihydrodiols and phenols as stable products. In our experiments the 9,10-, 4,5- and 7,8-dihydrodiols were identified along with the 9- and 3- phenols (Fig. 1). No other oxidizing system gave such a range of products for BP. Fenton's and HRP/H_2O_2 gave similar products, namely quinones with an additional peak in the diol range. TFA/H_2O_2 also gave peaks in the diol region as well as quinones. Oxidation with H_2SO_4 and TTFA/TFA (12) gave complex well resolved EPR signals similar to those previously reported in H_2SO_4 (18); the radical species in H_2SO_4 may well be the BP cation radical because only small amounts of products were detected by HPLC. In TTFA/TFA the species giving rise to the EPR spectrum may be the 6-trifluoroacetoxy derivative obtained by reaction of the initially produced cation radical with the trifluoroacetoxy anion. This parallels previously reported behavior (19,20) and is supported by the observation of a high yield reaction product assigned as the 6-trifluoroacetoxy derivative.

7-Methyl, 10-methyl and 7,10-dimethyl-BP were found to behave similarly in the oxidizing systems investigated. After microsomal oxidation 7-methyl-BP gave 3 major peaks on HPLC separation, one in the diol region (assigned as a 9,10-diol), one in the quinone region which may be due to a hydroxymethyl derivative (21), and one phenolic peak (Fig. 1). 10-Methyl-BP similarly gave major peaks in the diol region (probably 4,5- and 7,8-diols), the quinone region and the phenol region (Fig. 1). 7,10-Dimethyl BP gave only a number of small peaks in the diol region (not identified), a peak in the quinone region and one in the phenol region. Consistent with the appearance of quinones as stable products, EPR spectra were observed for 7-methyl, 10-methyl and more weakly for 7,10-dimethyl-BP which are assigned to the 6-oxy radicals of these compounds. Similar better resolved EPR signals were obtained from these compounds on oxidation with Fenton's reagent, HRP/H_2O_2 or TFA/H_2O_2 (12). Product analysis under these conditions also showed the presence of peaks identified as quinones. HRP/H_2O_2 (Fig. 1) and Fenton's gave 3 peaks in the quinone region for all 3 compounds, probably the 1,6-, 3,6- and 6,12-quinones; additionally a strong peak was found in the diol region for

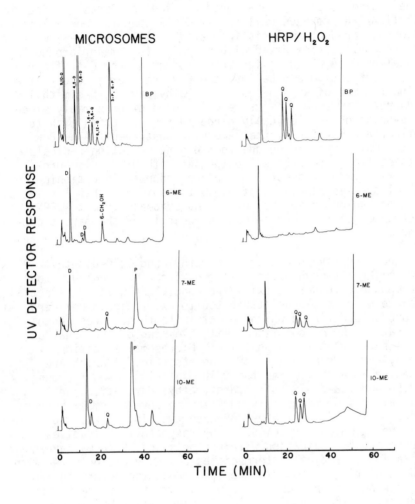

FIGURE 1. HPLC chromatograms obtained after oxidation with rat liver microsomes (microsomes) and horseradish peroxidase-hydrogen peroxide (HRP/H_2O_2) of benzo(a)pyrene (BP), 6-methyl-benzo(a)pyrene (6-ME), 7-methyl-benzo(a)pyrene (7-ME) and 10-methyl-benzo(a)pyrene (10-ME). Peaks marked with a D occur in the diol region, those with a Q in the quinone region and those with a P in the phenol region.

all 3 compounds. Products from TFA/H_2O_2 oxidation also included quinones plus several unidentified compounds. Oxidation with TTFA/TFA produced well resolved EPR spectra from all three compounds; product analysis indicated that 10-methyl-BP was almost completely oxidized within 5 mins to a major product. By analogy with BP we assign this product as the 6-trifluoroacetoxy derivative and believe the EPR spectrum is due to the cation radical of this product. 7-Methyl and 7,10-dimethyl-BP do not show such a major reaction product presumably since the 6-position is sterically hindered from reaction with the trifluoroacetoxy ion by the 7-methyl substituent. The EPR spectra may therefore be due to the genuine cation radicals of these two hydrocarbons. Oxidation with H_2SO_4 also produced resolved EPR spectra which are undoubtedly cationic radical species; analysis of products indicated that little reaction had occurred thus providing evidence that the EPR spectra are of the genuine cation radicals of the parent compounds.

The oxidative behavior of 6-methyl and 6,10-dimethyl-BP are similar although quite different from the previous group of compounds. Under no oxidizing conditions were EPR spectra obtained which could be attributed to oxy radicals nor were any products detected which we attribute to quinones, both observations being consistent with blocking of the usual oxidation pathway at the 6-position. Microsomal oxidation of 6-methyl-BP gave a variety of products (Fig. 1) which are tentatively identified as the 9,10-diol plus lesser amounts of 4,5- and 7,8-diols, a 6-hydroxymethyl derivative and small amounts of phenols in addition to several small unidentified peaks. 6,10-Dimethyl-BP gave two diol peaks, two small peaks which may be due to hydroxymethyl derivatives and a phenol peak. Oxidation with Fenton's and HRP/H_2O_2 gave no indication of radical intermediates and only a single reaction product eluting in the diol region. Reaction occurs in TFA/H_2O_2 for both compounds to produce a number of unidentified products. Oxidation with TTFA/TFA gave well resolved EPR spectra which are similar to those obtained on oxidation with H_2SO_4 and which are assigned as the cation radicals of the parent compounds.

Oxidation Products of Fluorinated BPs

7-,8-,9- and 10-Fluoro BPs behave quite similarly to each other and to 7- and 10-methyl BPs in the oxidizing systems investigated. Microsomal oxidations gave peaks in the regions of the chromatograms expected for dihydrodiol, quinone and phenolic products (Fig. 2). The quinone peaks

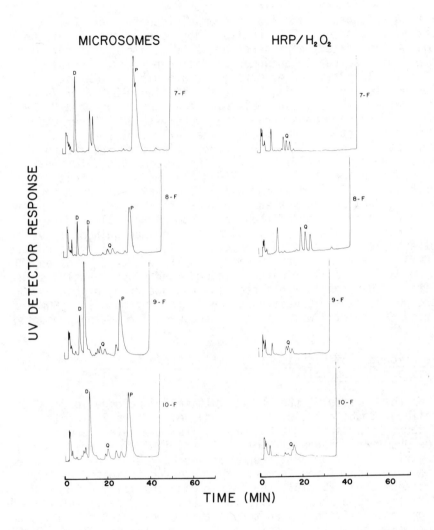

FIGURE 2. HPLC chromatograms obtained after oxidation with rat liver microsomes (microsomes) and horseradish peroxidase-hydrogen peroxide (HRP/H_2O_2) of 7-fluoro- (7-F), 8-fluoro- (8-F), 9-fluoro- (9-F) and 10-fluoro-benzo(a)pyrene (10 F). Peaks marked with a D occur in the diol region, those with a Q in the quinone region, and those with a P in the phenol region.

OXIDATION OF SUBSTITUTED BENZO(A)PYRENES

match similar peaks obtained on both HRP/H_2O_2 (Fig. 2) and Fenton's reagent oxidations presumably due to the 1,6-, 3,6- and 6,12-quinones. EPR spectra under both of the latter conditions indicate the presence of oxy-radicals consistent with the 6-oxy-radical to quinones oxidation pathway. In the microsomal system one or more major products in the diol region are found for 7-, 8-, 9- and 10-fluoro-BP. For 7- and 8-fluoro-BP there is an early diol peak which may be a 9,10-diol if the compounds behave analogously to BP. Other major diol peaks could be 4,5-diols. The early diol peak is absent for both 9- and 10-fluoro BP's suggesting that the fluoro substituent is an efficient blocking group. The later peaks are anticipated to be due to 7,8-diols and/or 4,5-diols. A peak in the diol region for all 4 fluoro-BPs is found for HRP/H_2O_2 and Fenton's reagent oxidations which does not appear to correspond to any of the diols from microsomal oxidation. Products from TFA/H_2O_2 oxidation also include quinones plus other major unidentified products particularly with the 9- and 10-fluoro BPs. Oxidation with TTFA/TFA produced resolved EPR spectra for the fluorinated compounds and product analysis indicated that considerable reaction had occurred, probably at the unsubstituted 6-position. Again, oxidation with H_2SO_4 produced EPR spectra with little in the way of reaction products.

Relationship of Microsomal Oxidation Products to Carcinogenicity and Mutagenicity

For BP itself, the primary ultimate carcinogenic form is believed to be a 7,8-diol-9,10-epoxide, and the 7,8-diol produced in microsomal metabolisms is its immediate precursor. The observed mutagenicity of BP is attributed primarily to contributions from BP-4,5-epoxide and BP-7,8-diol-9,10-epoxide, the formation of the former compound being indicated by the presence of a 4,5-diol, the latter compound arising from the 7,8-diol precursor. Thus, perhaps the most significant reaction products with regard to possible carcinogenicity or mutagenicity of the substituted BPs are also the diols. The reduced carcinogenicity (relative to BP) of all 7-,8-,9-, or 10-substituted BPs so far investigated may be due to the blocking effect of the substituents which probably reduce or eliminate formation of a 7,8-diol-9,10-epoxide. Thus 7-methyl-BP shows no major products in the region expected for a 7,8-diol, the only diol peak observed being consistent with a 9,10-diol, which may explain its low mutagenicity in the Ames test if, by analogy with BP, the 4,5-oxide and diol-epoxide are its most mutagenic metabolites. On the other hand both 7-fluoro and 8-fluoro BPs show other diol products in addition to products consistent with 9,10-diols which may be 4,5-diols. Both compounds are mutagenic (Table I), the 8-fluoro in particular being approximately 4 times more so than BP. The explanation for this effect must await mutagenicity testing of the individual metabolites of 8-fluoro-BP.

Just as 7- or 8-substitution appears to block the formation of a 7,8-diol, 9- or 10-substitution seems to prevent the formation of a 9,10-diol. If the elution order of diols remains the same as for BP (i.e. 4,5-diol before 7,8-diol), then 10-methyl-BP produces more 4,5-diol, whereas 9- and 10-fluoro BPs produce more 7,8-diol. This could explain the greater mutagenicity of 10-methyl-BP relative to BP and 9- or 10-fluoro BPs.

Microsomal oxidation of 6-methyl and 6,10-dimethyl-BP proceeds somewhat differently than for the previous compounds. Blocking the 6-position prevents the initial reaction at this position leading to quinones. For 6-methyl-BP the major products are diols (predominantly the 9,10-diol) and the 6-hydroxymethyl derivative. Since 6-hydroxymethyl-BP is found to be more mutagenic than 6-methyl-BP (Table I), the high mutagenicity of the latter compound may be due to side-chain oxidation rather than the diol-epoxide or epoxide. Such oxidation products have also been suggested as the ultimate carcinogenic forms of 6-methyl-BP (1) and may be formed via a cation radical intermediate. 6,10-Dimethyl-BP behaves similarly and the mutagenicity of this compound is probably due to a 6-hydroxymethyl derivative which is found to be much more mutagenic than 6,10-dimethyl-BP itself (Table I).

It is clear that further studies to more conclusively identify oxidation products are required to substantiate the interpretation of our observations.

ACKNOWLEDGEMENTS

We would like to thank Drs. E. Cavalieri, M. S. Newman, R. G. Harvey and G. H. Daub for kindly providing us with samples of various compounds and Drs. E. C. and J. A. Miller for communicating their results prior to publication. This work was supported by Grant No. CA-22209 from the National Cancer Institute of the National Institutes of Health.

REFERENCES

1. Cavalieri, E., Rogan, E., and Roth, R. (1981): Multiple mechanisms of activation in aromatic hydrocarbon carcinogenesis. In: Free Radicals and Cancer, edited by R. Floyd, Marcel Dekker, New York (in press).
2. Harvey, R. G., and Dunne, F. B. (1978): Multiple regions of metabolic activation of carcinogenic hydrocarbons. Nature, 273:566-568.
3. Levin, W., Wood, A. W., Wislocki, P. G., Chang, R. L., Kapitulnik, J., Mah, H. D., Yagi, H., Jerina, D. M., and Conney, A. H. (1978): Mutagenicity and carcinogenicity of benzo(a)pyrene and benzo(a)pyrene derivatives. In: Polycyclic Hydrocarbons and Cancer, Vol. 1: Environment, Chemistry and Metabolism, edited by H. V. Gelboin and P. O. P. Ts'o, pp 189-202, Academic Press, New York.
4. Wood, A. W., Levin, W., Chang, R. L., Yagi, H., Thakker, D. R., Lehr, R. E., Jerina, D. M., and Conney, A. H. (1979): Bay-region activation of carcinogenic polycyclic hydrocarbons. In: Polynuclear Aromatic Hydrocarbons, edited by P. W. Jones and P. Leber, pp 531-551, Ann Arbor Science, Ann Arbor, Michigan.
5. Schurch, O., and Winterstein, A. (1935): Uber die krebserregende Wirkung aromatischer Kohlenwasserstoffe. Z. Physiol. Chem., 236, 79-91.
6. Shear, M. J., and Perrault, A. (1939): Studies in carcinogenesis VII. Compounds related to 3,4-benzopyrene. Am. J. Cancer, 36, 211-228.
7. Lacassagne, A., Zajdela, F., Buu-Hoi, N. P., Chalvet, O., and Daub, G. H. (1968): Activité cancérogene elevée des mono-, di- et triméthylbenzo(a)pyrénes. Intl. J. Cancer, 3, 238-243.
8. Hecht, S. S., Hirota, N., and Hoffman, D. (1978): Comparative tumor initiating activity of methylated benzo(a)pyrene derivatives in mouse skin. Cancer Res., 40, 179-183.
9. Cavalieri, E., Roth, R., Grandjean, C., Althoff, J., Patil, K., Liakus, S., and Marsh, S. (1979): Carcinogenicity and metabolic profiles of 6-substituted benzo(a)pyrene derivatives on mouse skin. Chem. Biol. Interactions, 22, 53-67.
10. Sydnor, K. L., Bergo, C. H., and Flesher, J. W. (1980): Effect of various substituents in the 6-position on the relative carcinogenic activity of a series of benzo(a)pyrene derivatives. Chem. Biol. Interactions, 29, 159-167.
11. Iyer, R. P., Lyga, J. W., Secrist, J. A., Daub, G. H., and Slaga, T. J. (1980): Comparative tumor-initiating activity of methylated benzo(a)pyrene derivatives in mouse skin. Cancer Res., 40, 1073-1076.

12. Sullivan, P. D., Calle, L. M., Ocasio, I. J., Kittle, J. D. Jr., and Ellis, L. E. (1980): The effect of antioxidants on the mutagenicity of benzo(a)pyrene and derivatives. In: Polynuclear Aromatic Hydrocarbons: Chemistry and Biological Effects, edited by A. Bjorseth and A. J. Dennis, pp 163-175, Battelle Press, Columbus, Ohio.
13. Dewhurst, F., and Kitchen, D. A. (1972): Synthesis and properties of 6-substituted benzo(a)pyrene derivatives. J. Chem. Soc. Perkin Trans. I, 710-712.
14. Newman, M. S., and Kumar, S. (1977): New synthesis of benzo(a)pyrene, 7,10-dimethylbenzo(a)pyrene. J. Org. Chem., 42, 3284-3286.
15. Nebert, D. W., and Gielen, J. E. (1972): Genetic regulation of aryl hydrocarbon hydroxylase induction in the mouse. Fed. Proc., 31, 1315-1325.
16. Lesko, S., Caspary, W., Lorenzten, R., and Ts'o, P. O. P. (1975): Enzymic formation of 6-oxobenzo(a)pyrene radical in rat liver homogenates from carcinogenic benzo(a)pyrene. Biochemistry, 14, 3978-3984.
17. Ioki, Y., and Nagata, C. (1977): A fluorimetric and electron spin resonance study of the oxygenation of benzo(a)pyrene: an interpretation of the enzymic oxygenation. J. Chem. Soc. Perkin Trans. II, 1172-1175.
18. Menger, E. M., Spokane, R. B., and Sullivan, P. D. (1976): Free radicals derived from benzo(a)pyrene. Biochem. Biophys. Res. Communs., 71, 610-616.
19. Sullivan, P. D., Menger, E. M., Reddoch, A. H., and Paskovich, D. H. (1978): Oxidation of anthracene by thallium (III) trifluoroacetate. Electron Spin Resonance and structure of the product cation radicals. J. Phys. Chem., 82, 1158-1160.
20. Rogan, E. G., Roth, R., and Cavalieri, E. (1980): Manganic Acetate and horseradish peroxidase/hydrogen peroxide: in vitro models of activation of aromatic hydrocarbons by one-electron oxidation. In: Polynuclear Aromatic Hydrocarbons: Chemistry and Biological Effects, edited by A. Bjorseth and A. J. Dennis, pp 259-266, Battelle Press, Columbus, Ohio.
21. Wong, T. K., Chiu, P-L., Fu, F. F., and Yang, S. K. (1981): Metabolic study of 7-methylbenzo(a)pyrene with rat liver microsomes: separation by HPLC and characterization of metabolites. Chem-Biol. Interactions, 36, 153-166.

EFFECTS OF BUTYLATED HYDROXYANISOLE ON THE METABOLISM OF BENZO(A)PYRENE

WASYL SYDOR, JR., KATHERINE F. LEWIS, RENXIU PENG, AND CHUNG S. YANG
Department of Biochemistry, New Jersey Medical-CMDNJ, Newark, N.J. 07103.

INTRODUCTION

Butylated hydroxyanisole [BHA,2(3)-tert-butyl-4-methoxyphenol] is a relatively non-toxic phenolic antioxidant used extensively as a food preservative (1,2). When added to the diet at a level of 5 mg/g diet, BHA has been demonstrated to inhibit neoplasia in animals caused by a variety of chemicals (3-10). The anticarcinogenic activity of BHA was also shown by administering BHA to mice per os (p.o.) 4 hr before the administration of benzo(a)pyrene (BP) (5).

Several lines of investigation have dealt with the possible mechanisms of the anticarcinogenic action of BHA. Previously, we have observed that BHA is an inhibitor of the cytochrome P-450-containing monooxygenase system (11,12) an enzyme system responsible for the metabolic activation of most chemical carcinogens. Dietary BHA also induces a wide variety of enzymes, such as epoxide hydrase, glutathione S-transferase, and glucose-6-phosphate dehydrogenase (13-16). These enzymes are involved in the disposition of carcinogenic metabolites generated by the monooxygenase system. The maximal induction of these enzymes, however, usually requires several days. It is not known whether this mechanism contributes to the anticarcinogenic action of short-term BHA treatment. Lam and Wattenberg observed that BHA treatment altered the metabolism of BP in an in vitro incubation with liver microsomes. The extent of epoxidation, particularly at the 4,5-position, was decreased and the formation of 3-OH-BP was increased (17, 18).

In order to investigate the mechanisms of the anticarcinogenicity of BHA, the effects of dietary and short-term BHA treatment on the metabolism of BP were investigated in the present work.

MATERIALS AND METHODS

Chemicals. BP, NADP and $_3$BHA were obtained from Sigma Chemical Co., St. Louis, MO. [G-^{3}H]BP (17.4 Ci/nmol) was purchased from Amersham/Searle Corp., Arlington Heights, IL. 3,3,3-trichloropropylene oxide (TCPO) was from Aldrich Chem. Co., Milwaukee, WI. HPLC solvents were obtained from J.T. Baker, Phillpsburg, NJ.

Treatment of Animals. Female Swiss Webster mice (body weight 18-20 g) were obtained from Taconic Farms, Inc., Germantown, NY. All animals were placed on "Wayne Lab-Blox", Allied Mills, Chicago, IL. In the short term BHA experiments, a group of mice received 0.5 mg BHA/g body weight in 0.1 ml olive oil p.o. at 4 hr before sacrificing and the control group received only 0.1 ml olive oil. In the dietary BHA experiments, one group of mice received a 0.5% BHA supplement in the diet for 7 days while the control received no supplement. Liver microsomes were isolated by differential centrifugation as previously described (19).

Metabolic Studies and HPLC Analysis. The substrates, BP and BP-trans-7,8-diol were purified by HPLC. The in vitro incubation consisted of microsomes (5 mg protein), 50 mM Tris-HCl, pH 7.4, 3 mM MgCl$_2$, 0.13 mM NADP, 2 mM glucose-6-phosphate, and 0.5 unit glucose-6-phosphate dehydrogenase in a final volume of 5 ml. The reaction was initiated by the addition of 100 nmol BP or BP-7,8-diol and allowed to proceed for 10 min at 37°C. The reaction was terminated by the addition of 5 ml of acetone and 10 ml of ethyl acetate. The mixture was vortexed for 1 min to extract the substrate and metabolites into the organic phase which was then dehydrated with anhydrous MgSO$_4$ and dried under N$_2$. The compounds were redissolved in 0.1 ml methanol for HPLC analysis. A 60 μl aliquot was injected into a Waters HPLC system equipped with a C$_{18}$-Radial-Pak column (8 mm ID x 10 cm) and was eluted with a 20 min linear 65-100% methanol:water gradient at a flow rate of 0.8 ml/min. Quantitation of metabolites was by scintillation counting of fracions, integration of absorbance peaks (254 nm) and fluorescence detection (ex 338 nm, em 425 nm). The quantity of each metabolite was expressed as a percentage of the total metabolites. The metabolites were identified by comparing their HPLC retention times with authentic samples obtained from the National Cancer Institute Chemical Repository and by their absorption spectra.

RESULTS

Metabolism of BP by Microsomes from Control and BHA Treated Mice. With the present HPLC system, the major metabolites were clearly resolved from the parent compound BP (Fig. 1). The peaks were monitored by both absorbance and fluorescence. Although less quantitative than the absorbance detector, the fluorescence detector afforded higher sensitivity, being especially useful in monitoring the more polar metabolites of BP.

In the initial work, the metabolites were quantitated by radioactivity. Dietary BHA was found to affect the in vitro metabolism of BP with regioselectivity (Table 1). Significant inhibition was observed in the formation of 9-OH-BP but not of other metabolites. The total metabolism was enhanced by 49%. Because dietary BHA was found previously to cause a specific inhibition in the formation of BP-9,10-diol (collaborative work

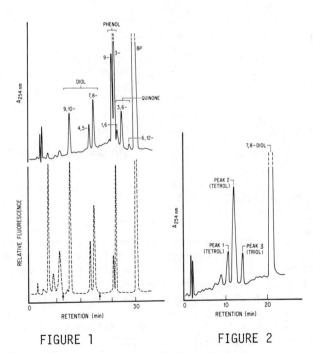

FIGURE 1.   HPLC profile of BP metabolites.
FIGURE 2.   HPLC profile of BP-7,8-diol metabolites.

with Drs. M. W. Chou and S. K. Yang of USUHS) and both the BP-9,10-diol and 9-OH-BP are reported to be derived from BP-9,10-oxide (20,21), the results suggest that the treatment caused a regiospecific inhibition of the oxygenation of BP at the 9,10-position. A similar pattern of inhibition was also observed when the metabolites were quantitated by absorbance. This method was not as laborious as the radioactive method, but the % metabolism was not measured accurately, therefore, the value is not shown in Table 1.

TABLE 1

EFFECT OF DIETARY BHA ON THE METABOLISM OF BENZO(A)PYRENE

Metabolite	Retention time, min	Control % CPM[a]	BHA[b]/Control	Control % area[c]	BHA[d]/Control
9,10-Diol	13.2	7.29	0.96 ± 0.20	4.48	1.06 ± 0.34
4,5-Diol	19.5	5.01	1.42 ± 0.30	8.31	1.16 ± 0.26
7,8-Diol	20.9	5.43	0.89 ± 0.31	6.76	0.99 ± 0.23
9-OH	26.9	21.35	0.58 ± 0.21	21.63	0.45 ± 0.12
1-,3- & 7-OH	27.7	50.68	1.22 ± 0.06	52.31	1.18 ± 0.07
1,6-Quinone	28.6	5.11	1.04 ± 0.15	2.95	1.08 ± 0.22
3,6-Quinone	30.1	3.90	2.12 ± 1.49	2.99	0.97 ± 0.24
6,12-Quinone	32.1	1.23	0.75 ± 0.27	0.57	0.78 ± 0.29
9,10-Diol + 9-OH		29.14	0.67 ± 0.13	26.26	0.59 ± 0.06
% of BP metabolized		19.49	1.49 ± 0.39	-	-

a. The data are from one representative experiment quantitated by radioactivity. Variability existed between experiments on the absolute quantity of the metabolites, but their ratios were reproducible.
b. The mean ± standard deviation of the ratios of % CPM from 4 experiments.
c. The data are from one representative experiment quantitated by absorbance at 254 nm.
d. The mean ± standard deviation of the ratios of % area from 8 experiments.

Short term BHA treatment decreased overall BP metabolism in vitro (by 23%). The 9-OH-BP peak was decreased by about 44% and the 3-OH-BP peak was increased by 12% (Table 2). The formation of 9,10-and 4,5-diol metabolites decreased slightly, but the changes are not statistically significant. This result is different from the inhibitory action of BHA when added in vitro which inhibited the formation of BP metabolites uniformly (12).

In order to assess the relationship between BP-9,10-diol and 9-OH-BP, the effect of TCPO, a potent inhibitor of epoxide hydrase (22) was studied. The addition of TCPO to the incubation completely eliminated the formation of the diol metabolites of BP by microsomes from either control or BHA intubated mice. On the other hand, the formation of 9-OH-BP was increased by TCPO. The results are consistent with the idea that the 9-OH-

TABLE 2

EFFECT OF SHORT TERM BHA ADMINISTRATION ON BENZO(A)PYRENE METABOLISM

Metabolite	Retention time, min	Control % area[a]	BHA p.o.[b] / Control
9,10-Diol	13.2	2.25	0.85 ± 0.26
4,5-Diol	19.5	3.75	0.84 ± 0.77
7,8-Diol	20.9	3.63	1.10 ± 0.16
9-OH	26.9	24.92	0.56 ± 0.09
1-,3- & 7-OH	27.7	52.60	1.12 ± 0.07
1,6-Quinone	28.6	6.16	1.28 ± 0.29
3,6-Quinone	30.1	5.90	1.12 ± 0.22
6,12-Quinone	32.1	1.79	1.14 ± 0.54
9,10-Diol + 9-OH		26.65	0.59 ± 0.07
% BP Metabolized		-	0.77 ± 0.15

a. Data are from one representative experiment.
b. Mean ± standard deviation of the ratios from 10 experiments.

BP is formed from BP-9,10-oxide by non-enzymatic rearrangement (20,21). Increased formation of 7-OH-BP from BP-7,8-oxide however, could not be clearly demonstrated because 7-OH-BP eluted at the same position as 3-OH-BP. 4-and 5-OH-BP, the possible products from the non-enzymatic rearragement of BP-4,5-oxide (21) were not detected in this experiment. The effect of BHA treatment was still evident even in the presence of TCPO (data not shown).

Effect of BHA on the in vitro Metabolism of BP-trans-7,8-diol. Several metabolites of the BP-trans-7,8-diol were detected by the HPLC analysis (Fig. 2). Dietary BHA treatment was found to decrease the metabolism of this substrate by 23% (Table 3). This is in contrast to the metabolism of BP which was enhanced 20% by dietary BHA treatment.

Effect of BHA on the Microsomal Monooxygenase System. Dietary BHA increased the specific content of cytochrome P-450 by 42%, the the activity of NADPH-cytochrome c reductase by 28%, and decreased the aryl hydrocarbon hydroxylase activity slightly (data not shown). The ethoxycoumarin o-dealkylase activity was markedly increased by 113%. Short-term BHA treatment did not alter the P-450 content and the reductase activity significantly and the slight decrease in P-450 content was not

TABLE 3

EFFECT OF DIETARY BHA ON THE METABOLISM OF THE TRANS-7,8-DIOL OF BENZO(A)PYRENE[a]

Metabolite	Retention time, min.	Control % area	BHA / Control
Peak 1 (tetrol)	10.57	24.65 ± 1.04	1.11 ± 0.01
Peak 2 (tetrol)	11.75	56.57 ± 0.72	1.03 ± 0.02
Peak 3 (triol)	13.38	18.78 ± 0.83	0.82 ± 0.04
% of BP-t-7,8-diol-metabolized		18.75 ± 0.69	0.77 ± 0.15

a. Conditions were the same as those described in Table 1 except that 20 μM trans-7,8-diol was used as a substrate.

reproduced in other experiments. The activities of aryl hydrocarbon hydroxylase and ethoxycoumarin o-dealkylase, however, were inhibited by the short-term BHA treatment. The decrease in these activities is probably due to the inhibitory action of the residual BHA in the microsomal sample.

Further investigation into the increased dealkylase activity revealed changed K_m and V_{max} values in the microsomes from mice receiving dietary BHA treatment. The values for K_m of the microsomes from control and dietary BHA treated mice were 123 and 37 µM, respectively and V_{max} values were 1.07 and 1.80 nmol/min/mg, respectively (Fig. 3).

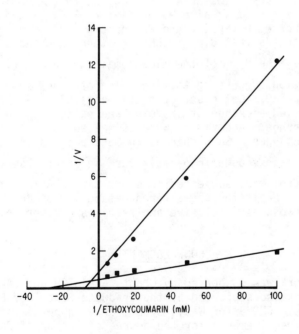

FIGURE 3. Double reciprocal plots of microsomal ethoxycoumarin o-dealkylase (nmol/min/mg). ■ and ●, microsomes from dietary BHA treated and control mice, respectively.

DISCUSSION

It is known that liver microsomes can catalyze the oxygenation at the 2,3-, 4,5-, 7,8-, and 9,10-positions of BP. The present work clearly demonstrated that dietary and short-term BHA treatment can cause a regiospecific inhibition in the oxygenation at the 9,10-position of BP. This differs from the the results by Lam and Wattenberg (17,18) which showed that BHA treatment specifically inhibited the formation of BP-4,5-oxide. The reasons for such a discrepancy is not clear, although differences in the strains and age of the mice and in the conditions of HPLC analysis are noted. The present results indicate that dietary BHA treatment also inhibits the further oxygenation of BP-trans-7,8-diol (Table 3). Based on this observation, it my be suggested that dietary BHA treatment specifically inhibits the formation of BP-7,8-diol-9,10-oxide, the commonly believed ultimate carcinogenic form of BP, but enhances the metabolism of BP via non-carcinogenic pathways.

Several lines of observations suggest that the microsomal monooxygenase system is altered by dietary BHA. In addition to those reported by Speier and Wattenberg (23), the present work demonstrates that dietary BHA treatment decreases the K_m and increases the V_{max} of the microsomal ethoxycoumarin o-dealkylase reaction suggesting the induction of a new species of cytochrome P-450. Gel electrophoresis analysis (data not shown) revealed a change of protein band pattern in the 42-60 kilodalton range. Additional work is needed to explore and define these changes. Such changes in band pattern and in the K_m and V_{max} of ethoxycoumarin o-dealkylase were not observed with microsomes from mice subjected to short-term BHA treatment. The enzymatic basis for the altered metabolism of BP (Table 2) by these microsomes is elusive and requires further investigation.

ACKNOWLEDGMENTS

This work was supported by Grants CA-16788 and CA-28298. This work is also supported by the Graduate School of Biomedical Sciences - CMDNJ for Wasyl Sydor, Jr., in partial fulfillment of the requirements for the degree of Doctor of Philosophy. We thank Drs. Shen K. Yang and Ming W. Chou (Uniformed Services Univeristy of the Health Sciences, Bethesda, MD) for their helpful advice.

REFERENCES

1. Stuckey, B. N. (1972): Antioxidants as food stabilizers. In: Handbook of Food Additives, 2nd edition, edited by T. E. Furia, pp. 185-223, Chemical Rubber Company, Cleveland, Ohio.
2. Branen, A. L. (1975): Toxicology and biochemistry of butylated hydroxyanisole, and butylated hydroxytoulene, J. Amer. Oil Chem. Soc., 52:59-63.
3. Wattenberg, L. W. (1973): Inhibition of chemcial carcinogen-induced pulmonary neoplasia by butylated hydroxyanisole, J. Natl. Cancer Inst., 50:1541-1544.
4. Wattenberg, L. W. (1978): Inhibition of chemical carcinogenesis, J. Natl. Cancer Inst., 60:11-18.
5. Speier, J. L., Lam, L. K. T., and Wattenberg, L. W. (1978): Effects of administration to mice of butylated hydroxyanisole by oral intubation on benzo(a)pyrene-induced pulmonary adenoma formation and metabolism of benzo(a)pyrene, J. Natl. Cancer Inst., 60:605-609.
6. Wattenberg, L. W., Jerina, D. M., Lam, L. K. T., and Yagi, H. (1979): Neoplastic effects of oral administration of (±)trans-7,8-dihydroxy-7,8-dihydrobenzo(a)pyrene and their inhibition by butylated hydroxyanisole, J. Natl. Cancer Inst. 62:1103-1106.
7. Wattenberg, L. W., Coccia, J. B., and Lam, L. K. T. (1980): Inhibitory effects of phenolic compounds on benzo(a)pyrene-induced neoplasia, Cancer Res. 40:2820-2823.
8. Wattenberg, L. W. (1978): Inhibitors of chemical carcinogenesis, Adv. Cancer Res., 26:197-226.
9. Wattenberg, L. W. (1972): Inhibition of carcinogenesis and toxic effects of polycyclic hydrocarbons by phenolic antioxidants and ethoxyquin, J. Natl. Cancer Inst., 48:1425-1430.
10. Wattneberg, L. W. (1980): Inhibition of chemical carcinogens, J. Environ. Pathol. Toxicol. 3:35-52.
11. Yang, C. S., Strickhart, F. S., and Woo, G. K. (1974): Inhibition of the monooxygenase system by butylated hydroxyanisole and butylated hydroxytoluene, Life Sci. 15:1496-1505.
12. Yang, C. S., Sydor, W., Jr., Martin, M. B., and Lewis, K. F. (1981): Effects of butylated hydroxyanisole on the aryl hydrocarbon hydroxylase of rats and mice, Chem. Biol. Interact. (in press).
13. Benson, A. M., Batzinger, R. P., Ou, S. Y. L., Bueding, E., Cha, Y. N., and Talady, P. (1978): Elevation of hepatic glutathione S-transferase activities and protection against mutagenic metabolites of benzo(a)pyrene by dietary antioxidants, Cancer Res. 38:4486-4495.

14. Benson, A. M., Cha, Y. N., Bueding, E., Heine, H. S., and Talady, P. (1979): Elevation of extrahepatic glutathione S-transferase and epoxide hydrase activities by 2(3)tert-butyl-4-hydroxyanisole, Cancer Res., 39:2971-2977.
15. Cha, Y. N., Martz, F., and Bueding, E. (1978): Enhancement of liver microsomal epoxide hydrase in rodents by treatment of 2(3)-tert-butyl-4-hydroxyanisole, Cancer Res., 38:4496-4498.
16. Cha, Y. N. and Bueding, E. (1979): Effect of 2(3)-tert-butyl-4-hydroxyanisole administration on the activities of several hepatic microsomal and cytoplasmic enzymes in mice, Biochem. Pharmacol., 28:1917-1921.
17. Lam, L. K. T., and Wattenberg, L. W. (1977): Effects of butylated hydroxyanisole on the metabolism of benzo(a)pyrene by mouse liver microsomes, J. Natl. Cancer Inst. 58: 413-417.
18. Lam, L. K. T., Fladmore, A. V., Hochalter, J. B., and Wattenberg, L. W., (1980): Short time interval effects of butylated hydroxyanisole on the metabolism of benzo-(a)pyrene, Cancer Res., 40:2824-2828.
19. Yang, C. S., and Strickhart, F. S. (1974): Inhibition of hepatic mixed function oxidase by propyl gallate, Biochem. Pharmacol, 23:3129-3188.
20. Sims, P., and Grover, P. L., (1974): Epoxides in polycyclic aromatic hydrocarbon metabolism and carcinogenesis, Adv. Cancer Res. 20:165-274.
21. Yang, S. K., Roller, P. P., and Gelboin, H. V., (1977): Enzymatic mechanism of benzo(a)pyrene conversion to diols and phenols and an improved high-pressure liquid chromatographic separation of benzo(a)pyrene derivatives, Biochemistry, 16:3680-3687.
22. Oesch, F., Kaubisch, N., Jerina, D. M. and Daly, J. W. (1971): Hepatic epoxide hydrase. Structure-activity relationships for substrates and inhibitor, Biochemistry, 10:4858-4866.
23. Speier, J. L., and Wattenberg, L. W. (1975): Alterations in microsomal metabolism of benzo(a)pyrene in mice fed butylated hydroxyanisole, J. Natl. Cancer Inst. 55:469-472.

BENZO[A]PYRENE METABOLISM IN HEPATIC S-9 FRACTIONS OF AROCLOR 1254-TREATED MULLET (MUGIL CEPHALUS).

BARRIE TAN[*], PAUL MELIUS[**]
*Department of Chemistry, University of Massachusetts, Amherst, Massachusetts, 01003; **Department of Chemistry, Auburn University, Auburn, Alabama 36849.

INTRODUCTION

Biotransformation of xenobiotic chemicals in aquatic species is less well known than that in mammalian species. Some example of benzo[a]pyrene (BaP) metabolites which have been detected using tissue preparations from marine fish include 3-hydroxybenzo[a]pyrene (3-OH-BaP) (1,2), BaP-3,6-quinone (2), and 7,8-dihydroxy-dihydrobenzo[a]pyrene (BaP-7,8-diol) (3). Recently, BaP-4,5-diol, BaP-7,8-diol, BaP-9,10-diol, 3-OH-BaP, 9-OH-BaP, and quinones were identified in trout (4,5), scup (6), and skate (7) when liver microsomes were incubated with BaP. Varanasi et al. (8) have identified the BaP metabolites in flounder, sole and salmon while we have separated these metabolites in mullet, sea catfish and gulf killifish (9,10). In order to ascertain the biotransformation pathways of xenobiotics, it is necessary to obtain profiles of metabolites and the factors that influence their formation. This paper presents the product patterns of phenols, diols, and quinones produced by the Aroclor-treated mullet (Mugil cephalus). Effects of incubation time, temperature, protein concentration and enzyme induction on the BaP metabolism were investigated.

MATERIALS AND METHODS

Preparation of Microsomes

Mullet (Mugil cephalus) weighing 250 to 350 g were collected from coastal waters in northwestern Florida and transferred to flow-through tanks (approximately 250L seawater/h; 18 to 22°C) located at the EPA Research Laboratory, Sabine Island, Gulf Breeze, Florida. The mullet was acclimated for seven days, after which Aroclor 1254 (200 mg Aroclor/kg body mass) in sterile corn oil (200 mg/mL) was administered as a single intraperitoneal injection. Control fish only received sterile corn oil. The animals were not fed for the eleven-day induction period before being killed. Livers were removed aseptically, rinsed twice with 20 mL aliquots of cold, sterile

0.15M KCl, and the masses livers transferred to beakers containing 0.15 M KCl (3 mL/g liver mass). Sterile scissors and a Potter-Elvehjem homogenizer were used to mince and macerate the livers, respectively. The homogenate was centrifuged at 9000 g for 25 min. (2^oC: r_{av} 10.8 cm). The resulting floating lipid layer was carefully removed and discarded while the supernatant fraction, S-9 was collected, stored at -20^oC and used in three weeks.

Spectrophotometric Determinations

Protein concentration was determined by the method of Lowry et al. (11), using bovine serum albumin as standard (fraction V; catalog number, A 4503) obtained from Sigma Chemical Company. The average protein concentration was found to be 130 mg/g liver mass (15 mg/mL S-9 protein) and no significant difference was observed between the control and treated fish. Cytochrome P_{450} content was determined by the carbon monoxide-bound reduced spectra using ε 91 $mM^{-1} cm^{-1}$ for the absorbance difference between 450 nm and 490 nm (12). The average hepatic content was 0.60 nmole cytochrome P_{450}/mg S-9 protein. The Gilford 250 spectrophotometer was used for the protein and cytochrome P_{450} determinations.

In vitro Metabolism of BaP

The in vitro metabolism of BaP was conducted as described by Selkirk et al.(13). The reaction mixture in a final volume of 1 mL contained 2 mg S-9 protein, 0.36 µmole NADPH, 3 µmole Mg $Cl_2.6H_2O$, 50 µmole Tris (pH 7.5) and 100 nmole BaP in 40 µL methanol. BaP was added last to initiate the reaction. All flasks were incubated in a Dubnoff metabolic shaking incubator under red light illumination and at pH 7.5. In the incubation time series, 2 mg protein each of the microsomes from control and Aroclor-treated animals were incubated at 37^oC for varying lengths of time from 1 min to 120 min. Similar series of experiments were repeated for treated animals at 25^oC and for incubation times from 1 min to 180 min. The protein concentration series were conducted at 37^oC for 30 min and over a protein concentration of 0.6 mg to 2.4 mg S-9 protein. The reaction was terminated by the addition of 1 mL acetone. The BaP and its lipophilic metabolites were extracted four times with 2 mL to 3 mL volumes of ethyl acetate. The ethyl acetate extracts were pooled, dried over anhydrous $MgSO_4$ (1g) and the ethyl acetate evaporated under nitrogen streams. BaP metabolites were redissolved in 0.5 mL methanol and then prepared for HPLC analysis.

Chromatographic Determinations

Chromatographic analyses were performed using a Micromeritics Model 7000B HPLC, fitted with a Lichrosorb RP-18 (Merck) packed stainless steel column (25 cm length & 0.4 cm diameter. Elution was achieved by 100% methanol (Mallinckrodt Nanograde) with a 0.3 or 0.6 mL/min flow rate and a column temperature and pressure of 50°C and 690 k Pa, respectively. The HPLC was modified to include a Schoeffel Model 970 spectrofluorometric detector (excitation: 263 nm; emission: 370 nm filter). A 10 µL sample was injected for each analysis. Metabolites were identified by comparison to the retention times of standards. No standards were available for 6-OH-BaP, quinones, triol or tetrol so concentrations of these metabolites were estimated using BaP as the standard. Each of the standards gave only one detectable peak by HPLC as did the BaP. Typical representative chromatograms of the effects of incubation time were shown in Figure 1.

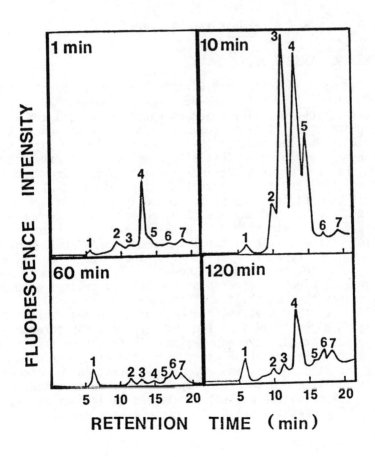

FIGURE 1. Changes of BaP Metabolites with Incubation Time. Metabolites, obtained form in vitro incubation (37°C, pH 7.5) with Aroclor-treated mullet hepatic S-9 protein (2mg) over various incubation times (1 min to 120 min), were separated by HPLC. Elution was achieved isocratically (100% methanol) with 0.3 mL/min flow rate and a column temperature and pressure of 50°C and 690 kPa, respectively. BaP metabolites: 1, triol/tetrol; 2, BaP-9, 10-diol; 3, BaP-4, 5-diol; 4, BaP-7, 8-diol; 5, quinones; 6, 9-OH-BaP; 7, 3-OH-BaP.

Biochemicals were purchased from Sigma Chemical Company. Organic solvents and Mallinckrodt Nanograde methanol was obtained from Fischer Scientific Products, Inc. The 3-OH-BaP 9-OH-BaP standards and Aroclor 1254 were provided by J. N. Keith (IIT Research Institute, Chicago, Ill.). Additionally, BaP-4,5-diol, BaP-7, 8-diol and BaP-9,10-diol standards were supplied by M. McCloud (Oak Ridge National Laboratory, Oak Ridge, Tenn.).

RESULTS AND DISCUSSION

Effects of Incubation Time

Figure 2 indicates the kinetic patterns of 5 BaP metabolites, namely 3-OH-BaP, 9-OH-BaP, BaP-4, 5-diol, BaP-7, 8-diol and BaP-9, 10-diol. The reaction rates showed complex kinetics which are different from what had been observed in the rat (14). For the initial 10-15 min, all the enzymatic rates increase linearly. The initial linearity of the rates suggests that BaP undergoes primary oxidation by mixed function oxidase (MFO) and the differential rates of the diols to the phenols suggest that the preferred positions of attack by the aryl hydrocarbon hydroxylase (AHH) enzyme are on carbon 4,5-, 7,8- and 9,10- rather than 3- or 9- of BaP. The rate of production of these metabolites sharply decreased at 30 min, followed by a 'trough' from 30-60 min. The rates of BaP-7, 8-diol, 3-OH-BaP, and 9-OH-BaP showed a second maximum at 90 min. The unusual kinetic patterns of the BaP metabolites were probably due to the possibility of conjugation reactions in S-9 protein. Furthermore, the presence of competing reactions utilizing NADPH could also account for the disappearance of metabolites at the 'trough' region. It is interesting to note that when microsomes were used, where conjugation and competing reactions were absent, the metabolite formation approached saturation kinetics instead (14). This is an important factor when considering the activated mutagens as in Ames tests, where S-9 proteins instead of microsomes are used.

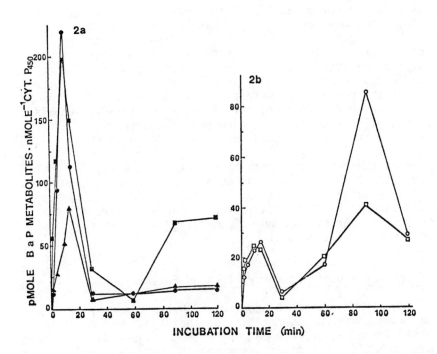

FIGURE 2. Time Course of Individual Phenol and Diol BaP Metabolites. BaP metabolites were formed by liver microsomes from Aroclor-treated mullet and measured by HPLC. Conditions were similar to those stated in Figure 1. In (a) formation of the diols:▲, BaP-9, 10-diol;●, BaP-4, 5-diol;■, BaP-7, 8-diol and in (b), formation of the phenols: ○, 9-OH-BaP; □, 3-OH-BaP.

Effects of Temperature

The effects of the reaction rates of BaP metabolite formation at 25°C and 37°C are shown in Figure 3. For the phenols and diols reactions, two maxima for each group of metabolites were found at 37°C, but only the phenols had one maximum (60 min) at 25°C. The conjugation reactions at 25°C might neither be optimal nor competing. James et al. (15), reported that the temperature optimum for in vitro MFO activity was higher in the Florida fish tested than for cold water acclimated marine or freshwater fish. This suggests that mullet, having been obtained from the same locality, has a much higher metabolism rate at 37°C than at 25°C. In the experiments where incubation time was varied, an early peak which eluted before any of the other metabolites. This metabolite or a group of metabolites was most polar and has relative retention times similar to triol and tetrol (16,17). The triol/tetrol did not appear to change significantly with temperature or with varying quantities of the other metabolites produced (see Figure 3b). Furthermore, this triol/tetrol was also independent of the protein concentration. Assuming that the triol/tetrol was hydrolyzed product of BaP-7, 8-diol-9,10-epoxide, this diol-epoxide might have undergone non-enzymatic hydrolysis to tetrol or reduction possibly by NADPH to triol (17,18).

On the basis of the relative retention times, as previously identified (9,13), the quinones were BaP-1, 6-quinone, BaP-3, 6-quinone, and BaP-6, 12-quinone. No quantification of quinones was made because these standards were not available, and that fluorescence intensities were greatly quenched by quinone functionalities.

Effects of Protein Concentration and Enzyme Induction

Aroclor-treated and control fish hepatic metabolism of BaP were compared. Quantities of individual metabolites were plotted against protein concentrations (Figure 4). When Aroclor-treated S-9 protein was used, the 9-OH-BaP, and quinones were 2-fold while BaP-4, 5-diol, and BaP-7, 8-diol were 10-fold higher than their corresponding metabolites using control S-9 protein (the ordinate scale for quinones was "fluorescence intensities" because no quinone standards were available). With the same treatment, the 3-OH-BaP, and BaP-9, 10-diol were increased drastically by greater than 15-fold. It is clear that an increase in protein concentration does not have a uniform effect on the formation of each metabolite. There is a striking similarity for the large excess of 3-OH-BaP and BaP-9, 10-diol formed by mullet compared to the rat (14,19), even though the inducing agent used in the rat was 3-methylcholanthrene. This suggests that mullet is capable of synthesizing the specific cytochrome P_{450} which was induced could uniquely mediate the formation of BaP to 3-OH-BaP and BaP-9,10-diol. Recently the separation of cytochrome P_{450c} (or P_{448}) has been elegantly established when rat (20) and trout (21) were treated with 3-methylcholanthrene, Aroclor or β-naphthoflavone. These data, taken with that of James *et al.* (15) confirm that hepatic MFO system of fish and mammalian species have a basic similarity in its mode of induction.

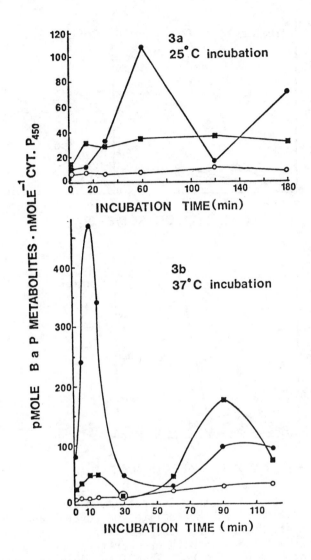

Figure 3. Kinetic Profile of Phenol, Diol, and Triol/Tetrol BaP Metabolites. Profiles of BaP metabolites were obtained by liver microsomes from Aroclor-treated mullet. Conditions were similar to those stated in Figure 1. In (a) at 25°C, and in (b) at 37°C incubation temperature; ■, phenols; ●, diols; ○, triol/tetrol.

The 'proximal carcinogen' BaP-7,8-diol, which though not the principal metabolite, increases proportionally to protein

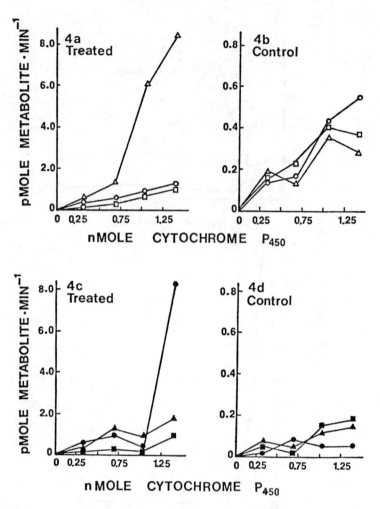

FIGURE 4. Effects of Protein Concentration and AHH Induction for the Formation of BaP Metabolites on Control and Aroclor-treated Mullet. Metabolites, obtained from in vitro incubation (37°C, pH 7.5) for 30 min with Aroclor-treated mullet hepatic S-9 protein (from 0.6-2.4 mg). Phenol and quinone metabolites for Acrolor-treated and control mullet are shown in (a) and (b), respectively. Diol metabolites for Aroclor-treated and control mullet are shown in (c) and (d), respectively. In (a) and (b) the phenols and quinones are: △ , 3-OH-BaP; ○ , 9-OH-BaP; □ , quinones. In (c) and (d) the diols are: ● , BaP-9,10-diol; ▲ , BaP-7,8-diol; ■ , BaP-4,5-diol. The ordinate scale for quinones (4a and 4b) is in "fluorescence intensities" because no quinone standards were available.

concentration as well as responding to a 10 fold increase with Aroclor induction (see Figure 4c and d). BaP-7, 8-diol is the precursor to the ultimate carcinogen BaP-7, 8-diol-9, 10-epoxide. Therefore, an increase in protein concentration and Aroclor induction must also implicate the increased production of this diol-epoxide necessary to bind covalently to macromolecules. When Ames tests were performed, incubating S-9 protein from Aroclor-treated mullet, sea catfish, or gulf killifish with BaP and Salmonella Typhimurium TA 98 mutagenic metabolites were formed (10).

The work performed at Auburn University was supported by a Contract CR806213010 from the EPA to P. Melius. One of us (B. Tan) would like to thank the EPA and the NCI for the funding of this collaborative carcinogen program. This work was done with the collaboration of W.P. Schoon at the EPA Laboratory, Sabine Island, Gulf Breeze, Florida.

REFERENCES

1. Pedersen, M.G., Hershberger, W.K., and Juchau, M.R.(1974): Metabolism of 3,4-benzo(a)pyrene in rainbow trout (Salmo gairdneri), Bull. Environ. Contam. Toxicol., 12: 481-486.
2. Ahokas, J.T., Pelkonen, O., and Karki, N.T. (1975): Metabolism of polycyclic hydrocarbons by a highly active aryl hydrocarbon hydroxylase system in the liver of a trout species, Biochem. Biophys. Res. Comm., 63:635-641.
3. Lee, R.F., Sauerheber, R., and Dobbs, G.H. (1972): Uptake, metabolism and discharge of polycyclic aromatic hydrocarbons by marine fish, Mar. Biol., 17:201-208.
4. Ahokas, J.T. (1979): Cytochrome P-450 in fish liver microsomes and carcinogen activation. In: Pesticide and Xenobiotic Metabolism in Aquatic Organisms, edited by M.A.Q. Khan, J.J. Lech, and J.J. Menn, pp 279-296, American Chemical Society, Washington, D.C.
5. Ahokas, J.T., Saarni, H., Nerbert, D.W., and Pelkonen, O. (1979): The in vitro metabolism and covalent binding of benzo(a)pyrene to DNA catalyzed by trout liver microsomes, Chem.-Biol. Interactions, 25:103-111.
6. Tjessum, K., and Stegeman, J.J. (1979): Improvement of reverse phase high pressure liquid chromatographic resolution of benzo(a)pyrene metabolites produced by fish, Anal. Biochem., 99:129-135.
7. Bend, J.R., Ball, L.M., Elmamlouk, T.H., James, M.O., and Philpot, R.M. (1979): Microsomal mixed-function oxidation in untreated and polycyclic aromatic hydrocarbon-treated marine fish. In: Pesticide and Xenobiotic Metabolism in Aquatic Organisms, edited by M.A.Q. Khan, J.J. Lech, and J.J. Menn, pp 297-318, American Chemical Society, Washington, D.C.

8. Varanasi, U., Gmur, D.J., and Krahn, M.M. (1980): Metabolism and subsequent binding of benzo(a)pyrene to DNA in pleuronectid and salmonid fish. In: Polynuclear Aromatic Hydrocarbon: Chemistry and Biological Effects, IV Symposium, edited by A. Bjorseth and A.J. Dennis, pp. 455-470, Battelle Press, Columbus, Ohio.
9. Tan, B., Kilgore, M.V., Elam, D.L., Melius, P., and Schoor, W.P. (1981): Metabolites of benzo(a)pyrene in Aroclor 1254-treated mullet. In: Aquatic Toxicology and Hazard Assessment, IV Symposium, edited by D.R. Brandson, and K.L. Dickson, pp 239-246, ASTM Press, Philadelphia,PA.
10. Melius, P., Elam, D., Kilgore, M., Tan, B., and Schoor, W.P. (1980): Mixed function oxidase inducibility and polyaromatic hydrocarbon metabolism in the mullet, sea catfish, and gulf killifish. In: Polynuclear Aromatic Hydrocarbon: Chemistry and Biological Effects, IV Symposium, edited by A. Bjorseth, and A.J. Dennis, pp. 1059-1075, Battelle Press, Columbus, Ohio.
11. Lowry, O.H., Rosebrough, N.J., Farr, A.L., and Randall, R.J. (1951): Protein measurement with Folin phenol reagent, J. Biol. Chem., 193:265-275.
12. Omura, T., and Sato, R. (1964): The carbon-monoxide binding pigment of liver microsomes, J. Biol. Chem., 239: 2379-2385.
13. Selkirk, J.K., Croy, R.G., and Gelboin, H.V. (1974): Benzo(a)pyrene metabolites: efficient and rapid separation by high pressure liquid chromatography, Science, 184: 169-171.
14. Yang, S.K., Selkirk, J.K., Plotkin, E.V., and Gelboin, H.V. (1975): Kinetic analysis of the metabolism of benzo(a)pyrene to phenols, dihydrodiols, and quinones by high-pressure liquid chromatography compared to analysis by aryl hydrocarbon hydroxylase assay, and the effect of enzyme induction, Can. Res., 35:3642-3650.
15. James, M.O., Khan, M.A.Q., and Bend, J.R. (1979): Hepatic microsomal mixed-function oxidase activities in several marine species common to coastal Florida, Comp. Biochem. Physiol., 62C:155-164.
16. Yang, S.K., Roller, P.P., and Gelboin, H.V. (1977): Enzymatic mechanism of benzo(a)pyrene conversion to phenols and diols and an improved high-pressure liquid chromatographic separation of benzo(a)pyrene derivatives, Biochemistry, 16:3680-3687.
17. Thakker, D.R., Yagi, H., Lu, A.Y.H., Levin, W., Conney, A.H., and Jerina, D.M. (1976): Metabolism of benzo(a)-pyrene: Conversion of ($\pm$)-trans-7,8-dihydroxy-7,8-dihydrobenzo(a)pyrene to highly mutagenic 7,8-diol-9,10-epoxides, Proc. Natl. Acad. Sci., 73:3381-3385.

18. Yang, S.K., McCourt, D.W., Roller, P.P., and Gelboin, H.V. (1976): Enzymatic conversion of benzo(a)pyrene leading predominantly to the diol-epoxide r-7, t-8-dihydroxy-t-9, 10-oxy-7,8,9,10-tetrahydrobenzo(a)pyrene through a single enantiomer of r-7, t-8-dihydroxy-7,8-dihydrobenzo(a)pyrene, Proc. Natl. Acad. Sci., 73:2594-2598.
19. Prough, R.A., Patrizi, V.W., Okita, R.T., Masters, B.S.S., and Jakobsson, S.W. (1979): Characteristics of benzo(a) pyrene metabolism in kidney, liver, and lung microsomal fractions from rodents and humans, Can. Res., 39:1199-1206.
20. Goldstein, J.A. (1979): The structure-activity relationships of halogenated biphenyls as enzyme inducers, Ann. N.Y. Acad. Sci., 320:164-178.
21. Elcombe, C.R., and Lech, J.J. (1979): Induction and characterization of hemoprotein(s) P-450 and monooxygenation in rainbow trout (Salmo gairdneri), Toxicol. App. Pharmacol., 49:437-450.

MULTICOMPONENT ISOLATION AND ANALYSIS OF POLYNUCLEAR AROMATICS*

B. A. TOMKINS, W. H. GRIEST, J. E. CATON, and R. R. REAGAN
Analytical Chemistry Division, Oak Ridge National Laboratory,
P. O. Box X, Oak Ridge, Tennessee 37830

INTRODUCTION

Polynuclear aromatics, including polycyclic aromatic hydrocarbons (PAH) and polycyclic aromatic amines (PAA), are major contributors (1-3) to the mutagenic and carcinogenic activity of synthetic fuels. Consequently, the progression of a safe and environmentally acceptable synthetic fuels industry will depend in part upon the continued development and application of improved analytical methodology for isolating and determining PAH and PAA.

Work performed at this laboratory (4,5) and at the National Bureau of Standards (6,7) has demonstrated that high-pressure liquid chromatography (HPLC) can rapidly produce isolates ranging in polarity from fairly nonpolar PAH to a weak acid such as phenol. These isolation procedures should be easily expanded to permit reproducible, rapid, and convenient sequential isolation of several chemical classes present in a sample. Such a method would be clearly superior to those using thin-layer chromatography (8), ambient-pressure liquid chromatography (9), or classical acid/base fractionation (10).

In this paper, we describe an HPLC fractionation procedure suitable for complex matrices such as synfuels or airborne particulate extracts. The procedure yields separate fractions bearing PAH and amines plus aza-arenes, in addition to fractions of saturates and simple alkylated phenols. Total turnaround time is two hours, including the time needed to regenerate the column. The fractions obtained are suitable for gas or analytical-scale liquid chromatography, without any further purification necessary.

---

*Research sponsored by the Office of Health and Environmental Research, U. S. Department of Energy under contract w-7405-eng-26 with Union Carbide Corporation.

MULTICOMPONENT ISOLATION

MATERIALS AND METHODS

Equipment

The details of the HPLC apparatus have been described elsewhere (4,5). The major changes were the addition of the Digital Programmer (Model DP 810) and the Automatic Stream Selection Valve (Model SSV-6), which were both purchased from Glenco Scientific Inc. (Houston, TX). The digital programmer also operated four three-way miniature teflon solenoid valves used to collect individual fractions (General Valve Corporation, East Hanover, NJ). When airborne particulate extracts were fractionated, the semi-preparative scale polar amino cyano (PAC) column described in reference 4 was used. When synfuels were fractionated, a 25 cm x 9.4 mm I.D. Lichrosorb-NH$_2$ (10 μ particles) column purchased from Altex Scientific, Inc. (Berkeley, CA) was substituted for the PAC column. The guard column, in both cases, consisted of an MPLC Guard Column equipped with a 3 cm "Amino" cartridge; both were purchased from Brownlee Labs, Inc. (Santa Clara, CA).

The gas chromatograph used to profile the synfuels isolates was a Sigma I (Perkin-Elmer Corporation, Norwalk, CT) equipped with a 1/8" O.D. x 10' glass column packed with 3.0% w/w Dexsil 400 on Supelcoport, 100/120 mesh. The oven was programmed from 90°C (hold for 8 minutes) to 320°C (hold for 15 min) at 2°C/min. The injector and detector temperatures were 300 and 320°C, respectively. The integrating and reporting features of the Sigma I were used to both tentatively identify and quantify peaks in the PAH-bearing isolate. A 30 m fused silica capillary coated with OV-101 was used with a similar temperature program on a Hewlett-Packard 5880 gas chromatograph for analysis of air particulate isolates.

Samples, Reagents, Sample Preparation

The solvents used were purchased from Burdick and Jackson Laboratories (Muskegon, MI), and were used as received. The coal oil, a "vacuum still overhead," was obtained from the U. S. Environmental Protection Agency/ Department of Energy, Fossil Fuel Materials Facility (11), Repository Sample Number 1310. The airborne particulate samples, which were collected in the vicinity of a coal gasifier, were supplied by B. R. Clark of this Laboratory on a Department of Energy site specific assessment program. The glass fiber filters bearing the airborne particulates were spiked with

^{14}C-labeled benzo(a)pyrene (BaP), shredded, and ultrasonically extracted with three 100 ml portions of benzene (3 minutes per extraction). The extracts were pooled and concentrated to 0.3 ml.

Typically, 1 g of coal oil 1310 was diluted to a final volume of 10 ml with "distilled in glass" grade methylene chloride. Consequently, each run of the HPLC system fractionated 25 mg of the coal oil.

Standards

A benzene solution containing 100 µg/ml each of phenanthrene and benzo(ghi)perylene was used to define the cutpoints for the PAH isolate in the airborne particulate extracts. An eleven-component standard was used to define cutpoints for four fractions in synfuels. The compounds included benzene (marker for dead volume of the column and the "alkane" fraction); phenanthrene and benzo(ghi)perylene ("PAH" fration); 2, 4, 6-trimethylphenol, 1-aminonaphthalene, acridine, quinoline, and 1-aminopyrene ("amines, aza-arenes, and alkylated phenols" fraction); and 2,4-dimethylphenol, p-cresol, and phenol ("simple phenols"). The eleven compounds were dissolved in methylene chloride in concentrations ranging from 20 µg/ml (acridine) to 4 mg/ml (phenol), in order to present a standard chromatogram in which each peak was approximately the same height.

Fractionation Procedures

Synfuels. The step gradient used was capable of dividing a synfuel into the four fractions defined by the eleven-component standard. The Digital Programmer was set as follows: (a) hexane, 6 min.; (b) 10% v/v methylene chloride (MeCl$_2$) in hexane, 30 min.; (c) 40% v/v MeCl$_2$ in hexane, 34 min.; (d) MeCl$_2$, 30 min.; and (e) hexane, 30 min. The programmer also permitted the last 24 minutes of solvent b, as well as all of solvents c and d, to be collected using independent solenoid valves.

Airborne Particulate Extracts. The PAH-bearing fraction was defined by the volume of 10% v/v MeCl$_2$ in hexane solvent required to elute the two-component standard described above from the PAC semi-preparative scale HPLC column. The remainder of the program was identical to that described in Reference 4.

RESULTS AND DISCUSSION

The order of elution of components in the four HPLC fractions is basically a function of polarity. Alkanes and alkylated naphthalenes elute within two column volumes, when either hexane or 10% v/v $MeCl_2$/hexane is the eluent. PAH are selectively eluted with the balance of the 10% $MeCl_2$/hexane solvent, while amines, aza-arenes, and heavily alkylated phenols are eluted with 40% $MeCl_2$/hexane. Phenol, p-cresol, and other simple alkylated phenols are eluted with $MeCl_2$. The polarity of a heavily-alkylated phenol is strongly affected by the electron-donor properties of the substituents. These substituted phenols therefore elute in an earlier fraction than would a simple cresol or phenol itself. The final 30 minute flush with hexane is sufficient to restore the column to its initial activity. This solvent program yields very reproducible separations. Under normal operating conditions, the cut points needed for each solvent have not varied by more than two minutes over a period of several months.

Aliphatic and aromatic acids are not eluted from the aminosilane column, even if the eluent is changed to acetonitrile. ^{14}C-labeled stearic or succinic acids are not recovered from the column. Consequently, the system is useful for compounds ranging in polarity from alkanes through phenols. The less polar amino cyano (PAC) column does elute such polar species and is preferred for use with highly polar samples such as air particulate extracts.

A fifty-two component PAH standard, which was nominally 20 µg/ml in each component, was used to test the accuracy of the isolation and analysis procedures. Two aliquots of the mixture, which contained PAH ranging from naphthalene to anthanthrene, were fractionated in the described manner. Both PAH- and amine/aza-arene -- bearing fractions were concentrated to 100 µl using dry, flowing nitrogen and vacuum and profiled using the Sigma I. The PAH fraction demonstrated excellent recovery (>90%) for compounds of four or more rings, and reasonable recoveries for three-ring PAH, as illustrated by the representative data in Table 1. Reduction in yield for the smaller PAH was due to either volatilization during solvent concentration or exclusion because of the fraction cutpoints. The amine/aza-arene fraction contained no peaks which could be positively identified as PAH.

TABLE 1

RECOVERIES OF INDIVIDUAL PAH USING THE HPLC FRACTIONATION PROCEDURE

Compound	% recovery[a]
Fluorene	8
Phenanthrene	53
2-methyl phenanthrene	57
1-methyl phenanthrene	~100
3,6-dimethyl phenanthrene	66
Fluoranthene	98
Pyrene	~100
Benzo(a)fluorene	~100
3-methyl pyrene	90
Chrysene/BAA	100
Dibenzanthracenes	77

[a]Result of duplicate determinations.

The utility of this procedure for the isolation of polynuclear aromatics in synfuels is illustrated by its application to a coal-derived oil. The gas chromatograms of the first three HPLC fractions of coal oil 1310 are shown in Figure 1. Detailed GC/MS examination of the PAH-bearing fraction (B) indicated compounds beginning with fluorene and increasing stepwise in mass to C_2-chrysene or C_2-benzanthracene, as shown in Table 2. No nitrogen-bearing species were detected in this fraction. Fraction C, on the other hand, contained exclusively nitrogen-containing species. The presence of aza-arenes, ranging from C_1-quinoline to C_3--azapyrene or C_3-azafluoranthene was confirmed using the ammonia/chemical ionization mass spectrometric technique described by Buchanan et al. (11). A detailed list of the specific compounds is given in Table 3. We conclude that this single-step HPLC procedure can be a powerful tool for the isolation of synfuel polynuclear aromatics.

Quantification has focused upon the PAH. The six major constituents of fraction B, sample 1310, were measured to evaluate the precision of this method. Nine separate PAH

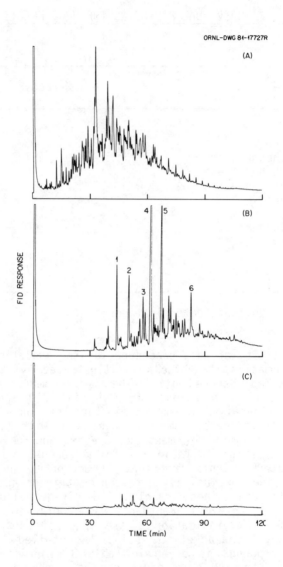

FIGURE 1. Chromatograms of the alkane/alkylnaphthalene (A), PAH (B), and aza-arene/amine (C) fractions of coal oil 1310. Identified peaks: 1, phenanthrene; 2, C_1-phenanthrene/anthracene; 3, phenylnaphthalene; 4, pyrene; 5, C_1-fluoranthene/pyrene; 6, C_1-chrysene/benzo(a)anthracene.

TABLE 2

PAH SPECIES IDENTIFIED IN COAL OIL 1310

Fluorene
C_1-fluorene (3 isomers)
*Phenanthrene/anthracene
C_2-fluorene/C_3-acenaphthylene (4 isomers)
*C_1-anthracene/phenanthrene (3 isomers)
C_3-fluorene/C_4-acenaphthylene (2 isomers)
*Phenylnaphthalene (2 isomers)
C_2-anthracene/phenanthrene (3 isomers)
Fluoranthene
C_4-fluorene
*Pyrene
C_3-phenanthrene/anthracene (3 isomers)
C_4-phenanthrene/anthracene (2 isomers)
*C_1-fluoranthene/pyrene (2 isomers)
C_1-benzylnaphthalenes (2 isomers)
C_2-pyrene/fluoranthene (2 isomers)
C_3-fluoranthene/pyrene (2 isomers)
C_3-phenylnaphthalene
Chrysene
C_4-pyrene/fluoranthene
*C_1-chrysene/BAA
C_2-chrysene/BAA (2 isomers)

---

*Denotes major species

fractions were prepared and analyzed for these six PAH in two sets of runs. The reproducibility, as shown in Table 4, is approximately 10% relative standard deviation, both for multiple aliquots analyzed in a single set of runs and for multiple sets of runs. These six components comprise approximately 6% w/w of the crude sample.

The fractionation system for synfuels can be readily adapted to airborne particulates extracts to facilitate the rapid and efficient analysis of airborne emissions. These extracts cannot be fractionated on the aminosilane column because the acid constituents adsorb to and seriously degrade the activity and performance of the aminosilane column. This difficulty, however, is not encountered with the less polar PAC column. Using this column, fractions enriched in PAH can

TABLE 3

AZA-ARENES AND AMINES IDENTIFIED IN COAL OIL 1310

C_1, C_2, C_3, C_4, and C_5 quinoline
C_1 indole
C_2 indole or N-C_1 aminoindene (2° amine)
C_3 indole or N-C_2 aminoindene (2° amine)
C_4 indole or N-C_3 aminoindene (2° amine)
C_2-aminoindene (1° amine)
C_5, C_6, and C_7 azaindane
C_6 and C_7 azaindene
C_2, C_3, C_4, and C_6 phenyl pyridine
Carbazole
C_1-carbazole
C_2 carbazole or N-C_1-aminofluorene
C_3 carbazole or N-C_2 aminofluorene
C_4 carbazole or N-C_3 aminofluorene
C_5 carbazole or N-C_4 aminofluorene
N-C_3 aminonaphthalene (2° amine)
C_0, C_1, and C_2 acridine
C_0, C_1, C_2, and C_3 azapyrene/azafluoranthene
N-C_1, C_2 aminoanthracene (2° amine)

be isolated with >95% recovery of the ^{14}C-BaP tracer. The fraction is sufficiently pure for analysis of PAH by capillary column GC. The GC profile in Figure 2 is of the PAH fraction obtained by HPLC of an extract of an ambient air particulate sample collected near a coal gasifier. Quantitative data for the major PAH identified in the sample are listed in Table 5. These data illustrate the range of PAH which can be determined using this approach. In the particular samples shown, the downwind air sample appears to contain elevated levels of several PAH.

SUMMARY

Semi-preparative scale HPLC using bonded normal-phase columns offers an attractive alternative to the classical solvent partition/adsorption column chromatographic procedure for obtaining a PAH or a PAA fraction of purity suitable for analysis by gas chromatography. Sample matrices ranging from fossil fuels to air particulate extracts can be fractionated successfully if the appropriate polarity column is chosen. The apparatus is low-cost, gives reproducible results, does

TABLE 4

REPRODUCIBILITY OF THE ANALYSIS OF SIX PAH IN COAL OIL 1310

COMPOUND	CONCENTRATION, mg/g of oil[a]	
	Aliquots 1-5	Aliquots 6-9
Phenanthrene	5.8 ± 0.5	6.6 ± 0.5
C-phenanthrene/anthracene	6.2 ± 0.3	7.3 ± 0.6
Phenyl naphthalene	3.8 ± 0.4	4.7 ± 0.4
Pyrene	29 ± 1.3	32 ± 2.0
C-fluoranthene/pyrene	14 ± 0.7	16 ± 1.1
C-chrysene/BAA	2.7 ± 0.1	3.1 ± 0.3

[a]Mean ± standard deviation using four or five replicates.

TABLE 5

CONCENTRATIONS OF AIR PARTICULATE PAH MEASURED UPWIND AND DOWNWIND OF A COAL GASIFIER

PAH	Upwind Air Concentration ng/m^3	Downwind Air Concentration, ng/m^3
Fluoranthene [A]	0.6	0.6
Pyrene [B]	0.9	0.7
C_2-pyrene [C]	0.5	0.6
Benzo(ghi)fluoranthene [D]	0.8	1.0
Benzo(a)anthracene	0.5	0.8
Chrysene/terphenyl [E]	1.3	2.4
Benzo (b,j, or k) fluoranthenes [F]	2.1	3.7
Benzo(e)pyrene [G]	0.8	2.7
Benzo(a)pyrene [H]	0.7	1.2
Benzo(ghi)perylene [I]	1.2	1.9

Bracketed letters refer to peaks in Figure 2.

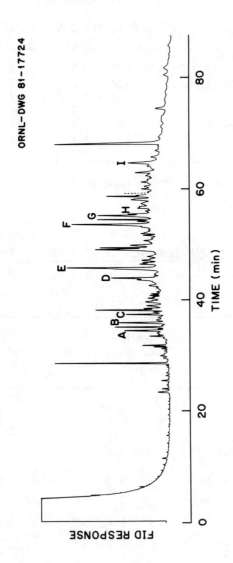

FIGURE 2. Gas chromatogram of an airborne particulate PAH fraction. Sample collected near a coal gasification plant. Peaks are identified in Table 5.

not require an expensive solvent gradient programmer, and can operate unattended.

ACKNOWLEDGEMENTS

The authors thank M. V. Buchanan and G. L. Kao for the mass spectrometric work reported in this paper.

REFERENCES

1. Dipple, A. (1976): Polynuclear aromatic carcinogens. In: Chemical Carcinogens, edited by C. Searle, pp. 245-315, American Chemical Society, Washington, D.C.
2. Guerin, M. R., Ho, C. -h., Rao, T. K., Clark, B. R., and Epler, J. L. (1980): Polycyclic aromatic primary amines as determinant chemical mutagines in petroleum substitutes, Env. Res., 23:42-53.
3. Wilson, B. W., Pelroy, R., and Cresto, J. T. (1980): Identification of primary aromatic amines in mutagenically active subfractions from coal liquifaction materials, Mutat. Res., 79:193-202.
4. Tomkins, B. A., Reagan, R. R., Caton, J. E., and Griest, W. H. (1981): Liquid chromatographic determination of benzo(a)pyrene in natural synthetic, and refined crudes, Anal. Chem., 53:1213-1217.
5. Tomkins, B. A., Ostrum, V. H., and Caton, J. E. (1981): A rapid screening procedure for 2-aminonapthalene in natural, synthetic, and refined crudes, Anal. Chim. Acta, (in press).
6. Hertz, H. S., Brown, J. M., Chesler, S. N., Guenther, F. R., Hilpert, L. R., May, W. E., Parris, R. M., and Wise, S. A. (1980): Determination of individual organic compounds in shale oil, Anal. Chem., 52:1650-1657.
7. Brown, J. M., Wise, S. A., and May, W. E. (1980): Determination of benzo(a)pyrene in recycled oils by a sequential HPLC method, J. Environ. Sci. Health, A15:613-623.
8. Daisey, J. M., and Leyko, M. A. (1979): Thin-layer gas chromatographic method for the determination of polycyclic aromatic and aliphatic hydrocarbons in airborne particulate matter, Anal. Chem., 51:24-26.

9. Lao, R. S., Thomas, R. S., Oja, H., and Dubois, L. (1973): Application of a gas chromatograph-mass spectrometer-data processor combination to the analysis of the polycyclic aromatic hydrocarbon content of airborne pollutants, Anal. Chem., 45:908-915.
10. Cautreels, W., and Van Cauwenberghe, K. (1976): Determination of organic compounds in airborne particulate matter by gas chromatography-mass spectrometry, Atmos. Environ., 10:447-457.
11. Griest, W. H., Coffin, D. L., and Guerin, M. R.: Fossil Fuels Research Matrix Program, ORNL/TM-7346, Oak Ridge National Laboratory, Oak Ridge, TN., 1980, 40 pp.
12. Buchanan, M. V., Ho, C. -h., Guerin, M. R., and Clark, B. R. (1981): Chemical characterization of mutagenic nitrogen-containing polycyclic aromatic hydrocarbon in fossil fuels. In: Chemical Analysis and Biological Fate: Polynuclear Aromatic Hydrocarbons, edited by M. Cooke and A. J. Dennis, pp. 133-145, Battelle Press, Columbus, Ohio.

EFFECTS OF TRANSPLACENTALLY ADMINISTERED POLYAROMATIC HYDROCARBONS ON THE GENOME OF DEVELOPING AND ADULT RATS MEASURED BY SISTER CHROMATID EXCHANGE

A. TURTURRO, N.P. SINGH, M.J.W. CHANG, AND R.W. HART
Dept. Health Human Services; Public Health Service; Food and Drug Administration; National Center for Toxicological Research, Jefferson, Arkansas 72079

INTRODUCTION

The polyaromatic hydrocarbon (PAH) 7,12-dimethylbenz(a)anthracene (DMBA) is a teratogen if given early in development (1), a neural carcinogen if administered late in gestation (2) and a mammary carcinogen in rats post-natally (3). Although it is not presently understood how these transitions occur, there is some evidence that modulation of genotoxicity is important (4). The induction of sister chromatid exchanges (SCE) is generally accepted as a biological indicator of genotoxicity (5) though its mechanism is obscure. For the PAH, SCE induction has been found to be correlated to the mutagenicity and carcinogenicity of various analogues (6,7). DMBA has been shown to increase the average number of SCE/cell in vitro with an adequate metabolic activating system (8) and in vivo post-natally in hamsters (9). To evaluate the relative genotoxic effects of DMBA on mother and fetus at stages when the fetus has very different biological responses to DMBA, the induction of SCE was measured at different stages of development in a transplacental system.

MATERIALS AND METHODS

Three to four month old female Sprague-Dawley derived CD rats were mated with males of the same age and strain. The presence of a sperm plug was noted as Day 0 of gestation. The details of this SCE procedure will be published elsewhere (N. P. Singh et al, manuscript in preparation). Briefly, the technique involves an interperitoneal infusion into pregnant rats of 500 mg bromodeoxyuridine (BUdR) (Sigma Co., St. Louis, MO.) /kg body weight in 13.5 ml of a 5% glucose (Fisher Co., Pittsburgh, PA.) solution in 23 hours to label the cells. After

2 hours of BUdR infusion, DMBA (Sigma) dissolved in dimethyl sulfoxide (DMSO (Sigma)) was injected through the infusion catheter at a concentration of 15 mg/kg body weight (approximately .3 ml), in treated animals and solvent only in control ones. After 23 hours, 10 mg /kg body weight colchicine (Sigma) was injected through the catheter.

Cell Harvest

Two hours after colchicine, animals were sacrificed by opening the chest after ether anesthesia and bone marrow taken from the tibia and femur or femur only. The marrow was suspended in 1 ml of balanced salt solution containing .1 microgram/ml colchicine. Early fetuses were separated from excised uteri in saline under a dissecting microscope. Late fetuses were removed from the uteri and the central region (around the liver and intestine) dissected out. Twelve fetuses were used for each Day 10 preparation, and ten and seven used on Days 12 and 13 respectively. Three fetuses were used for each determination on Day 20. Cells were suspended by cutting into small pieces and gently pipetting the tissue in 1 ml of the same solution as the bone marrow.

SCE Scoring

A modification of the fluorescence plus Giemsa procedure (10) visualized the SCE. They were scored using oil immersion (1000X) from metaphases having 2n (=42) chromosomes. Fifty cells were counted for bone marrow and thirty for each fetal preparation. In four mothers at 20 days, tibia was compared to femur. All determinations of the average SCE/cell used two preparations except 20 day fetus (three). All counts are expressed as the number of SCE per cell (2n chromosomes). All statistics was done by a two-tailed t-test.

RESULTS

The average number of spontaneous SCE in rat bone marrow was found to be dependent upon the origin of the marrow, the femur containing $2.8 \pm .2$ (mean $\pm$ standard error) and the tibia $3.3 \pm .2$ (n= 4) (different with $p \leq .01$). Since the femur is the most commonly reported source of marrow for SCE, only this marrow was used.

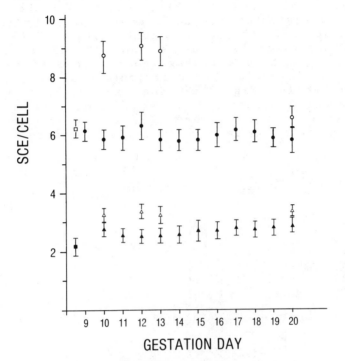

FIGURE 1.   SCE/cell at different gestation days. Closed circles are DMBA treated maternal bone marrow; open circles, the fetal treated values (plus standard error bars).   Triangles are respective control values.   The open box is non-pregnant treated, the closed, solvent. The abscissa is the gestation day, the ordinate is the SCE/cell. Each value is from 2 determinations except Day 20 fetus (3 determinations).

The average number of SCE did not vary significantly from the 9th to the 20th day of pregnancy in the mother and was approximately the same as in nonpregnant animals (Figure 1), both for control and DMBA induced animals. The DMBA treated animals had a level of SCE ($5.93 \pm .29$; n=20) (mean $\pm$ standard error) more than twice that of the controls ($2.66 \pm .2$; n=20)(significance $p<.01$). If the results from Days 10 to 20 are combined, the distribution of maternal bone marrow SCE/cell (Figure 2) seems to show two different types of cells, I (2-8 SCE/cell)(which appear distributed around a mean similarly to the distribution in the solvent control) and II (≥ 9 SCE/cell) where Type II comprise about 10% of the total. Type I cells have a average number of SCE/ cell of $5.39 \pm .2$ (n= 20). The distribution of the solvent control cells (Figure 2) does not appear to have this clear separation. If separated, type I cells have about twice as many SCE/cell as the solvent control cells, while type II cells have more.

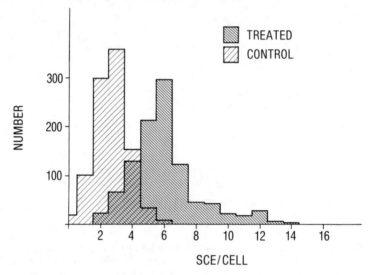

FIGURE 2. Distribution of maternal SCE/cell. Hatching from upper left to lower right indicated treated maternal and that from lower left to upper right, control values. The abscissa is SCE/cell and the ordinate is the number of cells with that value.

Fetal tissue from days 10, 12 and 13 fetuses had a similar average value of SCE/cell (Figure 1), with a mean of $8.89 \pm .4$ (n=6), which was about 3 times the fetal solvent control value of $3.2 \pm .3$ (n=6). The ratio of the difference between the treated and solvent values for the early fetus and the mother is 5.69/3.07, or approximately two.

Fetal cells from the central region of the 20 day old fetus had SCE/cell of $6.48 \pm .9$ (n=3) (Figure 1), which was not significantly different than the maternal bone marrow. However, if the distribution of the SCE in these cells is displayed, as in Figure 3, the number of cells which have a distribution similar to the type II cells in the mother seem to have a higher percentage (about 20%). The distribution of SCE/cell in the early fetus (Figure 3) is not so obviously separated into these types and there seems to be much more heterogeneity.

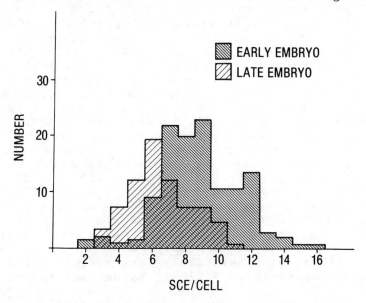

FIGURE 3. Distribution of fetal SCE/cell. Upper left to lower right hatching indicates early fetus and lower left to upper right, 20 day fetus. The axes are similar to Figure 2. Control distributions are similar in the two groups, and similar to the maternal control distribution in Figure 2.

DISCUSSION

Studies which include an <u>in vivo</u> measurement of the number of SCE in bone marrow do not always indicate the source of the marrow. These data show that this has an effect on the average number of SCE/cell perhaps related to the ratio of cell types. Tibia bone marrow contains about 65% of the hematopoeitic (11) and 75% of the reticuloendothelial (related to the myeloid series) (12) cells of the femur, indicating an enrichment of the latter.

There seems to be no significant variation of maternal SCE/cell with gestational age in either DMBA treated or DMSO control mothers, similar to the results of Kram et al (13) for cyclophosphamide and saline, although the treated rats have a higher average number. This dose of DMBA produces mammary tumors in about 55% of the mothers (2), supporting the results in hamster relating increased average SCE/cell in vivo and carcinogenicity of DMBA. The scoring of SCE, since it measures individual cells, has the capacity to identify different subpopulations. The distribution of the induced SCE/cell in the mother suggests that the effect of the chemical is not homogeneous on the marrow. SCE induction by DMBA seems higher in about 10% of the cells. The reason for this is unclear. The bone marrow is a heterogeneous tissue with cells in different stages of maturation which may differ in response to the genotoxic effect of DMBA (e.g. during maturation of a series in human bone marrow there is a decrease in inducible DNA repair (14)). Leukemia induced in rat by DMBA is almost always of myeloid origin (15) from stem cells (16), which is evidence of heterogeneity of the carcinogenic response of the blood to DMBA. It is tempting to speculate that these sensitive cells are involved in that phenomenon, especially since tibia, with relatively more myeloid cells, has a higher spontaneous SCE/cell value than femur.

The twofold higher induction of SCE/cell in the early fetus correlates well with the embryotoxic and teratogenic effect of DMBA (1). Decline with fetal age for DMBA induced SCE has also been reported for cyclophosphamide and mitomycin C (13). The cause for this change is not clear, but it does not seem to be related to metabolic activation since the former compound requires activation (as does DMBA) but the latter does not. Whether the monooxygenase system begins to produce genotoxic DMBA metabolites at birth in the rodent, as is generally accepted (17), or earlier (18), it is unlikely, as noted by Kram et al (13), that there is an actual decline in the activating systems near birth which would be required to explain the lower induction. One possibility, suggested by Kram et al (13), that placental transport for the agents changes with development, given the lipid solubility characteristics of DMBA, is also unlikely to explain any differential effect in development. Finally, Doerjer et al (4) have indicated that the 20 day fetus, on the average, has less DMBA binding to the DNA than in the mother. Yet the amount of SCE/cell at Day 20 is the same, if not higher, in the fetus. These considerations suggest that the important factor in the differential effect is not effective dose.

Since only cells which had undergone two divisions were counted in this assay, replication rate is probably not a factor either since only a subset of cells of fairly restricted rate was scored. Also the distribution of the SCE/cell in the early fetal cells has wide variations at a stage when, presumably, most cells are replicating quickly and fairly homogeneously. This broad distribution is not merely a result of tissue heterogeneity since the distribution of SCE/cell in 20 day old fetuses seem to be fairly similar to the adult even though it is derived from many different tissues.

The relative sensitivity of the genome of the early fetal cells to DMBA is probably, therefore, a result of some process(es) associated with differentiation. Cells in culture will decrease their SCE response to a compound in the later passages (19) as they "differentiate" (20). In the mouse, rat and hamster *in vitro* spontaneous SCE are higher than *in vivo* ones, i.e. in conditions where cells "dedifferentiate" the level of SCE increase (21). Differentiation seems to decrease the capacity of cells for SCE induction and, perhaps, may cause them to become more resistant to the other effects of DNA damaging agents. Cells which are not as far along the differentiation pathway as most other cells in an organ may be particularly sensitive to an insult leading to SCE and other biological effects of the PAH.

One possible mechanism for this is related to chromosome structure. Condensed heterochromatic regions in chromosomes have lower amounts of SCE than euchromatic ones (22). In model developing systems, chromatin becomes more condensed with differentiation (23). Condensation in the state of the chromatin in differentiation, as genes turn off, may also decrease the capacity for SCE induction. However, it is not clear how this explanation accounts for the increased induction of SCE in mutant cells derived from patients with Bloom's syndrome (24) or xeroderma pigmentosum (25). Some other factor associated with differentiation, such as the decrease in ultraviolet (UV) type excision repair (14,26) or the increase in X-ray type repair (27) may be more relevant since these cells are deficient in some repair system. The correlation of capacity for X-ray type or UV type excision repair to SCE induction is not a simple one (21), but it is clear that there are complicating factors since replication is a prerequisite for SCE and is unnecessary for excision repair.

Further studies on the relationship of differentiation and SCE and the use of SCE as a biological marker for genotoxicity in systems especially sensitive to the carcinogenic effects of DMBA, e.g. the neural system in transplacental carcinogenesis, are needed to help elucidate the mechanism by which expression of genetic damage caused by the PAH can result in teratogenesis or carcinogenesis in totally different organs depending on the stage of development at the time of insult.

ACKNOWLEDGEMENTS

We wish to thank Bob Loe and his staff for their assistance with the figure preparation and Dr. Peter Fu for purifying and recrystallizing the DMBA, all of the National Center for Toxicological Research.

REFERENCES

1. Currie, A.R., Crawford, A.M. and Bird, C.C. (1973): The embryopathic and adenocorticolytic effects of DMBA and its metabolites. In: Transplacental Carcinogenesis, WHO-IARC Sci. Publ. 4. Lyon, France pp. 149-153.
2. Napalkov, N.P., and Alexandrov, V.A. (1974): Neurotropic effect of 7,12-dimethylbenz(a)-anthracene in transplacental carcinogenesis, J. Nat. Cancer Inst., 52:1365-1366.
3. Huggins, C., Grand, L.C., and Brillantes, F.P. (1961): Mammary cancer induced by a single feeding of polynuclear hydrocarbons, and its suppression, Nature, 189:204-207.
4. Doerjer, G., Diessner, H., Bucheler, J., and Kleihues, P. (1978): Reaction of 7,12-dimethyl-benz(a)anthracene with DNA of fetal and maternal

rat tissue in vivo, Int. J. Cancer, 22:288-291.
5. Carrano, A.V., Thompson, L.H., Lindl, P.A., and Minkler, J.L. (1978): Sister chromatid exchanges as an indicator of mutagenesis, Nature, 271:551-553.
6. Roszinsky-Kocher, G., Basler, A., and Rohrborn, G. (1979): Mutagenicity of polycyclic hydrocarbons. V. Induction of sister chromatid exchanges in vivo, Mutat. Res. 66:65-67.
7. Pal, K., Grover, P.L., and Sims, P. (1980): The induction of sister chromatid exchanges in Chinese Hamster ovary cells by some epoxides and phenolic derivatives of benz(a)pyrene, Mutat. Res., 78:193-199.
8. Popescu, N.C., Turnbull, D., and DiPaolo, J.A. (1977): Sister chromatid exchange and chromosome aberration analysis with the use of several carcinogens and non-carcinogens, J. Nat. Cancer Inst., 59:289-292.
9. Bayer, U., and Bauknecht, T. (1977): The dose dependence of sister chromatid exchanges induced by 3 hydrocarbons in the in vivo bone marrow test with Chinese Hamsters, Experimentia, 15:25.
10. Perry, P., and Wolff, S (1974): A new Giemsa method for the differential staining of sister chromatids, Nature, 251:256-258.
11. Van dyke, D., Anger, H., and Pollycove, M. (1964): The effect of erythropoietic stimulation on marrow distribution in man, rabbit and rat as shown by ^{59}Fe and ^{52}Fe, Blood, 24:356-371.
12. Keene, W.R., and Jandl, J.H. (1965): Studies of reticuloendothelial mass and sequestering function of rat bone marrow, Blood, 26:157-175.
13. Kram, D., Bynum, G.D., Senula, G.C., Bickings, C.K., and Schneider, E.L. (1980): In utero analysis of sister chromatid exchange: Alteration in susceptibility to mutagenic damage as a function of fetal cell type and gestational age, Proc. Nat. Acad. Sci. 77:4784-4787.
14. Lewensohn, R., and Ringborg, U. (1979): Induction of unscheduled DNA synthesis in human bone marrow cells by bifunctional alkylating agents, Blood, 54:1320-1329.
15. Ioachim, H.L., Sabbath, M., Andersson, B., and Keller, S. (1971): Viral and chemical leukemia in the rat: Comparative study, J. Nat. Cancer Inst., 147:161-168.
16. Huggins, C., Grand, L., and Oka, H. (1970): Hundred day leukemia: Preferential induction in

rat by pulse doses of 7,8,12-trimethylbenz-(a)anthracene, J. Exp. Med., 131:321-330.
17. Sato, R., and Omura, T., editors (1978): Cytochrome P-450. Academic Press, N.Y., 233 pp.
18. Juchau, M.R., DiGiovanni, J., Namkung, M.J., and Jones, A.H. (1979): A comparison of the capacity of fetal and adult liver, lung and brain to convert polycyclic aromatic hydrocarbons to mutagenic and cytogenic metabolites in mice and rats, Tox. and Applied Pharm., 49:171-178.
19. Schneider, E.L., and Monticone, R.E. (1978): Aging and sister chromatid exchanges: II. The effect of the in vitro passage level of human fetal lung fibroblasts on baseline and mutagen induced sister chromatid exchange frequencies, Exp. Cell Res., 115:269-276.
20. Bell, E., Marek, L.F., Levinstone, D.S., Merrill, C., Sher, S., Young, I.T., and Eden, M. (1978): Loss of division potential in vitro: Aging or differentiation?, Science, 202:1158-1163.
21. Perry, P.E. (1980): Chemical mutagens and sister chromatid exchange. In: Chemical Mutagens: Principles and Methods for Their Detection, Vol. 6, edited by F.J. de Serres and A. Hollaender, pp. 1-39, Plenum Press, N.Y.
22. Kato, H. (1979): Preferential occurrence of sister chromatid exchanges at the heterochromatic-euchromatic junctions in the wallaby and hamster chromosomes, Chromosoma 74:307-316.
23. Arceci, R.J., and Gross, P.R. (1980): Histone variants and chromatin structure during sea urchin development, Develop. Biol., 80:186-209.
24. Chaganti, R.S.K., Schonberg, S., and German, J. (1974): A manyfold increase in sister chromatid exchanges in Bloom's syndrome lymphocytes, Proc. Nat. Acad. Sci., 71:4508-4512.
25. Wolff, S., Rodin, B., and Cleaver, J. (1977): Sister chromatid exchanges induced by mutagenic carcinogens in normal and xeroderma pigmentosum cells, Nature, 265:347-349.
26. Chang, A.C., Ng, S.K., and Walker, I.G. (1976): Reduced DNA repair during differentiation of a myogenic cell line, J. Cell Bio., 70:685-691.
27. Counis, M.F., Chaudun, E., Simonneau, L., and Courtois, Y. (1979): DNA repair in lens cells during chick embryo development, Biochim. Biophys. Acta, 561:85-98.

THE APPLICATION OF HIGH RESOLUTION PREPARATIVE LIQUID CHROMATOGRAPHY TO THE POLYCYCLIC AROMATIC HYDROCARBONS

F. L. VANDEMARK AND J. L. DiCESARE
Perkin-Elmer Corporation, Norwalk, Connecticut 06856

INTRODUCTION

Until very recently, preparative liquid chromatography has generally been considered a cumbersome, low-resolution technique, capable of only simple separations. Because of limited efficiency, little similarity could be expected between preparative and analytical scale chromatography when analytical separations were to be scaled up to the preparative level. With recent advances in column and instrument technology has come the development of high-resolution preparative scale columns. These columns enable the user to achieve very difficult separations, comparable to those with analytical columns, and yet obtain large quantities of material in a short time frame. The determination of polycyclic aromatic hydrocarbons (PAHs), which is difficult due to the large number of analytes, is further compounded by the complex sample matrices in which they are often found, and usually requires sample pretreatment prior to anlaysis. This paper reviews a variety of fractionation techniques used prior to analytical chromatography. These methods are compared with one which uses high-resolution preparative chromatography.

MATERIALS AND METHODS

Apparatus

<u>Instrument</u>. The liquid chromatograph was a Perkin-Elmer Series 3B, equipped with a Rheodyne 7125 injection valve, (with 20 µl or 2 ml loop), a Model LC-75 injection wavelength UV detector, and Model 650-10S fluorescence detector. A Model Sigma 15 chromatography data station was used for data aquisition and reduction.

<u>Columns</u>. Columns were: 1.) a 25 X 0.46 cm. column packed with 10-µm C18 reverse phase packing, and 2.) a 25 cm. X 22 mm. column packed with 10-µm C8 reversed phase packing.

Reagents and Standards

PAH standard materials were obtained in highest purity grade from Aldrich Chemical, Milwaukee, Wisconsin, Applied Science, State College, Pennsylvania, Chem. Services, West Chester, Pennsylvania, and Eastman Kodak, Rochester, New York. PAH standards were prepared as pure solutions in acetonitrile at a concentration of 1 mg/L. HPLC grade acetonitrile and tetrahydrofuran (THF) were purchased from MCB, Plainfield, New Jersey. Water for use as mobile phase was purified using a Millipore reverse-osmosis system, Bedford, Massachusettes.

SAMPLE PREPARATION

The petroleum samples and coal liquid samples were weighed, dissolved in THF, and filtered prior to analysis.

CHROMATOGRAPHY

All analyses on the high resolution analytical C-18 reverse phase column were performed using gradient elution from 40 to 100% acetonitrile, for 15 minutes in a linear step. The flowrate was set at 1.3 ml/min. The effluent of the column was monitored with a fluorescence detector, excitation 305 nm. emmission 430 nm. Petroleum samples analyzed on the preparative column utilized eluents of THF, acetonitrile in water using gradient elution from 50 to 100%, or an isocratic mobile phase of THF and water at 70:30. The preparative column was monitored with a UV detector set at 250 nm.

RESULTS AND DISCUSSION

The complex sample matrices in which PAHs are normally found, have posed problems for analysis by chromatographic techniques. Many workers have incorporated a variety of fractionation steps prior to the chromatographic analysis (1,2). Chromatography techniques have been widely used in this fractionation step because of their high selectivity (3,4). A limitation of these techniques is the very long sample preparation time required before analytical chromatograms may be obtained. One solution to this problem was a fractionation method which combined partition chromatography with size exclusion chromatography (5,6). In this method, the sample was first subjected to a low resolution separation on a C-18 bonded phase. The unretained material was collected on-line and subjected to size exclusion, where a discrete fraction of the eluent containing the PAHs was collected and then subjected to analysis by high-resolution reverse-phase techniques.

The importance of this two-step process is illustrated in Figure 1, which shows a chromatogram of a solvent refined coal not subjected to the coupled-column sample pretreatment.

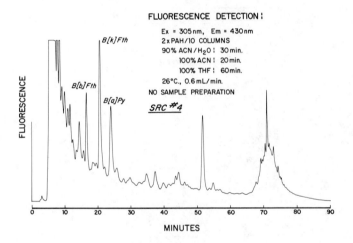

FIGURE 1. Chromatogram of Solvent Refined Coal without Pretreatment.

There is a large amount of polar material which elutes early in the chromatogram, obscuring peaks. In addition, there is a large amount of strongly retained material which requires high solvent strength to elute, with corresponding long analysis times.

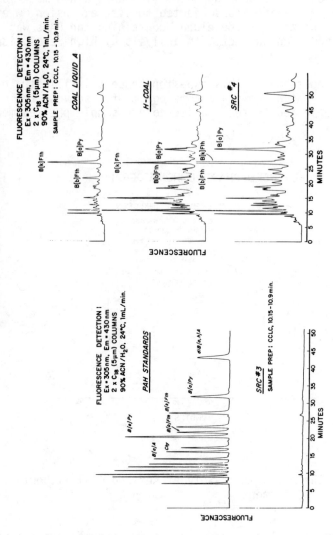

FIGURE 2. PAH Standards, Petroleum Samples Subjected to Coupled Column Chromatography Fractionation.

Figure 2 illustrates a chromatogram of PAH standards and several examples of similar samples pretreated using coupled-column techniques. Significant reduction of the early-eluting polar material and very highly-retained material has reduced the analysis time, as well as the number of unresolved peaks in the analytical chromatogram. The coupled-column techniques are easily automated, which effectively eliminates long external sample preparation times. However, it would be advantageous to be able to isolate larger quantities of fractionated material for bio-assays, off-line spectroscopic characterizations, or for alternative analytical evaluation. In an attempt to further simplify the fractionation process, as well as allow isolation of significant quantities of material, a high-resolution preparative LC column was evaluated for use as a means of fractionation. Figure 3 illustrates a mixture of PAH standards and a sample of solvent refined coal chromatographed at high solvent strength on the preparative column.

Comparison of the elution pattern of standards permits the establishment of times which define the desired PAH fraction. Analytical chromatograms from such a collected fraction of the solvent refined coal were shown to eliminate the highly retained materials; however, large amounts of early eluting polar material were still present. Reduction in the solvent strength used on the preparative column resulted in a partial resolution of PAH standards, and improvement in the separation of PAHs from matrix material in the SRC sample.

Figure 4 illustrates a 20 mg. sample of the diluted solvent refined coal injected directly onto the preparative column. Subsequent collection of the fraction containing PAHs and injection of a 20 µL aliquot onto the high resolution chromatogram results in chromatograms similar to those obtained using coupled column techniques.

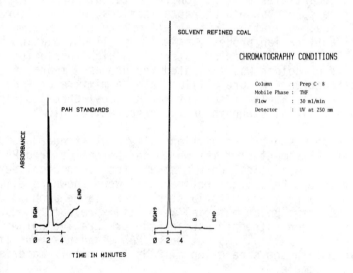

FIGURE 3. PAH Standards and SRC Sample Chromatographed in THF on C8 Preparative Column.

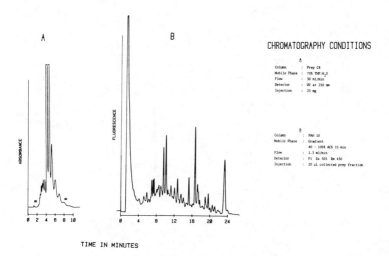

FIGURE 4. A. 20 mg SRC on Preparative Column.
B. Analytical Chromatogram Obtained from Injection of 20 µl Collected Aliquot.

Besides the utilization of preparative liquid chromatography to greatly facilitate isolation of PAHs in complex matrices, it has also proven useful for the rapid preparation of high purity standards which are essential for accurate calibration of analytical methods. Figure 5 illustrates an analytical chromatogram of a commercial chrysene standard, and reveals a number of significant contaminants present. Subsequent purification of a 100 mg sample on the preparative column and reinjection on the PAH column demonstrates that essentially pure material is obtained. Additional confirmation of purity was obtained from the stop flow spectrum run on the purified fraction, shown in Figure 5.

PREPARATIVE LIQUID CHROMATOGRAPHY

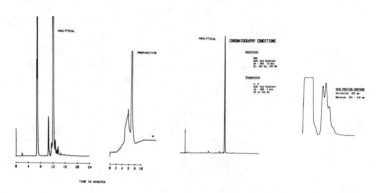

FIGURE 5. Analytical Chromatogram Chrysene, 100 mg. Sample Injection on Preparative Column for Collection. Analytical Chromatogram Obtained from Purified Product, Stopped Flow Emission scan Ex 295 Em 295 - 520.

CONCLUSION

Experiments with high-resolution preparative LC columns demonstrate that fairly pure fractions containing PAHs can be easily isolated from complex sample matrices in several minutes. These rapidly isolated fractions may then be directly analyzed by high-resolution analytical chromatography. The use of preparative techniques compliments the existing procedure using coupled-column chromatography. The initial results obtained with the preparative system can serve as an accurate assessment of whether a secondary size exclusion step would be required. The preparative sample fractionation system has the added advantage of enabling the isolating of large sample quantities suitable for other analytical tests. High resolution preparative techniques can be used for the rapid preparation of very high purity standards, important for reference purposes.

REFERENCES

1. Robbiwig, W.K., (1980): Solvent extractions of polynuclear aromatic hydrocarbons, In; Polynuclear Aromatic Hydrocarbons; Chemical and Biological Effects, edited by A. Bjorseth and A.J. Dennis, pp. 841-861, Battelle Press, Columbus, Ohio.
2. Smillie, R.D., Wang, D.T., Ibid., pp. 863-877.
3. Wise, S.A., Chesler, S.N., Hertz, H.S., Hilpert, L.R., and May, W.E., (1977): Chemically-bonded amino silane stationary phase for the high-performance liquid chromatographic separation of polynuclear aromatic compounds, Anal. Chem., 49: 2306-2310.
4. Erni, F., Frei, R.W., (1978): Two-dimensional liquid chromatographic technique for resolution of complex mixtures, Journal of Chromatography, 149: 501.
5. Ogan, K., and Katz, E. (1981): Analysis of complex samples by coupled-column chromatography, Anal. Chem., (in press).
6. Katz, E., and Ogan, K., (1981): The use of coupled-column and high-resolution chromatography in the analysis of petroleum and coal liquid samples, Chemical Analysis and Biological Fate: Polynuclear Aromatic Hydrocarbons, edited by M. Cooke and A.J. Dennis, Battelle Press, Columbus, OH, pp. 169-178.

DETERMINATION AND BIOCONCENTRATION OF POLYCYCLIC AROMATIC SULFUR IN HETEROCYCLES IN AQUATIC BIOTA

D. L. VASSILAROS, D. A. EASTMOND, W. R. WEST, G. M. BOOTH, and M. L. LEE
Departments of Chemistry and Zoology, Brigham Young University, Provo, Utah 84602

INTRODUCTION

It has long been known that the presence of polycyclic aromatic compounds (PAC) in the aqueous environment poses an elusive but theoretically real threat both to organisms which make up the food web and to man at the top of the food chain. Cognizance of this threat has encouraged efforts to improve the analytical chemistry of PAH in aquatic biota in order to increase the amount and reliability of knowledge of the concentrations and fates of these compounds. There is also a growing awareness of the potential toxicity and mutagenicity (1) of the polycyclic aromatic sulfur and nitrogen heterocycles (PASH, PANH), and toxicological studies using coal liquids and model compounds have been performed on various members of different aqueous trophic levels (2-5). The focus in this paper will be upon the behavior and presence of PASH in the aqueous ecosystem.

The structural similarities between the PAH and PASH suggest that the heterocyclic compounds will behave generally in a manner analogous to the PAH in the aquatic ecosystem, but the influence of the heteroatom on aqueous solubility, octanol/water partitioning ratio, and uptake, metabolism, and depuration rate of these compounds by invertebrates and vertebrates has not been fully investigated. There are strong indications that the heteroatom character of the PASH will lead to unique situations in the ecosystem. For example, of the macrofauna in the Galveston Bay, Texas, which were exposed to about 400,000 gallons of bunker C residual oil and No. 2 fuel oil spilled from a barge, the only species noted to have suffered a large mortality was the supralittoral isopod, *Lygia exotica* (6). Chromatographic investigation of contaminated animals demonstrated the presence of high levels of dibenzothiophene (DBT) and many of its alkylated derivatives either on the carapace or incorporated in the tissues of the isopods.

Warner (7) compared the relative levels of PASH and PAH in bunker C residual oil and the marine invertebrates described above (6). He concluded from sulfur-selective GC data that the PASH seemed to be concentrated to a greater extent than the PAH in the marine environment.

Studies on the uptake and depuration of benzo[b]thiophene, dibenzothiophene, and a few of their alkylated derivatives in eels and short-necked clams have been reported (8-10). Capillary gas chromatography with flame ionization detector (FID) and flame photometric detector (FPD), and gas chromatography-mass spectrometry (GC-MS), were used for the analysis of the PAH/PASH fractions. Although many details in the gas chromatograms were obscured by biogenic interfering materials which were not completely removed from the hydrocarbon fractions during the sample work-up procedure, it was clear that the clams and eels took up PAH and PASH from their environment and concentrated these compounds in their tissues. After 16 days of exposure to a suspension of crude oil in seawater, the ratio of the concentration of DBT in the eel tissue to the concentration in the ambient water was 463/1, and after a 16-day depuration period, the ratio was still around 100/1.

Joined with the concepts of potential toxicity and mutagenicity, and bioconcentration, is the observation by Berthou et al. (11) that the alkylated DBT seem to be the most persistent fossil fuel derived compounds in the environment. The PAH/PASH fractions from clams taken in the area of the Amoco Cadiz break-up off the northern coast of Brittany showed elevated levels of organosulfur compounds two years after the accident.

The purpose of this paper is to report some preliminary results on the study of the bioconcentration of PASH in fresh water and estuarine vertebrate fish. The PASH content in the fish and in related river sediments are compared, and the observations are discussed in terms of aqueous solubility, octanol/water partitioning ratio, and volatility.

METHODS AND MATERIALS

The brown bullhead catfish (Ictalurus nebulosus) and striped bass (Morone saxatillis) samples and sediment from the Black River were provided by the U.S. Fisheries and Wildlife Service, Columbia National Research Laboratory, Columbia, MO. The fish (composited whole fish) were analyzed by the procedure described elsewhere (12), and the sediment was extracted and the extract purified by a modification of the same method. GC-MS analyses were performed on a HP 5984 GC-MS-DS, using SE-52 coated fused silica columns prepared in this laboratory. Dual trace (FID/FPD) and single trace FPD gas chromatograms were obtained on a Perkin-Elmer Sigma 3 gas chromatograph, using the same capillary columns. A fused silica effluent splitter was used for the dual detector chromatography, and both detector outputs were attenuated to give identical response to DBT. Helium was used for the carrier at a linear velocity of 30 cm/sec, and the Sigma 3 oven was programmed from 50°C to 265°C at 4°C/min.

Studies following the disappearance of thiophene, benzo[b]thiophene, and DBT in water were conducted using 200 ml of filtered, autoclaved river water in 250 ml squat jars. A known concentration of a particular compound was added to a jar and replicate samples were taken over a maximum period of 96 hours. The UV absorbance of each sample was monitored using a Zeiss Model PM6KS spectrophotometer, and the concentration was determined by comparison with a standard curve. The jars were maintained at a constant temperature of 18±1°C.

The solubility of benzo[b]thiophene, DBT, and benzo[b]naphtho[1,2-d]thiophene was determined by a modified procedure of Hague and Schmedding (13). The water was passed through a C-18 Sep-Pak cartridge to remove organics. Stirring in 250 ml ground-glass stoppered jars was constant for 48 hours, at which time samples were taken to determine solubility. Solubility was determined by ultraviolet and fluorescence spectroscopy using standard curves. All other solubility data were taken from the literature (14,15).

The partition coefficients of thiophene, benzo[b]thiophene, DBT, benzene, naphthalene, and anthracene were taken from the literature (16,17). High performance liquid chromatography (18), radio-labelling (19), and UV spectroscopy were used to confirm the literature values. The partition coefficients of benzo[b]naphtho[1,2-d]thiophene and benz[a]anthracene were calculated according to Hansch and Leo (20).

RESULTS AND DISCUSSION

The numbered peaks in Figures 1-5 are identified and quantified in Table I, in which the amounts are given in ppb (ng/g dry sediment, ng/g wet fish).

Black River Samples

The PAH profile (Figure 1) of sediment taken from under a coking plant outfall on the Black River in Ohio conforms to the expected pattern for a pyrolytic source: high levels of parent compounds with respect to the alkylated derivatives.

Figure 1 shows that a wide boiling point range of PAH is present in the outfall sediment, starting with naphthalene and ending with a series of peri-condensed 6-ring compounds. The profile is dominated by phenanthrene, anthracene, fluoranthene, and pyrene. The sulfur fraction is very low in concentration compared with the PAH, with practically no alkylated DBT visible above the baseline.

The ratio of the amount of phenanthrene to the amount of DBT is used as an indicator of the relative level of the PASH, and the ratio of DBT to its C_1 and C_2 derivatives provides additional information. In the sediment fraction the phenanthrene/DBT ratio is 19/1. The ratios of DBT to the total

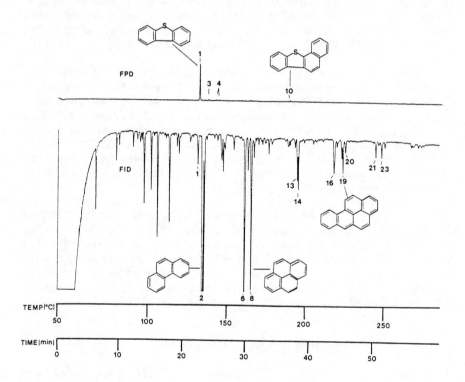

FIGURE 1. Dual detector (FID/FPD) GC trace of Black River Outfall Sediment PAH fraction.

methyl and C_2-dibenzothiophenes are 11/1 and 44/1, respectively.

TABLE 1

IDENTIFICATION AND QUANTIFICATION OF PEAKS IN FIGURES 1-5

Peak #	Compound Name	Amount[a] BROS[b]	BRNL[c]	SACR[d]
	Naphthalene	3.1×10^4	140	4.0
	2-Methylnaphthalene	1.5×10^4	320	2.0
	Biphenyl	9.7×10^3	120	
	Acenaphthylene	4.0×10^4	2400	
	Acenaphthene	3.6×10^4	258	3.0
	Dibenzofuran	6.4×10^4	1800	
1	Dibenzothiophene	2.2×10^4	700	2200
2	Phenanthrene	3.9×10^5	5700	2200
3	Naphtho[2,3-b]thiophene	7.0×10^3	28	e
4	Methyldibenzothiophenes	2.0×10^3	440	2000
5	C_2-Dibenzothiophenes	5×10^2	180	200
6	Fluoranthene	2.2×10^5	1900	350
7	Phenanthro[4,5-bcd]thiophene	4.8×10^3	120	400
8	Pyrene	1.4×10^5	1100	1200
9	Methylphenanthro[4,5-bcd]thiophenes			250
10	Benzo[b]naphtho[2,1-d]thiophene	7.6×10^3	3.6	460
	Benzo[ghi]fluoranthene	7.2×10^3		
	Benzo[c]phenanthrene	6.9×10^3		
11	Benzo[b]naphtho[1,2-d]thiophene	1.9×10^3		110
12	Benzo[b]naphtho[2,3-d]thiophene	2.9×10^3		88
	Cyclopenta[cd]pyrene	1.0×10^4		
13	Benz[a]anthracene	5.1×10^4	33	26
14	Chrysene/Triphenylene	5.1×10^4	83	130
15	Methyl 4-ring thiophenes	4.6×10^2		270
16	Benzo[b] & [k]fluoranthenes	7.5×10^4	32	35
17	Benzo[e]pyrene	2.8×10^4	21	29
18	Peri-condensed 5-ring thiophenes	$\sim 5.0 \times 10^2$		27
19	Benzo[a]pyrene	4.3×10^4	18	
20	Perylene	1.2×10^4	8	
21	Indeno[1,2,3-cd]pyrene	2.6×10^4		
	Dibenz[a,h] + [a,c]anthracenes	9.4×10^3		
	Benzo[b]chrysene	8.0×10^3		
22	cata-condensed 5-ring thiophenes	$\sim 5.5 \times 10^2$		
23	Benzo[ghi]perylene	2.4×10^4		5.0
	Anthanthrene	1.3×10^4		
	Coronene	4.8×10^3		
	Sum of 6-ring PAH isomers (mw = 302)	4.5×10^4		

[a] Amount given in ng/g dry sediment, ng/g wet tissue
[b] Black River Outfall Sediment
[c] Catfish from Black River
[d] Striped bass from Sacramento River
[e] Spaces indicate compound not detected

PASH IN FISH

The profile of the PAH/PASH fraction from a brown bullhead catfish caught near the coking plant outfall, shown in Figure 2, is very similar to that of the outfall sediment, but the levels of PAC beyond pyrene drop off dramatically. The PAH profiles of four different catfish from the same area are practically identical. The PASH content is remarkably different from the sediment fraction: there is a greater amount of DBT and of the alkylated DBT, relatively speaking. While the phenanthrene/DBT ratio in the sediment is 19/1, and the same ratio in a coal tar PAH fraction analyzed routinely in this laboratory is 15/1, the average ratio in the four fish from the Black River is 9/1, suggesting that the amount of DBT relative to the amount of phenanthrene doubled when compared to the sediment. In fact, the ratio of DBT/C_1-DBT in the fish

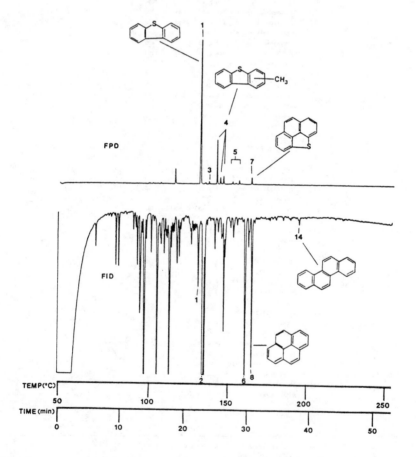

FIGURE 2. Dual detector (FID/FPD) GC trace of Black River catfish PASH fraction.

is 1.6/1 (11/1 in sediment), and the ratio of DBT/C_2-DBT is 4/1 (44/1 in sediment). The alkylated derivatives show a ten-fold concentration increase with respect to the parent compound in the fish over the sediment. The sediment DBT appears to be depleted while the fish DBT seems to be enhanced. The ratio of pyrene to phenanthro[4,5-bcd]thiophene in the two Black River samples shows the same relationship: 29/1 in the sediment, and 9.1/1 in the fish.

Figures 3 and 4 compare the PASH content of the outfall sediment and the catfish. Four 3-ring thiophene isomers are present in each sample, but the fish tissue does not contain the higher molecular weight PASH which are present in the sediment (but these fish did not appear to concentrate the higher molecular weight PAH, either); however, the alkylated species are much more abundant relative to DBT in the fish than in the sediment.

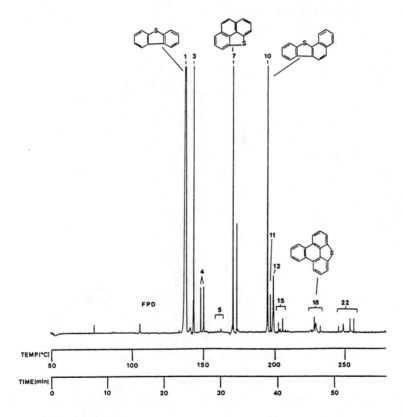

FIGURE 3. GC-FPD trace of Outfall Sediment PAH fraction.

PASH IN FISH

Sacramento sample

Figure 5 is the dual detector GC trace of the PAH/PASH fraction from a striped bass caught in the Sacramento River, CA. Like the catfish sample described above, this fish contains high levels of alkylated and parent PASH. On the other hand, there is not the paucity of high molecular weight PAC in the Sacramento fish as in the Black River fish. The ratio of phenanthrene to DBT is 1/1, and the ratio DBT/C_1-DBT is 1.1/1. Only one isomer (DBT) of the four three-ring thiophenes which are in the Black River samples is present, but higher levels of the 4-ring thiophenes and their methyl derivatives can be seen. For example, the ratio of amount of benzo[b]naphtho-[2,1-d]thiophene to chrysene is 3.5/1, and in the Black River fish it is 0.043/1. The ratio of pyrene to phenanthro[4,5-bcd]thiophene is 3/1. The relatively large concentration of alkylated species relative to the parent compounds suggests a nonpyrolytic source of the aromatics. A decrease in the striped bass fishing around San Francisco (21) may be attri-

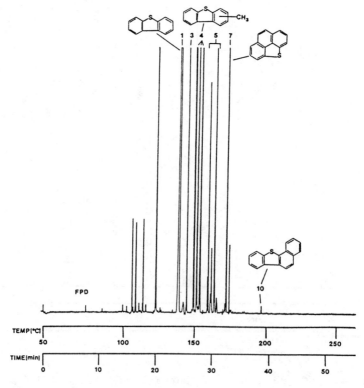

FIGURE 4. GC-FPD trace of Black River catfish PAH fraction.

butable to the toxic effects of elevated levels of PASH in the fish tissue.

Physico-Chemical Data

The conclusion that the PASH are being bioconcentrated is supported by the data in Table II, which show how the PASH could be transferred from the primary pollution source into the food web. The octanol/water partitioning coefficient (log P) is a measure of the tendency of the compound to partition into a lipid phase; and in this Table a comparison of the log P and aqueous solubility of several PASH and analogous PAH is made. The partition coefficients of both classes of compounds are practically identical. Since an inverse linear relation-

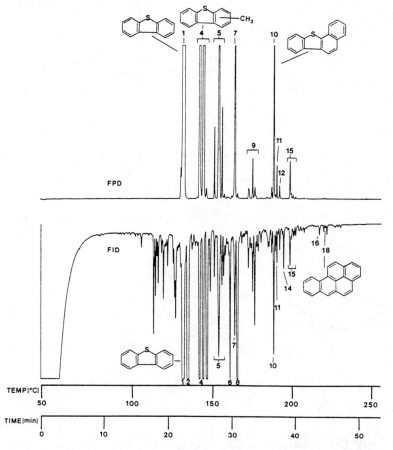

FIGURE 5. Dual detector (FID/FPD) GC trace of Sacramento River striped bass PAH fraction.

ship between the logarithm of the partition coefficient and logarithm of the aqueous solubility is typical, the aqueous solubilities of the PASH and PAH might be expected to be very similar, also. However, the PASH are somewhat more soluble in water than the analogous PAH: the aqueous solubility of the PAH decreases more rapidly with increasing molecular weight than does the solubility of the PASH. This discrepancy can be accounted for by considering the chemical constitution of the molecules. The lipophilicity of the compounds is determined by the hydrocarbon skeleton; thus the similarity in the partition coefficients may be attributed to the predominant lipophilic hydrocarbon nature of both the PASH and the PAH. However, the addition of a sulfur atom to the hydrocarbon skeleton slightly increases the polarity of the PASH, making them more water soluble without affecting the hydrophobic nature of the hydrocarbon skeleton. The result is that while anthracene and DBT may exhibit the same quantitative tendency to partition into a lipid phase, DBT, due to its greater aqueous solubility, will be present in relatively greater amounts in the aqueous phase, and this enrichment should be reflected by the relative amounts of PASH and PAH in the fish tissue. The data in Table I support the hypothesis of the bioconcentration of PASH.

TABLE 2

COMPARISON OF LOG P AND SOLUBILITY OF SELECTED PASH AND PAH

PASH	Log P	Water Solubility(ppm)	PAH	Log P	Water Solubility(ppm)
thiophene	1.8	1430	benzene	1.6	1515
benzo[b]thiophene	3.1	113	naphthalene	3.4	30
dibenzothiophene	4.4	1.7	anthracene	4.4	0.04
benzo-dibenzothiophene	5.6	0.06	benz[a]anthracene	5.6	0.01

Another parameter, the volatility, also plays an important role in determining the relative concentrations of PASH in the aqueous environment. The solubility of benzo[b]thiophene suggests that elevated levels of that compound should be found in water; however, the high volatility of the compound would account for its rapid depletion. Figure 6 compares the volatility of three homologous PASH.

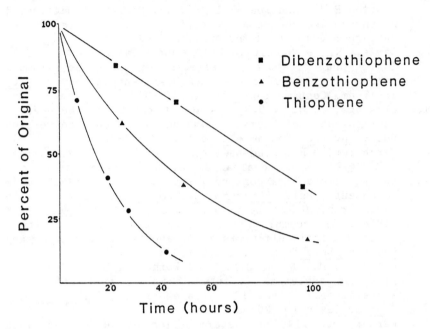

FIGURE 6. Disappearance of selected sulfur heterocycles at 18°C.

The aqueous solubility, lipid partitioning coefficient, and volatility of the PASH may account for the enhanced levels of PASH in the two fish samples discussed here. This study verifies that the sulfur heterocycles are indeed concentrated by the fish against the environmental concentration gradient. It remains to be determined whether fish exposed to chronically high levels of PASH will accumulate them throughout their lives and pass them upward through the food chain to man.

ACKNOWLEDGMENTS

The authors thank Mike Ribick and Paul Baumann for the samples and for suggestions during the preparation of the manuscript. This work was supported in part by a contract from the Columbia National Fisheries Research Laboratory, U.S. Fish and Wildlife Service, Columbia, MO.

REFERENCES

1. Pelroy, R.A., Stewart, D.L., Tominaga, Y., Iwao, M., Castle, R.N., and Lee, M.L. (1981): Microbial mutagenicity of three- and four-ring polycyclic aromatic heterocycles, Mut. Res. (submitted for publication).
2. Davis, K.R., Schultz, T.W., and Dumont, J.N. (1981): Toxic and teratogenic effects of selected aromatic amines on embryos of the amphibian Xenopus laevis, Arch. Environm. Contam. Toxicol., 10:371-391.
3. Giddings, J.M. (1979): Acute toxicity to Selenastrum capricornutum of aromatic compounds from coal conversion, Bull. Environ. Contam. Toxicol., 23:360-364.
4. Parkhurst, B.R., Bradshaw, A.S., Forte, J.L., and Wright, G.P. (1979): An evaluation of the acute toxicity to aquatic biota of a coal conversion effluent and its major components, Bull. Environm. Contam. Toxicol., 23:349-356.
5. Southworth, G.R., Beauchamp, J.J., and Schmieder, P.K. (1978): Bioaccumulation potential and acute toxicity of synthetic fuels effluents in freshwater biota: Azaarenes, Environ. Sci. Tech., 12: 1062-1066.
6. Dillon, T.M., Neff, J.M., and Warner, J.S. (1978): Toxicity and sublethal effects of No. 2 fuel oil on the supralittoral isopod Lygia exotica, Bull. Environm. Contam. Toxicol., 20:320-327.
7. Warner, J.S. (1975): Determination of sulfur-containing petroleum components in marine samples. In: Proc. Conf. Prev. Control Oil Poll., American Petroleum Institute, pp. 97-101, Washington, D.C.
8. Ogata, M., Miyake, Y., Fuhisawa, K., and Yoshida, Y. (1980): Accumulation and dissipation of organosulfur compounds in short-necked clam and eel, Bull. Environm. Contam. Toxicol., 25:130-135.
9. Ogata, M. and Miyake, Y. (1980): Gas chromatography combined with mass spectrometry for the identification of organic sulfur compounds in shellfish and fish, J. Chromatog. Sci., 18:594-605.
10. Ogata, M. and Miyake, Y. (1981): Identification of organic sulfur compounds and polycyclic hydrocarbons transferred to shellfish from petroleum suspension by capillary mass chromatography, Water Res., 15:257-266.
11. Berthou, F., Gourmelun, Y., Dreano, Y., and Friocourt, M.P. (1981): Application of gas chromatography on glass capillary columns to the analysis of hydrocarbon pollutants from the Amoco Cadiz oil spill, J. Chromatogr., 203: 279-292.

12. Vassilaros, D.L., Stoker, P.W., Booth, G.M., and Lee, M.L. (1981): Capillary gas chromatographic determination of polycyclic aromatic compounds in vertebrate fish tissue. Anal. Chem., 54 (in press).
13. Hague, R. and Schmedding, D. (1975): A method of measuring the water solubility of hydrophobic chemicals: solubility of five polychlorinated biphenyls, Bull. Environm. Sci. Tech., 14:13-18.
14. W.P. Fleiderer (1963): The Solubility of Heterocyclic Compounds. In: Physical Methods in Heterocyclic Chemistry, Vol. 1. p. 185, Academic Press, New York.
15. Lee, M.L., Novotny, M.V., and Bartle, K.D. (1981): Analytical Chemistry of Polycyclic Aromatic Compounds. Academic Press, New York, 462 pp.
16. Leo. A., Hansch, C., and Elkins, D. (1971): Partition coefficients and their uses, Chem. Rev., 71:525-616.
17. Means, J.C., Hassett, J.J., Wood, S.G., Banwart, W.L., Ali, S., and Khan, A. (1979): Sorption properties of polynuclear aromatic hydrocarbons and sediments: heterocyclic and substituted compounds. In: Polynuclear Aromatic Hydrocarbons: Chemistry and Biological Effects, edited by A.B. Bjørseth and A.J. Dennis, pp. 395-404, Battelle Press, Columbus, Ohio.
18. Nahum, A. and Horvath, C. (1980): Evaluation of octanol-water partition coefficients by using high performance liquid chromatography, J. Chromatogr., 192:315-322.
19. Means, J.C., Hassett, J.J., Wood, S.G., and Banwart, W.L. (1979): Sorption properties of energy-related pollutants and sediments. In: Carcinogenesis, Vol. 1, Polynuclear Aromatic Hydrocarbons, edited by P.W. Jones and P. Leber, pp. 327-340. Ann Arbor Science, Ann Arbor, Michigan.
20. Hansch, C. and Leo, A. (1979): Substituent Constants for Correlation Analysis in Chemistry and Biology. John Wiley and Sons, New York.
21. DeSalvo, L. (1976): Panel discussion on research needs. In: Fate and Effects of Petroleum Hydrocarbons in Marine Organisms and Ecosystems, edited by D.A. Wolfe, p. 458, Pergamon Press, New York.

STEREOSELECTIVITY IN THE METABOLISM, MUTAGENICITY AND TUMORIGENICITY OF THE POLYCYCLIC AROMATIC HYDROCARBON CHRYSENE

K. P. VYAS*, H. YAGI*, D. R. THAKKER*, R. L. CHANG**, A. W. WOOD*, W. LEVIN**, A. H. CONNEY**, AND D. M. JERINA*
*Laboratory of Bioorganic Chemistry, National Institute of Arthritis, Diabetes, and Digestive and Kidney Diseases, National Institutes of Health, Bethesda, MD 20205; **Department of Biochemistry and Drug Metabolism, Hoffmann-La Roche Inc., Nutley, NJ 07110.

INTRODUCTION

Polycyclic aromatic hydrocarbons comprise a large group of environmental contaminants which often produce cytotoxic, mutagenic, and carcinogenic responses in mammals (1). It is now well documented that biotransformation to reactive metabolites plays a key role in the production of these adverse effects. In accord with predictions of bay-region theory (2,3), direct and indirect evidence for over a dozen hydrocarbons has indicated that diol epoxides in which the epoxide group forms part of a bay-region of the hydrocarbon are ultimate carcinogenic metabolites. Bay-region diol epoxides are formed from the parent hydrocarbons by the combined action of the cytochrome P-450 dependent monooxygenases and epoxide hydrolase as shown in Fig. 1. In the first step, the cytochrome P-450 dependent monooxygenase system catalyses formation of the requisite arene oxide with a bay-region double bond, the arene oxide is then hydrolysed to a trans-dihydrodiol by epoxide hydrolase, and the trans-dihydrodiol is then oxidized by cytochrome P-450 to a pair of diastereomerically related bay-region diol epoxides-1 and -2 in which the benzylic hydroxyl group and the epoxide oxygen have either cis (isomer-1) or trans (isomer-2) relative stereochemistry. Both the cytochromes P-450 and epoxide hydrolase display a high degree of stereoselectivity in the formation of bay-region diol epoxides from the potent carcinogen benzo[a]pyrene (BaP) and the noncarcinogen phenanthrene (4-10). The present report describes the stereoselectivity observed in the metabolism of the weak carcinogen chrysene and compares the biological activity of its metabolites to those formed from BaP.

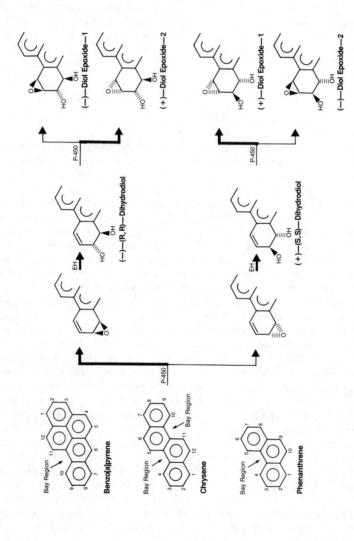

FIGURE 1. Stereoselectivity in the metabolism of polycyclic aromatic hydrocarbons to bay-region diol epoxides. Heavy arrows indicate major metabolic pathways. Signs of rotation for phenanthrene diol epoxides are assumed based on the other two hydrocarbons. Abbreviations are as follows: P-450, cytochrome P-450 dependent monooxygenanse system; EH, epoxide hydrolase.

MICROSOMAL METABOLISM OF CHRYSENE

The bay-region theory predicts that a 1,2-diol-3,4-epoxide should be an ultimate carcinogenic metabolite of chrysene. In order to determine the rate and extent of formation of the requisite 1,2-dihydrodiol, which is the immediate metabolic precursor of the 1,2-diol-3,4-epoxide, metabolism of chrysene by liver microsomes from control and induced rats has been investigated (8,11, cf. 12). Metabolism of chrysene by liver microsomes from control, phenobarbital- and 3-methylcholanthrene-treated rats is compared in Table 1. Dihydrodiols were the major metabolites formed by all three types of microsomes and accounted for 64-76% of the total metabolites. The major sites of attack by the cytochrome P-450 was at the 1,2- and 3,4-positions of the molecule. Very little K-region 5,6-dihydrodiol was formed. Chrysene 1,2-dihydrodiol with a bay-region 3,4-double bond was formed in substantial amounts (>25%) by all three types

TABLE 1

METABOLISM OF CHRYSENE BY LIVER MICROSOMES FROM CONTROL AND INDUCED RATS[a]

	Microsomes		
	Control	Phenobarbital	3-Methylcholanthrene
1,2-dihydrodiol	33.5[b]	28.5	26.1
3,4-dihydrodiol	34.5	30.4	46.8
5,6-dihydrodiol	2.0	5.7	3.5
Phenols	17.2	14.9	12.9
Unknowns	12.8	20.5	10.5
% Conversion[c]	3.0	4.2	24.1
Rate[d]	0.39	0.21	1.54

a Data taken from ref. 8
b Individual metabolites are given as a percent of total metabolites
c Substrate and protein concentrations used were 50 nmol and 0.5 mg per ml, respectively
d Rate of metabolism expressed as nmol of substrate metabolized/nmol cytochrome P-450/min.

of microsomes. Turnover numbers suggested that chrysene is a relatively poor substrate for the cytochrome P-450 dependent mixed-function oxidase system when compared to several other polycyclic aromatic hydrocarbons. Treatment of rats with 3-methylcholanthrene caused a four-fold induction in the rate of metabolism compared to microsomes from control rats, whereas phenobarbital-treatment reduced the rate by about 50% when the data were expressed per nmol of cytochrome P-450. The rates and extents of formation of benzo-ring dihydrodiols with bay-region double bonds from chrysene, BaP and phenanthrene by liver microsomes from rats induced with 3-methylcholanthrene are shown in Table 2. Compared to chrysene, benzo-ring dihydrodiols with bay-region double bonds at the 7,8-position of BaP and the 1,2-position of phenanthrene were formed in smaller amounts. Liver microsomes from rats induced with 3-methylcholanthrene metabolized all three hydrocarbons to corresponding benzo-ring dihydrodiols with 90-97% stereoselectivity. In each case, the (-)-dihydrodiols have (R,R) absolute configuration (Table 2).

TABLE 2

STEREOSELECTIVE METABOLISM OF POLYCYCLIC AROMATIC HYDROCARBONS TO BENZO-RING DIHYDRODIOLS WITH A BAY-REGION DOUBLE BOND BY LIVER MICROSOMES FROM 3-METHYLCHOLANTHRENE-TREATED RATS[a]

Compound	% of total metabolites	Rate[b]	% (-)-(R,R)-dihydrodiol
Chrysene 1,2-dihydrodiol	26.1	0.35	90
BaP 7,8-dihydrodiol	12.0	0.62	96
Phenanthrene 1,2-dihydrodiol	17.9	1.22	96.5

a Data taken from ref. 8 (chrysene 1,2- and phenanthrene 1,2-dihydrodiols) and ref. 4 (BaP 7,8-dihydrodiol).
b nmol dihydrodiol formed/nmol cytochrome P-450/min

METABOLISM OF (+)- AND (-)-CHYRSENE 1,2-DIHYDRODIOL TO 1,2-DIOL-3,4-EPOXIDES-1 AND -2

Metabolism of the optically pure (+)-(1S,2S) and (-)-(1R,2R)-enantiomers of chrysene 1,2-dihydrodiol was investigated (13) to determine the regio and stereoselectivity in the formation of their bay-region 1,2-diol-3,4-epoxides, ultimate carcinogenic metabolites of chrysene (14) as shown in Table 3. The extent to which diol epoxides are formed and the ratio of the two diastereomeric diol epoxides was highly dependent on treatment of the rats with inducing agents. Diol epoxides were formed to greater extents from the (-)-(1R,2R)-dihydrodiol compared to (+)-(1S,2S)-dihydrodiol regardless of the source of microsomes. In general, formation of diol epoxides was highly stereoselective. The (-)-(1R,2R) enantiomer was predominantly metabolized to the (+)-diol epoxide-2 diastereomer, whereas the (+)-diol epoxide-1 diastereomer was the major metabolite of the (+)-(1S,2S) enantiomer. Stereoselectivity was highest with microsomes from 3-methylcholanthrene-treated rats compared to microsomes from control or phenobarbital-treated rats. Microsomes from 3-methylcholanthrene-treated rats stereoselectively metabolized the (-)-(1R,2R)-dihydrodiol to the (+)-diol epoxide-2 diastereomer in a (+)-diol epoxide-2 to (-)-diol epoxide-1 ratio of 20:1, whereas (+)-(1S,2S)-dihydrodiol was metabolized to (+)-diol epoxide-1 and (-)-diol epoxide-2 in a ratio of 6:1. A similarly high degree of stereoselectivity was observed on metabolism of the BaP 7,8-(4) and phenanthrene 1,2-dihydrodiols (9). As shown in Table 4, liver microsomes from rats treated with 3-methylcholanthrene stereoselectively metabolized all three (-)-(R,R)-dihydrodiols to (+)-diol epoxide-2 diastereomers and all three (+)-(S,S)-dihydrodiols were metabolized to the (+)-diol epoxide-1 diastereomers. Diastereomeric bay-region diol epoxides were formed at almost equal rates from the (+)- and (-)-enantiomers of the phenanthrene 1,2-dihydrodiols, whereas the rate of diol epoxide formation from the chrysene 1,2- and BaP 7,8-dihydrodiols was almost twice as great from the (-)-enantiomers compared to the (+)-enantiomers. Since the metabolically formed dihydrodiols are predominantly the (-)-(R,R)-enantiomers, liver microsomes from 3-methylcholanthrene-treated rats stereoselectively metabolize chrysene, BaP and phenanthrene to their respective (+)-diol epoxide-2 diastereomers. These results are in good accord with the predictions of a stereochemical model for the catalytic binding site of cytochrome P-450c (15).

Pretreatment of rats with inducers of various isozymes of cytochrome P-450 failed to enhance the rate of microsomal

TABLE 3

METABOLISM OF (+)- AND (-)-CHRYSENE 1,2-DIHYDRODIOLS BY LIVER MICROSOMES FROM CONTROL, PHENO-BARBITAL-, AND 3-METHYLCHOLANTHRENE-TREATED RATS[a]

Isomer	Microsomes	Rate[b]	Diol epoxides as % of total metabolites	Ratio of diol epoxides formed isomer 1/isomer 2
(+)-(1S,2S)	Control	3.58	15.4	2.3/1
	Phenobarbital	1.47	23.6	1.4/1
	3-Methyl-cholanthrene	2.05	38.2	6.0/1
(-)-(1R,2R)	Control	2.60	29.5	1/5.3
	Phenobarbital	1.42	49.8	1/7.0
	3-Methyl-cholanthrene	2.25	65.8	1/19.7

a Data taken from ref. 13
b nmol dihydrodiol metabolized/nmol cytochrome P-450/min

TABLE 4

RATES AND RELATIVE AMOUNTS OF DIOL EPOXIDES FORMED FROM POLYCYCLIC AROMATIC HYDROCARBON DIHYDRODIOLS WITH A BAY-REGION DOUBLE BOND BY LIVER MICROSOMES FROM 3-METHYLCHOLANTHRENE-TREATED RATS[a]

Compound	Enantiomer	Rate of diol epoxide formation[b]	Relative amounts of diol epoxides	
			1	2
Chrysene 1,2-dihydrodiol	(−)-(1R,2R)	1.48	5	95
	(+)-(1S,2S)	0.78	86	14
BaP 7,8-dihydrodiol	(−)-(7R,8R)	1.00	14	86
	(+)-(7S,8S)	0.42	97	3
Phenanthrene 1,2-dihydrodiol	(−)-(1R,2R)	0.81	15	85
	(+)-(1S,2S)	0.73	85	15

a Data taken from ref. 4 (BaP 7,8-dihydrodiol), ref. 9 (phenanthrene 1,2-dihydrodiol), and ref. 13 (chrysene 1,2-dihydrodiol).
b nmol diol epoxides formed/nmol cytochrome P-450/min

metabolism of the enantiomeric chrysene 1,2-dihydrodiols (Table 3). Phenobarbital-treatment decreased the rate of microsomal metabolism of the (+)- and (−)-dihydrodiols by 59% and 45%, respectively. Similar reductions in the rate of metabolism per nmol of cytochrome P-450 after phenobarbital-treatment were also observed for the metabolism of chrysene 3,4- (16) and phenanthrene 1,2-dihydrodiols (9). Interestingly, 3-methylcholanthrene-treatment also decreased the rate of metabolism of the chrysene 1,2-dihydrodiols. This is the first example where 3-methylcholanthrene-treatment has decreased the rate of metabolism of a polycyclic aromatic dihydrodiol substrate.

BIOLOGICAL ACTIVITY OF CHRYSENE DIOL EPOXIDES

Four bay-region 1,2-diol 3,4-epoxides of chrysene are metabolically possible; the (+)- and (−)-enantiomers of the diastereomeric diol epoxides-1 and -2. These isomers have been synthesized in optically pure form (17) and were tested for mutagenicity and tumorigenicity in order to aid in the identification of ultimate carcinogenic metabolites of chrysene. Relative mutagenic activity of the four optically pure bay-region diol epoxides of chrysene in strains TA98 and TA100 of S. typhimurium and in Chinese hamster V79 cells is compared to the corresponding bay-region 7,8-diol-9,10-epoxides of BaP in Table 5. The metabolically predominant (+)-diol epoxide-2 isomer of chrysene was 5 to 25 times as

TABLE 5

MUTAGENIC ACTIVITY OF THE OPTICALLY PURE BAY-REGION DIOL EPOXIDES OF CHRYSENE IN BACTERIAL AND MAMMALIAN CELLS[a]

Compound	Relative Mutagenic Activity		
	S. typhimurium		Chinese hamster
	TA98	TA100	V79 cells
(+)-diol epoxide-1[b]	9 (75)[c]	4 (54)	2 (9)
(-)-diol epoxide-1	8 (100)	12 (100)	5 (15)
(+)-diol epoxide-2	19 (48)	100 (63)	100 (100)
(-)-diol epoxide-2	100 (35)	18 (10)	10 (6)

a Data taken from ref. 18. Mutagenicity for the chrysene diol epoxides is given relative to the most active compound in each test system; 377 his$^+$ rev/nmol in TA98; 3193 his$^+$ rev/nmol in TA100; and 9 mutant colonies/nmol/ 10^5 survivors in V79 cells.
b Note that the atoms of the diol epoxide rings for the two hydrocarbons are superimposable for a given diol epoxide isomer.
c The numbers in parentheses are values for the corresponding BaP isomers taken from ref. 19. Mutagenicity for the BaP diol epoxides is given relative to the most active compound in each test system; 5200 his$^+$ rev/nmol in TA98; 9500 his$^+$ rev/nmol in TA100; and 400 mutant colonies/nmol/ 10^5 survivors in V79 cells.

mutagenic as the other three chrysene diol epoxides in strain TA100 and 10-50-fold more active in V79 cells than the other three optical isomers (18). In strain TA98, however, (-)-diol epoxide-$\underline{2}$ was the most active isomer and had 5 to 10 times the activity of the other isomers. For the isomeric BaP 7,8-diol-9,10-epoxides, (+)-diol epoxide-$\underline{1}$ was the most active isomer in both strains of S. typhimurium (19). In mammalian V79 cells, however, (+)-diol epoxide-$\underline{2}$ was 7- to 20-fold more active compared to the other three diol epoxides of BaP. Thus, in V79 cells, the metabolically predominant (+)-diol epoxide-$\underline{2}$ isomer was the most potent mutagen for both hydrocarbons. No consistent trend for the relative mutagenic activity of the eight diol epoxides from the above two hydrocarbons in the two strains of S. typhimurium was evident.

Initiation-promotion studies on mouse skin have revealed that (-)-BaP (7R,8R)-dihydrodiol, the metabolic precursor of (+)-diol epoxide-$\underline{2}$, was 5-10-fold more tumorigenic than the (+)-enantiomer (20) and (+)-diol epoxide-$\underline{2}$ was the most potent tumorigenic compound among the four diol epoxides of BaP (21). In newborn Swiss-Webster mice, (-)-BaP (7R,8R)-dihydrodiol was found to be 20 times more tumorigenic than the (+)-BaP (7S,8S)-dihydrodiol (22) and the BaP (+)-diol epoxide-$\underline{2}$ isomer was highly tumorigenic whereas the other three isomers had little or no activity (23). Since (+)-diol epoxide-$\underline{2}$ was more active than the (-)-7,8-dihydrodiol which was in turn more active than BaP in newborn mice, (+)-diol epoxide-$\underline{2}$ qualifies as an ultimate carcinogen of BaP. Tumorigenicity studies of optically pure 1,2-diol-3,4-epoxides and 1,2-dihydrodiols of chrysene are currently in progress. Preliminary results indicate that (-)-chyrsene (1R,2R)-dihydrodiol is several fold more active than either chrysene or (+)-chyrsene (1S,2S)-dihydrodiol and that the metabolically predominant (+)-1,2-diol-3,4-epoxide-$\underline{2}$ isomer is the most active among the four diol epoxides. Thus it appears that (-)-chrysene (1R,2R)-dihydrodiol is proximate and (+)-chrysene 1,2-diol-3,4-epoxide-$\underline{2}$ is an ultimate carcinogenic metabolite of chrysene. For diol epoxides of both hydrocarbons, mutagenicity in Chinese hamster V79 cells accurately correlated with tumorigenic activity.

A particularly fascinating feature of the tumorigenic activity of chrysene is its dependence upon methyl substitution. Studies have shown that a single methyl group at the 5-position makes the resultant hydrocarbon about as tumorigenic as BaP (24). Furthermore, the 1,2-dihydrodiol of 5-methylchrysene seems to be the isomeric dihydrodiol which has the highest tumorigenic activity (25). Studies which establish the absolute configuration of this dihydrodiol and its further metabolism will be of considerable interest. Theoretical studies predict that this dihydrodiol, like the 1,2-dihydrodiol of chrysene, will be predominantly the (R,R)-enantiomer when formed by liver microsomes from 3-methylcholanthrene-treated rats (15).

SUMMARY

The data reported in the present manuscript clearly indicate that the cytochrome P-450 dependent monooxygenase system from 3-methylcholanthrene-treated rats stereoselectively metabolizes chrysene, BaP and phenanthrene to their respective (+)-diol epoxide-$\underline{2}$ isomers. Interestingly, for both the strong carcinogen Ba$\overline{P}$ and the weak carcinogen chrysene, the metabolically predominant (+)-diol epoxide-$\underline{2}$ isomer is also the most active mammalian cell mutagen and tumorigen among the four metabolically possible bay-region diol epoxides. The (+)-diol epoxide-$\underline{2}$ isomers of chrysene and BaP have the same absolute configuration and thus are superimposible when their bay regions are aligned, indicative of specific and highly chiral interactions in the elicitation of their tumorigenic response.

REFERENCES

1. Dipple, A. (1976): Polynuclear aromatic carcinogens. In: <u>Chemical Carcinogens, ACS Monograph Series, No. 173</u>, edited by C. E. Searle, pp. 245-314, American Chemical Society, Washington, D. C.
2. Nordqvist, M., Thakker, D. R., Yagi, H., Lehr, R. E., Wood, A. W., Levin, W., Conney, A. H., and Jerina, D. M. (1980): Evidence in support of the bay-region theory as a basis for the carcinogenic activity of polycyclic aromatic hydrocarbons. In: <u>Molecular Basis of Environmental Toxicity</u>, edited by R. S. Bhatnagar, pp. 329-357, Ann Arbor Science Publishers Inc., Ann Arbor, Michigan.
3. Jerina, D. M., Lehr, R. E., Yagi, H., Hernandez, O., Dansette, P. M., Wislocki, P. G., Wood, A. W., Chang, R. L., Levin, W., and Conney, A. H. (1976): Mutagenicity of benzo[a]pyrene derivitatives and the description of a quantum mechanical model which predicts the ease of carbonium ion formation from diol epoxides. In: <u>In Vitro Metabolic Activation in Mutagenesis Testing</u>, edited by F. J. de Serres, J. R. Fouts, J. R. Bend and R. M. Philpot, pp 159-177, Elsevier, Amsterdam.
4. Thakker, D. R., Yagi, H., Akagi, H., Koreeda, M., Lu, A. Y. H., Levin, W., Wood, A. W., Conney, A. H., and

Jerina, D. M. (1977): Metabolism of benzo[a]pyrene. VI. Stereoselective metabolism of benzo[a]pyrene and benzo[a]pyrene 7,8-dihydrodiol to diol epoxides. Chem. -Biol. Interact. 16:281-300.
5. Thakker, D. R., Yagi, H., Lu, A. Y. H., Levin, W., Conney, A. H. and Jerina, D. M. (1976): Metabolism of benzo[a]pyrene: Conversion of (±)-trans-7,8-dihydroxy-7,8-dihydrobenzo[a]pyrene to highly mutagenic 7,8-diol-9,10-epoxides. Proc. Natl. Acad. Sci. USA 73: 3381-3385.
6. Yang, S. K., McCourt, D. W., Roller, P. P., and Gelboin, H. V. (1976): Enzymatic conversion of benzo[a]pyrene leading predominantly to the diol-epoxide r-7,t-8-dihydroxy-t-9,10-oxy-7,8,9,10-tetrahydrobenzo[a]pyrene through a single enantiomer of r-7,t-8-dihydroxy-7,8-dihydrobenzo[a]pyrene. Proc. Natl. Acad. Sci. USA 73: 2594-2598.
7. Deutsch, J., Vatsis, K. P., Coon, M. J., Leutz, J. C., and Gelboin, H. V. (1979): Catalytic activity and stereoselectivity of purified forms of rabbit liver microsomal cytochrome P-450 in the oxygenation of the (-)- and (+)-enantiomers of trans-7,8-dihydroxy-7,8-dihydrobenzo[a]pyrene. Mol. Pharmacol. 16:1011-1018.
8. Nordqvist, M., Thakker, D. R., Vyas, K. P., Yagi, H., Levin, W., Ryan, D. E., Thomas, P. E., Conney, A. H., and Jerina, D. M. (1981): Metabolism of chrysene and phenanthrene to bay-region diol epoxides by rat liver enzymes. Mol. Pharmacol. 19:168-178.
9. Vyas, K. P., Thakker, D. R., Levin, W., Yagi, H., Conney, A. H., and Jerina, D. M. (1981): Stereoselective metabolism of the optical isomers of trans-1,2-dihydroxy-1,2-dihydrophenanthrene to bay-region diol epoxides by rat liver microsomes. Chem.-Biol. Interact. (in press).
10. Thakker, D. R., Levin, W., Yagi, H., Conney, A. H., and Jerina, D. M. (1982): Regio- and stereoselectivity of hepatic cytochrome P-450 toward polycyclic aromatic hydrocarbon substrates. In: Biological Reactive Intermediates, Vol. 2. Chemical Mechanisms and Biological Effects, edited by R. Snyder, D. V. Parke, J. Koesis, D. J. Jollow, and G. G. Gibson, Plenum Publishing Co., New York, (in press).
11. Sims, P. (1970): Qualitative and quantitative studies of the metabolism of a series of aromatic hydrocarbons by rat liver preparations. Biochem. Pharmacol. 19: 795-818.
12. MacNicoll, A. D., Grover, P. C., and Sims, P. (1980): The metabolism of a series of polycyclic hydrocarbons by mouse skin maintained in short-term organ culture. Chem.-Biol. Interact. 29:169-188.

13. Vyas, K. P., Levin, W., Yagi, H., Thakker, D. R., Conney, A. H., and Jerina, D. M. (1982): Stereoselective metabolism of (+)- and (-)-enantiomers of trans-1,2-dihydroxy-1,2-dihydrochrysene to bay-region diol epoxides by rat liver enzymes. (submitted).
14. Buening, M. K., Levin, W., Karle, J. M., Yagi, H., Jerina, D. M., and Conney, A. H. (1979): Tumorigenicity of bay-region epoxides and other derivatives of chrysene and phenanthrene in newborn mice. Cancer Res. 39:5063-5068.
15. Jerina, D. M., Michaud, D. P., Feldmann, R. J., Armstrong, R. N., Vyas, K. P., Thakker, D. R., Yagi, H., Thomas. P. E., Ryan, D. E., and Levin, W.: Stereochemical modeling of the catalytic site of cytochrome P-450c. Fifth International Symposium on Microsomes and Drug Oxidations (Microsomes, Drug Oxidations, and Drug Toxicity), edited by R. Sato and R. Kato, Japan Scientific Societies Press, Tokyo, (in press).
16. Vyas, K. P., Yagi, H., Levin, W., Conney, A. H., and Jerina, D. M. (1981): Metabolism of (-)-trans (3R,4R)-dihydroxy-3,4-dihydrochrysene to diol epoxides by liver microsomes. Biochem. Biophys. Res. Commun. 98:961-969.
17. Yagi, H., Vyas, K. P., Tada, M., Thakker, D. R., and Jerina, D. M. (1981): Synthesis of the enantiomeric bay-region diol epoxides of benz[a]anthracene and chrysene. J. Org. Chem. (in press).
18. Wood, A. W., Chang, R. L., Levin, W., Yagi, H., Vyas, K. P., Jerina, D. M., and Conney, A. H. (1982): Mutagenicity of the optical isomers of the diastereomeric bay-region chrysene 1,2-diol-3,4-epoxides in bacterial and mammalian cells. (submitted).
19. Wood, A. W., Chang, R. L., Levin, W., Yagi, H., Thakker, D. R., Jerina, D. M., and Conney, A. H. (1977): Differences in the mutagenicity of the optical enantiomer of the diastereomeric benzo[a]pyrene 7,8-diol-9,10-epoxides. Biochem. Biophys. Res. Commun. 77: 1389-1396.
20. Levin, W., Wood, A. W., Chang, R. L., Slaga, T. J., Yagi, H., Jerina, D. M., and Conney, A. H. (1977): Marked differences in the tumor-initiating activity of optically pure (+)- and (-)-trans-7,8-dihydroxy-7,8-dihydrobenzo[a]pyrene on mouse skin. Cancer Res. 37: 2721-2725.
21. Slaga, T. J., Bracken, W. M., Gleason, G., Levin, W., Yagi, H., Jerina, D. M., and Conney, A. H. (1979): Marked differences in the skin tumor-initiating activities of the optical enantiomers of the diastereomeric benzo[a]pyrene 7,8-diol-9,10-epoxides. Cancer Res. 39:67-71.

22. Kapitulnik, J., Wislocki, P. G., Levin, W., Yagi, H., Thakker, D. R., Akagi, H., Koreeda, M., Jerina, D. M., and Conney, A. H. (1978): Marked differences in the carcinogenic activity of optically pure (+)- and (-)-trans-7,8-dihydroxy-7,8-dihydrobenzo[a]pyrene in newborn mice. Cancer Res. 38:2661-2665.
23. Buening, M. K., Wislocki, P. G., Levin, W., Yagi, H., Thakker, D. R., Akagi, H., Koreeda, M., Jerina, D. M., and Conney, A. H. (1978): Tumorigenicity of the optical enantiomers of the diastereomeric benzo[a]pyrene 7,8-diol-9,10-epoxides in newborn mice. Exceptional activity of (+)-7β,8α-dihydroxy-9α,10α-epoxy-7,8,9,10-tetrahydrobenzo[a]pyrene. Proc. Natl. Acad. Sci. USA 75:5358-5361.
24. Hecht, S. S., Loy, M., Maronpot, R. R., and Hoffmann, D. (1976): Comparative carcinogenicity of 5-methylchyrsene, benzo[a]pyrene and modified chrysenes. Cancer Lett. 1:147-154.
25. Hecht, S. S., Rivenson, A., and Hoffmann, D. (1980): Tumor initiating activity of dihydrodiols formed metabolically from 5-methylchrysene. Cancer Res. 40:1396-1399.

MODEL FOR PREDICTING ADSORPTION OF PAH FROM AQUEOUS SYSTEMS

RICHARD W. WALTERS AND RICHARD G. LUTHY
Department of Civil Engineering, Carnegie-Mellon University, Pittsburgh, Pennsylvania 15213

INTRODUCTION

Polycyclic aromatic hydrocarbons (PAH) are an important class of organic compounds found in liquid effluents from many industrial facilities. A previous study by Walters and Luthy (1,2) investigated liquid/suspended solid phase partitioning of PAH in coal conversion and coking wastewaters. This study involved separation of field samples into liquid and suspended solid phases by filtering through 0.45 μm glass fiber filters and characterizing PAH associated with each phase by high performance liquid chromatography (HPLC). Eleven PAH ranging in molecular weight from 128 for naphthalene (NA) to 278 for dibenz(a,h)anthracene (DBahA) were evaluated in this study. The results of that investigation indicated several general trends: (i) for a given PAH, the fraction of the total wastewater concentration associated with the suspended solid phase increased with increasing molecular weight (or roughly with decreasing aqueous solubility, Cs, or increasing octanol-water partition coefficient, Kow) from approximately 50 percent to above 99 percent, (ii) partitioning on a per-mass-of-phase basis heavily favored the suspended solid phase, with the partition coefficients ranging from 10,000 to 10,000,000 and (iii) partitioning was qualitatively observed to increase with increasing organic carbon content of the suspended solid phase and decreasing concentration of dissolved organic materials in the liquid phase. This suspended solid phase partitioning has important implications on the selection, design and efficiency of treatment systems for the control of PAH in liquid effluents. These factors motivated an investigation of the adsorptive behavior of PAH in aqueous systems.

The present study focused on evaluating the adsorption of PAH from water onto Filtrasorb 400 activated carbon. Adsorption isotherm data at 25 °C for eleven PAH were evaluated by batch shake testing. PAH which were evaluated included one two-ring compound (NA), five three-ring compounds (acenaphthylene, acenaphthene, fluorene, anthracene and phenanthrene) and five four-ring compounds (fluoranthene, pyrene, triphenylene, benz(a)anthracene and chrysene). Ranges of physicochemical parameters for these compounds are molecular weight, 128 to 228 g/mol; Cs, 31.7 to 0.002 mg/l at 25 °C (3); and log(Kow), 3.36 to 5.91 (4,5). Cs generally decreases and Kow generally increases with increasing molecular weight. The Henry's Law, Freundlich, Langmuir, BET and Redlich-Peterson equations were fitted to the adsorption isotherm data and evaluated to determine which most appropriately represented relative adsorption. Relative adsorption was then correlated with several molecule-specific properties and with energy terms from the solvophobic (6), net adsorption energy (7) and Polanyi adsorption potential (8) theories. Details of the approach and results of this study are presented by Walters (9). A summary of this work is presented below.

MATERIALS AND METHODS

PAH Standards

PAH standards were used as received from Aldrich Chemical Company (Milwaukee, Wisconsin).

Activated Carbon

Filtrasorb 400 activated carbon was obtained from Calgon Corporation (Pittsburgh, Pennsylvania) and was mechanically ground to a powder and classified by passing through a 200 mesh U.S. standard sieve. Stock carbon slurries were prepared by dispersing the powdered carbon in water.

Solvents

Organic solvents were HPLC grade and were obtained from Fisher Scientific (Pittsburgh, Pennsylvania). All solvents were filtered through 0.5 μm teflon filters and deaerated by vacuum prior to use. Tap water was filtered, deionized and purified by an ion exchange and activated carbon bed system from Continental Water Company (El Paso, Texas). Organic-free, deionized water from this system was filtered through 0.45 μm cellulose filters and deaerated by vacuum prior to use.

Preparation of PAH Solutions

PAH solutions were prepared using the method of May et al. (10). This method involved pumping water through a column packed with 60 to 80 mesh glass beads precoated with the PAH being studied.

Adsorption Shake Tests

Batch adsorption shake tests were performed using 40 ml of solution in 50-ml silanized glass centrifuge tubes fitted with teflon lined screw caps. Centrifuge tubes containing various amounts of saturated solution, carbon slurry and dilution water were sealed and shaken for 24 to 36 hr to achieve equilibrium. Equilibrated samples were centrifuged at 8000 rpm for 5 min, and 5-ml aliquots of the water solution were withdrawn and added to 5 ml methanol prior to HPLC analysis.

Instrumentation and Analysis

A Perkin-Elmer (Norwalk, Connecticut) Series 3 liquid chromatograph was used with a 0.26 cm x 25 cm PAH 10 column and a model 204-S fluorescence detector equipped with variable excitation and emission settings. Optimum excitation and emission settings were determined for each compound studied. Sample analysis was performed using injection volumes ranging from 5 μl to 175 μl of the methanol-water solutions.

RESULTS

Results of this work are presented in detail by Walters (9). Results presented below are abstracted from separate articles which are in preparation regarding the experimental investigation and evaluation of the adsorption isotherm equations (11) and the correlation between relative adsorptive behavior and molecule-specific features and energy terms from adsorption theories (12).

Isotherm Equations

The five isotherm equations which were fitted to the adsorption isotherm data are shown in Table 1. Each of these equations have been used in previous adsorption studies and each contains a parameter which reflects relative adsorption. The Henry's Law equation, which contains just one parameter, indicates a linear relationship between concentration and capacity. This relationship is typically observed at low relative concentrations. The Langmuir and BET equations, which contain two parameters, both reduce to the Henry's Law equation at low concentration, but tend towards limiting capacities at high concentrations. The Langmuir equation levels off at this limiting capacity as concentration approaches saturation, while the BET equation rises near saturation. The Freundlich equation, which also contains two parameters, plots as a straight line on a log-log plot. The Redlich-Peterson equation contains three parameters. At low concentration, this equation reduces to the Henry's Law equation while at high concentration it reduces to the Freundlich equation. The Redlich-Peterson equation also reduces to the Langmuir equation when the exponent is zero.

Isotherm equations were fitted to the data by linear regression for the Henry's Law, Freundlich, Langmuir and BET equations and by non-linear regression for the Redlich-Peterson equation.

Adsorption Isotherm Data

Figure 1 shows a log-log plot of the adsorption isotherm data for NA in which equilibrium concentration in mg/l is plotted against equilibrium capacity in mg/g. Plots of the Freundlich, Langmuir and BET equations which fit the isotherm data also appear in Figure 1.

TABLE 1

ADSORPTION ISOTHERM EQUATIONS

Name	Equation[a]	Parameter(s)[b]	Capacity Term
Henry's Law	$q_e = k_h c_e$	k_h	k_h
Langmuir	$q_e = \dfrac{bq^o c_e}{1+bc_e}$	b, q^o	bq^o (low c_e) q^o (high c_e)
BET	$q_e = \dfrac{Bq^o c_e}{(c_s-c_e)(1+(B-1)(c_e/c_s))}$	B, q^o	Bq^o (low c_e) q^o (high c_e)
Freundlich	$q_e = k_f c_e^{1/n}$	$k_f, 1/n$	k_f
Redlich-Peterson	$\dfrac{1}{q_e} = \dfrac{1}{k_{hr} c_e} + \dfrac{1}{k_{fr} c_e^m}$	k_{hr}, k_{fr}, m	k_{hr} (low c_e) k_{fr} (high c_e)

a Variables in the equations are equilibrium capacity (q_e, mg/g), equilibrium concentration (c_e, mg/l) and aqueous solubility (c_s, mg/l).
b Parameters are evaluated by regression analysis.

The isotherm data in Figure 1 for NA are representative of the data for other PAH which were studied. These data show a curved isotherm in which adsorption at low equilibrium concentration is represented by a Henry's Law relationship while a limiting capacity is achieved at high concentrations. These characteristics impose limits on the Henry's Law, Freundlich and BET equations. In addition, results of the curve fitting techniques used to evaluate the parameters of the Redlich-Peterson equation indicated that this equation reduced to the Langmuir equation for more soluble compounds. For these reasons and others (11), the Langmuir equation was used to represent relative adsorption.

Two different linearized forms of the Langmuir equation were used to evaluate its parameters. One form weights data at high concentrations and thus provides parameters which apply at high

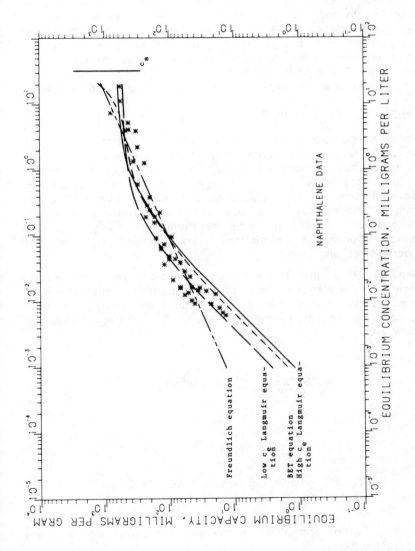

FIGURE 1. Log-log plot of naphthalene isotherm data.

concentration, while the other weights data at low concentration and provides parameters which apply at low concentration. Differences in relative adsorption at these different concentration extremes were observed; adsorption generally increased with decreasing solubility at low concentration, while relative capacity generally decreased with decreasing solubility at high concentration. Consequently, parameters from each of the linear forms were necessarily used to evaluate correlations between relative adsorption and molecule-specific parameters or energy terms. The product bq^o from the low concentration linear form was used to correlate adsorption at low concentration. The limiting capacity q^o from the high concentration linear form was used to correlate adsorption at high concentration.

Correlating Adsorptive Behavior

Correlations were attempted between the relative capacity values and various molecular and physicochemical properties. Molecular features included a limiting dimension ω (12), total surface area, the connectivity index and the shape parameter. Physicochemical properties included melting point, molar volume and the enthalpy and entropy of melting (ΔH_m^o and ΔS_m^o, respectively). Correlations with the energy terms mentioned previously were also attempted. Successful correlations were obtained for many of these parameters and terms.

Adsorption at low concentration correlated well with several parameters. Correlations were best in multiple regression with ΔH_m^o and ΔS_m^o and in single regression with log(Kow). These regressions are shown in Figures 2 and 3. Plotted regression equations are:

$$\log(bq^o) = -0.0188 \, \Delta H_m^o - 11.0 \, \Delta S_m^o + 5.45, \; r^2 = 0.67$$

and:

$$\log(bq^o) = 0.594 \, \log(Kow) + 1.49, \; r^2 = 0.70.$$

The units for ΔH_m^o and ΔS_m^o in the equation above are cal/g and cal/g·°K, respectively.

Adsorption at high concentration also correlated well with several parameters. Correlations were again best in multiple regression with ΔH_m^o and ΔS_m^o and in single regression with the Polanyi adsorption potential ϵ. These regressions are shown in Figures 4 and 5. Plotted regression equations are:

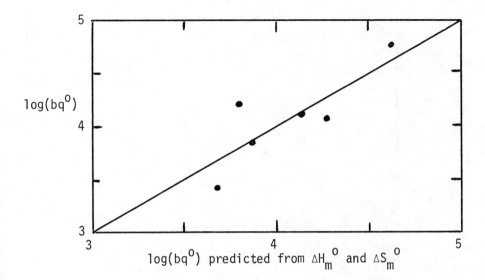

FIGURE 2. Correlation of $\log(bq^o)$ with ΔH_m^o and ΔS_m^o.

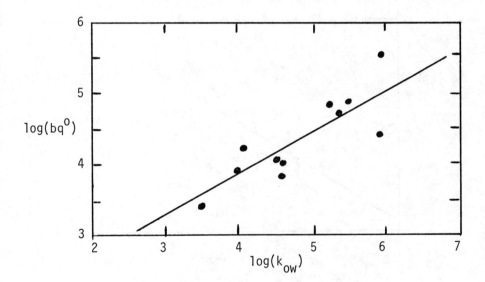

FIGURE 3. Correlation of $\log(bq^o)$ with $\log(k_{ow})$.

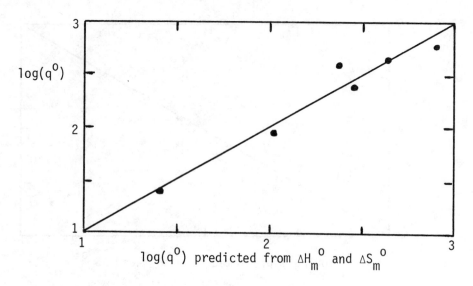

FIGURE 4. Correlation of $\log(q^o)$ with ΔH_m^o and ΔS_m^o.

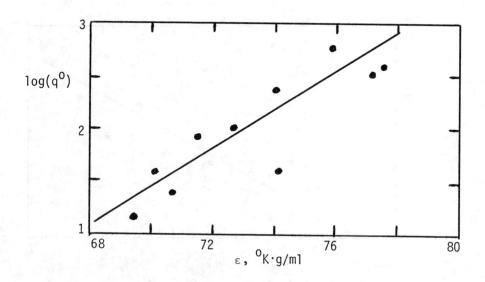

FIGURE 5. Correlation of $\log(q^o)$ with the Polanyi adsorption potential ε.

$$\log(q^\circ) = -0.121 \, \Delta Hm^\circ + 51.4 \, \Delta Sm^\circ + 2.04, \; r^2 = 0.92$$

and:

$$\log(q^\circ) = 0.178 \, \epsilon - 11.0, \; r^2 = 0.79.$$

The units for ΔHm° and ΔSm° are cal/g and cal/g·°K, respectively. The units for ϵ are °K·g/ml.

DISCUSSION

Observations made for data presented elsewhere (12) indicated that the adsorption of AN and BaA, the two most linear PAH studies, was anamolously low. Consequently, correlations tended to overestimate the adsorption of these compounds. This was true of correlations with most parameters with the exception of those related to melting. This observation suggests the fundamental importance of the melting process in the adsorption of these compounds. The importance of melting is also reflected in the correlation with the Polanyi adsorption potential at high concentration; the Polanyi theory inherently assumes that adsorption takes place as a result of precipitation of solute onto the adsorbent.

This study has indicated that relative adsorption can be predicted using several different parameters in regression equations in addition to those presented here. The choice of parameter to be use reflects a tradeoff between the desired reliability of the regression and the availability of ease of generating the correlating parameter. While this work has evaluated adsorption onto activated carbon, a similar approach is expected to be applicable to other adsorbent systems as well.

ACKNOWLEDGEMENTS

Helen M. Cordy assisted with the experimental portion of this study. This work was supported by the U.S. Department of Energy under Contract Number DE-FG22-80PC30246.

REFERENCES

1. Walters, R. W., and Luthy, R. G. (1981): Physicochemical

Behavior of PAH in Coal Conversion Liquid Effluents. In: Chemical Analysis and Biological Fate: Polynuclear Aromatic Hydrocarbons, edited by M. Cooke and A. J. Dennis, pp. 539-550, Battelle Press, Columbus, Ohio.

2. Walters, R. W., and Luthy, R. G. (1981): Liquid/Suspended Solid Phase Partitioning of Polycyclic Aromatic Hydrocarbons in Coal Coking Wastewaters. In preparation.

3. Mackay, D., and Shiu, W. Y. (1977): Aqueous Solubility of Polynuclear Aromatic Hydrocarbons, Journ. Chem. Eng. Data, 22:399-402.

4. Karickhoff, S. W., Brown, D. S., and Scott, T. A. (1979): Sorption of Hydrophobic Pollutants on Natural Sediments, Wat. Res., 13:241-248.

5. Yalkowsky, S. H., and Valvani, S. C. (1979): Solubilities and Partitioning 2. Relationships between Aqueous Solubilities, Partition Coefficients, and Molecular Surface Areas of Rigid Aromatic Hydrocarbons, Journ. Chem. Eng. Data, 24:127-129.

6. Belfort, G. (1979): Selective Adsorption of Organic Homologues onto Activated Carbon from Dilute Aqueous Solutions. Solvophobic Interaction Approach and Correlations of Molar Adsorptivity with Physicochemical Properties, Env. Sci. Tech., 13:939-946.

7. McGuire, M. J., and Suffet, I. H. (1980): The Calculated Net Adsorption Energy Concept. In: Activated Carbon Adsorption of Organics from the Aqueous Phase. Volume I., edited by I. H. Suffet and M. J. McGuire, pp. 91-115, Ann Arbor Science, Ann Arbor, Michigan.

8. Manes, M. (1980): The Polanyi Adsorption Potential Theory and its Applications to Adsorption from Water Solution Onto Activated Carbon. In: Activated Carbon Adsorption of Organics from the Aqueous Phase. Volume I., edited by I. H. Suffet and M. J. McGuire, pp. 43-64, Ann Arbor Science, Ann Arbor, Michigan.

9. Walters, R. W. (1981): Relationships Between Molecular and Physicochemical Properties of Polycyclic Aromatic Hydrocarbons and Adsorptive Behavior in Aqueous Systems. Ph.D. Dissertation, Carnegie-Mellon University, Pittsburgh, Pennsylvania.

10. May, W. E., Wasik, S. P., and Freeman, D. H. (1978): Determination of the Solubility Behavior of Some Polycyclic Aromatic Hydrocarbons in Water, Anal. Chem., 50:997-1000.

11. Walters, R. W., and Luthy, R. G. (1981): Adsorption of Polycyclic Aromatic Hydrocarbons From Water Onto Activated Carbon. In Preparation.

12. Walters, R. W., and Luthy, R. G. (1981): Predicting Adsorption of Polycyclic Aromatic Hydrocarbons From Physicochemical and Molecular Properties. In Preparation.

OCCURRENCE AND POTENTIAL UPTAKE OF POLYNUCLEAR AROMATIC HYDROCARBONS OF HIGHWAY TRAFFIC ORIGIN BY PROXIMALLY GROWN FOOD CROPS

DAVID T. WANG and OTTO MERESZ
Laboratory Services Branch, Ontario Ministry of the Environment, P.O. Box 213, Rexdale, Ontario M9W 5L1 Canada

INTRODUCTION

Polynuclear aromatic hydrocarbons (PAHs) are well recognized as ubiquitous contaminants of the environment, including air (1), soil (2), water (3) and certain foodstuffs (4). While a considerable amount of information has been reported on food PAH contents, it relates to specially processed items such as smoked or barbecued meat products (5) or is restricted to benzo(a)pyrene (BaP) alone (6).

Recent studies indicate that communities exposed to high density vehicular traffic show higher than average incidence of cancer (2). Although air pollution can be considered as the primary cause of higher cancer rates under such conditions, locally grown and consumed food crops may also contribute to the total PAH intake in these locations.

The objective of the present study was to explore whether the consumption of some popularly grown vegetables represents an additional load in PAH exposure. The study included analysis for 17 PAHs in three selected vegetables grown in a high PAH environment, about 50 meters south of one of Canada's busiest highways, Hwy. 401 in Toronto. This highway carries both urban and transcontinental traffic, an average of 250,000 vehicles per day, on fourteen lanes at the chosen location. The site was selected so that crops received not only airborne contaminants but also road run-off, being situated on a gentle slope ascending to the highway.

The selected area features predominantly clay soil, deficient in organic matter. It was analysed repeatedly and showed a relatively high PAH content (above 1 mg/kg total PAHs). The vegetables investigated initially were two root crops (beet and onion) and tomato. The latter was expected to take up airborne PAHs owing to the waxy skin of its fruit. Peanuts were also grown for analysis. This crop is of interest as a high oil content root crop. Unfortunately, because of its late maturing, analytical data on this produce is not yet available.

Soil and crops were analysed for the following seventeen PAHs: phenanthrene (Ph), anthracene (An), fluoranthene (Fl), pyrene (Py), benz(a)anthracene (BaA), chrysene (Ch), benzo(e)pyrene (BeP), benzo(b)fluoranthene (BbF), perylene (Per), benzo(k)fluoranthene (BkF), benzo(a)pyrene (BaP), benzo(g,h,i) perylene (BghiP), dibenz(a,h)anthracene (DBahA), indeno (1,2,3-cd)pyrene (IP), benzo(b)chrysene (BbCh), anthanthrene (AA) and coronene (Cor).

MATERIALS AND METHODS

All solvents used were of high purity and the extraction and concentration equipment was thoroughly cleaned prior to use. Samples were analyzed in parallel with blanks.

High Performance Liquid Chromatography (HPLC)

The analytical method for PAH determination utilized HPLC with a Vydac 5 μ C_{18} reversed phase column. The eluent was acetonitrile and water which was linear gradient programed over 60 min. from 70% (V/V) to 100% acetonitrile. The flow rate was 1 mL/min. A variable wavelength excitation (Ex) and emission (Em) fluorescence detector was used. The wavelengths used were Ex = 250 nm and EM > 370 nm for the first 42 minutes, then changed to Ex = 285 nm and EM > 418 nm. A loop injector equipped with a 25 μL sampling loop was employed and PAHs were identified by peak retention times.

Sample Preparation

Soil samples were taken at three depths: surface, 5 cm and 15 cm. The samples were first extracted with acetone by an ultrasonic device, then soxhlet extracted with methylene chloride for 8 hours. The extracts were combined, concentrated and prepared for HPLC analysis.

Vegetables were washed with tap water simulating household kitchen conditions before analysis. Peels were removed for separate analysis (8). For onions, three pieces of outer shell of each onion were removed. The skins of beets and tomatoes were removed by using a paring knife. Tomatoes were picked while they were still firm and green.

The peels were first soxhlet extracted with acetone for eight hours, then with methylene chloride for another eight hours. The peeled vegetables were chopped into small pieces and were similarly extracted with acetone and methylene chloride.

A cleaning up procedure was necessary for vegetable extracts. The acetone and methylene chloride extracts were combined, water was added and the mixture was extracted with methylene chloride. The methylene chloride extracts were transferred to cyclohexane for column chromatography on 5 g of 15% H_2O deactivated "Florisil". PAHs were eluted with 100 mL of cyclohexane. The cyclohexane solution was then concentrated and prepared for HPLC analysis.

RESULTS AND DISCUSSION

Analytical results are summarized in Tables 1, 2 and 3. For easier comparison and correlation, PAH profiles were prepared by defining BkF concentration as 1 and those of all other PAHs are given relative to BkF (8).

TABLE 1

PAHS IN SOIL

PAHs	at surface (ng/g)		at 5 cm depth (ng/g)		at 15 cm depth (ng/g)	
	A	B	A	B	A	B
Ph	51	106	31	72	32	75
An	50	70	29	62	22	53
Fl	141	170	74	97	73	103
Py	101	183	81	104	80	119
BaA	87	101	50	59	47	59
Ch	90	128	50	64	60	60
BeP	116	95	60	47	60	53
BbF	97	95	51	49	53	65
Per	34	71	20	19	23	32
BkF	54	62	31	37	31	42
BaP	108	108	58	65	54	87
BghiP	147	109	67	67	64	75
DBahA	19	29	15	16	11	19
IP	80	63	34	44	32	43
BbCh	16	20	13	16	9	16
AA	38	20	14	11	10	26
Cor	61	66	32	38	32	38

A: virgin soil (covered with grass)
B: cultivated soil

Soil

Two sets of soil samples were examined. Each set consisted of three samples collected at the surface, and at depths of 5 cm and 15 cm, respectively. The first set was taken at a location where the soil had not been cultivated. The second set was collected near the edge of a vegetable bed where the soil had been turned over in the previous year but had no vegetation growth.

In the first set of soil samples, PAH concentrations were higher at the surface than below the surface. PAH concentrations at the depth of 5 cm and 15 cm were approximately the same level. In the second set of soil samples, the PAH concentrations were slightly higher than in the first set. When PAH profiles of the soil samples at different depths were compared, (Figure 1) they showed little variation, indicating that PAHs in the soil originated from the same source.

Onions

Onions (Southport Red Globe) were picked when they were grown from seed to approximately 3 cm in width. Two groups of onions were analyzed. One was analyzed immediately after picking (fresh onion) and the other group was analyzed after winter storage (mature onion).

The fresh onions contained much lower concentrations of PAHs than the mature onions. In both cases, PAH concentrations were high in the peels.

It was observed (Figure 2) that the ratios of the low molecular weight, "light" PAHs were significantly higher in onions than in the soil. This may indicate that "light" PAHs such as phenanthrene are absorbed by onions more readily than high molecular weight, "heavy" PAHs such as BaP. It was also noticed (Figures 2 and 3) that fresh onion had higher ratios of "light" PAHs than mature onions. This may be explained by differences in contact time with soil and the degree of hydration of the outer skins.

Beets

Beets (Cylindra) were collected when they were grown to approximately 5 cm in diameter. They were analyzed immediately after picking.

The levels of PAHs in beets were lower than those in the mature onions but higher than those in the fresh onions. Again, the peels had higher PAH concentrations. It was also noticed (Figure 4) that the ratio

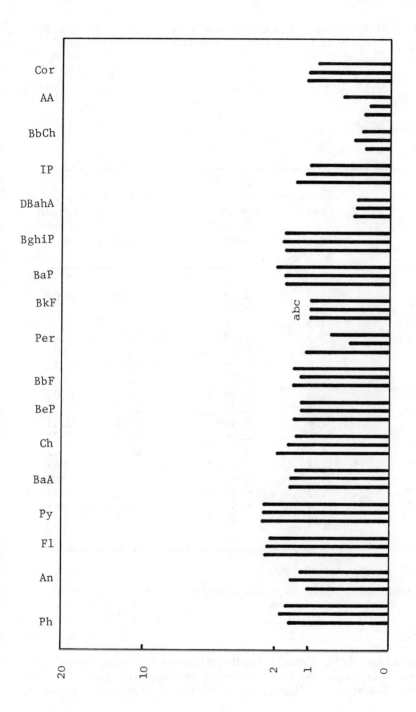

FIGURE 1. Soil PAH profiles (relative to BkF=1) a. surface soil b. soil at 5 cm depth c. soil at 15 cm depth

of "light" PAHs was higher in beets than in soil. This may also indicate that "light" PAHs were more readily absorbed by beets than "heavy" PAHs.

TABLE 2

PAHS IN ONIONS

PAHs	Mature Onion Peel (ng/g)	Mature Onion Without Peel (ng/g)	Fresh Onion Peel (ng/g)	Fresh Onion Without Peel (ng/g)
Ph	44.2	0.83	0.98	0.09
An	14.2	1.05	--	--
Fl	29.8	0.35	--	--
Py	29.8	0.48	0.34	0.04
BaA	4.96	ND	ND	ND
Ch	22.0	0.30	0.17	0.03
BeP	2.92	ND	ND	ND
BbF	2.64	0.09	0.05	0.01
Per	2.08	0.04	0.03	0.02
BkF	5.36	0.05	0.06	0.01
BaP	7.36	0.06	0.03	0.01
BghiP	16.6	0.14	0.07	ND
DBahA	1.72	0.07	ND	ND
IP	8.52	0.06	0.03	ND
BbCh	1.32	0.05	0.03	ND
AA	1.04	ND	ND	ND
Cor	8.20	0.07	0.07	ND

ND = Not detected
-- = Not measured due to peak uncertainty

Tomatoes

Tomatoes (O.M. Canada Northstar) were collected when they were near maturing, still firm and green and approximately 4 cm in diameter. Tomatoes had the lowest PAH content among the vegetables

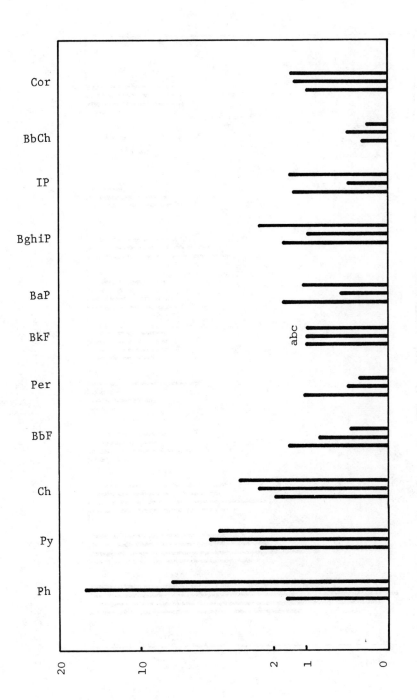

FIGURE 2. Onion peels PAH profiles (relative to BkF=1) a. surface soil b. fresh onion peel c. mature onion peel

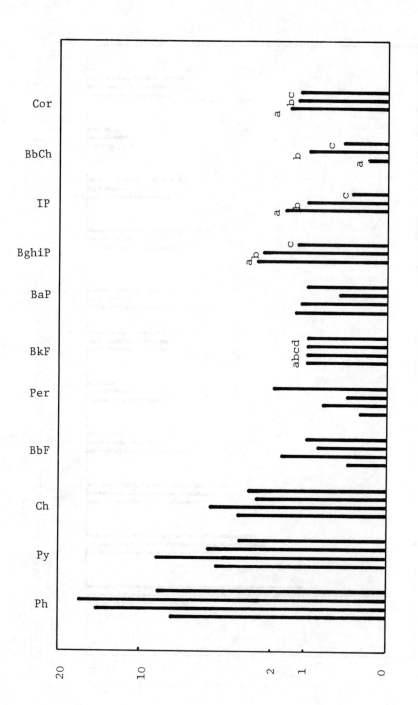

FIGURE 3. Onion PAH profiles (relative to BkF=1) a. mature onion without peel b. mature onion peel c. fresh onion peel d. fresh onion without peel

investigated. PAHs were only detected in the skins. The PAH profile indicated (Figure 5) that the ratios of "light" PAHs as well as coronene, were high. This may result from tomatoes only exposed to airborne deposition or the waxy coat on tomatoes has some special absorbancy property.

TABLE 3

PAHS IN BEETS AND TOMATOES

PAHs	Beets		Tomatoes	
	Peel (ng/g)	Without Peel (ng/g)	Peel (ng/g)	Without Peel
Ph	4.26	0.38	0.28	ND
An	1.48	0.06	0.08	ND
Fl	2.59	0.24	0.17	ND
Py	2.27	0.14	0.14	ND
BaA	0.44	ND	ND	ND
Ch	0.72	0.13	ND	ND
BeP	1.05	0.08	ND	ND
BbF	0.46	0.06	0.01	ND
Per	0.50	0.03	ND	ND
BkF	0.14	0.03	0.02	ND
BaP	0.21	0.02	0.01	ND
BghiP	0.38	0.04	0.02	ND
DBahA	ND	ND	ND	ND
IP	0.20	0.03	ND	ND
BbCh	0.08	ND	ND	ND
AA	0.05	ND	ND	ND
Cor	0.33	ND	0.22	ND

ND = Not Detected

It is not within the scope of this paper to make definitive statements concerning the absorbance of PAHs by vegetables. However, there are several observations worthy of note. As the data presented, it is obvious that vegetables having smooth skin and smaller

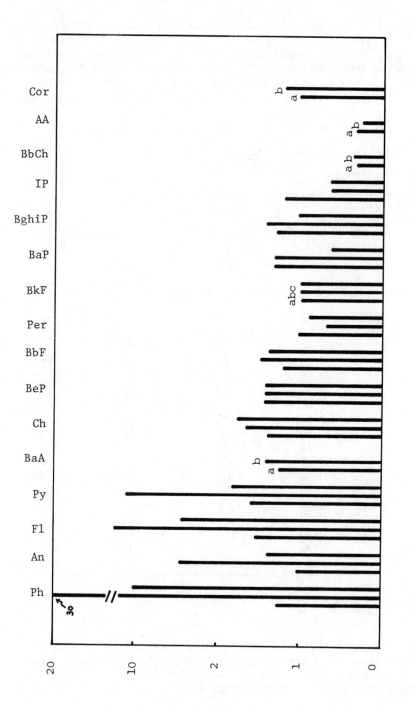

FIGURE 4. Beet PAH profiles (relative to BkF=1) a. surface soil b. beet peel
c. beet without peel

contacting surface areas contained less PAHs. The PAH profiles indicated that low molelcular weight, "light" PAH were absorbed by vegetables more readily than high molecular weight, "heavy" PAHs. It is also possible that the surface layers of the vegetables analysed have a higher retentive power for "light" PAHs than has the supporting soil. The level of PAHs also depended on the freshness of the vegetables. Most PAHs were detected in the outer shell or skin of the vegetables.

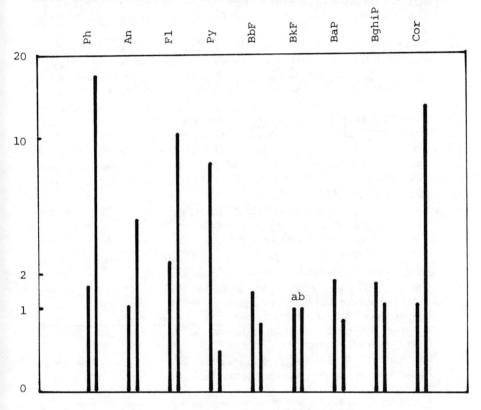

FIGURE 5. Tomato peel PAH profile (relative to BkF=1)
a. surface soil   b. tomato peel

Published estimates based on animal data suggested that the allowable daily intake of BaP for humans was 48 ng (9). From the results obtained in this study, one may conclude that when vegetables grown in a PAH polluted area are thoroughly washed and peeled before consuming, their contribution to the total PAH intake cannot be regarded as significant.

References

1. Katz, M. Sakuma, T. and Ho, A. (1978): Chromatographic and spectral analysis of polynuclear aromatic hydrocarbons: Quantitative distribution in air of Ontario cities. Environ. Sci. Technol. 12: 909-915.
2. Blumer, M., Blumer, W. and Reich, T. (1977): Polycyclic Aromatic hydrocarbons in soil of a mountain valley: Correlation with highway traffic and cancer incidence, Environ. Sci. Technol. 11: 1083.
3. Hase, A. and Hites, R. A. (1978): On the Origin of polycyclic aromatic hydrocarbons in the aqueous environment. In: Identification and Analysis of Organic Pollutants in Water, edited by L. H. Keith, pp. 205-214, Ann Arbor Science, Ann Arbor, Mich.
4. Lo, M. T. and Sandi, E. (1978): Polycyclic Aromatic Hydrocarbons in Foods. In: Residue Reviews, edited by F. A. Gunther and J. D. Gunther, Vol. 69, pp. 34-86, Springer-Verlag, New York.
5. Panalaks, T. (1976): Determination and Identification of polycyclic aromatic hydrocarbons in smoked and charcoal broiled food products by high pressure liquid chromatography. J. Environ. Sci., Health B, 11: 399-415.
6. Kolar, L. R., Ledvina, J. T. and Hanus, F. (1975): Contamination of soil, agriculture plants and vegetables by 3,4-benzopyrene in the Ceske Budejovice. Cesk. Hyg. 20: 135.
7. Grimmer, G. (1966): Cancerogeme Kohlenwasserstoffe in der Umgebung des Menschen. Erdoel Kohle 19: 578.
8. Grimmer, G., Nanjack, K. W. and Schneider, D. (1980): Changes in PAH Profiles in Different Areas of a City During the Year. In: Polynuclear Aromatic Hydrocarbons: Chemistry and Biological Effects, edited by A. Bjorseth and A. J. Dennis, pp. 107-125, Battelle Press, Columbus, Ohio.
9. Santodonato, J., Basu, D. and Howard, P. H. (1979): Multimedia Human Exposure and Carcinogenic Risk Assessment for Environmental PAH. In: Polynuclear Aromatic Hydrocarbons: Chemistry and Biological Effects, edited by A. Bjorseth and A. J. Dennis, pp. 435-454, Battelle Press, Columbus, Ohio.

FACTORS MODIFYING THE EFFECT OF AMBIENT TEMPERATURE ON TUMORIGENESIS IN MICE.[1]

HAROLD S. WEISS, JOHN F. O'CONNELL, JOSEPH F. PITT, RICHARD E. MILLER and DOUGLAS E. GRAFF.
Department of Physiology, College of Medicine, The Ohio State University, Columbus, Ohio 43210.

INTRODUCTION

Our findings that skin tumorigenesis is accelerated in cool acclimated mice and retarded in warm acclimated mice (18, 19) forms part of a generally inconsistent picture of the role of temperature in cancer. Apart from the possible beneficial effects of hyperthermia (7,11), the literature is highly variable in this area. With both cold and warm ambient temperature (Ta), reports range from stimulation to inhibition of tumors in species from man to fish (1-4,8,12-17). However, no two reports appear to have utilized precisely the same Ta, neoplasm or species.

The work reported here was undertaken to extend and expand our original observations, which were limited to benzo(a)pyrene (BaP) in toluene applied topically and repetetively to C3H male mice acclimated to Ta of $16°$, $23°$ and $32°C$. In particular, we attempted to extend the Ta range to which the mice were acclimated, to use the SENCAR mouse (5) in addition to the C3H and to utilize 7,12-dimethylbenz(a)anthracene (DMBA) in addition to BaP (6).

MATERIALS AND METHODS

Animals

C3H/HeJ males were purchased from Jackson Laboratories and SENCAR males from Oak Ridge Laboratories. Mice were obtained at 5-7 weeks of age. On arrival they were placed 5 to a plastic "shoe box" cage on ground corncob litter and given 1-2 wks to adjust to normal vivarium conditions. Prior to being placed in temperature controlled rooms for acclimation, they were ear punched for identification and the fur trimmed from the back by means of a fine-toothed electric clipper. Clipping of fur was repeated during the trials as needed. Feed (Purina Rodent Laboratory Chow, 5601), and water were <u>ad libitum</u>.

[1]Supported in part by EPA Grant CR-807342-01-0

TEMPERATURE AND TUMORIGENESIS

Temperature Rooms and Acclimation

Low Ta ranged from 16 to 10° and high Ta from 32 to 33°C. Normal vivarium Ta was around 23°C. These Ta's were maintained within an average of $\pm 1^\circ$C automatically and recorded continuously. Room temperature stability was checked additionally on max-min Hg thermometers. Lighting was clock controlled in all rooms, 12 hrs. on, 12 off. Animals were allowed approximately 2 wks to acclimate before treatment began. Acclimation (10) without undue stress or debilitation was assumed when food intake rose to a stable level in the cool Ta and fell to a stable level in the warm Ta while body weights and rectal temperatures were maintained near normal.

Chemicals and Dosing

The carcinogens were BaP (Aldrich Chemical Co., Milwaukee, Wis.) and DMBA (Eastman Kodak Co., Rochester, NY). Purity of original materials, obtained in powder form, was verified by spectrophotometry (18). The toluene used as vehicle for the carcinogen was of reagent grade. All stock solutions were stored in dark bottles in the refrigerator. Doses are described in amounts/wk (e.g., 0.5 mg/wk in 0.8 ml), but a wk's dose was always applied in two equal parts at least one day apart. A given day's application consisted of two drops of 0.02 ml each, separated by a few seconds to reduce runoff, for a total volume of 0.08 ml/wk. All applications were made by micropipette under yellow light to the clipped area of the back.

Data and Analysis

The basic information gathered was a weekly count of mice alive, of mice with tumors and a number of tumors per mouse. Body weights were obtained once a month and food intake sporadically. Number of mice with tumors was plotted against wks and presented as visually fitted incidence curves. Where incidence reached 100%, an average latency or mean tumor appearance time (MTAT $\pm$ S.E.) was calculated from the number of mice showing tumors for the first time at each weekly observation and used in statistical comparisons, usually by "t" test. Average number of tumors per mouse alive (ANT/M) was also plotted against wks, but fitted with a line derived by regression analysis started between avg values of 1/2 and one tumor/mouse and continued until the curve plateaued or became erratic, due apparently to mortality and/or coalescence of tumors. The slopes of these curves and their standard errors ($b \pm S_b$) were used in statistical comparisons, usually by "t" test.

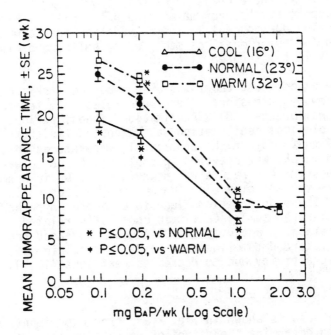

FIGURE 1. Summation of mean tumor appearance time (MTAT) for tumors induced by continuous BaP in C3H male mice acclimated to various Ta's.

General

Much of our previous work on the effect of Ta on incidence on skin tumors in C3H male mice is summarized in Fig. 1. Plotted in Fig. 1 are MTAT's for continuous doses of BaP ranging from 2.0 down to 0.1 mg/wk, all in 0.08 ml toluene/wk. It is evident that acclimation to 16°C accelerates and acclimation to 32°C retards tumorigenesis, compared to a normal Ta of 23°C. In general the accelerative effect of cool Ta is larger and more consistent than is the retardation due to warm Ta. The cool acceleration of tumorigenesis becomes more accentuated as the BaP dose decreases. On the other hand, total time to MTAT also increases with decreasing dose with the

MTAT increasing from about 9 wks at 1.0 mg/wk to 25 wks at 0.1 mg/wk. At the highest doses (and shortest MTAT) the Ta effect appears to be overwhelmed and obliterated. Similar acceleration due to cool Ta and retardation due to warm Ta can be shown for tumor mass, as measured by ANT/M, and also by mortality in the treated mice (19).

The predominant tumor was a squamous cell skin carcinoma which grew by extension. There were relatively few metastases to internal organs (18). Acclimation without serious physical or physiological debilitation was suggested for these mice by the fact that body weight and rectal temperatures were normal but food intake was elevated in cool, depressed in warm Ta. Although skin temperature was down in cool, up in warm, tissue O_2 uptake tended to be elevated in cool, depressed in warm (9,19). These observations lead to a general hypothesis that the accelerated tumorigenesis in cool Ta is related to a higher metabolic activity stimulated by increased heat loss and the retarded tumorigenesis in $32°C$ is related to a lower metabolic activity due to decreased heat loss (10,19).

Decreasing Temperature

The results shown in Fig. 2 are from experiments which sought to determine the effect of decreasing the cool Ta from $16°$ to $10°C$. The trial was run with C3H male mice treated continuously with 0.5 mg BaP in 0.08 ml toluene/wk, and is therefore generally comparable to our previous results [(19) and Fig. 1]. The tumor mass data suggest that there is an incremental acceleration of tumorigenesis with incremental lowering of Ta. From $16°$ to $10°$ there is an 18% increase in slope of ANT/M line, and from $23°$ to $16°$ the increase is 33%. The overall acceleration accomplished by $10°$ compared to the normal $23°$ is a highly significant 57%.

Incidence of tumors was also accelerated as Ta was lowered--i.e., the MTAT occurred 3.2 wks (25%) earlier at $10°$ than at $23°$ (highly significant). Whether this trend of accelerated tumorigenesis continues for temperatures below $10°C$ is presently unknown. The effect of extending the temperature range above $32°C$ will be discussed in subsequent sections. Food intake increased at each Ta below normal; in units of g/mouse/day it was for $10° = 9.0±0.27$, for $16° = 7.2±0.28$ and for $23° = 4.0±0.09$.

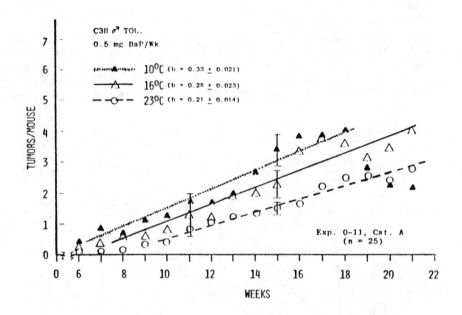

FIGURE 2. Acceleration by low temperature of single stage, BaP induced skin tumors in C3H male mice.

SENCAR mice

Since our initial work relating Ta to tumorigenesis was all based on the C3H mouse, it seemed desirable to extend the study to other strains. The SENCAR was thought to be a reasonable choice in view of its special selection for sensitivity to carcinogens (5). Accordingly, SENCAR male mice were treated with BaP in the same manner as C3H's, except for an extension of low Ta to $13°$ and warm Ta to $33°C$. These choices of acclimation Ta were based in part on the work shown in Fig. 2 where in the C3H mouse a lowering of acclimation Ta to $10°$ resulted in further acceleration of tumorigenesis over that at $16°$ and in part on a desire to keep the spread between warm and normal the same as between cool and normal (i.e., $\Delta = 10°C$). The upper limit was put at $33°$ because from earlier preliminary trials with C3H mice, we felt that higher Ta was likely to result in depressed body weights, possibly negating what we have considered to be an important factor in these studies, acclimation without undue stress (19).

Possibly because of their high sensitivity to BaP, results from SENCARS proved somewhat difficult to interpret. At doses down to 0.1 mg BaP/wk we failed to elicit a significant Ta effect on tumorigenesis, although the MTAT's all fell in a range where in C3H's at least (Fig. 1) a clear Ta effect had been observed (i.e., in SENCARS treated with 0.1 mg BaP/wk, the MTAT's were: cool = 14.8±0.7, normal = 14.8±0.8 and warm = 13.7±5). However, halving the BaP dose to 0.05 mg/wk clearly led to accelerated tumor incidence in cool Ta, although warm

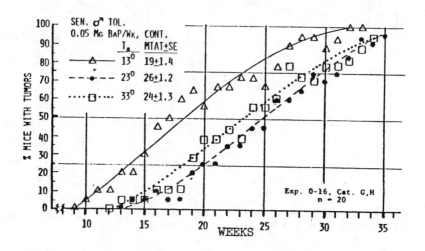

FIGURE 3. Effect of temperature acclimation (Ta) on single stage, BaP induced skin tumorigenesis in SENCAR male mice.

Ta was without effect (Fig. 3). The 7 wk earlier MTAT in cool Ta over normal Ta is highly significant statistically, and might perhaps be of practical value if rapid results were desired in a whole animal trial.

This acceleration of tumorigenesis in cool Ta was only partly supported by the tumor mass data where although the regression line for ANT/M was displaced 3-5 wks earlier in time, its slope was not significantly different from normal Ta (0.16± 0.01 vs. 0.18±0.01 tumor/mouse/wk). However, the slope of the warm Ta curve was significantly flatter than that for normal Ta (0.14±0.01 vs. 0.18±0.01 tumor/mouse/wk respectively), though not displaced later in time. Thus although the acceleration of tumorigenesis associated with acclimation to cool Ta and retardation due to warm Ta are not as clear cut and consistent in SENCARS as in C3H mice, the pattern of response is nevertheless generally similar.

DMBA as Carcinogen

To supplement all of our previous studies in which the Ta effect had been demonstrated on tumors induced by BaP, tests were made using DMBA. The procedure was otherwise the same as with BaP, except that the Ta range was extended to $13°C$ on the cool side and $33°C$ on the warm side, for reasons outlined in the previous section on SENCARS.

Fig. 4 illustrates that cool Ta can cause acceleration of skin tumors induced by DMBA in the male C3H mouse as it does for skin tumors induced by BaP [Fig. 1 and (19)]. Rate of increase in number of tumors per mouse is 1.8 times higher at $13°C$ than at $23°C$. The regression coefficients are highly reliable, fitting the weekly averages with correlations of better than 0.9. Incidence data corroborated tumor mass results in that cool MTAT occurred at 11±0.6 wks, a highly significant 3 wk earlier than the 14±0.9 wks of normals.

Fig. 4 also shows that warm Ta failed to retard tumorigenesis, in contrast to our experience with BaP in the C3H male mouse [Fig. 1 and (19)]. At present, we attribute this lack of response to warm Ta more to the fact that a higher Ta was used ($33°C$ rather than our usual $32°C$) than to the carcinogen being DMBA. We have other evidence (see e.g. section in SENCARS) that suggests that even with BaP, a $1-2°C$ difference at the upper end of the Ta scale may markedly diminish and even reverse the retardation in tumorigenesis usually seen with warm acclimation. Even at $32°C$ with BaP, it may be seen that the retardation associated with warm Ta was considerably less pronounced than the acceleration associated with cool Ta of $16°$ (Fig. 1).

The comparison of the DMBA results with the BaP data in Fig. 1 also supports the well known fact that DMBA is more potent than BaP, in this case on the order of 50 fold (i.e., normal Ta, a MTAT of 14 wks required a BaP dose of about 0.5 mg/wk compared to 0.01 mg/wk for DMBA).

In looking for physiological mechanisms that might help explain the Ta effect on tumorigenesis, we have used body weight and food intake as simple indications of physical status and metabolic activity in the mice. In the DMBA trials cool body weights were close to normals, showing that the $13°C$

mice have acclimated well to low temperature and have also tolerated the DMBA treatment well. The data suggest, however, that 33°C might be relatively more of a stress than 13° since body weights were around 10% below normal. In previous work at 32°, body weights were well maintained (19). Presumably this weight loss might bear on the failure of 33° Ta to elicit the typical retardation in tumorigenesis.

With respect to food intake, the pattern was the typical one of marked increase in cool Ta (close to 2 times normal) and marked decrease in warm Ta (about 1/2 of normal). This data supports a link, at least in cool Ta, between increased metabolic activity and accelerated carcinogenesis, but clearly other factors than depressed food intake can intervene in the warm Ta effect.

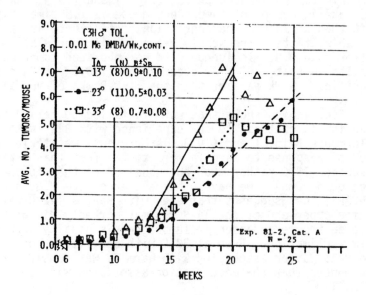

FIGURE 4. Effect of temperature acclimation (Ta) on DMBA induced skin tumorigenesis in C3H male mice.

SUMMARY

The previously found acceleration of BaP skin tumorigenesis by acclimation of C3H male mice to cool Ta (16°C) and retardation by acclimation to warm Ta (32°C) was amplified and extended as follows:

Lower and higher Ta. Acclimation to $10°C$ accelerated tumorigenesis beyond that seen at $16°$ in C3H male mice treated with continuous topical doses of 0.5 mg BaP in 0.08 ml toluene/wk. At $10°C$ compared to $23°C$, incidence (MTAT) was shortened by 25% and tumor mass (ANT/M) increased by 33%. However acclimation to a Ta of $33°$ tended to erase the retardation normally seen with $32°C$.

SENCAR mice. In general, male SENCARS responded to Ta as did C3H's but less consistently. With BaP in toluene at a dose of 0.05 mg/wk continuously, cool Ta of $13°C$ accelerated MTAT by a highly significant 7 wks, but did not alter slope of the ANT/M curve. SENCARS showed no change in incidence with acclimation to $33°$, but a flattening of the ANT/M curve did occur.

DMBA as carcinogen. Acceleration of tumorigenesis by a cool Ta of $13°$ was seen with DMBA as it had been with BaP. With 0.01 mg DMBA in 0.08 ml toluene/wk continuously in the C3H male mouse, the MTAT occurred 3 wks earlier and the slope of the ANT/M line was 1.8 times steeper than at $23°C$. An acclimation Ta of $33°C$, however, did not produce the expected retardation of tumorigenesis. Food intake showed the typical increase in cool Ta and decrease in warm Ta. Body weight was maintained in the $13°C$ mice but tended to be lower at $33°C$.

REFERENCES

1. Baker, D.G. (1977): Influence of a Chronic Environmental Stress on the Incidence of Methylcholanthrene-induced Tumors. Cancer Res. 37:3939-3944.
2. Baker, D.G. and Jahn, A. (1976): The Influence of a Cronic Environmental Stress on Radiation Carcinogenesis. Radiation Res. 68:449-458.
3. Blatteis, C.M., Cardoso, S.S., Narayanan, T.K., Hughes, M.H., and Morris, H.P. (1980): Depressed Growth of Morris Hepatomas in Altitude - and Heat-Stressed but not in Cold-Stressed Buffalo Rats. J. Nat'l. Cancer Inst. 61:1451-1458.
4. Freeman, R.G. and Knox, J.M. (1964): Influence of Temperature on Ultraviolet Injury. Arch. Derm. 89:858-864.
5. DiGiovanni, Slaga, T.J., Boutwell, R.K. (1980): Comparison of tumor-initiating activity of 7,12 dimethylbenz(a)anthracene and benzo(a)pyrene in female SENCAR and CD-1 mice. Carcinogenesis 1:381-389.
6. Gelboin, H.V. and Ts'o, P.O.P., editors, (1978): Polycyclic Hydrocarbons and Cancer. Vol. 1 & 2, Academic Press, New York, 860 pp.

7. Gillette, E.L. and Ensley, B.A. (1979): Effect of Heating Order on Radiation Response of Mouse Tumor and Skin. Int. J. Radiation Oncology Biol. Phys. 5:209-213.
8. Griffiths, J.D., Hoppe, E. and Cole, W.H. (1981): The Influence of Thermal Stress and Changes in Body Temperature on the Development of Carcinosarcoma 256 Walker in Rats After Inoculation of Cells. Cancer, 14:111-116.
9. Hakaim, A.G. and Weiss, H.S. (1981): Interaction Between Topical Toluene and Ambient Temperature on Tissue Oxygen Uptake. Res. Comm. Chem. Path. Pharm. 33:95-102.
10. Horwitz, B.A. symp. chr., (1979): Metabolic Aspects of Thermogenesis. Fed. Proc. 38:2147-2181.
11. Marmor, J.B., Hahn, N. and Hahn, G.M. (1977): Tumor Cure and Cell Survival After Localized Radiofrequency Heating. Cancer Res. 37:879-883.
12. Peters, G. and Peters, N. (1977): Temperature-Dependent Growth and Regression of Epidermal Tumors in the European Eel (Anguilla Anguilla L.) Annals New York Acad. of Sci., 245-260.
13. Sonstegard, R.A. (1976): Studies of the Etiology and Epizootiology of Lymphosarcoma in Esox. Prog. Exp. Tumor Res. 20:141-155.
14. Stich, H.F., Acton, A.B., Oishi, K., Yamazaki, R., Harada, T., Hibino, T. and Moser, H.G. (1977): Systematic Collaborative Studies on Neoplasm in Marine Animals as Related to the Environment. Annals New York Acad. Sci., 374-387.
15. Trubcheninova, L.P., Khutoryansky, A.A., Svet-Moldavsky, G.J., Kuznetsova, L.E., Sokolov, P.P. and Belianchykova, N.I. (1977): Body Temperature and Tumor Virus Infection. I. Tumorogenicity of Rous Sarcoma Virus for Reptiles. Neoplasma 24:1-19.
16. Turbiner, S., Shklar, G., and Cataldo, E. (1970): The Effect of Cold Stress on Chemical Carcinogenesis of Rat Salivary Glands. Oral Surg. 29:130-137.
17. Wallace, W., Wallace, H.M. and Mills, C.A. (1942): Effect of Climatic Environment Upon the Genesis of Subcutaneous Tumors Induced by Methylcholanthrene and Upon the Growth of a Transplantable Sarcoma in C3H Mice. J. Natl. Cancer Inst. 2:99-110.
18. Weisbrode, S.E. and Weiss, H.S. (1981): Effect of Temperature on Benzo(a)pyrene-Induced Hyperplastic and Neoplastic Skin Lesions in Mice. J. Nat'l Cancer Inst. 66: 975-978.
19. Weiss, H.S., Pitt, J.F., Kerr, K.M., Hakiam, A.G., Weisbrode, S.E. and Daniel, F.B. (1981): Sensitivity of Mice to Benzo(a)pyrene Skin Cancer Following Acclimation to Cool and Warm Temperature. Proc. Soc. Exptl. Biol. Med. 167:122-128.

COMPARATIVE ANALYSIS OF POLYCYCLIC AROMATIC SULFUR

HETEROCYCLES ISOLATED FROM FOUR SHALE OILS.

CHERYLYN WILLEY,* RICHARD A. PELROY** AND DOROTHY L. STEWART**
*Physical Sciences Department; **Biology Department,
Pacific Northwest Laboratory
Richland, Washington 99352

INTRODUCTION

Recently, a number of synthesized polycyclic aromatic sulfur heterocycles (PASH) have been tested for biological activity and have shown evidence of mutagenic and/or carcinogenic response (1,2). These compounds, based on the five-membered thiophene ring system, have been found in low concentrations in most polycyclic aromatic hydrocarbons (PAH) fractions of various fossil fuels such as coal liquids and shale oils (3,4). Although sulfur-selective detection can be used for preliminary screening for sulfur compounds, sulfur detection in complex mixtures often gives nonlinear response and is severely hampered by response quenching due to coeluting non-sulfur components (5). In view of this, an isolated PASH fraction is necessary for the detailed chemical characterization of the sulfur compounds through gas chromatography/mass spectrometry (GC/MS). In addition, separation is necessary to obtain mutagenicity data for this class of compounds.

This report describes the isolation of sulfur heterocycle fractions from four shale oils (Paraho, Geokinetics, Occidental, and Rio Blanco), the use of capillary column gas chromatography and mass spectrometry for the identification of individual mixture components, and a reverse (Ames) and forward mutation assay with Salmonella typhimurium to screen for possible health hazards.

MATERIALS AND METHODS

Samples

Four shale oils, all from the Green River oil shale formation in either Colorado or Utah, were chosen for this study. Paraho shale oil represented material formed above ground in constructed metal retorts. The crude oil was sampled on August 24, 1977 at Anvil Points, Colorado with an initial sampling temperature of 150°F. It contained 0.7 weight percent sulfur. The three other shale oils, Occidental,

Geokinetics, and Rio Blanco, were all produced below ground by in situ pyrolysis of oil shale kerogen. Occidental and Rio Blanco operate a vertical modified in situ (MIS) mode of shale oil production, while Geokinetics uses a horizontal in situ mode. The crude Occidental product oil was sampled on March 6, 1979 from the heater-treater unit of the MIS retort at Logan, Colorado at a sampling temperature between 150 to 170°F. It contained 0.7 weight percent sulfur. The crude Geokinetics product oil was sampled from the retort in Vernal, Utah on July 11, 1978 and July 12, 1978 at 130°F. It contained 0.6 weight percent sulfur. The crude Rio Blanco product oil was sampled at the retort on November 22, 1980 at approximately 130 to 150°F. It contained 1.9 percent sulfur. Many features of the retorting processes are proprietary, but a general explanation of the Paraho method can be obtained by reference to Sladek (6).

Chemical Fractionation

Ten-gram quantities of the shale oils were first fractionated into chemical classes by a modification of the method developed by Later et al. (7). The samples of the shale oils were adsorbed onto 25 grams of neutral alumina (Fischer, Brockman Activity 1, 80/200 mesh), and introduced onto a 100-gram neutral alumina column. The ambient pressure column was eluted with (a) 100 ml of hexane to obtain the aliphatic hydrocarbons, (b) 500 ml of benzene to obtain the neutral polycyclic aromatic compounds (including the PASH), (c) 1000 ml of chloroform:ethanol (99:1) to obtain the nitrogen- and oxygen-containing polycyclic aromatic compounds (N-PAC), and (d) 200 ml of methanol to obtain the phenolic compounds. Each elution from the neutral alumina column was then concentrated on a rotary evaporator at 40°C to a volume suitable for gas chromatography.

Isolation of PASH

The PASH were isolated from the neutral polycyclic aromatic compound (PAC) fraction according to the method described previously (3,4). Briefly, the aromatic fraction was oxidized overnight with an excess of 30 percent H_2O_2 in a 1:1 refluxing mixture of benzene:acetic acid. The organic phase was separated from the aqueous phase and was introduced to a silica gel (Baker Analyzed, 60/200 mesh) column and eluted first with benzene to obtain the oxidized PASH (the sulfones) and any oxidized PAH (the quinones). The oxidized portion was then reduced with $LiAlH_4$ in refluxing ether and passed through a silica gel column eluting with hexane to obtain the sulfur heterocycles and then with methanol to obtain the more polar hydroquinones (the reduction products of quinones).

The isolation method is shown pictorially in Figure 1 using a mixture of dibenzothiophene, anthracene, and chrysene as an example. After oxidation, the unoxidized anthracene and chrysene are separated from the more polar dibenzothiophene sulfone and anthraquinone. During reduction, the dibenzothiophene sulfone returns to its original state as dibenzothiophene and can then be separated from the more polar 9,10-dihydroxyanthracene. The PASH in the complex neutral PAC fraction follow a path similar to the dibenzothiophene to obtain an isolated sulfur heterocycle fraction.

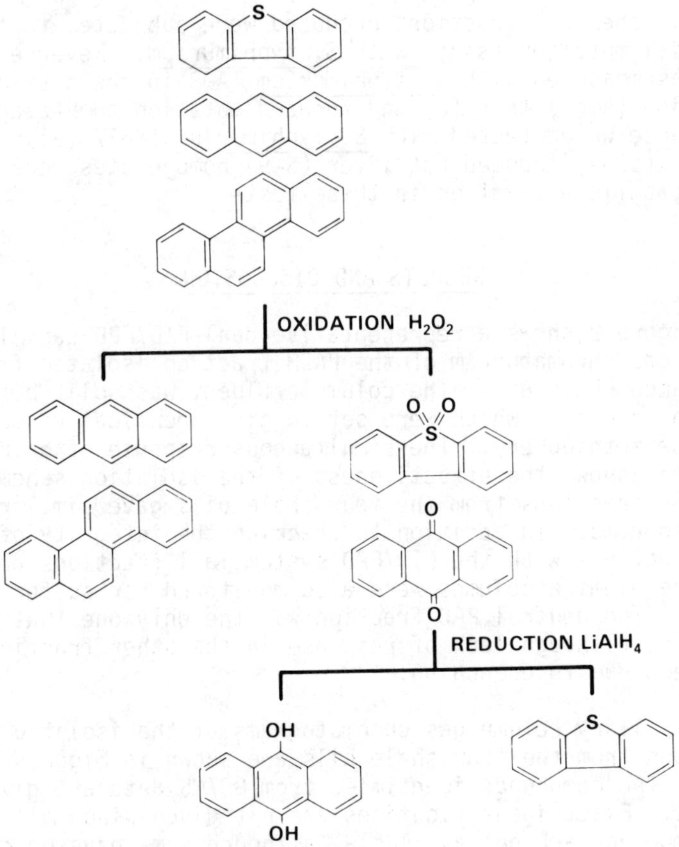

FIGURE 1. PASH isolation method.

Analysis of PASH

After the isolation of the PASH fraction, the integrity of the isolation scheme was checked through the use of a sulfur-specific Flame Photometric Detector (FPD). The column effluent was split between the universal Flame Ionization Detector (FID) and the FPD on a Perkin-Elmer Sigma 2 gas chromatograph. Along with the sulfur-specific detection, further identification of the sulfur compounds was achieved through the use of GC/MS and high resolution gas chromatography using fused silica capillary columns coated with SE-54. A Hewlett-Packard 5982 GC/MS system and a Hewlett-Packard 5880 gas chromatograph were used.

Mutagenesis Assays

All chemical fractions produced were subjected to two bacterial mutation assays with *S. typhimurium*. Reverse mutation was measured with *S. typhimurium* TA98 in the histidine reversion (Ames) test (8) and forward mutation to 8-azaguanine resistance was detected with *S. typhimurium* TM677 (9). Aroclor (1254) induced rat liver (S-9) homogenates were used for metabolic activation in these tests.

RESULTS AND DISCUSSION

Figure 2 shows a representative dual FID/FPD capillary column gas chromatogram of the PASH fraction isolated from the Rio Blanco shale oil. The column effluent was split between the two detectors which were set to give identical response for dibenzothiophene. The simultaneous response of both detectors shows the effectiveness of the isolation scheme. All PASH fractions from the four shale oils gave similar gas chromatograms. In addition to checking the integrity of the PASH fractions with the FID/FPD system, all fractions obtained from the alumina columns were also monitored for sulfur compounds. The neutral PAC fraction was the only one that gave response, although lack of response in the other fractions may have been due to quenching.

Capillary column gas chromatograms of the isolated PASH fractions from the four shale oils are shown in Figures 3 and 4. The compounds identified from GC/MS data are given in Table 1. Exact identifications are not given since all standard compounds are not available, although some have been recently synthesized (10). The PASH fractions are characterized mainly by the presence of benzothiophenes,

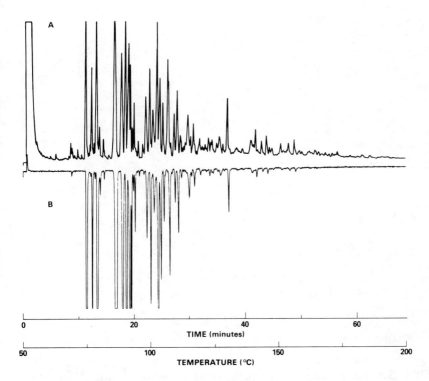

FIGURE 2. FID/FPD dual trace capillary column gas chromatogram of PASH fraction isolated from Rio Blanco shale oil. Conditions: 20 m x 0.25 mm fused silica column coated with SE-52, temperature programmed from 50 to 200°C at 2°C/min.

dibenzothiophenes, phenanthro[4,5-bcd]thiophene (the sulfur heterocycle homolog of pyrene), and several four-ring thiophene compounds. As can be seen in the table, there is a high degree of alkylation of the parent compounds. Partially hydrogenated benzo[b]thiophenes, formed under the harsh conditions of the $LiAlH_4$ reduction, are also present in the PASH fractions. In comparing the four shale oils, it should be noted that all gave very similar chromatograms suggesting the presence of many of the same compounds in all of the shale oils analyzed. Thus, even though the retorting processes for the four oils differ, the sulfur heterocycles present in the products are apparently very similar.

The total weight percent recovery of the PASH fractions based on original (unfractionated) samples for the Geokinetics, Rio Blanco, Occidental, and Paraho shale oils was

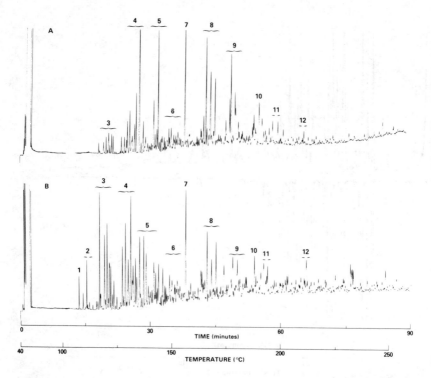

FIGURE 3. Capillary column gas chromatogram of the PASH fractions isolated from (A) Geokinetics and (B) Rio Blanco shale oils. Conditions: 30-m x 0.25-mm fused silica column coated with SE-54 temperature programmed from 40 to 100°C at 10°C/min and 100 to 260°C at 2°C/min. Numbers refer to identifications in Table 1

0.11 percent, 0.33 percent, 0.18 percent, and 0.10 percent, respectively. The only marked difference between these samples was that the Rio Blanco shale oil had two to three times more weight in the PASH fractions, consistent with the higher sulfur content of the unfractionated oil. The low recoveries, based on the sulfur content of the unfractionated materials, can be attributed to two main causes: First, the procedure entails many steps where losses could occur and the quantitative nature of the isolation procedure cannot be determined until adequate standards are available; and second, not all of the sulfur present in the shale oil is in the form of thiophene ring systems for which the isolation scheme was optimized. For example, inorganic sulfur is present as well as compounds such as thiols, mercaptans, thiazoles,

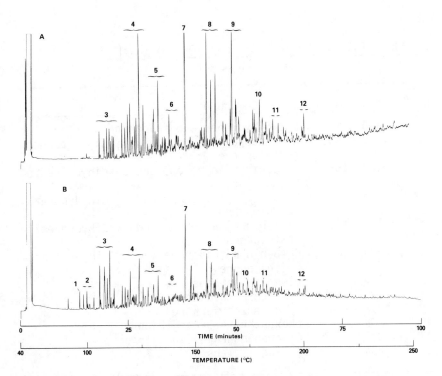

FIGURE 4. Capillary column gas chromatogram of the PASH fractions isolated from (A) Occidental and (B) Paraho shale oils. Conditions: same as in Figure 3. Numbers refer to identifications in Table 1.

thianthrenes, and other forms of organic sulfur which may not make it through the isolation procedure.

Figure 5 shows the results of both the reverse (Ames) and forward (8-azaguanine) mutation assays. From the Ames test it was found that most of the fractions from the alumina column chromatography fractionation, the unfractionated material, and the PASH fractions for all four shale oils were negative by the criteria used in the test (less than two-times increase in induced mutation response over background response) and they were therefore considered inactive. The only exceptions were the N-PAC fraction and the unfractionated Paraho shale oil. In quantitative terms, the unfractionated Paraho shale oil had a potency (based on dose response data) of 0.64 rev TA98/µg. The N-PAC fraction of the Paraho shale oil gave a dose response of 0.60 rev TA98/µg, nearly the same

TABLE 1

COMPOUNDS IDENTIFIED IN PASH FRACTIONS BY GC/MS

Peak No.[a]	Mol. Wt.	Compound
1	136	2,3-dihydrobenzo[b]thiophene
2	148, 150	C_1-benzo[b]thiophene and C_1-dihydrobenzo[b]thiophene
3	162, 164	C_2-benzo[b]thiophene and C_2-dihydrobenzo[b]thiophene
4	176, 178	C_3-benzo[b]thiophene and C_3-dihydrobenzo[b]thiophene
5	190, 192	C_4-benzo[b]thiophene and C_4-dihydrobenzo[b]thiophene
6	204, 206	C_5-benzo[b]thiophene and C_5-dihydrobenzo[b]thiophene
7	184	dibenzothiophene
8	198	C_1-dibenzothiophene
9	212	C_2-dibenzothiophene
10	208	phenanthro[4,5-bcd]thiophene
11	222	C_1-phenanthro[4,5-bcd]thiophene
12	234	4-ring isomers

a Numbers refer to peaks in Figures 3 and 4.

specific activity as the unfractionated material. These data are consistent with previously published results showing that Paraho shale oil was more active in the Ames test than the Geokinetics or Occidental shale oils (11).

The results of the forward mutation assay (a test which may respond to a somewhat greater array of mutagens than the standard Ames test) showed mutagenic activity in all the N-PAC fractions with the exception of the Rio Blanco shale oil. Moreover, by this test the neutral PAC fractions of the Geokinetics and Rio Blanco shale oils were also mutagenic contrasted to results in the Ames assay. It is of particular interest that the PASH fraction of the Rio Blanco shale oil showed nearly as large a specific response as did the neutral PAC fraction from which it was isolated. It was the only PASH fraction to show mutagenicity response, although chemical analyses thus far show no difference in the compounds present.

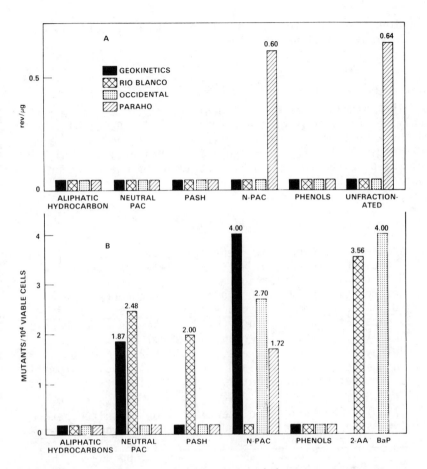

FIGURE 5. Results of (A) Ames and (B) forward mutation assays for the four shale oils. 2-AA is 2-amioanthracene and BaP is benzo[a]pyrene.

In summary, it was found that the major components of the PASH fractions for all four shale oils were two- to three-ringed parent and alkylated thiophene compounds. In all cases the PASH fractions showed no more specific mutagenic response than the neutral PAC fractions from which they were isolated. The only mutagenic response which was detected in the PASH fractions was for the Rio Blanco shale oil and showed specific mutagenic response similar to the neutral PAC fraction from which it was isolated. Finally, the forward mutation 8-azaguanine test was apparently more sensitive than the Ames histidine reversion test in detecting mutagenic activity for the chemical fractions from the shale oils.

ACKNOWLEDGEMENT

The authors thank J. S. Fruchter and D. S. Sklarew for supplying the samples of the four shale oils, C. L. Wilkerson for the weight percent sulfur determinations, and M. L. Lee for the use of the FPD at Brigham Young University. This work was prepared for the U.S. Department of Energy under Contract DE-AC06-76RLO 1830.

REFERENCES

1. Karcher, W., Neben, A., Depaus, R., van Eijk, J., Glaude, P., and Jacob, J. (1981): New results in the detection, identification and mutagenic testing of heterocyclic polycyclic aromatic hydrocarbons. In: Polynuclear Aromatic Hydrocarbons: Chemistry and Biological Effects, edited by W. M. Cooke and A. J. Dennis, pp. 317-327, Battelle Press, Columbus, Ohio.
2. Pelroy, R. A., and Stewart, D. L. (1982): Microbial mutagenicity of three- and four-ring polycyclic aromatic sulfur heterocycles, Mutat. Res., 91 (in press).
3. Lee, M. L., Willey, C., Castle, R. N., and White, C. M. (1980): Separation and identification of sulfur heterocycles in coal-derived products. In: Polynuclear Aromatic Hydrocarbons: Chemistry and Biological Effects, edited by A. Bjorseth and A. J. Dennis, pp. 59-73, Battelle Press, Columbus, Ohio.
4. Willey, C., Iwao, M., Castle, R. N., and Lee, M. L. (1981): Determination of sulfur heterocycles in coal liquids and shale oils, Anal. Chem., 53:400-407.
5. Grice, H. W., Yates, M. L., and David, D. J. (1970): Response characteristics of the melpar flame photometric detector, J. Chromatogr. Sci., 8:90-94.
6. Sladek, T. A. (1975): Recent trends in oil shale-part 2: Mining and shale oil extraction processes, Colo. Sch. Mines Miner. Ind. Bull., 18:1-21.

7. Later, D. W., Lee, M. L., Bartle, K. D., Kong, R. C., and Vassilaros, D. L. (1981): Rapid chemical class separation and detailed characterization of organic compounds in synthetic fuels, Anal. Chem., 53:1612-1620.
8. Ames, B. N., McCann, J., and Yamasaki, E. (1975): Methods for detecting carcinogens and mutagens with the Salmonella/mammalian-microsome mutagenicity test, Mutat. Res., 31:347-364.
9. Skopek, T. R., Liber, M. L., Krolowski, J. J., and Thilly, W. G. (1978): Quantitative forward mutation assay in Salmonella typhimurium using 8-azaquanine resistance as a genetic marker, Proc. Nat. Acad. Sci. USA, 75:410-414.
10. Iwao, M., Lee, M. L. and Castle, R. N. (1980): Synthesis of phenanthro[b]thiophenes, J. Heterocycl. Chem., 17:1259-1264.
11. Pelroy, R. A., Sklarew, D. S., and Downey, S. P. (1981): Comparison of the mutagenicities of fossil fuels, Mutat. Res., 90:233-245.

ANALYTICAL METHODS FOR THE DETERMINATION OF POLYCYCLIC AROMATIC HYDROCARBONS ON AIR PARTICULATE MATTER

STEPHEN A. WISE, SHARON L. BOWIE, STEPHEN N. CHESLER,
WILLIAM F. CUTHRELL, WILLIE E. MAY, and RICHARD E. REBBERT
Organic Analytical Research Division, Center for Analytical
Chemistry, National Bureau of Standards, Washington, DC 20234

INTRODUCTION

At the National Bureau of Standards (NBS) two samples of urban air particulate material [Standard Reference Material (SRM) 1648, "Urban Particulate Matter" (collected in St. Louis, MO) and Washington, DC urban particulate matter] have been used for the development and evaluation of analytical methods for the characterization and quantification of polycyclic aromatic compounds (PAC) on air particulate matter. SRM 1648, available in 2 g quantities, is certified only for inorganic constituents. However, a number of analysts have expressed interest in the organic constituents in this SRM for use in analytical methods evaluation, interlaboratory calibrations, and biological testing studies. As a result of this interest, NBS is evaluating the feasibility of issuing the Washington urban particulate material as an SRM certified for selected organic constituents. This SRM will be issued in 10 g quantities and be certified for the concentrations of several major polycyclic aromatic hydrocarbons (PAH). The analytical methods for the extraction, isolation, and measurement of the major PAH in this material are described briefly in this paper. The quantitative results obtained by gas chromatography (GC) and liquid chromatography (LC) are compared.

EXPERIMENTAL

The air particulate samples were collected in St. Louis, MO and Washington, DC using a baghouse especially designed for this purpose. These materials were collected over periods in excess of 12 months and, therefore, represent time-integrated samples. For each collection site the particulate material was removed from the filter bags by a specially designed vacuum cleaner and combined into a single lot. This material was subsequently screened through a fine mesh sieve to remove bag fibers and other extraneous material. The sieved material was then thoroughly mixed in a V-blender, bottled, and sequentially numbered. Randomly selected bottles were used for the analytical measurements. Sample aliquots of 1 g were extracted in a Soxhlet extractor for 48 h with a cycle time of about 20 min. Samples prepared for GC analysis were extracted with 450 mL of a 1:1 mixture of benzene/methanol, whereas samples for LC analysis were extracted with a similar volume of methylene chloride. An internal standard solution of 1-methylpyrene (for GC analysis) or 7-methylfluoranthene/perylene-d_{12} (for LC analysis) was added to the particulate samples prior to extraction. Prior to GC analysis the extract was concentrated in a rotary evaporator, redissolved in cyclohexane, and liquid-liquid partitioned between N,N-dimethylformamide (DMF) and water as described by Bjørseth (1). After the liquid-liquid partition the total PAH fraction was isolated by normal-phase LC on an aminosilane column (30 cm x 9 mm i.d., µBondapak NH_2, Waters Associates, Milford, MA) as described previously (2). The PAH fraction was collected, concentrated, solvent changed to toluene, and analyzed by GC on a 30 m x 0.25 mm i.d. fused silica column coated with a 0.25 µm film thickness of SE-52. The sample (1 µL sample plus 1 µL toluene solvent flush) was injected using an on-column injector (J & W Scientific, Rancho Cordova, CA) at 50 °C; the oven temperature was held for 1 min and then increased to 200 °C during a 3 min period and then programmed at 2 °C/min to 300 °C. The carrier gas was hydrogen with a head pressure of 14 psig (1418 Pa).

Sample clean-up for the LC analyses consisted of concentration, solvent exchange to cyclohexane, followed by a liquid-liquid partition between cyclohexane and nitromethane (4). The nitromethane solution was concentrated to approximately 1 mL and diluted with 1 mL tetrahydrofuran. The LC analyses were performed on a 5 µm C_{18} column (Vydac 201TP, The Separations Group, Hesperia, CA) with a solvent gradient from 40% acetonitrile in water to 100% acetonitrile in 45 min. A fluorescence detector was used which is capable of changing excitation and emission conditions (three sets of conditions are stored) during the chromatographic run.

RESULTS AND DISCUSSION

In the past twenty years numerous papers have reported the characterization of PAC in air particulate samples [see Bartle et al. (3) and references 36-68 therein]. The majority of these papers deal with the characterization of the PAH components. Several recent studies illustrate the use of GC (1,5) and LC (6-8) for the quantitation of PAH extracted from air particulate matter.

The extraction of air particulate matter with an organic solvent results in a complex mixture of organic constituents from which the PAH must be isolated (e.g., by column chromatography, normal-phase HPLC, or a liquid-liquid partition scheme) prior to identification and quantification. The PAH mixtures isolated from air particulate extracts are extremely complex because of the presence of numerous alkylated PAH as well as the numerous isomeric parent compounds. The complexity of these mixtures necessitates the use of high resolution gas and liquid chromatographic techniques to achieve separation, identification, and quantification of individual components. For the characterization of the St. Louis and Washington air particulate materials, three chromatographic approaches have been used: [1] multi-dimensional chromatographic techniques (i.e., normal-phase HPLC to isolate specific PAH fractions followed by GC, gas chromatography/mass spectrometry (GC/MS), or reversed-phase LC); [2] LC with selective fluorescence detection; and [3] high resolution capillary GC with flame ionization detection (FID). The first approach is used primarily for the characterization of both the major (unsubstituted PAH) and trace components (generally alkyl-substituted PAH). The second and third approaches are generally adequate for the measurement of the major PAH components after appropriate sample cleanup.

Qualitative Analysis by Multi-dimensional LC

For the qualitative characterization of these two air particulate samples, a multi-dimensional chromatographic approach was employed. The major PAH constituents in the extracts were found to be the unsubstituted PAH with smaller amounts (< 10 percent of the parent PAH) of the alkyl-substituted PAH. In order to isolate and identify these minor components in the complex PAH mixture, a normal-phase HPLC procedure on an aminosilane column was used to separate the PAH according to the number of aromatic carbons as previously described (2,9,10). In this procedure the alkyl-substituted PAH elute in the same region as the parent compound. These normal-phase fractions were then analyzed by GC/MS and reversed-phase LC. The reversed-phase liquid chromatograms of these fractions (including a total PAH fraction) are shown in Figure 1 to illustrate the usefulness of this approach for the analysis of complex PAH mixtures. Using UV absorption detection at 254 nm (generally considered as a "universal" detector for PAH), quantification of even the major PAH would be difficult due to the complexity of the mixture. However, the normal-phase LC preseparation based on the number of aromatic carbons provides fractions suitable for analysis even with the "universal" UV detector.

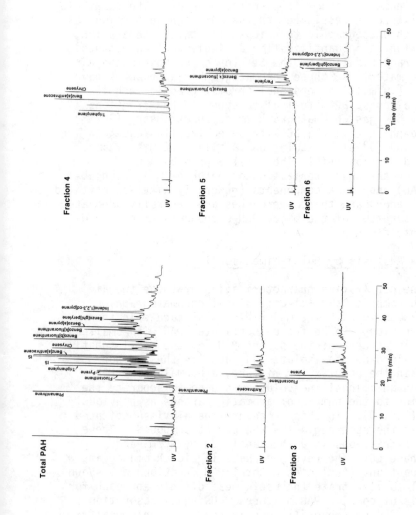

FIGURE 1. Reversed-phase LC separations of fractions obtained from normal-phase LC fractionation of an extract of Washington urban particulate matter.

Quantification by GC and LC

Quantification of the major PAH in the air particulate extract was achieved using both GC and LC. In order to provide two "independent and reliable methods" (10,11) for the certified concentration values, different sample extraction solvents and cleanup steps were employed prior to the chromatographic quantification. Prior to analysis by GC, the PAH were isolated from the complex extract using a liquid-liquid partition with cyclohexane/DMF/water to remove aliphatic hydrocarbons, followed by a normal-phase LC separation to isolate the total PAH fraction (2). A gas chromatogram of the PAH isolated from the Washington air particulate sample is shown in Figure 2.

In the LC analyses fluorescence detection was used to achieve the selectivity necessary to quantify the individual PAH components in the complex mixture without preseparation. Because of the selectivity for PAH of LC fluorescence detection compared to the universal FID for GC, a less rigorous PAH isolation procedure was used for the LC analyses than for the GC analyses (i.e., cyclohexane/nitromethane partition vs. cyclohexane/DMF/water partition and normal-phase LC). The selectivity of the LC analysis with fluorescence detection is illustrated in Figure 3 for the Washington sample. The various excitation and emission wavelength combinations, which are used to optimize the selectivity and/or sensitivity for the various PAH, are summarized in Table 1. Four liquid chromatographic runs, each with three sets of wavelength conditions (one in each run specific for the internal standard) were used for the quantification of 12 PAH. A detailed description of the LC method of quantitation of the PAH on these air particulate samples will be reported elsewhere (12).

Preliminary results of the LC and GC analyses of the Washington and St. Louis air particulate samples are summarized in Table 2. The majority of the samples were analyzed by LC because of the reduced sample preparation time. Table 2 contains LC results using two different internal standards, 7-methylfluoranthene and perylene-d_{12}. For the LC analyses using perylene-d_{12} as internal standard, the relative precision (1σ) varied from 2 to 6% for most of the PAH with only benzo[ghi]perylene at greater than 10%. For the GC analyses the relative precision (1σ) varied from 3 to 12%. Recovery studies indicated that the internal standards successfully accounted for any losses during the extraction, isolation, and analyses.

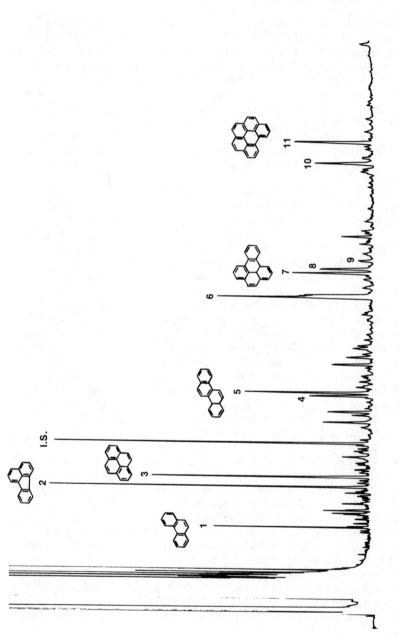

FIGURE 2. Gas chromatographic separation of PAH fraction from Washington urban particulate matter. Peaks identified as: (1) phenanthrene, (2) fluoranthene, (3) pyrene, (I.S.) 1-methylpyrene, (4) benz[a]anthracene, (5) chrysene/triphenylene, (6) benzofluoranthenes, (7) benzo[e]pyrene, (8) benzo[a]pyrene, (9) perylene, (10) indeno[1,2,3-cd]pyrene, and (11) benzo[ghi]perylene.

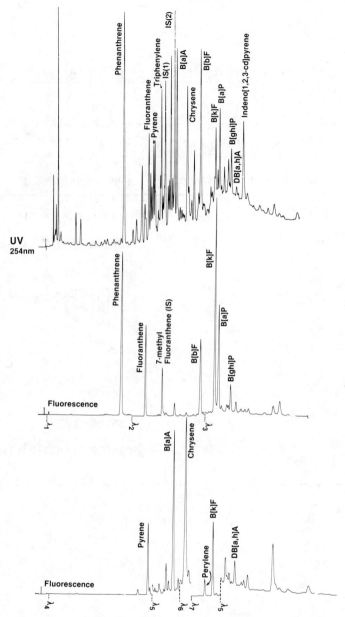

FIGURE 3. Reversed-phase liquid chromatograms of total PAH fraction from Washington particulate matter. Upper chromatograms: UV detection at 254 nm; middle and lower chromatograms: fluorescence detection at conditions described in Table 1.

TABLE 1

FLUORESCENCE CONDITIONS FOR LC DETERMINATION OF PAH IN AIR PARTICULATE SAMPLES

	Wavelengths (nm) (see Figure 3)		PAH Quantified
	excitation	emission	
λ_1	250	360	phenanthrene
λ_2	285	450	fluoranthene, 7-methyl-fluoranthene, benzo[b]fluoranthene
λ_3	295	400	benzo[k]fluoranthene, benzo[a]pyrene, benzo[ghi]perylene
λ_4	335	385	pyrene
λ_5	285	390	benz[a]anthracene, dibenz[a,h]anthracene
λ_6	270	360	chrysene
λ_7	400	440	perylene, benzo[k]fluoranthene

TABLE 2

DETERMINATION OF SELECTED PAH IN URBAN AIR PARTICULATE MATTER (µg/g)

	Washington, DC			St. Louis, MO
	GC [4][a]	LC [18][b]	LC [9][c]	LC [3][c]
Phenanthrene	-	4.8 ± 0.3 (17)	4.5 ± 0.3[d]	-
Fluoranthene	7.3 ± 0.2 (8)	7.4 ± 0.8 (36)	6.9 ± 0.4	7.9 ± 0.6
Pyrene	-	6.3 ± 0.4 (17)	6.2 ± 0.2	7.4 ± 0.2
Chrysene	4.6 ± 0.2 (8)[e]	3.5 ± 0.1 (6)	3.7 ± 0.2	6.6 ± 0.1
Benz[a]anthracene	2.4 ± 0.1 (8)	2.5 ± 0.3 (13)	2.4 ± 0.1	2.8 ± 0.1
Perylene	0.84± 0.09(8)	0.80 ± 0.04 (20)	0.65 ± 0.02	0.65 ± 0.02
Benzo[b]fluoranthene	-	6.2 ± 0.4 (31)	-	-
Benzo[k]fluoranthene	-	2.0 ± 0.2 (33)	2.1 ± 0.1 (18)	3.3 ± 0.1
Benzo[a]pyrene	2.9 ± 0.3 (8)	2.6 ± 0.4	2.6 ± 0.1	2.6 ± 0.2
Benzo[e]pyrene	3.3 ± 0.2 (8)	-	-	-
Benzo[ghi]perylene	4.7 ± 0.2 (8)	3.9 ± 0.8 (12)[f]	5.2 ± 0.6	5.5 ± 0.8
Dibenz[a,h]anthracene	-	0.54 ± 0.04	0.41 ± 0.07[g]	-
Indeno[1,2,3-cd]pyrene	3.3 ± 0.3 (8)	3.4 ± 0.4 (16)	3.6 ± 0.2	4.8 ± 0.2

[a] 1-methylpyrene used for internal standard, numbers in [ ] indicate number of samples extracted; numbers in ( ) indicate number of determinations if different from number of samples, uncertainties are ± 1σ from the mean.
[b] 7-methylfluoranthene used as internal standard; fluorescence wavelengths as in Table 1 except as noted.
[c] perylene-d_{12} used as internal standard; fluorescence wavelengths as in Table 1 except as noted.
[d] Determined at excitation 290 nm and emission 360 nm to achieve more selectivity.
[e] Chrysene and triphenylene coelute in the GC analyses.
[f] Determined at λ_2 conditions.
[g] Determined at excitation 290 nm and emission 395 nm to achieve more selectivity.

ACKNOWLEDGMENTS

The authors acknowledge partial support from the U. S. Environmental Protection Agency.

Identification of any commercial product does not imply recommendation or endorsement by the National Bureau of Standards, nor does it imply that the material or equipment identified is necessarily the best available for the purpose.

REFERENCES

1. Bjørseth, A. (1977): Analysis of polycyclic aromatic hydrocarbons in particulate matter by glass capillary gas chromatography, Anal. Chim. Acta, 94:21-27.
2. Wise, S.A., Bonnett, W.J., and May W.E. (1980): Normal- and reverse-phase liquid chromatographic separation of polycyclic aromatic hydrocarbons. In: Polynuclear Aromatic Hydrocarbons: Chemistry and Biological Effects, edited by A. Bjørseth and A. J. Dennis, pp. 791-806, Battelle Press, Columbus, Ohio.
3. Bartle, K.D., Lee, M.L., and Wise, S.A. (1981): Modern analytical methods for environmental polycyclic aromatic compounds, Chem. Soc. Rev., 10:113-158.
4. Novotny, M., Lee, M.L., and Bartle, K.D. (1974): The methods for fractionation, analytical separation, and identification of polynuclear aromatic hydrocarbons in complex mixtures, J. Chromatogr. Sci., 12:606-612.
5. Giger, W., and Schaffner, C. (1978): Determination of polycyclic aromatic hydrocarbons in the environment by glass capillary gas chromatography, Anal. Chem., 50: 243-249.
6. Dong, M., Locke, D.C., and Ferrand, E. (1976): High pressure liquid chromatographic method for routine analysis of major parent polycyclic aromatic hydrocarbons in suspended particulate matter, Anal. Chem., 48:368-372.
7. Lankmayr, E.P., and Müller, K. (1979): Polycyclic aromatic hydrocarbons in the environment: High-performance liquid chromatography using chemically modified columns, J. Chromatogr., 170:139-146.
8. Fechner, D., and Seifert, B. (1979): Determination of selected polynuclear aromatic hydrocarbons in settled dust by high-performance liquid chromatography with multi-wavelength detection. In: Polynuclear Aromatic Hydrocarbons, edited by P.W. Jones and P. Leber, pp. 191-199, Ann Arbor Science Publishers, Inc., Ann Arbor, Michigan.

9. Wise, S. A., Chesler, S.N., Hertz, H.S., Hilpert, L.R., and May, W.E. (1977): Chemically-bonded aminosilane stationary phase for the high performance liquid chromatographic separation of polynuclear aromatic compounds. Anal. Chem., 49:2306-2310.
10. May, W.E., Brown-Thomas, J., Hilpert, L. R., and Wise, S.A. (1981): The certification of selected polynuclear aromatic hydrocarbons in Standard Reference Material 1580, "Organics in Shale Oil". In: Chemical Analysis and Biological Fate: Polynucelar Aromatic Hydrocarbons, edited by M. Cooke and A.J. Dennis, pp. 1-16, Battelle Press, Columbus, Ohio.
11. Uriano, G.A., and Gravatt, C.C. (1977): The Role of Reference Materials and Reference Methods in Chemical Analysis. In: CRC Critical Reviews in Analytical Chemistry, 6:361-411.
12. May, W.E., Cuthrell, W.F., and Wise, S.A.: Quantitation by liquid chromatography of selected polycyclic aromatic hydrocarbons from air particulate matter, in preparation.

METABOLISM OF BAY REGION TRANS-DIHYDRODIOLS TO VICINAL DIHYDRODIOL EPOXIDES

SHEN K. YANG*, MING W. CHOU*[1], AND PETER P. FU**
*Department of Pharmacology, School of Medicine, Uniformed Services University of the Health Sciences, Bethesda, Maryland 20814; **Division of Carcinogenesis, National Center for Toxicological Research, Jefferson, Arkansas 72079 USA

INTRODUCTION

It has been shown that little if any of the bay region trans-dihydrodiols of benzo[a]pyrene (BaP) (Fig. 1, structure 5) and of benzo[e]pyrene (BeP) (Fig. 1, structure 1) are converted to the corresponding vicinal dihydrodiol epoxides (1,8, 9,11). These results were interpreted to be due to the directing effect of the quasidiaxial hydroxyl groups which shift the metabolic oxidation away from the vicinal double bond (8,11) and these findings were thought to be a general phenomenon for bay region trans-dihydrodiols (6). Metabolic studies of other bay region and "bay-like" region trans-dihydrodiols of polycyclic aromatic hydrocarbons (PAHs) indicated that the metabolic formation of vicinal dihydrodiol epoxides depends not only on the dihydrodiol conformation (quasidiequatorial vs. quasidiaxial) but also on the geometric location (bay region vs. non-bay region) of the vicinal double bond (14). Other studies also indicated that the extent of metabolism of the bay region dihydrodiols to vicinal dihydrodiol epoxides may depend on the source of microsomal enzymes as well as the presence of cosubstrate (2,7). This paper reports the metabolic study of racemic bay region trans-dihydrodiols of benz[a]anthracene (BA), dibenz[a,h]anthracene (DBA), and chrysene (Fig. 1, structures 2, 3, and 4 respectively). In each case, 1,2,3,4-tetrahydrotetrols are the major products indicating that the vicinal double bond of the dihydrodiol is the major metabolic site. The available results (3,10,13, 14) thus indicate that quasidiaxial hydroxyl groups do not direct metabolism toward distant end of the molecule.

---

[1]Present address: Division of Carcinogenesis, National Center for Toxicological Research, Jefferson, Arkansas 72079 USA

METABOLISM OF BAY REGION TRANS-DIHYDRODIOLS

MATERIALS AND METHODS

DBA (±)trans-1,2-dihydrodiol and chrysene (±)trans-3,4-dihydrodiol were synthesized by the method of Karle et al. (5). The racemic BA trans-1,2-dihydrodiol and BA trans-1,2-dihydrodiol anti- and syn-3,4-epoxides were obtained from the Chemical Repository of the National Cancer Institute. Dihydrodiols were purified prior to use by reversed-phase HPLC on a DuPont Zorbax ODS column (6.2 x 250 mm) with methanol/water (3:1 or 3:2, v/v) as the elution solvent at a flow rate of 1.2 ml/min. Liver microsomes were prepared from 3-methylcholanthrene (MC)-pretreated (i. p. injections; 0.5 ml of corn oil containing 0.25 mg of MC for each of 3 consecutive days) male Sprague-Dawley rats weighing approximately 100 g.

In Vitro Incubations

The bay region dihydrodiol (2 μmole dissolved in 2 ml of methanol or acetone) was added to a 48-ml incubation mixture (pH 7.5) containing 2.5 mmol of Tris-HCl, 5 mg of NADP$^+$, 0.15 mmol of MgCl$_2$, 29.3 mg of glucose-6-phosphate (G-6-P, monosodium salt), 5 units of G-6-P dehydrogenase (type II, Sigma), and 50 mg of protein equivalent of rat liver microsomes. The reaction mixture was incubated at 37°C for 30 min and was

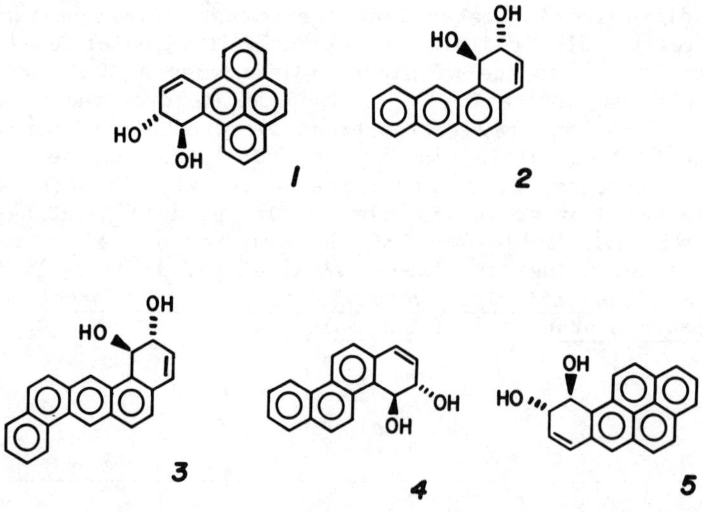

FIGURE 1. Structures of some bay region trans-dihydrodiols whose hydroxyl groups are predominantly in quasidiaxial conformations.

stopped by addition of 50 ml of acetone. The substrate and metabolites were extracted with 100 ml of ethyl acetate. The solvent of the organic phase was removed and the residue was dissolved in tetrahydrofuran/methanol (1:1, v/v) for HPLC analysis.

HPLC

Reversed-phase HPLC analysis was performed on a Spectra-Physics model 3500B liquid chromatograph fitted with a DuPont Zorbax ODS column (4.6 x 250 mm). The sample was injected via a Valco model N60 loop injector (Valco Instruments, Houston, Texas). Metabolite mixtures of DBA trans-1,2-dihydrodiol and chrysene trans-3,4-dihydrodiol were eluted with a 30-min linear gradient of methanol/water (1:1, v/v) to methanol at a solvent flow rate of 0.8 ml/min. Metabolite mixture of BA trans-1,2-dihydrodiol was eluted with methanol/water (1:1, v/v) for 5 min followed by a 40-min linear gradient of methanol/water (1:1, v/v) to methanol at a solvent flow rate of 0.8 ml/min. The uv absorbance of the eluent was monitored at 254 nm.

Identification of Metabolites

All HPLC-purified metabolites were characterized by analysis of their uv-vis absorption and mass spectra. Uv-vis absorption spectra were measured on a Cary 118C spectrophotometer. Mass spectral analysis was performed on a Finnigan model 4000 gas chromatograph/mass spectrometer: data system with a solid probe by electron impact at 70 eV and 250°C ionizer temperature.

RESULTS AND DISCUSSION

The reversed-phase HPLC profile of DBA (±)trans-1,2-dihydrodiol and its metabolites is shown in Fig. 2A. The uv-vis absorption spectra of the two most abundant metabolites are identical and have characteristics of a BA nucleus (Fig. 2B). Mass spectral analysis indicate that these two metabolites have molecular ions at m/z 346 and characteristic fragment ions at m/z 328 and 310 (loss of water molecules). Based on these data, the two most abundant metabolites of DBA (±)trans-1,2-dihydrodiol are identified as the 1,2,3,4-tetrahydrotetrols of DBA. It is apparent that the 3,4-double bond of DBA trans-1,2-dihydrodiol is the predominant metabolic site by rat liver microsomes.

The HPLC profile of BA (±)trans-1,2-dihydrodiol and its

METABOLISM OF BAY REGION TRANS-DIHYDRODIOLS

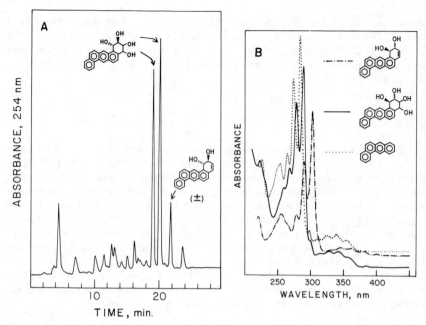

FIGURE 2. Reversed-phase HPLC separation (A) and uv-vis absorption spectra (B) of DBA (±)trans-1,2-dihydrodiol and its metabolites.

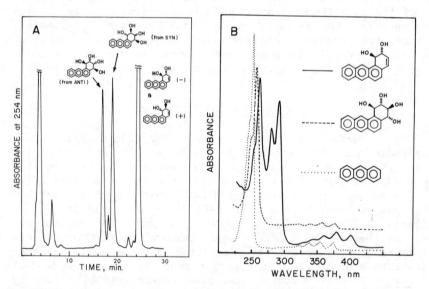

FIGURE 3. Reversed-phase HPLC separation (A) and uv-vis absorption spectra (B) of benz[a]anthracene (±)trans-1,2-dihydrodiol and its metabolites.

metabolites is shown in Fig. 3A. The two most abundant metabolites have uv-vis absorption spectra characteristic of an anthracene nuleus (Fig. 3B). Mass spectral analysis (M⁺ at m/z 296 and fragment ions at m/z 278 and 260) indicate that both metabolites are 1,2,3,4-tetrahydrotetrols of BA. As shown in Fig. 2A, one of the two tetrol metabolites is formed by stereoselective metabolism of (-)trans-1,2-dihydrodiol via an anti-3,4-epoxide and the other is formed stereoselectively from (+)trans-1,2-dihydrodiol via a syn-3,4-epoxide (4,13). Uv-vis absorption spectral analysis of the peaks which elute before the tetrols indicates that they are resulted from material extracted from liver microsomes in the incubation mixture. The results thus indicate that the 3,4-double bond of either (-)- or (+)-BA trans-1,2-dihydrodiol is the predominant metabolic site by rat liver microsomes.

The HPLC profile of chrysene (±)trans-3,4-dihydrodiol and its metabolites is shown in Fig. 4A. The metabolites contain in peaks T1 and T2 of Fig. 4A are similarly identified as

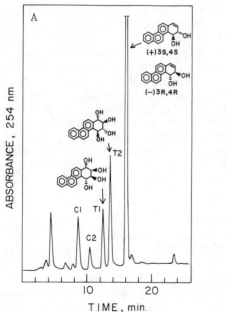 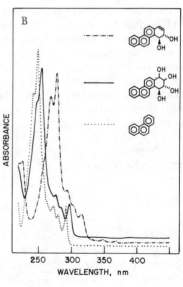

FIGURE 4. Reversed-phase HPLC separation (A) and uv-vis absorption spectra (B) of chrysene (±)trans-3,4-dihydrodiol and its metabolites.

1,2,3,4-tetrahydrotetrols. Their uv-vis absorption spectra are characteristic of a phenanthrene nucleus (Fig. 4B). Mass spectral analysis indicate that the metabolites contain in these two peaks are tetrahydrotetrols with molecular ions at m/z 296 and characteristic fragment ions at m/z 278 and 260. Tetrol T1 is formed by trans addition at C_1 of an anti-1,2-epoxide intermediate in which the epoxide ring is trans to the 4-hydroxyl group (10). Tetrol T2 is formed by trans addition at C_1 of a syn-1,2-epoxide intermediate in which the epoxide ring is cis to the 4-hydroxyl group (10). Mass spectral analysis of metabolites contain in peaks C1 and C2 (Fig. 4A) indicate that they are also tetrahydro derivatives. However, their uv-vis absorption spectra (3) indicate that they do not have a phenanthrene nucleus and are therefore not 1,2,3,4-tetrahydrotetrols. Thus peaks C1 and C2 are products resulting by epoxidation and hyration at positions other than the 1,2-double bond of chrysene trans-3,4-dihydrodiols.

The results reported here indicate that the quasidiaxial hydroxyl groups of DBA trans-1,2-dihydrodiol, BA trans-1,2-dihydrodiol, and chrysene trans-3,4-dihydrodiol do not shift the rat liver microsomal metabolism away from the corresponding vicinal double bonds of these bay region dihydrodiols. Thus the earlier suggestion of the directing effects by the quasidiaxial hydroxyl groups obtained in the metabolic studies of BaP trans-9,10-dihydrodiol and BeP trans-9,10-dihydrodiol (6,8,9,11) may be exceptions rather than a general phenomenon. This regioselective metabolism is a property of the drug-metabolizing enzyme system toward its substrate molecules. To date there is no apparent trend for a substituent such as a methyl group in certain methylbenz[a]anthracenes (14-16) and in 7-methylbenzo[a]pyrene (12) or a hydroxyl group in a dihydrodiol to "direct" the cytochrome P-450-containing microsomal mixed-function oxidases toward metabolism at a particular portion of the substrate molecules.

REFERENCES

1. Booth, J., and Sims, P. (1976): Different pathways involved in the metabolism of the 7,8- and 9,10-dihydrodiols of benzo[a]pyrene. Biochem. Pharmacol. 25:979-980
2. Cerniglia, C. E., and Gibson, D. T. (1980): Fungal oxidation of (±)9,10-dihydroxy-9,10-dihydrobenzo(a)pyrene: formation of diastereomeric benzo(a)pyrene 9,10-diol 7,8-epoxide. Proc. Natl. Acad. Sci., USA, 77: 4554-4558
3. Chou, M. W., Fu, P. P., and Yang, S. K. (1981): Metabolic conversion of dibenz[a,h]anthracene (±)trans-1,2-dihydrodiol and chrysene (±)trans-3,4-dihydrodiol to vicinal

dihydrodiol epoxides. Proc. Natl. Acad. Sci., USA, 78:4270-4273
4. Chou, M. W., Fu, P. P., and Yang, S. K. (1982): Stereoselective metabolism of optically pure (-) and (+)trans-1,2-dihydroxy-1,2-dihydrobenz[a]anthracene to vicinal 1,2-dihydrodiol 3,4-epoxides. submitted for publication
5. Karle, J. M., Mah, H. D., and Jerina, D. M., and Yagi, H. (1977): Synthesis of dihydrodiols from chrysene and dibenz-[a,h]anthracene, Tetrahedron Lett., 4021-4024
6. Lehr, R. E., Taylor, C. W., Kumar, S., Levin, W., Chang, R. L., Wood, A. W., Conney, A. H., Thakker, D. R., Yagi, H., Mah, H. D., and Jerina, D. M. (1979): Differences in metabolism provide a basis for the low mutagenicity and carcinogenicity of benzo[e]pyrene compared to benzo[a]pyrene. In: Polynuclear Aromatic Hydrocarbons, edited by P. W. Jones and P. Leber, pp. 37-49, Ann Arbor Science Publishers, Inc., Ann Arbor, Michigan.
7. Thakker, D. R., Levin, W., Buening, M., Yagi, H., Lehr, R. E., Wood, A. W., Conney, A. H., and Jerina, D. M. (1981): Species-specific enhancement by 7,8-benzoflavone of hepatic microsomal metabolism of benzo[e]pyrene 9,10-dihydrodiol to bay-region diol epoxides. Cancer Res., 41:1389-1396
8. Thakker, D. R., Nordqvist, M., Yagi, H., Levin, W., Ryan, D., Thomas, P., Conney, A. H., and Jerina, D. M. (1979): Comparative metabolism of a series of polycyclic aromatic hydrocarbons by rat liver microsomes and purified cytochrome P-450. In: Polynuclear Aromatic Hydrocarbons, edited by P. W. Jones and P. Leber. pp.455-472, Ann Arbor Science Publishers, Inc., Ann Arbor, Michigan.
9. Thakker, D. R., Yagi, H., Lehr, R. E., Levin, W., Buening, M., Lu, A. Y. H., Chang, R. L., Wood, A. W., Conney, A. H., and Jerina, D. M. (1978): Metabolism of trans-9,10-dihydroxy-9,10-dihydrobenzo[a]pyrene occurs primarily by arylhydroxylation rather than formation of a diol epoxide. Mol. Pharmacol. 14:502-513
10. Vyas, K. P., Yagi, H., Levin, W., Conney, A. H., and Jerina, D. M. (1981): Metabolism of (-)trans-(3R,4R)-dihydroxy-3,4-dihydrochrysene to diol epoxides by liver microsomes. Biochem. Biophys. Res. Commun., 98:961-969
11. Wood, A. W., Levin, W., Thakker, D. R., Yagi, H., Chang, R. L., Ryan, D. E., Thomas, P. E., Dansette, P. M., Whittaker, N., Turujman, S., Lehr, R. E., Kumar, S., Jerina, D. M., and Conney, A. H. (1979): Biological activity of benzo[e]pyrene. An assessment based on mutagenic activities and metabolic profiles of the polycyclic hydrocarbon and its derivatives. J. Biol. Chem., 254:4408-4415
12. Wong, T. K., Chiu, P.-L., Fu, P.P., and Yang, S. K. (1981): Metabolic study of 7-methylbenzo[a]pyrene with rat liver microsomes: separation by reversed-phase and normal-phase

high performance liquid chromatography and characterization of metabolites. Chem.-Biol. Interac., 36:153-166
13. Yang, S. K., and Chou, M. W. (1980): Metabolism of the bay-region trans-1,2-dihydrodiol of benz[a]anthracene in rat liver microsomes occurs primarily at the 3,4-double bond. Carcinogenesis, 1:803-805
14. Yang, S. K., Chou, M. W., and Fu, P. P. (1981): Metabolic and structural requirements for the carcinogenic potencies of unsubstituted and methyl-substituted polycyclic aromatic hydrocarbons. In: Carcinogenesis: Fundamental Mechanisms and Environmental Effects, edited by B. Pullman, P. O. P. Ts'o, and H. Gelboin. pp. 143-156, D. Reidel Publishing Co., Dordrecht-Holland.
15. Yang, S. K., Chou, M. W., and Fu, P. P. (1981): Microsomal oxidations of methyl-substituted and unsubstituted aromatic carbons of monomethylbenz[a]anthracenes. In: Polynuclear Aromattic Hydrocarbons, edited by M. Cooke and A.J. Dennis, pp. 253-264, Battelle Press, Columbus, Ohio
16. Yang, S. K., Chou, M. W., Weems, H.B., and Fu, P.P. (1979): Enzymatic formation of an 8,9-diol from 8-methylbenz[a]anthracene, Biochem. Biophys. Res. Commun., 90: 1136-1141

Author Index

Abrams, L.D. 439
Alsberg, T. 73,765
Andrews, P.A. 667
Austin, A. 449

Balfanz, E. 305
Bancsi, J.J. 471
Barbella, R. 83
Barnes, R.H. 229
Baumann, P.C. 93
Beland, F.A. 287
Beltz, P. 687
Benedek, A. 471
Benson, J.M. 103
Beretta, F. 83
Bergstrom, J.G.T. 109
Bickers, D.R. 121
Blau, L. 133
Booth, G.M. 845
Boparai, A.S. 347
Bowie, S.L. 919
Bruce, C. 491
Brune, H. 335
Buchan, R.M. 567
Burbaum, J.J. 515
Burton, R. 449
Butler, M.A. 325

Calle, L.M. 779
Casciano, D.A. 287
Caton, J.E. 813
Cavalieri, E. 145
Cazer, F.D. 537
Cerniglia, C.E. 157
Chang, M.J.W. 39,167,825
Chang, R.L. 859
Chen, D.J-C. 177,773
Chesler, S.N. 919
Chiu, P-L. 183,193
Chou, M.W. 287,931
Chuang, C.C. 641
Ciajolo, A. 83
Claxton, L. 449,695
Colmsjo, A.L. 201
Conney, A.H. 21,859
Cooper, C.S. 211

Crooks, C.S. 597
Cuthrell, W.F. 919

D'Alessio, A. 83
Daniel, F.B. 221
Davis, R.C. 229
DePaus, R. 405
Dettbarn, G. 335
Deutsch-Wenzel, R. 335
DiCesare, J.L. 237,835
Dommen, J. 515
Dong, M.W. 237
Drum, M.A. 221
DuBois, J. 405
Dunn, B.P. 247

Easterling, R. 695
Eastmond, D.A. 845
Eisenhut, W. 255
Eklund, G. 109
El-Bayoumy, K. 263
Ellis, L.E. 779
Evans, D.L. 325
Evans, F.E. 287

Facklam, T.J. 229
Fennelly, P.F. 367
Finkelmann, H. 275
Fisher, G. 551,615
Fordham, R.J. 405
Fox, D.G. 567
Fu, P.P. 39,157,183,287,931
Fulford, J.E. 297
Funcke, W. 305

Geacintov, N.E. 311
Giammarise, A.T. 325
Glaude, Ph. 897
Gold, A. 695
Gower, W.R. 537
Graff, D.E. 897
Griest, W.H. 813
Griffin, G.D. 725
Grimmer, G. 335,383
Grover, P.L. 211
Gschwend, P.M. 357
Gusten, H. 133

Hall, R.R. 367
Hardy, R. 491
Hart, R.W. 39,167,825
Haugen, D.A. 347
Hayes, T.L. 641
Hecht, S.S. 1,263
Heflich, R.H. 287
Hewer A. 211
Hill, J.O. 103
Hites, R.A. 357
Hoffmann, D. 1
Holko, A.P. 471
Hoyt, M. 367
Hsieh, D. 615
Hunt, G.T. 367

Inbasekaran, M. 537
Ivancic, W.A. 229

Jaasma, D.R. 655
Jacob, J. 383
Jerina, D.M. 21,859
Joyce, N.J. 221
Jungers, R. 449

Kadlubar, F.F. 287
Kandaswami, C. 389
Karcher, W. 405
Katz, M. 687
Kaur, P. 633
Kindya, R.J. 367
Kiriazides, D.K. 325
Konig, J. 305
Kowalczyk, P.J. 529
Kumar, S. 575

Lane, D.A. 297
Langer, E. 255
Larsson, B. 417
Later, D.W. 427
Laub, R.J. 275
Lavoie, E.J. 1
Lee, M.L. 427,845
Lehr, R.E. 21
Levin, W. 21,859
Levine, S.P. 439
Lewtas, J. 449,603
Lewis, K.F. 791
Lichtenstein, I.E. 461
Lowe, J.P. 743
Lu, C. 667
Luthy, R.G. 873

Mackerer, C.R. 705
Mackie, P.R. 491
MacLeod, M.C. 725
MacNicoll, A.D. 211
Malaiyandi, M. 471
Marsh, D. 325
Mason, T.O. 537
Mast, T. 615
May, W.E. 755,919
Mays, D. 551
McCalla, D.R. 667
McCarry, B.E. 667
McElroy, F.C. 673
McGill, A.S. 491
McMillan, L. 633
Melius, P. 801
Meresz, O. 885
Mermelstein, R. 325
Messier, F. 667
Meyer, C. 255
Michl, F. 501
Miller, K.J. 515,529
Miller, R.E. 897
Milo, G.E. 537
Misfeld, J. 335
Mosberg, A. 551
Mukhtar, H. 121
Mumford, J. 551
Murphy, B.P. 597
Murphy, C.B. 325
Murphy, D.J. 567
Murray, R.W. 575

Matusch, D.F.S. 713
Naujack, K.-W. 335
Nelen, A. 405
Nesnow, S. 585,695
Neue, U.D. 597
Neville, A.M. 211
Nishioka, M.G. 603

O'Brien, P.O. 389
Ocasio, I.J. 779
O'Connell, J.F. 897
Ogan, K. 237
O'Hare, M. 211
Okinaka, R.T. 177,773
Olsen, H. 615
Ostman, C.E. 201

Pal, K. 211
Palmer, III, A.G. 439
Parsons, E. 491
Peak, M.J. 347
Pederson, T.C. 623
Pelroy, R.A. 427,907
Peng, R. 791
Pereira, M.A. 633
Petersen, B.A. 603,641
Pierce, T.L. 655
Pitt, J.F. 897

Quilliam, M.A. 667

Rannug, U. 73,765
Reagan, R.R. 813
Rebbert, R.E. 919
Ribick, M. 93
Riggin, R. 551
Robbins, W.K. 673
Roberts, W.L. 275
Rogan, E. 145
Romanowski, T. 305
Rokosh, D.A. 735
Royer, R.E. 103

Sabharawal, P.S. 633
Sahlberg, G. 417
Sakuma, T. 297,735
Salamone, M.F. 687,735
Sangaiah, R. 695
Schmoldt, A. 383
Schreiner, C.A. 705
Schure, M.R. 551,713
Schuresko, D.D. 725
Segmuller, W. 529
Seiber, J. 615
Selkirk, J.K. 725
Shushan, B.I. 735
Siak, J-S. 623
Silverman, B.D. 743
Sims, P. 211
Singh, N.P. 167,825
Skewes, L.M. 439
Slaga, T.J. 585
Smith, C.A. 275
Smith, W.D. 93
Sonnefeld, W.J. 755
Stamoudis, V.C. 347

Stenberg, U. 73,765
Stewart, D.L. 907
Strniste, G.F. 177,773
Subrahmanyam, V.V. 389
Sullivan, P.D. 779
Sundvall, A. 73,765
Sydor, Jr., W. 791

Tan, B. 801
Thakker, D.R. 21,859
Tomkins, B.A. 813
Toney, G.E. 695
Toney, S.H. 695
Trayser, D.A. 641
Triplett, L.L. 585
Trzcinski, K. 109
Turturro, A. 167,825

Unruh, L.E. 287

Vandemark, F.L. 835
Vassilaros, D.L. 845
Vyas, K.P. 859

Walters, R.W. 873
Wang, D.T. 885
Weeks, G.H. 501
Weiss, H.S. 897
West, W.R. 845
Westerholm, R. 765
Willey, C. 907
Wilson, B.W. 427
Wise, S.A. 755,919
Witiak, D.T. 537
Wong, T.K. 183
Wood, A.W. 21,859
Woodrow, J. 615

Yagi, H. 21,859
Yang, C.S. 791
Yang, S.K. 121,157,183,
  193,287,931
Yee, J. 615

Zoller, W.H. 755

Key Word Index

Absorption spectroscopy, 311
Acclimation, 897
Acenaphthylene, 83
Acridine, 229
Activated carbon, 873
Adduct, 229
Adsorption, 167,873
Air particulate matter, 919
Airborne particulate, 713
Airborne particulate extract, 813
Airborne particulate matter, 305
Aluminum smelter, 417
Alveolar macrophages, 389
Ambient particulate, 449
Ames, 325
Ames assay, 427,695,907
Ames test, 73,183,193,765
Ames testing, 103,367
Amino polycyclic aromatic compound, 427
Amosite, 389
Anthophyllite, 389
Anthracene, 83
APCI/MS/MS, 297
Aroclor 1254, 801
Aromatic amine, primary, 347
Aryl hydrocarbon hydroxylase, 801
Aryl hydrocarbon monooxygenese (AHM), 725
Asbestos, 167,389
Asphalt volatiles, 471
Automobile exhaust, 335,765
Azaarene, 347,813

Bacon, 491
BaP, 21,39,83,121,167,177,183,193,229,
    247,389,515,585,633,735,791,897
BaP dioepoxide, 39,515
BaP-7,8-dihydrodiol, 183
BaP-7,8-dihydrodiol-9,10-epoxide, 183
BaP-7,8-diol, 791
BaP-7,8-diol-9,10-epoxide, 311
BaP-4,5-epoxide, 183
BaP-9,10-oxide, 311
BaP metabolism, 211
BaP metabolites, 801

Batch shake tests, 873
Bay region, 193,743
Bay region dihydrodiol, 931
Bay region diol epoxide, 145,859
Beet PAH levels, 885
Benz(a)anthracene, 21,931
Benz(a)anthracene-5,6-dione, 735
Benz(a)anthracene trans-1,2-
    dihydrodiol, 931
Benz(a)anthracene trans-1,2-
    dihydrodiol 3,4-epoxide, 931
Benz(a)anthracene-1,2,3,4-
    tetrahydrotetrol, 931
Benz(a)anthracene metabolism, 211
Benzo(c)phenanthrene, 21
Benzo(e)pyrene, 21
5,6-Benzoflavone, 687
Benzo(ghi)fluoranthene, 73
Benzo(j)aceanthrylene, 695
Benzophenone, 575
Bioaccumulation, 357
Bioconcentration, 845
Bovine, 615
7-Bromomethylbenz(a)anthracene, 39
Brown bullhead, 93
Butylated hydroxyanisole (BHA), 791

Capillary GC, 765,845
Capillary GC-MS, 73,305
Carbon black, 73,201,325
Carbonyl oxides, 575
Carcinogenesis, 585,897
Carcinogenic, 743
Carcinogenicity, 515,779
Cell mediated activation, 177
Certification, 405
Chemical oxidation, 779
Chinese hamster V79 cells, 21
Cholangiomas, 93
Chlorobenzene, 325
CHO, 177,773
Chrysene, 859,931
Chrysene-1,2,3,4-tetrahydrotetrol, 931
Chrysene-trans3,4-dihydrodiol, 931

943

Chrysene-trans-3,4-dihydrodiol-
  1,3-epoxide, 931
Chrysotile, 167,389
Coal combustion, 109
Coal fired power plant, 713
Coal fired stove, 335
Coal gasification, 103
Coal tar pitch volatiles, 471
Cocarcinogens, 1
Coke, 255
Coke oven emissions, 585
Coke plant, 93
Coking plant outfall, 845
Collision induced dissociation, 735
Comparison, 335
Complete carcinogenesis, 585
Complex mixture, 585,773
Computer graphics, 529
Coronene, 83
Correlation, 73
Crankcase oil, 335
Crocidolite, 389
Cryo gradient sampling system, 765
Cyclopenta PAH, 695
Cyclopenteno(cd)pyrene, 73,145
Cytotoxicity, 773

Deuterated internal standards, 641
Diaxial conformation, 743
Dibenz(ah)anthracene, 931
Dibenz(ah)anthracene-1,2,3,4-
  tetrahydrotetrol, 931
Dibenz(ah)anthracene trans-1,2-
  dihydrodiol-3,4-epoxide, 931
Dibenzo(ah)pyrene, 21
Dibenzo(ai)pyrene, 21
Diesel engine emissions, 297,439,585
Diesel exhausts, 765
Diesel exhaust emissions, 633
Diesel exhaust particulate, 641
Dihydrodiols, 211
Dihydrodiol epoxide, 931
4,5-Dihydroxyphenanthrene, 735
7,12-Dimethylbenz(a)anthracene (DMBA),
  39,221,825,897
DMBA metabolism, 211
Diol epoxide, 743
Diol epoxide formation, 21
Diphenyldiazomethane, 575
Direct acting mutagens, 603
Dispersion modeling, 567
Distribution coefficients, 673
DNA, 229
DNA adduct, 39,247,311

DNA adduct removal, 39
DNA binding, 221,247,515
DNA complexes, noncovalent, 311
DNA hydrocarbon adducts, 211
DNA repair, 39
DNA unwinding, 515,529
Dose response, 247

Electric linear dichroism, 311
Enzymatic oxidations, 779
Ether-soluble metabolites, 211
Environmental contaminants, 93
Extraction, 325
Extrapolation, 39

Fish, 93,491
Flame analysis, 83
Flame ionization detector (FID), 907
Flame photometric detector (FPD), 907
Fluoridated BP, 779
Fluidized bed, 109
Fluidized bed combustion (FBC), 367
Fluorene, 83
Fluorescence, 229,367,667,873
Fluorescence detection, 919
Fluorescence spectroscopy, 221,311
Flyash, 713
Food crops, 885
Forward mutation, 907
Fractionation, 347,439
Fused silica capillary columns, 305

Gas chromatography, 83,275,305,919
Gas chromatography/mass spectrometry
  (GC/MS), 83,347,367,615,641,907
Gasification, 347
Gasoline engine combustion, 705
Gasoline engine emissions, 585
Gasoline exhaust, 765
Gas phase trapping, 765
Gel permeation chromatography, 615
Gravimetry, 367

Hamster, 633
Hamster embryo fibroblasts (HEF), 725
Hamster embryonic cells, 725
Hepatoma rate, 83
HGPRT, 177,773
High performance liquid chromatography
  (HPLC), 83,177,183,193,221,237,325,
  575,597,667,791,801,873,919,931
HPLC analysis, 885
HPLC fractionation, 367
HPLC ion exchange, 347

High resolution gas chromatography, 907
High volume air sampling, 615
Horseradish peroxidase, 145
Hot gas streams, 297
Human cancer, 211
Human cell transformation, 537
Human mammary epithelial cells, 211
Human mammary fibroblasts, 211
Hydrocarbon-deoxyribonucleoside adducts, 221
3-Hydroxy-6-nitrobenzo(a)pyrene, 287
6-Hydroxybenzo(a)pyrene, 735
Hydroxylation, 735
7-Hydroxymethylbenz(a)anthracene, 157
7-Hydroxymethyl-12-methylbenz(a)-anthracene, 221
1-Hydroxy-6-nitrobenzo(a)pyrene, 287
1-Hydroxypyrene, 735
2-Hydroxypyrene, 735
4-Hydroxypyrene, 735

Industrial hygiene, 255
Initiators, 1
Intercalation, 311,515
In utero, 633
Isotherms, 873

K-region, 193
K-region epoxide, 193
Kinetic product pattern, 801

Laser, 229
Laser excited fluorescence, 83
Lettuce, 417
Liquefaction, 347
Liquid chromatography, 919
Liquid chromatography/mass spectrometry, 439
Liquid crystal phase, 275
Liver microsomal oxidation, 383
Low temperature enrichment, 765
Lowest unoccupied molecular orbital (LUMO), 743
Lung microsomal oxidation, 383

Magnetic circular, 501
Mammalian enzymes, 623
Mammalian microsomes, 389
Mammary cancer, 145,211
Marine, 357
Mass spectrometry, 287,304,765,931

MS/MS, 735
Melting point, 873
Metabolic activation, 177,735
Metabolism, 39,183,193,211,383,667
Methyl group substitution, 743
Methylation, 743
7-Methylbenz(a)anthracene, 157
1-Methylbenzo(a)pyrene, 193
2-Methylbenzo(a)pyrene, 193
3-Methylbenzo(a)pyrene, 193
4-Methylbenzo(a)pyrene, 193
5-Methylbenzo(a)pyrene, 193
6-Methylbenzo(a)pyrene, 193
7-Methylbenzo(a)pyrene, 183,193
9-Methylbenzo(a)pyrene, 193
10-Methylbenzo(a)pyrene, 193
11-Methylbenzo(a)pyrene, 193
12-Methylbenzo(a)pyrene, 193
1-Methylbenzo(a)pyrene 7,8-dihydrodiol, 193
4-Methylbenzo(a)pyrene 7,8-dihydrodiol, 193
7-Methylbenzo(a)pyrene 7,8-dihydrodiol, 183
7-Methylbenzo(a)pyrene 7,8-dihydrodiol 9,10-epoxide, 183
3-Methylcholanthrene derivatives, 145
Mice, 633
Mice, C3H, 897
Mice, SENCAR, 585,897
Microbial mutagenicity, 449
Micronucleus, 633
Microsomal metabolism, 121,263
Microsomal monooxygenase, 791
Mixed function oxidase, 801
Molecular features, 873
Molecular spectra, 405
Monomethylbenzo(a)pyrenes, 193
Monomethyl isomers, 743
Monooxygenase induction, 383
Motor oils, 705
Mouse, 247,335
Mouse, SENCAR, 585,897
Mouse, Swiss Webster, 791
Mouse skin painting, 335
Mullett (mugil cephalus), 801
Multicomponent isolation, 813
Mutagenesis assay, 347
Mutagenic response, 615
Mutagenicity, 145,183,193,287, 623,687,773,779
Mutagens, bases, 347

Naphthalene, 83
Near ultraviolet light, 773
Nitroacenaphthene, 263
Nitroaromatics, 603,623
6-Nitrobenzo(a)pyrene, 287
6-Nitrobenzo(a)pyrene-1,9-
  hydroquinone, 287
6-Nitrobenzo(a)pyrene-3,9-
  hydroquinone, 287
Nitronaphthalene, 263
Nitro-PAH, 287,297,427
Nitroreductase, 667
Nitropyrene, 325,439,667
Nonalternant PAH, 695
Nonlinear poisson statistical
  model, 585
Normal human fibroblast, 167
Normal phase LC, 439,919
Nuclear magnetic resonance
  spectroscopy, 287

Occupational exposure to PAH, 471
Oil combustion, 109
Oil shale, 773
On-line multidimensional system, 755
One electron oxidation, 145
Onion PAH levels, 885
Oxidation microbial, 157
Oxygenated metabolites, 537
Oxygenated PAH, 439
Oxy-PAH, 73
Outfall, 93

P450, 695
PAH, 83,305,537,615,765,845
PAH analysis sampling cartridge
  modification, 471
PAH fractions, 335
PAH residue levels, 93
Parent PAH, 93
Paving operations, 471
Peroxidases, 145
Phagocytic ability, 615
Phenanthrene, 21,83,575
Phenanthrene, 9-10, oxide, 575
Phenanthridine, 103
9-Phenanthrol, 575
Photodecomposition, 133
Polycyclic aromatic amines, 813
Polycyclic aromatic sulfur
  heterocycles (PASH), 845,907
Predicting adsorption, 873

Preparative LC, 439,835
Probit statistical model, 585
Protein content, 167
Pulmonary alveolar macrophage, 615
Pyrene chromophore, 311

Quantum yields, 133

Radical cations, 145
Radiolabeled material, 167
Rats, 825
Rat liver metabolism, 221
Rat liver microsomes, 931
Rat mammary epithelial cells, 211
Rat mammary fibroblasts, 211
Rat mammary gland, 145
Reactivity, 743
Real time analysis, 297
Recovery studies, 641
Regioselective metabolism, 931
Residential wood combustion, 567,655
Respirable particulate, 449
Retort process waters, 773
Reversed phase chromatography, 835
Reversed phase LC, 919
Rice straw smoke, 615
Risk assessment, 247
Roofing activity, 471
Roofing tar emissions, 585
Rotary evaporation, 367

S9 optimization, 73
Salmonella, 325
Salmonella mutagenesis, 145
Salmonella/mutagenicity assay, 615
Salmonella typhimurium, 21,183,193,
  287,705,907
Sample preparation, 673
Sediments, 357,845
Selective excision, 39
Semi-preparative HPLC, 73
Shale oil, 907
Shpol'skii spectra, 201
Silica fractionation, 73
Silica gel column, 439
Singlet oxygen, 575
Sister chromatid exchange, 633,825
Skin microsomes, 121
Smoked foods, 491
Soil, 201
Soil analysis, 885
Solvent exchange, 367

Solvent refined coal, 427
Soot, 83
Source assessment sampling system (SASS), 367
Soxhlet extraction, 73,325,367
Spectrophotometry, 931
Spot test, 655
Sprague-Dawley rats, 221,287,779,931
Spray combustion, 83
Standard Reference Material (SRM), 919
Stereoselectivity, 515,859
Steric fit, 515,529
Stomach carcinogenesis, 247
Structure activity, 1
Substituted BaP, 779
Sulfur heterocycles, 907
Sunlight, 773
Surface degradation, 133
Synfuels, 813
Synthetic fuels, 427
Synthetic motor lubricant, 705
Syrian hamster embryonic cell, 177,221

Target tissues, 145
Telluride Colorado, 567
Temperature, 713
Temperature, cool, 897
Temperature, warm, 897

Tenax GC packing, 471
7,8,9,10-Tetrahydroxytetrahydrobenzo(a)pyrene, 311
Thiophenes, 93
Thymodine incorporation, 167
Toluene, 325
Tomato PAH levels, 885
Traffic, environmental impact, 417
Transplacental carcinogenesis, 825
1,1,1-Trichloropropene oxide, 183,193
Triphenylene, 21
Triple quadrupole MS, 297
Tumor initiation, 145,585
Tumorigenesis, 585,897
Tumorigenicity, 1

Ultimate carcinogens, 859
Ultraviolet-visible absorption spectra 931

Vacuum sublimation, 73
Vegetables, 491
Vertebrate fish, 845
Viable cell count, 167
Vicinal dihydrodiol epoxide, 931

Water soluble metabolites, 211
Wood combustion, 109